Operations Research
Models and Methods

Operations Research Models and Methods

Paul A. Jensen and Jonathan F. Bard

John Wiley & Sons, Inc.

Acquisitions Editor   *Wayne Anderson*
Marketing Manager   *Katherine Hepburn*
Senior Production Editor   *Valerie A. Vargas*
Designer   *Kevin Murphy*
Production Management Services   *Argosy Publishing*

This book was set in Times Roman by *Argosy Publishing* and printed and bound by *Hamilton Printing*. The cover is printed by *Phoenix Color*.

This book is printed on acid-free paper.

ISBN 0-471-38004-0

Printed in the United States of America

10 9 8 7 6 5 4 3 2 1

This book is dedicated to our faculty colleagues and students from the Operations Research/Industrial Engineering Program in the Mechanical Engineering Department at the University of Texas at Austin.

Preface

Over the years, a traditional approach to the organization and presentation of introductory material on operations research (OR) has evolved. The basic progression is to introduce a topic, such as linear programming or Markov chains, with a simple example; follow the example with a brief discussion of the theory; use the theory to derive a solution methodology; and conclude with additional examples to demonstrate the computations. Virtually all OR textbooks take this approach. What has occurred recently, however, has been a broadening of the topics covered at the expense of theory and even at the expense of algorithms. The rationale for this shift has been based in part on the dwindling need for analysts to write their own software. In some minds, this means that it is less important for students to develop an in-depth knowledge of the underlying theory and methods.

To some extent, we are sympathetic with this view, but what we see as lacking in current OR texts is an in-depth treatment of modeling. Rather than broaden the topics covered, we have chosen to increase the number of examples presented and to emphasize modeling issues. We do not feel that much has been sacrificed in this approach, because we have also included chapters that describe the theories and computations associated with the various models. In addition, we present a wealth of supplementary material on the accompanying CD. Those topics that an instructor would most likely cover in a typical two-course sequence can be found in the text; advanced topics and material related to OR applications, such as inventory theory, are presented on the CD. Most instructors would agree that it is just about impossible to survey the entire field of OR in two or even three semesters. We feel that many topics, including nonlinear programming theory, decision analysis, and reliability, are best suited either for more advanced classes or for parallel classes such as fundamentals of industrial engineering or operations management.

The realization that there was a need for more emphasis on modeling arose in the early 1980s, when the first-named author of this book was asked to write a student's guide to the classical Hillier and Lieberman text. Around the same time, this author began developing microcomputer-based software that could be used to solve a wide range of OR models. This work has culminated in an extensive collection of Excel add-ins that are included on the CD. About 10 years ago, the second-named author began writing a series of books that included sections on basic topics in optimization. These two efforts provided the foundation for this book.

As anyone who has undertaken a writing project of this nature knows there are countless individuals who make contributions of all shapes and sizes. Although it is impossible to recognize them all, we would first and foremost like to thank Nathan Jensen whose artistry and imagination are at the heart of the book's web page. We are also indebted to our faculty colleagues who used materials from this manuscript in their classes and offered helpful suggestions and criticisms. Especially, we note the contributions of J. W. Barnes who authored the first drafts of the nonlinear programming chapters. Our thanks also go to the steady stream of high-caliber students who have taken our classes at The University of Texas and have provided feedback on all aspects of this book. We hope that in some way our efforts have helped to shape the minds of those students and have instilled in them an appreciation of the richness of, and an abiding love for, the field of operations research.

Paul A. Jensen
Jonathan F. Bard

Contents

Chapter 1

Problem Solving with Operations Research

The field of operations research (OR) comprises a rich collection of analytic techniques that have been developed over the last 50 years to solve complex problems arising in all aspects of human activity. The scope of such problems can be as broad as designing the layout of a new manufacturing facility or as narrow as selecting a battery technology for an electric vehicle. To gain an appreciation for the OR field, it is necessary to study problem formulation and modeling as well as the analytic techniques themselves. This premise has motivated the way this book is organized. Rather than group all the material on a specific topic together in a single chapter, we provide separate chapters on specific models and separate chapters on their associated solution methods.

To some extent, this arrangement may contribute to the view that operations research is a collection of diverse, mostly unrelated computational techniques for solving hypothetical problems. It would be most unfortunate if the reader jumped to this conclusion and dismissed the possibility of application. In fact, the field of operations research was born entirely for the purpose of solving real problems. The original idea was to study a system or organization with the goal of improving its operations. The use of models and mathematics was only one of the means, not the end, of the study. Although the development of quantitative techniques for solving large mathematical models is certainly a valuable and worthy activity for academics and others, from the operations research point of view it is the application of these methods that is the goal. Of course, application is not possible without knowledge, and providing an introduction to this body of knowledge has been our aim in writing this book.

The purpose of this chapter is to provide perspective and focus. In particular, the models and methods to be studied in this book are but one part of the problem-solving process. The process, described in the next section, typically involves a number of individuals, with the operations research professional contributing primarily by creating and solving quantitative models. When applying the ideas and techniques studied in subsequent chapters, the reader should always remember that the nontangible and qualitative aspects of a problem are important in every problem-solving situation, and the results of a quantitative analysis are at best only guides to the persons entrusted with making a decision.

1.1   THE PROBLEM-SOLVING PROCESS

It is a mark of modern society that decisions must often be made under circumstances characterized by conflicting goals, changing conditions, limited resources, complex interpersonal dynamics, uncertainty, and unyielding deadlines. Operations research is a discipline explicitly devoted to aiding decision makers under such circumstances. Closely related fields include management science, decision science, operations management, systems engineering, and combinations thereof. The goal of operations research is to provide a framework for constructing models of decision-making problems, finding the best solutions with respect to given measures of merit, and implementing those solutions in an attempt to solve the problems. To this end, several mathematical techniques have been developed that are exercised primarily with the help of a computer. The purpose of this text is to provide an introduction to these techniques.

An educational program in operations research typically covers prescriptive models and algorithms for optimizing deterministic systems as well as normative models and methods for understanding stochastic systems. Topics include mathematical programming, queuing analysis, and simulation, to name a few. Studies range from the theoretical to the computational to applications of textbook size. The danger in this approach is that the student, and sometimes even the professor, may forget that a particular analytic technique is a means rather than an end in solving a problem. Practitioners have long claimed that the selection and application of an analytic technique constitute a small component of the overall solution process. This process begins with the formulation stage where the problem is recognized, and continues through the implementation stage where some procedure or system is installed. A broad outline of the information flow and the analytic steps needed to arrive at a solution is presented in Figure 1.1.

In practice, these steps may not be well defined and may not be executed in a strict order from start to finish. Rather, there are many loops in the process, with experimentation and observation at each step suggesting modifications to decisions made earlier. The process rarely terminates with all the loose ends tied up. Work continues after a solution has been proposed and implemented. Parameters and conditions change over time requiring constant review of the solution and continuing repetition of portions of the process.

The Situation

Decision making begins with a situation in which a problem is recognized (see Figure 1.2). The problem may be actual or abstract, it may involve current operations or proposed expansions or contractions resulting from expected market shifts, it may become apparent through consumer complaints or employee suggestions, and it may be a conscious effort to improve efficiency or a response to an unexpected crisis. It would be impossible to circumscribe the entire breadth of circumstances that might be appropriate for this discussion, because problem situations that are amenable to objective analysis arise in every area of human activity.

To provide context, we identify the *organization* as the society in which the problem arises or for which the solution is important. The organization may be a corporation, a branch of government, a department within a firm, a group of employees, or perhaps even a household or an individual.

The situation will probably suggest that some action is appropriate to alleviate, correct, improve, or otherwise address the problem. An individual, committee, or organization is identified as being interested and capable of proposing and implementing the necessary actions. We call this individual or collection of persons the *decision maker*.

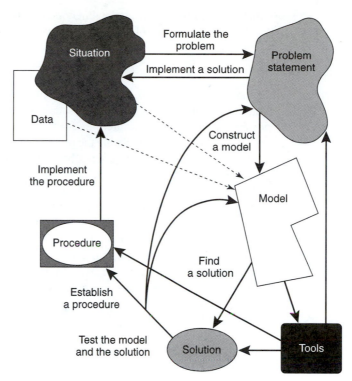

Figure 1.1   Outline of the problem-solving process

The individual or group called on to aid the decision maker in the problem-solving process is called the *analyst*. The decision maker may also play the part of the analyst, but typically the latter has special skills in modeling, mathematics, data gathering, and computer implementation. The analyst may, in fact, be a team of individuals whose members bring unique skills and perspectives to the problem-solving process.

When the situation is portrayed in the figures of this chapter, for example, in Figure 1.2, it is shown with a vague and irregular outline because most problems are poorly defined in their original conception. Historical data describing organizational operations and performance may be present in various forms. The data may be immediately relevant to the

Figure 1.2   The situation

situation or further investigations may determine the need for additional data collection. The first analytical step of the solution process is to formulate the problem in precise terms.

Formulating the Problem

At the formulation stage, statements of objectives, constraints on solutions, appropriate assumptions, descriptions of processes, data requirements, alternatives for action, and metrics for measuring progress are introduced. Because of the ambiguity of the perceived situation, the process of formulating the problem is extremely important. The analyst is usually not the decision maker and may not be part of the organization, so care must be taken to get agreement on the exact character of the problem to be solved from those who perceive it. There is little value in either a poor solution to a correctly formulated problem or a good solution to a problem that has been incorrectly formulated.

It is at this stage that the boundaries of the system under consideration should be identified. Because decisions usually have impacts beyond the perhaps narrow interests of the decision maker and the organization, the boundaries should encompass these primary effects. Generally, we use the term *systems approach* to show that the analysis has breadth beyond the local effects of an action; however, every study has boundaries, and these boundaries must be clearly stated.

Figure 1.3 shows a loop between the situation and the problem statement, implying that a careful examination of a problem often leads to solutions without complex mathematics. Since this book is mostly concerned with the other activities outlined in Figure 1.1, we might ask when it is necessary to continue around the loop.

When we step outside of our familiar surroundings, we often find it difficult to cope with the resulting complexity, variability, and uncertainty without technical assistance. For example, problems involving gate assignments at airports, machine setups for assembly operations, or vehicle routing are simple in concept but complex in solution given the many alternatives that need to be evaluated. Variability can be unsettling because it is difficult to conceive of solutions that will respond effectively to a rapidly changing environment. Examples include determination of replenishment policies for inventory systems and development of manufacturing plans for seasonal markets. It is a continuing challenge to derive robust solutions in the face of an uncertain future. Airlines must publish their flight schedules months in advance with only rough estimates of demand and weather, and military commanders must commit available forces before the intentions of the enemy are known. Operations research has been particularly attentive to such problems, and the reader will find many examples throughout this book.

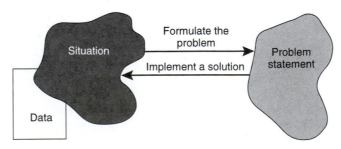

Figure 1.3   Formulation stage

In Figure 1.3, we show the problem statement with slightly more definition than the situation; however, greater simplification is still necessary before a computer-based analysis can be performed. This is achieved by constructing a model.

Constructing a Model

Computers allow objective methods such as statistical analysis, simulation, optimization, and expert systems to be used in the solution of a problem. First, however, the problem must be translated from verbal, qualitative terms to logical, quantitative terms (see Figure 1.4). A logical model is a series of rules, usually embodied in a computer program, by which the impact of alternative decisions can be predicted and evaluated. A mathematical model is a collection of functional relationships by which allowable actions are delimited and evaluated.

Although the analyst would hope to study the broad implications of the problem using a systems approach, a model cannot include every aspect of a situation. A model is always an abstraction that is, by necessity, simpler than the reality. Elements that are irrelevant or unimportant to the problem are to be ignored, hopefully leaving sufficient detail so that the solution obtained with the model has value with regard to the original problem. The statements of the abstractions introduced in the construction of the model are called the *assumptions*. It is important to observe that assumptions are not necessarily statements of belief, but are descriptions of the abstractions used to arrive at a model. The appropriateness of the assumptions can be determined only by subsequent testing of the model's validity. Models must be both tractable (capable of being solved) and valid (representative of the true situation). These dual goals are often contradictory and are not always attainable.

Model construction is primarily the function of the analyst; however, to ensure acceptance, this activity should be carried out in cooperation with the decision maker and other stakeholders. It is important that the decision maker as well as the full complement of users understand and agree to the assumptions underlying the model. Every model is an abstraction that embodies a set of assumptions. Without abstraction, little useful analysis is possible.

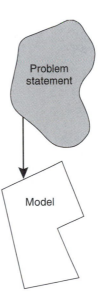

Figure 1.4 Model construction

The analyst is often confronted with an array of qualitative factors and perplexing constraints that cannot be captured in the model. These factors, however, may have a strong influence on the decision maker's thinking when it comes time to implement a solution, and may far outweigh the analytic results. This can be very frustrating for the analyst who may see things predominantly in quantitative terms. The analyst must accept that since the model is an abstraction, its results can only be an approximation to reality, and the nontangible aspects of a problem will sometimes be more relevant than the quantitative results.

We have intentionally represented the model with well-defined boundaries to indicate its relative simplicity. The next step in the process is to solve the model to obtain a solution to the problem. It is generally true that the most powerful solution methods can be applied to the simplest, or most abstract, model.

Deriving a Solution

In this step, tools available to the analyst, many of which are described in this book, are used to derive a solution to the mathematical model. Some methods can prescribe optimal solutions, whereas others only evaluate candidates, thus requiring a trial-and-error approach to finding an acceptable course of action.

To solve the model, the analyst must have a broad knowledge of the available solution methodologies (see Figure 1.5). It may be necessary to develop new techniques specifically tailored to the problem at hand. A model that is impossible to solve may have been formulated incorrectly or burdened with too much detail. Such a case signals a return to the previous step for simplification or perhaps the postponement of the study if no acceptable, tractable model can be found.

The variety of effective tools is constantly growing as research provides new insights and methodologies for general classes of problems. Access to these tools is increasing rapidly with the proliferation of inexpensive microcomputers and workstations. Most spreadsheet programs have built-in statistical and optimization routines. Stand-alone software for simulation, decision analysis, and almost every other subject described in this book is relatively inexpensive and readily available. As the analytic tools become widespread, it is important

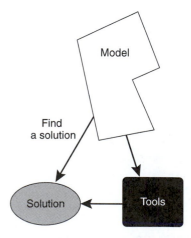

Figure 1.5   Finding solutions

that some member of the study team have the professional knowledge necessary to understand their applicability and limitations.

The solution obtained in this step prescribes a set of actions to resolve the problem under consideration. The solution, however, is a product of the model. Before it is imposed on the organization, it is necessarily to test its validity with respect to the problem formulation, the modeling assumptions, the intangible factors, and the original situation.

Testing the Model and the Solution

The testing step involves determination of the validity of the model and the solution obtained. Are the computations being performed correctly? Does the model have relevance to the original problem? Do the assumptions used to obtain a tractable model render the solution useless? These questions must be answered before the solution is implemented in the field. There are several ways to do this. The simplest way is to determine whether the solution makes sense to the decision maker. Solutions obtained by quantitative studies may not be predictable, but they are not often very surprising. Other testing procedures include sensitivity analysis, the use of the model under a variety of conjectured conditions including a range of parameter values, and the use of the model with historical data.

If the testing determines that the solution or model is inappropriate, the process may return to the formulation step to derive a more complex model embodying details of the problem formerly eliminated through abstractions. This may, of course, render the model intractable, and it may be necessary to conclude that an acceptable quantitative analysis is out of reach. It may also be possible to construct a less abstract model and accept less powerful solution methods. In many cases, finding a good or acceptable solution is almost as satisfactory as obtaining an optimal one. This is particularly true when the quality of the input data is low or when important parameters cannot be specified with certainty.

Testing may also return the process to the problem formulation stage. Perhaps the wrong problem has been solved. The decisions obtained may violate constraints not previously recognized or require actions not available to the decision maker or organization.

Of course, the solution provided by the computer is only a proposal. An analysis does not promise a solution but only guidance to the decision maker. Choosing a solution to implement is the responsibility of the decision maker and not the analyst. The decision maker may modify the solution to incorporate practical or intangible considerations not reflected in the model. Once a solution has been accepted, a procedure must be designed to retain control of the implementation effort.

Establishing the Control Procedure

Problems are usually ongoing rather than unique. Solutions are implemented as procedures to be used repeatedly in an almost automatic fashion under perhaps changing conditions (see Figure 1.6). Control may be achieved with a set of operating rules, a job description, laws or regulations promulgated by a government body, or computer programs that accept current data and prescribe actions.

Once a procedure has been established (and implemented), the analyst and perhaps the decision maker are ready to tackle new problems, leaving the procedure to handle the required tasks. But what if the situation changes? An unfortunate result of many analyses is a remnant procedure designed to solve a problem that no longer exists or a remnant procedure that burdens an organization with restrictions that are limiting and no longer appropriate. Therefore, it is important to establish controls that recognize a changing situation and signal the need to modify or update the solution.

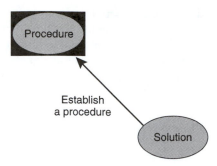

Figure 1.6   Ongoing control

Implementing the Solution

A solution to a problem usually implies changes for some individuals in the organization. Because resistance to change is a common human trait, the implementation of solutions is perhaps the most difficult part of a problem-solving exercise (see Figure 1.7). Some say it is the most important part. Although solution implementation is not strictly the responsibility of the analyst, the solution process itself can be designed to smooth the way for implementation. The people who are likely to be affected by the changes brought about by a solution should take part, or at least be consulted, during the various stages involving problem formulation, solution testing, and the establishment of the procedure.

Relevance of the Approach

The problem-solving process illustrated in Figure 1.1 and discussed in the associated text is only a model. Different organizations have different ways of approaching problems, and many do not admit quantitative techniques or analysts as part of the process. Today

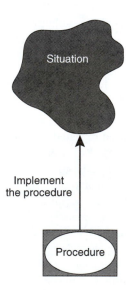

Figure 1.7   Implementation

more than ever before, however, it is important to note that problems do arise and decisions must be made (even inaction is a decision made by default). Many problems are solved in the first step of our process, but there will be cases in which complexity, variability, or uncertainty suggests that further analysis is necessary. In these cases, the models and methods described in this book will be of assistance in problem solving and decision making.

The process described here is a representative abstraction of what happens in many organizations. One of the most serious difficulties associated with its use is that it is most often applied to controversial issues. When a problem is perceived in an organization, it is usually perceived differently by different parties. For example, particular solutions may call for the replacement of workers by automated equipment, the restricted use of water resources when sufficient quantities are not available to all, the disposal of nuclear wastes in rural settings, or the construction of low-income housing throughout a community. All of these solutions are clearly of interest to some members of an organization or group but are perhaps anathema to others.

Like mathematics and other sciences, the models and methods of operations research do not embody ethical judgments. They are appropriate in a wide range of contexts for problems of interest to a wide variety of organizations and decision makers. One of the benefits of an objective analytical approach is that the models, with their associated assumptions, can be clearly and explicitly stated. The ideal model allows alternative assumptions that should be helpful in resolving controversial issues.

Terminology

Operations research, like every other discipline, has its own vocabulary. Throughout this book we introduce terms that are related to the specific topics under consideration. Following are terms related to the process of modeling and analysis.

OPERATIONS   The activities carried out in an organization that are related to the attainment of its goals and objectives.

RESEARCH   The process of observation and testing characterized by the scientific method. The steps of the process include observing the situation and formulating a problem statement, constructing a mathematical model, hypothesizing that the model represents the important aspects of the situation, and validating the model through experimentation.

MODEL   An abstract representation of reality. As used here, a representation of a decision problem related to the operations of the organization. The model is usually presented in mathematical terms and includes a statement of the assumptions used in the functional relationships. A model can also be physical or narrative, or can be a set of rules embodied in a computer program.

SYSTEMS APPROACH   An approach to analysis that attempts to ascertain and include the broad implications of decisions for the organization at each stage in the process. Both quantitative and qualitative factors are included in the analysis.

OPTIMAL SOLUTION   A solution to the model that optimizes (maximizes or minimizes) some objective measure of merit over all feasible solutions—the best solution among all alternatives given the existing organizational, physical, and technological constraints.

TEAM   A group of individuals bringing various skills and viewpoints to a problem. Historically, operations research has used the team approach in order to prevent the solution from being limited by past experience or too narrow in its focus. A team also provides a collection of specialized skills that are rarely found in a single individual.

OPERATIONS RESEARCH TECHNIQUES   A collection of general mathematical models, analytic procedures, and optimization algorithms that have been found to be useful in quantitative studies, including linear programming, integer programming, network programming, nonlinear programming, dynamic programming, statistical analysis, probability theory, queuing theory, stochastic processes, simulation, inventory theory, reliability, decision analysis, and others. Operations research professionals have created some of these fields whereas others have been derived from allied disciplines.

1.2 PROBLEMS, MODELS, AND METHODS

Real-world problems are typically steeped in conflict and skewed by politics, passions, and vested interests. The goal of OR is to abstract from the morass of reality (1) an objective statement of the decisions to be made, (2) the criteria for measuring the effectiveness of alternative solutions, and (3) the restrictions under which the decision maker must operate. The ultimate aim is to help the organization and its managers make better decisions. Whether or not this approach is successful and appropriate in a particular context depends on the complexity of the problem, the skill of those who apply the methods, the receptivity of the decision maker, and the environment in which the decisions are to be applied. Success is not guaranteed, but the goal of adding rationality to the process should have a strong appeal for those who must make difficult decisions and for those who must live with the consequences. Figure 1.8 identifies the three components that are paramount in the history, practice, and application of operations research: problems, models, and methods.

Problems

The OR approach is to simplify or abstract a real-world situation to a level at which analysis is possible. The result is a model of the situation that can be studied and manipulated in order to measure the effects of alternative decisions before they are actually implemented.

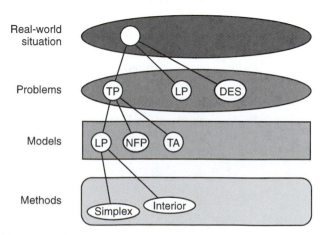

Figure 1.8   Problems, models, and methods

To this end, several general problem classes have been defined that provide forms into which the modeler may attempt to fit a given situation. The variety of these problems, partially enumerated in this book, is rich and growing, and thus many situations are likely to fit one or another. There is no requirement, however, that every problem be modeled as a member of the general set. Reality continually inspires creation, study, and solution of new problem classes.

The top level in Figure 1.8 identifies a situation arising from the real world. There may be several problem classes that would fit elements of the problem depending on which parts of the situation are eliminated or modified through abstraction. Perhaps some components of the situation involve a large number of decisions constrained by a limited set of resources. This might suggest that the problem could be modeled as a linear program, as indicated by the LP node in the second level of the figure. Additional simplifications might reduce the problem to a general transportation problem (TP). A completely different route of analysis might suggest that the stochastic elements of the situation require that a discrete-event simulation (DES) model be constructed. There is no reason why several different analytic techniques could not be used in a particular situation, each providing different insights regarding behavior and suggestions for possible courses of action.

Problem classes arise in a variety of ways. Certainly, reality breeds the creation of new classes. The process of abstracting a real problem to a tractable one has many steps. Perhaps the abstraction must go too far to obtain a computable solution, and thus the model ceases to be representative of the situation and the solution has no value. It is then necessary to step back and consider a new, less abstract problem. If this problem is solved in some fashion, it may be of sufficient general interest to be included in the class of problems available to other users.

Researchers designing new solution methodologies often add to the list of general problems through invention. Also, new problems can be added through generalization or simplification of a known problem class. Generalization adds complicating factors to the problem that make it more difficult, hence requiring new solution strategies. Simplification removes one or more elements to allow the use of more efficient solution methods. Although these efforts often occur without the driving and constraining influences of reality, they add to the collection of problems that might be useful in the future.

Models

The model is the middle ground of the operations researcher. It is the model that is manipulated, analyzed, and solved using quantitative methods. It is the result of this analysis that suggests the solution to be imposed on the real-world situation providing the impetus for the study. Important activities must take place on both sides of the model. On the side of reality, we create the model as an abstraction of the original problem, gather data to estimate its parameter values, test the model for usefulness, and interpret the solution with respect to the original situation. On the side of analysis, we apply known methodologies to derive solutions. If the current collection of these methodologies is not effective, new approaches must be devised. Also on the analysis side, we use our theoretical knowledge and computational power to investigate the sensitivity of solutions to changes in the model's structure, parameters, and coefficients.

Figure 1.8 indicates that there may be several models that are appropriate for a particular problem. It happens that the transportation problem can be described by a linear programming (LP) model, a network flow programming (NFP) model, or a matrix usually called the transportation array (TA). The selection of a model determines the applicability of solution methods.

Some of the models take their names from, and are closely related to, general problem classes. For example, the linear programming problem is usually described with the help of the linear programming model. The terms "model" and "problem" are equally appropriate in many instances.

We also find a hierarchy among model types, as illustrated by the following models listed in order of decreasing generality: nonlinear programming, linear programming, generalized network flow programming, pure network flow programming, transportation, and assignment. The assignment problem can be modeled using any of the forms on the list.

Models are also thought of as being general or specific. A general model has general parameters usually identified by letters. A specific instance of a general model has real numbers assigned to parameters and relates to a particular situation.

Methods

Solution methods are associated with specific models. In most cases, there are several methods appropriate for a given model. Figure 1.8 indicates the existence of two methods that can be applied to the linear programming model. The first is the simplex algorithm that has been available for many years, and the second includes the more recently discovered interior point approaches. Researchers are constantly searching for more efficient methods of solving general models.

The application of operations research requires constant attention to the available methods. Whether a particular model can be solved with reasonable effort depends on the method that is used. Researchers have identified classes of problems that will be difficult under any circumstances and other classes that have the potential for easy solution—a distinction that is critical to successful analysis.

Most OR methods are implemented with software. As computer power has grown and become more available, these methods have become more convenient to apply. It is likely that computer technology will be a continuing factor in the growth and relevance of the OR approach.

Team Approach

One of the difficulties in successfully applying OR to real-world problems is that it is hard for any one person to combine an understanding of the practical situation, knowledge of the broad class of general problems, skill at modeling, and the technical abilities needed for implementation. Individuals frequently get trapped in one area of expertise, such as mathematical programming or simulation, and rely too heavily on it at the expense of all the others. The team approach is a means of overcoming the limitations that any single individual brings to a problem.

Whether one is working independently or as a member of a team, it is important to have at least a rudimentary understanding of all three components described above. We have attempted to provide this understanding throughout the book. In each chapter, we identify many important problem classes and offer instruction in creating models that can be solved with available methodologies.

1.3   ABOUT THIS BOOK

This book is designed to bridge the gap between theory and practice by presenting the quantitative tools that are most suited for modern operations research. A principal goal is to give analysts, engineers, and decision makers a greater appreciation of their roles by defining a common terminology and by explaining the interfaces among the underlying methodologies.

Theoretical aspects are covered at a level appropriate for a senior undergraduate course or a first-year graduate course in either an engineering or an MBA program. Special attention is paid to the use and evaluation of specific methods with respect to their real-world applicability. Subjects are covered at a survey level with ample illustrations and examples. When appropriate, insights from the latest research are cited to give perspective. As is the case with all introductory texts, one of our primary goals is to provide a starting point from which interested readers can pursue a more thorough study of the material in advanced texts, journals, and courses. Whether this book is adopted for a course or is read by practitioners who want to learn the "tools of the trade," we have tried to present the material in a concise and fully integrated manner.

Topics are structured along functional lines and span mathematical programming, stochastic processes, and simulation. The presentation is designed to give a full picture of the relationships that exist among modeling, analysis, computations, algorithmic implementation, and decision making. For most topics, separate chapters are included for models and methods. This allows the instructor to tailor the material to the aims of the class. If one wishes to emphasize applications rather than algorithms, the methods chapters can be either skipped or selectively assigned. By virtue of this formal separation of models and methods, those whose main interests do not lie in the mathematics of operations research can study modeling techniques without intimidation. For those who have the motivation or need to understand the mathematics, a simple but rigorous development of OR methods is provided. Although this will be difficult material for some, we feel that it is presented in a style that will encourage and simplify learning by the majority of students.

Advances in computer technology have driven much of the recent progress in operations research. This has given OR more visibility at the upper levels of management and has created a new component of the software industry dedicated to the implementation of OR techniques. On the CD-ROM accompanying this book, we provide computational support for each chapter with a collection of Microsoft Excel™ add-ins. Each is programmed in Visual Basic for Applications (VBA), the Microsoft macro language for the Office suite of programs, and works with both the Windows and Macintosh operating systems.

The algorithms embodied in the add-ins are quite robust and, according to the feedback we have received, extremely easy to use. In general, they reflect the procedures outlined in the text. When the user enters the data and selects a solution procedure, results are automatically computed and presented on the Excel™ spreadsheet. Very little instruction is necessary to input and solve complex problems. A prime advantage of the spreadsheet medium is that most people are familiar with its basic operations. In addition, data analysis can be performed in the same file that contains the model and the solution, experimentation with several different OR methods can be carried out using a common interface, and data can be easily modified for "what-if" analysis. The CD-ROM contains detailed instructions for the use of each program. The book's website (http://www.ormm.net) provides updates to the software, animated demonstrations, teaching aids, and additional examples and exercises.

Virtually all of the examples and exercises in this text were solved with the Excel add-ins. Nevertheless, there are limitations on the size of problems that they can handle, so it would be naïve to think that every real-world problem can be solved with this type of software. Spreadsheets have their advantages and conveniences but are not a substitute for creativity and special-purpose codes. Many optimization problems found in scheduling and logistics, for example, will quickly eat up the resources of any spreadsheet software package. In such cases, advanced or custom-tailored techniques, such as decomposition, are likely to be required to obtain solutions. An insightful look at issues related to spreadsheet use in the classroom can be found in Gass *et al.* [2000].

At the end of each chapter there is a series of exercises designed to stimulate thought and to test the readers' grasp of the material. In some cases, the intent is to explore supplementary topics in a more open-ended manner. This is one mechanism we occasionally use to introduce models and algorithms that are a bit beyond the elementary level. A bibliography of important works is also provided at the end of each chapter. The interested reader can further his or her understanding of a particular topic by consulting these references. For those who adopt the book, a solution manual is available from the publisher, and Power Point slides for some of the chapters are available from the authors.

Material on the CD

The CD contains chapter supplements covering more advanced subjects and areas of application. In creating these supplements, our goal was to limit the length of the hardbound text without sacrificing content. For the most part, we wanted to include in print only material that would likely be covered in a two-course sequence. Individual instructors, of course, have their own preferences, so we recognize that not everyone will be satisfied with our selection.

The pedagogical contents of the CD, including additional chapters and appendices, the Excel add-ins, example spreadsheets, and additional exercises, are as follows:

- Minimizing piecewise linear convex functions
- Goal programming
- Minimum-cost network flow algorithms
- Implicit enumeration for integer programming
- Constrained nonlinear programming
- Dynamic programming models and methods
- Time series and forecasting
- Statistical analysis of simulation output
- Decision analysis
- Game theory
- Inventory theory
- Reliability
- Excel add-ins
- Excel workbooks with solutions to textbook examples
- Additional exercises

The contents of the CD are organized under five general headings: models, methods, computation, problems, and supplements. The models section contains brief discussions on how decision problems can be expressed in a form that is amenable to analysis, along with additional examples. The methods section contains write-ups that explain the theoretical constructs behind the solution methods. The computation section provides instructions for the Excel add-ins that can be used to solve the models. The problems section presents modeling and algorithmic-oriented problems that are too long to be included in the text. The supplements section provides an annotated and linked index to the supplementary text materials. In addition, the CD includes what might be considered a horizontal structure in the form of tours. A tour spans a specific topic, such as integer programming or simulation, and links to sections of the CD contents that pertain to that subject. An internet site, reached through a link from the CD, includes a portion of this material as well as access to updates and activities of ongoing interest.

Student versions of commercial software for optimization and simulation are provided on the CD illustrating professional level tools used by practicing OR analysts. See the instructions on the CD for installing and operating these packages.

Notation and Terminology

As the field of operations research has matured, notation and terminology have become increasingly standardized. We have done our best to adhere to these standards and to be consistent in the use of symbols and style throughout the text. In several of the chapters that introduce an abundance of unique terms, a separate section is provided to highlight relevant definitions and vocabulary.

Several rules have been followed regarding the use of mathematical notation and the construction of equations. In general, boldface font is used to signify vectors and matrices, boldface-italics to signify sets, and plain italics to represent scalars. Decision variables, model coefficients, and system parameters may be written in either uppercase or lowercase, depending on the context. Unless otherwise stated, vectors and gradients assume a column orientation. For the most part, the transpose symbol "T" is used only when it is necessary to distinguish the orientation of a vector or matrix, such as in the quadratic expression $\mathbf{x}^T\mathbf{Q}\mathbf{x}$. It is always used in equations to ensure that the algebraic operations make sense.

Subscripts have their usual interpretation as components of a vector or matrix [e.g., $\mathbf{x} = (x_1, \ldots, x_n)$] except when they are attached to a scalar. In that case, they usually denote an index that counts the number of iterations of an algorithm. Superscripts on vectors similarly refer to an iteration index. In the expression $\mathbf{x}^{k+1} = \mathbf{x}^k + t_k\mathbf{d}^k$, for example, the symbol k is an index, $\mathbf{x}$ and $\mathbf{d}$ are vectors, and t is a scalar. A superscript can also refer to an element of a set—e.g., $\mathbf{x}^1 \in \mathbf{S}$, where the symbol $\in$ means "belongs to" and $\mathbf{S}$ is a set of appropriate dimension. Braces are used for sets, whereas parentheses and square brackets are used for both vectors and matrices. The set of points in n-dimensional Euclidean space that satisfies several linear inequalities is denoted by $\{\mathbf{x} \in \Re^n : \mathbf{A}\mathbf{x} \le \mathbf{b}\}$, where the colon means "such that." In the expression $k \leftarrow k + 1$, the symbol $\leftarrow$ instructs an algorithm to replace the contents of the storage location holding the value of k with $k + 1$. More specialized notation will be introduced as needed.

The Future

It goes without saying that the huge body of knowledge in the area of operations research cannot be condensed into a single book. Over the last 10 years alone much has been written on the subject in technical journals, textbooks, and trade publications. Graduates are in top demand, and all indications are that they will have a profound impact on the new economy. Currently, the hottest areas of application include financial engineering, supply-chain management, and revenue management. A revolution in education is also taking place, including distance learning and web-based instruction. We hope that our internet site will be a resource for web-based instruction. We expect that over time and with constructive feedback we will be able to expand the focus of the web presentation, improve its interactive capabilities, and allow the user more control over the pace of instruction.

EXERCISES

1. Why is the object labeled "Situation" in Figure 1.1 depicted with ambiguous borders?

2. In Figure 1.1, why is the object labeled "Model" drawn with straight lines?

3. In Figure 1.1, why is the object labeled "Procedure" drawn as an oval with a box around it?

4. What are the assumptions associated with a model? Why is it necessary to make assumptions?

5. Why may it be necessary to go back and change the model after it has been solved?

6. What is the meaning of the term "systems approach"?

7. What is meant by the term "optimal solution"?

8. Why is implementation of an operations research solution sometimes difficult?

9. In the title of this book, what is meant by the term "methods"?

10. How would you define an organization in the context of problem solving? Who are its members?

11. What is an abstraction? Why does one make abstractions in the modeling process?

12. What must happen if the solution to a model turns out to violate some important constraints not previously stated by the decision maker?

13. Why do operations research studies often involve teams rather than single individuals?

14. Explain how the conflicting goals of tractability and validity may cause the modeling process to fail.

15. How is a solution different from a model?

16. In what two activities of the problem-solving process should the decision maker play a large role?

17. Why would an analyst make assumptions that he or she did not necessarily believe when formulating a model?

18. What is the purpose of a control procedure?

19. How should the value of a model be determined?

BIBLIOGRAPHY

Gass, S.I., D.S. Hirshfeld, and E.A. Wasil, "Model World: The Spreadsheeting of OR/MS," *Interfaces*, Vol. 30, No. 5, pp. 72–81, 2000.

Hall, R., "What's So Scientific about MS/OR?," *Interfaces*, Vol. 15, No. 2, pp. 40–45, 1985.

Hillier, F.S. and G.J. Lieberman, *Introduction to Operations Research*, Seventh Edition, McGraw-Hill, New York, 2001.

Horner, P. (editor), *OR/MS Today, Special Issue on Executives' Guide to Operations Research*, The Institute for Operations Research and the Management Sciences, Vol. 27, No. 3, 2000.

Labe, R.R., Jr. and S.C. Graves, "Franz Edeleman Award for Management Science Achievement," *Interfaces*, Vol. 30, No. 1, pp. 1–6, 2000.

Murthy, D.N.P., N.W. Page, and E.Y. Rodin, *Mathematical Modelling: A Tool for Problem Solving in Engineering, Physical, Biological and Social Sciences*, Pergamon, New York, 1990.

Ragsdale, C.T., *Spreadsheet Modeling and Decision Analysis: A Practical Introduction to Management Science*, Second Edition, South-Western, Cincinnati, 1998.

Savage, S., "Weighing the PROS and CONS of Decision Technology in Spreadsheets," *OR/MS Today*, Vol. 24, No. 1, pp. 42–45, 1997.

Winston, W.L., *Operations Research, Applications and Algorithms*, Third Edition, Duxbury, Belmont, CA, 1994.

Chapter 2

Linear Programming Models

The branch of operations research that deals with the optimal allocation of scarce resources among competing activities is known as *mathematical programming,* of which *linear programming* (LP) is a special case. Here, the word "programming" is a synonym for "planning" and should not be confused with the term "computer programming." The adjective "mathematical" indicates that mathematics is the primary mechanism by which problems are represented. The fact that this chapter is concerned with *modeling* indicates that we are abstracting reality rather than studying it directly.

A typical mathematical program consists of a single objective function, representing either a profit to be maximized or a cost to be minimized, and a set of constraints that circumscribe the decision variables. In the case of a linear program, we will see that the constraints define a polyhedron. Although some compromise is usually called for when modeling a problem as a linear program, the power and importance of such formulations cannot be overstated. Countless real-world applications have been successfully modeled and solved using LP techniques. This has produced an ongoing revolution in the way decisions are made throughout all sectors of the economy.

Because planning is such an important topic in our increasingly complex world, mathematical programming has gained widespread acceptance as one of the most practical model forms in operations research. It has achieved much of its stature from advances in computer technology and modeling languages, a combination that has put the solution of formidable problems within easy reach of the analyst. Typical applications include the scheduling of airline crews, the distribution of products through a manufacturing supply chain, and production planning in the petrochemical industry.

In contrast to the models introduced in chapters 11 through 18, we generally treat mathematical programs as deterministic rather than stochastic in nature—that is, all model parameters are assumed to be known constants rather than random variables with accompanying probability distributions. This assumption is designed to keep the text at an elementary level, recognizing that most real problems contain both deterministic and stochastic elements. In addition to LP, the field includes integer programming, network flow programming, nonlinear programming, and dynamic programming. Each of these topics is covered in chapters 2 through 10 and in the supplements on the CD.

For the linear program, all functional relationships are required to be linear. The objective function, for example, is typically written as $z = f(\mathbf{x}) = c_1 x_1 + c_2 x_2 + \cdots + c_n x_n$. In such expressions we use subscripts to identify the individual terms and use j for the general term. The jth decision variable is x_j and c_j is the jth unit cost or profit coefficient. An additional restriction is that the decision variables x_j are required to be continuous rather than discrete. Variable names are commonly derived from the context of the problem, but for the generic case, z and x_j are the names traditionally used for the objective value and the jth decision variable, respectively. The number of decision variables is n. A complete list of relevant definitions can be found later in the chapter.

At first glance, these restrictions would seem to limit the scope of the LP model, but this is hardly the case. Because of the simplicity of the LP model, software has been developed that is capable of solving problems containing millions of variables and tens of thousands of constraints. Computer implementations are widely available for most mainframes, workstations, and microcomputers. Moreover, special modeling techniques expand the usefulness of LP to situations that do not, at first, appear to fit the underlying assumptions. A variety of problems with nonlinear functions, multiple objectives, uncertainties, or multiple decision makers, such as those arising in game theory, can be modeled as linear programs. Consequently, before proposing complex mathematical models, the analyst should first determine whether or not an abstraction can be created that fits the LP format.

LP is at the center of virtually all optimization. Many common models, such as those arising in network flows and personnel scheduling, are special cases. Thus, understanding LP models and methods provides the foundations for understanding the special cases. When the assumptions associated with the special cases fail to hold, standard LP is the natural replacement. On the other hand, algorithms developed to solve nonlinear, integer, and stochastic programming models usually use an LP code as a component in the solution process.

This chapter focuses on the development of the skills required to formulate optimization problems as linear programs. As in every modeling activity, there is often more art than science, so practice and experience are the best teachers. Our goal is to foster a level of understanding that will allow the reader to pursue LP applications in a realistic setting.

2.1   A MANUFACTURING EXAMPLE

The basic components of an LP model will be highlighted with the help of the following manufacturing example. An operations manager is trying to determine a production plan for the next week. The data for this problem are enumerated in Tables 2.1 and 2.2, and are arranged with respect to machine processing requirements in Figure 2.1. The company produces three products: P, Q, and R. The unit revenues and maximum sales for the week are indicated in the boxes at the top of Figure 2.1. Storage from one week to the next is not

Table 2.1   Machine Data

Machine	Unit processing time (min)			Availability (min)
	Product P	Product Q	Product R	
A	20	10	10	2400
B	12	28	16	2400
C	15	6	16	2400
D	10	15	0	2400
Total processing time	57	59	42	

Table 2.2   Product Data

Item	P	Q	R
Revenue per unit	$90	$100	$70
Material cost per unit	$45	$40	$20
Profit per unit	$45	$60	$50
Maximum sales	100	40	60

permitted. The rectangles below the product boxes indicate the operations used to manufacture the products, and the circles describe the raw material requirements. Each operation square has a machine assignment and the time required for the operation. For example, the square labeled "D" for product P indicates that each unit of product P is processed on machine D for 10 minutes.

To illustrate the notation used in Figure 2.1, consider product P, which consists of two components. Component 1 is manufactured from one unit of raw material RM1 by processing it on machines A and C. Component 2 is made from one unit of raw material RM2 by processing it on machines B and C. The components are assembled with a purchased part at machine D and then sold as a complete product. We observe from Figure 2.1 that products P and Q use component 2, whereas products Q and R use component 3. The purchased part is required only for product P.

Each of the four machines (A, B, C, and D) performs a unique process. There is one machine of each type, and each machine is available for 2400 minutes per week. The operating expenses associated with the plant consist of labor and overhead costs amounting to $6000 each week.

There are several questions that are appropriate for the manufacturing process, but here we seek the "optimal" product mix—that is, the amount of each product that should be manufactured during the present week in order to maximize profits. Constructing the model requires several steps.

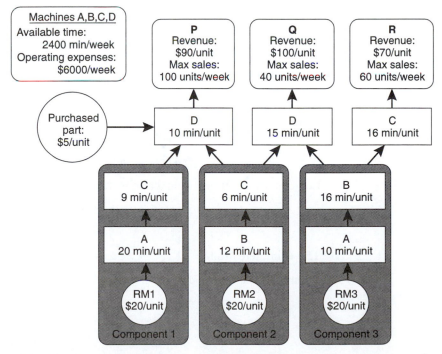

Figure 2.1   Manufacturing example

Defining the Variables

We are trying to select the optimal product mix, so we define three decision variables as follows:

$$P = \text{number of units of product P to produce during the week}$$
$$Q = \text{number of units of product Q to produce during the week}$$
$$R = \text{number of units of product R to produce during the week}$$

Choosing an Objective

A measure is necessary to compare alternative solutions. Here we choose to maximize profit. The unit contribution of each product to profit is the difference between unit revenue and unit raw material cost. Assuming that there are no volume discounts, the total profit per week is proportional to the number of units manufactured and sold, and thus

$$\text{Profit} = (90 - 45)P + (100 - 40)Q + (70 - 20)R - 6000$$
$$= 45P + 60Q + 50R - 6000$$

Because the operating costs are not a function of the variables in the problem, we can drop the $6000 term from the profit function. We typically use the variable Z or z to denote the value of the objective.

$$Z = 45P + 60Q + 50R$$

Identifying the Constraints

The amount of time a machine is available and the maximum sales potential for each product restrict the quantities to be manufactured. Since we know the unit processing times for each machine, the corresponding constraints can be written as linear inequalities, again assuming no economies of scale.

$$\text{Machine A: } 20P + 10Q + 10R \leq 2400$$
$$\text{Machine B: } 12P + 28Q + 16R \leq 2400$$
$$\text{Machine C: } 15P + 6Q + 16R \leq 2400$$
$$\text{Machine D: } 10P + 15Q + 0R \leq 2400$$

Observe that the unit for these constraints is minutes per week. Both sides of an inequality must be in the same unit.

The market limitations are written as simple upper bounds.

$$\text{Market constraints: } P \leq 100, \, Q \leq 40, \, R \leq 60$$

Logic indicates that we should also include nonnegativity restrictions on the variables.

$$\text{Nonnegativity constraints: } P \geq 0, \, Q \geq 0, \, R \geq 0$$

The Complete Model

Combining the objective function and the constraints provides the complete LP model.

$$\text{Maximize } Z = 45P + 60Q + 50R$$

$$
\begin{aligned}
\text{subject to} \quad & 20P + 10Q + 10R \le 2400 && \text{Machine A} \\
& 12P + 28Q + 16R \le 2400 && \text{Machine B} \\
& 15P + 6Q + 16R \le 2400 && \text{Machine C} \\
& 10P + 15Q + 0R \le 2400 && \text{Machine D} \\
& P \le 100, \, Q \le 40, \, R \le 60 && \text{Market constraints} \\
& P \ge 0, \, Q \ge 0, \, R \ge 0 && \text{Nonnegativity constraints}
\end{aligned}
$$

This is the mathematical statement of the model for selecting the optimal product mix. It is called a linear programming model because the functions representing the objective and each constraint are linear in form.

Obtaining the Optimal Solution

To obtain the optimal product mix, we must select values for the variables P, Q, and R that maximize the objective variable Z but do not violate any of the constraints. There are indeed an infinite number of choices of P, Q, and R that satisfy the constraints, but in this case there is only one choice that maximizes Z. Using the software that comes with the book,[1] we determine that the optimal solution is

$$P = 81.82, \, Q = 16.36, \, R = 60$$

with the corresponding objective value $Z = \$7664$. To compute the profit for the week, we reduce this value by $6000 for operating expenses and get $1664.

By setting the production quantities to the amounts specified by the solution, we find machine usage. These values are shown in Table 2.3. The capacities of machines A and B are fully used by the solution, whereas machines C and D are idle during part of the week. Only product R is manufactured at a level that matches its market limit; products P and Q are manufactured in quantities below their respective maximum market limits.

Table 2.3   Solution to the Manufacturing Example

Machine	Available usage (min)	Actual usage (min)
A	2400	2400
B	2400	2400
C	2400	2285
D	2400	1064

Product	Maximum sales	Units produced
P	100	81.82
Q	40	16.36
R	60	60

[1] The problems in this chapter were modeled using the Math Programming Excel add-in and were solved with the LP/IP Solver add-in.

In the solution, the availability constraints for machines A and B and the maximum sales constraint for product R are satisfied as equalities. Such constraints, which are called *active* or *tight*, represent the bottlenecks in the manufacturing process. They are particularly important because they limit profit for the company. If management can do something to increase the availabilities of machines A and B, or increase the market for product R, additional profit can be realized. Efforts to expand the capacities of machines C and D or the markets for products P and Q would not be worthwhile for the given model data.

Abstractions

Although we have obtained a solution to the manager's problem, it should be recognized that several assumptions and abstractions have been made to arrive at a tractable model, the most prominent of which are discussed in the subsections that follow. Some are generic for the LP model whereas others are specific to the situation at hand.

Divisibility

If the products are discrete items, the solution is not realizable because it prescribes the production of fractional amounts of products P and Q. A reasonable approximation in this case is to round the production quantities of products P and Q to the next lower integer values, yielding $P = 81$, $Q = 16$. This approach, however, does not in general result in feasible, much less optimal, solutions. The assumption of linear programming is that the variables are real values and that units are arbitrarily divisible.

The imposition of integrality requirements leads to an integer programming model that is considered in Chapter 7. For larger problems, this model is much harder to solve than the LP model which does not require integrality so we will often overlook the need for integrality in our pursuit of tractable models. When solved as an integer program, the current example yields the solution $P = 82$, $Q = 16$, $R = 60$ and the objective value $Z = \$7650$.

Proportionality

A major abstraction in LP is that the objective function and the constraints are formulated by summing individual terms that are proportional to the values of the variables. There are many instances, however, in which such formulations are not valid. There might be economies of scale that allow one to purchase raw materials at lower unit prices as volume increases, or there might be diseconomies of scale that require price discounts for products sold at higher volumes. When nonlinearities are present, they must be approximated by or converted to linear forms if LP is to be used.

Models with nonlinear objective functions can be approached directly with nonlinear programming methods, a subject that will be considered in Chapter 10. As in integer programming, however, we will find that the solution algorithms are much less effective than the algorithms for solving linear models.

Certainty

In reality, many parameters and coefficients in a model are only estimates rather than known constants. Actual values are likely to be uncertain; for example, demand estimates are affected by a variety of unknowns, machines often fail and reduce the available capacity, and processing times are only averages of perhaps highly variable quantities. The standard LP model ignores the changing dynamics and temporal fluctuations that are present in a system. All parameters and coefficients are assumed to be constant and known with certainty.

Situation-Dependent Abstractions

In our formulation of the manufacturing model, we have ignored many features that must be addressed at the operational level. Paramount among these features is the issue of scheduling. The products share several machines, and each machine can work on only one product at a time. Is there a schedule of machine operations that can accommodate the optimal product mix? We have identified machines A and B as bottlenecks. If the optimal mix is to be produced, these machines must be active 100% of the time. Finding a feasible schedule is often very difficult.

Other factors that have been neglected include inventory considerations, machine setup times, process time variability, and material movement, to name the most obvious. Nevertheless, the analyst should always bear in mind that the value of a model should not be judged by its fidelity to reality but by the insights it provides and by its usefulness in predicting system behavior.

Relevance of the Solution

There will always be abstractions in the creation of a model. Decision making at the level addressed in this book is often accompanied by factors that are unknown, uncertain, too complicated to model, or too difficult to resolve. In some cases we might handle complicating issues by creating a more comprehensive model, but always at the expense of increasing the computational burden.

Returning to the manufacturing example, it is fair to ask whether the solution has any value. Although we have made a variety of abstractions in the development of the model, it still represents an interesting aspect of the problem—namely, the allocation of the scarce resources (machine time) among competing alternatives (products). But even this problem is difficult, and most people would be hard-pressed to find an optimal solution without the help of a computer.

Given a solution in the form of a specific production plan does not mean that implementation can be achieved in a straightforward manner. The factors that were abstracted in the formulation stage will make themselves known during implementation and may produce unforeseen difficulties. For example, market estimates may be wrong, machines might fail, and a feasible schedule might be impossible to achieve.

In any case, the solution does suggest strategies that can aid in the implementation. Such strategies might include (1) making sure that product R is manufactured to meet its demand, (2) encouraging the production of product P over product Q, (3) keeping machines A and B busy, and (4) not trying to increase the efficiencies of machines C and D. Strategies often incorporate important features of the solution rather than specific numbers. We can also collect data as the week progresses and perhaps modify and re-solve the model as better data become available.

2.2 COMPUTATIONAL CONSIDERATIONS

Although all mathematical programming models appear to be similar, with an objective function to be optimized and a set of constraints circumscribing the decision variables, they are very different with respect to the computational effort required to obtain an optimal solution.

Compared with nonlinear and integer programming models, the LP model is by far the easiest to solve. Large-scale LP codes are available for most mainframes and workstations as well as for microcomputers, with only limited restrictions.

Alternatively, models with discrete variables or nonlinear functions can be very difficult to solve. Although general statements should be viewed with caution, it is fair to say that a model with no special structure and containing only a few hundred integer variables may be impossible to solve even with the latest technology. Moreover, many nonlinear programs with only a few variables may be challenging. For some integer and nonlinear problems, however, it is often possible to derive special solution procedures from the unique characteristics of the model. This may allow the solution of considerably larger formulations. Thus, although we find only a few algorithms for LP, a large number of integer and nonlinear programming algorithms are available in the scientific community, each designed to tackle a small class of problems.

One method used to solve large integer or nonlinear programs is to abstract the model by, respectively, dropping the integrality constraints or approximating nonlinear functions by linear ones and thereby obtaining a linear program. The resultant LP is called a *relaxation* of the original model.

Regarding the current state of the art in algorithms, there are two messages for the modeler. First, great economy in defining the variables and enumerating the constraints of a linear program is not often necessary. The first priority in modeling should be clarity rather than reducing the number of variables and constraints. LP codes are powerful enough to allow some excess for the sake of understanding. Second, a model with a large number of integer variables or nonlinear functions will be difficult if not impossible to solve. Integer variables and nonlinear functions are appealing because of the modeling richness they allow, but in most cases the analyst must be willing to sacrifice accuracy in favor of computational tractability.

Terminology

DECISION VARIABLES   Decision variables are represented by algebraic variables with names such as $x_1, x_2, x_3, \ldots, x_n$. The number of decision variables is n, and x_j is the name of the jth variable. In a specific situation, it is often convenient to use other names such as x_{ij}, y_k, or $z(i,j)$. In computer models, we use names such as FLOW1 or AB_5 to represent specific problem-related quantities. An assignment of values to all variables in a problem is called a *solution.*

OBJECTIVE FUNCTION   The objective evaluates some quantitative criterion of immediate importance, such as cost, profit, utility, or yield. The general linear objective function can be written as

$$z = c_1 x_1 + c_2 x_2 + \cdots + c_n x_n = \sum_{j=1}^{n} c_j x_j$$

Here, c_j is the coefficient of the jth decision variable. The criterion selected can be either maximized or minimized.

CONSTRAINTS   A constraint is an inequality or equality placing restrictions on decisions. Constraints arise from a variety of sources, such as limited resources, contractual obligations, or physical laws. In general, an LP is said to have m linear constraints that can be stated as

$$\sum_{j=1}^{n} a_{ij} x_j \begin{Bmatrix} \leq \\ = \\ \geq \end{Bmatrix} b_i, \ \ i = 1, \ldots, m$$

One of the three relations shown in the braces must be chosen for each constraint. The number a_{ij} is called a *technological coefficient,* and the number b_i is called the *right-hand-side* (RHS) value of the ith constraint. Strict inequalities (<, >, and $\neq$) are not permitted. We sometimes identify constraints written in this form as *structural constraints* to distinguish them from the nonnegativity restrictions and simple upper bounds defined below.

NONNEGATIVITY RESTRICTIONS   In most practical problems, the variables are required to be nonnegative—i.e.,

$$x_j \geq 0, \ \ j = 1, \ldots, n$$

This special kind of constraint is called a *nonnegativity restriction.* Sometimes variables are required to be nonpositive or may even be unrestricted (allowing any real value).

SIMPLE UPPER BOUND   Associated with each variable x_j may be a specified quantity called the *simple upper bound u_j.* A simple upper bound limits the value of x_j from above—i.e.,

$$x_j \leq u_j, \ \ j = 1, \ldots, n$$

When a simple upper bound is not specified for a variable, the variable is said to be *unbounded from above.*

COMPLETE LINEAR PROGRAMMING MODEL   Combining the aforementioned components into a single statement yields

$$\text{Maximize or minimize } z = \sum_{j=1}^{n} c_j x_j$$

$$\text{subject to} \quad \sum_{j=1}^{n} a_{ij} x_j \begin{Bmatrix} \leq \\ = \\ \geq \end{Bmatrix} b_i, \ \ i = 1, \ldots, m$$

$$x_j \leq u_j, \ \ j = 1, \ldots, n$$
$$x \geq 0, \ \ j = 1, \ldots, n$$

The constraints, including nonnegativity restrictions and simple upper bounds, define the *feasible region* of a problem.

PARAMETERS   The collection of coefficients (c_j, a_{ij}, b_i, u_j) for all values for the indices i and j are called the *parameters* of the model. For the model to be completely determined, all parameter values must be specified.

2.3   SOLUTION CHARACTERISTICS

Linear programming models with two decision variables can be illustrated graphically. Using a two-dimensional grid and simple examples, we now discuss a variety of solution properties that may be exhibited by more general models.

The Manufacturing Example in Two Variables

To facilitate the presentation, we simplify the manufacturing example in Section 2.1 by eliminating the variable R. Recall that the optimal solution called for as much of product R as possible, so in the following model R takes on its maximum value of 60. The revised model includes only the P and Q variables. When the machine capacities are reduced by the time required to produce 60 units of product R, we have

$$\text{Maximize } Z = 45P + 60Q$$

subject to	$20P + 10Q \leq 1800$	Machine A
	$12P + 28Q \leq 1440$	Machine B
	$15P + 6Q \leq 1440$	Machine C
	$10P + 15Q \leq 2400$	Machine D
	$P \leq 100, Q \leq 40$	Market constraints
	$P \geq 0, Q \geq 0$	Nonnegativity constraints

Because there are only two decision variables, one can plot the feasible region defined by the constraints on a graph with the axes corresponding to P and Q. Figure 2.2 shows the region that is feasible for the constraint imposed by machine A and the nonnegativity restrictions. The line $20P + 10Q = 1800$ is drawn by connecting the intercepts on the P and Q axes. The shaded region, including the line, encompasses the nonnegative points that satisfy the constraint $20P + 10Q \leq 1800$. These points are said to be *feasible* for the constraint. Points above the line violate the constraint and so are infeasible. Points outside the positive quadrant violate the nonnegativity restrictions and are also *infeasible*.

Feasible points must satisfy all the constraints of the problem. Figure 2.3 shows the line corresponding to each inequality, as well as the overall feasible region. For a particular constraint, the feasible region will be on one side of the line or the other. To determine which, simply pick a point in the plane and see whether or not it satisfies the inequality. In most cases, the easiest point to work with is the point $(0, 0)$.

The feasible region for the problem is the intersection of the feasible regions of the individual constraints—the shaded area in Figure 2.3. It is easy to see that only the constraints related to the times on machines A and B and the upper bound on production of product Q affect the feasible region. The other constraints are said to be redundant. The scale is expanded in Figure 2.4. In general, the geometric shape associated with any LP feasible region is called a *polyhedron*.

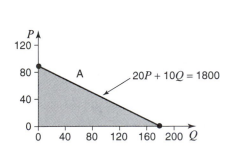

Figure 2.2   Points that are feasible for machine A

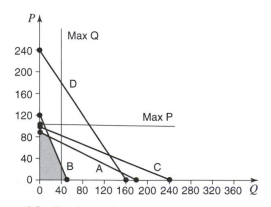

Figure 2.3   Feasible region for the revised example

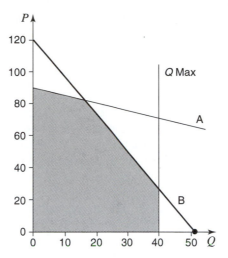

Figure 2.4   Magnified feasible region

Figure 2.5 depicts the feasible region with a plot of two objective functions for fixed values $Z = \$3600$ and $Z = \$4664$. The corresponding lines are called *isovalue contours*. Any point on an isovalue contour gives the same objective function value. Because we are working with linear functions, all isovalue contours are parallel. As can be seen in Figure 2.5, if the value of the objective is increased from 3600 to 4664, the line moves higher on the graph. Since the goal is to maximize, we want to move the line as far as possible so that it just intersects the feasible region. For this example, we can increase the objective function to \$4664. The corresponding isovalue contour is tangent to the feasible region at the point $(Q, P) = (16.36, 81.82)$, which represents the intersection of the constraint lines defined by machines A and B. This is the optimal solution.

The graphical analysis has illustrated an important theoretical result for linear programming. In particular, when an optimal solution exists, there will always be an optimal solution at an *extreme point* of the feasible region. An extreme point is a vertex of the polyhedron and is created by the intersection of two or more of the constraint lines. In general,

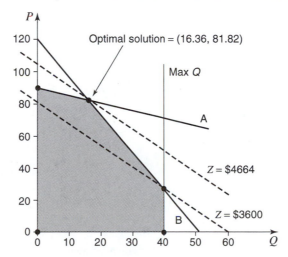

Figure 2.5   Isovalue contours and optimal solution

for an n-dimensional problem with m inequality constraints and n nonnegativity restrictions, at least n of the constraints or nonnegativity restrictions must be active or tight to create each extreme point. Five feasible extreme points can be observed in Figure 2.5.

Additional Examples

We now present several two-dimensional examples to illustrate other features that may be observed when modeling or solving LPs. Figure 2.6 shows the situation in which the optimal isovalue contour *exactly* coincides with a constraint boundary as it leaves the feasible region. In this case, there are an infinite number of *alternative optima,* each lying precisely on the constraint boundary. Two of the optimal solutions, however, are extreme points of the feasible region.

Figure 2.7 depicts a situation in which the feasible region is unbounded. This means that there is no circle with finite radius that can hold the feasible region. Although it is possible that the objective function may also be unbounded, this is not the case here. The optimal solution is finite.

Figures 2.8 and 2.9 illustrate the conditions under which the optimal solution is unbounded. In Figure 2.8, the variable x_1 is bounded; in Figure 2.9, either variable can get infinitely large. Of course, if explicit upper bounds exist on all the variables, the feasible region is necessarily bounded.

Figures 2.10 and 2.11 depict two situations in which the constraints of the problem allow no feasible solutions. In Figure 2.10, the structural constraints are inconsistent. In Figure 2.11, the structural constraints are consistent but have no feasible intersection in the positive quadrant.

Model for Figure 2.6:

Maximize $z = 3x_1 - x_2$

subject to $15x_1 - 5x_2 \leq 30$

$10x_1 + 30x_2 \leq 120$

$x_1 \geq 0,\ x_2 \geq 0$

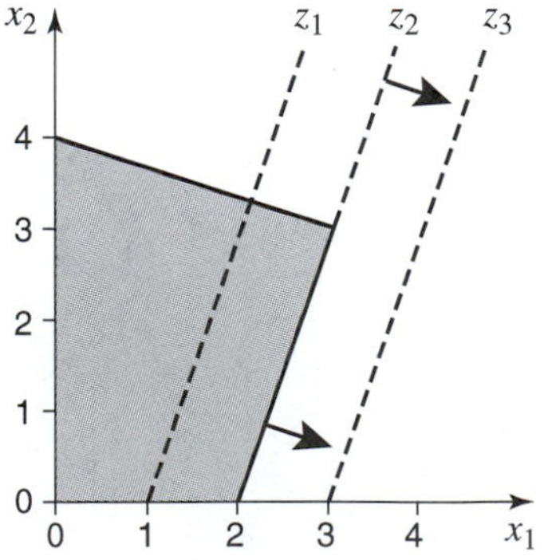

Model for Figure 2.7:

Maximize $z = -x_1 + x_2$

subject to $-x_1 + 4x_2 \leq 10$

$-3x_1 + 2x_2 \leq 2$

$x_1 \geq 0,\ x_2 \geq 0$

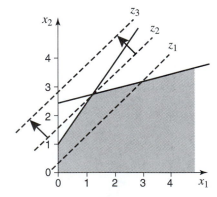

Figure 2.6 Example with alternative optimal solutions

Figure 2.7 Bounded objective function with an unbounded feasible region

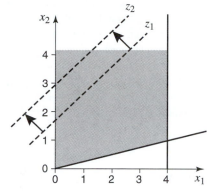

Figure 2.8 Unbounded objective function with x_1 bounded

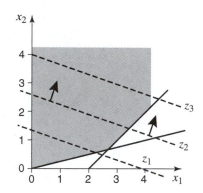

Figure 2.9 Unbounded objective function with unbounded variables

Model for Figure 2.8:

Maximize $z = -x_1 + x_2$

subject to $\quad -x_1 + 4x_2 \geq 0$

$\qquad\qquad x_1 \leq 4$

$\qquad\qquad x_1 \geq 0, x_2 \geq 0$

Model for Figure 2.9:

Maximize $z = x_1 + 3x_2$

subject to $\quad -x_1 + 4x_2 \geq 0$

$\qquad\qquad x_1 - x_2 \leq 2$

$\qquad\qquad x_1 \geq 0, x_2 \geq 0$

Should any of the conditions illustrated in Figures 2.8 through 2.11 occur, it is likely that an error has been made in the development of the model. Problems arising in practice are generally feasible and routinely bounded by resource limits. Omitted constraints, formulation errors, and typing errors account for the majority of such anomalies.

Model for Figure 2.10:

Maximize $z = x_1 + x_2$

subject to $\quad 3x_1 + x_2 \geq 6$

$\qquad\qquad 3x_1 + x_2 \leq 3$

$\qquad\qquad x_1 \geq 0, x_2 \geq 0$

Model for Figure 2.11:

Maximize $z = x_1 + x_2$

subject to $\quad x_1 - 2x_2 \geq 0$

$\qquad\qquad -x_1 + x_2 \geq 1$

$\qquad\qquad x_1 \geq 0, x_2 \geq 0$

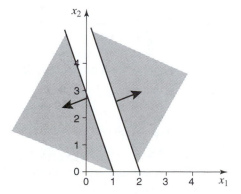

Figure 2.10 Inconsistent constraint system

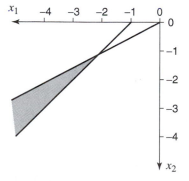

Figure 2.11 Constraint system allowing only nonpositive values of x_1 and x_2

Of course, most LP models are not as easy to visualize as the ones in this section. When more than three variables are present, it is not possible to draw the complete feasible region. The concepts and language introduced here, however, are still appropriate for describing the geometry of a problem and the characteristics of the solutions. In higher dimensions, the lines defining the boundaries of the feasible region become hyperplanes and the feasible regions become n-dimensional polyhedra. Regardless of the dimensions, optimal solutions can still be found at extreme points of the feasible region, a particular problem may have a finite or unbounded solution, and models may have no feasible regions.

Summary of Concepts

SOLUTION   An assignment of values to the decision variables is a solution to the LP model. Given a solution, the expressions describing the objective function and the constraints can be evaluated. A solution is *feasible* if all the constraints, the non-negativity restrictions, and the simple upper bounds are satisfied. If any one of these restrictions is violated, the solution is said to be *infeasible*.

OPTIMAL SOLUTION   A feasible solution that maximizes or minimizes the objective function (depending on the criterion) is called the optimal solution. The purpose of an LP algorithm is to find the optimal solution or to determine that no feasible solution exists.

ALTERNATIVE OPTIMA   If there is more than one optimal solution (solutions that yield the same value of z), the model is said to have multiple or alternative optimal solutions. Many practical problems have alternative optima.

NO FEASIBLE SOLUTION   If there is no specification of values for the decision variables that satisfies all the constraints, the problem is said to have no feasible solution. In practical problems, it is possible that the set of constraints does not allow for a feasible solution. Such a situation might result from a mistake in the problem statement or an error in data entry. Redundant equality constraints or nearly identical inequality constraints in the problem formulation may lead to a false indication that no feasible solution exists. Although a set of equalities may have a solution in theory, rounding errors inherent in computer computations may make the simultaneous satisfaction of these equalities (and sometimes inequalities) impossible.

UNBOUNDED MODEL   If there are feasible solutions for which the objective function can achieve arbitrarily large values (if maximizing) or arbitrarily small values (if minimizing), the model is said to be unbounded. When all variables are restricted to be nonnegative and have finite simple upper bounds, this condition is impossible. If no bounds are specified for some variables, the model may have an unbounded solution. However, since most decisions must take into account limitations on resources and laws of nature, such a model is probably a poor representation of the real problem.

2.4 SOLUTIONS AND SENSITIVITY ANALYSIS

In practical situations, there are three important aspects of solving problems with mathematical programming: (1) creating the model, (2) finding the solution, and (3) interpreting the solution in terms of the original problem. In this chapter, we provide examples and exercises with the hope that the student will gain modeling skills. The computer has the job of finding the solution. There are many good LP algorithms, including those provided with this book, that will find optimum solutions, so this is really not a problem for the analyst. An often neglected part of modeling instruction, but perhaps the most important part of practical application, is the task of interpreting the solution.

In this section, we discuss several solution characteristics that are helpful in interpreting results obtained by algorithms designed to solve linear programs. The mathematical details are presented in Chapter 3. Of particular interest here is sensitivity analysis. This analysis describes the consequences of changing certain parameters of the model, such as cost coefficients or RHS values, over specified ranges. From a practical viewpoint, this type of analysis is important because model parameters are normally outside the control of the decision maker and are rarely known with certainty.

Basic Solution

The first step in solving a linear program with any of the common algorithms is to rewrite the model so that all structural constraints appear as equations rather than as inequalities. Inequalities are transformed into equations with the help of nonnegative *slack variables (or slacks)* denoted by s_1 through s_4 in the following revised model.

Model with Inequalities:

Maximize $Z = 45P + 60Q$
subject to
Machine A: $20P + 10Q \leq 1800$
Machine B: $12P + 28Q \leq 1440$
Machine C: $15P + 6Q \leq 1440$
Machine D: $10P + 15Q \leq 2400$
Market constraints: $P \leq 100, Q \leq 40$
Nonnegativity restrictions: $P \geq 0, Q \geq 0$

Revised Model with Slacks and Equalities:

Maximize $Z = 45P + 60Q$
subject to
C1: $20P + 10Q + s_1 = 1800$
C2: $12P + 28Q + s_2 = 1440$
C3: $15P + 6Q + s_3 = 1440$
C4: $10P + 15Q + s_4 = 2400$
Simple upper bounds: $P \leq 100, Q \leq 40$
Nonnegativity restrictions: $P \geq 0, Q \geq 0,$
$$s_i \geq 0, i = 1, \ldots, 4$$

Both forms of the model have exactly the same feasible region with respect to the original variables. That is, for any values of P and Q that satisfy the original constraints, it is always possible to find accompanying values of $s_1, \ldots s_4$ such that the revised model is also satisfied at these values. The reverse is also true. This result follows from the fact that the slack variables are restricted to be nonnegative. By convention, we say that the number of structural constraints in the equality form of the model is m and that the number of variables, including the slacks, is n, where $n \geq m$. For this example,

$$m = 4 \text{ and } n = 6$$

Notice that we do not add slacks to the simple upper bound constraints or count them as constraints in the determination of m. If we did add slack variables to the upper bound constraints, they would be trivially determined from the values of P and Q.

For reasons that will become clear in Chapter 3, it is generally assumed that the m equality constraints are linearly independent—that is, none can be written as a linear combination of the others. The original variables, P and Q here, are called the *structural variables*, whereas the variables added to form the equalities are the slacks.

Figure 2.12 shows the feasible region for the manufacturing problem. Constraints 3 and 4 do not affect the feasible region and are not shown. The optimal solution is indicated by the large black dot. Although the slack variables do not appear explicitly in the figure, they are proportional to the perpendicular distances from the associated constraint lines. When the structural variables and all slacks are nonnegative, the solution is feasible. When

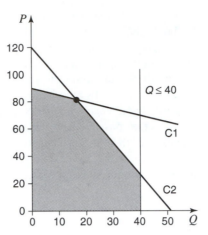

Figure 2.12   Manufacturing example revisited

a slack variable is zero, the solution must lie on the line representing the associated constraint. Table 2.4 shows the variable values at the optimal solution.

When the slack variables are zero, the corresponding constraints are said to be *tight*— restricting the solution. Alternatively, slack variables that are positive indicate constraints that are *loose*. In the example, the tight constraints are C1 and C2 with s_1 and s_2 equal to zero, whereas the remaining constraints are loose.

For every model, there exists an optimal solution with at most m variables that are not at their upper or lower bounds. These variables are termed the *basic* variables, and the corresponding solution is called a *basic* solution. (Technically speaking, a basic solution consists of exactly m variables, some of which may be at their bounds. If $k < m$ variables are between their bounds, it is always possible to find an additional $m - k$ variables, from the remaining $n - k$ variables, with which to construct a basic solution.)

As we will show later, every linear program for which an optimal solution exists has an optimal solution that is basic. This is illustrated by the solution shown in Table 2.4, which has four basic variables (P, Q, s_3, and s_4) and two *nonbasic* variables (s_1 and s_2). In general, a problem with m structural constraints will have m basic variables and $n - m$ nonbasic variables.

Sensitivity Analysis for Decision Variables

Two kinds of sensitivity analyses are reported by most linear programming codes. The first concerns the structural variables, and the second concerns the constraints. We illustrate variable sensitivity results in Table 2.5.

Table 2.4   Variable Values for the Optimal Solution

Structural variables	Slack variables	Objective value
$P = 81.82$	$s_1 = 0$	$Z = 4363.6$
$Q = 16.36$	$s_2 = 0$	
	$s_3 = 114.5$	
	$s_4 = 1336.4$	

Table 2.5   Variable Analysis for the Manufacturing Example

Number	Name	Value	Status	Reduced cost	Objective coefficient	Range lower limit	Range upper limit
1	P	81.82	Basic	0	45	25.71	120
2	Q	16.36	Basic	0	60	22.50	105

Value, Status, and Reduced Costs

The sensitivity analysis has relevance only at the optimal solution. The variable values at optimality appear in the third column of Table 2.5.

A linear programming solution has as many basic variables as the number of structural constraints. For this problem, with four constraints and six variables, we expect to have four basic variables. The "Status" column shows that both P and Q are basic. Other possible entries in this column are "Lower" and "Upper," indicating that the corresponding variable is nonbasic with its value at its lower or upper bound, respectively.

One might ask, "If P and Q are basic, where are the remaining two basic variables?" The answer is that they are the slacks associated with the loose constraints, s_3 and s_4 in the example.

The *reduced cost* of a variable represents the marginal amount the objective function would change if we were to force that variable to enter the basis. Since P and Q are already basic, their reduced costs are both zero. When the objective is to minimize, at optimality all reduced costs of nonbasic variables at their lower bounds are nonnegative, whereas all reduced costs of nonbasic variables at their upper bounds are nonpositive.

Simple Ranging of the Objective Function Coefficients

This kind of sensitivity analysis provides the range over which an objective function coefficient—call it c_j ($j = 1, \ldots, n$)—may vary (holding all other parameter values constant) while the current solution remains optimal. The value of the objective function may change, but the values of the variables at the optimal solution do not.

This type of information is important, because the actual values of the coefficients may not be known precisely. A large range indicates that the solution is insensitive to variations in the data, so we don't have to worry about uncertainty. A small range indicates that the value may be critical in determining the true optimum.

The last three columns in Table 2.5 show the current values of the objective coefficients, the lower limit of the range, and the upper limit of the range. For example, the value used in the analysis for the contribution of product P to profit was 45. Table 2.5 indicates the surprising result that this value can range between 25.71 and 120 while the same values for P and Q remain optimal. These results were obtained with a few simple calculations described in Chapter 3, and are a natural by-product of all computer codes designed to solve linear programs. They suggest that the solution is not very sensitive to the unit profits (objective function coefficients) in the model because the ranges are relatively large. In such cases, the analyst can feel assured that the solution is accurate despite any uncertainty regarding the objective coefficients.

Changing a single objective coefficient within the ranges specified by the sensitivity analysis causes the objective function line to change slope and pivot around the optimal extreme point. The value of the objective function will vary linearly as the coefficient changes. This is illustrated in Figure 2.13 for the objective coefficient c_2, the unit profit for product Q. When $c_2 = 22.5$, the smallest value in the range, the isovalue contour $Z = 45P + 22.5Q$, becomes parallel to constraint C1. The solution $(Q, P) = (16.36, 81.82)$ remains optimal, but now all the points on the line C1 between $(0, 90)$ and $(16.36, 81.82)$ are also optimal. Thus, at the limit of the range, we have alternative optima. If we decrease the unit profit for product Q below 22.5, the extreme point $(Q, P) = (0, 90)$ becomes the unique optimum for the problem.

At the other extreme, the objective function is

$$Z = 45P + 105Q$$

The corresponding isovalue contour is now coincident with constraint C2, so all points on C2 between $(16.36, 81.82)$ and $(40, 26.67)$ are alternative optima.

Sensitivity Analysis for the Constraints

The sensitivity of the solution to variations in the right-hand-side vector **b** are described in Table 2.6.

Value and Status

The "Value" column in Table 2.6 indicates the value of the left-hand side of each structural constraint at the optimal solution. In the example, each constraint measures the number of hours of machine usage during the week.

The "Status" column classifies the constraint into one of four categories. "Basic" means that the constraint is loose and the corresponding slack variable is a basic variable. As expected, two of the constraints have this characteristic, with slack variables s_3 and s_4 being basic.

For some LP models, a particular constraint may have two limits—a lower limit on value and an upper limit on value. When the constraint has a status of "Upper," the upper limit is tight. This is the case for this problem, in which the times for machines A and B are restricting the solution. A status of "Lower" means that the lower limit is tight. This is

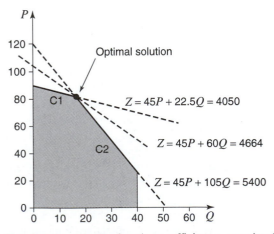

Figure 2.13   Effect of varying the objective function coefficient c_2 associated with product Q

Table 2.6   Constraint Analysis for the Manufacturing Example

Number	Name	Value	Status	Shadow price	Constraint limit	Range lower limit	Range upper limit
1	Machine A	1800	Upper	1.23	1800	933	1945
2	Machine B	1440	Upper	1.70	1440	1080	1960
3	Machine C	1325	Basic	0	1440	1325	—
4	Machine D	1063	Basic	0	2400	1063	—

not possible for the current example, which has only upper limits. The fourth value for status, "Equality" is used for equality constraints.

Dual Variables

Associated with each constraint is a *dual variable,* or *shadow price* as it is called in the economics literature. This information is presented in the "Shadow price" column in Table 2.6. An important theoretical result states that the dual variable gives the marginal rate of change of the objective function value with respect to the RHS parameter of the constraint (while all other problem parameters are held fixed at their original values). For constraint i, let π_i be the corresponding dual variable and let Δb_i be a small change (positive for increase or negative for decrease) in the RHS constant b_i. Accordingly, Δb_i produces a change in the objective function $\Delta z = \pi_i \Delta b_i$.

The dual variables for loose constraints are always zero, since changing the RHS of a loose constraint by a small amount does not affect the intersecting boundaries that determine the optimal solution. For our example, we see that increasing the time availability of machine A has the marginal value of $1.23 per minute, which means that for each additional minute that machine A is available the company increases its profit by $1.23. That is, increasing b_1 from 1800 minutes to 1801 minutes produces an increase in profit of $1.23. The marginal value of increasing the time for machine B is $1.70 per minute, a greater benefit than that for machine A. For nonbinding constraints 3 and 4, the shadow price is zero, so there is no gain (or loss) in increasing (or decreasing) b_3 or b_4.

Because this problem is linear, there is some range of the RHS parameter over which the rate of change remains constant. Ranging information is an output of sensitivity analysis, which will be discussed in the next subsection.

Simple Ranging of the Right-Hand-Side Parameter

We now describe the effect of changing one RHS parameter, b_i ($i = 1, \ldots, m$), while keeping all others constant. Although it might be of interest to consider changing several parameters at once, the accompanying analysis is much more cumbersome and not easily summarized. The goal here is to find the range over which an RHS constant can be varied without affecting some characteristic of the optimal solution. The last three columns in Table 2.6 display this ranging information.

The "Constraint limit" column gives the current RHS value. The lower and upper limits of the desired range are shown in the next two columns. The dash indicate a limit that is unbounded in the associated direction. Altering any one of the four parameters in either direction from its current value will induce a change in the value of one or more of the basic variables. (The nonbasic variables are not affected and remain at their lower or upper bounds). As long as the value of the parameter stays within its specified range, the set of basic variables will remain the same. Once this range is exceeded, the optimal set of basic variables will change and the results in Table 2.6 will no longer be valid.

For the loose constraints (i.e., the time limits on machines C and D), the RHS parameter may be increased or decreased over its specified range without inducing a change in the value of any of the basic variables except the slack variable associated with the loose constraint. Consider, for example, constraint C3 with RHS parameter $b_3 = 1440$. In Table 2.6, we see that the time available on machine C can be reduced to 1325 without affecting the solution[2] except that the slack variable s_3 will decrease to zero as b_3 approaches the limit. Any further reduction in b_3 will cause a different set of basic variables to become optimal.

Figure 2.14 graphically shows the effect of changing the RHS parameter for constraint 3: $15P + 6Q \leq b_3$. With $b_3 = 1440$, this constraint is loose and thus does not affect the optimal solution. When b_3 is decreased to 1325, the constraint line remains parallel to its original orientation but moves down until it passes directly through the optimal solution. Further reduction makes the solution infeasible.

When a constraint is tight, changing its RHS value within the specified range does not change the identities of the basic variables; however, the values of the basic variables change in a linear fashion over the range. The objective function also changes at the rate indicated by the value of the corresponding dual variable. The range for C2 given in Table 2.6 shows that the number of hours available on machine B can be varied between 1080 and 1960 while the values of P, Q, s_3, and s_4 remain basic. Changing the number of hours to a value outside this range will lead to a different set of basic variables at optimality.

Figure 2.15 shows what happens as the RHS parameter for C2—call it b_2—is reduced to 1080. The optimal solution remains at the intersection of constraints C1 and C2, but as b_2 is reduced, the line representing C2 moves parallel to its original orientation, causing the optimal solution to move also. When b_2 reaches 1080, the variable Q has gone to zero and the basic variables must change in order for the solution to remain feasible. In this case, Q would leave the basis and s_1 would take its place—that is, C1 would no longer be tight.

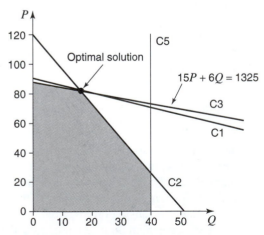

Figure 2.14   Effect of changing the right-hand-side constant of a loose constraint

---

[2] The numbers in the limit columns have been rounded to integers. The value to two decimal places is 1325.45.

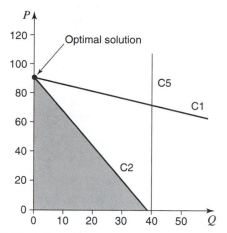

Figure 2.15   Effect of decreasing the right-hand-side constant of C2

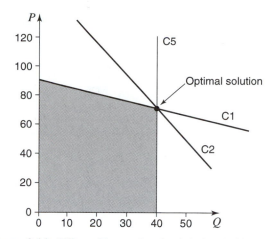

Figure 2.16   Effect of increasing the right-hand-side constant of C2

Figure 2.16 shows the effect of increasing b_2. Again the solution remains at the intersection of C1 and C2 as b_2 is increased. When b_2 reaches a value of 1960, the basis must change in order for the solution to remain feasible. What would happen is that C2 would become loose and Q would reach its upper bound, implying that s_2 would enter the basis and Q would leave the basis. Now Q would be a nonbasic variable at its upper bound.

2.5   PROBLEM CLASSES

Operations research is characterized by many problem classes. At the most general level, we identify problems as essentially deterministic or stochastic, although most situations contain both deterministic and stochastic elements. One way to learn how to model is to learn the modeling techniques associated with the various classes, and when a new situation arises try to place it into one of these molds—at least initially. In this section we introduce several problem classes and show how they fit the LP model using representative examples. Our aim is to illustrate the surprising variety of problems that can be modeled as LPs. Of course, the list is not exhaustive; other LP models can be found throughout the book. Moreover, the integer programming models discussed in Chapter 7 provide an extension of the discussion in this chapter because the integrality requirement is sometimes dropped to reduce the computational burden of finding an optimal solution when a nearly optimal one will do.

Resource Allocation Problem

The type of problem most often identified with the application of LP is the problem of distributing scarce resources among alternative activities. In this example, similar to the one in Section 2.1, we consider a manufacturing facility that produces five different products using four machines. The scarce resources are the times available on the machines, and the alternative activities are the individual production volumes. The machine requirements in hours per unit are shown for each product in Table 2.7. These are the technological coefficients a_{ij}. With the exception of product 4, which does not require machine 1, each product must pass through all four machines. The unit profits c_j are also shown in the table.

Table 2.7 Machine Data and Processing Requirements (Hours Per Unit)

Machine	Quantity	Product 1	Product 2	Product 3	Product 4	Product 5
M1	4	1.2	1.3	0.7	0	0.5
M2	5	0.7	2.2	1.6	0.5	1.0
M3	3	0.9	0.7	1.3	1.0	0.8
M4	7	1.4	2.8	0.5	1.2	0.6
Unit profit ($)	—	18	25	10	12	15

The facility has four machines of type 1, five of type 2, three of type 3, and seven of type 4. Each machine operates 40 hours per week. The problem is to determine the optimum weekly production quantities for the products. The goal is to maximize total profit. In constructing a model, the first step is to define the decision variables; the next step is to write the constraints and objective function in terms of these variables and the given data. In the problem statement, phrases such as "at least," "no greater than," "equal to," "less than or equal to," and "must satisfy" imply one or more constraints.

Model

Variable Definitions

P_j: quantity of product j to manufacture each week, $j = 1, \ldots, 5$

Machine Availability Constraints

The number of hours available per week for each type of machine is 40 times the number of machines of that type. All the constraints are in hours. For machine 1, for example, we have four machines at 40 hours per machine for a total of 160 hours. In writing out the constraints, it is customary to provide a column in the model for each variable. Each coefficient a_{ij} represents the number of hours required by product j on machine i.

$$\text{M1: } 1.2P_1 + 1.3P_2 + 0.7P_3 \qquad\quad + 0.5P_5 \leq 160$$
$$\text{M2: } 0.7P_1 + 2.2P_2 + 1.6P_3 + 0.5P_4 + 1.0P_5 \leq 200$$
$$\text{M3: } 0.9P_1 + 0.7P_2 + 1.3P_3 + 1.0P_4 + 0.8P_5 \leq 120$$
$$\text{M4: } 1.4P_1 + 2.8P_2 + 0.5P_3 + 1.2P_4 + 0.6P_5 \leq 280$$

Nonnnegativity

$P_j \geq 0$ for $j = 1, \ldots, 5$

Objective Function

The unit profit coefficients c_j ($j = 1, \ldots, 5$) are given in the bottom row of Table 2.7. Assuming proportionality, the profit maximization criterion can be written as

$$\text{Maximize } Z = 18P_1 + 25P_2 + 10P_3 + 12P_4 + 15P_5$$

Solution

The optimal solution to this LP model has the objective value of 2989. The variable values and sensitivity analysis are shown in Table 2.8. The optimal values for the decision variables are listed in the "Value" column.

Table 2.8   Variable Sensitivity Analysis for the Resource Allocation Problem

Number	Name	Value	Status	Reduced cost	Objective coefficient	Range lower limit	Range upper limit
1	P_1	58.96	Basic	0.	18	13.26	24.81
2	P_2	62.63	Basic	0.	25	23.84	41.82
3	P_3	0	Lower	−13.53	10	—	23.53
4	P_4	10.58	Basic	0.	12	11.29	17.92
5	P_5	15.64	Basic	0	15	9.91	15.36

Because this problem has four constraints, there will be four basic variables selected from the five structural and four slack variables. In this example, the basic variables are all structural, and since there are five of them, no more than four will be nonzero whereas at least one will be zero—in this case, product 3 ($P_3 = 0$). If it were desirable to produce at least some of every product, it would be necessary to specify nonzero simple lower bounds for the P_j variables.

The "reduced cost" column has nonzero values only for nonbasic variables, P_3 in this case. The reduced cost indicates that for each unit of product 3 produced, the profit will decrease by $13.53. It is clearly not beneficial to produce product 3. Because of the problem's linearity, the loss of $13.53 will be realized for every unit of product 3 over some range, but that range is not indicated in Table 2.8.

The objective function sensitivity analysis shown in the last three columns indicates the ranges of the unit profit coefficients over which the solution remains optimal. The lower bound on c_4 and the upper bound on c_5 are close to the current values. This implies that the solution is sensitive to these values.

The constraint analysis presented in Table 2.9 indicates that at the optimal solution, all machine capacity is used. If one could increase the available time of one of the machines by 1 hour, the values of the dual variables (shadow prices) suggest that the choice should be machine 3. This would lead to a marginal gain in profit of almost $9. The fact that the current values of the constraint limits b_i ($i = 1, \ldots, 4$) are relatively distant from their bounds implies that the optimal basis is insensitive to minor changes in the constraint limits.

Blending Problem

Another classic problem that can be modeled as a linear program concerns blending or mixing of ingredients to obtain a product with certain properties. We illustrate this class with the problem of determining the optimal amounts of three ingredients to include in an animal

Table 2.9   Constraint Analysis for the Resource Allocation Program

Number	Name	Value	Status	Shadow price	Constraint limit	Range lower limit	Range upper limit
1	M1	160	Upper	4.82	160	99.35	173.00
2	M2	200	Upper	5.20	200	184.69	230.36
3	M3	120	Upper	8.96	120	101.13	237.5
4	M4	280	Upper	0.36	280	259.32	295.64

feed mix. The final product must satisfy several nutrient requirements. The possible ingredients, their nutrient contents (as a proportion of the ingredient), and the unit costs are shown in Table 2.10.

The mixture must meet the following restrictions.

- Calcium: at least 0.8% but not more than 1.2%
- Protein: at least 22%
- Fiber: at most 5%

The problem is to find the composition of the feed mix that satisfies these constraints while minimizing cost.

Model

Variable Definitions

L, C, S: proportions of limestone, corn, and soybean meal, respectively, in the mixture

Constraints

The constraints all represent proportions and therefore are unitless. By definition, the proportions of the ingredients must sum to 1. Observe that the proportion of calcium has both a lower limit and an upper limit.

$$\text{Minimum calcium:}\ \ 0.38L + 0.001C + 0.002S \geq 0.008$$
$$\text{Maximum calcium:}\ \ 0.38L + 0.001C + 0.002S \leq 0.012$$
$$\text{Minimum protein:}\ \ \quad\quad\quad 0.09C + 0.50S \geq 0.22$$
$$\text{Maximum fiber:}\ \ \quad\quad\quad 0.02C + 0.08S \leq 0.05$$
$$\text{Conservation:}\ \ \quad\quad L + \quad C + \quad S = 1$$

Nonnegativity

$$L,\ C,\ S \geq 0$$

Objective Function

Because each decision variable is defined as a fraction of a kilogram, the objective is to minimize the cost of providing 1 kg of feed mix.

$$\text{Minimize } Z = 10L + 30.5C + 90S$$

Solution

The solution shown in Table 2.11 results in a unit cost of about $0.50 ($0.4916) for the feed, made up of approximately two-thirds corn and one-third soybean meal. Limestone, the least expensive ingredient, constitutes roughly 3% of the mixture. Its low cost does not compensate for the fact that it contains no protein and no fiber.

Table 2.10   Nutrient Contents and Costs of Ingredients

Ingredient	Calcium	Protein	Fiber	Unit cost (cents/kg)
Limestone	0.38	0.0	0.0	10.0
Corn	0.001	0.09	0.02	30.5
Soybean meal	0.002	0.50	0.08	90.0

Table 2.11   Variable Analysis for the Blending Problem

Number	Name	Value	Status	Reduced cost	Objective coefficient	Range lower limit	Range upper limit
1	L	0.03	Basic	0	10.0	—	17.44
2	C	0.65	Basic	0	30.5	24.40	90.21
3	S	0.32	Basic	0	90.0	30.45	123.89

From the objective ranging analysis, we see that none of the unit cost coefficients is near its bounds. This suggests that the solution is not very sensitive to the current values, so minor uncertainties or perturbations should be of little consequence.

The constraint analysis presented in Table 2.12 indicates that the upper limit on calcium is binding whereas the lower limit is not. The maximum fiber specification is also not binding, and would have to be reduced to 0.0388, or 3.88%, before it would become active. The maximum calcium and minimum protein specifications determine the solution. The dual values indicate that if the minimum protein restriction were reduced by 0.01 (1%) to 21%, the unit cost of the feed would decrease by

$$145.17 \times 0.01 = 1.45 \text{ cents}$$

The conservation constraint "Conserve" has the status of "Equality." The sensitivity analysis for this constraint does not have much meaning since the product volume must be made up entirely of these components.

Car Rental Problem

A Delaware company has just set up a new manufacturing plant in Austin, Texas. Several corporate employees are planning to visit the facility during the next week, and the travel office must arrange rental cars for all of them. Table 2.13 shows the daily demand for cars.

There are several rental plans available with different prices.

1. Daily cost on Saturday or Sunday $35
2. Daily cost for weekdays $50
3. Three-day plan (three consecutive weekdays) $125
4. Weekend plan (Saturday and Sunday) $60
5. All-weekdays plan (Monday through Friday) $180
6. All-week plan (Saturday through Friday) $200

Table 2.12   Constraint Analysis for the Blending Problem

Number	Name	Value	Status	Shadow price	Constraint limit	Range lower limit	Range upper limit
1	Calcium	0.01	Upper	−19.62	0.01	0.0080	0.2137
2	Protein	0.22	Lower	145.17	0.22	0.0874	0.2963
3	Fiber	0.04	Basic	0	0.05	0.0388	—
4	Conserve	1	Equality	17.45	1	0.4693	1.2876

Table 2.13   Daily Requirements for Rental Cars

Day	Saturday	Sunday	Monday	Tuesday	Wednesday	Thursday	Friday
Cars	2	5	10	9	16	7	11

A car is used by one person at a time, but when a car is rented for more than one day it may used by different people on different days. How should the travel office arrange for car rentals to cover all requirements at minimum cost?

Variable Definitions

In the definitions for x_j and y_k the subscripts represent day numbers where Saturday is day 1, Sunday is day 2, and so on.

x_j: number of cars rented on day j, $j = 1, \ldots, 7$

y_k: number of cars rented for three consecutive weekdays, $k = 3, 4, 5$. For example, y_3 is the number of cars rented on Monday and kept until Wednesday.

w_e: number of cars rented for the weekend

w_d: number of cars rented for Monday through Friday

w: number of cars rented for entire week

Constraints

The constraints require that at least b_i cars be available on each day i. This type of restriction leads to what are termed *covering constraints*, which typically arise whenever resources must be scheduled to meet demand. Assigning aircraft to flight legs and assigning crews to aircraft are two other examples.

We have chosen to represent the model in matrix or detached coefficient form in Table 2.14 in order to convey the problem structure more clearly. The matrix gives the constraints as rows. The coefficients of each variable, listed in the respective columns, are given as the matrix entries. The relation and RHS parameter values are depicted in the last two columns. For instance, the first row corresponds to the constraint

$$x_1 + w_e + w \geq 2$$

Table 2.14   Constraint Coefficients for Car Rentals

	x_1	x_2	x_3	x_4	x_5	x_6	x_7	y_3	y_4	y_5	w_e	w_d	w		b
Saturday	1	0	0	0	0	0	0	0	0	0	1	0	1	$\geq$	2
Sunday	0	1	0	0	0	0	0	0	0	0	1	0	1	$\geq$	5
Monday	0	0	1	0	0	0	0	1	0	0	0	1	1	$\geq$	10
Tuesday	0	0	0	1	0	0	0	1	1	0	0	1	1	$\geq$	9
Wednesday	0	0	0	0	1	0	0	1	1	1	0	1	1	$\geq$	16
Thursday	0	0	0	0	0	1	0	0	1	1	0	1	1	$\geq$	7
Friday	0	0	0	0	0	0	1	0	0	1	0	1	1	$\geq$	11
Cost	35	35	50	50	50	50	50	125	125	125	60	180	200		

which states that the number of cars rented for Saturday only plus the number rented for the weekend plan plus the number rented on the all-week plan must cover the demand for the persons requiring a car on Saturday.

Objective Function

The objective function is the total cost of the rentals. Using the data given for the cost of the various plans, we have

$$\text{Minimize } z = 35x_1 + 35x_2 + 50x_3 + \cdots + 200w$$

Solution

The solution is shown in Table 2.15. A curious characteristic of the optimal solution is that it is an all-integer solution although this condition was not included in the problem statement. In general, linear programming models do not have integer solutions. To achieve this condition, it is usually necessary to perform a great deal more work. This problem, however, exhibits the *total unimodularity* property that guarantees an integral solution whenever the RHS parameter values are integral. This property is partially due to the fact that the matrix describing the constraints consists of all 0's and 1's although this is not a sufficient condition in and of itself. The particular arrangement of nonzero a_{ij} coefficients must also be considered.

Sensitivity analysis reveals some insights about rental costs. The solution is stable if these costs do not change by at least $5 in either direction. Although the weekday plan is not used by the current solution, the analysis indicates that the solution would change if the cost were reduced below $175. In particular, lowering the cost to $174 in the model would shift the solution so that two cars would be rented with that plan.

From the "Status" column in Table 2.15, we note that five of the variables are basic. Since there are seven constraints, we expect that there should be seven basic variables. We see from the constraint analysis in Table 2.16 that the first and the fourth constraints (Saturday

Table 2.15 Variable Analysis for the Car Rental Problem

Number	Name	Value	Status	Reduced cost	Objective coefficient	Range lower limit	Range upper limit
1	Sat	0	Lower	35	35	0	—
2	Sun	0	Lower	10	35	25	—
3	M	5	Basic	0	50	40	55
4	Tu	0	Lower	50	50	0	—
5	W	9	Basic	0	50	25	75
6	Th	0	Lower	25	50	25	—
7	F	4	Basic	0	50	25	75
8	MThW	0	Lower	25	125	100	—
9	TuWTh	0	Lower	50	125	75	—
10	WThF	2	Basic	0	125	115	130
11	Weekend	0	Lower	35	60	25	—
12	Weekday	0	Lower	5	180	175	—
13	Week	5	Basic	0	200	175	210

Table 2.16   Constraint Analysis for the Car Rental Problem

Number	Name	Value	Status	Shadow price	Constraint limit	Range lower limit	Range upper limit
1	Sat	5	Basic	0	2	—	5
2	Sun	5	Lower	25	5	3	7
3	M	10	Lower	50	10	5	—
4	Tu	5	Basic	0	3	—	5
5	W	16	Lower	50	16	7	107
6	Th	7	Lower	25	7	5	11
7	F	11	Lower	50	11	7	107

and Tuesday) are loose. The corresponding slacks are the remaining basic variables. The shadow prices indicate the cost of adding one more traveler each day. Of course, Saturday and Tuesday have excess capacity, so additions can be made with no cost until the excess is zero.

Aggregate Planning Problem

A company wants a high-level, aggregate production plan for the next 6 months. Projected orders for the company's product are listed in Table 2.17. Over the 6-month period, units may be produced in one month and stored in inventory to meet demand at some later time. Because of seasonal factors, the cost of production is not constant, as shown in the table.

The cost of holding an item in inventory is $4 per unit per month. Items produced and sold in the same month are not placed in inventory. The maximum number of units that can be stored is 250. The initial inventory level at the beginning of the planning horizon is 200 units; the final inventory level at the end of the planning horizon is to be 100 units. The problem is to determine the optimal amount to produce in each month so that demand is met while minimizing total production cost and inventory holding cost. Shortages are not permitted.

Model

The aggregate planning problem is interesting not only because it represents an important application of linear programming but also because it illustrates how multiperiod planning problems are approached. It also contains aspects of network flow models.

Table 2.17   Aggregate Planning Data

Month	Demand (units)	Production cost ($/unit)
1	1300	100
2	1400	105
3	1000	110
4	800	115
5	1700	110
6	1900	110

Variable Definitions

The measure for all variables is the number of product units. The index t ranges from 1 to 6.

$$P_t: \text{production level in month } t$$
$$I_t: \text{inventory level at the end of month } t$$

Parameters

The problem statement identifies the following parameters.

$$D_t: \text{demand in month } t$$
$$I_0: \text{initial inventory level}$$
$$I_6: \text{final inventory level}$$

Constraints

Conservation of flow: A basic requirement in production planning problems is that product or material must be conserved. In our case, this leads to the balance constraint for each month

$$I_{t-1} + P_t - I_t = D_t, t = 1, \ldots, 6$$

which states that the demand in month t must be met by the production in month t plus the net change in inventory. Notice that we describe six constraints by a general expression that depends on the parameter t. This is common practice for describing LP models.

Maximum inventory: This is simply an upper bound constraint on the inventory levels for each month.

$$I_t \leq 250, t = 1, \ldots, 6$$

Initial and final conditions:

$$I_0 = 200, I_6 = 100$$

Although I_0 and I_6 have constant values because of these constraints, we leave them as variables in the model. Aggregate planning models, as well as many others, are meant to be solved over and over again as time progresses and as parameters change. It is easier to treat initial and final values as constraints rather than replace the two variables by their equivalent values.

Nonnegativity

$$I_t \geq 0, P_t \geq 0 \text{ for all } t$$

Objective Function

$$\text{Minimize } Z = 100P_1 + 105P_2 + 110P_3 + 115P_4 + 110P_5 + 110P_6$$
$$+ 4I_1 + 4I_2 + 4I_3 + 4I_4 + 4I_5 + 4I_6$$

Solution

The variable analysis shown in Table 2.18 provides the solution. Although production takes place in all 6 months, it is considerably greater than required in the first 3 months because

Table 2.18   Variable Analysis for the Aggregate Scheduling Problem

Number	Name	Value	Status	Reduced cost	Objective coefficient	Range lower limit	Range upper limit
1	P_1	1350	Basic	0	100	4	101
2	P_2	1400	Basic	0	105	104	106
3	P_3	1000	Basic	0	110	109	111
4	P_4	550	Basic	0	115	114	—
5	P_5	1700	Basic	0	110	106	119
6	P_6	2000	Basic	0	110	0.0	114
7	I_0	200	Upper	−96	4	—	100
8	I_1	250	Upper	−1	4	—	5
9	I_2	250	Upper	−1	4	—	5
10	I_3	250	Upper	−1	4	—	5
11	I_4	0	Lower	9	4	−5	—
12	I_5	0	Lower	4	4	0	—
13	I_6	100	Upper	110	0	—	110

of the prevailing price structure. For example, it is better to produce a unit in month 3 at a cost of $110 and hold it for 1 month at a cost of $4 than to produce the unit in month 4 for $115. The trade-off between production cost and inventory cost favors the former in the early months. As a consequence, inventory levels in months 1 through 3 are at their maximum levels, which implies that it is not possible to increase production in these months. Another observation that can be made is that there is no advantage in overproducing in month 4 and holding the excess units in inventory to meet the demand in month 5 or 6. The same can be said for overproducing in month 5.

Note that when this model was solved, the maximum inventory level constraints were treated as simple upper bounds and thus were not included as explicit structural constraints. As we have stated previously, it is always possible to take this approach with upper bound constraints of this type. From a computational point of view, implicit rather than explicit treatment of bounds is much more efficient because it reduces the number of constraints in the model. For the aggregate planning problem, we are left with only the six conservation-of-flow constraints, so there should be at most six nonzero basic variables in the solution. The six production variables P_1 through P_6 comprise the optimal basic set. Although the inventory variables I_1, I_2, and I_3 are nonzero, they are at their upper bounds and are not considered to be basic. The general rule, again, for a model with m constraints is that there can be at most m positive variables not at their bounds in a basic set.

From the objective coefficient ranges in Table 2.18, it can be seen that the current values of the production costs are very close to one or both bounds. The optimal solution is particularly sensitive to changes. A decrease in the cost for P_2 of only $2 per unit, for example, would induce a change in the optimal solution.

Because the conservation-of-flow constraints are written as equalities, the model has no slack variables. The shadow prices in Table 2.19 indicate that adding one unit of demand in any period will increase the overall cost by the production cost in that period. At first glance, it seems that this would seem always be true—increasing demand by one unit means that we must incur the cost of producing one additional unit. However, that unit need not be produced in the period in which the demand is increased. Although the pricing structure

Table 2.19   Constraint Analysis for the Aggregate Scheduling Problem

Number	Name	Value	Status	Shadow price	Constraint limit	Range lower limit	Range upper limit
1	M1	1300	Equality	100	1300	−50	—
2	M2	1400	Equality	105	1400	0	—
3	M3	1000	Equality	110	1000	0	—
4	M4	800	Equality	115	800	250	—
5	M5	1700	Equality	110	1700	0	—
6	M6	1900	Equality	110	1900	−100	—

here ensures that this will be the case, it may be optimal, with different inventory and production costs, to produce the unit in a prior period and hold it in inventory until needed.

Power Distribution Problem

Consider a regional power system with three generating stations A, B, and C each serving its own local area. Three outlying areas X, Y, and Z are also served by the system and have demands of 25 MW (megawatts), 50 MW, and 30 MW, respectively, which must be satisfied. The maximum generating capacity beyond local requirements and the costs of power generation at the three stations are shown in Table 2.20.

Power can be transmitted directly from each generating station to a subset of the outlying areas but with a 10% loss in each case. There are lines from stations A and C to area X, from stations B and C to area Y, and from stations A and B to area Z. Power can also be transmitted between any two generating stations, but 5% of the transmitted power is lost. Our goal is to construct an LP model to find the optimal plan for power distribution between generating stations and to outlying areas.

This type of problem can be represented by a special class of linear programming models known as *generalized network flow* models. Their special structure permits the development of highly efficient solution algorithms, and so they are often studied separately. Chapter 5 provides a detailed discussion of these models. Another advantage associated with network models is that they can be described graphically. Whereas a series of algebraic expressions is commonly used to represent a general LP model, a diagram can be used to represent the important elements of a network problem. Although a particular problem might not fully admit a network structure, quite often many of its major components do.

Table 2.20   Data for Power Distribution Problem

Generating station	Capacity (MW)	Unit cost of generation ($/MW)
A	100	500
B	75	475
C	100	400

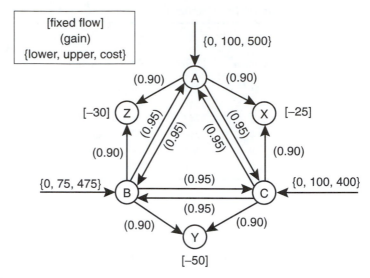

Figure 2.17   Network flow model of power distribution problem

Model

The network model for this problem consists of a set of nodes joined by a set of directed arcs, as shown in Figure 2.17. The numbers in square brackets adjacent to the nodes represent either flow entering the network from an external source or flow leaving the network en route to an external destination. Positive numbers indicate incoming flows, whereas negative numbers denote outgoing flows. For this problem, all such flows correspond to demand at the outlying areas X, Y, and Z and thus are negative.

Power enters at the generating stations A, B, and C. Incident to each of these nodes is an arc originating outside the network with three numbers in braces adjacent to it. The first number is the flow that *must* enter at the node, the second number is an upper bound on the flow that *may* enter, and the third number is the unit cost of the entering flow. In our example, the entering flow corresponds to power generated. At node A, this flow must lie between 0 and 100 MW and costs \$500 per MW to generate.

Power leaves the network at nodes X, Y, and Z. The numbers in square brackets in Figure 2.17 indicate the flow that is to be withdrawn. The numbers in parentheses on some of the arcs are called *gains*. A gain factor multiplies the flow at the beginning of the arc to obtain the flow that leaves the arc, and can have any positive value. For this problem, the gain factors represent the losses of power in the lines and so will all be less than 1. A feasible solution to the network model is an assignment of flows to the arcs such that all node requirements are satisfied and flow is conserved at the nodes. The optimal solution minimizes cost.

Linear Programming Model

The measure for all variables and constraints in this problem is megawatts. The objective function is expressed in dollars.

Variable Definitions

$x_{AB}, x_{AC}, x_{BA}, x_{BC}, x_{CA}, x_{CB}$: power transmitted between stations

$y_{AX}, y_{AZ}, y_{BY}, y_{BZ}, y_{CX}, y_{CY}$: power transmitted from a station to an outlying area

P_A, P_B, P_C: power generated at the stations to meet outlying area demands

Constraints

The principal constraints in a network model require that flow be conserved at each node. In general, flow can enter or leave a node on one or more arcs. Moreover, it can enter from external sources and leave to fulfill demand at external destinations. To achieve balance, however, any flow that enters a node must equal the flow that leaves the node.

In formulating the constraints, we use positive coefficients to express flow out of a node to another node and negative coefficients to express flow into a node from another node. RHS constants are positive if they represent flow from external sources and negative if they correspond to demand. For example, the coefficient for the y_{AX} term in the station A constraint is +1 signifying flow from station A to area X, whereas the coefficient for the y_{AX} term in the area X constraint is –0.9, signifying flow to area X from station A. The area constraints for X, Y, and Z all have negative RHS values, indicating that these are demand nodes. These sign conventions are standard in the network literature.

Conservation Constraints at Stations

$$\text{Station A: } x_{AB} + x_{AC} + y_{AX} + y_{AZ} - 0.95x_{BA} - 0.95x_{CA} - P_A = 0$$
$$\text{Station B: } x_{BA} + x_{BC} + y_{BY} + y_{BZ} - 0.95x_{AB} - 0.95x_{CB} - P_B = 0$$
$$\text{Station C: } x_{CA} + x_{CB} + y_{CX} + y_{CY} - 0.95x_{AC} - 0.95x_{BC} - P_C = 0$$

Conservation Constraints at Outlying Areas

$$\text{Area X: } - 0.9y_{AX} - 0.9y_{CX} = -25$$
$$\text{Area Y: } - 0.9y_{BY} - 0.9y_{CY} = -50$$
$$\text{Area Z: } - 0.9y_{AZ} - 0.9y_{BZ} = -30$$

Maximum Generation at Stations

These constraints are simple upper bounds: $P_A \leq 100$, $P_B \leq 75$, $P_C \leq 100$.

Nonnegativity

All variables ≥ 0

Objective Function

The goal is to satisfy the power demands of the outlying areas at minimum cost. Considering generation costs only yields

$$\text{Minimize } Z = 500P_A + 475P_B + 400P_C$$

Solution

The optimal solution to the model is depicted in Figure 2.18, where the flows are given directly on the arcs. Arcs with zero flow are not shown. As can be seen, station C, the least expensive source, generates up to its limit of 100 MW. This power serves the needs of areas X and Y, with the remaining 16.66 MW being transshipped to station A. The latter quantity is then transmitted directly to area Z incurring an additional 10% loss on the way. Evidently, station A serves only its local area as a result of its high operating costs.

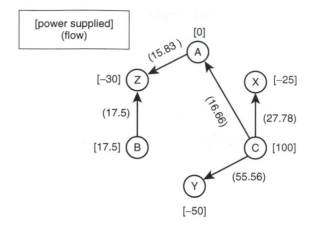

Figure 2.18 Optimal flows for power distribution problem, $Z = \$48{,}313$

Table 2.21 shows the analysis for the six equality constraints for this problem. The shadow price for an area constraint indicates how much the objective function would increase if the RHS value for that constraint were increased by 1 unit (1 MW). At first, we are surprised to see negative values as the shadow prices for the area constraints. These negative values result from the negative RHS values used in the model for these constraints. Increasing the RHS value for area X means changing its value from −25 to −24. The shadow price indicates that the objective value will go up by −501.39 (or down by 501.39). This is reasonable, because reducing the requirement by 1 should also reduce the cost of power distribution.

The range information associated with the constraint analysis also provides some useful insights. For example, the range for area X indicates that the demand at that location can vary from 8.42 to 40 MW without affecting the marginal cost of power, which will remain at $501.39.

Network flow models have the interesting characteristic that the coefficients of the conservation-of-flow equations have at most two nonzero entries for each variable. This can be seen in the matrix shown in Table 2.22, where there is one column per variable. The gain factors appear with negative signs in each column. For the *pure network problem* in which all arc gains are 1, this arrangement is a sufficient condition for total unimodularity, so the optimal solution is guaranteed to be an all-integer solution.

Table 2.21 Constraint Analysis for Power Distribution Problem

Number	Name	Value	Status	Shadow price	Constraint limit	Range lower limit	Range upper limit
1	A	0	Equality	−475	0	−15.83	17.5
2	B	0	Equality	−475	0	−57.5	17.5
3	C	0	Equality	−451.25	0	−16.67	18.42
4	X	0	Equality	−501.39	−25	−40	−8.42
5	Y	0	Equality	−501.39	−50	−65	−33.42
6	Z	0	Equality	−527.78	−30	−81.75	−14.25

Table 2.22   Constraint Coefficients for Power Generating Problem

	x_{AB}	x_{AC}	x_{BA}	x_{BC}	x_{CA}	x_{CB}	y_{AX}	y_{AZ}	y_{BY}	y_{BZ}	y_{CX}	y_{CY}
Station A	1	1	−0.95	0	−0.95	0	1	1	0	0	0	0
Station B	−0.95	0	1	1	0	−0.95	0	0	1	1	0	0
Station C	0	−0.95	0	−0.95	1	1	0	0	0	0	1	1
Area X	0	0	0	0	0	0	−0.90	0	0	0	−0.90	0
Area Y	0	0	0	0	0	0	0	0	−0.90	0	0	−0.90
Area Z	0	0	0	0	0	0	0	−0.90	0	−0.90	0	0

EXERCISES

1. Write the constraints of the LP model whose feasible region is shown in the diagram. The axis intercepts, (x_1, x_2) of the lines are

   Line 1: (0, 1) and (2, 0)
   Line 2: (0, 2.5) and (−5, 0)
   Line 3: (0, −5) and (4, 0)
   Line 4: (0, 8) and (6, 0).

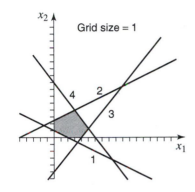

2. The figure shows the graphical model of a linear program. The large numbers on the right (1, 2, and 3) indicate the constraints. The feasible region is shown in white, and the infeasible region is shaded. The variables are restricted to nonnegative values. The small numbers (0, 1, 2, and 3) indicate four feasible corner points: 0, 1, 2, 3. Three objective functions are under consideration, as indicated by the three lines labeled **A**, **B**, and **C**. The arrows represent the directions of increasing objective function. Objective **B** is parallel to constraint **3**. In each case, specify the location of the optimal solution. If there is more than one optimal solution, characterize all of them.

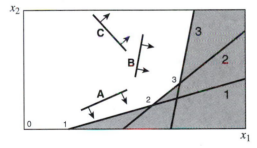

   (a) Maximize **A**
   (b) Maximize **B**
   (c) Maximize **C**
   (d) Minimize **A**
   (e) Minimize **B**
   (f) Minimize **C**
   (g) Drop $x_1 \geq 0$ and minimize **C**

3. Consider the following LP model.

$$\text{Minimize } z = 2x_1 + x_2$$

$$\begin{array}{llr}
\text{subject to} & -x_1 + 2x_2 \le 10 & (1) \\
& x_1 - 2x_2 \le 4 & (2) \\
& x_1 + x_2 \ge 8 & (3) \\
& x_1 + 2x_2 \ge 20 & (4) \\
& x_1 \ge 0, x_2 \ge 0 & (5)
\end{array}$$

(a) Graph the feasible region using a two-dimensional graph. Identify the lines on the graph with the constraint numbers assigned in the model. Show an isovalue contour for the objective function and indicate the direction of decrease. Identify the optimal solution on the graph.

(b) Graphically perform a sensitivity analysis for each of the objective function coefficients and each of the right-hand-side constants.

Exercises 4 through 10 refer to the material in Section 2.5.

4. For the resource allocation problem, show that the dual variables can be used to estimate the effects of changes in the right-hand sides of the constraints. Add one more machine of type 3 with a corresponding increase in availability of 40 hours. Solve the resulting linear program using the software that accompanies this book or any LP code available to you. Show that the objective function value obtained is the same as that predicted from the dual variables associated with the optimal solution given in the text.

5. For the blending problem, show that the ranges given in the objective function analysis suggest how the solution will change as the price of corn changes. Solve the problem for corn prices of 24, 25, 90, and 91 cents per kilogram. Compare the four solutions with the solution given in the text for a corn price 30.5 cents per kilogram.

6. Change the car rental problem to add a Monday–Wednesday–Friday plan that costs $105 for the three days and a Tuesday–Thursday plan that costs $64 for the two days. What is the new optimal rental plan? If one of these new plans is not used, from sensitivity analysis determine how much the cost of the plan should be reduced to have the plan adopted.

7. Use an approach similar to the car rental problem to solve this workforce scheduling problem. A bus company is scheduling drivers for its buses. The requirements for drivers vary by time of day, as shown in the table. These requirements repeat every day of the week.

Hours	Midnight to 4 A.M.	4 A.M. to 8 A.M.	8 A.M. to noon	Noon to 4 P.M.	4 P.M. to 8 P.M.	8 P.M. to midnight
Requirement	4	8	10	7	12	4

Drivers are hired for 8-hour shifts that start at midnight, 4 A.M., 8 A.M., noon, 4 P.M., and 8 P.M. That is, a driver starting at midnight will work until 8 A.M. Drivers that start at 8 P.M. will work until 4 A.M. the next morning. Find a schedule of drivers that will minimize the number of drivers necessary to meet the daily requirements. Note that some drivers may be idle for portions of their shifts.

8. Solve the workforce scheduling problem described in Exercise 7 for each of the given work rules. Use the same requirements as in Exercise 7. Comment on the integrality of the solution for each case.

(a) Each driver works a 12-hour shift.

(b) Drivers brought in at a particular time work for 4 hours, break for 4 hours, and then work for another 4 hours.

9. How must the model for the aggregate planning problem be modified to account for costs incurred when production levels change? In particular, if production goes up from one month to the next, there is an additional cost of $2 per unit of change. If production goes down from one month to the next, there is a cost of $1 per unit of change. The current production level is 1000 and the production level following month 6 is to be 1000. (*Hint:* Introduce variables E_t and F_t in the model to represent, respectively, the increase or decrease in production in month t.)

10. For the power distribution problem, assume that a generator failure occurs at station B so that it can no longer serve any outlying areas and falls 25 MW short of serving its own area. That is, the demand at B becomes 25 MW. Set up and solve the LP model to find the optimal distribution of power in this circumstance.

11. Consider the following linear program.

$$\text{Maximize } z = 6x_1 + 3x_2$$

$$\text{subject to} \quad x_1 + 2x_2 \geq 10$$

$$2x_1 + x_2 \leq 20$$

$$x_1 - x_2 \leq 10$$

$$-x_1 + x_2 \leq 3$$

Sketch the feasible region and several isovalue contours for the objective function in the (x_1, x_2) space. Show the optimal solution on the graph.

12. Ten jobs are to be completed by three workers during the next week. Each worker has a 40-hour work week. The times for the workers to complete the jobs are shown in the table. The values in the cells assume that each job is completed by a single worker; however, jobs can be shared, with completion times being determined proportionally. If no entry exists in a particular cell, it means that the corresponding job cannot be performed by the corresponding worker. Set up and solve an LP model that will determine the optimal assignment of workers to jobs. The goal is to minimize the total time required to complete all the jobs.

Man \ Task	1	2	3	4	5	6	7	8	9	10
A	—	7	3	—	—	18	13	6	—	9
B	12	5	—	12	4	22	—	17	13	—
C	18	—	6	8	10	—	19	—	8	15

13. A company has two manufacturing plants (A and B) and three sales outlets (I, II, and III). Shipping costs from the plants to the outlets are in $/unit and are as follows:

	Outlet		
Plant	I	II	III
A	4	6	8
B	7	4	3

The company wants to plan production, shipping, and sales for the next two periods. The manufacturing and demand data for the two periods are shown in the following tables. The plants can store products produced in one period for sale in the next. The maximum storage at each plant is 50 units and the inventory cost is $1 per unit. Find the solution that maximizes profit.

	Manufacturing Data				
	Plant A		Plant B		
Period	Unit cost ($/unit)	Capacity	Unit cost ($/unit)	Capacity	
1	8	175	7	200	
2	10	150	8	170	

	Demand Data					
	Selling price ($/unit)			Maximum sales		
Period	I	II	III	I	II	III
1	15	20	14	100	200	150
2	18	17	21	150	300	150

14. A company is planning its aggregate production schedule for the next 3 months. Units may be produced on regular time or overtime. The relevant costs and capacities are shown in the table. The demand for each month is also shown. There are three ways of meeting this demand: inventory, current production, and back-orders. Units produced in a particular month may be sold in that month or kept in inventory for sale in a later month. There is a $1 cost per unit for each month an item is held in inventory. Initially, there are 15 units in inventory. Also, sales can be back-ordered at a cost of $2 per unit per month. Back-orders represent production in future months to satisfy demand in past months, and hence incur an additional cost. Initially there are no back-orders. There should be no inventory or back-orders after month 3.

	Capacity (units)		Production cost ($/unit)		
Month	Regular time	Overtime	Regular time	Overtime	Demand
1	100	20	14	18	60
2	100	10	17	22	80
3	60	20	17	22	140

(a) Develop a model that when solved will yield the optimal production plan. Use only one variable for the inventory and one variable for back-orders for each month. Find the solution with a computer program.

(b) How would the model change if the inventory cost depended on the total time an item was stored? Let the cost be $1 per unit for items kept in inventory for 1 month, $3 for items kept for 2 months, and $5 for items kept for 3 months. Assume that the initial inventory has been in storage for 1 month. You will need more than one inventory variable for each month. Solve the model given these conditions.

15. A shuttle bus system operates at a university from 6 A.M. to midnight, a period of 18 hours. To meet student demands for service, the following schedule has been determined for the number of bus drivers required during each hour of the day. Times are stated with respect to a 24-hour clock.

Time Period:	6–7	7–8	8–9	9–10	10–11	11–12	12–13	13–14	14–15
Number of drivers:	5	20	25	20	15	12	25	20	15

Time Period:	15–16	16–17	17–18	18–19	19–20	20–21	21–22	22–23	23–24
Number of drivers:	15	12	20	20	18	15	10	5	3

During an 8-hour day, each driver receives two 1-hour breaks. The 8-hour period is divided between work and breaks, as shown in the following table.

Eight-Hour Work Schedule							
1	2	3	4	5	6	7	8
Work	Work	Break	Work	Work	Break	Work	Work

The company would like to determine how many drivers to call in at each hour of the day. Drivers begin their shifts on the hour (at 6, 7, 8, . . . , 16). No drivers are called after hour 16. There must be at least enough drivers scheduled to cover the hourly requirements under the condition that each follows the 8-hour pattern given in the table. The goal is to minimize the total number of drivers used. Formulate and solve an LP model for this problem.

BIBLIOGRAPHY

Brooke, A., D. Kendrik, and A. Meeraus, *GAMS: A User's Guide*, GAMS Development Corp., Washington, DC, 2000.

Dantzig, G.B., "The Diet Problem," *Interfaces*, Vol. 20, No. 4, pp. 43–47, 1990.

Dantzig, G.B. and M.N. Tapia, *Linear Programming: Introduction*, Springer-Verlag, New York, 1997.

Epstein, F., R. Morales, J. Seron, and A. Weintraub, "Use of OR Systems in the Chilean Forest Industry," *Interfaces*, Vol. 29, No. 1, pp. 7–29, 1999.

Fourer, R., D.M. Gay, and B.W. Kernighan, *AMPL: A Modeling Language for Mathematical Programming*, Scientific Press, South San Francisco, 1993.

Gass, S.I., *Linear Programming: Methods and Applications*, Fourth Edition, McGraw-Hill, New York, 1975.

Glassy, C.R., "Dynamic LP's for Production Scheduling," *Operations Research*, Vol. 19, pp. 45–56, 1971.

Hesse, R., *Management Science and Operations Research*, McGraw-Hill, New York, 1996.

Hillier, F.S., M.S. Hillier, and G.J. Lieberman, *Introduction to Management Science: A Model & Case Studies Approach*, McGraw-Hill, New York, 2000.

Infanger, G., *Planning under Uncertainty: Solving Large-Scale Stochastic Linear Programs*, Boyd and Fraser, Danvers, MA, 1994.

Johnson, R.B., A.J. Svoboda, C. Greif, A. Vojdani, and F. Zhuang, "Positioning for a Competitive Electric Industry with PG&E's Hydro-Thermal Optimization Model," *Interfaces*, Vol. 28, No. 1, pp. 53–74, 1998.

Liebman, J.S., L. Lasdon, L. Shrage, and A. Waren, *Modeling and Optimization with GINO*, Boyd and Fraser, Danvers, MA, 1986.

McCarl, B.A., "Repairing Misbehaving Mathematical Programming Models: Concepts and a GAMS-Based Approach," *Interfaces*, Vol. 28, No. 5, pp. 124–138, 1998.

Chapter **3**

Linear Programming Methods

The development of the simplex algorithm for solving linear programs in the 1940s was undoubtedly one of the epic occurrences in operations research. Subsequent decades have brought great improvement in computer implementations to the point where today's codes are easy to use and capable of solving the most challenging problems.

Research continues on a variety of different fronts. Whereas the simplex method searches for the optimal solution at the boundary of the feasible region, the latest developments are based on the idea of following a path through the interior of the feasible region until the optimum is reached. These so-called *interior point methods* are of great theoretical importance because they provide a bound on the computational effort required to solve a problem that is a polynomial function of its size. No polynomial bound is available for the simplex algorithm. Nevertheless, simplex codes have proven to be highly efficient in practice and remain at the center of virtually all commercial optimization packages. In our view, it is not possible to study modern optimization techniques without studying the simplex method.

We begin this chapter by presenting the tableau form of the algorithm. This approach emphasizes the algebraic nature of the model; however, it is computationally inefficient, serving primarily to illustrate underlying concepts, to allow the solution of small problems by hand, and to prepare the student for the more rigorous treatment that follows. Beginning with Section 3.10, the focus shifts to the matrix form of the model, in which linear algebra is used to find solutions. An introduction to interior point methods is presented in Chapter 4.

3.1 STANDARD FORM OF THE MODEL

Solution algorithms for the linear program (LP) require that a problem instance be put into equality form in which all structural constraints are written as equalities and all variables are restricted to be nonnegative with zero lower bounds. We will see presently that it is easy to write any LP in this manner with a few minor adjustments and additions to the original formulation. The standard form of the model, sometimes referred to as the *canonical* form, is written as

$$\text{Maximize} \quad z = \sum_{j=1}^{n} c_j x_j \tag{1}$$

$$\text{subject to} \quad \sum_{j=1}^{n} a_{ij} x_j = b_i, \quad i = 1, \ldots, m \tag{2}$$

$$x_j \geq 0, \qquad j = 1, \ldots, n \tag{3}$$

where all the right-hand-side parameters b_i are nonnegative and $n > m$. We often use the n-dimensional vector $\mathbf{x}$ to represent the decision variables—i.e., $\mathbf{x} = (x_1, \ldots, x_n)$. It is assumed that this vector contains the original problem variables as well as any slack variables added to inequalities of the form $a_{i1}x_1 + \cdots + a_{in}x_n \leq b_i$ or $a_{i1}x_1 + \cdots + a_{in}x_n \geq b_i$ to convert them to equations. Also, some variables might have simple upper bounds, such as $x_j \leq u_j$. In such cases, it is assumed that a slack variable has been added to the inequality, and the resultant equation is included as one of the m constraints in Equation (2). For convenience, we sometimes use vector notation to write these equations. Letting $\mathbf{a}_i$ be an n-dimensional row vector, the ith constraint in Equation (2) can be written as $\mathbf{a}_i\mathbf{x} = b_i$.

With regard to the simplex algorithm, the reason why we wish to deal only with equality constraints is that the computations require that we identify a subset of the variables and solve for them in terms of the remaining variables. To do this, it is necessary to perform elementary row operations on the structural constraints in Equation (2). Such operations are much more cumbersome when there is a mix of equalities and inequalities. When interior point methods are used, the computations are more involved, but the same principles apply.

The feasible region of the problem is comprised of constraints in the form of Equations (2) and (3), which define a polyhedron in geometric terms. Although a polyhedron contains an infinite number of points, except for the degenerate case, and may be unbounded, linear programming theory tells us that if a finite optimal solution exists, there is an optimal solution at one of the extreme points or vertices of the feasible region. To be more explicit in our terminology, we define an extreme point as follows.

Definition 1: Let P be a polyhedron in n-dimensional space, written as $P \subset \Re^n$. A vector $\mathbf{x} \in P$ is an *extreme point* of P if we cannot find two vectors $\mathbf{y}, \mathbf{z} \in P$, both different from $\mathbf{x}$, and a scalar $\lambda \in (0, 1)$, such that $\mathbf{x} = \lambda\mathbf{y} + (1 - \lambda)\mathbf{z}$.

In other words, $\mathbf{x}$ is an extreme point if it doesn't lie on the line between *any* two points in P. This concept is illustrated in Figure 3.1. The vector $\mathbf{w}$ is not an extreme point, because we can find two vectors $\mathbf{u}$ and $\mathbf{v}$ such that $\mathbf{w}$ is on the line joining them. The same claim cannot be made for $\mathbf{x}$. For our purposes, "vertex" and "extreme point" have the same meaning.

Definition 2: Let $\mathbf{a}$ be a nonzero row vector in n-dimensional space, written as $\mathbf{a} \in \Re^n$, and let b be a scalar.

   a. The set $\{\mathbf{x} \in \Re^n : \mathbf{a}\mathbf{x} = b\}$ is called a *hyperplane*.
   b. The set $\{\mathbf{x} \in \Re^n : \mathbf{a}\mathbf{x} \geq b\}$ is called a *halfspace*.

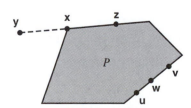

Figure 3.1   Some points in a polyhedron

Note that a hyperplane is the boundary of a corresponding halfspace. For a polyhedron in n-dimensional space, an extreme point is the intersection of $n - 1$ non-coplanar hyperplanes. For the linear programming problem defined by Equations (1) to (3), some of these hyperplanes will be of the form $x_j = 0$.

3.2 PREPARING THE MODEL

We now describe the steps that can be taken to put a linear program into standard form. In particular, the model must have the following characteristics.

- The objective must be to maximize.
- The objective function must be linear in the variables and must not contain any constant terms.
- All variables must be restricted to be nonnegative.
- Each constraint must be written as a linear equation with the variables on the left of the equal sign and a positive constant on the right.

If any of these characteristics are not present in the original model, the following simple linear transformations can be used to put the model into the required form.

Transformations

Objective Is to Minimize

Although the simplex algorithm can be readily adapted to handle minimization problems directly, most implementations optimize in only one direction. In our case, the objective is to maximize. A minimization problem can be transformed into an equivalent maximization problem by changing the signs of all terms in the objective function—that is,

$$\text{Minimize } \{c_1x_1 + c_2x_2 + \cdots + c_nx_n\} \Leftrightarrow \text{Maximize } \{-c_1x_1 - c_2x_2 - \cdots - c_nx_n\}$$

After optimization, we can recover the optimal value for the original problem by multiplying the optimal value of the new problem by -1.

A Constant Term in the Objective

Often, a model will have a constant term in the objective function. For example, when profits are being maximized, revenues might be fixed, with only the costs varying with the decision variables. A constant term may also appear when the lower bound transformation described in the next subsection is used. Because the optimal values of the decision variables are not affected by such a term in the objective, it may be dropped from the model during optimization. To get the true objective function value, the constant must be reinstated after the optimal solution is found.

Nonzero Lower Bounds on Variables

If a variable has a nonzero lower bound such as

$$x_j \geq l_j$$

the variable x_j may be replaced everywhere by the expression

$$x_j = \hat{x}_j + l_j$$

to obtain a new model in which $\hat{x}_j \geq 0$. This transformation is necessary if l_j is negative; however, most computer codes automatically make this transformation even if l_j is positive, because it eliminates a constraint.

Nonpositive Variables

If a variable x_j is restricted to be nonpositive—i.e.,

$$x_j \leq 0$$

it may be replaced everywhere by

$$x_j = -\hat{x}_j$$

and a new model obtained in which $\hat{x}_j \geq 0$.

Unrestricted Variables

An unrestricted or *free* variable is one that may assume both positive and negative values between $+\infty$ and $-\infty$. The simplex algorithm does not permit this situation, so one of two modifications is necessary to obtain a model that satisfies the inequalities in Equation (3).

The most effective way to handle unrestricted variables is to eliminate them by using one of the structural equations. With the model in equality form, any equation that includes an unrestricted variable can be solved for that variable in terms of the others. The resultant expression can then be substituted into the other constraints and the objective function, thus eliminating the variable. This will lead to a model with one fewer variable and one fewer constraint.

Sometimes it is convenient to keep unrestricted variables in the model to simplify the constraint equations or to report some quantity as part of the solution. In such cases, the most common approach is to replace the unrestricted variables with the difference between two nonnegative new variables. For x_j, a free variable, we let

$$x_j = x_j^+ - x_j^-$$

where $x_j^+ \geq 0$ and $x_j^- \geq 0$. In an optimal solution, at most one of these new variables will be positive.

In an alternative approach, only one additional variable is used no matter how many unrestricted variables are present. If x_j is unrestricted, the idea is to replace it everywhere in the model with the expression

$$x_j = x_j^+ - x^-, \text{ where } x_j^+ \geq 0 \text{ and } x^- \geq 0$$

Although x_j^+ and x^- must assume nonnegative values, the equivalent expression for x_j can be both positive and negative. When more than one variable is unrestricted, x^- can be used for all of them and has the interpretation of being the absolute value of the most negative original variable for a given solution. Assuming x_1 through x_k are unrestricted variables, $x^- = \max\{-x_1, \ldots, -x_k, 0\}$. In this approach, x^- is the constant amount by which x_j^+ exceeds x_j in any feasible solution.

Constraints Not in Correct Form

Inequality constraints are transformed into equations by adding or subtracting slack variables, as described previously. A linear term in x_j appearing on the right-hand side of an equation should be moved to the left by subtracting the term from both sides of the equation.

A constant term appearing on the left should be moved to the right by subtracting it from both sides of the equation. After these transformations are complete, if the equation has a negative constant on the right-hand side, it must be converted to a positive number by multiplying both sides of the constraint by -1.

Example

To illustrate some of these transformations, we consider the following example, whose feasible region is shown in Figure 3.2.

$$\text{Minimize } z = x + y$$
$$\text{subject to} \quad x + y \leq 20$$
$$x + y \geq -20$$
$$x - y \leq 20$$
$$x - y \geq -20$$
$$x \text{ and } y \text{ unrestricted}$$

To put the model into standard form, we first convert the four inequalities to equations by adding slack variables s_1, s_2, s_3, and s_4 to the constraints, with the appropriate signs. We then make the substitutions

$$x = x^+ - x^- \text{ and } y = y^+ - x^-$$
$$x^+ \geq 0,\ y^+ \geq 0,\ x^- \geq 0$$

At this point, the model is

$$\text{Minimize } z = x^+ + y^+ - 2x^-$$
$$\text{subject to} \quad x^+ + y^+ - 2x^- + s_1 \qquad\qquad = 20$$
$$x^+ + y^+ - 2x^- \qquad - s_2 \qquad\quad = -20$$
$$x^+ - y^+ \qquad\qquad\qquad + s_3 \quad = 20$$
$$x^+ - y^+ \qquad\qquad\qquad\qquad - s_4 = -20$$
$$x^+ \geq 0,\ y^+ \geq 0,\ x^- \geq 0,\ s_i \geq 0,\ i = 1, \ldots, 4$$

We convert the objective to maximization by multiplying its coefficients by -1. To obtain nonnegative right-hand-side constants, we multiply both sides of the second and fourth constraints by -1.

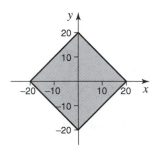

Figure 3.2 Feasible region for transformation example

$$\text{Maximize } z = -x^+ - y^+ + 2x^-$$

$$
\begin{array}{llr}
\text{Subject to} & x^+ + y^+ - 2x^- + s_1 & = 20 \\
& -x^+ - y^+ + 2x^- \quad + s_2 & = 20 \\
& x^+ - y^+ \qquad\qquad + s_3 & = 20 \\
& -x^+ + y^+ \qquad\qquad\quad + s_4 & = 20 \\
\end{array}
$$

$$x^+ \geq 0,\ y^+ \geq 0,\ x^- \geq 0,\ s_i \geq 0,\ i = 1, \ldots, 4$$

Solving the transformed problem yields the following nonzero values for the decision variables and the objective function.

$$x^- = 10,\ s_1 = 40,\ s_3 = 20,\ s_4 = 20,\ z = 20$$

In terms of the original variables, $x = -10$, $y = -10$, and $z = -20$. Note that this problem has alternative optima. Any point on the line $x + y = -20$ yields the objective value $z = -20$.

3.3   GEOMETRIC PROPERTIES OF A LINEAR PROGRAM

When solving two-dimensional problems graphically, we saw that solutions were always found at extreme points of the feasible region. These solutions are called *basic solutions*. In a problem with n variables and m constraints, a basic solution is determined by identifying m variables as basic, setting the remaining $n - m$ (nonbasic) variables equal to zero, and solving the resultant set of simultaneous equations. In order for these equations to have a unique solution and hence correspond to an extreme point, care must be taken in choosing which variables to make basic. The following discussion clarifies this point.

Linear Independence

Let us consider a system of m linear equations in m unknowns.

$$a_{11}x_1 + a_{12}x_2 + \cdots + a_{1m}x_m = b_1$$
$$a_{21}x_1 + a_{22}x_2 + \cdots + a_{2m}x_m = b_2$$
$$\vdots$$
$$a_{m1}x_1 + a_{m2}x_2 + \cdots + a_{mm}x_m = b_m$$

Using vector notation, this system can be written compactly as $\mathbf{Ax} = \mathbf{b}$, where $\mathbf{A}$ is an $m \times m$ matrix and $\mathbf{b}$ is an m-dimensional column vector. The components of $\mathbf{A}$ are denoted by a_{ij}. For this system to have a unique solution, the matrix $\mathbf{A}$ must be *invertible* or *nonsingular*— that is, there must exist another $m \times m$ matrix $\mathbf{B}$ such that $\mathbf{AB} = \mathbf{BA} = \mathbf{I}$, where $\mathbf{I}$ is the $m \times m$ identity matrix. Such a matrix $\mathbf{B}$ is called the *inverse* of $\mathbf{A}$ and is unique. It is denoted by $\mathbf{A}^{-1}$.

A concept intimately related to invertibility is linear independence.

Definition 3: Let $\mathbf{A}_1, \ldots, \mathbf{A}_k$ be a collection of k column vectors, each of dimension m. We say that these vectors are *linearly independent* if it is not possible to find k real numbers $\alpha_1, \alpha_2, \ldots, \alpha_k$ not all zero such that $\sum_{j=1}^{k} \alpha_j \mathbf{A}_j = \mathbf{0}$, where $\mathbf{0}$ is the k-dimensional null vector; otherwise, they are called *linearly dependent*.

An equivalent definition of linear independence requires that none of the vectors is a linear combination of the remaining vectors. Using linear algebra, it is then possible to prove the following results.

Theorem 1: Let $\mathbf{A}$ be a square matrix. The following statements are equivalent.

    a. The matrix $\mathbf{A}$ is invertible, as is its transpose $\mathbf{A}^\mathrm{T}$.

    b. The determinant of $\mathbf{A}$ is nonzero.

    c. The rows and columns of $\mathbf{A}$ are linearly independent.

    d. For every vector $\mathbf{b}$, the linear system $\mathbf{A}\mathbf{x} = \mathbf{b}$ has a unique solution.

Assuming that $\mathbf{A}^{-1}$ exists, a popular approach to solving the system $\mathbf{A}\mathbf{x} = \mathbf{b}$ and thus obtaining the unique solution $\mathbf{x} = \mathbf{A}^{-1}\mathbf{b}$ is Gauss–Jordan elimination. The simplex algorithm is based on this approach.

Basic Solutions

In working toward a solution of a linear program, it is convenient to start with an analysis of the system of equations $\mathbf{A}\mathbf{x} = \mathbf{b}$ given in Equation (2), where once again $\mathbf{x}$ is an n-dimensional vector, $\mathbf{b}$ is an m-dimensional vector, and $\mathbf{A}$ is an $m \times n$ matrix. Suppose that from the n columns of $\mathbf{A}$ we select a set of m linearly independent columns (such a set exists if the rank of $\mathbf{A}$ is m). For notational simplicity, assume that we select the first m columns of $\mathbf{A}$ and denote the corresponding $m \times m$ matrix by $\mathbf{B}$. The matrix $\mathbf{B} = (\mathbf{A}_1, \mathbf{A}_2, \ldots, \mathbf{A}_m)$ is then nonsingular, and we can uniquely solve the equation

$$\mathbf{B}\mathbf{x}_\mathrm{B} = \mathbf{b}$$

for the m-dimensional vector $\mathbf{x}_\mathrm{B} = (x_1, \ldots, x_m)$. By setting $\mathbf{x} = (\mathbf{x}_\mathrm{B}, \mathbf{0})$—that is, by setting the first m components of $\mathbf{x}$ equal to those of $\mathbf{x}_\mathrm{B}$ and the remaining $n - m$ components equal to zero—we obtain a solution to $\mathbf{A}\mathbf{x} = \mathbf{b}$. This leads to the following definition.

Definition 4: Given a set of m simultaneous linear equations [Equation (2)] in n unknowns, let $\mathbf{B}$ be any nonsingular $m \times m$ matrix made up of columns of $\mathbf{A}$. If all the $n - m$ components of $\mathbf{x}$ not associated with columns of $\mathbf{B}$ are set equal to zero, the solution to the resulting set of equations is said to be a *basic solution* to Equation (2) with respect to the basis $\mathbf{B}$. The components of $\mathbf{x}$ associated with columns of $\mathbf{B}$ are called *basic variables*.

In some instances, Equation (2) may have no basic solution. However, to avoid nonessential difficulties, several elementary assumptions regarding the nature of $\mathbf{A}$ will be made. These assumptions have already been mentioned: The first is that the number of x variables exceeds the number of equality constraints ($n > m$); the second is that the rows of $\mathbf{A}$ are linearly independent. A linear dependence among the rows of $\mathbf{A}$ would imply either a redundancy in the m equations that could be eliminated or contradictory constraints and hence no solution to Equation (2).

Given the assumption that $\mathbf{A}$ has full row rank, the system $\mathbf{A}\mathbf{x} = \mathbf{b}$ always has a solution and, in fact, will always have at least one basic solution; however, the basic variables in a solution are not necessarily all nonzero. This is noted in the following definition.

Definition 5: A *degenerate basic solution* is said to occur if one or more of the basic variables in a basic solution have values of zero.

Thus, in a nondegenerate basic solution, the basic variables, and hence the basis $\mathbf{B}$, can be immediately identified from the nonzero components of the solution. The same cannot be said for a degenerate basic solution, because a subset of the zero-value basic and nonbasic variables can be interchanged. This implies some ambiguity but does not cause any difficulties.

So far in this discussion, we have treated the equality constraints of the linear program only to the exclusion of the nonnegativity constraints on the variables. We now want to consider the full set of constraints given by Equations (2) and (3) for an LP in standard form.

Definition 6: A vector $\mathbf{x} \in S = \{\mathbf{x} \in \mathfrak{R}^n : \mathbf{Ax} = \mathbf{b}, \mathbf{x} \geq \mathbf{0}\}$ is said to be *feasible* to the linear programming problem in standard form; a feasible solution that is also basic is said to be a *basic feasible solution* (BFS). If this solution is degenerate, it is called a *degenerate basic feasible solution*.

Foundations

In this section, we establish the relationship between optimality and basic feasible solutions in the fundamental theorem of linear programming. The results tell us that when seeking a solution to an LP it is necessary to consider only BFSs.

Theorem 2: Given a linear program in standard form [Equations (1) to (3)], where $\mathbf{A}$ is an $m \times n$ matrix of rank m,

   a. if there is a feasible solution, there is a BFS

   b. if there is an optimal feasible solution, there is an optimal BFS.

In general, if there are m constraints and n variables, the number of basic solutions will be bounded by the number of ways to choose m basic variables (or, equivalently, the $n - m$ nonbasic variables) from the n variables—i.e.,

$$\binom{n}{m} = \frac{n!}{m!(n-m)!} \tag{4}$$

This is only an upper bound, because not all selections of the basic variables may yield a set of equations that has a solution.

Example

As an example consider the following linear program in standard form, in which slacks x_3, x_4, and x_5 have been added to the three constraints.

$$\text{Maximize } z = 2x_1 + 3x_2$$

C1:   subject to $-x_1 + x_2 + x_3 \qquad = 5$

C2: $x_1 + 3x_2 \qquad + x_4 \qquad = 35$

C3: $x_1 \qquad\qquad + x_5 = 20$

$$x_1 \geq 0, x_2 \geq 0, x_3 \geq 0, x_4 \geq 0, x_5 \geq 0$$

The equivalent of the feasible region in the (x_1, x_2) space appears in Figure 3.3. Each slack variable can be viewed as the distance from any point in the shaded polyhedron to the constraint associated with that variable.

Not including z, the model has $n = 5$ variables and $m = 3$ constraints. A basic solution is found by selecting three variables as basic, with the remaining $n - m = 2$ variables as nonbasic. For instance, let's try x_1, x_2, and x_3 as basic, and x_4 and x_5 as nonbasic. Setting the nonbasic variables equal to zero, we obtain the equations

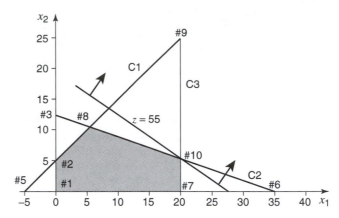

Figure 3.3   Geometry of example

$$-x_1 + x_2 + x_3 = 5$$
$$x_1 + 3x_2 = 35$$
$$x_1 = 20$$

which are easily solved to obtain $x_1 = 20$, $x_2 = 5$, and $x_3 = 20$. This is evidently a BFS, since all the variables are positive. It is identified in Figure 3.3 as extreme point #10.

Although only the structural variables are plotted in Figure 3.3, the slacks also have meaning. The example shows that those slacks that are zero identify the constraint lines that intersect at the solution. Thus, Constraints C2 and C3, associated with slack variables x_4 and x_5, are satisfied exactly by the solution. Recall that these constraints are said to be *tight* or *binding*. Those slacks that are basic and nonzero identify the constraint lines that do not intersect at the solution [Constraint C1 in the example]. Such constraints are said to be *loose* or *nonbinding*. For cases in which a structural variable is nonbasic, the nonnegativity conditions are *tight*, indicating that the structural variable is at zero.

In the example, the upper bound on the number of basic solutions is computed to be $\binom{n}{m} = \binom{5}{3} = 10$. The possible selections of three variables as basic are listed in Table 3.1. Because the boundaries of Constraint C3 and the nonnegativity restriction $x_1 \geq 0$ are parallel, only one of them can be tight at a time. Consequently, x_1 and x_5 cannot be zero simultaneously, so (x_2, x_3, x_4), denoted by #4 in the table, cannot be a basic solution. A review of the data shows that of the five BFSs, #10 produces the largest objective value and so is the optimal solution.

From the preceding analysis, it is possible to deduce the first part of the following result. The second part is not as obvious.

Theorem 3: For any linear program, there is a unique extreme point in the feasible region $S = \{\mathbf{x} \in \mathfrak{R}^n : \mathbf{Ax} = \mathbf{b}, \mathbf{x} \geq \mathbf{0}\}$ corresponding to a BFS. In addition, there is at least one BFS corresponding to each extreme point in S.

When more than one BFS maps into a specific extreme point, at least one of the binding constraints at that extreme point is redundant. It could be either one of the structural constraints or one of the nonnegativity restrictions. In either case, the BFSs would all be degenerate. This issue is taken up in Section 3.7.

Table 3.1   Basic Solutions to Example

Point	Basic variables	Nonbasic variables	Solution $(x_1, \ldots, x_5)$	Feasibility	Objective value
#1	(3, 4, 5)	(1, 2)	(0, 0, 5, 35, 20)	Yes	0
#2	(2, 4, 5)	(1, 3)	(0, 5, 0, 20, 20)	Yes	15
#3	(2, 3, 5)	(1, 4)	(0, 11.7, −6.7, 0, 20)	No	—
#4	(2, 3, 4)	(1, 5)	No solution	—	—
#5	(1, 4, 5)	(2, 3)	(−5, 0, 0, 40, 25)	No	—
#6	(1, 3, 5)	(2, 4)	(35, 0, 40, 0, −15)	No	—
#7	(1, 3, 4)	(2, 5)	(20, 0, 25, 15, 0)	Yes	40
#8	(1, 2, 5)	(3, 4)	(5, 10, 0, 0, 10)	Yes	40
#9	(1, 2, 4)	(3, 5)	(20, 25, 0, −60, 0)	No	—
#10	(1, 2, 3)	(4, 5)	(20, 5, 20, 0, 0)	Yes	55

One additional geometric concept is required before we present the details of the simplex method. Because we are interested only in examining BFSs, we need to develop a procedure for systematically transitioning from one to another. This involves the concept of adjacency.

Definition 7: For a linear program in standard form with m constraints, two basic feasible solutions are said to be *adjacent* if they have $m - 1$ basic variables in common.

For example, referring to Figure 3.3 we see that points #2 and #8 are adjacent in a geometric sense. Given that $m = 3$, these points should have $3 - 1 = 2$ basic variables in common to be adjacent according to Definition 7. Table 3.1 confirms that x_2 and x_5 are in both basic feasible solutions. Point #7, whose basic vector is (x_1, x_3, x_4), is not adjacent to either #2 or #8. Geometrically, two nondegenerate BFSs are adjacent if they both lie on the same edge of the boundary of the feasible region.

An important property of adjacency concerns optimality. For all but the smallest problem instances, it is impractical to identify and evaluate the objective function at all BFSs. For the example problem, point #10 yielded the largest objective function value and so was declared the optimal solution. The value at #10 is $z_{10} = 55$. The objective values at the two BFSs adjacent to #10 are $z_7 = 40$ and $z_8 = 40$. Considering only BFSs adjacent to #10, we have $z_{10} \geq \{z_7, z_8\}$, which is no surprise, since we have already said that #10 is the optimal solution. However, this simple observation leads to a critical result regarding optimal solutions.

Theorem 4: For a linear program in standard form [Equations (1) to (3)], let $\mathbf{x}^l$ be a BFS and let $E(l)$ be the index set of all BFSs adjacent to $\mathbf{x}^l$. Define $z_l \equiv z(\mathbf{x}^l)$. If

$$z_l \geq z_k \text{ for all } k \in E(l)$$

then $\mathbf{x}^l$ is an optimal BFS.

With this background in mind, we now outline how the simplex method finds an optimal BFS with a maximization objective.

Step 1:   Find an initial BFS to the linear program. As the algorithm progresses, the most recent BFS will be called the *incumbent.*

Step 2:   Determine if the incumbent is optimal. If not, move to an adjacent BFS that has a larger z value.

Step 3:   Return to Step 2, using the new BFS as the incumbent.

An algorithm that searches only BFSs has the benefit of finiteness. Because the number of basic solutions is bounded by the factorial expression in Equation (4), the effort of finding an optimal BFS is bounded from above by the value of that expression. Unfortunately, the value is very large for even moderate-size problems. For $m = 10$ and $n = 20$, we have

$$\binom{n}{m} = \binom{20}{10} = 184{,}756$$

potential BFSs. Nevertheless, the simplex algorithm is seen in practice to be highly efficient. Vanderbei [1996] has empirically demonstrated that in the absence of degeneracy approximately $0.5(m + n)$ BFSs have to be examined before optimality is confirmed (when degeneracy is present, this number doubles). For the preceding data, this means $0.5(10 + 20) = 15$ iterations compared with the upper bound of 184,756. The exponential growth in the computation effort implied by Equation (4) is a worst-case scenario; linear growth is the norm for real-world applications.

3.4   SIMPLEX TABLEAU

As we have seen, a linear program can always be written as a collection of equations in nonnegative variables. In this section, we describe what is called the *simplex form* of these equations. For reasons that will become apparent, equations written in this form are very useful for determining whether a particular BFS is optimal and, if not, suggesting alternative adjacent BFSs that are better. The equations can be written in the usual algebraic fashion or in an equivalent tabular format known as a *tableau*. Our first goal is to show how the tableau is constructed and interpreted.

The Simplex Form

The example introduced in the preceding section will be used for illustrative purposes. It is rewritten below in slightly modified form.

$$\text{Maximize } z$$

E0:	$z - 2x_1 - 3x_2$	$= 0$
E1:	$-x_1 + x_2 + x_3$	$= 5$
E2:	$x_1 + 3x_2 + x_4$	$= 35$
E3:	$x_1 + x_5$	$= 20$
	$x_j \geq 0, \quad j = 1, \ldots, 5$	

In addition to the three structural constraints, we have inserted a fourth "constraint" [Equation E0] that describes how the objective value z varies with the decision variables. These four equations have the characteristic that all variables appear on the left and all constants appear on the right. Except for the nonnegativity conditions, the model is now a set of four equations in six variables. This system admits an infinite number of solutions. [This is true in general as long as the system $\mathbf{Bx} = \mathbf{b}$ is consistent—that is, as long as the rank of the $m \times m$ matrix $\mathbf{B}$ is equal to the rank of the $m \times (m + 1)$ matrix $(\mathbf{B}, \mathbf{b})$.] Any solution can be found by setting two of the variables equal to arbitrary values and solving for the remaining four.

An important property of linear equations is that they can be represented in an infinite number of ways, each having the same set of solutions. Linear operations (elementary row operations) on the equations, such as multiplying one equation by any number other than zero, or replacing one equation with the sum of two others, result in an entirely equivalent model. In the simplex form, m variables are identified as basic and $n - m$ variables as nonbasic, where m is the number of equations not including Equation E0 and n is the number of variables not including z.

Conditions for the Simplex Form

- Every basic variable appears with a nonzero coefficient in one and only one equation and does not appear in Equation E0.
- In the equation in which the basic variable does appear, it has the coefficient 1.
- Each equation includes only one basic variable.
- z appears only in Equation E0 with the coefficient 1.

The preceding example happens to be in the simplex form if we identify the basic variables as x_3, x_4, and x_5.

One advantage of this form is that a solution to the model is immediately available. In particular, we simply set the nonbasic variables equal to zero and the basic variables equal to the constants on the right-hand side's of the equations. For our example, when $x_1 = x_2 = 0$, the solution $z = 0$, $x_3 = 5$, $x_4 = 35$, $x_5 = 20$ is easily determined by observation. Since the decision variables are all nonnegative, this is a BFS corresponding to the origin (point #1) in Figure 3.3.

An alternative representation of the equations with, for example, x_1 basic, can be obtained by a series of linear operations. There are several different ways to proceed. The first step is to multiply Equation E1 by -1 to get a $+1$ coefficient for x_1. This yields Equation E1′. The remaining row operations have the purpose of eliminating x_1 (obtaining 0 coefficients) in the remaining three equations. The new system is

$$
\begin{array}{llll}
(E0') = (E0) + 2(E1') & z & -5x_2 - 2x_3 & = -10 \\
(E1') = -(E1) & x_1 - x_2 - x_3 & & = -5 \\
(E2') = (E2) - (E1') & 4x_2 + x_3 + x_4 & & = 40 \\
(E3') = (E3) - (E1') & x_2 + x_3 & + x_5 & = 25
\end{array}
$$

These equations are entirely equivalent to the original system in that they have the same set of solutions. Note that

$$(x_1, x_2, x_3, x_4, x_5) = (0, 0, 5, 35, 20) \text{ and } z = 0$$

is still a solution to the new system. By design, the equations are again in the simplex form, in which we identify the basic variables as x_1, x_4, and x_5. Thus, the basic solution is

$$(x_1, x_4, x_5) = (-5, 40, 25), (x_2, x_3) = (0, 0), z = -10$$

which is evidently infeasible because x_1 is negative. The solution is shown as #5 in Figure 3.3.

General Simplex Form

During the simplex computations, we will use $\bar{a}_{ij}$ to denote the current coefficient of x_j in Equation Ei, for $i = 1, \ldots, m$ and $j = 1, \ldots, n$. We use $\bar{c}_j$ in Equation E0 to denote the current value of the jth objective function coefficient. When the problem objective is minimization, these values are known as *reduced costs* or *relative costs*. They are the unit costs relative to the current set of nonbasic variables in the reduced model in which the basic variables are fixed.

The general form of the model is

E0:
$$z + \sum_{j=1}^{n} \overline{c}_j x_j = \overline{z}$$

Ei:
$$\sum_{j=1}^{n} \overline{a}_{ij} x_j = \overline{b}_i, \quad i = 1, \ldots, m$$

The "bars" on the coefficients $\overline{a}_{ij}$, $\overline{b}_i$, and $\overline{c}_j$ indicate that these values are not in general the same as those in the original model [Equations (1) to (3)]. Because linear operations are performed on the constraints in Equation (2), different representations are realized.

When the equations are in the simplex form, we can associate a basic variable with each constraint. Let B(i) be the index of the basic variable for the ith constraint so that $x_{B(i)}$ is basic for $i = 1, \ldots, m$. Then, from the definition of the simplex form, $\overline{a}_{i,B(i)} = 1$ and all other coefficients in column B(i) are zero. This is true for Equation E0 as well, so $\overline{c}_{B(i)} = 0$. Thus, the basic variables can be viewed as slacks in the modified model.

Tableau Description of Equations

The simplex form of the linear programming model plays an important role in the algorithms to follow. To organize the computations, it is common to convert the algebraic model to a tableau, as illustrated in Table 3.2 for the example problem.

Algebraic Model

E0: $z - 2x_1 - 3x_2 = 0$
E1: $-x_1 + x_2 + x_3 \qquad = 5$
E2: $x_1 + 3x_2 \qquad + x_4 \qquad = 35$
E3: $x_1 \qquad\qquad\qquad + x_5 = 20$

The tableau shows the coefficients of the variables and the right-hand-side (RHS) constants without repeating the variable names or equality signs. The "Row" column identifies the constraints [Equation Ei], with row 0 being the objective function [Equation E0]. The "Basic" column identifies the basic variable associated with each equation. By examining the columns associated with z, x_3, x_4, and x_5, we see that the conditions for the simplex form of the model are satisfied. Although we will use the tableau in the following discussion, it is important to remember that it is simply a representation of the equations constituting the linear programming model.

Table 3.2   Tableau Form of Example Problem Equations (E0) to (E3)

Row	Basic	Coefficients						RHS
		z	x_1	x_2	x_3	x_4	x_5	
0	z	1	−2	−3	0	0	0	0
1	x_3	0	−1	1	1	0	0	5
2	x_4	0	1	3	0	1	0	35
3	x_5	0	1	0	0	0	1	20

The tableau is used to describe alternative forms of the equations. Table 3.3 gives the tableau for the updated set of Equations E0′ to E3′.

The tableau in Table 3.4 gives the general form for a problem with n variables and m constraints. It is assumed that m variables have been singled out as basic. The column of the kth basic variable is as follows:

$$
\begin{pmatrix}
\bar{c}_k \\
\bar{a}_{1k} \\
\vdots \\
\bar{a}_{k,k} \\
\bar{a}_{k+1,k} \\
\vdots \\
\bar{a}_{mk}
\end{pmatrix}
=
\begin{pmatrix}
0 \\
0 \\
\vdots \\
1 \\
0 \\
\vdots \\
0
\end{pmatrix}
\tag{5}
$$

Interpretation of the Coefficients

The simplex form, in either the algebraic or tableau representation, is useful because the coefficients of the variables describe the marginal effect of increasing a nonbasic variable from zero. Consider a single nonbasic variable x_k for some selection of the basic variables. Assume that x_k is increased from zero while all other nonbasic variables are held at zero.

Table 3.3 Tableau for Equations (E0′) to (E3′)

Row	Basic	z	x_1	x_2	x_3	x_4	x_5	RHS
				Coefficients				
0	z	1	0	−5	−2	0	0	−10
1	x_1	0	1	−1	−1	0	0	−5
2	x_4	0	0	4	1	1	0	40
3	x_5	0	0	1	1	0	1	25

Table 3.4 General Form of Tableau

Row	Basic	z	x_1	x_2	$\cdots$	x_n	RHS
				Coefficients			
0	z	1	$\bar{c}_1$	$\bar{c}_2$	$\cdots$	$\bar{c}_n$	$\bar{z}$
1	$x_{B(1)}$	0	$\bar{a}_{11}$	$\bar{a}_{12}$	$\cdots$	$\bar{a}_{1n}$	$\bar{b}_1$
.	.	.	.	.	.	.	.
.	.	.	.	.	.	.	.
.	.	.	.	.	.	.	.
m	$x_{B(m)}$	0	$\bar{a}_{m1}$	$\bar{a}_{m2}$	$\cdots$	$\bar{a}_{mn}$	$\bar{b}_m$

The basic variables must be adjusted so that the equations describing the model still hold. Written in terms of the basic variables and x_k, the equations in the simplex form are

E0: $$z + \bar{c}_k x_k = \bar{z}$$

Ei: $$x_{B(i)} + \bar{a}_{ik} x_k = \bar{b}_i, \; i = 1, \ldots, m$$

When x_k is increased by one unit, the value of z must *decrease* by $\bar{c}_k$ units (of course, when $\bar{c}_k < 0$, z will increase by $-\bar{c}_k$ units). Similarly, when x_k is increased by one unit, the value of $x_{B(i)}$ must *decrease* by $\bar{a}_{ik}$. Thus, the marginal effects of increasing x_k can be described by the following partial derivatives.

$$\frac{\partial z}{\partial x_k} = -\bar{c}_k \text{ and } \frac{\partial x_{B(i)}}{\partial x_k} = -\bar{a}_{ik} \text{ for } i = 1, \ldots, m \tag{6}$$

These results play a critical role in the simplex algorithm.

Referring to the example problem, when the basic variables are x_3, x_4, and x_5, as given in Table 3.2, the coefficients in the column for the nonbasic variable x_2 indicate the following effects when x_2 is increased by one unit: z increases by 3, x_3 decreases by 1, x_4 decreases by 3, and x_5 does not change.

3.5   INFORMATION FROM THE SIMPLEX FORM

For each set of basic variables, the equations can be written in the simplex form and placed in a tableau. These equations describe how the basic variables change with different values of the nonbasic variables. Letting Q be the set of nonbasic variables, each row in Table 3.4 can be written as $x_{B(i)} = \bar{b}_i - \sum_{j \in Q} \bar{a}_{ij} x_j$. The basic solution is found by setting the nonbasic variables equal to zero. To obtain a new basic solution, one of the nonbasic variables must be increased from zero. At each stage in the process, a number of questions regarding the current solution must be answered. If the solution is feasible, is it optimal? If the solution is feasible but not optimal, how can a better solution be obtained? How do we ensure that each solution is basic? The answers to these questions and others that may arise in the computations are easily available from the equations written in the simplex form. The answers lead to the simplex algorithm.

Feasible Region from Different Views

This section provides a graphical argument that the equations in the simplex form, in addition to providing an easy method for computing a basic solution, really describe the entire feasible region just as accurately as does the original statement of the model. The selection of a set of basic variables and the conversion of the equations for that basis to the simplex form simply provide a different view of the feasible region. A room in a house will appear differently when viewed from various locations, but no matter what the view, the room remains the same. Similarly, different selections of basic variables provide different views of the feasible region, but the region remains the same. When the number of degrees of freedom of the model is two (or at most three), this is easily visualized with a graph; for more degrees of freedom, graphs are no longer possible, but the same insights are available.

As an example, we use the following linear program along with its graphical representation shown in Figure 3.4.

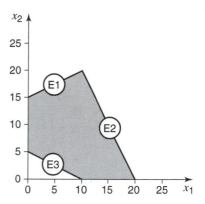

Figure 3.4   Feasible region

$$\text{Maximize } z = 3x_1 + 2x_2$$
$$\text{subject to} \quad -x_1 + 2x_2 \le 30$$
$$2x_1 + x_2 \le 40$$
$$x_1 + 2x_2 \ge 10$$
$$x_1 \ge 0, x_2 \ge 0$$

Adding slacks, we obtain the equality form of the model. After multiplying the third equation by -1, we obtain the simplex form with basic variables x_3, x_4, and x_5.

E1:	$-x_1 + 2x_2 + x_3 \qquad\qquad = 30$
E2:	$2x_1 + x_2 \qquad + x_4 \qquad = 40$
E3:	$-x_1 - 2x_2 \qquad\qquad + x_5 = -10$

These equations make up three of the boundary lines in the feasible region shown in Figure 3.4.

An alternative way to view the solution space is shown in Figure 3.5. Here, each boundary of the feasible region is numbered to correspond to the *variable* that is zero for all solutions that fall on the boundary. For example, the x_2 axis is numbered 1 because all solutions on this boundary have $x_1 = 0$. The line labeled 3 was derived from Equation E1, but now we recognize that all solutions on this line have $x_3 = 0$.

At an extreme point, two of the variables have a value of zero, indicating that two of the constraints or nonnegativity conditions of the original problem are tight. The numbered extreme points in Figure 3.5 correspond to the following solutions.

$$\mathbf{x}^1 = (10, 0, 40, 20, 0)$$
$$\mathbf{x}^2 = (0, 5, 20, 35, 0)$$
$$\mathbf{x}^3 = (0, 15, 0, 25, 20)$$
$$\mathbf{x}^4 = (10, 20, 0, 0, 40)$$
$$\mathbf{x}^5 = (20, 0, 50, 0, 10)$$

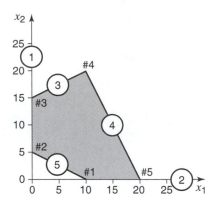

Figure 3.5   Boundaries and extreme points

Take particular note of the origin in Figure 3.5, where x_1 and x_2 are zero. This point corresponds to a basic solution with basic variables $(x_3, x_4, x_5) = (30, 40, -10)$. Because x_5 is negative, the solution is not feasible. This is evident Figure 3.5, which shows that the origin is outside the crosshatched feasible region.

The optimal solution occurs at point #4, which is obtained by setting $(x_1, x_2) = (10, 20)$. The corresponding values of the remaining variables are $(x_3, x_4, x_5) = (0, 0, 40)$. Although we know that this point is a BFS, this is not obvious when we simply look at Equations E1 to E3, because this solution has been reached by assigning nonzero values to the nonbasic variables (x_1, x_2). In fact, every solution of the set of equations, feasible or infeasible, is the result of assigning some values to (x_1, x_2).

The graphical representation of the model can be drawn with respect to any two variables. Say we choose x_2 and x_5 as the nonbasic variables and the axes of our graph. By performing linear operations on Equations E1, E2, and E3, x_2 and x_5 can be rewritten strictly in terms of x_1, x_3, and x_4.

The simplex form of the new BFS is

$$
\begin{aligned}
z \quad + 4x_2 \quad\quad\quad - 3x_5 &= 30 \\
4x_2 + x_3 \quad\quad - x_5 &= 40 \\
-3x_2 \quad\quad + x_4 + 2x_5 &= 20 \\
x_1 + 2x_2 \quad\quad - x_5 &= 10
\end{aligned}
$$

The feasible region can now be viewed as a function of x_2 and x_5, as shown in Figure 3.6. We have numbered the boundaries and extreme points to correspond to their equivalents in Figure 3.5. Both figures illustrate the same model, but from different points of view. Figure 3.5 shows the feasible region in terms of x_1 and x_2, whereas Figure 3.6 shows it in terms of x_2 and x_5.

When the nonbasic variables are zero (origin in Figure 3.6), the basic variables assume the values $(x_1, x_3, x_4) = (10, 40, 20)$. This is a BFS and corresponds to extreme point #1.

For this example, the feasible region can, in fact, be represented by the selection of any two nonbasic variables whose columns are linearly independent. Each selection provides a view of the feasible region from a different corner or extreme point. Setting the nonbasic variables to zero results in a basic solution, but that solution may not be optimal or even feasible. In each of these cases, we want to know what changes in the nonbasic variables will drive the solution toward optimality or feasibility. These issues are taken up in the remainder of this section.

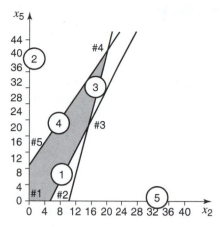

Figure 3.6 Feasible region in terms of x_2 and x_5

Table 3.5 gives the equations for the example in terms of the basic variables (x_1, x_3, x_4). In general terms, the equations are written as

E0:
$$z + \sum_{j \in Q} \overline{c}_j x_j = \overline{z}$$

Ei:
$$x_{B(i)} + \sum_{j \in Q} \overline{a}_{ij} x_j = \overline{b}_i, \quad i = 1, \dots, m$$

where Q is the set of nonbasic variables. For our example, B(1) = 3, B(2) = 4, and B(3) = 1.

What Is the Basic Solution?

Setting the nonbasic variables to zero, we see that the values of the basic variables are simply the right-hand sides of the equations. Thus, for our example, the basic solution is

$$\mathbf{x} = (10, 0, 40, 20, 0) \text{ with } z = 30$$

From the general expression of the simplex form we read the objective function value and the values of the basic variables as

$$z = \overline{z}, \ x_{B(i)} = \overline{b}_i \text{ for } i = 1, \dots, m, \text{ and } x_j = 0 \text{ for } j \in Q$$

Table 3.5 Tableau Corresponding to the Origin in Figure 3.6

Row	Basic	Coefficients						RHS
		z	x_1	x_2	x_3	x_4	x_5	
0	z	1	0	4	0	0	−3	30
1	x_3	0	0	4	1	0	−1	40
2	x_4	0	0	−3	0	1	2	20
3	x_1	0	1	2	0	0	−1	10

Is the Basic Solution Feasible?

Because the equations are obviously satisfied, the issue of determining feasibility reduces to whether or not the basic variables are nonnegative. The values of the basic variables are equal to the right-hand sides of the equations, so the solution is feasible if the RHS column is nonnegative. An examination of Table 3.5 indicates that the example solution is feasible. In general, a basic solution is feasible if

$$\overline{b}_i \geq 0 \text{ for } i = 1, \ldots, m$$

Is the Basic Feasible Solution Optimal?

This question is of concern only if the basic solution is feasible. To answer the question, we examine the effect of increasing one of the nonbasic variables from zero while holding all others at zero. This is the approach we use to investigate adjacent BFSs. Since the equations are linear, any change involving more than one nonbasic variable can be expressed as a linear combination of the individual changes, so examining changes in single variables is sufficient to determine optimality.

The objective function equation for the example problem is

E0: $z + 4x_2 - 3x_5 = 30$

For the current BFS, both x_2 and x_5 are zero, so decreasing either one causes the solution to become infeasible. We therefore restrict our attention to increases.

When one of the nonbasic variables is held at zero, only z and the other nonbasic variable play a role in Equation E0.

$$\text{For } x_2 \text{ active, } z = 30 - 4x_2$$
$$\text{For } x_5 \text{ active, } z = 30 + 3x_5$$

Increasing x_2 causes the objective function to decrease, whereas increasing x_5 causes the objective function to increase. Given that we are maximizing, a solution better than the current BFS can be obtained by increasing x_5, so we are not at the optimal solution.

As we have seen in Equation (6), the coefficient of the nonbasic variable x_j in Equation E0 is the negative of the marginal change in the objective function with respect to x_j. In the general case, if any of these coefficients is negative, the marginal change is positive and the objective value can be increased. This observation leads to a test for optimality.

Optimality conditions: $\overline{c}_j \geq 0$ for all $j \in Q$

This condition is easily tested by examining Equation E0 in the simplex form or in row 0 of the tableau.

How Can a Nonoptimal Feasible Solution Be Improved?

When the test for optimality indicates that a BFS is not optimal, a change in the solution is suggested. The current solution can be improved by increasing *any* nonbasic variable with a negative coefficient in Equation E0. For the example problem, if we move away from the origin in Figure 3.6 in the x_5 direction, we will obtain a larger objective function value.

In the general case, all nonbasic variables x_j with $\overline{c}_j < 0$ are candidates for conversion to basic variables. Some algorithms choose the variable with the most negative $\overline{c}_j$. This is called the *steepest ascent* rule. The idea is that the rate of increase will be greatest in the direction of change; however, the total gain will be determined by how much the variable can be increased before another variable becomes negative. This issue is addressed in the next subsection. For our purposes, the selection of the nonbasic variable to increase in a nonoptimal situation is an optional feature of the solution algorithm.

There is no immediate reason why two or more variables could not be increased simultaneously. However, this would mean moving into the feasible region rather than remaining on the boundary. Virtually all simplex-type algorithms prohibit this option because it complicates other computational steps.

How Much Should the Nonbasic Variable Be Changed?

Given a BFS that is not optimal, we have determined that it can be improved by increasing a nonbasic variable. How much should that variable be increased? Considering only the objective value, the variable should be increased as much as possible. Because the objective function is linear, the value of z will increase by $-\overline{c}_j$ per unit of x_j indefinitely. The question of how much to increase it is governed by the constraints.

For the example problem, when we increase x_5 along the axis, we move it either toward or away from the line defining each constraint. To illustrate this point consider Figure 3.7. Increasing x_5 causes the solution to move from the origin toward the second constraint, labeled 4 in the figure, and away from constraint 5. Our goal is to move it as far as possible while still keeping it feasible. In this case, increasing x_5 to 10 will cause the solution to lie on the line marked 4. Any further increase will cause the solution to become infeasible—that is, will force x_4 to become negative.

Once again, this fact is easily determined from the simplex form of the equations without recourse to the graph. When one nonbasic variable is changed from zero while holding all others at zero, only two terms are active in each of the constraint equations. For our example, we have

$$x_3 = 40 + x_5$$
$$x_4 = 20 - 2x_5$$
$$x_1 = 10 + x_5$$

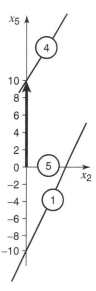

Figure 3.7   Effect of increasing x_5

These equations indicate that for each unit increase in x_5, x_1 and x_3 will increase by 1 and x_4 will decrease by 2. Increasing x_5 to 10 will cause x_4 to become zero, whereas any greater increase will drive x_4 negative. Since x_4 is the slack variable for the second constraint, changing x_5 to 10 will cause the second constraint to become tight, a fact that has already been illustrated graphically.

In general, $\overline{b}_i$ is the right-hand side of equation i, $x_{B(i)}$ is the basic variable for constraint i, and $\overline{a}_{ij}$ is the marginal decrease in $x_{B(i)}$ with respect to an increase in nonbasic variable x_j. Then the ratio $\overline{b}_i / \overline{a}_{ij}$ is the amount that x_j must increase to drive $x_{B(i)}$ to zero. When the current basic solution is feasible, the values of $\overline{b}_i$ will be nonnegative, so when $\overline{a}_{ij} < 0$, the ratio is negative and x_j must actually decrease to drive the corresponding basic variable to zero. This is illustrated by the ratios for x_1 and x_3 in the example, which are –10 and –40, respectively.

When $\overline{a}_{ij} = 0$, an increase in x_j will not cause a change in $x_{B(i)}$ (and the ratio is undefined), so row i does not restrict the value of the nonbasic variable.

When $\overline{a}_{ij} > 0$, a positive value of x_j will drive the corresponding basic variable to zero. When more than one row has this characteristic, the nonbasic variable can increase only to the value of the minimum positive ratio if feasibility is to be guaranteed. Thus the nonbasic variable may increase by

$$\theta = \text{minimum} \left\{ \frac{\overline{b}_i}{\overline{a}_{ij}} : \ \overline{a}_{ij} > 0, \ i = 1, \ldots, m \right\}$$

This is called the *minimum ratio test* and is an integral component of the simplex algorithm. The row (or equation) in which the minimum occurs is called the *pivot row* and will be denoted by r. When a tie occurs, implying that two or more rows have ratio θ, any one of them can be chosen.

The basic variable $x_{B(r)}$ is driven to zero by increasing x_j to θ. For the example problem, if we choose to increase x_5, only $\overline{a}_{25}$ will be positive, so the pivot row must be $r = 2$. The value of θ determined by the ratio for that row is 20/2 = 10.

There is a possibility that no positive value of $\overline{a}_{ij}$ exists for some nonbasic variable x_k such that $\overline{c}_k < 0$. This implies that x_k and z can be increased indefinitely without the solution becoming infeasible. When this occurs, there is no finite optimum, so the solution is said to be *unbounded*.

How Is a New Simplex Form Obtained?

When a nonbasic variable is increased by the value θ, one (or more) of the basic variables goes to zero. Now we have a situation in which a nonbasic variable is positive while a basic variable is zero. The fact that this is not the BFS corresponding to the current basic variables is easily remedied by replacing the zero basic variable by the nonzero nonbasic variable in the basic set. This process is called a *pivot* and describes how we move from one BFS to an adjacent BFS.

For the example problem, the basic set identified in Table 3.5 is x_3, x_4, and x_1. When x_5 is selected as the variable to increase, we discover that the pivot row is row 2. The maximum increase for x_5 is computed to be 10, causing the basic variable for the second row, x_4, to go to zero. To obtain a new BFS, we replace x_4 with x_5, yielding the new basic variables (x_3, x_5, x_1). The tableau in Table 3.5, however, is not in the simplex form for this BFS. We must create a new tableau with x_5 basic in row 2. In the transformed equations, the column of coefficients for x_5 must resemble column k in Equation (5), where each element is zero except for $\overline{a}_{kk}$ which is 1. The process of obtaining the new tableau is called *pivoting* or *pricing out* the entering column. It is simply a series of linear operations on the equations that results in the simplex form of the new set of basic variables. To illustrate these operations, we repeat the initial tableau for the example, and then show the pivot operations in two steps, in Table 3.6. The third tableau in Table 3.6 is in the simplex form for the basic variables (x_3, x_5, x_1). The BFS is easily seen to be

$$\mathbf{x} = (20, 0, 50, 0, 10) \text{ with } z = 60$$

As expected, the objective value has improved by $-\bar{c}_5 \times \theta = 3 \times 10 = 30$ to a value of 60. We observe from row 0 that the solution is still not optimal.

In general, the pivot operations for obtaining the new tableau can be carried out in two steps. Let the variable entering the basis be x_s and let the pivot row be r. Then proceed as follows.

1. Divide row r by the pivot element $\bar{a}_{rs}$.
2. For each row i $(i = 0, \ldots, m, i \neq r)$, multiply the new row r by $-\bar{a}_{is}$ and add it to row i.

Table 3.6   Original and Subsequent Tableaus

Row	Basic		Coefficients					RHS
		z	x_1	x_2	x_3	x_4	x_5	
0	z	1	0	4	0	0	−3	30
1	x_3	0	0	4	1	0	−1	40
2	x_4	0	0	−3	0	1	2	20
3	x_1	0	1	2	0	0	−1	10

Divide row 2 by 2 to obtain a new row 2:

Row	Basic		Coefficients					RHS
		z	x_1	x_2	x_3	x_4	x_5	
0	z	1	0	4	0	0	−3	30
1	x_3	0	0	4	1	0	−1	40
2	x_4	0	0	−1.5	0	0.5	①	10 ←
3	x_1	0	1	2	0	0	−1	10

Multiply the new row 2 by 3 and add it to row 0 to obtain a new row 0, multiply the new row 2 by 1 and add it to row 1 to obtain a new row 1, and multiply the new row 2 by 1 and add it to row 3 to obtain a new row 3:

Row	Basic		Coefficients					RHS
		z	x_1	x_2	x_3	x_4	x_5	
0	z	1	0	−0.5	0	1.5	0	60
1	x_3	0	0	2.5	1	−0.5	0	50
2	x_5	0	0	−1.5	0	0.5	1	10
3	x_1	0	1	0.5	0	−0.5	0	20

This procedure tells us how to move from one BFS to another and obtain an improved objective function value. All the information required is immediately observable from the tableau or is easily computed by the ratio test. The next section provides a complete statement of the primal simplex method that uses this information to find the optimal solution of a linear program.

3.6 SIMPLEX METHOD USING TABLEAUS

For an LP in simplex form, the objective function equation provides a direct indication of the optimality of the basic solution. If it is not optimal, the reduced costs point to one or more nonbasic variables that can be increased to improve the current solution. When one of these variables is selected, a simple calculation using the ratio test determines the amount the variable can be increased without making the solution infeasible. It also indicates which basic variable goes to zero. Finally, the basis is changed by deleting that variable from the basic set and adding the selected nonbasic variable. The equations are rewritten in the simplex form for the new basis, and the process is repeated.

This is a rough statement of the primal simplex method for linear programming. We use the term "primal" to distinguish this method from the dual simplex algorithm to be described in Section 3.9. In this section, we provide a formal description of the algorithm and a complete illustration. The algorithm assumes that a starting BFS is available. This is easily accomplished if the linear programming model has all "less than or equal to" constraints with positive right-hand sides. In Section 3.8, we describe how an initial BFS is obtained when the model is not in this form.

Primal Simplex Algorithm

Step 1:   (*Initialization*) *Express the linear program in the equation form.* Choose a subset of the variables that yields a BFS. Transform the equations into the simplex form. Create the first tableau using the coefficients from these equations. If row 0 has all nonnegative coefficients, stop with the optimal solution.

Step 2:   (*Iteration k*)

  a. *Select the variable that will enter the basis.* Find the nonbasic variable that has the most negative reduced cost in row 0 (we are using the steepest ascent rule here). Let that variable be the entering variable, and call the corresponding column of the tableau the pivot column. Let the index of the entering variable be s.

  b. *Find the variable that must leave the basis.* Compute the ratios between the right-hand-side $\bar{b}_i$ and the positive elements of the pivot column $\bar{a}_{is} > 0$. Find the row with the minimum ratio and call it the pivot row. Let the index of the pivot row be r.

  c. *Change the basis.* Let x_s become the basic variable for row r, and call $\bar{a}_{rs}$ the pivot element. Let $\bar{\mathbf{a}}_i$ be the vector of coefficients for the ith row in the current tableau with right-hand-side $\bar{b}_i$. Let $\bar{\mathbf{a}}_i^{\text{new}}$ be the vector of coefficients for the ith row in the new tableau with right-hand-side $\bar{b}_i^{\text{new}}$. Create a new tableau by performing the following operations:

  For row r (divide the pivot row by the pivot element)

  $$\bar{\mathbf{a}}_r^{\text{new}} = \bar{\mathbf{a}}_r \, / \, \bar{a}_{rs} \ \text{ and } \ \bar{b}_r^{\text{new}} = \bar{b}_r \, / \, \bar{a}_{rs}$$

For row $i = 0,1, \ldots, m$ and $i \neq r$ (multiply the new pivot row by the negative of the ith coefficient in the pivot column and add the result to row i),

$$\overline{\mathbf{a}}_i^{\text{new}} = \left(-\overline{a}_{is} \times \overline{\mathbf{a}}_r^{\text{new}}\right) + \overline{\mathbf{a}}_i \text{ and } \overline{b}_i^{\text{new}} = \left(-\overline{a}_{is} \times \overline{b}_r\right) + \overline{b}_i$$

Step 3: (*Optimality Test*) If all reduced costs in row 0 are nonnegative, stop with the optimal solution. Otherwise, return to Step 2.

Example

We will use as an example the problem considered in Section 3.4. The tableau in Table 3.2 is repeated in Table 3.7 with slacks x_3, x_4, and x_5.

Step 1: (*Initialization*)

Note that the initial tableau is in the simplex form. Choosing the slacks x_3, x_4, and x_5 as the basic variables yields a BFS. This is a direct result of the original model having all "less than or equal to" constraints and positive constants on the right-hand side. In cases such as this, the slack variables always define a BFS and provide a convenient starting basis for the simplex method. Thus, the initialization step is accomplished for the example. From row 0 we see that the solution is not optimal.

Iteration 1

Step 2: (*Iterate and Update Tableau*)

a. Using the steepest ascent rule, we choose x_2 as the entering variable ($s = 2$).

b. We compute the ratios $\overline{b}_i / \overline{a}_{is}$ for positive $\overline{a}_{is}$. For convenience, we list them in a column to the right of the tableau, as in Table 3.8. The index of the pivot row is $r = 1$.

c. The new tableau, which was created using the prescribed linear operations, appears in Table 3.9. The value of the pivot element $\overline{a}_{12}$ is 1. To economize on tables, we show both the results of the current iteration and the selection of the entering and leaving variables for the next iteration in the same tableau. For example, Table 3.9 shows the results of iteration 1 and the selections for iteration 2.

Step 3: (*Optimality Test*) One of the reduced costs is negative, so we return to Step 2.

Table 3.7 Initial Tableau for Simplex Example

Row	Basic	Coefficients						RHS
		z	x_1	x_2	x_3	x_4	x_5	
0	z	1	-2	-3	0	0	0	0
1	x_3	0	-1	1	1	0	0	5
2	x_4	0	1	3	0	1	0	35
3	x_1	0	1	0	0	0	1	20

Table 3.8   Tableau with Ratios Computed

Row	Basic	z	x_1	x_2	x_3	x_4	x_5	RHS	Ratio
0	z	1	−2	−3	0	0	0	0	—
1	x_3	0	−1	⓵	1	0	0	5	5 ←
2	x_4	0	1	3	0	1	0	35	11.67
3	x_5	0	1	0	0	0	1	20	—

Table 3.9   Tableau after Iteration 1

Row	Basic	z	x_1	x_2	x_3	x_4	x_5	RHS	Ratio
0	z	1	−5	0	3	0	0	15	—
1	x_2	0	−1	1	1	0	0	5	—
2	x_4	0	④	0	−3	1	0	20	5 ←
3	x_5	0	1	0	0	0	1	20	20

Iteration 2

Step 2:   (*Iterate and Update Tableau*)

   a. Only x_1 has a negative coefficient in row 0, so it is selected to enter the basis ($s = 1$).

   b. From the ratios, we discover that the pivot row is $r = 2$ and the pivot element $\bar{a}_{21} = 4$.

   c. Using the prescribed linear operations, we create the new tableau, which is shown in Table 3.10.

Step 3:   (*Optimality Test*) One of the coefficients in row 0 is still negative, so we return to Step 2.

Iteration 3

Step 2:   (*Iterate and Update Tableau*)

   a. Now x_3 has a negative coefficient in row 0, so we select it to enter the basis ($s = 3$).

   b. From the ratio test, we discover that the pivot row is $r = 3$ and the pivot element $\bar{a}_{33} = 0.75$.

Table 3.10   Tableau after Iteration 2

Row	Basic	z	x_1	x_2	x_3	x_4	x_5	RHS	Ratio
0	z	1	0	0	−0.75	1.25	0	40	—
1	x_2	0	0	1	0.25	0.25	0	10	40
2	x_1	0	1	0	−0.75	0.25	0	5	—
3	x_5	0	0	0	⓪.75	−0.25	1	15	20 ←

c. Using the prescribed linear operations, we create the new tableau which is presented in Table 3.11.

Step 3: (*Optimality Test*) All reduced costs are nonnegative, so the current solution is optimal.

If done correctly, the tableaus should always be in the simplex form, the right-hand sides should remain nonnegative, and the objective function should never decrease from one iteration to the next. The solution prescribed by the tableau should also satisfy the equations in the original model. If at some iteration one or more of these conditions is violated, a computational or logical error has been made.

3.7   SPECIAL SITUATIONS

There are several situations that may occur during the execution of the simplex algorithm that provide alternative paths or signal some special characteristics of the problem. Some of these situations, such as alternative optimal solutions and unbounded solutions, have important implications for the decision maker, whereas others, such as degeneracy and cycling, have important implications with respect to computational efficiency. Each of these situations will now be discussed.

Tie for the Entering Variable

When the steepest ascent rule is used for selecting the variable that will enter the basis, if there is a tie for the most negative value, any one of the tied variables may be chosen arbitrarily. The route to the optimal solution will be affected, but the simplex method will ultimately find it.

The rule for selecting the entering variable does not affect the finiteness of the simplex algorithm but may affect its computational efficiency as measured by the time it takes to reach the optimal solution. The only absolute requirement is that the variable that enters must have a negative value in row 0 of the tableau. We will discuss alternative selection strategies in later subsections.

Tie for the Leaving Variable

At times, the ratio test may indicate a tie for the row of the leaving variable. In the simplest implementation, the basic variable for any one of the tied rows may be chosen arbitrarily to leave the basis. Since the ratio indicates the change in the entering variable that will cause a basic variable to go to zero, if more than one row has the same positive ratio θ more than one basic variable will go to zero at the next iteration. The new tableau will show zero for the right-hand sides of each of the tied rows not selected as the pivot row,

Table 3.11   Final Tableau

Row	Basic	Coefficients						RHS
		z	x_1	x_2	x_3	x_4	x_5	
0	z	1	0	0	0	1	1	55
1	x_2	0	0	1	0	0.333	−0.333	5
2	x_1	0	1	0	1	0	1	20
3	x_3	0	0	0	1	−0.333	1.333	20

yielding a degenerate BFS. Degeneracy is associated with the problem of cycling, which will be discussed later in this section.

More sophisticated implementations of the simplex method may include mechanisms for choosing the variable that will leave the basis in the event of a tie so that cycling does not occur. A standard procedure is to choose either the first or last occurrence of the minimum ratio. A method that provides greater numerical stability is to choose the tied row with the largest pivot element.

Ties in the leaving variable often result because of similarity among the constraints in the linear program. For instance, a network model with conservation of flow constraints will have many right-hand-side values of zero, so it is likely that the minimum ratio will be zero and thus there will be several candidates for the leaving variable. Pure network models frequently have degenerate BFSs.

Degenerate Basic Solution

According to Definition 5, the m columns associated with a basic solution must be linearly independent. This does not preclude some of the basic variables from being zero and giving rise to a degenerate solution. In such a case, the solution identifies a point in the decision space at which the selection of basic variables is not unique. It is always possible to exchange a basic variable at zero value for one of the nonbasic variables and satisfy the linear independence requirement.

From a geometric perspective, we have an alternative to Definition 4.

Definition 8: A basic solution $\mathbf{x} \in \Re^n$ is said to be *degenerate* if the number of structural constraints and nonnegativity conditions active at $\mathbf{x}$ is greater than n.

In two dimensions, a degenerate basic solution occurs at the intersection of three or more lines; in three dimensions, a degenerate solution occurs at the intersection of four or more planes.

The effect of degeneracy is to interrupt the steady progress of the simplex algorithm toward optimality. When the column of the entering variable has a positive coefficient in a degenerate row, the ratio θ will be zero for that row. The entering variable need not increase at all to drive the basic variable associated with the degenerate row to zero (it is already zero). The set of basic variables will change at the next iteration, but the solution will not, and the next tableau will also be degenerate. The only way to escape from a degenerate point is to find some entering variable that has nonpositive coefficients in all degenerate rows. Then the ratio will be positive and the solution will change.

Example with Degeneracy

We start with a revised version of the problem given in Table 3.7. In the tableau displayed in Table 3.12, we choose x_2 to enter the basis, which occasions a tie in the ratio test. We arbitrarily choose row 1 as the pivot row.

After pivoting to obtain the tableau in Table 3.13 after iteration 1, we note that a degenerate solution has been realized as a result of the tie in the ratio test.

$$(x_1, x_2, x_3, x_4, x_5) = (0, 5, 0, 0, 20)$$

Now we must select x_1 to enter the basis. The ratio is minimum in the degenerate row, and thus row 2 is the pivot row and x_4 leaves the basis.

After pivoting again, we obtain the tableau in Table 3.14 after iteration 2. The solution is the same, but the basis has changed. The entering variable is x_3. Table 3.15 shows the

Table 3.12   Initial Tableau for Degeneracy Example

Row	Basic	z	x_1	x_2	x_3	x_4	x_5	RHS	Ratio
				Coefficients					
0	z	1	−3	−4	0	0	0	0	—
1	x_3	0	−1	(1)	1	0	0	5	5 ←
2	x_4	0	1	3	0	1	0	15	5
3	x_5	1	0	0	0	1	1	20	—

Table 3.13   Tableau after Iteration 1

Row	Basic	z	x_1	x_2	x_3	x_4	x_5	RHS	Ratio
				Coefficients					
0	z	1	−7	0	4	0	0	20	—
1	x_2	0	−1	1	1	0	0	5	—
2	x_4	0	(4)	0	−3	1	0	0	0 ←
3	x_5	0	1	0	0	0	1	20	20

Table 3.14   Tableau after Iteration 2

Row	Basic	z	x_1	x_2	x_3	x_4	x_5	RHS	Ratio
				Coefficients					
0	z	1	0	0	−1.25	1.75	0	20	—
1	x_2	0	0	1	0.25	(0.25)	0	5	20 ←
2	x_1	0	1	0	−0.75	0.25	0	0	—
3	x_5	0	0	0	0.75	−0.25	1	20	26.67

Table 3.15   Optimal Solution

Row	Basic	z	x_1	x_2	x_1	x_4	x_5	RHS
				Coefficients				
0	z	1	0	5	0	3	0	45
1	x_3	0	0	4	1	1	0	20
2	x_1	0	1	3	0	1	0	15
3	x_5	0	0	−3	0	−1	1	5

next solution. It is not degenerate because in Table 3.14 the coefficient of x_3 in row 2, the degenerate row, is negative (−0.75), and there is no tie in the minimum ratio.

Cycling

When degeneracy occurs, the possibility arises that the simplex algorithm will return to a previously encountered basis. In a deterministic algorithm (always following the same sequence of basic solutions from a given starting point), this is the signal that a cycle has occurred. The solution will never leave the current extreme point. The code continues to perform degenerate pivots and never terminates.

Although cycling is possible, most implementations of the simplex algorithm do not make special provisions to prevent its occurrence. To do so would require, on average, the computation of a majority of the reduced costs. This can be very expensive for large-scale problems. The common supposition is that rounding errors in the computations will cause most computer codes eventually to move away from degenerate points without introducing special procedures. Nevertheless, cycling can be eliminated by modifying the rules used to select the entering and leaving variables. The following rules are based on the indices of the problem variables x_j ($j = 1, \ldots, n$) and are attributable to Bland [1977].

> *Rule for selecting the entering variable:* From all the variables having negative coefficients in row 0, select as the entering variable the one with the smallest index.

> *Rule for selecting the leaving variable:* Use the standard ratio test to select the leaving variable, but if there is a tie in the ratio test, select from the tied rows the variable having the smallest index.

Example with Cycling

The following problem exhibits cycling in a situation in which the steepest ascent rule is used for selecting the entering variable.

$$\text{Maximize } z = 10x_1 - 57x_2 - 9x_3 - 24x_4$$
$$\text{subject to} \quad 0.5x_1 - 5.5x_2 - 2.5x_3 + 9x_4 \leq 0$$
$$0.5x_1 - 1.5x_2 - 0.5x_3 + x_4 \leq 0$$
$$x_1 \qquad\qquad\qquad \leq 1$$
$$x_j \geq 0 \text{ for } j = 1, \ldots, 4$$

Let x_5, x_6, x_7 be the slacks for the three constraints. Starting with these variables as basic, the sequence of solutions is given in Table 3.16. At iteration 7, the BFS is the same as the BFS at iteration 1, so the sequence will continue indefinitely.

When Bland's rules are used instead of the steepest ascent rule, the first six iterations are the same but iteration 7 is different and iteration 8 provides the optimal solution. The final two iterations in the sequence are shown in Table 3.17. The reader is encouraged to develop the corresponding tableaus to confirm the results.

A phenomenon related to cycling occurs when the simplex algorithm goes through an exceedingly long (although finite) sequence of degenerate pivots whose number may be

Table 3.16   Cycling Results

Iteration	Basic variables	Basic values	Objective value
1	(x_5, x_6, x_7)	$(0, 0, 1)$	0
2	(x_1, x_6, x_7)	$(0, 0, 1)$	0
3	(x_1, x_2, x_7)	$(0, 0, 1)$	0
4	(x_3, x_2, x_7)	$(0, 0, 1)$	0
5	(x_3, x_4, x_7)	$(0, 0, 1)$	0
6	(x_5, x_4, x_7)	$(0, 0, 1)$	0
7	(x_5, x_6, x_7)	$(0, 0, 1)$	0

Table 3.17 Elimination of Cycling

Iteration	Basic variables	Basic values	Objective value
7	(x_5, x_1, x_7)	$(0, 0, 1)$	0
8	(x_5, x_1, x_3)	$(2, 1, 1)$	1

exponential in the size (m and n) of the problem. This phenomenon is known as *stalling*, because with increasing problem size the algorithm can spend an enormous amount of time at a degenerate point before verifying optimality or moving on. In any implementation, one would like to preclude stalling by guaranteeing that the length of a sequence of degenerate pivots is bounded from above by a polynomial in m and n.

Alternative Optima

If the optimal tableau shows one or more nonbasic variables with zero coefficients in row 0, there are alternative optimal solutions. Recall that the coefficient of a nonbasic variable in row 0 indicates the marginal change in the objective function with respect to a small increase in that variable. When the marginal change is zero, that variable can enter the basis without altering the objective value. For an optimal nondegenerate solution, this implies that there is another distinct solution that has the same value of the objective function.

Although the simplex method automatically stops when the first optimal BFS is reached, others can be found by forcing a nonbasic variable with zero reduced cost to enter the basis. When there are $\hat{m}$ zeros in row 0 and the corresponding columns of the **A** matrix are linearly independent, there are

$$\frac{\hat{m}!}{m!(\hat{m}-m)!}$$

ways to choose the m basic variables from the $\hat{m}$ variables whose reduced costs are zero. These are alternative basic solutions with the same objective value. Not all are necessarily feasible. When more than one optimal BFS exists, any weighted average of them is also optimal, where the weights are nonnegative and sum to 1. Such a weighted linear average is called a *convex combination*. In particular, if $\mathbf{x}^1, \ldots, \mathbf{x}^s$ are optimal BFSs, then so is $\hat{\mathbf{x}}$, where

$$\hat{\mathbf{x}} = \sum_{k=1}^{s} \lambda_k \mathbf{x}^k, \quad \sum_{k=1}^{s} \lambda_k = 1, \ \lambda_k \geq 0, \ k = 1, \ldots, s$$

Alternative optima may be useful for decision makers. The linear programming model rarely encompasses all factors of importance, and the existence of multiple solutions allows the decision maker to select the one that best meets nonquantifiable goals.

Unbounded Solution

When there is a nonbasic variable with a negative coefficient in row 0 and all the other coefficients in that column are nonpositive, the value of that variable can be increased indefinitely without driving any basic variable to zero. The objective function can then be made arbitrarily large. This condition signals an unbounded solution, and there is no optimal solution for the model. The algorithm must stop if such an event occurs. To illustrate, consider the following problem.

$$\text{Maximize } z = 6x_1 + 3x_2$$
$$\text{subject to} \qquad x_2 \leq 5$$
$$-4x_1 + 3x_2 \leq 15$$
$$2x_1 - 3x_2 \geq 2$$
$$x_1 \geq 0, \, x_2 \geq 0$$

After several iterations, the tableau in Table 3.18 is obtained. Variable x_5 is slated to enter the basis, but it is not possible to perform the ratio test.

An unbounded solution is probably the result of a poorly specified model. An LP model usually involves linear approximations of inherently nonlinear phenomena. The approximations are valid over a relatively small range of variation in the decision variables. An unbounded solution will certainly be outside the range of any reasonable approximations and is therefore a signal that the assumptions of the analyst need to be reviewed.

Because of the additional work involved, computer implementations rarely *look for* columns that describe an unbounded solution. It may be that nonbasic variables that are candidates to enter the basis and that exhibit unboundedness are present but are not chosen. Eventually, however, a problem that is unbounded will be revealed under any rule for selecting the entering variable.

3.8   THE INITIAL BASIC FEASIBLE SOLUTION

Given an initial BFS, the primal simplex algorithm systematically moves through the extreme points of the feasible region until it finds the optimal solution or discovers that the problem is unbounded. When the model has all "less than or equal to" inequalities with nonnegative right-hand-side values, the initial solution is easy to find; the slack variables constitute the initial BFS, and the equality form of the model is the simplex form. When the original model contains "greater than or equal to" inequalities or equations, a BFS is not immediately available. In this section, we show how to find an initial solution by solving an augmented linear program as the first phase of a two-phase procedure. The second phase involves solving the original problem using the BFS obtained in the first phase as the starting point.

The Expanded Model

To provide a starting solution to a linear programming model with constraints that are equalities or "greater than or equal to" inequalities and with nonnegative RHS values, we will expand the model by adding *artificial variables*. The equality form of the model has all constraints written as equations with nonnegative constants on the right. The "less than or

Table 3.18   Example of Unboundedness

Row	Basic	Coefficients						RHS	Ratio
		z	x_1	x_2	x_3	x_4	x_5		
0	z	1	0	0	12	0	−3	66	—
1	x_2	0	0	1	1	0	0	5	—
2	x_4	0	0	0	3	1	−2	34	—
3	x_1	0	1	0	1.5	0	−0.5	8.5	—

↑

equal to" constraints have slack variables with +1 coefficients that will take the role of basic variables in the first solution. The "greater than or equal to" constraints also have slack variables (sometimes called *surplus variables* in this context), but their coefficients are −1 and so they cannot be basic in the first solution. Equality constraints do not have slack variables.

The required number of basic variables is obtained by adding an artificial variable to each constraint without a positive slack. For constraint i, the variables will be called α_i. As an example, consider the following model.

$$\text{Maximize } z = -7x_1 + 3x_2$$
$$\text{subject to} \quad 4x_1 + 8x_2 = 30$$
$$-2x_1 + x_2 \geq 3$$
$$x_1 \geq 0, x_2 \geq 0$$

The equality form of the constraints with slacks and artificial variables added is

$$4x_1 + 8x_2 + \alpha_1 = 30$$
$$-2x_1 + x_2 - x_3 + \alpha_2 = 3$$
$$x_1 \geq 0, x_2 \geq 0, x_3 \geq 0, \alpha_1 \geq 0, \alpha_2 \geq 0$$

Because the artificial variables are designated as basic, the equations are in the simplex form, so the computations can begin. The expanded model, however, is not equivalent to the original model. Only when the artificial variables are zero does a feasible solution to the new formulation also represent a feasible solution to the original problem.

We now describe phase 1 of the algorithm, in which a search is made for a BFS to the expanded model in which all artificial variables are zero. If such a BFS cannot be found, it must be that no feasible solution exists for the original problem. If such a solution is found, the artificial variables are eliminated and the simplex algorithm proceeds to find the optimum of the original problem.

Algorithm for the Two-Phase Method

Initial Step

Put the problem into equality form by adding slack variables. For each constraint that has at least one variable with a +1 coefficient that does not appear in any other constraint, include one of these variables in the initial selection of basic variables. If m basic variables are assigned, go directly to phase 2. Otherwise, introduce artificial variables for each constraint that does not have this characteristic, and proceed to phase 1. Let F be the set of constraints that have artificial variables.

Phase 1

The objective is to minimize the sum of the artificial variables created in the initial step, subject to the original constraints augmented by the artificial variables. That is,

$$\text{Minimize } w = \left\{ \sum_{i \in F} \alpha_i : \sum_{j=1}^{n} a_{ij} x_j + \alpha_i = b_i, i = 1, \ldots, m \right\}$$

where α_i is present in constraint i only if $i \in F$. The simplex method is used to solve this problem. If at optimality some artificial variables remain in the basis at a nonzero level, the linear program has no feasible solution. Stop and indicate that the original problem is infeasible. If some artificial variables remain in the basis at a zero level, the constraints for which they are basic are redundant and can be deleted.

Phase 2

Delete the artificial variables from the problem and revert to the original objective function:

$$\text{Maximize } z = \sum_{j=1}^{n} c_j x_j$$

Starting from the BFS obtained at the termination of phase 1 (or in the initial step if phase 1 was skipped), use the simplex method to solve the original problem to optimality.

Example

Placing the preceding LP in simplex form by adding slacks and artificial variables, we get the following model. Both the phase 1 and phase 2 objectives are included.

Phase 1: Maximize $w =$ $\quad\quad\quad\quad\quad -\alpha_1 - \alpha_2$

Phase 2: Maximize $z = -7x_1 + 3x_2$

subject to $\quad 4x_1 + 8x_2 \quad\quad + \alpha_1 \quad\quad = 30$

$\quad\quad\quad\quad -2x_1 + x_2 - x_3 \quad\quad + \alpha_2 = 3$

$\quad\quad\quad\quad x_1 \geq 0, x_2 \geq 0, x_3 \geq 0, \alpha_1 \geq 0, \alpha_2 \geq 0$

In constructing the tableau in Table 3.19, we introduce row 0′ for the phase 1 objective. Row 0′ is not in the simplex form, because the coefficients of the basic variables α_1 and α_2 are not zero. To price out these coefficients we subtract row 1 and row 2 from row 0′ with the result shown in row 0, which is now used for the phase 1 computations.

Examining the reduced costs in row 0 leads us to select x_2 as the entering variable. The ratio test indicates that α_2 is the leaving variable. After pivoting, one artificial variable remains in the basis. Using the steepest ascent rule in the tableau in Table 3.20, we select x_1 as the entering variable and α_1 as the leaving variable. The new tableau, shown in Table 3.21, indicates that an optimal solution has been found for the phase 1 objective function. Because the artificial variables are not in the basis, they can be deleted for phase 2.

Phase 2 is initiated with the basic solution $(x_1, x_2, x_3) = (0.3, 3.6, 0)$ found in phase 1. As shown in Table 3.22, we put the phase 2 objective function in row 0′. Again, we find that the tableau is not in the simplex form, so we price out the nonzero coefficients in the columns for the basic variables. The result is labeled row 0.

Examining the coefficients in row 0, we discover that x_3 enters the basis and x_1 leaves. The optimal solution for phase 2, and the original problem, are shown in Table 3.23.

Table 3.19   Initial Tableau for Phase 1

					Coefficients				
Row	Basic	w	x_1	x_2	x_3	α_1	α_2	RHS	Ratio
0′	w	1	0	0	0	1	1	0	
0	w	1	−2	−9	1	0	0	−33	—
1	α_1	0	4	8	0	1	0	30	3.75
2	α_2	0	−2	①	−1	0	1	3	3 ←

Table 3.20   Iteration 1

Row	Basic	w	x_1	x_2	x_3	α_1	α_2	RHS	Ratio
0	w	1	−20	0	−8	0	9	−6	—
1	α_1	0	⬭20	0	8	1	−8	6	0.3 ←
2	x_2	0	−2	1	−1	0	1	3	—

↑

Table 3.21   Optimal Solution for Phase 1

Row	Basic	w	x_1	x_2	x_3	α_1	α_2	RHS	Ratio
0	w	1	0	0	0	1	1	0	—
1	x_1	0	1	0	0.4	0.05	−0.4	0.3	—
2	x_2	0	0	1	−0.2	0.1	0.2	3.6	—

Table 3.22   Initial Solution for Phase 2

Row	Basic	z	x_1	x_2	x_3	RHS	Ratio
0′	z	1	−7	3	0	0	
0	z	1	0	0	−3.4	8.7	—
1	x_1	0	1	0	⬭0.4	0.3	0.75 ←
2	x_2	0	0	1	−0.2	3.6	—

↑

Table 3.23   Optimal Solution for Phase 2

Row	Basic	z	x_1	x_2	x_3	RHS	Ratio
0	z	1	8.5	0	0	11.25	—
1	x_3	0	2.5	0	1	0.75	—
2	x_2	0	0.5	1	0	3.75	—

3.9   DUAL SIMPLEX ALGORITHM

In the tableau implementation of the primal simplex algorithm, the right-hand-side column is always nonnegative, so the basic solution is feasible at every iteration. For the purposes of this section, we will say that the basis for the tableau is *primal feasible* if all elements of the right-hand side are nonnegative. Alternatively, when some of the elements are negative, we say that the basis is *primal infeasible*. Up to this point, we have always been concerned with primal feasible bases.

For the primal simplex algorithm, some elements in row 0 will be negative until the final iteration when the optimality conditions are satisfied. In the event that all elements of row 0 are nonnegative, we say that the associated basis is *dual feasible*. Alternatively, if some of the elements of row 0 are negative, we have a *dual infeasible* basis.

As described previously, the primal simplex method works with primal feasible but dual infeasible (nonoptimal) bases. At the final (optimal) solution, the basis is both primal and dual feasible. Throughout the process, we maintain primal feasibility and drive toward dual feasibility.

The variant of the primal approach that we now consider works in just the opposite fashion, and is known as the *dual simplex method*. Until the final iteration, each basis examined is primal infeasible (there are some negative values on the right-hand side) and dual feasible (all elements in row 0 are nonnegative). At the final (optimal) iteration, the solution is both primal and dual feasible. Throughout the process, we maintain dual feasibility and drive toward primal feasibility. For a given problem, both the primal and dual simplex algorithms terminate at the same solution but arrive there from different directions.

The dual simplex algorithm is best suited for problems in which an initial dual feasible solution is easily available. It is particularly useful for reoptimizing a problem after a constraint has been added or some parameters have been changed so that the previously optimal basis is no longer feasible.

We will have much more to say about duality and the relationship between primal and dual solutions in Chapter 4; however, in this section we are principally concerned with the mechanics of implementing the dual simplex method in the tableau format. We shall see that the dual simplex algorithm is very similar to the primal simplex algorithm.

Algorithm

With reference to the tableau, the algorithm must begin with a basic solution that is dual feasible, so all the elements of row 0 must be nonnegative. The iterative step of the primal simplex algorithm first selects a variable to enter the basis and then finds the variable that must leave so that primal feasibility is maintained. The dual simplex method does the opposite; it first selects a variable to leave the basis and then finds the variable that must enter the basis to maintain dual feasibility. This is the principal difference between the two methods. The following algorithm assumes that a basic solution is described by a tableau.

Step 1: (*Initialization*)

Start with a dual feasible basis and let $k = 1$. Create a tableau for this basis in the simplex form. If the right-hand-side entries are all nonnegative, the solution is primal feasible, so stop with the optimal solution.

Step 2: (*Iteration k*)

a. *Select the leaving variable.* Find a row (call it r) with a negative right-hand-side constant—i.e., $\bar{b}_r < 0$. Let row r be the pivot row and let the leaving variable be $x_{B(r)}$. A common rule for choosing r is to select the most negative RHS value—i.e.,

$$\bar{b}_r = \min\{\bar{b}_i : i = 1, \ldots, m\}$$

b. *Determine the entering variable.* For each negative coefficient in the pivot row, compute the negative of the ratio between the reduced cost in row 0 and the structural coefficient in row r. If there is no negative coefficient $\bar{a}_{rj} < 0$, stop; there is no feasible solution. Otherwise, let the column with the minimum ratio, designated by the index s, be the pivot column, and let x_s be the entering variable. The pivot column is determined by the following ratio test.

$$\frac{-\bar{c}_s}{\bar{a}_{rs}} = \min\left\{\frac{-\bar{c}_j}{\bar{a}_{rj}} : \bar{a}_{rj} < 0, j = 1, \ldots, n\right\}$$

c. *Change the basis.* Replace $x_{B(r)}$ by x_s in the basis. Create a new tableau by performing the following operations (the same operations as for the primal simplex algorithm). Let $\bar{\mathbf{a}}_i$ be the vector of the ith row of the current tableau, and let $\bar{\mathbf{a}}_i^{\text{new}}$ be the ith row in the new tableau. Let $\bar{b}_i$ be the RHS for row i in the current tableau, and let $\bar{b}_i^{\text{new}}$ be the RHS of the new tableau. Let $\bar{a}_{is}$ be the element in the ith row of the pivot column s. The pivot row in the new tableau is

$$\bar{\alpha}_r^{\text{new}} = \bar{\alpha}_r / \bar{a}_{rs} \quad \text{and} \quad \bar{b}_r^{\text{new}} = \bar{b}_r / \bar{a}_{rs}$$

The other rows in the new tableau are

$$\bar{\mathbf{a}}_i^{\text{new}} = \left(-\bar{a}_{is} \times \bar{\mathbf{a}}_r^{\text{new}}\right) + \bar{\mathbf{a}}_i$$

and

$$b_i^{\text{new}} = \left(-\bar{a}_{is} \times \bar{b}_r^{\text{new}}\right) + \bar{b}_i \quad \text{for } i = 0, 1, \ldots, m, i \neq r$$

(These operations have the effect of pricing out the pivot column. Its replacement will have a single 1 in row r and zeros in all other rows, as required by the simplex form.)

Step 3: (*Feasibility Test*)

If all entries on the right-hand side are nonnegative, the solution is primal feasible, so stop with the optimal solution. Otherwise, let $k \leftarrow k + 1$ and return to Step 2.

Examples

An Easy Dual Feasible Starting Solution

The simplest situation arises when there is an obvious dual feasible basis that can be used to initialize the algorithm. Consider the following problem.

$$\text{Maximize } z = -5x_1 - 35x_2 - 20x_3$$
$$\text{subject to} \quad x_1 - x_2 - x_3 \leq -2$$
$$-x_1 - 3x_2 \qquad \leq -3$$
$$x_1 \geq 0, x_2 \geq 0, x_3 \geq 0$$

Step 1: Adding slack variables x_4 and x_5 leads to the first tableau, shown in Table 3.24, which is primal infeasible but dual feasible.

Table 3.24 Initial Dual Feasible Tableau

Row	Basic		Coefficients					RHS
		z	x_1	x_2	x_3	x_4	x_5	
0	z	1	5	35	20	0	0	0
1	x_4	0	1	−1	−1	1	0	−2
2	x_5	0	−1	−3	0	0	1	−3

Iteration 1: Row 2 is selected as the pivot row, so x_5 leaves the basis. The ratio test indicates that x_1 is to enter the basis. The tableau in Table 3.25 shows the ratio calculations.

After the pricing-out operations to obtain the simplex form for x_1, we get the tableau depicted in Table 3.26. The feasibility test at Step 3 fails because the basis is not yet primal feasible, so we return to Step 2. The row and column selection steps for the next iteration are also identified.

Iteration 2: Row 1 is selected as the pivot row, so x_4 leaves the basis and x_2 enters. This leads to the tableau shown in Table 3.27, which still has a negative RHS value, so we return to Step 2.

Table 3.25 Selecting the Leaving and Entering Variables

Row	Basic	z	x_1	x_2	x_3	x_4	x_5	RHS
				Coefficients				
0	z	1	5	35	20	0	0	0
1	x_1	0	1	−1	−1	1	0	−2
2	x_2	0	−1	−3	0	0	1	−3 ←
	Ratio	—	5	11.67	—	—	—	

Table 3.26 Tableau for Iteration 2

Row	Basic	z	x_1	x_2	x_3	x_4	x_5	RHS
				Coefficients				
0	z	1	0	20	20	0	5	−15
1	x_4	0	0	−4	−1	1	1	−5 ←
2	x_1	0	1	3	0	0	−1	3
	Ratio	—	—	5	20	—	—	

Table 3.27 Tableau for Iteration 3

Row	Basic	z	x_1	x_2	x_3	x_4	x_5	RHS
				Coefficients				
0	z	1	0	0	15	5	10	-40
1	x_2	0	0	1	0.25	−0.25	−0.25	1.25
2	x_1	0	1	0	−0.75	0.75	−0.25	−0.75 ←
	Ratio	—	—	—	20	—	40	

Table 3.28   Optimal Tableau

Row	Basic	z	x_1	x_2	x_3	x_4	x_5	RHS
					Coefficients			
0	z	1	20	0	0	20	5	−55
1	x_2	0	0.333	1	0	0	−0.333	1
2	x_3	0	−1.333	0	1	−1	0.333	1

Iteration 3:   Row 2 is selected as the pivot row, so x_1 leaves the basis and x_3 enters. The updated tableau, in shown in Table 3.28, is both primal and dual feasible, which indicates that the optimal solution has been obtained. The algorithm terminates.

Restarting After Changing the Right-Hand-Side Constants

A primary use of the dual simplex algorithm is to reoptimize a problem after it has been solved and one or more of the RHS constants have been changed. This is illustrated by the following problem. The optimal tableau is shown in Table 3.29, with x_{s1}, x_{s2}, and x_{s3} as slacks.

$$\text{Maximize } z = 2x_1 + 3x_2$$

$$\text{subject to} \quad -x_1 + x_2 \le 5$$

$$x_1 + 3x_2 \le 35$$

$$x_1 \quad\quad \le 20$$

$$x_1 \ge 0, x_2 \ge 0$$

Changing the RHS constants will change only the entries in the last column of the tableau. In particular, if we change b_2 from 35 to 20 and b_3 from 20 to 26 in the original problem statement, the RHS vector in the tableau shown in Table 3.29 for the current basis **B** becomes

$$\mathbf{x}_B = \overline{\mathbf{b}} = \mathbf{B}^{-1}\mathbf{b}^{\text{new}} = \begin{pmatrix} -1 & 1 & 1 \\ 1 & 3 & 0 \\ 1 & 0 & 0 \end{pmatrix}^{-1} \begin{pmatrix} 5 \\ 20 \\ 26 \end{pmatrix} = \begin{pmatrix} 26 \\ -2 \\ 33 \end{pmatrix} \text{ with } z = 46$$

Consequently, when $\mathbf{b}^{\text{new}} = (5, 20, 26)^{\text{T}}$ replaces $\mathbf{b}^{\text{old}} = (5, 35, 20)^{\text{T}}$, we get the tableau shown in Table 3.30, which is a candidate for the dual simplex algorithm. Note that $\overline{b}_2$ appears in

Table 3.29   Optimal Tableau before the Change

Row	Basic	z	x_1	x_2	x_{s1}	x_{s2}	x_{s3}	RHS
					Coefficients			
0	z	1	0	0	0	1	1	55
1	x_2	0	0	1	0	0.33	−0.3	5
2	x_1	0	1	0	0	0	1	20
3	x_{s1}	0	0	0	1	−0.3	1.33	20

Table 3.30   Tableau Showing the Changed RHS

Row	Basic	z	x_1	x_2	x_{s1}	x_{s2}	x_{s3}	RHS
0	z	1	0	0	0	1	1	55
1	x_2	0	0	1	0	0.33	(−0.3)	−2 ←
2	x_1	0	1	0	0	0	1	26
3	x_{s1}	0	0	0	1	−0.3	1.33	33
	Ratio	—	—	—	—	—	3	

row 1 and $\overline{b}_1$ appears in row 2 in the tableau, because row 2 corresponds to x_1 and row 1 to x_2.

Changing the RHS values does not affect the reduced costs, so the entries in row 0 remain nonnegative; however, a negative value for $\overline{b}_2$ indicates that the basic solution is now infeasible. It is clear from the tableau that x_2 will leave the basis and x_{s3} will enter at the next iteration. The result is shown in Table 3.31.

Adding a Constraint

Using the preceding problem, we now add the constraint $x_2 \geq 10$. The solution in the optimal tableau, $x_1 = 20$ and $x_2 = 5$, does not satisfy this constraint, so action must be taken to incorporate it into the tableau. First we subtract a slack variable x_{s4} to get the equality

$$x_2 - x_{s4} = 10$$

and then multiply it by −1 to achieve the correct form. A row corresponding to this constraint and a column corresponding to the slack variable are added to the current tableau, resulting in the modified tableau shown in Table 3.32.

To regain the simplex form for column x_2, we must add row 1 to row 4. Now the tableau shown in Table 3.33 is in the simplex form. As expected, the solution is dual feasible but not primal feasible. The only negative RHS value appears in row 4, so x_{s4} must leave the basis. The entering variable is x_{s3}, the only candidate with a negative entry in the pivot row. The optimal tableau is shown in Table 3.34.

Table 3.31   Tableau after a Dual Simplex Iteration

Row	Basic	z	x_1	x_2	x_{s1}	x_{s2}	x_{s3}	RHS
0	2	1	0	3	0	2	0	40
1	x_{s3}	0	0	−3	0	−1	1	6
2	x_1	0	1	3	0	1	0	20
3	x_{s1}	0	0	4	1	1	0	25

Table 3.32   Tableau with an Added Constraint

Row	Basic				Coefficients				RHS
		z	x_1	x_2	x_{s1}	x_{s2}	x_{s3}	x_{s4}	
0	z	1	0	0	0	1	1	0	55
1	x_2	0	0	1	0	0.333	−0.333	0	5
2	x_1	0	1	0	0	0	1	0	20
3	x_{s1}	0	0	0	1	−0.333	1.333	0	20
4	x_{s4}	0	0	−1	0	0	0	1	−10

Table 3.33   Tableau in the Simplex Form

Row	Basic				Coefficients				RHS
		z	x_1	x_2	x_{s1}	x_{s2}	x_{s3}	x_{s4}	
0	z	1	0	0	0	1	1	0	55
1	x_2	0	0	1	0	0.333	−0.333	0	5
2	x_1	0	1	0	0	0	1	0	20
3	x_{s1}	0	0	0	1	−0.333	1.333	0	20
4	x_{s4}	0	0	0	0	0.333	(−0.333)	1	−5 ←
	Ratio	—	—	—	—	—	3	—	

Table 3.34   Reoptimized Tableau

Row	Basic				Coefficients				RHS
		z	x_1	x_2	x_{s1}	x_{s2}	x_{s3}	x_{s4}	
0	z	1	0	0	0	2	0	3	40
1	x_2	0	0	1	0	0	0	−1	10
2	x_1	0	1	0	0	1	0	3	5
3	x_{s1}	0	0	0	1	1	0	4	0
4	x_{s3}	0	0	0	0	−1	1	−3	15

3.10   SIMPLEX METHOD USING MATRIX NOTATION

For many purposes it is more convenient to describe the linear programming model and steps of the computational procedure using matrices rather than the algebraic format of the previous sections. The matrix statement of the problem is much simpler to contemplate and only the elements of the tableaus that are necessary for immediate algorithmic decisions are computed. This leads to much greater efficiency. Since the model is really a collection of arrays describing the objective function coefficients, decision variables, constraint coefficients, and right-hand-side constants, the coding of the algorithm is closer to the matrix description than the algebraic description. Virtually all computer implementations of the simplex method are based on the approach presented in this and the next section.

Linear Programming Model

In the development of the matrix form of the simplex method, we will work with the equality form of the model, assuming that all modifications required to put it into that form have been made. A linear program in n nonnegative variables and m constraints is thus completely described by the following vectors and matrices.

$$\begin{aligned}
&\textit{Decision variables:} && \mathbf{x} = (x_1, \ldots, x_n)^T \\
&\textit{Objective coefficients:} && \mathbf{c} = (c_1, \ldots, c_n) \\
&\textit{Right-hand-side constants:} && \mathbf{b} = (b_1, \ldots, b_m)^T \\
&\textit{Structural coefficients:} && \mathbf{A} = \begin{pmatrix} a_{11} & a_{12} & \cdots & a_{1n} \\ a_{21} & a_{22} & \cdots & a_{2n} \\ \vdots & \vdots & & \vdots \\ a_{m1} & a_{m2} & \cdots & a_{mn} \end{pmatrix}
\end{aligned}$$

Making use of this notation, Equations (1) to (3) can be rewritten as

$$\text{Maximize } z = \mathbf{cx}$$
$$\text{subject to} \quad \mathbf{Ax} = \mathbf{b}$$
$$\mathbf{x} \geq \mathbf{0}$$

As an example, consider the following model expressed in algebraic form.

$$\begin{aligned}
\text{Maximize } z = {}& 2x_1 + 1.25x_2 + 3x_3 \\
\text{subject to} \quad 2x_1 + {}& x_2 + 2x_3 \leq 7 \\
3x_1 + {}& x_2 \qquad\;\; \leq 6 \\
& x_2 + 6x_3 \leq 9 \\
& x_1 \geq 0, x_2 \geq 0, x_3 \geq 0
\end{aligned}$$

With x_4, x_5, and x_6 as slacks, the matrices and vectors defining the equality form of the model are

$$\mathbf{x} = (x_1, x_2, x_3, x_4, x_5, x_6)^T$$
$$\mathbf{c} = (2, 1.25, 3, 0, 0, 0)$$
$$\mathbf{A} = \begin{pmatrix} 2 & 1 & 2 & 1 & 0 & 0 \\ 3 & 1 & 0 & 0 & 1 & 0 \\ 0 & 1 & 6 & 0 & 0 & 1 \end{pmatrix} \text{ and } \mathbf{b} = \begin{pmatrix} 7 \\ 6 \\ 9 \end{pmatrix}$$

Computing a Basic Solution

Suppose we now assume that the n variables are permuted so that the basic variables are the first m components of $\mathbf{x}$. Then we can write $\mathbf{x} = (\mathbf{x}_B, \mathbf{x}_N)$, where $\mathbf{x}_B$ and $\mathbf{x}_N$ refer to the basic and nonbasic variables, respectively. The matrix $\mathbf{A}$ can also be partitioned similarly into $\mathbf{A} = (\mathbf{B}, \mathbf{N})$, where $\mathbf{B}$ is the $m \times m$ basis matrix and $\mathbf{N}$ is $m \times (n - m)$. The equation $\mathbf{Ax} = \mathbf{b}$ can thus be written as

$$(\mathbf{B}, \mathbf{N}) \begin{pmatrix} \mathbf{x}_B \\ \mathbf{x}_N \end{pmatrix} = \mathbf{Bx}_B + \mathbf{Nx}_N = \mathbf{b}$$

Multiplying through by $\mathbf{B}^{-1}$ yields

$$\mathbf{x}_B + \mathbf{B}^{-1}\mathbf{Nx}_N = \mathbf{B}^{-1}\mathbf{b}$$

Solving this vector equation allows us to express the basic variables in terms of the non-basic variables.

$$\mathbf{x}_B = \mathbf{B}^{-1}(\mathbf{b} - \mathbf{N}\mathbf{x}_N)$$

The objective function can be written as the sum of two terms, one for the basic variables and the other for the nonbasic variables:

$$z = \mathbf{c}_B\mathbf{x}_B + \mathbf{c}_N\mathbf{x}_N = \mathbf{c}_B\mathbf{B}^{-1}(\mathbf{b} - \mathbf{N}\mathbf{x}_N) + \mathbf{c}_N\mathbf{x}_N \tag{7}$$

A basic solution is found by setting the nonbasic variables equal to zero, $\mathbf{x}_N = \mathbf{0}$, so

$$\mathbf{x}_B = \mathbf{B}^{-1}\mathbf{b} \text{ and } z = \mathbf{c}_B\mathbf{B}^{-1}\mathbf{b}$$

For convenience and later use in the section on the revised simplex method, we introduce the m-dimensional row vector of *dual variables*, $\boldsymbol{\pi}$, and define it as

$$\boldsymbol{\pi} \equiv \mathbf{c}_B\mathbf{B}^{-1}$$

so $z = \boldsymbol{\pi}\mathbf{b}$.

In the example, when $\mathbf{x}_B = (x_1, x_3, x_5)^T$, the basic solution is

$$\mathbf{x}_B = \mathbf{B}^{-1}\mathbf{b} = \begin{pmatrix} 0.5 & 0 & -0.166 \\ 0 & 0 & 0.166 \\ -1.5 & 1 & 0.5 \end{pmatrix}\begin{pmatrix} 7 \\ 6 \\ 9 \end{pmatrix} = \begin{pmatrix} 2 \\ 1.5 \\ 0 \end{pmatrix}$$

with objective value

$$z = \mathbf{c}_B\mathbf{x}_B = (2,3,0)\begin{pmatrix} 2 \\ 1.5 \\ 0 \end{pmatrix} = 8.5$$

and dual solution

$$\boldsymbol{\pi} = (\pi_1, \pi_2, \pi_3) = \mathbf{c}_B\mathbf{B}^{-1}$$

$$= (2,3,0)\begin{pmatrix} 0.5 & 0 & -0.166 \\ 0 & 0 & 0.166 \\ -1.5 & 1 & 0.5 \end{pmatrix} = (1,0,1/6))$$

Marginal Information Concerning the Objective

The simplex algorithm must have sufficient information to determine if a particular basic solution is optimal and, if not, to determine variables that are candidates for entry into the basis. This information is easily found using the matrix equations already derived. From Equation (7) we see that the objective function value for a given basis can be written as

$$z = \mathbf{c}_B\mathbf{B}^{-1}\mathbf{b} + (\mathbf{c}_N - \mathbf{c}_B\mathbf{B}^{-1}\mathbf{N})\mathbf{x}_N$$

For consistency with the material in the sections on the simplex tableau, we will work with the preceding equation in the form

$$z = \mathbf{c}_B\mathbf{B}^{-1}\mathbf{b} - (\mathbf{c}_B\mathbf{B}^{-1}\mathbf{N} - \mathbf{c}_N)\mathbf{x}_N$$

When the nonbasic variables are set to zero, the objective value for the basic solution is $z = \mathbf{c}_B\mathbf{B}^{-1}\mathbf{b}$, but now we want to hold all the nonbasic variables to zero except one, x_k, and observe the effect. The column for x_k in N is the same as the column for x_k in **A**; call it $\mathbf{A}_k$. The objective value as a function of x_k alone is

$$z = \mathbf{c}_\mathrm{B}\mathbf{B}^{-1}\mathbf{b} - (\mathbf{c}_\mathrm{B}\mathbf{B}^{-1}\mathbf{A}_k - c_k)x_k$$

The first term on the right in this expression is a constant for a given basis. The coefficient of x_k in the second term (without the – sign) is the marginal change in the objective function for a unit increase in x_k. We define, for the general nonbasic variable x_k, the marginal change

$$\overline{c}_k = \mathbf{c}_\mathrm{B}\mathbf{B}^{-1}\mathbf{A}_k - c_k = \boldsymbol{\pi}\mathbf{A}_k - c_k \tag{8}$$

where $\overline{c}_k$ is what we have called the reduced cost of x_k.

A sufficient condition for a given basis to be optimal is that increasing any nonbasic variable will cause the object to decrease or stay the same. Mathematically, the basic solution is optimal if

$$\overline{c}_k \geq 0 \text{ for all } k \in Q$$

where Q is the set of nonbasic variables. If this condition is not satisfied, every nonbasic variable with a negative marginal value is a candidate for entry into the basis.

For the example problem, the nonbasic variables are $\mathbf{x}_\mathrm{N} = (x_2, x_4, x_6)$ with $\mathbf{c}_\mathrm{N} = (1.25, 0, 0)$ and $\boldsymbol{\pi} = (1, 0, 1/6)$. The reduced cost for each of the nonbasic variables is computed using Equation (8) as follows.

For x_2 we have $\overline{c}_2 = (1, 0, 1/6) (1, 1, 1)^\mathrm{T} - 1.25 = -1/12$
For x_4 we have $\overline{c}_4 = (1, 0, 1/6) (1, 0, 0)^\mathrm{T} - 0 = 1$
For x_6 we have $\overline{c}_6 = (1, 0, 1/6) (0, 0, 1)^\mathrm{T} - 0 = 1/6$

Because the reduced cost for x_2 is negative, the solution can be improved. Hence, the current basis is not optimal.

Marginal Information Concerning the Basic Variables

When some nonbasic variable is chosen to enter the basis, the ratio test is performed to determine which variable is to leave the basis. Recall that the leaving variable is associated with the row that minimizes the ratio $\overline{b}_i/\overline{a}_{ik}$, for positive values of $\overline{a}_{ik}$. The value $\overline{a}_{ik}$ is the marginal decrease in the value of $x_{\mathrm{B}(i)}$ per unit increase in the nonbasic variable x_k. The values of the marginal change are easily computed using matrix computations. For a particular basis **B**, we have

$$\mathbf{x}_\mathrm{B} = \mathbf{B}^{-1}(\mathbf{b} - \mathbf{N}\mathbf{x}_\mathrm{N})$$

When we set all the nonbasic variables equal to zero except x_k, this expression becomes

$$\mathbf{x}_\mathrm{B} = \mathbf{B}^{-1}\mathbf{b} - \mathbf{B}^{-1}\mathbf{A}_k x_k$$

The first term on the right is a vector of constants corresponding to the current basic solution. The coefficient of the second term is a vector that describes the marginal decreases in the values of the basic variables per unit increase in x_k. Thus, we identify the vector $\overline{\mathbf{A}}_k$ as the vector of marginal decreases in the basic variables per unit increase in x_k.

$$\overline{\mathbf{A}}_k = \mathbf{B}^{-1}\mathbf{A}_k = \begin{pmatrix} \overline{a}_{1k} \\ \overline{a}_{2k} \\ \vdots \\ \overline{a}_{mk} \end{pmatrix}$$

This is exactly the information required for the ratio test.

For the example problem, we start with the basic solution

$$\mathbf{x}_B = \begin{pmatrix} x_1 \\ x_3 \\ x_5 \end{pmatrix} = \mathbf{B}^{-1}\mathbf{b} = \overline{\mathbf{b}} = \begin{pmatrix} 2 \\ 3/2 \\ 0 \end{pmatrix}$$

Allowing x_2 to enter the basis, we compute

$$\overline{\mathbf{A}}_2 = \mathbf{B}^{-1}\mathbf{A}_2 = \mathbf{B}^{-1}\begin{pmatrix} 1 \\ 1 \\ 1 \end{pmatrix} = \begin{pmatrix} 1/3 \\ 1/6 \\ 0 \end{pmatrix}$$

The minimum ratio is $\theta = 6$ for the first equation, so x_1 must leave the basis.

Matrix Representation of the Simplex Tableau

The preceding developments show how all the operations we had previously performed with the simplex tableau can be accomplished using matrix computations. Given a basis and a basis inverse, the matrix representation of a linear program can be written as

$$\begin{pmatrix} 1 & \mathbf{c}_B\mathbf{B}^{-1}\mathbf{A} - \mathbf{c} \\ 0 & \mathbf{B}^{-1}\mathbf{A} \end{pmatrix}\begin{pmatrix} z \\ \mathbf{x} \end{pmatrix} = \begin{pmatrix} \mathbf{c}_B\mathbf{B}^{-1}\mathbf{b} \\ \mathbf{B}^{-1}\mathbf{b} \end{pmatrix} \tag{9}$$

which is equivalent to Equations E0, Ei, $i = 1, \ldots, m$, in Section 3.4. The tableau form of Equation (9) comes from the individual terms. The matrix on the right is the "RHS" column. The matrix on the left forms the remainder of the numerical portion of the tableau.

Table 3.35 shows the tableau form of the example problem for the current basis. The data in the table can be readily confirmed by working out the matrix elements in Equation (9). Note that the basis inverse $\mathbf{B}^{-1}$ appears in the columns under the slack variables (x_4, x_5, x_6) in rows 1, 2, and 3. This is a direct consequence of the columns of the slack variables forming an identity matrix in $\mathbf{A}$, and the fact that $\mathbf{I}^{-1} = \mathbf{I}$.

3.11   REVISED SIMPLEX METHOD

The tableau form of the simplex method is most useful for instructional purposes because it embodies the essential elements of the technique and can be organized for hand calculations. Nevertheless, the tableau method requires an excessive amount of memory and an excessive number of unnecessary calculations, primarily those associated with updating of

Table 3.35   Tableau for the Example

Row	Basic	z	x_1	x_2	x_3	x_4	x_5	x_6	RHS
				Coefficients					
0	z	1	0	−0.083	0	1	0	0.1667	8.5
1	x_1	0	1	0.3333	0	0.5	0	−0.1667	2
2	x_3	0	0	0.1667	1	0	0	0.1667	1.5
3	x_5	0	0	0	0	−1.5	1	0.5	0

nonbasic columns. These inefficiencies cause the method to bog down for large problems. The principal impediment is the size of the tableau, which contains m rows and $m + n$ columns, in general, after adding slack and artificial variables. A fairly small problem with 100 constraints and 1000 variables would require more than 100,000 words of computer memory.

To get around this problem, virtually all implementations are based on the revised simplex method, which does not update and store the entire tableau but only those data elements needed to construct the current basis inverse and to reproduce the matrices describing the original problem. Because most real problems are sparse, having only a handful of variables in each constraint, the original **A** matrix is stored in compressed form rather than as an $m \times n$ array. This leads to much greater efficiencies in the computations.

Matrix Equations

The revised simplex method performs the same steps as the tableau method but does not keep the tableau as an aid. Rather, whenever the algorithm requires a number from the tableau it is computed from one of several matrix equations, often involving the inverse of the basis. The data for the algorithm are the matrices **A**, **c**, and **b** defining the original problem, the numbers of variables and constraints (n and m), and a record of the current basic and nonbasic variables. There are several ways to store this information that affect computational efficiency; however, this discussion assumes that the matrices are stored in explicit form as arrays of numbers and that an explicit representation of $\mathbf{B}^{-1}$ is available. Commercial codes do not store $\mathbf{B}^{-1}$ as an $m \times m$ matrix but use an implicit approach such as *LU*-decomposition to reconstruct it as needed. In this approach, the **B**-matrix is decomposed into an upper triangular matrix, **U**, and a lower triangular matrix, **L**, such that $\mathbf{B} = \mathbf{LU}$. This representation facilitates the computation of the four vectors listed below by eliminating the need to invert **B** explicitly. Instead, two triangular systems of equations are solved by back-substitution. This is further discussed in Section 4.3 in the subsection on Implementation Issues where Cholesky factorization is outlined. The context is slightly different but the procedure is the same. Full details can be found in most linear algebra texts, such as the one by Strang [1988].

The components of the revised simplex algorithm require computation of the following information.

Primal variables: $\mathbf{x}_B = \mathbf{B}^{-1}\mathbf{b}$

Dual variables: $\boldsymbol{\pi} = \mathbf{c}_B\mathbf{B}^{-1}$

Marginal cost for x_k: $\overline{c}_k = \boldsymbol{\pi}\mathbf{A}_k - c_k$ ($\mathbf{A}_k$ is the kth column of **A**)

Pivot column: $\overline{\mathbf{A}}_k = \mathbf{B}^{-1}\mathbf{A}_k$

An optional feature of the algorithm is the procedure used to select the variable that will enter the basis. In the example problem, we compute the reduced costs of all nonbasic variables and choose the most negative as the entering variable. More efficient methods will be discussed in the next subsection.

Statement of the Algorithm

We present the algorithm in parallel with the example introduced at the beginning of this section. As always, the algorithm must start with a basic feasible solution. Although the usual choice is the slack variables, we will start with the BFS given in Table 3.24—i.e.,

$$\mathbf{x}_B = (x_1, x_3, x_5)^T, \; \mathbf{c}_B = (2, 3, 0),$$

$$\mathbf{B} = \begin{pmatrix} 2 & 2 & 0 \\ 3 & 0 & 1 \\ 0 & 6 & 0 \end{pmatrix}, \quad \mathbf{B}^{-1} = \begin{pmatrix} 1/2 & 0 & -1/6 \\ 0 & 0 & 1/6 \\ -3/2 & 1 & 1/2 \end{pmatrix}$$

In the steps below, symbolic representations are given in the lefthand boxes, and computations are given in the righthand boxes.

Step 1: *Compute the basic solution.*

For the current basis $\mathbf{B}$, compute $\mathbf{B}^{-1}$ and the primal and dual solutions $$\mathbf{x}_B = \mathbf{B}^{-1}\mathbf{b}, \; \pi = \mathbf{c}_B \mathbf{B}^{-1}$$	For the example problem, using $\mathbf{B}^{-1}$, we compute $$\mathbf{x}_B = (x_1, x_3, x_5)^T = (2, 3/2, 0)^T$$ $$\pi = (\pi_1, \pi_2, \pi_3) = (1, 0, 1/6)$$

Step 2: *Select the variable that will enter the basis.*

For each nonbasic variable, compute the reduced cost $$\bar{c}_j = \pi \mathbf{A}_j - c_j$$ If each of these values is nonnegative, stop with the optimal solution and compute $z^* = \mathbf{c}_B \mathbf{x}_B$. Otherwise, select the variable with the most negative reduced cost for entry into the basis. The entering variable is x_s.	As we have seen, for $x_2, \bar{c}_2 = (1, 0, 1/6)(1, 1, 1)^T - 1.25 = -1/12$ for $x_4, \bar{c}_4 = (1, 0, 1/6)(1, 0, 0)^T - 0 = 1$ for $x_6, \bar{c}_6 = (1, 0, 1/6)(0, 0, 1)^T - 0 = 1/6$ Since the reduced cost for x_2 is negative, the solution can be improved by allowing x_2 to enter the basis ($s = 2$).

Step 3: *Compute the pivot column.*

Let column s of the matrix $\mathbf{A}$ be the vector $\mathbf{A}_s$. Compute the pivot column $$\bar{\mathbf{A}}_s = \mathbf{B}^{-1}\mathbf{A}_s$$	For $s = 2$, the pivot column is $$\bar{\mathbf{A}}_2 = \mathbf{B}^{-1}(1, 1, 1)^T = (1/3, 1/6, 0)^T$$

Step 4: *Find the variable that will leave the basis.*

Find the row r for which the minimum ratio is obtained—i.e., $$\theta = x_{B(r)} / \bar{a}_{rs}$$ $$= \min\left\{ \frac{x_{B(i)}}{\bar{a}_{is}} : \bar{a}_{is} > 0 \right\}$$ If there are no rows that have $\bar{a}_{is} > 0$, the solution is unbounded. Otherwise, the basic variable $x_{B(r)}$ leaves the basis; θ is the amount that the variable x_s needs to be increased in order to drive $x_{B(r)}$ to zero.	At the current iteration, $$\theta = \min\left\{ \frac{2}{1/3}, \frac{3/2}{1/6} \right\} = 6$$ with the minimum obtained for row 1. Thus, x_1 will leave the basis as x_2 increases from 0 to 6. Note that x_1 is the basic variable for row 1. It is important to keep track of this correspondence.

Step 5: *Change the basis.*

Replace $x_{B(r)}$ by x_s in the set of basic variables and replace $c_{B(r)}$ by c_s in $\mathbf{c}_B$. Update the basis inverse (this is equivalent to pivoting in the simplex tableau). Return to Step 1.	For the new basis, $$\mathbf{x}_B = (x_2, x_3, x_5)^{\mathrm{T}}, \ \mathbf{c}_B = (1.25, 3, 0)$$ The new basis inverse is $$\mathbf{B}^{-1} = \begin{pmatrix} 3/2 & 0 & -1/2 \\ -1/4 & 0 & 1/4 \\ -3/2 & 1 & 1/2 \end{pmatrix}$$ We now go to Step 1.

The algorithm continues until the optimality condition is met at Step 2 or an unbounded solution is discovered at Step 4. For the example problem, the new BFS is

$$\mathbf{x}_B = (x_2, x_3, x_5)^{\mathrm{T}} = (6, \ 1/2, \ 0)^{\mathrm{T}} \text{ with } z = \mathbf{c}_B\mathbf{x}_B = 9$$

which is an improvement over the initial basic BFS, whose objective function value is $z = 8.5$, as can be seen in Table 3.24.

Alternatives for Selecting the Entering Variable

The rule at Step 2 for selecting the entering variable can be implemented in several ways. Although the algorithm as written uses the steepest ascent rule, the only requirement is that the entering variable have a negative reduced cost. In fact, it is highly inefficient to compute all the reduced costs. When n is large, this operation would require excessive computation.

An alternative is to compute the reduced costs in the order of the variable indices. This rule selects the entering variable as the *first* one found with a negative reduced cost. Say x_s is selected at some iteration. At the next iteration, the search begins with the nonbasic variable with the next-higher index than s. When the search reaches x_n, it cycles back to x_1 if no negative reduced cost is found. Of course, optimality is reached only when $\overline{c}_j \geq 0$ for all $j \in Q$.

Candidate lists are used in more sophisticated implementations. At the first iteration, a candidate list of variables with negative reduced costs is created. Subsequent iterations choose the entering variables from this list until no more variables with negative reduced costs remain. Then another list is constructed. Because the marginal values of the variables change from iteration to iteration, the reduced cost of a member of the list must be recomputed before that variable is considered for entry into the basis. If the reduced cost is no longer positive, the variable is deleted from the list.

An adaptive approach is one in which the search strategy changes during the course of the algorithm. At the beginning, when the solution is far from optimal, many variables are likely to be candidates for the entering variable, and so a candidate list approach should be effective. As we converge on the optimal solution, however, only a few variables will have negative reduced costs, and so switching to a *select first* strategy might be called for. More complex approaches for choosing the entering variable, such as "devex" pricing, are quite popular in commercial codes. These and other issues related to efficient implementation are discussed in many of the references listed at the end of this chapter.

EXERCISES   1. Consider the following linear program.

$$\text{Maximize } z = 5x_1 + 3x_2$$
$$\text{subject to} \quad 3x_1 + 5x_2 \leq 15$$
$$5x_1 + 2x_2 \leq 10$$
$$-x_1 + x_2 \leq 2$$
$$x_2 \leq 2.5$$
$$x_1 \geq 0, x_2 \geq 0$$

(a) Show the equality form of the model.

(b) Sketch the graph of the feasible region and identify the extreme point solutions. From this representation, find the optimal solution.

(c) Analytically determine all solutions that derive from the intersection of two constraints or nonnegativity restrictions. Identify whether or not these solutions are feasible, and indicate the corresponding objective function values. Which one is optimal?

(d) Let the slack variables for the first two constraints, call them x_3 and x_4, be the axes of the graph, and sketch the geometric representation of the model. Show an objective isovalve contour in these variables, and from it determine the optimal solution.

2. You are given the following linear program.

$$\text{Maximize } z = 8x_1 + 4x_2 + 7x_3 - 3x_4$$
$$\text{subject to} \quad 2x_1 - 2x_2 + 3x_3 - 4x_4 = 12$$
$$3x_1 + 8x_2 - x_3 + 7x_4 = 18$$
$$x_j \geq 0, \ j = 1, \ldots, 4$$

(a) Select x_3 and x_4 as the axes and sketch the feasible region and an objective isovalue contour in terms of these variables. Identify the extreme points of the region.

(b) Analytically determine the set of all solutions that are intersections of two constraints or nonnegativity restrictions. Identify whether or not the solutions are feasible, and from the feasible subset select the optimal solution by evaluating the objective function at each one.

3. Consider the linear program

$$\text{Maximize } z = 10x_1 + 5x_2 + 8x_3 - 3x_4$$
$$\text{subject to} \quad -2x_1 + x_2 + 2x_3 - 3x_4 \geq 12$$
$$x_1 + x_2 + x_3 + x_4 \leq 20$$
$$x_j \geq 0, \ j = 1, \ldots, 4$$

(a) Construct the equality form of the model by introducing slack variables x_5 and x_6 for the two constraints.

(b) What is an upper bound on the number of basic solutions?

(c) Analytically determine the set of basic solutions by listing all possible selections of the basic variables. Identify which solutions are feasible and which are infeasible. Compute the objective function values for the feasible solutions, and select the optimal solution.

4. Given the linear program

$$\text{Maximize } z = x_1 + 2.5x_2 + x_3$$
$$\text{subject to} \quad x_1 + x_2 \geq 10$$
$$x_2 + x_3 \geq 10$$
$$x_j \geq 0, \ j = 1, \ldots, 3$$

Algebraically determine the set of all basic solutions and identify whether or not each solution is feasible. Determine the optimal solution by evaluating the objective function at each feasible point.

5. The following tableau is not in simplex form.

| | | Coefficients | | | | | | |
Row	Basic	z	x_1	x_2	x_3	x_4	x_5	RHS
0	z	1	1	−1	−1	0	2	20
1	—	0	1	5	−1	1	12	12
2	—	0	0	8	1	2	16	16

(a) Write the set of equations described by the tableau as it stands.

(b) Using linear operations, convert the tableau into simplex form by making x_1 and x_4 basic. What is the solution corresponding to the new tableau?

(c) From the new tableau, predict the effects of increasing x_5 by 1, by 0.5, and by 2.

(d) From the new tableau, predict the effects of increasing x_3 by 1, by 0.5, and by 2.

6. Starting with the tableau found in Exercise 5(b), consider the three cases that follow. Construct the new tableau in the simplex form, write the basic solution obtained, and use the marginal information available from the tableau to comment on any characteristics that the solution exhibits. Each part of this problem refers to the original basis.

(a) Allow x_2 to replace x_4 as a basic variable.

(b) Allow x_5 to replace x_4 as a basic variable.

(c) Allow x_5 to replace x_1 as a basic variable.

7. For the example problem in Section 3.5, start with the basic solution labeled #1 in Figure 3.5. Sequentially change the basis by allowing variables to enter and leave so that the basic solutions are #2, #3, #4, and #5, in that order. Show the equations in the simplex form for each of the four cases, and sketch the view of the feasible region described for each basis.

8. This is an exercise designed to illustrate the pivoting process. In each case listed, start from the following tableau and let the specified variable enter the basis. Specify which variable should leave the basis, predict the new basic solution, and predict the new value of the objective function. Indicate whether the objective value increases or decreases.

(a) x_1 enters.

(b) x_5 enters.

(c) x_6 enters.

(d) x_7 enters.

(e) x_{10} enters.

| | | Coefficients | | | | | | | | | | |
Row	Basic	z	x_1	x_2	x_3	x_4	x_5	x_6	x_7	x_8	x_9	x_{10}	RHS
0	z	1	0	0	0	0	−0.2	0.2	−0.2	0	0	0.6	11
1	x_3	0	2	0	1	0	1	0	−1	0	0	2	5
2	x_4	0	−2	0	0	1	2	1	2	0	0	−4	0
3	x_8	0	0	0	0	0	0	1	1	1	0	−3	12
4	x_9	0	0	0	0	0	0	2	1	0	1	−9	9
5	x_2	0	1	1	0	0	1	−1	−1	0	0	3	3

9. The following tableau has been found at an intermediate stage of the simplex algorithm for a maximization problem. In each part of this exercise, start from the tableau and perform the suggested pivot operation. Show the complete tableau obtained and comment on any special characteristics that it exhibits. The parts are not cumulative.

Row	Basic	z	x_1	x_2	x_3	x_4	x_5	RHS
					Coefficients			
0	z	1	0	−2	2	0	−2	−2
1	x_1	0	1	1	−1	0	4	4
2	x_4	0	0	4	1	1	8	8

(a) Let x_2 enter the basis and let x_1 leave the basis.
(b) Let x_2 enter the basis and let x_4 leave the basis.
(c) Let x_3 enter the basis and let x_1 leave the basis.
(d) Let x_3 enter the basis and let x_4 leave the basis.
(e) Let x_5 enter the basis and let x_1 leave the basis.
(f) Let x_5 enter the basis and let x_4 leave the basis.

10. You are given the following tableau.

Row	Basic	z	x_1	x_2	x_3	x_4	x_5	x_6	RHS
						Coefficients			
0	z	1	−3	0	0	−3	−2	0	10
1	x_3	0	4	0	1	5	−3	0	4
2	x_2	0	2	1	0	−3	1	0	2
3	x_6	0	1	0	0	2	−2	1	2

Starting from this tableau in each case, perform the indicated operations to derive a new tableau.

(a) Let x_1 enter the basis.
(b) Let x_4 enter the basis.
(c) Let x_5 enter the basis.
(d) Which of the three operations results in the greatest increase in the objective function? Referring to the primal simplex algorithm in Section 3.6, rewrite Step 2(a) so that the variable chosen to enter the basis is the one that produces the greatest increase in the objective function. Would you suggest that this procedure replace the steepest ascent rule? Explain.

11. Using the tableau given in Exercise 10, if x_1 enters the basis, there is a tie for the variable that leaves the basis. Starting from the original tableau, consider the two possibilities and determine what the solution will be in each case.

(a) Let x_3 leave the basis.
(b) Let x_2 leave the basis.

12. (a) What special property does the following tableau exhibit?

Row	Basic	z	x_1	x_2	x_3	x_4	x_5	x_6	RHS
						Coefficients			
0	z	1	−3	2	0	0	−1	0	4
1	x_3	0	2	0	1	0	0	0	13
2	x_4	0	−2	1	0	1	−7	0	7
3	x_6	0	4	2	0	0	−2	1	2

(b) Use the most negative rule to select the entering variable. Proceed until the algorithm terminates.

13. Find all alternative optima in the following tableau.

Row	Basic				Coefficients				
		z	x_1	x_2	x_3	x_4	x_5	x_6	RHS
0	z	1	0	0	0	0	2	4	10
1	x_3	0	4	0	1	5	−3	2	4
2	x_2	0	2	1	0	−3	1	3	2

14. Convert the following problem into simplex form and solve with the Math Programming Excel Add-in. What are the values of the original problem variables?

$$\text{Minimize } z = 3x_1 + 2x_2$$
$$\text{subject to} \quad x_1 - x_2 = -11$$
$$-x_1 - 2x_2 \le -10$$
$$x_1 \ge -3$$
$$x_1 \text{ and } x_2 \text{ are unrestricted}$$

15. Drop the second constraint from the example problem in Section 3.2 and solve it. Comment on any special characteristics of the optimal solution.

16. Add the simple lower bound constraints $x \ge -5$ and $y \ge -5$ to the example problem in Section 3.2. Solve the problem with these constraints explicitly included and comment on any special characteristics of the optimal solution.

17. Repeat Exercise 16, but use a transformation to eliminate the lower bounds on x and y from the constraint set (i.e., do not treat x and y as unrestricted variables).

18. Add the constraint $x_1 - x_2 \ge 0$ to the example problem in Section 3.8, and solve it using the two-phase method.

19. For the following problem, use the phase 1 procedure to find a feasible solution. Set up the tableau to begin phase 2.

$$\text{Maximize } z = 2x_1 - 3x_2 + x_3 - 4x_4$$
$$\text{subject to} \quad 2x_1 - x_2 + 3x_3 - 5x_4 \le 20$$
$$x_1 + 2x_2 - x_3 + 4x_4 \ge 2$$
$$x_4 \le 20$$
$$x_j \ge 0, j = 1, \ldots, 4$$

20. Consider the linear program

$$\text{Maximize } z = 2x_1 + 3x_2 + x_3 + 4x_4$$
$$\text{subject to} \quad x_1 - x_3 + x_4 \le 5$$
$$-x_1 + 2x_2 + x_4 \le 6$$
$$x_2 + 2x_3 + 0.5x_4 \le 8$$
$$0 \le x_j \le 1, j = 1, \ldots, 4$$

After several pivots, the simplex tableau appears as follows.

Row	Basic					Coefficients				
		z	x_1	x_2	x_3	x_4	x_5	x_6	x_7	RHS
0	z	1	0	−1.5	0	−1.25	2	0	1.5	22
1	x_1	0	1	0.5	0	1.25	1	0	0.5	9
2	x_6	0	0	2.5	0	2.25	1	1	0.5	15
3	x_3	0	0	0.5	1	0.25	0	0	0.5	4

Here, x_5, x_6, and x_7 are the slack variables for the three constraints.

(a) What is the basic solution described by the tableau? Give the values of all variables and the objective function.

(b) Without performing a pivot, predict the solution that would be obtained if x_5 were to enter the basis. Use the usual rule to determine the leaving variable.

(c) Without performing a pivot, predict the solution that would be obtained if x_5 were to enter the basis. Let the leaving variable be x_6.

(d) Write the equation described by row 3 in the tableau.

(e) What variable should enter the basis to obtain the greatest total increase in the objective at the next iteration?

(f) Using the most negative rule to determine the entering variable, what variables will enter and leave the basis at the next iteration?

(g) Let x_7 enter the basis and select the leaving variable that will ensure that the next solution will be feasible. Perform a pivot and show the new tableau.

21. Consider the linear program

$$\text{Minimize } z = 2x_1 + x_2$$
$$\text{subject to } \quad -x_1 + 2x_2 \leq 10$$
$$x_1 - 2x_2 \leq 4$$
$$x_1 + x_2 \geq 8$$
$$x_1 + 2x_2 \leq 20$$
$$x_1 \geq 0, x_2 \geq 0$$

(a) Write the equality form of the model.

(b) Give an upper bound on the number of basic solutions for this problem.

(c) What selection of basic variables will cause the first and third constraints to be tight? Assign values to the variables for this basic solution.

(d) For which set of basic variables is there no solution to the model in equality form?

(e) Given that the two-phase method is to be used to find a solution, construct the first tableau for the computations. The tableau should be in simplex form.

(f) Based on this tableau, select the variables that should enter and leave the basis. You don't have to determine the next tableau.

22. Starting from the following tableau, perform at least two iterations of the dual simplex algorithm. Comment on the solution obtained after constructing the second tableau.

Row	Basic	z	x_1	x_2	x_3	x_4	RHS
				Coefficients			
0	z	1	0	2	0	3	−5
1	x_3	0	0	2	1	6	4
2	x_1	0	1	−1	0	−2	−4

23. The tableau that follows shows the optimal solution for the problem

$$\text{Maximize } z = -3x_1 + 2x_2$$
C1: $\quad\quad$ $\text{subject to }\quad\quad 6x_1 + 2x_2 \geq 4$
C2: $\quad\quad\quad\quad\quad\quad\quad\quad 2x_1 + x_2 \leq 4$
$$x_1 \geq 0, x_2 \geq 0$$

with x_3 and x_4 slacks.

Row	Basic	z	x_1	x_2	x_3	x_4	RHS
				Coefficients			
0	z	1	7	0	0	2	8
1	x_2	0	2	1	0	1	4
2	x_3	0	-2	0	1	2	4

(a) Reoptimize after the original problem is changed such that the right-hand side of constraint C1 is 6 and the right-hand side of constraint C2 is 2.

(b) Reoptimize after the original problem is changed such that the objective coefficient of x_1 is 2 and the objective coefficient of x_2 is -3.

(c) Reoptimize using the dual simplex method after adding the constraint $x_2 \leq 1$ to the original problem.

24. Consider the following problem for which the tableau indicates the optimal solution.

$$\text{Maximize } z = 2x_1 + 5x_2 + x_3 + 4x_4$$
$$\text{subject to} \quad 3x_2 + 2x_3 - x_4 \leq 5$$
$$-x_1 + 2x_2 + 2x_3 \leq 6$$
$$x_1 + x_2 + x_3 + x_4 \leq 8$$
$$x_j \geq 0, j = 1, \ldots, 4$$

Row	Basic	z	x_1	x_2	x_3	x_4	x_5	x_6	x_7	RHS
						Coefficients				
0	z	1	1.5	0	4	0	0	0.5	4	35
1	x_2	0	-0.5	1	1	0	0	0.5	0	3
2	x_4	0	1.5	0	0	1	0	-0.5	1	5
3	x_5	0	3	0	-1	0	1	-2	1	1

(a) Change the right-hand sides of the three constraints to 1, 10, and 3, respectively. Show the updated tableau and state the primal solution.

(b) Starting from the tableau found in part (a), perform two iterations of the dual simplex method. Use the most negative rule to choose the pivot row. State the solution obtained in the second iteration.

25. Answer the following questions regarding the simplex algorithm. Each question is independent of the others.

(a) How could you change the algorithm to solve a minimization problem directly without changing it to an equivalent maximization problem?

(b) If the algorithm always chose as the entering variable the one with the negative entry having the largest value, how would the effectiveness of the algorithm be changed?

(c) If the algorithm always chose as the leaving variable the one having the greatest ratio, how would the effectiveness of the algorithm be changed?

26. Convert the following problem into the matrix format. Use x_{si} as the slack variable for constraint i.

$$\text{Maximize } z = x_1 + x_2 + 1.2x_3 + 1.2x_4 + 0.8x_5$$
$$\text{subject to} \quad 2x_1 + 2x_3 + x_4 + x_5 \leq 12$$
$$2x_2 + x_3 + 2x_4 + x_5 \leq 15$$
$$x_1 + x_2 + x_5 \leq 15$$
$$5x_1 + 7x_2 + 4x_3 + 5x_4 + 6x_5 \leq 60$$
$$x_1 + x_2 + x_3 + x_4 + x_5 \leq 10$$
$$x_j \geq 0, j = 1, \ldots, 5$$

27. Referring to Exercise 26, for $\mathbf{x}_B = (x_2, x_3, x_{s2}, x_{s3}, x_{s4})$, the basis inverse is

$$\mathbf{B}^{-1} = \begin{pmatrix} -0.5 & 0 & 0 & 0 & 1 \\ 0.5 & 0 & 0 & 0 & 0 \\ 0.5 & 1 & 0 & 0 & -2 \\ 0.5 & 0 & 1 & 0 & -1 \\ 1.5 & 0 & 0 & 1 & -7 \end{pmatrix}$$

The rows and columns are in the order in which they appear in $\mathbf{x}_B$.

(a) Find the primal and dual basic solutions corresponding to this information.

(b) Compute the reduced cost associated with each basic variable.

(c) Perform two iterations of the revised simplex procedure starting from this basis.

28. For the problem in Exercise 26, at optimality $\mathbf{x}_B = (x_2, x_3, x_4, x_{s3}, x_{s4})$ and the basis inverse is

$$\mathbf{B}^{-1} = \begin{pmatrix} -1 & -1 & 0 & 0 & 3 \\ 0 & -1 & 0 & 0 & 2 \\ 1 & 2 & 0 & 0 & -4 \\ 1 & 1 & 1 & 0 & -3 \\ 2 & 1 & 0 & 1 & -9 \end{pmatrix}$$

Construct the simplex tableau showing this optimal solution. Note the order of the basic variables given by $\mathbf{x}_B$.

29. Consider the problem

$$
\begin{aligned}
\text{Maximize } z = x_1 + {}& x_2 \\
\text{subject to } \quad x_1 + {}& 3x_2 \geq 15 \\
2x_1 + {}& x_2 \geq 10 \\
x_1 + {}& 2x_2 \leq 40 \\
3x_1 + {}& x_2 \leq 60 \\
x_1 \geq 0, {}& x_2 \geq 0
\end{aligned}
$$

Solve using the revised simplex method and comment on any special characteristics of the optimal solution. Sketch the feasible region for the problem as stated above and show on the figure the solutions at the various iterations.

BIBLIOGRAPHY

Bard, J.F., *Practical Bilevel Optimization: Algorithms and Applications*, Kluwer Academic, Boston, 1998.

Bazaraa, M.S., J.J. Jarvis, and H.D. Sherali, *Linear Programming and Network Flows*, Wiley, New York, 1990.

Bertsimas, D. and J.N. Tsitsiklis, *Introduction to Linear Optimization*, Athena Scientific, Belmont, MA, 1997.

Bixby, R.E., "Progress in Linear Programming," *ORSA Journal on Computing,* Vol. 6, No. 1, pp. 15–22, 1994.

Bland, R.C., "New Finite Pivoting Rules for the Simplex Method," *Mathematics of Operations Research*, Vol. 2, No. 1, pp. 103–107, 1977.

Dantzig, G.B. and M.N. Tapia, *Linear Programming: Introduction*, Springer-Verlag, New York, 1997.

Gass, S.I., *Linear Programming: Methods and Applications*, Fourth Edition, McGraw-Hill, New York, 1975.

Harris, P.M.J., "Pivot Selection Methods for the Devex LP Code," *Mathematical Programming,* Vol. 5, pp. 1–28, 1973.

Hooker, J.N., "Karmarkar's Linear Programming Algorithm," *Interfaces*, Vol. 16, No. 4, pp. 75–90, 1986.

Lustig, I.J., R.E. Marsten, and D.F. Shanno, "Computational Experience with a Primal-Dual Interior Point Method," *Linear Algebra and Its Applications*, Vol. 152, pp. 191–222, 1991.

Marsten, R.E., R. Subramanian, M. Saltzman, I.J. Lustig, and D.F. Shanno, "Interior Point Methods for Linear Programming: Just Call Newton, Lagrange, and Fiacco and McCormick," *Interfaces*, Vol. 20, No. 4, pp. 105–116, 1990.

Murty, K.G., *Linear Programming*, Wiley, New York, 1983.

Nering, E.D. and A.W. Tucker, *Linear Programs and Related Problems*, Academic Press, New York, 1992.

Strang, G., *Linear Algebra and Its Applications*, Third Edition, Harcourt Brace Jovanovich, San Diego, 1988.

Vanderbei, R.J., *Linear Programming: Foundations and Extensions*, Kluwer Academic, Boston, 1996.

Ye, Y., *Interior Point Algorithms: Theory and Analysis*, Wiley, New York, 1997.

Chapter **4**

Sensitivity Analysis, Duality, and Interior Point Methods

When a practitioner develops a linear programming model for a problem and finds the optimal solution, it rarely happens that the results are accepted without further analysis. Usually there will be some factors that were omitted from the original formulation and that require additional consideration, or it might be necessary to update estimates of parameter values because of the arrival of new information. For example, the decision maker might be interested not only in the maximum profit attainable under the present constraints but also in how much the maximum profit would change if a new constraint were added to reflect more realistic limits on a critical resource. Moreover, the decision maker might not know the exact value of a particular coefficient c_j but might only be able to place it within a certain range. The decision maker might then want to know whether or not the same basic solution would remain optimal if the true value of the coefficient were any number within the range. Unit profit contributions often fail to capture the uncertainties associated with selling prices and with the costs of items such as wages, raw materials, and shipping. Such investigations fall under the heading of *sensitivity analysis* and are the subject of the first section of this chapter.

Throughout this presentation, we make extensive use of the m-dimensional dual vector $\boldsymbol{\pi}$, whose components can be calculated directly from the current basis matrix $\mathbf{B}$ and the corresponding objective coefficients $\mathbf{c}_B$. In Section 4.2 we will provide the context in which these variables arise. We will show that for every linear program (LP) there is a corresponding dual LP. Both problems are constructed from the same data but in such a way that if one is a maximization problem, the other is a minimization problem, and the optimal objective values of both, if finite, are equal. The variables of the dual problem can be interpreted as prices associated with the constraints of the original or *primal* problem. Through this association it is possible to provide an economic characterization of the dual problem whenever there is such a characterization of the primal problem.

We conclude this chapter by highlighting the primal–dual path following algorithm. This is perhaps the most effective *interior point method* available for solving LPs and, as its name implies, addresses both the primal and dual problems simultaneously. Recall that the simplex method solves an LP by visiting extreme points on the boundary of the feasible region, each time improving the objective function. Although the simplex method works well in practice, it suffers from the theoretical liability that in the worst case it is possible to construct instances whose running times are proportional to an exponential function of their size. A measure of size is the number of bits required to encode an instance. In the mid-1980s, a new class of algorithms was proposed—algorithms that traverse the interior of the feasible region and thus are called interior point methods. These methods have revolutionized the field of optimization, reducing solution times from hours (or longer for some applications) to minutes. In addition, their worst-case running times are only proportional to a polynomial function of their size. This in itself was a remarkable theoretical breakthrough. Nevertheless, the determination of whether a variant of the simplex method or an interior point method performs better depends on the problem as well as on the instance, with experience suggesting that for large, sparse problems, interior point methods are often superior.

4.1   SENSITIVITY ANALYSIS

Generally speaking, the basic assumption that all the coefficients of a linear programming model are known with certainty rarely holds in practice. Moreover, it may be expedient to simplify causal relationships and to omit certain variables or constraints at the beginning of the analysis to achieve tractability. We have already explained how the dual simplex method can be used to reoptimize a model when a new constraint is added to the formulation. Adding a new variable can also be handled efficiently by simply pricing out the new column and seeing if its reduced cost is nonnegative. If so, the new variable has no effect on the optimal solution; if not, it becomes the entering variable and the algorithm continues until all reduced costs are nonnegative.

In this section, we deal implicitly with the issue of uncertainty in the data elements c_j, a_{ij}, and b_i by determining the bounds over which each such element can range without effecting a change in either the optimal solution or the optimal basis. Such investigations fall under the heading of sensitivity analysis. For the most part, we deal with the simple case of perturbing one coefficient at a time. This allows us to derive closed-form solutions. When two coefficients are varied simultaneously, the analysis is much more complex because the "range" is described by a two-dimensional polyhedron rather than by an interval on the real line. When k coefficients are varied simultaneously, a k-dimensional polyhedron results. The only exception is *proportional ranging,* which allows all elements of the original vector to vary simultaneously but in fixed proportion as defined by a second vector. This topic falls under the more general heading of *postoptimality analysis*.

Sensitivity to Variation in the Right-Hand Side

We have seen that for every basis **B** associated with an LP, there is a corresponding set of m dual variables $\pi = (\pi_1, \ldots, \pi_m)$, one for each row. The optimal values of the dual variables can be interpreted as prices. In this section, this interpretation is explored in further detail starting with an LP in standard equality form.

$$\text{Maximize } \{\mathbf{cx} : \mathbf{Ax} = \mathbf{b}, \mathbf{x} \geq \mathbf{0}\}$$

Suppose that the optimal basis is **B** with solution $(\mathbf{x}_B, \mathbf{0})$, where $\mathbf{x}_B = \mathbf{B}^{-1}\mathbf{b}$ and where $\pi = \mathbf{c}_B\mathbf{B}^{-1}$ is unrestricted in sign. Now, assuming nondegeneracy, small changes ($\Delta\mathbf{b}$) in the vector **b** will not cause the optimal basis to change. Thus, for $\mathbf{b} + \Delta\mathbf{b}$, the optimal solution is

$$\mathbf{x} = (\mathbf{x}_B + \Delta\mathbf{x}_B, \mathbf{0})$$

where $\Delta\mathbf{x}_B = \mathbf{B}^{-1}\Delta\mathbf{b}$. Therefore the corresponding increment in the objective function is

$$\Delta z = \mathbf{c}_B \Delta \mathbf{x}_B = \pi \Delta \mathbf{b}$$

This equation shows that π gives the sensitivity of the optimal payoff with respect to small changes in the vector $\mathbf{b}$. In other words, if a new problem were solved with $\mathbf{b}$ changed to $\mathbf{b} + \Delta \mathbf{b}$, the change in the optimal value of the objective function would be $\pi \Delta \mathbf{b}$.

For a maximization problem, this interpretation says that π_i directly reflects the change in profit owing to a change in the ith component of the vector $\mathbf{b}$. Thus, π_i may be viewed equivalently as the *marginal price* of b_i, because if b_i is changed to $b_i + \Delta b_i$, the value of the optimal solution changes by $\pi_i \Delta b_i$. When the constraints $\mathbf{Ax} = \mathbf{b}$ are written as $\mathbf{Ax} \le \mathbf{b}$, the dual variables are nonnegative, implying that for π_i positive, a positive change in b_i will produce an increase in the objective function value. In economic terms, it is common to refer to the dual variables as *shadow prices*.

Example 1

The shadow prices are associated with constraints but are often used to evaluate prices or cost coefficients associated with variables of the primal problem. As an example, we have a matrix $\mathbf{A}$ representing the daily operation of an oil refinery, and a particular variable x_j representing the purchase of crude oil feedstock, with a cost of \$22.65 per barrel ($c_j = 22.65$). The refinery wants to minimize its costs. There is an upper limit on the purchase of this oil of 50, 000 barrels per day at this price. This limit is represented by the constraint

$$x_j + x_s = 50,\ 000$$

where x_s is the associated slack variable. Assume that at optimality x_s has a reduced cost of –\$2.17 per barrel (because we are minimizing, the reduced cost at optimality must be $z \le 0$): What does this value mean?

As we shall see, the shadow price on the constraint is also –\$2.17 per barrel. This does not mean, however, that we should pay only \$2.17 for another barrel of crude. It means that we should be prepared to pay another \$2.17 per barrel for an opportunity to purchase extra supplies given that any further purchases would cost \$22.65 per barrel—i.e., the objective function will decrease by \$2.17 for every extra barrel we can purchase at the price c_j already in the cost row. This means that we should be prepared to bid up to \$22.65 + \$2.17 = \$24.82 per barrel on the spot market for extra supplies of crude. Note that \$24.82 per barrel is the *break-even price,* in that we decrease our objective function z if we can purchase a barrel for less than this price, increase z if we purchase a barrel for more, and make no change in z if we purchase crude for exactly \$24.82 per barrel.

Reduced Cost

The reduced cost of a nonbasic variable at its lower bound is often referred to as the *opportunity cost* of that variable. If management makes the (nonoptimal) decision of increasing that nonbasic variable from its lower bound, for a maximization problem the reduced cost gives the decrease in z per unit increase in the variable (for a certain range). This represents the opportunity loss in departing from the optimal solution.

Ranging

For reasons that practitioners understand implicitly, it is often said that sensitivity analysis is the most important part of the LP calculations. The majority of the coefficients that appear in an LP are rarely known with certainty and so have to be estimated from historical or empirical data. Under these circumstances, we would like to know the range of variation of these coefficients for which the optimal solution remains optimal—i.e., the basis does not change. We will now investigate three categories below: objective coefficients c_j, right-hand-side terms b_i, and matrix coefficients a_{ij}.

Changes in the Objective Row

Nonbasic variable. The change in the objective coefficient of a nonbasic variable affects the reduced cost of that variable only. Letting Q be the index set of nonbasic variables, if δ is a perturbation associated with the original objective coefficient c_q for some $q \in Q$, then at optimality we can write the reduced cost coefficient of nonbasic variable x_q as $\bar{c}_q(\delta) = \pi\mathbf{A}_q - (c_q + \delta)$. In order for the current basis $\mathbf{B}$ to remain optimal, we must have $\bar{c}_q(\delta) \geq 0$. This means

$$\delta \leq \pi\mathbf{A}_q - c_q = \bar{c}_q$$

Not surprisingly, there is no lower bound on δ. Reducing the value of an objective coefficient associated with a nonbasic variable in a maximization problem cannot make the variable more attractive.

The reduced costs of all the other variables are independent of c_q and so will remain nonnegative. If a δ is chosen that violates the inequality, above x_q be identified as the entering variable and we continue the application of the simplex method until an optimal basis for the modified problem found.

It is worth mentioning that in most commercial LP codes a second range is given as well—the range over which x_q can be increased from zero before a change of basis occurs. When $\delta = \bar{c}_q$, the reduced cost is zero, implying that x_q can be increased without affecting the value of the objective function; alternatively, it implies that there are multiple optima. The maximum value that x_q can take without effecting a change in basis is given by $\min_i\{\bar{b}_i/\bar{a}_{iq}: \bar{a}_{iq} > 0\}$, which is the minimum ratio test in Step 4 of the revised simplex algorithm in Section 3.11.

Basic variable. A change in the objective coefficient of a basic variable may affect the reduced cost of all the nonbasic variables. Let $\mathbf{e}_i$ be the ith unit vector of length m and suppose we increment the objective coefficient of the ith basic variable $x_{B(i)}$ by δ—i.e., $\mathbf{c}_B \leftarrow \mathbf{c}_B + \delta\mathbf{e}_i^T$. This yields $\pi(\delta) = (\mathbf{c}_B + \delta\mathbf{e}_i^T)\mathbf{B}^{-1}$, so the dual vector is an affine function of δ. The reduced cost of the qth nonbasic variable is now

$$\begin{aligned}\bar{c}_q(\delta) &= (\mathbf{c}_B + \delta\mathbf{e}_i^T)\mathbf{B}^{-1}\mathbf{A}_q - c_q \\ &= \mathbf{c}_B\mathbf{B}^{-1}\mathbf{A}_q + \delta\mathbf{e}_i^T\mathbf{B}^{-1}\mathbf{A}_q - c_q \\ &= \bar{c}_q + \delta\bar{a}_{iq}\end{aligned}$$

where $\bar{a}_{iq} = (\mathbf{B}^{-1}\mathbf{A}_q)_i$ is the ith component of the updated column of $\mathbf{A}_q$. This value can be found for the nonbasic variable x_q by solving $\mathbf{B}^T\mathbf{y} = \mathbf{e}_i$ for $\mathbf{y}$, then computing $\bar{a}_{iq} = \mathbf{y}^T\mathbf{A}_q$. (Obviously, if $\bar{a}_{iq} = 0$ for any x_q, the reduced cost does not change.)

For a solution to remain optimal, we must have $\bar{c}_q(\delta) \geq 0$, or

$$\bar{c}_q + \delta\bar{a}_{iq} \geq 0 \text{ for all } q \in Q \tag{1}$$

where $\bar{c}_q$ is the reduced cost at the current optimal solution. This constraint produces bounds on δ. For a basic variable, the *range* over which c_i can vary and the current solution remain optimal is given by $c_i + \delta$, where

$$\max_{q \in Q}\left\{\frac{-\bar{c}_q}{\bar{a}_{iq}}: \bar{a}_{iq} > 0\right\} \leq \delta \leq \min_{q \in Q}\left\{\frac{-\bar{c}_q}{\bar{a}_{iq}}: \bar{a}_{iq} < 0\right\}$$

since this is the range for which Equation (1) is satisfied. If there is no $\bar{a}_{iq} > 0$, then $\delta > -\infty$; likewise, if there is no $\bar{a}_{ij} < 0$, then $\delta < \infty$.

Note that perturbing the value c_i for the ith basic variable has no effect on the reduced costs of any of the basic variables. All reduced costs remain zero. This can be seen from the definition $\overline{\mathbf{c}}_B \equiv \mathbf{c}_B \mathbf{B}^{-1} \mathbf{B} - \mathbf{c}_B = \mathbf{0}$ at optimality, so if any of the components of $\mathbf{c}_B$ are perturbed by δ, the effect cancels itself out.

Example 2

Suppose we have an optimal solution to an LP given in tableau form with attached variables:

$$\text{Maximize } z = 4.9 - 0.1x_3 - 2.5x_4 - 0.2x_5$$
$$\text{subject to } x_1 = 3.2 - 0.5x_3 - 1.0x_4 - 0.6x_5$$
$$x_2 = 1.5 + 1.0x_3 + 0.5x_4 - 1.0x_5$$
$$x_6 = 5.6 - 0.5x_3 - 2.0x_4 - 1.0x_5$$

The index set of nonbasic variables $Q = \{3, 4, 5\}$, so the current basis remains optimal as long as $\delta \leq \overline{c}_q$ for all $q \in Q$. When $q = 3$, for instance, this means that $\delta \leq 0.1$. If the original coefficient $c_3 = 1$, then the current basis remains optimal for $c_3 \leq 1.1$.

If the objective coefficient of the basic variable x_2 becomes $c_2 + \delta$, the reduced costs of the nonbasic variables become

$$x_3: \overline{c}_3(\delta) = 0.1 + \delta(-1.0)$$
$$x_4: \overline{c}_4(\delta) = 2.5 + \delta(-0.5)$$
$$x_5: \overline{c}_5(\delta) = 0.2 + \delta(+1.0)$$

Note that $x_{B(i)} = \overline{b}_i - \sum_{j \in \{3,4,5\}} \overline{a}_{ij} x_j$ for $i = 1, 2, 6$, so $\overline{a}_{ij}$ is the negative of the number appearing in the preceding equations. The range that δ can assume is given by

$$\max\left\{\frac{-0.2}{1.0}\right\} \leq \delta \leq \min\left\{\frac{-2.5}{-0.5}, \frac{-0.1}{-1.0}\right\}$$
$$-0.2 \leq \delta \leq 0.1$$

When δ assumes one of the limits of its range, a reduced cost becomes zero. In this example, for $\delta = 0.1$, the reduced cost of x_3 is zero, so that if the objective coefficient of x_2 increases by more than 0.1, it becomes advantageous for x_3 to become active. The minimum ratio test, $\min\left\{\frac{3.2}{0.5}, \frac{5.6}{0.5}\right\} = 6.4$, indicates that x_3 can be increased to 6.4 before x_1 becomes zero and a change of basis is required. Analogously, for $\delta = -0.2$, the reduced cost for x_5 is zero and any further decrease in c_2 will require a basis change to remain optimal. In this case, the ratio test indicates that x_2 will be the leaving variable.

The preceding analysis can be generalized without too much difficulty to allow proportional changes in the vector $\mathbf{c}$ rather than changes in only one coefficient at a time. To perform proportional ranging, we must stipulate an n-dimensional row vector $\mathbf{c}^*$ and consider the new vector $\mathbf{c}(\delta) = \mathbf{c} + \delta \mathbf{c}^*$. The analysis would proceed as previously discussed under "Basic variable" but with $\mathbf{e}_i$ replaced by $\mathbf{c}^*$ (see Exercise 3).

Changes in the Right-Hand-Side Vector

We wish to investigate the effect of a change $b_i \leftarrow b_i + \delta$ for some $1 \leq i \leq m$. It is usual to consider the case in which b_i is the right-hand side of an *inequality* constraint, which therefore has a slack variable associated with it. The goal is to determine the range over which the current solution remains optimal. If the constraint is an equality, it can be analyzed by regarding its associated artificial variable as a positive slack (which must be nonbasic for a feasible solution).

Basic slack variable. If the slack variable associated with the ith constraint is basic, the constraint is not binding at the optimal solution. The analysis is simple: The value of the slack gives the range over which the right-hand-side b_i can be reduced for a "less than or equal to" constraint or increased for a "greater than or equal to" constraint. The solution remains feasible and optimal for the range $b_i + \delta$, where

$$-\hat{x}_s \leq \delta \leq \infty \qquad \text{for a "$\leq$" constraint}$$

$$-\infty \leq \delta \leq \hat{x}_s \qquad \text{for a "$\geq$" constraint}$$

where $\hat{x}_s$ is the value of the associated slack variable.

Nonbasic slack variable. If a slack variable is nonbasic at zero, then the original inequality constraint is binding at the optimum. At first glance, it would seem that, because the constraint is binding, there is no possibility of changing the right-hand-side term, particularly in *decreasing* the value of b_i (for "less than or equal to" constraints). It turns out that by changing the vector $\mathbf{b}$, we also change $\mathbf{x}_B$ ($= \mathbf{B}^{-1}\mathbf{b} = \bar{\mathbf{b}}$) so there is a range over which $\mathbf{x}_B$ remains nonnegative. For the associated values, we still retain an optimal feasible solution in the sense that the basis does not change. (Note that both $\mathbf{x}_B$ and $z = \mathbf{c}_B\mathbf{x}_B$ change values.)

Consider the kth constraint

$$a_{k1}x_1 + a_{k2}x_2 + \cdots + a_{kn}x_n + x_s = b_k$$

where x_s is the slack variable. If the right-hand side becomes $b_k + \delta$, rearranging this equation yields

$$a_{k1}x_1 + a_{k2}x_2 + \cdots + a_{kn}x_n + (x_s - \delta) = b_k \tag{2}$$

so that $(x_s - \delta)$ replaces x_s. Thus, if x_s is nonbasic at zero in the final tableau, we have the expression

$$\mathbf{x}_B = \bar{\mathbf{b}} - \bar{\mathbf{A}}_s(-\delta)$$

where $\bar{\mathbf{A}}_s$ is the updated column in the tableau corresponding to x_s. Because $\mathbf{x}_B$ must remain nonnegative, we have $\bar{\mathbf{b}} + \delta\bar{\mathbf{A}}_s \geq \mathbf{0}$, which is used to solve for the range over which δ can vary.

$$\max_i\left\{\frac{\bar{b}_i}{-\bar{a}_{is}} : \bar{a}_{is} > 0\right\} \leq \delta \leq \min_i\left\{\frac{\bar{b}_i}{-\bar{a}_{is}} : \bar{a}_{is} < 0\right\}$$

If there is no $\bar{a}_{is} > 0$, then $\delta > -\infty$; if there is no $\bar{a}_{is} < 0$, then $\delta < \infty$.

For "greater than or equal to" constraints, δ changes sign. This follows, because we can analyze $\sum_{j=1}^n a_{ij}x_j \geq b_i$ in the form $-\sum_{j=1}^n a_{ij}x_j \leq -b_i$, so that $-(x_s + \delta)$ replaces $(x_s - \delta)$ in Equation (2). Another way of seeing this is to consider the change in the right-hand side of the form

$$\mathbf{b}(\delta) = \mathbf{b} + \delta\mathbf{e}_k$$

Thus, the new value of $\mathbf{x}_B$ is given by

$$\mathbf{x}_B(\delta) = \mathbf{B}^{-1}\mathbf{b}(\delta) = \mathbf{B}^{-1}\mathbf{b} + \delta\mathbf{B}^{-1}\mathbf{e}_k$$
$$= \bar{\mathbf{b}} + \delta\mathbf{B}^{-1}\mathbf{e}_k$$

However,

$$\bar{\mathbf{A}}_s = \mathbf{B}^{-1}\mathbf{e}_k \text{ for a "$\leq$" constraint}$$

and

$$\overline{\mathbf{A}}_s = -\mathbf{B}^{-1}\mathbf{e}_k \text{ for a "}\geq\text{" constraint}$$

since the column corresponding to the slack variable is $+\mathbf{e}_k$ for a "less than or equal to" constraint and $-\mathbf{e}_k$ for a "greater than or equal to" constraint. Thus, we have

$$\overline{\mathbf{b}} - \overline{\mathbf{A}}_s(-\delta) \geq \mathbf{0} \text{ for a "}\leq\text{" constraint}$$
$$\overline{\mathbf{b}} - \overline{\mathbf{A}}_s(+\delta) \geq \mathbf{0} \text{ for a "}\geq\text{" constraint}$$

Example 3

Consider Example 2 again and suppose that x_4 represents a slack variable for constraint 1 (originally written as a "$\leq$" constraint). If the coefficient b_1 is varied by an amount δ for

$$\overline{\mathbf{b}} = (3.2, 1.5, 5.6)^\mathrm{T} \text{ and } \overline{\mathbf{A}}_s = \overline{\mathbf{A}}_4 = (1.0, -0.5, 2.0)^\mathrm{T}$$

we have

$$x_1(\delta) = 3.2 - 1.0(-\delta)$$
$$x_2(\delta) = 1.5 + 0.5(-\delta)$$
$$x_6(\delta) = 5.6 - 2.0(-\delta)$$

Thus,

$$x_1(\delta) \geq 0 \text{ or } 3.2 - 1.0(-\delta) \geq 0 \left(\text{that is, } \delta \geq \frac{3.2}{-1.0} \right)$$

$$x_2(\delta) \geq 0 \text{ or } 1.5 + 0.5(-\delta) \geq 0 \left(\text{that is, } \delta \leq \frac{1.5}{0.5} \right)$$

$$x_6(\delta) \geq 0 \text{ or } 5.6 - 2.0(-\delta) \geq 0 \left(\text{that is, } \delta \geq \frac{5.6}{-2.0} \right)$$

Therefore, δ can vary in the following range.

$$\max \left\{ \frac{3.2}{-1.0}, \frac{5.6}{-2.0} \right\} \leq \delta \leq \min \left\{ \frac{1.5}{0.5} \right\}$$

$$-2.8 \leq \delta \leq 3.0$$

When it is desirable to perform proportional ranging on the vector $\mathbf{b}$, the preceding analysis is the same, but the unit vector $\mathbf{e}_k$ is replaced with $\mathbf{b}^*$, yielding $\mathbf{b}(\delta) = \mathbf{b} + \delta\mathbf{b}^*$. The current basis $\mathbf{B}$ will remain optimal as long as $\mathbf{x}_\mathrm{B}(\delta) = \mathbf{B}^{-1}\mathbf{b}(\delta) \geq \mathbf{0}$. This inequality gives rise to a range on δ within which primal feasibility is preserved (see Exercise 4).

Changes in Matrix Coefficients

The structural coefficients a_{ij} are usually known with much more certainty than the objective or right-hand-side coefficients, since they customarily represent some physical interaction between variables and are not subject to the same market fluctuations as are costs and demands. We shall consider changes in the coefficients of nonbasic variables only; changes in basic variables coefficients alter the basis matrix $\mathbf{B}$ and are rather complicated to analyze (see Murty [1983]).

Consider the jth nonbasic variable with corresponding column $\mathbf{A}_j$. If the ith element of $\mathbf{A}_j$ is changed by an amount δ, this affects the reduced cost $\bar{c}_j$ as follows.

If
$$\mathbf{A}_j(\delta) = \mathbf{A}_j + \delta \mathbf{e}_i$$

then
$$\bar{c}_j(\delta) = \pi\left(\mathbf{A}_j + \delta\mathbf{e}_i\right) - c_j$$

$$= \bar{c}_j + \delta\pi\mathbf{e}_i$$

$$= \bar{c}_j + \delta\pi_i$$

where π $(= \mathbf{c}_\mathrm{B}\mathbf{B}^{-1})$ is the dual vector. Thus, the solution remains optimal as long as $\bar{c}_j(\delta) \geq 0$. The corresponding range for δ is

$$\delta \geq -\frac{\bar{c}_j}{\pi_i} \text{ for } \pi_i > 0$$

$$\delta \leq -\frac{\bar{c}_j}{\pi_i} \text{ for } \pi_i < 0$$

4.2 THE DUAL LINEAR PROGRAM

When a solution is obtained for a linear program with the revised simplex method, the solution to a second model, called the dual problem, is readily available and provides useful information for sensitivity analysis, as we have just seen. There are several benefits to be gained from studying the dual problem, not the least of which is that it often has a practical interpretation that enhances the understanding of the original model. Moreover, it is sometimes easier to solve than the original model, and likewise provides the optimal solution to the original model at no extra cost. Duality also has important implications for the theoretical basis of mathematical programming algorithms. In this section, the dual problem is defined, the properties that link it to the original problem (called the primal problem) are listed, and the procedure for identifying the dual solution from the tableau is presented.

Definition of the Dual LP Model

In discussing duality, it is common to depart from the standard equality form of the LP given in Section 4.1 in order to highlight the symmetry of the primal–dual relationships. The dual model is derived by construction from the standard inequality form of the linear programming model as shown in Tables 4.1 and 4.2. All constraints of the primal model are written as "less than or equal to," and right-hand-side constants may be either positive or negative. In the primal model, it is assumed that there are n decision variables and m constraints,

Table 4.1 Matrix Definitions of the Primal and Dual Problems

Primal problem	Dual problem
Maximize $z_\mathrm{P} = \mathbf{cx}$	Minimize $z_\mathrm{D} = \pi\mathbf{b}$
subject to $\mathbf{Ax} \leq \mathbf{b}$	subject to $\pi\mathbf{A} \geq \mathbf{c}$
$\mathbf{x} \geq \mathbf{0}$	$\pi \geq \mathbf{0}$

Table 4.2   Algebraic Definitions of the Primal and Dual Problems

Primal problem	Dual problem
Maximize $z_P = c_1 x_1 + c_2 x_2 + \cdots + c_n x_n$	Minimize $z_D = b_1 \pi_1 + b_2 \pi_2 + \cdots + b_m \pi_m$
subject to $a_{11}x_1 + a_{12}x_2 + \cdots + a_{1n}x_n \le b_1$	subject to $a_{11}\pi_1 + a_{21}\pi_2 + \cdots + a_{m1}\pi_m \ge c_1$
$a_{21}x_1 + a_{22}x_2 + \cdots + a_{2n}x_n \le b_2$	$a_{12}\pi_1 + a_{22}\pi + \cdots + a_{m2}\pi_m \ge c_2$
$a_{m1}x_1 + a_{m2}x_2 + \cdots + a_{mn}x_n \le b_m$	$a_{1n}\pi_1 + a_{2n}\pi + \cdots + a_{mn}\pi_m \ge c_n$
$x_1 \ge 0, x_2 \ge 0, \ldots, x_n \ge 0$	$\pi_1 \ge 0, \pi_2 \ge 0, \ldots, \pi_m \ge 0$

thus **c** and **x** are n-dimensional vectors. The matrix of structural coefficients, **A**, has m rows and n columns. The dual model uses the same arrays of coefficients but arranged in a symmetric fashion. The dual vector π has m components.

Example 4

Maximize $z_P = 2x_1 + 3x_2$
subject to
$-x_1 + x_2 \le 5$
$x_1 + 3x_2 \le 35$
$x_1 \le 20$
$x_1 \ge 0, x_2 \ge 0$

Minimize $z_D = 5\pi_1 + 35\pi_2 + 20\pi_3$
subject to
$-\pi_1 + \pi_2 + \pi_3 \ge 2$
$\pi_1 + 3\pi_2 \ge 3$
$\pi_1 \ge 0, \pi_2 \ge 0, \pi_3 \ge 0$

The optimal solution to the primal problem, including slacks, is $\mathbf{x}^* = (20, 5, 20, 0, 0)^T$, with $z_P = 55$. The corresponding dual solution, including slacks, is $\pi^* = (0, 1, 1, 0, 0)$, with $z_D = 55$. Note that $z_P = z_D$. This is always the case, as will be shown presently.

From an algorithmic point of view, solving the primal problem with the dual simplex method is equivalent to solving the dual problem with the primal simplex method. When written in inequality form, the primal and dual models are related in the following ways.

- When the primal model has n variables and m constraints, the dual model has m variables and n constraints.
- The constraints for the primal model are all "less than or equal to," while the constraints for the dual model are all "greater than or equal to."
- The objective for the primal model is to maximize, while the objective for the dual model is to minimize.
- All variables for either problem are restricted to be nonnegative.
- For every primal constraint, there is a dual variable. Associated with the ith primal constraint is dual variable π_i. The dual objective function coefficient for π_i is the right-hand side of the ith primal constraint b_i.
- For every primal variable, there is a dual constraint. Associated with primal variable x_j is the jth dual constraint whose right-hand side is the primal objective function coefficient c_j.
- The number a_{ij} is, in the primal model, the coefficient of x_j in the ith constraint, whereas in the dual model, a_{ij} is the coefficient of π_i in the jth constraint.

Modifications of the Inequality Form

It is rare that a linear program is given in inequality form. This is especially true when the model has definitional constraints that are introduced for convenience, or when it has been prepared for the tableau simplex method, in which all RHS constants must be positive. Nevertheless, no matter how the primal model is stated, the corresponding dual model can always be found by first converting the primal model to the inequality form in Table 4.1 and then writing the dual model accordingly. For example, given an LP in standard equality form

$$\text{Maximize } z_\text{P} = \mathbf{c}\mathbf{x}$$
$$\text{subject to} \quad \mathbf{A}\mathbf{x} = \mathbf{b}, \mathbf{x} \geq \mathbf{0}$$

we can replace the constraints $\mathbf{A}\mathbf{x} = \mathbf{b}$ with two inequalities, $\mathbf{A}\mathbf{x} \leq \mathbf{b}$ and $-\mathbf{A}\mathbf{x} \leq -\mathbf{b}$, so that the coefficient matrix becomes $\begin{pmatrix} \mathbf{A} \\ -\mathbf{A} \end{pmatrix}$ and the right-hand-side vector becomes $\begin{pmatrix} \mathbf{b} \\ -\mathbf{b} \end{pmatrix}$. Introducing a partitioned dual row vector (γ_1, γ_2) with $2m$ components, the corresponding dual model is

$$\text{Minimize } z_\text{D} = \gamma_1 \mathbf{b} - \gamma_2 \mathbf{b}$$
$$\text{subject to} \quad \gamma_1 \mathbf{A} - \gamma_2 \mathbf{A} \leq \mathbf{c}$$
$$\gamma_1 \geq \mathbf{0}, \gamma_2 \geq \mathbf{0}$$

Letting $\pi = \gamma_1 - \gamma_2$, we can simplify the representation of this problem as shown in Table 4.3. This is the asymmetric form of the duality relation. Similar transformations can be worked out for any linear program by first putting the primal model into inequality form, constructing the dual model, and then simplifying the latter to account for special structure. We say that two LPs are *equivalent* if one can be transformed into the other so that feasible solutions, optimal solutions, and corresponding dual solutions are preserved—e.g., the inequality form in Table 4.1 and the equality form in Table 4.3 are equivalent primal–dual representations. This suggests the following result, which can be proven by constructing the appropriate models.

Proposition 1: Dual models of equivalent problems are equivalent. Let (P) refer to an LP and let (D) be its dual model. Let $(\overline{\text{P}})$ be an LP that is equivalent to (P). Let $(\overline{\text{D}})$ be the dual model of $(\overline{\text{P}})$. Then $(\overline{\text{D}})$ is equivalent to (D)—that is, they have the same optimal objective function values or they are both infeasible.

Table 4.4 describes more general relations between the primal and dual models that can be easily derived from the standard definition. They relate the sense of constraint i in the primal model with the sign restriction of π_i in the dual model, and the sign restriction of x_j in the primal model with the sense of constraint j in the dual. Note that when these alternative definitions are allowed, there are many ways to write the primal and dual problems; however, they are all equivalent.

Table 4.3   Equality Forms of the Primal and Dual Models

Primal problem	Dual problem
Maximize $z_\text{P} = \mathbf{c}\mathbf{x}$ subject to $\mathbf{A}\mathbf{x} = \mathbf{b}$ $\mathbf{x} \geq \mathbf{0}$	Minimize $z_\text{D} = \pi\mathbf{b}$ subject to $\pi\mathbf{A} \geq \mathbf{c}$

Table 4.4   Modifications in the Primal–Dual Formulations

Primal model	Dual model
Constraint i is $\leq$	$\pi_i \geq 0$
Constraint i is $=$	π_i is unrestricted
Constraint i is $\geq$	$\pi_i \leq 0$
$x_j \geq 0$	Constraint j is $\geq$
x_j is unrestricted	Constraint j is $=$
$x_j \leq 0$	Constraint j is $\leq$

Example 5

Maximize $z_P = -3x_1 - 2x_2$
subject to
$$-x_1 - x_2 = 8$$
$$x_1 + 2x_2 \geq 13$$
$$x_1 \geq 0, x_2 \text{ unrestricted}$$

Minimize $z_D = 8\pi_1 + 13\pi_2$
subject to
$$-\pi_1 + \pi_2 \geq -3$$
$$-\pi_1 + 2\pi_2 = -2$$
$$\pi_1 \text{ unrestricted}, \pi_2 \leq 0$$

Relations Between Primal and Dual Objective Function Values

There are several relationships between solutions to the primal and dual problems that are interesting to theoreticians, useful to algorithm developers, and important to analysts for interpreting solutions. We present these relationships as theorems and prove them for the primal–dual pair in Table 4.1; however, they are true for all primal–dual formulations. In what follows, $\mathbf{x}$ refers to any feasible solution of the primal problem; and $\boldsymbol{\pi}$ to any feasible solution of the dual problem; $\mathbf{x}^*$ and $\boldsymbol{\pi}^*$ are the respective optimal solutions, if they exist.

Theorem 1 (*Weak Duality*): In a primal–dual pair of LPs, let $\mathbf{x}$ be a primal feasible solution and let $z_P(\mathbf{x})$ be the corresponding value of the primal objective function that is to be maximized. Let $\boldsymbol{\pi}$ be a dual feasible solution and let $z_D(\boldsymbol{\pi})$ be the corresponding dual objective function that is to be minimized. Then, $z_P(\mathbf{x}) \leq z_D(\boldsymbol{\pi})$.

Theorem 1 states that the objective value for a feasible solution to the dual problem will always be greater than or equal to the objective function for a feasible solution to the primal problem. The following sequence demonstrates this result.

1. The primal solution is feasible by hypothesis: $\mathbf{Ax} \leq \mathbf{b}$

2. Premultiply both sides by $\boldsymbol{\pi}$: $\boldsymbol{\pi}\mathbf{Ax} \leq \boldsymbol{\pi}\mathbf{b}$

3. The dual solution is feasible by hypothesis: $\boldsymbol{\pi}\mathbf{A} \geq \mathbf{c}$

4. Post-multiply both sides by $\mathbf{x}$: $\boldsymbol{\pi}\mathbf{Ax} \geq \mathbf{cx}$

5. Combine the results of Steps 2 and 4: $\mathbf{cx} \leq \boldsymbol{\pi}\mathbf{Ax} \leq \boldsymbol{\pi}\mathbf{b}$ or $z_P(\mathbf{x}) \leq z_D(\boldsymbol{\pi})$

There are a several useful relationships that can be derived from Theorem 1. In particular,

- The value of $z_P(\mathbf{x})$ for any feasible $\mathbf{x}$ is a lower bound to $z_D(\boldsymbol{\pi}^*)$.

- The value of $z_D(\boldsymbol{\pi})$ for any feasible $\boldsymbol{\pi}$ is an upper bound to $z_P(\mathbf{x}^*)$.

- If there exists a feasible $\mathbf{x}$ and the primal problem is unbounded, there is no feasible $\boldsymbol{\pi}$.

- If there exists a feasible $\boldsymbol{\pi}$ and the dual problem is unbounded, there is no feasible $\mathbf{x}$.

- It is possible that there is no feasible $\mathbf{x}$ and no feasible $\boldsymbol{\pi}$.

The last point is demonstrated by the following example.

$$\begin{array}{ll}
\text{Maximize } z = x_1 + 3x_2 & \text{Minimize } z_D = 3\pi_1 - 5\pi_2 \\
\text{subject to} \quad x_1 - x_2 \leq 3 & \text{subject to} \quad \pi_1 - \pi_2 = 1 \\
\quad\quad\quad -x_1 + x_2 \leq -5 & \quad\quad\quad -\pi_1 + \pi_2 = 3 \\
\quad\quad\quad x_1, x_2 \text{ unrestricted} & \quad\quad\quad \pi_1 \geq 0, \pi_2 \geq 0
\end{array}$$

Theorem 2: (*Sufficient Optimality Criterion*): In a primal–dual pair of LPs, let $z_P(\mathbf{x})$ be the primal objective function and let $z_D(\boldsymbol{\pi})$ be the dual objective function. If $(\hat{\mathbf{x}}, \hat{\boldsymbol{\pi}})$ is a pair of primal and dual feasible solutions satisfying $z_P(\hat{\mathbf{x}}) = z_D(\hat{\boldsymbol{\pi}})$, then $\hat{\mathbf{x}}$ is an optimal solution of the primal problem and $\hat{\boldsymbol{\pi}}$ is an optimal solution of the dual problem.

The proof is presented seen in the following sequence:

1. Definition of optimality for the primal solution: $z_P(\hat{\mathbf{x}}) \leq z_P(\mathbf{x}^*)$
2. Feasible dual solution bound on z_P: $z_P(\mathbf{x}^*) \leq z_D(\boldsymbol{\pi}^*)$
3. Definition of optimality for the dual solution: $z_D(\boldsymbol{\pi}^*) \leq z_D(\hat{\boldsymbol{\pi}})$
4. Combine the results of Steps 1, 2, and 3: $z_P(\hat{\mathbf{x}}) \leq z_P(\mathbf{x}^*) \leq z_D(\boldsymbol{\pi}^*) \leq z_D(\hat{\boldsymbol{\pi}})$
5. Objectives are equal by hypothesis: $z_P(\hat{\mathbf{x}}) = z_D(\hat{\boldsymbol{\pi}})$
6. Combine Steps 4 and 5: $z_P(\hat{\mathbf{x}}) = z_P(\mathbf{x}^*) = z_D(\boldsymbol{\pi}^*) = z_D(\hat{\boldsymbol{\pi}})$

Therefore, $\hat{\mathbf{x}}$ and $\hat{\boldsymbol{\pi}}$ are optimal.

Theorem 2 states that equality of objective values implies optimality; moreover, we have

- Given feasible solutions $\mathbf{x}$ and $\boldsymbol{\pi}$ for a primal–dual pair, if the objective values are equal, they are both optimal.
- If $\mathbf{x}^*$ is an optimal solution to the primal problem, a finite optimal solution exists for the dual problem with objective value $z_P(\mathbf{x}^*)$.
- If $\boldsymbol{\pi}^*$ is an optimal solution to the dual problem, a finite optimal solution exists for the primal problem with objective value $z_D(\boldsymbol{\pi}^*)$.

Taking these results one step farther leads to the *Fundamental Duality Theorem* that is sometimes called *strong duality*.

Theorem 3 (*Strong Duality*): In a primal–dual pair of LPs, if either the primal problem or the dual problem has an optimal feasible solution, then the other problem does also and the two optimal objective values are equal.

We will prove the result for the primal and dual problems given in Table 4.3. Solving the primal problem by the simplex algorithm yields an optimal solution $\mathbf{x}_B = \mathbf{B}^{-1}\mathbf{b}$, and $\mathbf{x}_N = \mathbf{0}$ with $\overline{c}_j = \mathbf{c}_B\mathbf{B}^{-1}\mathbf{N} - \mathbf{c}_N \geq \mathbf{0}$, which can be written $[\mathbf{c}_B, \mathbf{c}_N - \mathbf{c}_B\mathbf{B}^{-1}(\mathbf{B}, \mathbf{N})] = \mathbf{c}_B\mathbf{B}^{-1}\mathbf{A} - \mathbf{c} \geq \mathbf{0}$. Now, if we define $\boldsymbol{\pi} = \mathbf{c}_B\mathbf{B}^{-1}$, we have $\boldsymbol{\pi}\mathbf{A} \geq \mathbf{c}$ and $z_P(\mathbf{x}) = \mathbf{c}_B\mathbf{x}_B = \mathbf{c}_B\mathbf{B}^{-1}\mathbf{b} = \boldsymbol{\pi}\mathbf{b} = z_D(\boldsymbol{\pi})$. By the sufficient optimality criterion (Theorem 2), $\boldsymbol{\pi}$ is a dual optimal solution. This completes the proof when the primal and dual problems are as stated.

In general, every LP can be transformed into an equivalent problem in standard equality form. This equivalent problem is of the same type as the primal problem in Table 4.3, hence the proof applies. Also, by Proposition 1, the dual model of the equivalent problem in standard form is equivalent to the dual model of the original problem. Thus, the theorem must hold for it, too.

Complementary Solutions

For the purposes of this section, it is helpful to repeat the definitions of the primal and dual problems given in Table 4.1 in a slightly different but equivalent form. Table 4.5 contains the modified representation, where $\mathbf{I}_m$ is an $m \times m$ identity matrix and $\mathbf{x}_s = (x_{s1}, \ldots, x_{sm})^T$ is an m-dimensional vector of slack variables.

Each structural variable x_j is associated with the dual constraint j, and each slack variable x_{si} is associated with the dual variable π_i. Recall that a basic solution is found by selecting a set of basic variables, constructing the basis matrix $\mathbf{B}$, and setting the nonbasic variables equal to zero. This yields the primal solution

$$\mathbf{x}_B = \mathbf{B}^{-1}\mathbf{b} \text{ with } z_P = \mathbf{c}_B\mathbf{B}^{-1}\mathbf{b}$$

The complementary dual solution associated with the basis is defined as

$$\pi = \mathbf{c}_B\mathbf{B}^{-1} \text{ with } z_D = \pi\mathbf{b} = \mathbf{c}_B\mathbf{B}^{-1}\mathbf{b}$$

Every basis defines complementary primal and dual solutions with identical objective function values.

Theorem 4 (*Optimality of Feasible Complementary Solutions*): Given the solution $\mathbf{x}_B$ determined from the basis $\mathbf{B}$, when $\mathbf{x}_B$ is optimal for the primal problem the complementary solution $\pi = \mathbf{c}_B\mathbf{B}^{-1}$ is optimal for the dual problem.

The proof of Theorem 4 is presented in the following sequence.

1. Primal and dual objective values are equal by construction:

$$z_P(\mathbf{x}_B) = \mathbf{c}_B\mathbf{x}_B = \mathbf{c}_B\mathbf{B}^{-1}\mathbf{b}$$
$$z_D(\pi) = \pi\mathbf{b} = \mathbf{c}_B\mathbf{B}^{-1}\mathbf{b}$$

2. The primal objective value for a basic solution when nonbasic variable x_k is allowed to increase is

$$z_P = \mathbf{c}_B\mathbf{B}^{-1}\mathbf{b} - (\pi\mathbf{A}_k - c_k)x_k$$

3. Since the primal solution is optimal:

$$\bar{c}_k = \pi\mathbf{A}_k - c_k \geq 0 \text{ or}$$
$$\pi\mathbf{A}_k \geq c_k$$

4. From Step 3, when x_k is a structural variable, dual constraint k is satisfied:

$$\pi\mathbf{A}_k \geq c_k$$

5. From Step 3, when the nonbasic variable is a slack variable x_{si}, π_i is nonnegative:

$$c_{si} = 0 \text{ and } \mathbf{A}_{si} = \mathbf{e}_i, \text{ so } \pi_i \geq 0$$

Table 4.5 Equivalent Primal–Dual Pair

Primal problem	Dual problem
Maximize $z_P = \mathbf{cx}$	Minimize $z_D = \pi\mathbf{b}$
subject to $\quad (\mathbf{A}_1, \ldots, \mathbf{A}_n)\mathbf{x} + \mathbf{I}_m\mathbf{x}_s = \mathbf{b}$	subject to $\quad \pi\mathbf{A}_j \geq c_j, j = 1, \ldots, n$
$\mathbf{x} \geq 0, \mathbf{x}_s \geq 0$	$\pi \geq 0$

6. For basic variables:

$$\mathbf{c}_B - \pi\mathbf{B} = \mathbf{c}_B - \mathbf{c}_B\mathbf{B}^{-1}\mathbf{B} = \mathbf{0}$$

7. From Step 6, when x_k is a structural variable and basic, the kth dual constraint is satisfied as an equality:

$$c_k - \pi\mathbf{A}_k = 0$$

8. From Step 6, when x_{si} is a slack variable and basic, π_i is zero:

$$c_{si} = 0 \text{ and } \mathbf{A}_{si} = \mathbf{e}_i, \text{ so } \pi_i = 0$$

All constraints are satisfied, so π is feasible; by Theorem 2, it must be optimal.

From Step 3 of the proof, it can be inferred that the reduced cost $\bar{c}_k$, for the primal variable x_k is equivalent to the dual slack π_{sk} for dual constraint k. Moreover, Steps 7 and 8 illustrate an important property known as *complementary slackness*. Given the primal–dual pair in Table 4.1, we have the following.

Complementary solutions property: For a given basis, when a primal structural variable is basic, the corresponding dual constraint is satisfied as an equality (the dual slack variable is zero), and when a primal slack variable is basic (the primal constraint is loose), the corresponding dual variable is zero.

This property holds whether or not the primal and dual solutions are feasible. We have already seen this in the simplex tableau—that is, when a primal structural variable is basic, its reduced cost (dual slack) is zero, and when a primal slack variable is basic, the corresponding structural dual variable is zero (Step 8 of proof).

Illustration of Complementary Solutions

Tables 4.6 and 4.7, respectively, show the 10 basic solutions for the primal and dual problems given in Example 4. In equality form, the primal problem has five variables and three constraints, whereas the dual problem has five variables and two constraints. Thus there are $\binom{n}{m} = \binom{5}{2} = 10$ potential bases in each case. The numbers in the leftmost column of each table identify the complementary solutions—e.g., solution 1 in Table 4.6 is complementary to solution 1 in Table 4.7. Note that for Number 7 in both tables, no solution exists because the columns associated with the variables (x_{s1}, x_{s2}, x_2), as well as the columns associated with (π_{s1}, π_3), do not form a basis.

The conditions derived in this section are illustrated by the data in Tables 4.6 and 4.7. Figure 4.1 shows the objective function values for the solutions that are feasible for the primal and dual problems. As can be seen, the objective value for every primal feasible solution provides a lower bound for the optimal dual objective ($z_D = 55$), and the objective value for every dual feasible solution provides an upper bound for the optimal primal objective ($z_P = 55$). These bounds converge to the optimal solution as might be expected; however, $z_P = z_D$ for all points.

The complementary solutions property is similarly exhibited by all points. For example, consider solution 9. Here, x_1 is basic in the primal problem and π_{s1} is nonbasic in the dual problem, so the first dual constraint is satisfied as an equality. Also, x_{s2} is basic and π_2, the corresponding dual variable, is nonbasic. These two observations can be written mathematically as $x_1\pi_{s1} = 0$ and $x_{s2}\pi_2 = 0$. In the next subsection, we provide a general statement of this result for all complementary pairs.

Table 4.6   Basic Solutions for Primal Problem in Example 4

Number	Basic variables	Nonbasic variables	x_1	x_2	x_{s1}	x_{s2}	x_{s3}	z_P	Primal status
1	x_{s1}, x_{s2}, x_{s3}	x_1, x_2	0	0	5	35	20	0	Feasible
2	x_{s1}, x_{s2}, x_{s3}	x_2, x_{s1}	−5	0	0	40	25	−10	Infeasible
3	x_{s2}, x_{s2}, x_{s3}	x_1, x_{s1}	0	5	0	20	20	15	Feasible
4	x_{s1}, x_{s1}, x_{s3}	x_2, x_{s2}	35	0	40	0	−15	70	Infeasible
5	x_{s1}, x_{s2}, x_{s3}	x_1, x_{s2}	0	11.67	−6.67	0	20	35	Infeasible
6	x_{s1}, x_{s2}, x_1	x_2, x_{s3}	20	0	25	15	0	40	Feasible
7	x_{s1}, x_{s2}, x_2	x_1, x_{s3}			No solution				
8	x_1, x_2, x_{s3}	x_{s1}, x_{s2}	5	10	0	0	15	40	Feasible
9	x_1, x_{s2}, x_2	x_{s1}, x_{s3}	20	25	0	−60	0	115	Infeasible
10	x_{s1}, x_1, x_2	x_{s2}, x_{s3}	20	5	20	0	0	55	Feasible

Table 4.7   Basic Solutions for Dual Problem in Example 4

Number	Basic variables	Nonbasic variables	π_{s1}	π_{s2}	π_1	π_2	π_3	z_D	Dual status
1	π_{s1}, π_{s2}	π_1, π_2, π_3	−2	−3	0	0	0	0	Infeasible
2	π_{s2}, π_1	π_{s1}, π_2, π_3	0	−5	−2	0	0	−10	Infeasible
3	π_{s1}, π_1	π_{s2}, π_2, π_3	−5	0	3	0	0	15	Infeasible
4	π_{s2}, π_2	π_1, π_{s1}, π_3	0	3	0	2	0	70	Feasible
5	π_{s1}, π_2	π_1, π_{s2}, π_3	−1	0	0	1	0	35	Infeasible
6	π_{s2}, π_3	π_1, π_2, π_{s1}	0	−3	0	0	2	40	Infeasible
7	π_{s1}, π_3	π_1, π_2, π_{s2}			No solution				
8	π_1, π_2	$\pi_{s1}, \pi_{s2}, \pi_3$	0	0	−0.75	1.25	0	40	Infeasible
9	π_1, π_3	$\pi_{s1}, \pi_2, \pi_{s2}$	0	0	3	0	5	115	Feasible
10	π_2, π_3	$\pi_1, \pi_{s1}, \pi_{s2}$	0	0	0	1	1	55	Feasible

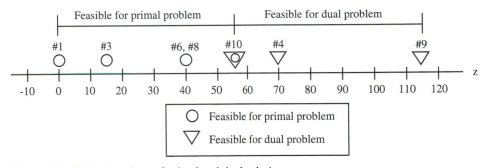

Figure 4.1   Objective values of primal and dual solutions

Finding Complementary Solutions for Standard Inequality Forms

The simplex algorithm solves the primal and dual problems simultaneously. This is obvious for the revised simplex method, which uses the complementary dual solution directly in the computations. When the primal and dual problems are in the standard inequality form given in Table 4.1, the tableau method provides all dual values in row 0. Table 4.8 shows row 0 of the general tableau. Note that the x's are labels but the entries in row 0 are numbers corresponding to the values of the dual variables. The dual slacks appear under the primal structural variable labels, and the dual structural variables appear under the primal slack variable labels.

To illustrate, solution 3 from Tables 4.6 and 4.7 is displayed in the tableau shown in Table 4.9. The primal solution is shown as the RHS vector. The complementary dual solution is given in row 0. In particular, $\pi_{s1} = -5$, $\pi_1 = 3$ and $z_D = 15$, whereas all other dual variables are zero.

The optimal tableau for example is shown in Table 4.10. From this tableau we can read both the primal and dual solutions. For the dual problem, the optimal solution is $\pi_2^* = \pi_3^* = 1$, and $\pi_1^* = \pi_{s1}^* = \pi_{s2}^* = 0$.

Finding Complementary Solutions for Nonstandard Forms

When the primal linear programming problem is in a nonstandard form with equality or "greater than or equal to" constraints, the dual variables do not appear directly in the tableau. For an equality in the primal model, there is no slack variable in the tableau; however, if a unit vector is inserted in phase 1 to represent the artificial variable for that constraint in the initial tableau, the dual variable will appear in row 0 under that artificial variable (assuming the artificial variable is given a zero objective coefficient in phase 2). For a "greater than or equal to" constraint, the dual variable associated with that constraint is the nega-

Table 4.8   Dual Variables Shown in the Simplex Tableau

Row	Basic variables	z	x_1	x_2	$\cdots$	x_n	x_{s1}	x_{s2}	$\cdots$	x_{sm}	RHS
0	z	1	π_{s1}	π_{s2}	$\cdots$	π_{sn}	π_1	π_2	$\cdots$	π_m	$z_D = z_P$

Table 4.9   Solution 3 from Tables 4.7 and 4.8

Row	Basic variables	z	x_1	x_2	x_{s1}	x_{s2}	x_{s3}	RHS
0	z	1	−5	0	3	0	0	15
1	x_2	0	−1	1	1	0	0	5
2	x_{s2}	0	4	0	−3	1	0	20
3	x_{s3}	0	1	0	0	0	1	20

Table 4.10   Solution 10 from Tables 4.6 and 4.7

Row	Basic variables	z	x_1	x_2	x_{s1}	x_{s2}	x_{s3}	RHS
0	z	1	0	0	0	1	1	55
1	x_2	0	0	1	0	1/3	−1/3	5
2	x_1	0	1	0	0	0	1	20
3	x_{s1}	0	0	0	1	−1/3	4/3	20

tive of the value appearing in row 0 of the column of the slack variable for the constraint. Finally, because the primal simplex method requires that all variables be nonnegative, nonstandard forms that contain unrestricted variables or variables constrained to be nonpositive are not allowed.

The foregoing developments are neatly summarized in the following theorem.

Theorem 5 (*Necessary and Sufficient Optimality Conditions*): Consider a primal–dual pair of LPs. Let $\mathbf{x}$ and $\boldsymbol{\pi}$ be the primal and dual variables and let $z_P(\mathbf{x})$ and $z_D(\boldsymbol{\pi})$ be the corresponding objective functions. If $\mathbf{x}$ is a primal feasible solution, it is optimal if and only if there exists a dual feasible solution $\boldsymbol{\pi}$ satisfying $z_P(\mathbf{x}) = z_D(\boldsymbol{\pi})$.

The "if" (sufficient) part of the proof follows directly from the sufficient optimality conditions of Theorem 2. The "only if" (necessary) part follows from the optimality of complementary solutions stated in Theorem 4. It also follows from the Fundamental Duality Theorem (Theorem 3).

Complementary Slackness

We have observed the complementary slackness property of complementary basic solutions, which holds for all bases whether optimal or not. To present this property in mathematical terms, we first recast the primal and dual problems given in Table 4.1 by introducing slack variables. Table 4.11 defines the revised models.

The vector of slacks for the primal problem is $\mathbf{x}_s = (x_{s1}, x_{s2}, \ldots, x_{sm})$, with x_{si} being the slack variable for the ith constraint. The vector of slacks for the dual problem is $\boldsymbol{\pi}_s = (\pi_{s1}, \pi_{s2}, \ldots, \pi_{sn})$, with π_{sj} being the slack variable for the jth constraint. $\mathbf{I}_m$ and $\mathbf{I}_n$ are identity matrices of sizes m and n, respectively. Both problems in equality form have $n + m$ variables.

The primal and dual variables are linked by identifying $n + m$ pairs with one variable in the pair from each problem.

Pair (x_j, π_{sj}) for $j = 1$ to n: The primal variable x_j is paired with the slack variable π_{sj} associated with the jth dual constraint.

Pair (x_{si}, π_i) for $i = 1$ to m: The dual variable π_i is paired with the slack variable x_{si} associated with the ith primal constraint.

Complementary slackness is the condition that at least one member of each pair is zero. For a particular pair (x_j, π_{sj}), the property implies that either x_j is zero or the corresponding dual constraint is satisfied as an equality. For the pair (x_{si}, π_i), it implies that either primal constraint i is satisfied as an equality or the corresponding dual variable is zero.

Theorem 6 (*Complementary Slackness*): The pairs $(\mathbf{x}, \mathbf{x}_s)$ and $(\boldsymbol{\pi}, \boldsymbol{\pi}_s)$ of primal and dual feasible solutions are optimal for their respective problems if and only if whenever a slack variable in one problem is strictly positive, the value of the associated nonnegative variable of the other problem is zero.

Table 4.11   Primal and Dual Problems with Slack Variables Added

Primal problem	Dual problem
Maximize $z_P = \mathbf{c}\mathbf{x}$	Minimize $z_D = \boldsymbol{\pi}\mathbf{b}$
subject to $\quad \mathbf{A}\mathbf{x} + \mathbf{I}_m\mathbf{x}_s = \mathbf{b}$	subject to $\quad \boldsymbol{\pi}\mathbf{A} - \boldsymbol{\pi}_s\mathbf{I}_n = \mathbf{c}$
$\quad\quad\quad \mathbf{x} \geq \mathbf{0}, \mathbf{x}_s \geq \mathbf{0}$	$\quad\quad\quad \boldsymbol{\pi} \geq \mathbf{0}, \boldsymbol{\pi}_s \geq \mathbf{0}$

For the primal–dual pair in Table 4.11, the theorem has the following interpretation. Whenever

$$x_{si} = b_i - \sum_{j=1}^{n} a_{ij}x_j > 0, \qquad \text{we have } \pi_i = 0 \qquad (3)$$

and whenever

$$\pi_{sj} = \sum_{i=1}^{m} a_{ij}\pi_i - c_j > 0, \qquad \text{we have } x_j = 0 \qquad (4)$$

Alternatively, we have

$$\pi_i x_{si} = \pi_i \left(b_i - \sum_{j=1}^{n} a_{ij}x_i \right) = 0, \qquad i = 1, \ldots, m \qquad (5)$$

$$x_j \pi_{sj} = x_j \left(\sum_{i=1}^{m} a_{ij}\pi_i - c_j \right) = 0, \qquad j = 1, \ldots, n \qquad (6)$$

The proof of this theorem is left as an exercise. In vector notation, Equations (5) and (6) can be written collectively as $\pi x_s = 0$ and $x\pi_s = 0$, respectively. Equations (3) and (5) require only that if $x_{si} > 0$, then $\pi_i = 0$. They do not require that if $x_{si} = 0$, then π_i must be > 0—that is, both x_{si} and π_i could be zero and the conditions of the theorem would be satisfied. Moreover, Equations (3) and (5) automatically imply that if $\pi_i > 0$, then $x_{si} = 0$. The same is true for Equations (4) and (6). For instance, if $x_j > 0$, then $\pi_{sj} = 0$.

The Complementary Slackness theorem does not say anything about the values of unrestricted variables (corresponding to equality constraints in the other problem) in a pair of optimal feasible solutions. This is the situation, for example, when the primal solution is written in equality form as in Table 4.4. It is concerned only with nonnegative variables of one problem and the slack variables corresponding to the associated inequality constraints in the other problem.

Economic Interpretation

Consider the primal–dual pair given in Table 4.1. Assume that the primal problem represents a chemical manufacturer that is under direction to limit its output of toxic wastes. Suppose that over a given period of time it produces n different types of chemicals that return a unit profit of c_j each for $j = 1, \ldots, n$, and that no more than b_i units of toxic waste i can result from the manufacturing process, $i = 1, \ldots, m$. Let a_{ij} be the amount of byproduct i generated by the manufacture of one unit of chemical j. The problem is to decide how many units of j to produce, denoted by x_j, so that no toxic waste levels are exceeded. These constraints can be written as $\sum_{j=1}^{n} a_{ij}x_j \leq b_i$ for all i.

To derive an equivalent dual problem, let $\pi_i \geq 0$ be the unit contribution to profit associated with byproduct i. Thus, π_i can be interpreted as the amount the company should be willing to pay to be allowed to generate one unit of toxic waste i. Consequently, the term $\sum_{i=1}^{m} \pi_i a_{ij}$ represents the implied contribution to profit associated with the current mix of toxic wastes when one unit of chemical j is produced. Because the same mix of wastes could probably be generated in other ways as well, no alternative use should be considered if it is less profitable than chemical j. This leads to the constraint $\sum_{i=1}^{m} \pi_i a_{ij} \geq c_j$ for all j; however, if a strict inequality holds for some j giving $\sum_{i=1}^{m} \pi_i a_{ij} > c_j$, better use of the permissible toxic waste levels can be found so it is optimal to set x_j equal to 0. If $x_j > 0$, the implied value of the toxic waste mix should be just equal to the unit profit for chemical j, yielding $\Sigma_{i=1}^{m} \pi_i a_{ij} = c_j$.

Similarly, if $\Sigma_{j=1}^{n} a_{ij} x_j < b_i$ for some i, the marginal contribution to profit associated with the ith toxic waste limit is zero, so we should set $\pi_i = 0$. If $\pi_i > 0$, the manufacturer should be generating as much byproduct i as permissible, implying that $\Sigma_{j=1}^{n} a_{ij} x_j = b_i$. These conditions are nothing more than complementary slackness in Equations (3) and (4). When they are satisfied, there is no incentive for the manufacturer to alter its production plan or to change the implied price structure.

The objective of the dual problem is to minimize pollution costs, which can be interpreted as minimizing the total implicit value of toxic wastes generated in the manufacturing process. At optimality, the minimum cost incurred is exactly the maximum revenue realized in the primal problem. This results in an economic equilibrium in which cost equals revenue, or

$$\sum_{i=1}^{m} \pi_i b_i = \sum_{j=1}^{n} c_j x_j$$

4.3   INTERIOR POINT METHODS

All forms of the simplex method reach the optimal solution by traversing a series of basic solutions. Because each basic solution represents an extreme point of the feasible region, the path followed by the algorithm moves around the boundary of the feasible region. In the worst case, it may be necessary to examine most if not all of the extreme points. This can be cripplingly inefficient given that the number of extreme points grows exponentially with n for m fixed.

The running time of an algorithm as a function of the problem size is known as its *computational complexity*. In practice, the simplex method works surprisingly well, exhibiting linear complexity—i.e., its running time is proportional to $m + n$. Researchers, however, have long tried to develop solution algorithms whose worst-case running times are polynomial functions of the problem size. The first success was attributed to the Russian mathematician Khachian, who proposed the *ellipsoid method* with a running time proportional to n^6 (see Bertsimas and Tsisiklis [1997] for a full discussion of this approach). Although the ellipsoid method is theoretically efficient, code developers were never able to realize an implementation that matched the performance of concurrent simplex codes.

Just about the time when interest in the ellipsoid method was waning, a new technique for solving linear programs was proposed by Karmarkar [1984]. His idea was to approach the optimal solution from the strict interior of the feasible region. This led to a series of *interior point methods* (IPMs), which combined the advantages of the simplex algorithm with the geometry of the ellipsoid algorithm. IPMs are of interest from a theoretical point of view because most have polynomial complexity, and from a practical point of view because they have produced solutions to many industrial problems that previously were intractable.

There are at least three major types of IPMs: (1) the potential reduction algorithm, which most closely embodies the constructs of Karmarkar; (2) the affine scaling algorithm, which is perhaps the simplest to implement; and (3) path following algorithms, which arguably combine excellent behavior in theory and practice. In this section, we highlight a member of the third category known as the *primal–dual path following algorithm,* which has become the method of choice in large-scale implementations.

The primal–dual path following algorithm is an example of an IPM that operates simultaneously on the primal and dual linear programming problems. Consider the following example, in which the slack variables in the dual model are denoted by the vector $\mathbf{z}$ to correspond to the notation in the literature.

Primal model	Dual model

Primal model

Maximize $z_P = 2x_1 + 3x_2$

subject to
$$2x_1 + x_2 + x_3 \qquad = 8$$
$$x_1 + 2x_2 \qquad + x_4 = 6$$
$$x_j \geq 0, \quad j = 1, \ldots, 4$$

Dual model

Minimize $z_D = 8\pi_1 + 6\pi_2$

subject to
$$2\pi_1 + \pi_2 - z_1 \qquad\qquad = 2$$
$$\pi_1 + 2\pi_2 \qquad - z_2 \qquad\qquad = 3$$
$$\pi_1 \qquad\qquad - z_3 \qquad = 0$$
$$\pi_2 \qquad\qquad - z_4 = 0$$
$$z_j \geq 0, \quad j = 1, \ldots, 4$$

Figures 4.2 and 4.3 show the progress of such an algorithm starting at point #1, which is in the interior of the feasible region. In general, the algorithm iteratively modifies the primal and dual solutions until the optimality conditions of primal feasibility, dual feasibility, and complementary slackness are satisfied to a sufficient degree. This event is signaled when the duality gap—the difference between the primal and dual objective functions—is sufficiently small. For the purposes of the discussion, "interior of the feasible region" is taken to mean that the values of the primal and dual variables are always strictly greater than zero.

The Primal and Dual Problems

In developing the algorithm, we will work with the primal and dual problems as defined in Table 4.12. The primal problem is assumed to consist of m nonredundant constraints in n variables and is given in equality form. This means that the n dual variables π are unrestricted in sign. The m-dimensional vector of nonnegative slack variables $\mathbf{z}$ transforms the dual inequalities into equations.

Basic Ideas

The use of path following algorithms to solve linear programs is based on the following three ideas.

1. The application of the Lagrange multiplier method of classical calculus to the transformation of an equality constrained optimization problem into an unconstrained one.

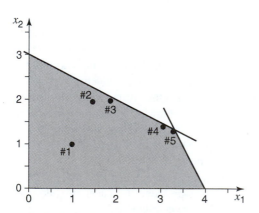

Figure 4.2   Path of primal solution for interior point algorithm

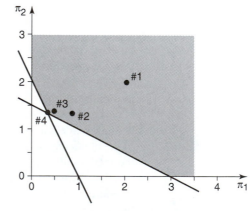

Figure 4.3   Path of dual solution for interior point algorithm

Table 4.12 Primal and Dual Problems for IPMs

Primal problem	Dual problem
Maximize $z_P = \mathbf{cx}$	Minimize $z_D = \boldsymbol{\pi}\mathbf{b}$
subject to $\quad \mathbf{Ax} = \mathbf{b}$	subject to $\quad \boldsymbol{\pi}\mathbf{A} - \mathbf{z} = \mathbf{c}$
$\mathbf{x} \geq 0$	$\mathbf{z} \geq 0$

2. The transformation of an inequality constrained optimization problem into a sequence of unconstrained problems by incorporating the constraints in a logarithmic barrier function that imposes a growing penalty as the boundary ($x_j = 0$, and $z_j = 0$ for all j) is approached.

3. The solution of a set of nonlinear equations using Newton's method, thereby arriving at a solution of the unconstrained optimization problem.

In the solution sequence of unconstrained problems, as the strength of the barrier function is decreased, the optimal solution follows a well-defined path (hence the term "path following") that ends at the optimal solution of the original problem. In subsequent subsections, we outline the three components of the algorithm.

The Lagrangian Approach

A well-known procedure for determining the minimum or maximum of a function subject to equality constraints is the Lagrange multiplier method. The details are given in Chapter 10, Nonlinear Programming Methods. For now, consider the general problem

$$\text{Maximize } f(\mathbf{x})$$
$$\text{subject to } g_i(\mathbf{x}) = 0, \ i = 1, \ldots, m$$

where $f(\mathbf{x})$ and $g_i(\mathbf{x})$ are scalar functions of the n-dimensional vector $\mathbf{x}$. The Lagrangian for this problem is

$$\mathscr{L}(\mathbf{x}, \boldsymbol{\pi}) = f(\mathbf{x}) - \sum_{i=1}^{m} \pi_i g_i(\mathbf{x})$$

where the variables $\boldsymbol{\pi} = (\pi_1, \pi_2, \ldots, \pi_m)$ are the Lagrange multipliers.

Necessary conditions for a stationary point (maximum or minimum) of the constrained optimization of $f(\mathbf{x})$ are zero values of the partial derivatives of the Lagrangian with respect to the components of $\mathbf{x}$ and $\boldsymbol{\pi}$—i.e.,

$$\frac{\partial \mathscr{L}}{\partial x_j} = 0, \ j = 1, \ldots, n \quad \text{and} \quad \frac{\partial \mathscr{L}}{\partial \pi_i} = 0, \ i = 1, \ldots, m$$

For linear constraints ($\mathbf{a}_i\mathbf{x} - b_i = 0$), the conditions are sufficient for a maximum if the function $f(\mathbf{x})$ is concave and sufficient for a minimum if $f(\mathbf{x})$ is convex.

The Barrier Approach

For the primal–dual pair of LPs in Table 4.12, the only essential inequalities are the nonnegativity conditions. The idea of the barrier approach, as developed by Fiacco and McCormick [1968], is to start from a point in the strict interior of the inequalities ($x_j > 0$, and $z_j > 0$ for all j) and construct a barrier that prevents any variable from reaching a boundary (e.g., $x_j = 0$). Adding $\log(x_j)$ to the objective function of the primal problem, for example, will cause the objective to decrease without bound as x_j approaches 0. The difficulty with this idea is that if the constrained optimal solution is on the boundary (that is, one or more $x_j^* = 0$, which is always the case in linear programming), then the barrier will prevent us from reaching it. To get around this difficulty, we use a barrier parameter μ that balances the contribution of the true objective function with that of the barrier term. The modified problems are presented in Table 4.13.

The parameter μ is required to be positive and controls the magnitude of the barrier term. Because function $\log(x)$ takes on very large negative values as x approaches zero from above, as long as the problem variables $\mathbf{x}$ and $\mathbf{z}$ remain positive the optimal solutions to the barrier problems will be interior to the nonnegative orthants ($x_j > 0$ and $z_j > 0$ for all j). The barrier term is added to the objective function for a maximization problem and subtracted for a minimization problem. The new formulations have nonlinear objective functions with linear equality constraints, and can be solved with the Lagrangian technique for $\mu > 0$ fixed. The solutions to these problems will approach the solution to the original problem as μ approaches zero.

Table 4.14 shows the development of the necessary optimal conditions for the barrier problems. These conditions are also sufficient, because the primal Lagrangian is concave and the dual Lagrangian is convex. Note that the dual variables $\boldsymbol{\pi}$ are the Lagrange multipliers of the primal problem, and the primal variables $\mathbf{x}$ are the Lagrange multipliers of the dual problem.

Thus, the optimal conditions are nothing more than primal feasibility, dual feasibility, and complementary slackness satisfied to within a tolerance of μ. Theory tells us that when μ goes to zero we get the solution to the original problem; however, we can't just set μ equal to zero, because that would destroy the convergence properties of the algorithm.

To facilitate the presentation, we define two $n \times n$ diagonal matrices containing the components of $\mathbf{x}$ and $\mathbf{z}$, respectively—i.e.,

$$\mathbf{X} = \text{diag}\{x_1, x_2, \ldots, x_n\}$$
$$\mathbf{Z} = \text{diag}\{z_1, z_2, \ldots, z_n\}$$

Also, let $\mathbf{e} = (1, 1, \ldots, 1)^{\mathrm{T}}$ be a column vector of size n. Using this notation, the necessary and sufficient conditions derived in Table 4.14 for the simultaneous solution of both the primal and dual barrier problems can be written as

Primal feasibility:	$\mathbf{Ax} - \mathbf{b} = \mathbf{0}$	(m linear equations)
Dual feasibility:	$\mathbf{A}^{\mathrm{T}}\boldsymbol{\pi}^{\mathrm{T}} - \mathbf{z} - \mathbf{c}^{\mathrm{T}} = \mathbf{0}$	(n linear equations)
μ-Complementary slackness:	$\mathbf{XZe} - \mu\mathbf{e} = \mathbf{0}$	(n nonlinear equations)

We must now solve this set of nonlinear equations for the variables $(\mathbf{x}, \boldsymbol{\pi}, \mathbf{z})$.

Table 4.13   Primal and Dual Barrier Problems

Primal problem	Dual problem
Maximize $B_{\mathrm{P}}(\mu) = \mathbf{cx} + \mu \sum_{j=1}^{n} \log(x_j)$	Minimize $B_{\mathrm{D}}(\mu) = \boldsymbol{\pi}\mathbf{b} - \mu \sum_{j=1}^{n} \log(z_j)$
subject to $\qquad \mathbf{Ax} = \mathbf{b}$	subject to $\qquad \boldsymbol{\pi}\mathbf{A} - \mathbf{z} = \mathbf{c}$

Table 4.14 Necessary Conditions for the Barrier Problems

Primal problem	Dual problem
Lagrangian	*Lagrangian*
$\mathscr{L}_P = \mathbf{c}\mathbf{x} + \mu \sum_{j=1}^{n} \log\left(x_j\right) - \boldsymbol{\pi}(\mathbf{A}\mathbf{x} - \mathbf{b})$	$\mathscr{L}_D = \boldsymbol{\pi}\mathbf{b} - \mu \sum_{j=1}^{n} \log\left(z_j\right) - (\boldsymbol{\pi}\mathbf{A} - \mathbf{z} - \mathbf{c})\mathbf{x}$

Primal problem	Dual problem
$\dfrac{\partial \mathscr{L}_P}{\partial x_j} = 0$	$\dfrac{\partial \mathscr{L}_D}{\partial z_j} = 0$
$c_j - \sum_{i=1}^{m} a_{ij}\pi_j + \dfrac{\mu}{x_j} = 0$	$-\dfrac{\mu}{z_j} + x_j = 0$
$-z_j + \dfrac{\mu}{x_j} = 0$	$z_j x_j = \mu, \, j = 1, \ldots, n$
$z_j x_j = \mu, \, j = 1, \ldots, n$	(μ-complementary slackness)
(μ-complementary slackness)	
$\dfrac{\partial \mathscr{L}_P}{\partial \pi_i} = 0$	$\dfrac{\partial \mathscr{L}_D}{\partial \pi_i} = 0$
$\sum_{j=1}^{n} a_{ij}x_j = b_i, \, i = 1, \ldots, m$	$\sum_{j=1}^{n} a_{ij}x_j = b_i, \, i = 1, \ldots, m$
(primal feasibility)	(primal feasibility)
	$\dfrac{\partial \mathscr{L}_D}{\partial x_j} = 0$
	$\sum_{i=1}^{m} a_{ij}\pi_j - z_j = c_j, \, j = 1, \ldots, n$
	(dual feasibility)

Stationary Solutions Using Newton's Method

Newton's method is an iterative procedure for numerically solving a set of nonlinear equations. To motivate the discussion, consider the single variable problem of finding y to satisfy the nonlinear equation

$$f(y) = 0$$

where f is once continuously differentiable. Let y^* be the unknown solution. At some point y^k, one can calculate the functional value $f(y^k)$ and the first derivative $f'(y^k)$. Using the derivative as a first-order approximation of how the function changes with y, we can predict the amount of change $\Delta = y^{k+1} - y^k$ required to bring the function to zero. This idea is illustrated in Figure 4.4.

Taking the first-order Taylor series expansion of $f(y)$ around y^k yields

$$f\left(y^{k+1}\right) \approx f\left(y^k\right) + \Delta f'\left(y^k\right)$$

Setting the approximation of $f(y^{k+1})$ equal to zero and solving for Δ yields

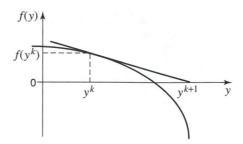

Figure 4.4 Newton's method for a function of a single variable

$$\Delta = \frac{-f\left(y^k\right)}{f'\left(y^k\right)}$$

The point $y^{k+1} = y^k + \Delta$ is an approximate solution of the equation. It can be shown that if one starts at a point y^0 sufficiently close to y^*, then $y^k \to y^*$ as $k \to \infty$.

Newton's method can be extended to multidimensional functions. Consider the general problem of finding the r-dimensional vector $\mathbf{y}$ that solves the set of r equations

$$f_i(\mathbf{y}) = 0, \, i = 1, \ldots, r$$
$$\text{or } \mathbf{f}(\mathbf{y}) = \mathbf{0}$$

Let the unknown solution to the equations be $\mathbf{y}^*$. The $r \times r$ Jacobian matrix describes the first-order variations of these functions with the components of $\mathbf{y}$. The Jacobian at $\mathbf{y}^k$ is

$$\mathbf{J}(\mathbf{y}^k) = \begin{pmatrix} \dfrac{\partial f_1}{\partial y_1} & \dfrac{\partial f_1}{\partial y_2} & \cdots & \dfrac{\partial f_1}{\partial y_r} \\[2ex] \dfrac{\partial f_2}{\partial y_1} & \dfrac{\partial f_2}{\partial y_2} & \cdots & \dfrac{\partial f_2}{\partial y_r} \\[2ex] \vdots & \vdots & & \vdots \\[2ex] \dfrac{\partial f_r}{\partial y_1} & \dfrac{\partial f_r}{\partial y_2} & \cdots & \dfrac{\partial f_r}{\partial y_r} \end{pmatrix}$$

All the partial derivatives are evaluated at $\mathbf{y}^k$. Now, taking the first-order Taylor series expansion around the point $\mathbf{y}^k$ and setting it equal to zero yields

$$\mathbf{f}(\mathbf{y}^k) + \mathbf{J}(\mathbf{y}^k)\mathbf{d} = \mathbf{0}$$

where $\mathbf{d} = \mathbf{y}^{k+1} - \mathbf{y}^k$ is an r-dimensional vector whose components represent the change in position for the $(k+1)$th iteration. Solving for $\mathbf{d}$ leads to

$$\mathbf{d} = -\mathbf{J}(\mathbf{y}^k)^{-1}\mathbf{f}(\mathbf{y}^k)$$

The point $\mathbf{y}^{k+1} = \mathbf{y}^k + \mathbf{d}$ is an approximation of the solution of the set of equations. Once again, if one starts at an initial point $\mathbf{y}^0$ sufficiently close to $\mathbf{y}^*$, the value of $\mathbf{y}^k$ will approach $\mathbf{y}^*$ for large values of k.

Newton's Method for Solving Barrier Problems

We are now ready to use Newton's method to solve the optimality conditions for the barrier problems given in Table 4.14 for a fixed value of μ. For $\mathbf{y} = (\mathbf{x}, \boldsymbol{\pi}, \mathbf{z})$ and $r = 2n + m$, the corresponding equations and Jacobian are

$$\left.\begin{array}{l} \mathbf{Ax - b = 0} \\ \mathbf{A}^{\mathrm{T}}\boldsymbol{\pi}^{\mathrm{T}} - \mathbf{z} - \mathbf{c}^{\mathrm{T}} = \mathbf{0} \\ \mathbf{XZe} - \mu\mathbf{e} = \mathbf{0} \end{array}\right\} => \mathbf{J(y)} = \begin{pmatrix} \mathbf{A} & \mathbf{0} & \mathbf{0} \\ \mathbf{0} & \mathbf{A}^{\mathrm{T}} & \mathbf{-I} \\ \mathbf{Z} & \mathbf{0} & \mathbf{X} \end{pmatrix}$$

Where $\mathbf{e}$ is an n-dimensional column vector of 1's. Assume that we have a starting point $(\mathbf{x}^0, \boldsymbol{\pi}^0, \mathbf{z}^0)$ satisfying $\mathbf{x}^0 > \mathbf{0}$ and $\mathbf{z}^0 > \mathbf{0}$, and denote by

$$\delta_{\mathrm{P}} = \mathbf{b} - \mathbf{Ax}^0$$
$$\delta_{\mathrm{D}} = \mathbf{c}^{\mathrm{T}} - \mathbf{A}^{\mathrm{T}}(\boldsymbol{\pi}^0)^{\mathrm{T}} + \mathbf{z}^0$$

the primal and dual residual vectors at this starting point. The optimality conditions can be written as

$$\mathbf{J(y)d = -f(y)} \tag{7}$$

$$\begin{pmatrix} \mathbf{A} & \mathbf{0} & \mathbf{0} \\ \mathbf{0} & \mathbf{A}^{\mathrm{T}} & \mathbf{-I} \\ \mathbf{Z} & \mathbf{0} & \mathbf{X} \end{pmatrix}\begin{pmatrix} \mathbf{d}_x \\ \mathbf{d}_\pi \\ \mathbf{d}_z \end{pmatrix} = \begin{pmatrix} \delta_{\mathrm{P}} \\ \delta_{\mathrm{D}} \\ \mu\mathbf{e} - \mathbf{XZe} \end{pmatrix}$$

where the $(2n + m)$-dimensional vector $\mathbf{d} \equiv (\mathbf{d}_x, \mathbf{d}_\pi, \mathbf{d}_z)^{\mathrm{T}}$ in Equations (7) is called the New-ton direction. We must now solve for the individual components of $\mathbf{d}$.

In explicit form, the preceding system is

$$\mathbf{Ad}_x = \delta_{\mathrm{P}}$$
$$\mathbf{A}^{\mathrm{T}}\mathbf{d}_\pi - \mathbf{d}_z = \delta_{\mathrm{D}}$$
$$\mathbf{Zd}_x + \mathbf{Xd}_z = \mu\mathbf{e} - \mathbf{XZe}$$

The first step is to find $\mathbf{d}_\pi$. With some algebraic manipulation, which is left as an exercise, we get

$$(\mathbf{AZ}^{-1}\mathbf{XA}^{\mathrm{T}})\mathbf{d}_\pi = -\mathbf{b} + \mu\mathbf{AZ}^{-1}\mathbf{e} + \mathbf{AZ}^{-1}\mathbf{X}\delta_{\mathrm{D}}$$

or

$$\mathbf{d}_\pi = (\mathbf{AZ}^{-1}\mathbf{XA}^{\mathrm{T}})^{-1}(-\mathbf{b} + \mu\mathbf{AZ}^{-1}\mathbf{e} + \mathbf{AZ}^{-1}\mathbf{X}\delta_{\mathrm{D}}) \tag{8}$$

Note that $\mathbf{Z}^{-1} = \mathrm{diag}\{1/z_1, 1/z_2, \ldots, 1/z_n\}$ and is trivial to compute. Further multiplications and substitutions lead to

$$\mathbf{d}_z = -\delta_{\mathrm{D}} + \mathbf{A}^{\mathrm{T}}\mathbf{d}_\pi \tag{9}$$

and

$$\mathbf{d}_x = \mathbf{Z}^{-1}(\mu\mathbf{e} - \mathbf{XZe} - \mathbf{Xd}_z) \tag{10}$$

From these results, we can see in part why it is necessary to remain in the interior of the feasible region. In particular, if either $\mathbf{Z}^{-1}$ or $\mathbf{X}^{-1}$ does not exist, the procedure breaks down.

Once the Newton direction has been computed, $\mathbf{d}_x$ is used as a search direction in the x space and $(\mathbf{d}_\pi, \mathbf{d}_z)$ as a search direction in the (π, z)-space. That is, we move from the current point $(\mathbf{x}^0, \boldsymbol{\pi}^0, \mathbf{z}^0)$ to a new point $(\mathbf{x}^1, \boldsymbol{\pi}^1, \mathbf{z}^1)$ by taking a step in the direction $(\mathbf{d}_x, \mathbf{d}_\pi, \mathbf{d}_z)$. The step sizes α_{P} and α_{D} are chosen in the two spaces to preserve $\mathbf{x} > \mathbf{0}$ and $\boldsymbol{\pi} > \mathbf{0}$. This requires a ratio test similar to that performed in the simplex algorithm. The simplest approach is to use

$$\alpha_{\mathrm{P}} = \gamma \min_{j} \left\{ \frac{-x_j}{(\mathbf{d}_x)_j} : (\mathbf{d}_x)_j < 0 \right\}$$

$$\alpha_{\mathrm{D}} = \gamma \min_{j} \left\{ \frac{-z_j}{(\mathbf{d}_z)_j} : (\mathbf{d}_z)_j < 0 \right\}$$

where γ is the step size factor that keeps us from actually touching the boundary. Typically, $\gamma = 0.995$. The notation $(\mathbf{d}_x)_j$ refers to the jth component of the vector $\mathbf{d}_x$. The new point is

$$\mathbf{x}^1 = \mathbf{x}^0 + \alpha_{\mathrm{P}} \mathbf{d}_x$$
$$\boldsymbol{\pi}^1 = \boldsymbol{\pi}^0 + \alpha_{\mathrm{D}} \mathbf{d}_\pi$$
$$\mathbf{z}^1 = \mathbf{z}^0 + \alpha_{\mathrm{D}} \mathbf{d}_z$$

which completes one iteration. Ordinarily, we would now resolve Equations (8) to (10) at $(\mathbf{x}^1, \boldsymbol{\pi}^1, \mathbf{z}^1)$ to find a new Newton direction and hence a new point. Rather than iterating in this manner until the system converges for the current value of μ, it is much more efficient to reduce μ at every iteration. The primal–dual method itself suggests how to update μ.

It is straightforward to show that the Newton step reduces the duality gap—the difference between the dual and primal objective values at the current point. Assume that $\mathbf{x}^0$ is primal feasible and that $(\boldsymbol{\pi}^0, \mathbf{z}^0)$ is dual feasible (Lustig et al. [1994] discuss the more general case). Let "gap(0)" denote the current duality gap.

$$\begin{aligned} \mathrm{gap}(0) &= \boldsymbol{\pi}^0 \mathbf{b} - \mathbf{c}\mathbf{x}^0 \\ &= \boldsymbol{\pi}^0(\mathbf{A}\mathbf{x}^0) - (\boldsymbol{\pi}^0\mathbf{A} - \mathbf{z}^0)^{\mathrm{T}}\mathbf{x}^0 \qquad \text{(primal and dual feasibility)} \\ &= (\mathbf{z}^0)^{\mathrm{T}}\mathbf{x}^0 \end{aligned}$$

If we let $\alpha = \min\{\alpha_{\mathrm{P}}, \alpha_{\mathrm{D}}\}$, then

$$\mathrm{gap}(\alpha) = (\mathbf{z}^0 + \alpha \mathbf{d}_z)^{\mathrm{T}} (\mathbf{x}^0 + \alpha \mathbf{d}_x)$$

and, with a little algebra, it can be shown that $\mathrm{gap}(\alpha) < \mathrm{gap}(0)$ as long as

$$\mu < \frac{\mathrm{gap}(0)}{n} \tag{11}$$

In our computations, we have used

$$\mu^k = \frac{\mathrm{gap}(\alpha^k)}{n^2} = \frac{(\mathbf{z}^k)^{\mathrm{T}}\mathbf{x}^k}{n^2}$$

which indicates that the value of μ^k is proportional to the duality gap. Termination occurs when the gap is sufficiently small—say, less than 10^{-8}.

Iterative Primal–Dual Interior Point Algorithm

We now summarize the basic steps of the algorithm. The following inputs are assumed.

1. The data of the problem $(\mathbf{A}, \mathbf{b}, \mathbf{c})$, where the $m \times n$ matrix $\mathbf{A}$ has full row rank
2. Initial primal and dual feasible solutions $\mathbf{x}^0 > \mathbf{0}$, $\mathbf{z}^0 > \mathbf{0}$, $\boldsymbol{\pi}^0$
3. The optimality tolerance $\varepsilon > 0$ and the step size parameter $\gamma \in (0, 1)$

Step 1: (*Initialization*) Start with some feasible point $\mathbf{x}^0 > \mathbf{0}$, $\mathbf{z}^0 > \mathbf{0}$, $\boldsymbol{\pi}^0$ and set the iteration counter $k = 0$.

Step 2: (*Optimality Test*) If $(\mathbf{z}^k)^T\mathbf{x}^k < \varepsilon$, stop; otherwise, go to Step 3.

Step 3: (*Compute Newton Directions*) Let

$$\mathbf{X}^k = \mathrm{diag}\left\{x_1^k, x_2^k, \ldots x_n^k\right\}, \quad \mathbf{Z}^k = \mathrm{diag}\left\{z_1^k, z_2^k, \ldots z_n^k\right\}, \quad \mu^k = \frac{(\mathbf{z}^k)^T\mathbf{x}^k}{n^2}$$

Solve the following linear system equivalent to Equation (7) to get $\mathbf{d}_x^k$, $\mathbf{d}_\pi^k$, and $\mathbf{d}_z^k$.

$$\mathbf{A}\mathbf{d}_x^k = \mathbf{0}$$
$$\mathbf{A}^T\mathbf{d}_\pi^k - \mathbf{d}_z^k = \mathbf{0}$$
$$\mathbf{Z}\mathbf{d}_x^k + \mathbf{X}\mathbf{d}_z^k = \mu\mathbf{e} - \mathbf{X}\mathbf{Z}\mathbf{e}$$

Note that $\delta_P = \mathbf{0}$ and $\delta_D = \mathbf{0}$ as a result of the feasibility of the initial point.

Step 4: (*Find Step Lengths*) Let

$$\alpha_P = \gamma \min_j\left\{\frac{-x_j^k}{(\mathbf{d}_x^k)_j} : (\mathbf{d}_x^k)_j < 0\right\}, \quad \alpha_D = \gamma \min_j\left\{\frac{-z_j^k}{(\mathbf{d}_z^k)_j} : (\mathbf{d}_z^k)_j < 0\right\}$$

Step 5: (*Update Solution*) Take a step in the Newton direction to get

$$\mathbf{x}^{k+1} = \mathbf{x}^k + \alpha_P\mathbf{d}_x^k$$
$$\boldsymbol{\pi}^{k+1} = \boldsymbol{\pi}^k + \alpha_D\mathbf{d}_\pi^k$$
$$\mathbf{z}^{k+1} = \mathbf{z}^k + \alpha_D\mathbf{d}_z^k$$

Put $k \leftarrow k + 1$ and go to Step 2.

Implementation Issues

From a theoretical point of view, it has been shown that for a slightly different choice of μ^k and step length, the algorithm takes about

$$\sqrt{n} \log(\varepsilon^0/\varepsilon)$$

iterations to reduce the duality gap from ε^0 to ε in the worst case, where $\varepsilon^0 = \mathrm{gap}(0)$. The observed average behavior is closer to $O[\log(n)\log(\varepsilon^0/\varepsilon)]$. In fact, extensive testing indicates that IPMs are surprisingly insensitive to problem size. Convergence usually takes from 20 to 80 iterations. Most of the work is at Step 2, where the system of linear equations has to be solved. The most computationally intensive component of this step involves forming the matrix $(\mathbf{A}\mathbf{Z}^{-1}\mathbf{X}\mathbf{A}^T)$ and then inverting it to get $\mathbf{d}_\pi$, as indicated in Equation (8). This consumes about 90% of the overall effort. It should be mentioned that one iteration of an IPM is much more time consuming than one iteration of a simplex-type algorithm as measured by a pivot.

Virtually all interior point methods, whether path following, affine scaling, or potential reduction, require the solution of a linear system similar to that in Step 2. Much research has gone into figuring out how to do this efficiently. In today's codes, the matrix $\mathbf{A}\mathbf{Z}^{-1}\mathbf{X}\mathbf{A}^T$ is never inverted explicitly; Cholesky factorization and back-substitution are used to solve a series of related triangular linear systems. Specifically, this involves finding the lower triangular $n \times n$ matrix $\mathbf{L}$ such that $\mathbf{AZ}^{-1}\mathbf{XA}^{\mathrm{T}} = \mathbf{LL}^{\mathrm{T}}$ and then solving

$$\mathbf{LL}^{\mathrm{T}}\mathbf{d}_\pi = \mathbf{r}$$

where $\mathbf{r}$ is the right-hand side of Equation (8). This can be done by first solving $\mathbf{Lg} = \mathbf{r}$ for $\mathbf{g}$ and then solving $\mathbf{L}^{\mathrm{T}}\mathbf{d}_\pi = \mathbf{g}$ for $\mathbf{d}_\pi$, both by simple substitution. This approach can be implemented in a highly efficient manner when $\mathbf{AZ}^{-1}\mathbf{XA}^{\mathrm{T}}$ is a large, sparse matrix. Although the matrix changes from iteration to iteration, the pattern of zeros remains the same, which implies that the calculations can be streamed by eliminating almost all multiplications by zero.

In the presentation of the path following algorithm, it was assumed that an initial feasible point was available. Surprisingly, this need not be the case for the primal–dual approach. The algorithm can be started at any interior point with the values of δ_P and δ_D in Step 2 set appropriately. Other IPMs, including the primal (only) path following method, require the equivalent of a phase 1 step to find a feasible point. Rather than explicitly solving a phase 1 problem, however, such methods generally take a big-M approach that involves the introduction of one artificial variable x_{n+1} in the primal problem.

To conclude this section, we offer a few comments on the relative performance of simplex-based methods versus interior point methods. Although the determination of which is better depends on the problem and the particular instance, some qualitative insights are available. In particular, simplex methods tend to bog down when solving large, massively degenerate problems. Such problems typically arise in transportation and scheduling applications that have network structures at their core. The accompanying formulations are often very sparse, thus making them prime candidates for IPMs.

When the formulations are dense, or in some cases when only a few columns of the $\mathbf{A}$ matrix are dense, IPMs are not likely to perform well. Such problems give rise to dense Cholesky factors $\mathbf{L}$, which nullify the efficiencies associated with solving triangular systems. Dense problems, however, are not usually degenerate, so simplex-type methods are the better choice.

Example 6

To illustrate the computations, consider the problem presented at the beginning of this section, which is in the desired form (see Table 4.12). In particular, we have $n = 4$ and $m = 2$, with data

$$\mathbf{c} = (2, 3, 0, 0), \quad \mathbf{A} = \begin{pmatrix} 2 & 1 & 1 & 0 \\ 1 & 2 & 0 & 1 \end{pmatrix}, \quad \mathbf{b} = \begin{pmatrix} 8 \\ 6 \end{pmatrix}$$

and variable definitions

$$\mathbf{x} = \begin{pmatrix} x_1 \\ x_2 \\ x_3 \\ x_4 \end{pmatrix}, \quad \mathbf{z} = \begin{pmatrix} z_1 \\ z_2 \\ z_3 \\ z_4 \end{pmatrix}, \quad \boldsymbol{\pi} = (\pi_1, \pi_2)$$

For an initial interior solution, we take $\mathbf{x}^0 = (1, 1, 5, 3)$, $\mathbf{z}^0 = (4, 3, 2, 2)$, and $\boldsymbol{\pi}^0 = (2, 2)$, as shown in Figure 4.5[1]. The primal residual vector δ_P and the dual residual (feasible)

---

[1] The tables in this section were constructed by the Interior Point demonstration in the Teach LP add-in.

vector $\boldsymbol{\delta}_\mathrm{D}$ are both zero, since $\mathbf{x}^0$ and $\mathbf{z}^0$ are interior and feasible to their respective problems. The column labeled "Comp. slack" is the μ-complementary slackness vector computed as

$$\mathbf{XZe} - \mu\mathbf{e}$$

Since this vector is not zero, the solution is not optimal.

We solve the set of equations in Step 3 of the algorithm to find the Newton directions and the step sizes. The results are shown in Figure 4.6. The first three columns refer to the directions $\mathbf{d}_\pi, \mathbf{d}_z$, and $\mathbf{d}_x$, and the last two columns are the ratios that determine the step sizes α_D and α_P, respectively. Recall that these values are used to ensure that the new point $(\mathbf{z}^1, \mathbf{x}^1)$ remains interior to the feasible region. For the calculations here, we use $\gamma = 0.8$ to avoid getting too close to the boundaries. As indicated in Step 4, the minimum of each ratio is multiplied by γ to find the step sizes for the dual and primal variables. The cell labeled "Z step" is the value of α_D, and the cell labeled "X step" is α_P.

Figure 4.7 shows next solution, obtained by multiplying $\mathbf{d}_\pi$ and $\mathbf{d}_z$ by α_D and multiplying $\mathbf{d}_x$ by α_P, as indicated in Step 5, to get the vectors $\boldsymbol{\pi}^1$, $\mathbf{z}^1$ and $\mathbf{x}^1$. The primal and dual solutions remain interior, and the gap is reduced from 23 to 6.36.

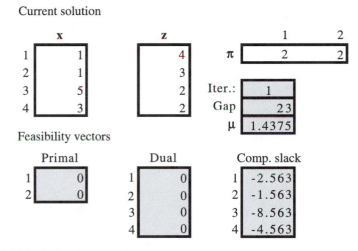

Figure 4.5   Initial solution for example

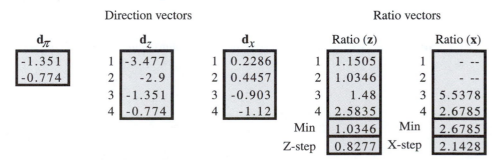

Figure 4.6   Direction and ratio vectors for initial solution

Current solution

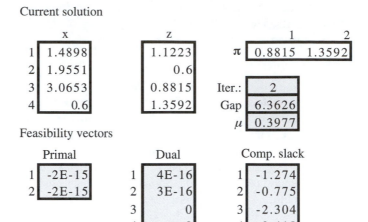

Figure 4.7   The solution at the second iteration

Table 4.15 shows the nine iterations necessary to obtain a gap of 0.0001. Observe that z_P increases to 10.667 from its initial value whereas z_D decreases to 10.667 from its initial value. The optimal solution is at the intersection of the two constraints, where the slack variables x_3 and x_4 are both zero. These values are reached only in the limit, because they are at the boundary of the feasible region, but we can get as close as desired (arbitrarily close) by adjusting the stopping criterion.

Table 4.15   Sequence of Solutions

Iteration	Primal solution					Dual solution		
	z_P	x_1	x_2	x_3	x_4	z_D	π_1	π_2
1	5	1	1	5	3	28	2	2
2	8.845	1.4898	1.9551	3.0653	0.6000	15.208	0.8815	1.3592
3	9.764	1.8872	1.9964	2.2292	0.1200	11.761	0.3964	1.4317
4	10.419	3.0608	1.4325	0.4458	0.0741	11.025	0.3111	1.4228
5	10.598	3.2836	1.3437	0.0892	0.029	10.744	0.3276	1.3538
6	10.651	3.3201	1.3370	0.0227	0.0058	10.685	0.3328	1.3371
7	10.664	3.3306	1.3341	0.0046	0.0012	10.670	0.3332	1.3341
8	10.666	3.3328	1.3335	0.0009	0.0002	10.667	0.3333	1.3335
9	10.667	3.3332	1.3334	0.0002	10^{-5}	10.667	0.3333	1.3334

EXERCISES

1. The following linear program has optimal solution $\mathbf{x}^* = (5, 0, 0, 0, 2)$ with $z^* = 60$ and $\mathbf{x}_B = (x_1, x_2, x_5)$.

$$\text{Maximize } z = 12x_1 + 6x_2 + 2x_3$$

$$\text{subject to} \quad 2x_1 + 4x_2 - x_3 + x_4 \quad = 10$$

$$x_1 + \quad 3x_3 \quad = 5$$

$$x_1 + 2x_2 \quad - x_5 = 3$$

$$x_j \geq 0, \, j = 1, \dots, 5$$

Compute simple ranges over which the basis does not change for each objective function coefficient and right-hand-side constant.

2. Consider the linear program

$$\text{Maximize } z = 5x_1 - 3x_2 + 4x_3 - x_4$$

$$\text{subject to} \quad 3x_1 + 4x_2 + 3x_3 - 5x_4 \leq 13$$

$$5x_1 - 3x_2 + x_3 + 2x_4 \leq 5$$

$$x_j \geq 0, j = 1, \ldots, 4$$

At optimality, x_3 and x_4 are basic, with basis inverse

$$\mathbf{B}^{-1} = \begin{pmatrix} 2/11 & 5/11 \\ -1/11 & 3/11 \end{pmatrix}$$

Perform the following ranging analyses.

(a) Find simple ranges for each of the objective function coefficients.

(b) Find simple ranges for each of the RHS constants.

(c) Find the range of parameter variation when for every unit increase in c_1 we have a unit decrease in c_3 (see Exercise 3).

(d) Find the range of parameter variation when for every unit increase in b_1 we have a unit decrease in b_2 (see Exercise 4).

(e) Find the range of parameter variation when for every unit increase in a_{11} there is a 0.5 decrease in a_{21}.

3. *Parametric Objective Coefficient Problem: Proportional Ranging.* Let $\mathbf{c}^*$ be an n-dimensional row vector and let λ be a scalar parameter. Consider the following parameterized linear program.

$$z_{\text{OF}}(\lambda) = \text{Maximize } z = (\mathbf{c} + \lambda \mathbf{c}^*)\mathbf{x}$$

$$\text{subject to} \quad \mathbf{Ax} = \mathbf{b}, \mathbf{x} \geq \mathbf{0}$$

For optimal basis $\mathbf{B}$, all reduced costs must be nonnegative—i.e.,

$$\bar{c}_j + \lambda \bar{c}_j^* \geq 0 \text{ for all } j$$

Determine the range over which λ can vary and $\mathbf{B}$ remain optimal. You must find $\underline{\lambda}_B$ and $\bar{\lambda}_B$ such that $\underline{\lambda}_B \leq \lambda \leq \bar{\lambda}_B$.

4. *Parametric Right-Hand-Side Problem: Proportional Ranging.* Let $\mathbf{b}^*$ be an m-dimensional vector and let λ be a scalar parameter. Consider the following parameterized linear program.

$$z_{\text{RHS}}(\lambda) = \text{Maximize } z = \mathbf{cx}$$

$$\text{subject to} \quad \mathbf{Ax} = \mathbf{b} + \lambda \mathbf{b}^*, \mathbf{x} \geq \mathbf{0}$$

Let $\mathbf{B}$ be the optimal basis for some value of λ, and let $\bar{\mathbf{b}} = \mathbf{B}^{-1}\mathbf{b}$ and $\bar{\mathbf{b}}^* = \mathbf{B}^{-1}\mathbf{b}^*$. For $\mathbf{B}$ to remain optimal, primal feasibility must be maintained—i.e.,

$$\bar{b}_i + \lambda \bar{b}_i^* \geq 0 \text{ for all } i$$

Find the lower and upper bounds $\underline{\lambda}_B$ and $\bar{\lambda}_B$, respectively, such that $\underline{\lambda}_B \leq \lambda \leq \bar{\lambda}_B$ and $\mathbf{B}$ is optimal.

5. Let $z_{\text{OF}}(\lambda)$ be the optimal objective function value for the parametric objective coefficient problem in Exercise 3. Prove that $z_{\text{OF}}(\lambda)$ is a convex function of λ. *Hint:* Consider two values λ^1 and λ^2 and corresponding solutions $\mathbf{x}^1$ and $\mathbf{x}^2$. Now use the definition of convexity.

6. Let $z_{\text{RHS}}(\lambda)$ be the optimal objective function value for the parametric RHS problem in Exercise 4. Consider two values λ^1 and λ^2 and corresponding solutions $\mathbf{x}^1$ and $\mathbf{x}^2$.

(a) Prove that $z_{RHS}(\lambda)$ is a concave function of λ by writing the dual formulation of the parametric RHS problem and making use of the results of Exercise 5.

(b) Prove that $z_{RHS}(\lambda)$ is a concave function of λ by making use of the definition of convexity and not resorting to the dual formulation.

7. Consider the parametric objective function LP

$$\text{Maximize } z = -3x_1 - (4+8\lambda)x_2 - 7x_3 + x_4 - (1-3\lambda)x_5$$
$$\text{subject to} \quad 5x_1 \quad - 4x_2 + 14x_3 - 2x_4 + \quad x_5 = 20$$
$$x_1 \quad - x_2 + 5x_3 - x_4 + \quad x_5 = 8$$
$$x_j \geq 0, \quad j = 1, \ldots, 5$$

For what range of values of λ is the basic vector (x_2, x_3) optimal for this problem? Determine the optimal BFSs for all $\lambda \geq 0$.

8. Let x_5, x_6, and x_7 be slack variables for the LP given below, and let $\mathbf{x}_B = (x_1, x_2, x_3)$. Compute $\mathbf{B}^{-1}$.

$$\text{Maximize } z = 2x_1 + 4x_2 + x_3 + x_4$$
$$\text{subject to} \quad x_1 + 3x_2 \quad + x_4 \leq 8 \quad \text{(raw material 1)}$$
$$2x_1 + x_2 \quad \leq 6 \quad \text{(raw material 2)}$$
$$x_2 + 4x_3 + x_4 \leq 6 \quad \text{(raw material 3)}$$
$$x_j \geq 0, \quad j = 1, \ldots, 4$$

(a) If the availability of only one of the raw materials can be marginally increased, which should it be? Explain.

(b) For what range of values of b_1 (the amount of raw material 1 available) does the basis $\mathbf{B}$ remain optimal? What is an optimal solution to the problem if $b_1 = 20$?

(c) If seven more units of raw material 1 can be made available (in addition to the current amount of eight), what is the maximum you should pay for them? Explain.

9. Write the dual model of the primal LP

$$\text{Maximize } z = 5x_1 - 3x_2 + 4x_3 - x_4$$
$$\text{subject to} \quad 3x_1 + 4x_2 + 3x_3 - 5x_4 \leq 13$$
$$5x_1 - 3x_2 + x_3 + 2x_4 \leq 15$$
$$x_j \geq 0, \quad j = 1, \ldots, 4$$

10. The following tableau shows an intermediate solution for the problem in Exercise 9. Use the information in the tableau to find the primal and dual solutions and to illustrate the various primal–dual relationships.

Row	Basic variables	Coefficients							RHS
		z	x_1	x_2	x_3	x_4	x_{s1}	x_{s2}	
0	z	1	0	0	-3	3	0	1	15
1	x_{s1}	0	0	5.8	2.4	-6.2	1	-0.6	4
2	x_1	0	1	-0.6	0.2	0.4	0	0.2	3

11. The following tableau shows another intermediate solution for the problem in Exercise 9. Use the information in the tableau to find the primal and dual solutions and to illustrate the various primal–dual relationships.

Row	Basic variables	Coefficients							RHS
		z	x_1	x_2	x_3	x_4	x_{s1}	x_{s2}	
0	z	1	5.18	1.64	0	0	0.82	1.55	33.82
1	x_3	0	2.82	-0.64	1	1	0.18	0.46	9.18
2	x_4	0	1.09	-1.18	0	0	-0.09	0.27	2.91

12. Consider the LP

$$
\begin{aligned}
\text{Minimize } z = 5x_1 + 3x_2 - x_3 \\
\text{subject to} \quad x_1 - 2x_2 + x_3 \geq 2 \\
-2x_1 - x_2 - 3x_3 \geq -10 \\
x_1 + x_2 + x_3 = 5 \\
x_j \geq 0, \quad j = 1, \ldots, 3
\end{aligned}
$$

  (a) Write the dual problem.

  (b) Solve the primal problem by hand using the two-phase method, and identify the optimal primal and dual solutions.

  (c) Solve the dual problem by hand directly using the two-phase method, and identify the optimal primal and dual solutions.

13. Consider the LP

$$
\begin{aligned}
\text{Maximize } z = 2x_1 + x_2 - 3x_3 + 5x_4 \\
\text{subject to} \quad 3x_1 - x_2 + x_3 + 2x_4 \leq 8 \\
x_1 + x_2 + 4x_3 - x_4 \leq 6 \\
2x_1 + 3x_2 - x_3 + x_4 \leq 10 \\
x_1 + x_3 + x_4 \leq 7 \\
x_j \geq 0, \quad j = 1, \ldots, 4
\end{aligned}
$$

Call the slack variables for the four constraints x_{s1}, x_{s2}, x_{s3}, and x_{s4}, respectively.

For each of the constraints, let the basic variables be x_4, x_{s2}, x_2, and x_{s4} in this order. Thus, x_4 is the basic variable for constraint 1, x_{s2} the basic variable for constraint 2, and so on. The basis inverse for the optimal solution under these assumptions is

$$
\mathbf{B}^{-1} = \begin{pmatrix}
3/7 & 0 & 1/7 & 0 \\
4/7 & 1 & -1/7 & 0 \\
-1/7 & 0 & 2/7 & 0 \\
-3/7 & 0 & -1/7 & 1
\end{pmatrix}
$$

Answer each of the following parts independently of each other using the information given above.

  (a) Show the basis matrix.

  (b) Compute the basic solution and corresponding dual solution.

  (c) Describe the marginal effects of allowing x_1 to enter the basis. (How much will the objective function and each of the other variables change for every unit increase in x_1?)

  (d) How much should x_1 increase to obtain the adjacent basic solution?

  (e) Perform the primal simplex iteration that will allow x_1 to enter the basis. Show the new basis inverse.

  (f) How many basic feasible solutions are adjacent to the solution found in part (b)?

  (g) Describe the marginal effects of changing the right-hand side of the fourth constraint on the objective function and the variables, assuming that the current solution is to remain basic.

  (h) What is the maximum amount that the RHS of constraint 4 can change in either direction without causing the current solution to become infeasible?

  (i) What if the RHS of constraint 4 were made one unit greater than the amount found in part (h)? Using the dual simplex algorithm to regain feasibility, what variables should enter and leave the basis at the next iteration?

  (j) Say a new constraint is added to the problem: $x_1 + x_3 \geq 6$. Starting from the original basic solution, how will this change the basis inverse? Without performing any computations, explain how the next step of the solution process is to be accomplished.

(k) Say that x_3 has an upper bound of 1.5, and that the problem is being solved by the upper bounded simplex method. What is the current basic solution? Find the variables that will enter and leave the basis in the next step. Assume that all the variables except x_3 are unbounded from above.

14. Give an economic interpretation of duality when the primal problem is a minimization problem with inequality constraints and the dual problem is a maximization problem with inequality constraints, both in nonnegative variables. Begin by writing out the two LPs. Assume that the primal problem represents a government agency that uses a set of n resources with fixed unit costs to provide m different services, each with a fixed demand. The agency's objective is to minimize its costs. Assume that the dual problem represents a consulting company that can provide the same set of m services. Define all model coefficients. Also, show how complementary slackness arises and interpret its meaning in the context of your models.

15. Consider the LP

$$\text{Maximize } z = -x_1 - x_2$$
$$\text{subject to } \quad x_1 + x_2 + x_3 = 1$$
$$x_1 \geq 0, x_2 \geq 0, x_3 \geq 0$$

(a) Formulate the primal barrier problem and find closed-form solutions for $x_1(\mu)$, $x_2(\mu)$, and $x_3(\mu)$.

(b) Show that as μ decreases to zero, the vector $(x_1(\mu), x_2(\mu), x_3(\mu))$ converges to the unique optimal solution.

(c) Solve the problem using the primal–dual path following algorithm starting at the point $\mathbf{x}^0 = (1/3, 1/3, 1/3)$.

(d) Solve the problem using the primal–dual path following algorithm starting at the point $\mathbf{x}^0 = (1/2, 1/2, 1/2)$.

16. The data for a primal LP in the form: $\max\{\mathbf{cx} : \mathbf{Ax} = \mathbf{b}, \mathbf{x} \geq \mathbf{0}\}$ are given below. Using the initial primal and dual solutions $\mathbf{x}^0 = (1, 1, 1, 1)^T$ and $\boldsymbol{\pi}^0 = (1, 1)$, take three steps of the primal–dual path following algorithm. Note that the dual solution is only partially specified, so you must find the complete solution in order to begin. Use a step size factor of $\gamma = 0.8$.

$$\mathbf{c} = (1,2,3,4), \quad \mathbf{A} = \begin{pmatrix} 1 & 2 & 2 & 3 \\ 2 & 1 & 3 & 2 \end{pmatrix}, \quad \text{and} \quad \mathbf{b} = \begin{pmatrix} 8 \\ 8 \end{pmatrix}$$

17. The optimality conditions for the primal–dual path following interior point algorithm are given in Equation (7). Perform the algebra necessary to obtain the Newton directions $\mathbf{d}_x$, $\mathbf{d}_\pi$, and $\mathbf{d}_z$ given in Equations (8) to (10).

18. Work through the algebra leading to the result in Equation (11), which says that the duality gap will shrink at each iteration of the primal–dual path following algorithm if the barrier parameter $\mu < \text{gap}(0)/n$. Begin with the equation for $\text{gap}(\alpha)$.

BIBLIOGRAPHY

Adler, I., N. Karmarkar, M. G. Resende, and G. Veiga, "Data Structures and Programming Techniques for the Implementation of Karmarkar's Algorithm," *ORSA Journal on Computing*, Vol. 1, No. 2, pp. 84–106, 1989.

Bertsimas, D. and J.N. Tsitsiklis, *Introduction to Linear Optimization*, Athena Scientific, Belmont, MA, 1997.

Bixby, R.E., "Progress in Linear Programming," *ORSA Journal on Computing*, Vol. 6, No. 1, pp. 15–22, 1994.

Bixby, R.E. and M.J. Saltzman, "Recovering the Optimal LP Basis from an Interior Point Solution," *Operations Research Letters*, Vol. 15, No. 4, pp. 169–178, 1994.

Chow, I.C., C.L. Monma, and D.F. Shanno, "Further Development of a Primal-Dual Interior Point Method," *ORSA Journal on Computing*, Vol. 2, No. 4, pp. 304–311, 1990.

Deif, A., *Sensitivity Analysis in Linear Systems*, Springer-Verlag, Berlin, 1986.

Fiacco, A.V. and G.P. McCormick, *Nonlinear Programming: Sequential Unconstrained Minimization Techniques*, Wiley, New York, 1968.

Greenberg, H.J., "The Use of the Optimal Partition in a Linear Programming Solution for Postoptimal Analysis," *Operations Research Letters*, Vol. 15, No. 4, pp. 179–186, 1994.

Hooker, J.N., "Karmarkar's Linear Programming Algorithm," *Interfaces*, Vol. 16, No. 4, pp. 75–90, 1986.

Karmarkar, N., "A New Polynomial-Time Algorithm for Linear Programming," *Combinatorica*, Vol. 4, pp. 373–395, 1984.

Khachin, L.G., "A Polynomial Algorithm in Linear Programming," *Soviet Mathematics Doklady*, Vol. 20, pp. 191–194, 1979.

Kojima, M., N. Megiddo, and S. Mizuno, "A Primal-Dual Infeasible-Interior-Point Algorithm for Linear Programming," *Mathematical Programming*, Vol. 61, pp. 263-280, 1993.

Lustig, I.J., R.E. Marsten, and D.F. Shanno, "Computational Experience with a Primal-Dual Interior Point Method," *Linear Algebra and Its Applications*, Vol. 152, pp. 191–222, 1991.

Lustig, I.J., R.E. Marsten, and D.F. Shanno, "Interior Point Methods: Computational State of the Art," *ORSA Journal on Computing*, Vol. 6, No. 1, pp. 1–14, 1994.

Marsten, R.E., R. Subramanian, M. Saltzman, I.J. Lustig, and D.F. Shanno, "Interior Point Methods for Linear Programming: Just Call Newton, Lagrange, and Fiacco and McCormick," *Interfaces*, Vol. 20, No. 4, pp. 105–116, 1990.

Murty, K.G., *Linear Programming*, Wiley, New York, 1983.

Vanderbei, R.J., *Linear Programming: Foundations and Extensions*, Kluwer Academic, Boston, 1996.

Ye, Y., *Interior Point Algorithms: Theory and Analysis*, Wiley, New York, 1997.

Chapter 5

Network Flow Programming Models

$\mathbf{A}$ network is a set of nodes joined by a set of arcs, much like a highway system linking major cities or a power grid that supplies electricity to a community over high-tension wires. We consider in this chapter networks in which the arcs carry flow and the total flow entering each node must equal the total flow leaving each node. One optimization problem that is defined on a network is to determine the assignment of flows to the arcs such that the total cost of the flow is minimized. This is called the *network flow programming* (NFP) problem or, alternatively, the *minimum-cost flow* problem. The NFP problem has a *linear programming* (LP) model, so NFP can be considered a special case of LP. We use an equivalent graphical model in this chapter that we will call the NFP model.

Problems that can be modeled with NFP include some of the classical problems of OR such as the assignment problem, the shortest path problem, the maximum flow problem, the pure minimum-cost flow problem, and the generalized minimum-cost flow problem. NFP is important because many aspects of real-world applications are readily recognized as networks. The network is a compact graphical representation that allows the analyst to quickly visualize the essence of the problem. When a situation can be modeled entirely with NFP, very efficient algorithms are available for solving the associated optimization problem, many times more efficient than the standard simplex method with respect to computational effort and storage requirements. In this chapter, we present a wide range of applications that can be modeled directly using NFP. With a little imagination, the reader should be able to extend these examples to many different situations.

As an introduction to network flow programming, we provide a graphical representation of a typical model in Figure 5.1. The situation associated with the model involves the distribution of a single commodity across the country and is described in Section 5.4. Definitions of common network terms follow shortly and are illustrated throughout this chapter. When more than one commodity flows in a network, the underlying problem becomes much more difficult to solve, especially when integer solutions are required. For the most part, we leave the subject of multicommodity networks to more advanced texts.

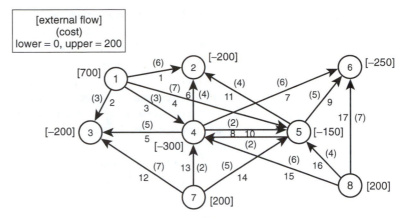

Figure 5.1 Example of a network flow programming model

Terminology

NODES AND ARCS   The network flow model consists of nodes and arcs. In the context of modeling a problem, each node, shown as a circle, represents some aspect of the problem, such as a physical location, an individual worker, or a point in time. For modeling purposes, it is often convenient to assign names to the nodes. Arcs are directed line segments that generally pass from an *origin node* to a *terminal node*, although in the case of external flow, an arc may be incident to only one node. If an arc does not have a direction, it is sometimes referred to as an *edge*. The number of nodes is denoted by *m*, and the number of arcs by *n*.

ARC FLOW   Flow is associated with the network, entering and leaving at the nodes and passing through the arcs. The flow in arc k is x_k. When flow is conserved at the nodes, the total flow entering a node must equal the total flow leaving the node. The arc flows are the decision variables for the network flow programming model.

UPPER AND LOWER BOUNDS ON FLOW   Flow is limited in an arc by lower and upper bounds. Sometimes the term *capacity* refers to the upper bound on flow. We use l_k and u_k for the lower and upper bounds of arc k, respectively.

COST   The criterion for optimality is cost. Associated with each arc k is the cost per unit of flow c_k. Negative values of c_k correspond to revenues.

GAIN   The arc gain g_k multiplies the flow at the beginning of the arc to obtain the flow at the end of the arc. When a flow x_k is assigned to an arc, this flow leaves the origin node of the arc. The flow entering the terminal node is $g_k x_k$. The arc's lower bound, upper bound, and cost all refer to the flow at the beginning of the arc. Gains less than 1 model losses, such as evaporation or spoilage. Gains greater than 1 model growth in flow.

A network in which all arcs have unit gains is called a *pure* network. The optimal solution for a pure network with integer parameters always has integer flows. If some gains have values other than 1, the network is a *generalized* network, and the solution is not usually integral.

ARC PARAMETERS   In a network diagram, the arc parameters are shown adjacent to arcs and are enclosed in parentheses (e.g., lower bound, upper bound, cost, gain). When a parameter is not shown, it assumes its default value. Default values are 0 for the lower bound, ∞ for the upper bound, 0 for the cost, and 1 for the gain.

EXTERNAL FLOWS   The external flow at node i, denoted by b_i, is the flow that *must* enter node i from sources, or leave node i for destinations outside the network. A positive external flow enters the node, and a negative external flow leaves the node. In a network diagram, the external flow is shown in square brackets adjacent to the node.

FEASIBLE FLOW   When an assignment of flows to the arcs satisfies conservation of flow for each node and is within the bounds of each arc, the flow is said to be *feasible.*

OPTIMAL FLOW   The feasible flow that minimizes the sum of all arc costs is the optimal flow.

5.1   CLASSICAL PROBLEMS

When one is learning to build network models, it is helpful to recognize several special cases such as the models for the transportation, shortest path, and maximum flow problems. These models differ primarily in the set of parameters that are relevant, or in the way the nodes and arcs are arranged in the flow diagram. In this section, we describe each of the aforementioned cases in terms of the subset of the parameters that is used in the NFP formulation. All nonrelevant parameters will assume their default values defined as 0 for the lower bound, M (∞) for the upper bound, 0 for the cost, and 1 for the gain.

Using the default values, all special cases discussed in this section can be solved with algorithms designed for the more general problem. There are, however, a variety of special-purpose algorithms that do not consider the nonrelevant parameters in their computations and hence are more efficient. Figure 5.2 shows the relationships among the various problems that have NFP models and the general LP problem. Starting from the left and moving to the right, the problems become increasingly more general. Thus, all problems to the left of the generalized minimum-cost flow problem can be solved with an algorithm designed for the generalized problem, which itself is a special case of the LP problem. Finally, we should mention that there are several network problems that are not based on the conservation-of-flow principle and hence are not included in this or the next chapter. A prominent example is the minimal spanning tree problem, which is introduced in the discussion of greedy algorithms in Chapter 8.

Transportation Problem

The components of a typical transportation problem are shown in Figure 5.3. This problem deals with a set of sources, labeled S1, S2, and S3, at which supplies of some commodity are available, and a set of destinations, labeled D1, D2, and D3, where the commodity is demanded. Shipping costs are specified in the 3×3 matrix, and the supply and demand data are shown outside the matrix.

The classic statement of the transportation problem uses a matrix or tableau with the rows representing sources and the columns representing destinations. The algorithms for solving the problem are based on this matrix representation. The cost c_{ij} of shipping one unit of the commodity from source i to destination j is indicated by the appropriate entry in the matrix. If shipment is not possible between a given source and a given destination, an M, representing a large cost, is entered. This discourages solutions that use such cells.

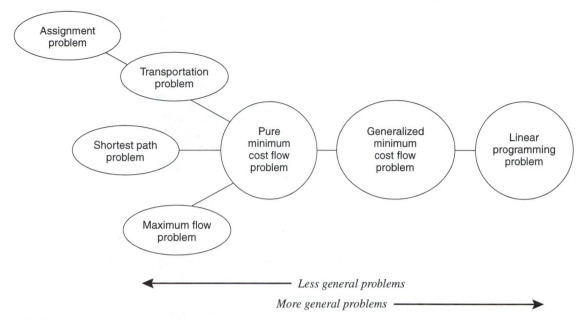

Figure 5.2   Relationships among the classical problems and the LP problem

Supply and demand data are shown along the margins of the matrix and are denoted symbolically as s_i and d_j, respectively, where $i = 1, \ldots , m$ and $j = 1, \ldots , n$. In the statement of the classical transportation problem, balance is present, which means that the total supply always equals the total demand—i.e., $\Sigma_{i=1}^m s_i = \Sigma_{j=1}^n d_j$. This is the case in our example and is a sufficient condition for feasibility. Instances of the problem can always be put into this form with the addition of a dummy supply point (call it $m + 1$) when demand exceeds supply, or a dummy demand point (call it $n + 1$) when supply exceeds demand. If Δ denotes the excess, we set $s_{m+1} = \Delta$ or $d_{n+1} = \Delta$, depending on the imbalance. All unit costs, $c_{m+1,j}$ or $c_{i,n+1}$, associated with the corresponding shipments are set equal to zero.

The NFP model for this example is shown in Figure 5.4. Sources are denoted by the nodes on the left and destinations by the nodes on the right. Allowable shipping links are shown as arcs whereas impermissible or nonexistent links are omitted. For example, the cost entry in cell (S3, D1) in Figure 5.3 indicates that it is not permissible to ship from source 3 to destination 1, and so arc (S3, D1) is omitted from the diagram.

	D1	D2	D3	*Supply*
S1	3	1	M	5
S2	4	2	4	7
S3	M	3	3	3
Demand	7	3	5	

Figure 5.3   Matrix model of a transportation problem

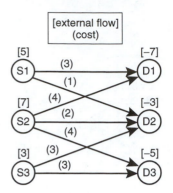

Figure 5.4 NFP model of the transportation problem

Only arc costs are included in the NFP model, because these costs are the only relevant arc parameters. All other parameters, such as upper and lower bounds, are set to their default values. The network has a special form important in graph theory. It is called a *bipartite* network, which means that the nodes can be divided into two subsets with all arcs going from one subset to the other.

On each supply node in Figure 5.4, the positive number in brackets indicates supply flow entering the network. On each destination node, the quantity in brackets is negative, indicating that this amount of flow must leave the network. The optimal solution for this example is shown in Figure 5.5, in which the matrix entries correspond to flow rather than to shipping costs. We see that five arcs have positive flow in the solution. The minimum cost associated with these values is $z^* = 46$.

Variations of the classical transportation problem are easily handled by modifications of the NFP model. If links have finite capacity, the arc upper bounds can be made finite. If supplies represent raw materials that are transformed into products at the sources and the demands are in units of product, gain factors can be used to represent transformation efficiency at each source. If some minimal flow is required on certain links, arc lower bounds can be set to nonzero values.

Assignment Problem

A typical assignment problem, presented in the classic form, is shown in Figure 5.6. Here there are five workers that are to be assigned to five jobs. The numbers in the matrix indicate the cost c_{ij} of worker i doing job j. Jobs with costs of M are disallowed assignments. The problem is to match workers and jobs at minimum cost.

	D1	D2	D3	*Supply*
S1	2	3	0	5
S2	5	0	2	7
S3	0	0	3	3
Demand	7	3	5	

Figure 5.5 Optimal solution, $z^* = 46$

	J1	J2	J3	J4	J5
W1	M	8	6	12	1
W2	15	12	7	M	10
W3	10	M	5	14	M
W4	12	M	12	16	15
W5	18	17	14	M	13

Figure 5.6   Matrix model of the assignment problem

The NFP model is given in Figure 5.7. It is very similar to the transportation model except that the external flows are all +1 and −1. The only relevant parameter for the assignment model is the arc cost (not shown in the figure to avoid clutter); all other parameters should be set to their default values. Note that this model also has the bipartite structure.

The decision variable for the assignment problem is denoted by x_{ij}, which takes the value 1 if worker i is assigned to job j, and 0 otherwise. As shown in Figure 5.8, the solution that minimizes total cost has, not coincidentally, a total flow of 1 in every column and row.

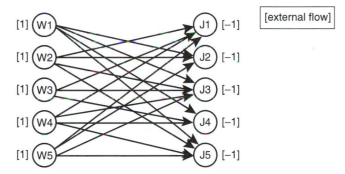

Figure 5.7   NFP model of the assignment problem

	J1	J2	J3	J4	J5
W1	0	0	0	0	1
W2	0	0	1	0	0
W3	0	0	0	1	0
W4	1	0	0	0	0
W5	0	1	0	0	0

Figure 5.8   Solution to the assignment problem, $z^* = 51$

Shortest Path Problem

This problem uses a general network structure in which the only relevant arc parameter is cost. A typical case is shown in Figure 5.9. The length of a path is the sum of the arc costs along the path. The problem is to find the shortest path from some specified node, node A for example, to some other node or perhaps to all other nodes. The latter problem is called the *shortest path tree problem* because the collection of all shortest paths from a specified node forms a graph structure called a *tree*. Since it is not much more difficult to find the paths to all nodes than it is to find the path to one node, the shortest path tree problem is usually solved.

The NFP equivalent of the shortest path tree problem is formed by equating arc distance to arc cost, assigning a fixed external flow of $m - 1$ (m is the number of nodes) to the source node, and assigning fixed external flows of -1 to every other node. This is illustrated in Figure 5.9 when node A is the source node. In solving this flow problem, the algorithm assigns flow from the source to each node by the least cost path, because there are no bounds on arc flows. The shortest path tree will consist of those arcs with nonzero flow in the optimal solution. Figure 5.10 depicts the solution to the problem.

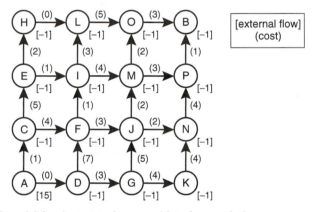

Figure 5.9   NFP model for shortest path tree problem from node A

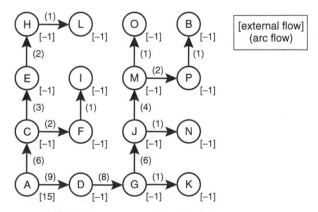

Figure 5.10   Solution of shortest path tree problem, $z^* = 48$

Maximum Flow Problem

The maximum flow problem again fits the NFP structure. With one exception, only the arc capacities or upper bounds are the relevant parameters. The problem is to find the maximum possible flow from a given source node s to a given sink node t. An NFP model with node 1 as the source and node 8 as the sink is depicted in Figure 5.11. All arc costs are zero with the exception that the cost on the arc leaving the sink is set to -1. Since the objective is to minimize cost, the maximum possible flow is delivered to the sink node. By convention, we assign a large number M to the upper bound on flow into the source and out of the sink node.

The solution to the example is given in Figure 5.12. The maximum flow from node 1 to node 8 is 30. The individual arc flows that yield this result are shown in parentheses. The bold arcs in the figure, $\{(1, 3), (2, 6), (2, 7), (5, 8)\}$, are called the *minimal cut*. In general, a cut is a set of arcs such that if they are removed from the graph, no flow can pass from the source to the sink. A cut partitions the nodes in a graph into two disjoint subsets or subgraphs. The arcs in the minimal cut are the bottlenecks that are restricting the maximum flow. The fact that the sum of the capacities of the arcs on the minimal cut equals the maximum flow in this example is not a coincidence. It is always the case and is a famous primal–dual result in network theory called the *max-flow min-cut theorem*. The arcs comprising the minimal cut can be identified using sensitivity analysis.

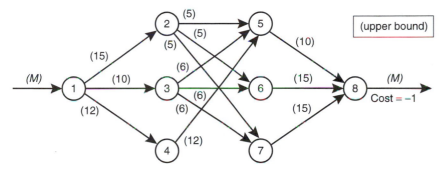

Figure 5.11   NFP model for the maximum flow problem

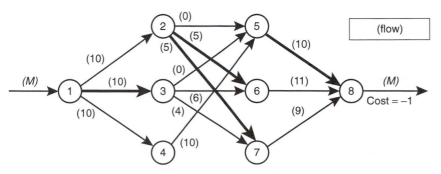

Figure 5.12   Maximum flow (minimal cut) = 30

5.2 EXTENSIONS OF THE BASIC MODELS

As illustrated by some of the special cases discussed in the preceding section, NFP models often involve a bipartite graph but are certainly not limited to this form. Some of the extensions that we now present show that a variety of complex problems have a network structure.

Transportation Problem with Costs and Revenues

Consider a company that has two plants (S1 and S2) and four customers (D1, D2, D3, and D4). The costs of shipping products from the plants to the customers are given in the upper left matrix shown in Figure 5.13. Also included are the minimum and maximum shipments, the unit manufacturing costs for each plant, the minimum and maximum receipts, and the unit revenues for each customer. The problem faced by the plant manager is to find the optimal shipping pattern that maximizes the net profit—that is, revenue less shipping costs less manufacturing costs.

From a modeling point of view, this problem closely resembles the standard transportation problem. The only difference is that the quantities available at the sources and the amounts demanded by the customers are no longer fixed but are given by a range of values. Figure 5.14 shows how these ranges can be included in the NFP model. We have added arcs to carry the extra supplies and demands allowed. The quantities in square brackets above the nodes represent either the minimum number of units that must be shipped from a source or the minimum number of units that must be received at a destination. Manufacturing costs appear on the arcs entering the source nodes, and revenues are the negative costs shown on the arcs leaving the customer nodes.

	Shipping costs				Shipping restrictions		
	Customer D1	Customer D2	Customer D3	Customer D4	Minimum shipment	Maximum shipment	Unit cost
Plant S1	10	15	6	13	10	25	8
Plant S2	3	6	7	10	10	25	10
Minimum receipt	5	5	5	5			
Maximum receipt	15	15	15	15			
Unit revenue	12	14	16	18			

Figure 5.13   Parameters for an extended transportation problem

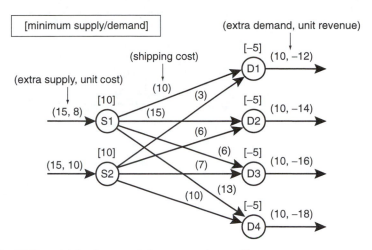

Figure 5.14   NFP model for the extended transportation problem

Multiperiod Operations

Now consider the scenario in which the company needs to establish a shipping schedule for the next two months. The demands for each customer are 15 units in the first month and 20 units in the second month. These demands must be met. Plant S1 has a manufacturing capacity of 30 units per month and plant S2 has a capacity of 50 units per month. In the first month, the production costs are $8 per unit at plant S1 and $10 per unit at plant S2. In the second month, the unit costs are $9 per unit at both plants. Products can be stored at the customer sites from one month to the next at a cost of $1 per unit. Products cannot be stored at the plants. Shipping costs are as given in the preceding example except that the shipping company is offering a discount of $1 per unit on all routes during the first month. The goal is to minimize total production costs, shipping costs, and inventory costs over the two-month period. Not all production capacity needs to be utilized.

The NFP model for this situation is shown in Figure 5.15. Because there are no longer any minimum shipping requirements, there are no square brackets above the source nodes. The first number in parentheses on an arc entering a source node represents an upper bound on the available supply. Also, demand is fixed at each destination—i.e., there is no extra demand.

Interconnected transportation subnetworks represent each period. Arcs that go from one period to the next correspond to inventories. With this construction, the size of the network is proportional to the number of periods. This same approach is useful for a variety of multi-period situations.

Transshipment Problem

An important extension of the classical transportation problem includes the introduction of nodes that can act as transshipment points. Such locations may or may not be supply and demand points as well. For example, transshipment often occurs in the distribution systems of national retail chains. It is common for companies such as Sears and Wal-Mart to have regional warehouses that ship to smaller district warehouses, which in turn supply retail

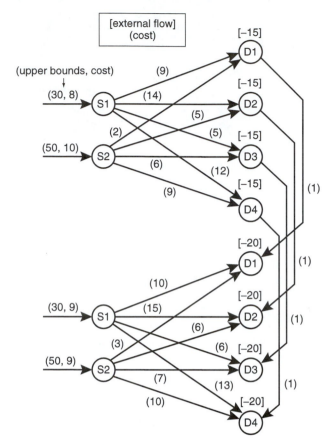

Figure 5.15 Two-period NFP model

outlets. In this scenario, the district warehouses are transshipment points. The underlying model is a useful tool for a company wishing to decide on the optimal number and locations of its warehouses, since the network analysis can provide a minimum cost routing plan for a specific configuration of warehouses and retail outlets.

The transshipment problem has an NFP model that is not bipartite. In particular, this model contains intermediate nodes with nonnegative supply or demand requirements and with arcs between pairs of sources or pairs of destinations. An example with three sources, three transshipment points, and three destinations is depicted in Figure 5.16.

Alternatively, using the following procedure, a transshipment problem can be transformed into a balanced transportation problem. The associated NFP model is bipartite. To begin, we assume that there are m pure supply points, n pure demand points, and p transshipment points. For an NFP model, a pure supply point would have only leaving arcs whereas a pure demand point would have only entering arcs. We let s_T be the total supply available for the problem—i.e., $s_T = \Sigma_{i=1}^{m+p} s_i$—and we let t_k be the net stock position at transshipment node k for $k = 1, \ldots, p$. If stock is supplied at node k, then t_k is positive, and if stock is demanded at node k, then t_k is negative. In the example, $m = n = p = 3$, with $s_T = 17$.

Step 1: If necessary, put the problem into balanced form by adding either a dummy supply point to meet the excess demand or a dummy demand point to absorb the excess supply. Shipments to or from the dummy node will have zero shipping cost.

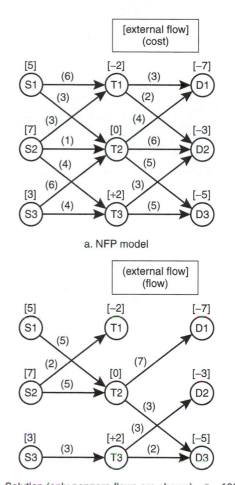

a. NFP model

b. Solution (only nonzero flows are shown), $z^* = 100$

Figure 5.16 Transshipment problem

Step 2: Construct a transportation tableau with $m + p$ rows—one for each supply point and transshipment point—and $n + p$ columns—one for each demand point and transshipment point. Each pure supply point i will have a supply equal to its original value s_i, and each pure demand point j will have a demand equal to its original value d_j. Each transshipment point l will have a supply equal to $s_k = t_k + s_T$ and a demand equal to $d_k = s_T$.

In the construction of the tableau in Step 2, we assign a large cost M to shipments that are not permissible. To facilitate modeling, we allow a shipment from a transshipment point to itself and assign it a unit transportation cost of zero—that is, we include x_{kk} in the model and set c_{kk} equal to zero for $k = 1, \ldots, p$. The transformed model for the example is shown in Figure 5.17 and the solution is given in Figure 5.18. Note that the solutions in Figures 5.16b and 5.18 are the same.

The decision variable x_{kk} corresponds to a fictitious shipment. For each transshipment point k that has a nonnegative quantity to be supplied ($t_k \geq 0$), the net difference between

	D1	D2	D3	T1	T2	T3	*Supply*
S1	M	M	M	6	3	M	5
S2	M	M	M	3	1	4	7
S3	M	M	M	M	6	4	3
T1	3	2	M	0	M	M	15
T2	4	6	5	M	0	M	17
T3	M	3	5	M	M	0	19
Demand	7	3	5	17	17	17	

Figure 5.17   Transportation representation for the transshipment example

	D1	D2	D3	T1	T2	T3	*Supply*
S1	0	0	0	0	5	0	5
S2	0	0	0	2	5	0	7
S3	0	0	0	0	0	3	3
T1	0	0	0	15	0	0	15
T2	7	0	3	0	7	0	17
T3	0	3	2	0	0	14	19
Demand	7	3	5	17	17	17	

Figure 5.18   Solution of the transportation model, $z^* = 100$

the amount demanded ($d_k = s_T$) and the fictitious shipment x_{kk} represents the transshipment quantity. Similarly, for each transshipment point k that has a negative quantity ($t_k < 0$) indicating the number of units demanded, the net difference between the amount supplied ($s_k = d_k + s_T$) and the fictitious shipment x_{kk} represents the transshipment quantity. Because the fictitious shipment variables x_{kk} have $c_{kk} = 0$, their levels do not contribute to the total cost of transportation.

To achieve the conversion to a standard transportation model, we have introduced s_T, which can be thought of as a fictitious buffer stock at each transshipment node k. Because s_T has been included in both s_k and d_k, the sum of all the s_i remains equal to the sum of all the d_j. Thus, balance is maintained. The total amount of stock transshipped through point k is $s_T - x_{kk}$ if $t_k \geq 0$, and $t_k + s_T - x_{kk}$ if $t_k < 0$.

Transformation of Flow

A company makes three products at four plants. Because of differences in labor skills, the amount of time required to manufacture the products varies among the plants. The time requirements (in minutes per unit) are shown in the matrix in Figure 5.19. A dash indicates that the specified product cannot be made at the specified plant. The demands for the three products are also shown. The total time available at each plant is 25 hours per week, but it is not necessary that all of it be used. The hourly charges for labor are $10, $12, $9, and

Time requirements (minutes per unit)

	Plant 1	Plant 2	Plant 3	Plant 4	Demand
Product A	30	25	–	–	70
Product B	28	–	25	22	80
Product C	–	30	30	29	40

Figure 5.19   Parameters for the transformation of flow problem

$13 for plants 1, 2, 3, and 4, respectively. There is no labor charge for time not used. The problem is to construct and solve a network model that can be used to minimize total manufacturing costs while meeting all demand.

This example illustrates the need for a gain factor to transform one type of flow into another. This situation is like a transportation problem except that the plant capacities are measured in hours whereas the demand is measured in product units. We use the gain factor to transform the flow, as indicated in Figure 5.20.

The requirements data are given in minutes per unit, so to compute the appropriate gain factors it is necessary to invert the quantities in Figure 5.19 and then multiply by 60 to find units per hour. The gain for plant 1 producing product A is 60 (min/hr) / 30 (min/unit) = 2 units per hour. The complete NFP model and the solution are shown in Figures 5.21 and 5.22, respectively.

Convex Costs and Concave Revenues

Consider an arc that carries flow between two nodes. Suppose that the first 10 units of flow have a unit cost of $5, the next five units have a unit cost of $8, and any additional units have a unit cost of $10. When unit cost is increasing with flow, the arc cost is a convex

Figure 5.20   Using the gain factor to transform flow

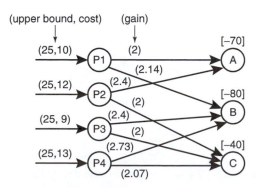

Figure 5.21   NFP model with gains

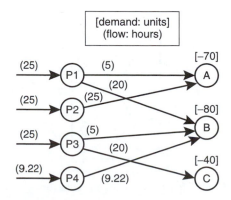

Figure 5.22   Solution

function of the flow and can be modeled with multiple arcs. Figure 5.23 depicts a linear model that has an arc for each cost level. In cases in which the cost function is continuous, we can use a piecewise linear approximation with the number of arcs determining the accuracy of the approximation. Nonlinear concave revenue functions can be handled in a similar manner. We simply represent the revenues with negative arc costs.

Unfortunately, concave cost functions and convex revenue functions cannot be transformed into linear models. To deal with such functions, it would be necessary to introduce integer variables into the formulation to ensure that arcs were selected in the proper order—i.e., higher-cost arcs first in the case of concave costs. Section 7.8 describes models for this circumstance.

Undirected Edges

Some situations will allow flow between pairs of nodes in either direction, where the cost per unit flow from node i to node j is the same as the cost per unit flow from j to i. This is conveniently represented with undirected edges, as in Figure 5.24. Since the flow model requires directed arcs, the transformation shown in Figure 5.25 leads to an equivalent model with each edge replaced by a pair of oppositely directed arcs. The two arcs in Figure 5.25

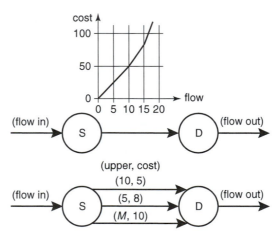

Figure 5.23   Representation of nonlinear convex costs

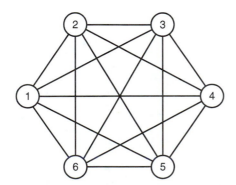

Figure 5.24   Model with nodes and edges

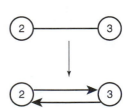

Figure 5.25   Transformed model with directed arcs

will each have zero lower bounds, upper bounds that provide appropriate limits to the directional flow, and equal unit costs.

Elimination of Arc Lower Bounds

Equivalent models with zero lower bounds on flow can replace models with nonzero lower bounds. Figure 5.26 shows a general arc with the parameters lower bound (l), upper bound (u), cost (c), and gain (g). An equivalent representation of the arc with a zero lower bound is also shown. The transformation requires modification of the external flows at both ends of the arc. When the upper and lower bounds are equal, the affected arc may be eliminated because its upper bound on flow will be zero.

Determining Economic Life

Consider an investment in a machine with an initial purchase price of $1000. The yearly operating costs and salvage value of the machine depend on its age, as shown in Table 5.1. It is anticipated that the machine will be used far into the future. Given that the salvage value is decreasing and operating costs are increasing, there must be some optimal time at which the machine should be replaced. The optimal replacement time is called the *economic life* of the machine.

Investment analysis recognizes that money spent or earned in the future has less value when viewed from the present. This is called the *time value of money principle*. Specifically, the present value of an amount c_n received n years from now is computed as follows.

$$P = \frac{c_n}{(1+i)^n}$$

The quantity i is a percentage expressed as a decimal and is variously called the interest rate, the discount rate, or the minimum acceptable rate of return. The term $1/(1+i)^n$ is the *discount factor*. When i is a positive quantity, the discount factor is less than 1.

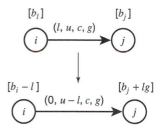

Figure 5.26   Elimination of lower bounds

Table 5.1   Cost Data for Machine

Year of ownership	Operating cost during year	Salvage value at year's end
1	$1000	$500
2	1200	300
3	1500	100
4	1900	0
5	2500	0

The network representation of decision problems involving the time value of money is not obvious. In this section, we describe the problem of determining the *economic life* of an investment. The approach is illustrative of a wide variety of situations that can be handled using NFP.

The investment can be viewed as a series of cash flows over time, as illustrated in Figure 5.27. Let the present time be 0 and assume that we keep the machine for 3 years. The operating costs are expended at the end of each year, and the salvage value is a revenue received at the end of the life of the machine—3 years in this case. In Figure 5.27, costs are shown as positive values and revenues as negative values.

To find a general expression for the present value of the cash flow as a function of the time of disposal, let c_n be the cost expended at time n, let N be the life of the machine, and let s_N be the amount realized when the machine is disposed of at time N. The net present value for a lifetime of N years, P_N, is then

$$P_N = \sum_{n=0}^{N} \frac{c_n}{(1+i)^n} - \frac{s_N}{(1+i)^N}$$

For the specific case for $N = 3$ and assuming that i is 20%, we have

$$P_3 = 1000 + 833.33 + 833.33 + 810.19 = \$3476.85$$

Note that the fourth term in this expression is the present value of the combined operating cost and salvage value in year 3.

The NFP representation of this calculation uses gains to model the discount factor, as shown in Figure 5.28 for the general case with $N = 3$. The net cash flow in year 3, considering both operating cost and salvage value, is denoted by c_3'. The gain factor g is equal to $1/(1 + i) = 0.8333$.

There is no optimization in the network shown in Figure 5.28. The fixed flow at node 0 is forced on the network, resulting in the arc flows

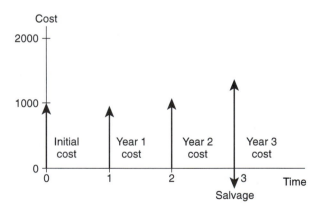

Figure 5.27 Cash flows for a 3-year life

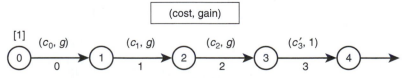

Figure 5.28 NFP model for present value calculation

$$x_0 = 1,\ x_1 = 0.833,\ x_2 = 0.833^2,\ x_3 = 0.833^3,\ P_3 = 3476.85$$

The cost associated with the network is the present value of the cash flows for the machine.

The fact that the present value can be computed as the cost of a generalized network allows us to address various decision issues associated with investments. We might ask for the present value of a series of replacements over an infinite period for the cash flows depicted in Figure 5.29.

Without proving this result, we observe that the NFP model of a continuous series of replacements can be represented by the cycle shown in Figure 5.30.

The economic life of a machine is that life that minimizes the present value of ownership over an infinite time horizon. To find this value, an optimization problem must be solved. The NFP model for our example is depicted in Figure 5.31. There are five cycles or alternative paths representing replacement after years 1 through 5, respectively.

The solution to this problem, shown in Figure 5.32, has positive flows on only one cycle. The result indicates that the machine should be replaced every 2 years. The cost of the optimal flow, $8045.46, is the present value of the costs incurred by following the optimal policy forever.

Arbitrage Example

A second example that makes use of gains arises in currency markets and concerns arbitrage. In general terms, arbitrage is the process whereby one simultaneously buys and sells the same or equivalent securities with the aim of profiting from price discrepancies. In our case, the securities being traded are five currencies. Table 5.2 lists the currencies along with their exchange rates r_{ij}, $i, j = 1, \ldots, 5$. For example, $r_{13} = 1.45$ in row 1, column 3 means that 1 US dollar will purchase 1.45 Swiss francs (CHF). The other currencies are Japanese yen, German Deutsche marks (D-Mark), and British pounds (Brit-£).

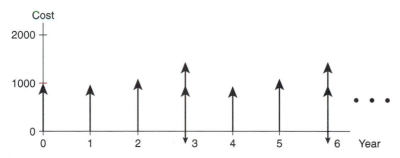

Figure 5.29   Cash flows for an infinite series of replacements

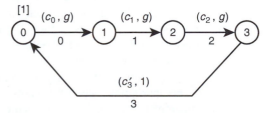

Figure 5.30   NFP model for an infinite series of replacements

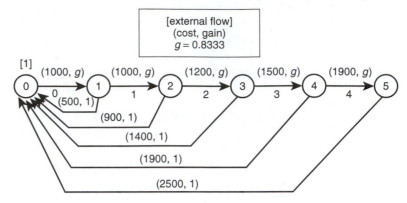

Figure 5.31   NFP model of the economic life problem

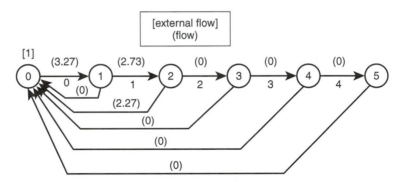

Figure 5.32   Optimal solution of the machine replacement problem

Table 5.2   Rates Among Various Currencies

	US $	Yen (100)	CHF	D-Mark	Brit-£
US $	1	1.05	1.45	1.72	0.667
Yen (100)	0.95	1	1.41	1.64	0.64
CHF	0.69	0.71	1	1.14	0.48
D-Mark	0.58	0.61	0.88	1	0.39
Brit-£	1.5	1.56	2.08	2.56	1

When buying and selling currencies, traders must pay a transaction fee to a broker. Let us assume that a 1% fee is charged on each exchange. The problem then is to determine if there is a series of exchanges that will be profitable—that is, to exchange one currency for a second currency, exchange the second currency for a third, and so on, and after the final exchange wind up with more of the original currency than you started with.

A network flow model with gains can be used to find a solution to this problem if one exists. The appropriate network is shown in Figure 5.33 and has one node for each currency and directed arcs between each pair of nodes. Outgoing arcs are associated with selling

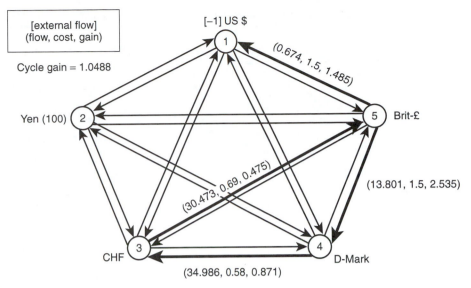

Figure 5.33   Arbitrage network

the currency and incoming arcs are associated with buying it. The gain on arc (i, j) describes the process of converting one unit of currency i for r_{ij} units of currency j.

$$g_{ij} = r_{ij} / 1.01$$

The denominator 1.01 in this expression accounts for the transaction fee.

The arc costs are the dollar equivalents of a currency and can be obtained from the first column in Table 5.2. For example, arc (5, 1) has the unit cost of $1.50. This is the dollar equivalent of 1 British pound. Note that all costs on arcs leaving node 5 (Brit-£) are 1.50. At node 1 we place a demand of $1. The model tries to reduce the total investment required to obtain $1 of gain in US dollars.

The flows in the network are in units of currency and depend on the originating node of an arc. For example, all flows leaving node 2 are in hundreds of yen. The flow on arc (5, 4) represents the conversion of British pounds to Deutsche marks but is measured in pounds.

To determine if arbitrage is possible, we must find a cycle in the network with a gain greater than 1. The gain of a cycle is the product of the arc gains on the cycle. Therefore, we will attempt to minimize the sum of the dollar equivalents of the exchanged currencies. This is the same as minimizing the total amount invested multiplied by the number of exchanges. Other objective functions are possible, but, for example, putting a cost of 1 on every arc would be equivalent to adding different currencies and so would not be valid. With a flow requirement of 1 at the US $ node, any objective function will yield a feasible solution as long as there exists a cycle with a gain greater than 1. If there are only break-even cycles or cycles with gains less than 1, the problem will not be feasible.

The solution is shown in Table 5.3 and was obtained by solving the generalized network formulation for the problem. As can be seen, the following cycle was found: British pounds → Deutsche marks → Swiss francs → British pounds. The gain around this cycle is $(2.535) \times (0.871) \times (0.475) = 1.0488$, which indicates that for every British pound invested in the arbitrage, a 4.88% gain is realized after the transfers on the cycle are complete. In particular, the solution involves first changing 13.801 pounds into Deutsche marks. This

Table 5.3 Solution Obtained for Arbitrage Problem, $z^* = 62.088$

Arc	Flow	Originating node	Terminal node	Cost	Gain*	Node	Name	Fixed
1	0	1	2	1	1.040	1	US $	−1
2	0	1	3	1	1.436	2	Yen (100)	0
3	0	1	4	1	1.703	3	CHF	0
4	0	1	5	1	0.660	4	D-Mark	0
5	0	2	1	0.95	0.941	5	Brit-£	0
6	0	2	3	0.95	1.396			
7	0	2	4	0.95	1.624			
8	0	2	5	0.95	0.634			
9	0	3	1	0.69	0.683			
10	0	3	2	0.69	0.703			
11	0	3	4	0.69	1.129			
12	30.473	3	5	0.69	0.475			
13	0	4	1	0.58	0.574			
14	0	4	2	0.58	0.604			
15	34.986	4	3	0.58	0.871			
16	0	4	5	0.58	0.386			
17	0.6734	5	1	1.5	1.485			
18	0	5	2	1.5	1.545			
19	0	5	3	1.5	2.059			
20	13.801	5	4	1.5	2.535			

*Note that the gain factors have been rounded to three significant digits. The solution is for the rounded values.

yields 34.986 Deutsche marks, which is then exchanged for 30.473 Swiss francs. Next, the 30.473 Swiss Francs is exchanged for 14.475 British pounds. Finally, 0.674 pound is exchanged for 1 US dollar. This leaves 13.801 pounds in the system to be circulated again.

The optimal objective value $z^* = 62.088$ is the approximate dollar equivalent of all flows in the solution based on the rates used to convert the flows to dollars. These rates do not include the 1% exchange fee. It is not surprising that the objective contribution of the flow form Brit-£ to US $ has the value 1. The other nonzero flows have the approximate value of $20.70, which is the amount invested in the cycle. It is counted three times since the cycle has three arcs.

The cycle gain of 1.0488 means that at node 5 a 4.88% profit will be realized every time the cycle is traversed regardless of the amount invested. The problem statement does not consider the time over which the transactions take place or address the question of how to start the process; the solution assumes that a steady state has been reached. If such a steady state did exist, however, it wouldn't be for long, because the price setters would quickly adjust the exchange rates to ensure that no cycle had a gain greater than 1.

5.3 LINEAR PROGRAMMING MODEL

Every NFP model has an LP equivalent. Although we usually start off with an NFP formulation when possible, it is often useful to construct the associated LP model. This is especially true when the NFP model has side constraints that prevents the use of specialized algorithms for finding solutions.

Consider a network model with n arcs and m nodes. For each node, we identify $\boldsymbol{K}_{O(i)}$ as the set of arcs originating at node i and $\boldsymbol{K}_{T(i)}$ as the set of arcs terminating at node i. We let b_i be the external flow at node i, and we let c_k, l_k, u_k, and g_k represent, respectively, the cost, lower bound, upper bound, and gain for general arc k. The decision variable in terms of flow for arc k is denoted by x_k. The LP model consists of the objective function, the conservation-of-flow constraints, and the lower and upper bound limits on the flow.

Objective function:

$$\text{Minimize } \sum_{k=1}^{n} c_k x_k \tag{1}$$

Conservation of flow:

$$\sum_{k \in \boldsymbol{K}_{O(i)}} x_k - \sum_{k \in \boldsymbol{K}_{T(i)}} g_k x_k = b_i, \quad i = 1, 2, \ldots, m \tag{2}$$

Lower and upper bounds on flow:

$$l_k \leq x_k \leq u_k, \quad k = 1, 2, \ldots, n \tag{3}$$

To illustrate the LP formulation, consider the network model of the economic life problem shown in Figure 5.31. Figure 5.34 depicts the network with the six nodes and 10 arcs renumbered to correspond to the notation used in Equations (1) to (3). The LP equivalent consists of the objective function, six conservation-of-flow constraints, and simple lower bounds on flow.

Objective function:

$$\text{Minimize } z = 1000x_1 + 1000x_2 + 1200x_3 + 1500x_4 + 1900x_5$$
$$+ 500x_6 + 900x_7 + 1400x_8 + 1900x_9 + 2500x_{10}$$

Conservation of flow:

Node 1: $x_1 - x_6 - x_7 - x_8 - x_9 - x_{10} = 1$

Node 2: $x_2 + x_6 - 0.833x_1 = 0$

Node 3: $x_3 + x_7 - 0.833x_2 = 0$

Node 4: $x_4 + x_8 - 0.833x_3 = 0$

Node 5: $x_5 + x_9 - 0.833x_4 = 0$

Node 6: $x_{10} - 0.833x_5 = 0$

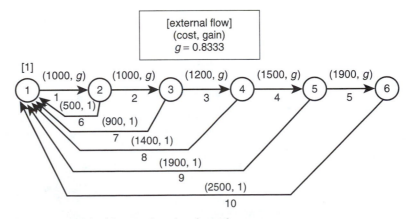

Figure 5.34   NFP model with renumbered nodes and arcs

Lower bounds on flow (the upper bounds are infinite):

$$x_k \geq 0, \quad k = 1, 2, \ldots, 10$$

Solving this model with a standard LP code yields the solution

$$x_1^* = 3.2727, \ x_2^* = 2.2727, \ x_7^* = 2.2727,$$
$$x_3^* = x_4^* = x_5^* = x_6^* = x_8^* = x_9^* = x_{10}^* = 0, \ z^* = 8045.5$$

This is the same solution obtained using an NFP code.

We should note that it is somewhat surprising that an LP model is appropriate for this discrete problem of finding the economic life of an investment. As in this case, it is often much easier to model a problem using network constructs and then find the equivalent LP model through the transformation presented in this section.

5.4 MINIMUM-COST FLOW PROBLEM

We now take a more expansive view of NFP models and describe a generic case that is often referred to as the *minimum-cost flow problem*. The following example is used to illustrate concepts of which many have already appeared in the models presented in the previous sections.

Suppose that the logistics manager of a company is faced with the problem of shipping a homogeneous commodity located at several plants to various customer sites around the country. The situation is illustrated in Figure 5.35. The plants are in Phoenix, Austin, and Gainesville (the gray nodes in the figure). The quantities available for shipping are indicated in the brackets adjacent to the nodes, with positive numbers indicating supplies of the commodity. Customers are in Chicago, Los Angeles, Dallas, Atlanta, and New York. The negative numbers adjacent to these nodes are the demands. Possible air shipping links are represented by the directed arcs between the nodes. The numbers adjacent to the arcs are unit shipping costs. The flow on each link is restricted to a maximum of 200 units. Dallas and Atlanta are hubs that, in addition to having their own demands, can transship items to other customers. The problem is to determine an optimal distribution plan that minimizes the total cost of shipping while meeting all customer demands with available supplies.

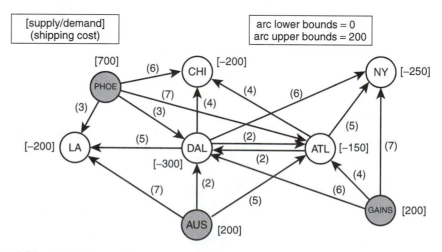

Figure 5.35   Distribution problem

Pure Minimum-Cost Flow Model

This distribution problem is ready-made for an NFP model, so we use it to describe the basic building blocks. In fact, it may seem to be nothing more than a transshipment problem as described earlier. A closer reading of the description reveals that the only difference between the two is that we have now placed finite upper bounds on the links.

In Figure 5.35, the nodes correspond to the cities, and the directed arcs correspond to transportation links. An origin node and a terminal node identify an arc. One arc in Figure 5.35, for example, originates at the Phoenix node and terminates at the Chicago node, implying that flow is one way between these two cities. If we wanted to consider flow from Chicago to Phoenix, another directed arc could be added to the graph joining these nodes.

Flow in the network enters and leaves at the nodes and passes through the arcs. In general, each network diagram is accompanied by one or more legends that describe the data in the problem. Flows entering or leaving a node from external sources are called *external flows* and are shown adjacent to the nodes in square brackets. A positive external flow is a supply—a flow that enters the network. A negative external flow is a demand—a flow that leaves the network. As we saw in Equation (2), flow is conserved at each node, implying that the total flow entering a node, either from arcs or from external sources, must equal the total flow leaving the node, either to arcs or to external demand points. The arc flows are the decision variables in the model.

Flow is limited in an arc by lower and upper bounds that are either implicit or explicit in the problem statement. In this example, we specify 0 as the lower bound and 200 as the upper bound for all arcs. As mentioned, the term *capacity* is used to refer to the upper bound on flow. Within the restrictions imposed by conservation of flow for each node and the bounds on flow for each arc, there are usually many feasible flows (in the context of a particular problem, this means the assignment of the variable quantity to each arc). The problem is to find an optimal flow with respect to the given objective function from the set of feasible flows assuming the feasible region is nonempty.

The objective we use here is cost minimization. Associated with each arc is a cost per unit of flow (the number in parentheses in Figure 5.35). If a flow x_k passes through arc k with unit cost c_k, a cost $c_k x_k$ is incurred. The total cost for the network is the sum of the arc costs $z = \Sigma_{k \in A} c_k x_k$, where A is the set of all arcs. The goal is to find the feasible flow that minimizes this measure.

We call the models in this section *pure minimum-cost flow model*s, because the flow entering an arc at its originating node is equal to the flow leaving the arc at its terminal node. The pure minimum-cost flow model is contrasted later in the chapter with the *generalized network flow model*, which does not have this requirement. The pure model has the important property of integral optimal solutions. Whenever all node external flows and all arc upper and lower bounds are integer valued, the solution to the pure model is also integer valued. As we shall see, this property has critical ramifications.

Linear Programming Model

Every minimum-cost flow model can be formulated as a linear program—i.e., with algebraic linear expressions describing the objective function and constraints, as indicated by Equations (1) to (3). We now develop the LP model for the distribution problem shown in Figure 5.35. For this purpose, it is convenient to number the nodes and arcs for reference as in Figure 5.1, which is repreated here as Figure 5.36. There are eight nodes and 17 arcs. The decision variables x_k represent the flow on arc k for all $k \in A$. The cost function is denoted by z.

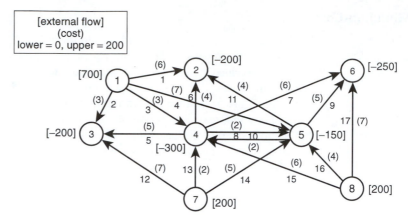

Figure 5.36   NFP representation with node and arc numbers

The objective is to minimize total cost:

$$z = 6x_1 + 3x_2 + 3x_3 + 7x_4 + \cdots + 7x_{17}$$

The main constraints require conservation of flow at each node:

Node 1: $x_1 + x_2 + x_3 + x_4 = 700$

Node 2: $-x_1 - x_6 - x_{11} = -200$

Node 3: $-x_2 - x_5 - x_{12} = -200$

Node 4: $x_5 + x_6 + x_7 + x_8 - x_3 - x_{10} - x_{13} = -300$

and similarly for nodes 5 through 8

A "+" sign indicates flow out of a node and a "–" sign indicates flow into a node. Simple lower and upper bounds are $0 \le x_k \le 200$ for $k = 1, \ldots, 17$.

A network flow model may be solved with either special-purpose network programming algorithms or with general-purpose linear programming codes. The solution shown in Figure 5.37 demonstrates the integrality property. The total cost $z^* = \$5300$.

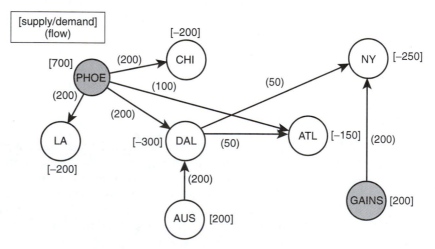

Figure 5.37   Optimal flow solution, $z^* = \$5300$

Variable External Flows

Further generalizing the distribution problem, we assume that the supply and demand values are not fixed as originally stated but are more dynamic. The following additional information must be factored into the analysis by the logistics manager.

- Phoenix: This plant is to be discontinued, and the entire inventory of 700 units must be shipped or sold for scrap. The scrap value is $5 per unit.

- Chicago: Minimum demand is 200 units. If available, however, an additional 100 units could be sold with a revenue of $20 per unit.

- Los Angeles: Demand is 200 units and must be met.

- Dallas: Contracted demand is 300 units. An additional 100 units may be sold with a revenue of $20 each.

- Atlanta: Demand is 150 units and must be met.

- New York: One hundred units are left over from previous shipments. There is no firm demand, but up to 250 units can be sold at $25 each.

- Austin: Maximum production is 300 units with a manufacturing cost of $10 per unit.

- Gainesville: Work rules require that all regular-time production of 200 units be shipped. An additional 100 units can be produced using overtime at a cost of $14 per unit.

Some of these qualifications represent variable external flows—that is, flows that can enter or leave the network at specified nodes in variable amounts. We handle such situations by adding arcs that either originate or terminate at one particular node. Table 5.4 lists the set of *external flow* arcs added to the network in Figure 5.36. The dashes indicate that the corresponding arcs do not originate (or terminate, depending on the case) at a node in the system.

Each of these new arcs affects a conservation of flow constraint at a single node only. Three negative values appear in the "Cost" column, representing the revenues at the cities with variable demand. In general, negative costs are equivalent to revenues. The lower bounds on all the new arcs are zero, but the upper bounds depend on the additional information.

The optimal solution for the example is shown in Figure 5.38 with the six external flow arcs. Reviewing the solution, we note that Phoenix ships its total supply of 700 units, whereas Austin ships 300 units and Gainesville ships its minimum amount, 200 units. All customers receive the maximum permitted. The objective function is negative ($z^* = -\$1600$), indicating a net profit. The fixed external flows at the nodes do not contribute to the profit figure, because no costs or revenues are associated with fixed flows.

Table 5.4   External Flow Arcs Added to Distribution Network

External flow arcs	Origin node	Terminal node	Lower bound	Upper bound	Cost
Phoenix scrap	1	—	0	700	5
Chicago extra demand	2	—	0	100	−20
Dallas extra demand	4	—	0	100	−20
New York demand	6	—	0	250	−25
Austin extra supply	—	7	0	300	10
Gainesville extra supply	—	8	0	100	14

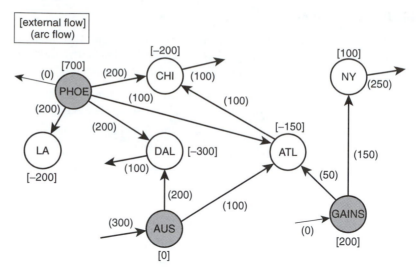

Figure 5.38   Solution with variable node flows, $z^* = -\$1600$

Generalized Minimum-Cost Flow Problem

Continuing with the distribution example, the manager notes that there is a possibility of shipping losses. As an approximation, let us assume that every transportation link loses 5% of the amount shipped.

We introduce an arc gain parameter to handle losses or gains that occur along a link. The gain parameter multiplies the flow at the beginning of the arc to obtain the flow at the end of the arc. Figure 5.39 illustrates the effect of the gain on flow. We model the 5% loss from Phoenix to Chicago with a gain of 0.95 on the arc. Thus, 200 units leave Phoenix, but only 190 units arrive at Chicago.

The solution with losses is shown in Figure 5.40 and is significantly different than before. The flows are no longer integer valued, the extra demand at Chicago is no longer satisfied, Austin does not produce to its full capacity, more arcs are used to provide merchandise lost during shipping, and the profit is considerably reduced. The optimal objective function value is now $z^* = -\$494.30$.

Gains are a very useful modeling device. When all arc gains are 1, a pure NFP model results and all solutions are guaranteed to be integer valued. When some gains are other than 1, we have a generalized NFP model with no assurance of integer solutions.

Side Constraints

Returning to the model without losses for this illustration, we find that the logistics manager is receiving complaints from the Gainesville plant. Whereas Austin is producing and shipping at full capacity, the Gainesville plant is producing at only two-thirds of its capac-

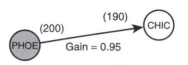

Figure 5.39   Modeling losses with gain factor

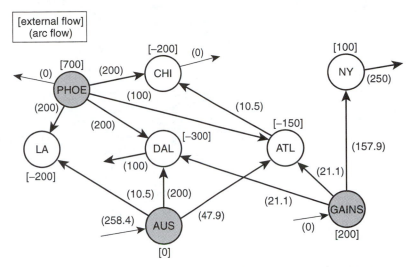

Figure 5.40 Solution with shipping losses represented by gain factors, $z^* = -\$494.30$

ity. Because both plants can manufacture up to 300 units, the manager wants to see the effect of a constraint that requires both plants to ship the same proportion of capacity. One way to write this constraint is to require that the total flow leaving Austin be equal to the total flow leaving Gainesville. Using the arc numbers in Figure 5.36, we have

$$x_{12} + x_{13} + x_{14} - x_{15} - x_{16} - x_{17} = 0$$

This equation is called a *side constraint* and is not compatible with Equation (2).

Although the essence of the problem can still be described using a network flow model, the new requirement cannot be handled within the NFP structure. The addition of side constraints precludes the use of special-purpose NFP algorithms; however, we can simply add the side constraint to the LP model and solve it with a general-purpose LP code. In this case, as shown in Figure 5.41, the optimal solution calls for a reduction in the output at Austin, which has the effect of eliminating the flows on the links from Austin to Atlanta and from Atlanta to Chicago and of not meeting the optional demand at Chicago. The corresponding profit is now $1500, down $100, so forcing equality between the two plants is not advisable.

Node Restrictions and Costs

Continuing with the distribution example, the manager learns that there is a problem with transshipments through Atlanta. The airport has instituted a fee for each unit transferred and limits transfers to 100 units per week. In addition, there is a spoilage of 10% of the units that are handled.

The model in Figure 5.36 does not allow this information to be included, because there is no way to identify on a single arc the flow that represents the amount transshipped. Restrictions can be placed on an arc but not on the material passing through a node. We allow for these possibilities by dividing the Atlanta node into two nodes, as shown in Figure 5.42, and adding an arc between them. The flow that actually passes through Atlanta now flows on arc 18, and any information related to that flow is described by the parameters of the arc.

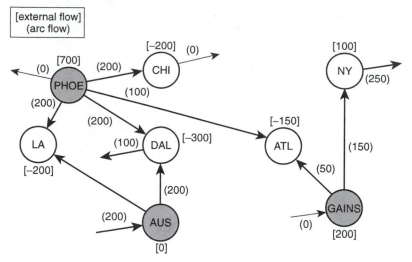

Figure 5.41   Forcing equality between Austin and Gainesville, $z^* = -\$1500$

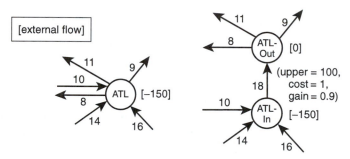

Figure 5.42   Modification of the model to include node restrictions

EXERCISES   *Provide an NFP model and then find the optimal solution for each of the following problems using the accompanying Excel add-ins or any available software package.*

1. Find the shipping plan between sources (S1 to S3) and destinations (D1 to D4) that minimizes total shipping cost while meeting the demands. The numbers in the matrix represent unit shipping costs.

	D1	D2	D3	D4	*Supply*
S1	10	10	6	15	10
S2	5	15	10	12	15
S3	11	8	7	21	8
Demand	5	3	8	17	

2. Using the shipping costs given in Exercise 1 and the following production and demand data, find the production schedule that maximizes profit.

Source	Production data			Destination	Demand data		
	Minimum number of units	Maximum number of units	Cost per unit		Minimum sales	Maximum sales	Unit revenue
S1	10	15	$10	D1	5	5	$20
S2	15	15	12	D2	3	10	25
S3	0	8	13	D3	0	8	22
				D4	10	20	30

3. (*Multiperiod Production Scheduling*) A company has one manufacturing plant and three sales outlets (A, B, and C). Unit shipping costs from the plant to outlets A, B, and C are $4, $6, and $8, respectively. The company wishes to develop a production, shipping, and sales plan for three periods. The corresponding data are as follows.

Period	Manufacturing data			Selling price			Maximum sales		
	Unit cost ($)	Capacity (units)	Outlet	Outlet A	Outlet B	Outlet C	Outlet A	Outlet B	Outlet C
1	8	175		$15	$20	$14	50	100	75
2	10	200		18	17	21	75	150	75
3	11	150		15	18	17	20	80	50

The plant has the capability of storing up to 100 units of product stored from one period to the next. The storage cost is $1 per unit per period. Find a plan that maximizes profit.

4. Five crews are available to do three jobs. The table shows the time in hours required for each crew to do each job completely, but a crew can do a portion of any job in a corresponding fraction of the time. The crews are paid only for the time they work, and they can work no more than 20 hours. Their hourly wage rates are also shown in the matrix. Find the assignment of jobs to crews that minimizes the total cost of completing the jobs.

Job	Crew 1	Crew 2	Crew 3	Crew 4	Crew 5
1	15	20	40	35	45
2	20	22	30	50	35
3	25	32	50	48	60
Cost / hr	150	130	100	80	70

5. Solve the 4 × 4 assignment problem given in the table.

	E	F	G	H
A	3	2	14	3
B	21	20	17	5
C	14	14	16	4
D	14	3	14	4

6. A company has three workers. On a particular day, six jobs are scheduled to be completed. The cost for each worker to do each job is shown in the table.

Worker	Job 1	Job 2	Job 3	Job 4	Job 5	Job 6
A	3	2	2	6	4	6
B	4	3	7	5	3	3
C	9	9	7	9	7	6

(a) What is the minimum cost assignment when each worker can do two jobs?

(b) What is the minimum cost assignment to do any three jobs when each worker can do only one job?

(c) What is the minimum cost assignment when each worker can do any number of jobs?

7. Find the shortest path tree rooted at node 1 for the following network.

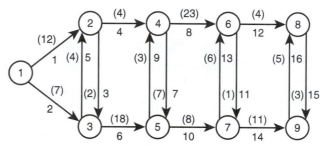

8. You are given the following directed network.

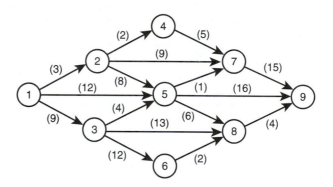

(a) Let the numbers on the arcs represent distances and find the shortest path tree rooted at node 1.

(b) Let the numbers represent flow capacity and find the maximum flow from node 1 to node 9.

9. A company has just purchased a new machine for $4000. The operating cost and resale value of the machine varies with age according to the schedule in the table. The discount rate for time value of money calculations is 10%.

Year of ownership	Operating cost during year	Resale value at year's end
1	$800	$2000
2	800	800
3	1000	400
4	1500	200
5	2000	0

(a) Draw the generalized NFP model that could be used to evaluate the policy "buy a new machine every 3 years." Show only the relevant arcs and the corresponding parameters. Use this model to compute the present worth of following this policy for an infinite period of time.

(b) Show the single additional arc that would be required to determine if replacing the machine every 2 years was superior to replacing it every 3 years. Which is the better policy?

(c) The economic life of a piece of equipment is the age of replacement that minimizes the lifetime cost of replacement. Considering the possibility of replacement in any one of the 5 years, develop an NFP model whose solution will determine the economic life of the machine. How sensitive is the solution to the costs in your model?

10. You have two alternatives for providing yourself with the use of a new Honda. In alternative (1) you can purchase a contract for $100,000 that provides you with the car for 10 years. There are no other costs associated with this option. For purposes of analysis, assume that the same contract will be available 10 years from now, and forever after. In alternative (2) you can purchase the Honda for $20,000 and keep it for either 2 years, 4 years, or 6 years. The annual operating costs and salvage values are listed in the table. Resale values for odd years are not shown because they are not included as options. Again, it is assumed that this or a similar car will be available forever.

Year of ownership	Operating cost during year	Salvage value at year's end
1	$500	—
2	500	$10,000
3	1000	—
4	2000	6000
5	3000	—
6	5000	2000

The selection depends on your minimum acceptable rate of return, which is 25% per year.

(a) Determine whether alternative (1) or alternative (2) is the optimal choice. The goal is to minimize the present worth of the cost of ownership for an infinite time period.

(b) Find the optimal resale time when alternative (2) is chosen.

11. The figure shows a generalized NFP model. All arc costs are 1, and all arc gains are 2. All arcs have lower bounds of zero and unlimited upper bounds on flow. Write the LP model for this network and find the solution. The objective is to minimize the total flow cost.

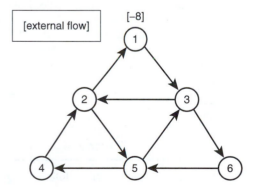

12. Draw an arc that describes a plant that transforms wood into chairs. A gain parameter should be used to reflect this transformation. Flow entering the arc is measured in pounds of wood. Flow leaving the arc is to be in units of chairs. Twenty pounds of wood is required for each chair. The plant can process no more than 1000 chairs at a cost of $40 per chair.

13. The following arc has a nonzero lower bound. Show an equivalent arc that has a zero lower bound.

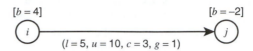

14. A company offers volume discounts for its products. The first 20 units are sold for $5 per unit, the next 10 units for $2 per unit, and any number of units beyond 30 for $1 per unit. Show the arc structure that describes this revenue function for a minimum-cost NFP model.

15. A warehouse facility in a transportation system has several incoming and outgoing links. At least 100 units but no more than 1000 units must pass through the facility per unit time. A cost of $15 is charged for each of the first 500 units and $20 for each additional unit. Develop a network model for this facility.

16. A mining company has discovered a mineral reserve with 400 tons of recoverable ore. Engineers estimate that the cost of mining the first 200 tons will be $150 per ton, mining the next 100 tons will cost $200 per ton, and mining the final 100 tons will cost $300 per ton. The raw ore can be stored at the mine site for a cost of $50 per ton per year. The equipment available allows a maximum of 50 tons to be mined each year.

   After being mined, the ore will be processed into ingots at a cost of $2 per ingot. Each ton will make 100 ingots. The processing capacity is 40 tons (or 4000 ingots) per year. Ingots may be stored after they are processed, but because of impurities in the atmosphere, 10% of the amount stored in any year will be lost. The cost of storing an ingot for a year is $0.60.

   The demand and the selling price for ingots for the next 10 years are shown in the table. The demand figures indicate the maximum sales. The actual sales may be less than the stated demands.

Year	1	2	3	4	5	6	7	8	9	10
Demand	0	0	2000	3000	4000	5000	6000	7000	8000	9000
Price ($)			4.5	5.0	5.5	6.0	6.0	6.5	6.5	6.0

   Set up and solve the NFP model to determine the optimal policy for mining ore, storing ore, processing ore into ingots, storing ingots, and selling ingots. The solution should determine each of these quantities for each of the next 10 years.

17. A company makes two items labeled A and B. Management wants to obtain a production plan for the next 3 months. The demand for the two items over this period is given in the table. The company does not have to sell the amount demanded; rather, the numbers given are upper bounds on sales. If the company provides less than the amount demanded, the difference is lost. The unit revenues for the items in each of the next 3 months are also shown.

Month	Demand A	Revenue A	Demand B	Revenue B
1	100	$50	75	$100
2	200	55	135	102
3	176	49	82	110

   Several additional restrictions follow.

   • The total sales of both items in any month cannot exceed 250 units.
   • The amount of labor is fixed at 15 workers, whose individual salaries are $600 per month when the workers are productively employed. When working full time on item A, each worker can produce 12 units per month. When working full time on item B, each worker can produce eight units per month. A worker can divide his or her time between the two items, with the amount of each

item produced being equal to the proportion of time spent on that item. The worker can also remain idle for part of the time. If such is the case, the worker's salary is reduced to $450 per month for the proportion of the time the worker is idle.

- Items may be produced in one month for sale in another. The cost of storing one unit from one month to the next is $5; however, there is spoilage of items in inventory. Of the amount stored at the beginning of a month, 10% must be discarded. In addition, no more than 20 units of item A, and no more than 10 units of item B, can be stored in any month. Items produced and sold in the same month do not enter inventory.

The problem is to determine the allocation of workers to the manufacture of items in each month so that profit is maximized. The sales quantities and inventory policy for each item are also to be determined. Solve using an NFP model.

18. A small university has four buildings, A through D, and six faculty parking lots, I through VI.

    (a) Using the data in the table, determine the parking lots to which the building occupants should be assigned so as to minimize the total walking distance for the faculty.

Building	Number of faculty	Distance from building to parking lot (blocks)					
		I	II	III	IV	V	VI
A	15	0	1	3	2	1	4
B	21	1	4	3	2	5	3
C	13	2	1	1	0	2	3
D	7	6	5	5	4	5	7
Capacity of lot		5	12	15	10	3	15

    (b) Say that five faculty members in each building are full professors, two are associate professors, and the remainder are assistant professors. Use a goal programming approach to ensure that the full professors have priority over all others and that the associate professors have priority over the assistant professors in the assignment of parking lots. (*Hint:* To give priority to the higher ranks, multiply each associate professor's travel cost by 10 and each full professor's travel cost by 100.)

19. For this distribution problem, supply and demand data, direct shipping costs from suppliers to customers, and transshipment costs to and from a terminal are given in the matrix.

		Customer				Supply
		5	6	7	Terminal	
Supplier	1	5	4	3	2	15
	2	1	2	3	2	15
	3	5	5	4	3	15
	4	6	1	4	3	15
Terminal		3	1	2		
Demand		17	15	20		

In addition, the following restrictions must be met.

- All demands must be satisfied.
- Suppliers 1 and 2 together can ship no more than a total of 10 units to customer 5 directly.
- No shipping link can handle more than seven units.
- The transshipment terminal can handle no more than 15 units.

Set up and solve the NFP model for this problem.

20. A developing country has two mines labeled M1 and M2 from which coal can be obtained, two power plants labeled P1 and P2 that burn coal to produce electricity, and three cities labeled C1, C2, and C3 that use the electricity. The government's energy department wants to determine the optimal plan for producing and delivering the electricity and for mining the coal over a given period of time. The demand for electricity, which is measured in megawatt hours (MWh), is 1000 MWh for city C1, 500 MWh for city C2, and 3000 MWh for city C3. Power is transmitted from the plants to the cities over lines that undergo losses as a result of electrical resistance. The size of each loss depends on the origin and destination of the power transmitted and is given as a percentage for each plant–city pair, as follows.

	Power loss (%)		
Plant	City C1	City C2	City C3
P1	1	5	2
P2	8	4	9

The efficiency of a plant is measured in megawatt hours produced per ton of fuel input. Each plant is limited in terms of the maximum input of fuel allowed during the period. Processing costs for the fuel are given in the following table.

Plant	Efficiency (MWh/ton)	Maximum input (tons)	Processing cost per ton ($/ton)
P1	1	4000	$10
P2	0.9	3000	$ 9

	Shipping cost ($/ton)	
Mine	Plant P1	Plant P2
M1	6	12
M2	3	8

The coal must be shipped from the mines to the plants. Shipping costs, in dollars per ton, are as follows.

During the time period covered by the plan, a limited amount of coal may be taken from each mine. For mine M1, 1200 tons may be extracted at a cost of $9 per ton and an additional 1500 tons can be extracted at a cost of $12 per ton. For mine M2, 1500 tons can be extracted at a cost of $13 per ton with an additional 2000 tons available at $15 per ton. Develop in detail an NFP model that can be used to solve this planning problem. Find the optimal solution.

21. A company has two plants, P1 and P2, that manufacture three products, A, B, and C. The number of labor hours needed to produce a unit of each product, and the cost of raw materials for each product, are as follows. The weekly demands for the three products are 1000, 500, and 700, respectively. There are 1500 labor hours available at each plant. Set up an NFP model with the objective of minimizing raw material costs, and find the solution that will meet the demand for the products without exceeding the labor constraints. (*Hint*: The network should be bipartite, with flow going from plants

	Labor hours per unit of product			Raw material cost ($) per unit of product		
Plant	Product A	Product B	Product C	Product A	Product B	Product C
P1	0.5	2.0	1.5	10	20	30
P2	0.75	1.2	2.0	15	20	35

to products. Each arc should have two parameters: cost per hour and a gain factor in terms of products per hour. The units of flow in the network should be hours.)

BIBLIOGRAPHY   Argüello, M.F., J.F. Bard, and G. Yu, "Models and Methods for Managing Airline Irregular Operations," in G. Yu (editor), *Operations Research in the Airline Industry*, pp. 1–45, Kluwer Academic, Boston, 1998.

Ball, M.O., T.L. Magnanti, C.L. Monma, and G.L. Nemhauser (editors), *Network Models; Handbooks in Operations Research and Management Science*, Vol. 7, Elsevier Science, Amsterdam, 1995.

Dembo, R.S., J.M. Mulvey, and S.A. Zenios, "Large-Scale Nonlinear Networks and Their Applications," *Operations Research*, Vol. 37, pp. 353–372, 1989.

Dror, M., P. Trudeau, and S.P. Ladany, "Network Models for Seat Allocation on Flights," *Transportation Research*, Vol. 22B, pp. 239–250, 1988.

Glover, F., D. Klingman, and N. Philips, *Network Models and Their Applications in Practice*, Wiley, New York, 1992.

Lawler, E., *Combinatorial Optimization: Networks and Matroids*, Holt, Rinehart & Winston, New York, 1976.

Magnanti, T.L. and R.T. Wong, "Network Design and Transportation Planning: Models and Algorithms," *Transportation Science*, Vol. 18, pp. 1–55, 1984.

Mulvey, J.M., "Strategies for Modeling: A Personnel Scheduling Example," *Interfaces*, Vol. 9, No. 3, pp. 66–76, 1979.

Chapter 6

Network Flow
Programming Methods

The focus of this chapter is on the development of algorithms for solving NFP problems. We begin with a discussion of several special cases, including the transportation problem, the shortest path problem, and the maximum flow problem, and their variants, and conclude with a presentation of the primal simplex algorithm for the pure minimum-cost flow problem. Although virtually all of the special cases are instances of the minimum-cost flow problem, a great deal can be learned by studying them separately. The individual algorithms provide insights into different ways of solving problems and have the benefit of being extremely efficient. In addition, many applications of the minimum-cost flow problem embody features of the special cases. From a modeling perspective, it is helpful to know how these features can be incorporated into broader formulations.

6.1 TRANSPORTATION PROBLEM

The transportation problem (TP) is concerned with finding an optimal distribution plan for a single commodity. A given supply of the commodity is available at several sources, the demand for the commodity at each of several destinations is specified, and the transportation cost for each source–destination pair is known. In the simplest case, the unit transportation cost is constant. The problem is to find the optimal distribution plan for shipments from sources to destinations that minimizes the total transportation cost.

Matrix Model

The traditional way to describe a TP is with a matrix, as in Figure 6.1. The m sources that provide the commodity are identified by name at the left side of the matrix, and the n destinations to which the commodity is to be shipped are arrayed along the top. The quantities available at the sources (supplies) are shown as numbers at the right of the matrix, with s_i being the supply at source i. The quantities required by the destinations (demands) are shown as numbers along the bottom, with d_j being the demand at destination j. The numbers in the body of the matrix are the unit shipping costs, with c_{ij} representing the cost of shipping from source i to destination j. If it is not possible to ship from a given source to a given destination, a large cost of M is entered in the appropriate cell.

We observed in Chapter 5 that the TP can also be represented as a bipartite network comprising m supply nodes and n destination nodes. When it is feasible to ship from source i to destination j, a directed arc between these two nodes is included with cost coefficient

Destination

Source	D1	D2	...	Dn	*Supplies*
S1	c_{11}	c_{12}	...	c_{1n}	s_1
S2	c_{21}	c_{22}	...	c_{2n}	s_2
⋮	⋮	⋮		⋮	⋮
Sm	c_{m1}	c_{m2}	...	c_{mn}	s_m
Demands	d_1	d_2	...	d_n	

Figure 6.1   Matrix model of the transportation problem

c_{ij}. The term "bipartite" refers to the division of the nodes into two disjoint sets. Transitions are permitted only between sets and not within sets, as is the case with the TP.

A requirement of most solution algorithms is that total supply equal total demand—i.e., $\Sigma_i s_i = \Sigma_j d_j$. This is known as the *feasibility property* and gives rise to the *balanced* TP. All instances of the traditional TP can be modified so that this requirement is satisfied by simply adding either a dummy source with index $m + 1$ if demand exceeds supply, or a dummy destination with index $n + 1$ if supply exceeds demand. Letting Δ be the excess, we set $s_{m+1} = \Delta$ in the first case and $d_{n+1} = \Delta$ in the second case. In addition, the corresponding cost coefficients are set equal to zero—i.e., either $c_{m+1, j} = 0$ for all j or $c_{i, n+1} = 0$ for all i.

A solution to the model is an assignment of flows to the cells of the matrix. In general, we call x_{ij} the flow in the cell representing shipments from source i to destination j. For a feasible solution, the sum of the flows in a row of the matrix must equal the supply at the associated source, and the sum of the flows in a column of the matrix must equal the demand at the associated destination. A numerical example is given in Figure 6.2a. The optimal solution is displayed in Figure 6.2b.

The TP has a linear programming (LP) representation, so it can be solved using the simplex method described in Chapter 3. Its special structure, however, allows considerable simplification of the standard simplex approach. In the following discussion, we assume that the feasibility property holds and thus the total supply equals the total demand.

	D1	D2	D3	*Supply*
S1	9	12	10	15
S2	8	15	12	15
S3	13	17	19	15
Demand	10	20	15	

	D1	D2	D3	*Shipped*
S1	0	5	10	15
S2	10	0	5	15
S3	0	15	0	15
Received	10	20	15	

a. Transportation model (c_{ij})   b. Optimal solution (x_{ij}^*)

Figure 6.2   Instance of the transportation problem in matrix format

Example

Throughout this section, we use the example depicted in Figure 6.3 in matrix (tableau) format. Note that the transportation tableau is different from the simplex tableau commonly employed to solve linear programs (LPs). This example has three sources and five destinations. The supplies at the sources are shown at the right, and the demands at the destinations are shown along the bottom.

A cell corresponds to each supply–demand combination; the unit cost of shipping is the boxed number in the upper left corner of the cell, and the boldface number in each cell is the flow assigned to that cell. For example, $x_{12} = 10$ implies that there is a flow of 10 units from S1 to D2. Because total supply equals total demand, the flows in each row sum to the supply at the corresponding source and the flows in each column sum to the demand at the corresponding destination. The flows shown in Figure 6.3 are feasible but not optimal. The goal of this section is to describe an algorithm that can be used to find the optimal flow—that is, the flow that minimizes total cost.

Theory

Primal Linear Programming Model

To formulate the transportation problem as an LP, we let x_{ij} be the flow from source i to destination j. The objective is to minimize the total shipping costs—that is,

$$\text{Minimize } \sum_{i=1}^{m}\sum_{j=1}^{n} c_{ij}x_{ij}$$

subject to the following constraints.

1. The supply at each source must be used:

$$\sum_{j=1}^{n} x_{ij} = s_i, \quad i = 1, \dots, m \tag{1}$$

2. The demand at each destination must be met:

$$\sum_{i=1}^{m} x_{ij} = d_j, \quad j = 1, \dots, n \tag{2}$$

	D1	D2	D3	D4	D5	*Supply*
S1	15	15	16	11	11	
	5	**10**				15
S2	13	11	15	9	6	
		0	**15**			15
S3	8	12	11	7	8	
			0	**5**	**10**	15
Demand	5	10	15	5	10	

Figure 6.3 Transportation example

3. The cell flows must be nonnegative:

$$x_{ij} \geq 0, \quad i = 1, \ldots, m \text{ and } j = 1, \ldots, n$$

Although we have an equality constraint for every source and destination, one of these equalities is redundant. This can be seen by separately adding the m supply constraints and the n demand constraints and observing from the feasibility property that the two sums are equal. Thus, the $m + n$ equations are linearly dependent. As we shall see, the transportation algorithm exploits this redundancy.

Basic Solutions

The primal problem has $n + m$ constraints, one of which is redundant. A basic solution for this problem is determined by a selection of $n + m - 1$ independent variables. The basic variables assume values that satisfy the supply and demand equations, whereas the nonbasic variables are zero.

For our example, $m = 3$ and $n = 5$, so the number of basic variables is 7. Figure 6.3 shows a basic solution. The flows in the seven basic cells uniquely satisfy the supplies and demands. The fact that some of the basic flows are zero indicates that this basic solution is degenerate.

The cells in the tableau are said to be dependent and do not form a basis if it is possible to trace a closed path through them. Such a path would consist of a series of horizontal and vertical moves turning only at the designated cells. Figure 6.4 depicts a collection of dependent cells marked by **x** and the closed path that connects them. The variables associated with those cells are $x_{12}, x_{15}, x_{22}, x_{23}, x_{33}$, and x_{35}. If we wrote out the corresponding equations—Equation (1) for $i = 1, 2, 3$, and Equation (2) for $j = 2, 3, 5$, we would see that they are linearly dependent.

A basis is any collection of $n + m - 1$ cells that does not include a dependent subset. The basic solution is the assignment of flows to the basic cells that satisfies the supply and demand constraints. The solution is feasible if all the flows are nonnegative. From the theory of linear programming, we know that there is an optimal solution that is a basic feasible solution (BFS).

	D1	D2	D3	D4	D5	Supply
S1	15	15	16	11	11	15
S2	13	11	15	9	6	15
S3	8	12	11	7	8	15
Demand	5	10	15	5	10	

Figure 6.4 Dependent cells

Dual Linear Programming Model

To formulate the dual formulation of the transportation LP model, we introduce a variable for each constraint in the primal formulation.

Let u_i be the dual variable corresponding to Equation (1).

Let v_j be the dual variable corresponding to Equation (2).

Then, using the procedures in Section 4.2, the dual problem is

$$\text{Maximize } \sum_{i=1}^{m} s_i u_i + \sum_{j=1}^{n} d_j v_j$$

$$\text{subject to } \quad u_i + v_j \leq c_{ij}, \quad i = 1, \ldots, m \text{ and } j = 1, \ldots, n \tag{3}$$

The values of u_i and v_j are unrestricted in sign for all i and j.

Optimality Conditions

From duality theory, we know that a complementary pair of primal and dual feasible solutions is optimal to their respective problems. To simplify the discussion, we define the slack variable for the general dual constraint (3) to be w_{ij}, where

$$w_{ij} = c_{ij} - u_i - v_j \tag{4}$$

In fact, w_{ij} is the reduced cost associated with the primal variable x_{ij} and thus measures the change in the objective function with a unit change in x_{ij}. The complementary slackness property states that for every basic primal variable the corresponding dual constraint must be satisfied as an equality—that is,

$$u_i + v_j = c_{ij} \text{ or } w_{ij} = 0 \text{ for every basic variable } x_{ij} \tag{5}$$

Because of the redundant primal constraint, one of the dual variables can be arbitrarily set equal to zero. The others can be determined using Equation (5).

By construction, we ensure that the primal solution is feasible. If all the dual constraints are satisfied for the dual solution, both the primal and dual solutions are optimal. The condition for dual feasibility is

$$u_i + v_j \leq c_{ij} \text{ or } w_{ij} \geq 0 \text{ for all nonbasic cells} \tag{6}$$

This result is the foundation of the solution algorithm described in this section.

In the numerical example, we use the matrix structure in Figure 6.5 to depict the dual variables and the corresponding dual slacks. The u_i variables are shown along the right margin, and the v_j variables are shown along the bottom. The dual slacks variables w_{ij} are shown in the upper right corners of the nonbasic cells. For example, $w_{14} = -4$. In the case of the basic cells, these values are zero and are omitted from the matrix.

For any BFS **x**, a value of zero is arbitrarily assigned to one of the u_i or v_j variables. For hand computations, we assign the zero value to the row or column with the most basic cells. This simplifies the process of determining the other dual values. Given the basic cells, it is a simple matter to compute the values of the remaining $n + m - 1$ dual variables. This is done by solving the $n + m - 1$ simultaneous equations given by Equation (5) using back-substitution. Note that there is no requirement that u_i or v_j be nonnegative. Once the dual variables have been determined, the values of w_{ij} for the nonbasic cells are computed using Equation (4).

Figure 6.5 shows the complete tableau for the current BFS. Given the basic cells, there is a unique assignment of flows that satisfies the supplies and demands. Once one of the

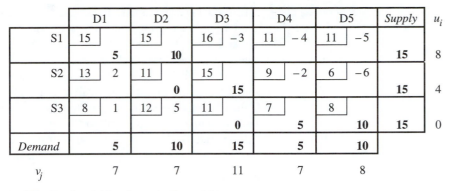

Figure 6.5   Dual variables shown in the matrix

dual variables has been assigned arbitrarily to zero, the others are uniquely determined. The value of w_{ij} is, in fact, the marginal benefit of allowing x_{ij} to enter the basis. Evidently, the basis picked for Figure 6.5 is not optimal, because some of the w_{ij} entries are negative. The value of $w_{25} = -6$ indicates that the objective function will be reduced by six units for every unit increase in x_{25}. The primal simplex algorithm proceeds by allowing x_{25} to enter the basis, as described in the complete solution to the example that follows the statement of the algorithm.

The Simplex Algorithm

Step 1: *Construct the initial tableau.*

Prepare the initial tableau showing problem parameters. Start with some selection of independent basic cells for which the primal solution is feasible. Determine the unique values of the basic variables (cell flows) for this basis, and place them in the cells.

Step 2: *Compute the dual variables for the current basis and check for optimality.*

a. Assign a value of zero to a dual variable in some row or column. Determine the values of the other dual variables such that the complementary slackness condition ($w_{ij} = 0$ for basic cells) is satisfied. Place these values at the boundaries of the tableau.

b. For each nonbasic cell, compute the dual slack variable w_{ij} and place it in the upper right portion of the cell.

c. If all the values of w_{ij} are nonnegative (that is, if all reduced costs are nonnegative), stop with the optimal solution; otherwise, go to Step 3.

Step 3: *Change the basis.*

a. (*Find the cell to enter the basis*) Select the entering cell as the one with the most negative value of w_{ij}.

b. (*Find the cell to leave the basis: ratio test*) Identify the (unique) cycle that starts with the entering cell, passes through only basic cells, and returns to the entering cell. Show the cycle in the tableau as a sequence of alternating vertical and horizontal lines connecting pairs of cells in the cycle. Mark the entering cell with a plus sign (+). Mark the first basic cell in the cycle with a minus sign (−). Alternately mark each basic cell in the cycle with a plus sign or a minus sign until the path returns to the entering cell. Find the basic cell marked with a minus sign that has the smallest flow value: This is the cell that is to leave the basis. If there is a tie, select any one of the tied cells arbitrarily. The cell(s) not chosen will remain basic but will go to zero during the pivot operation, indicating a degenerate BFS. Let δ be the value of the basic variable for the leaving cell.

c. (*Change the basic solution: pivoting*) For every cell in the cycle marked with a plus sign, increase the basic variable by δ. For every cycle cell marked with a minus sign, decrease the basic variable by δ. Remove the leaving cell from the basis. Add the entering cell to the basis with flow equal to δ.

d. Return to Step 2.

Solution to Example

We illustrate how the algorithm works with the 3×5 example starting from the basic solution shown in Figure 6.5. Any iteration of the algorithm can be completely described with a tableau similar to the one shown in Figure 6.6. The basic cells are those with boldface flow values. The dual slack variables are computed for the nonbasic cells and are shown at the upper right. A cell with a negative w_{ij} is selected to enter the basis, as indicated by a plus sign. The most negative value is $w_{25} = -6$, so x_{25} is chosen as the entering variable. We find the unique cycle formed by this cell and a subset of cells comprising the current basis. The vertical and horizontal arrows indicate the cells in the cycle: $(2, 5) \rightarrow (3, 5) \rightarrow (3, 3) \rightarrow (2, 3) \rightarrow (2, 5)$. The signs adjacent to the arrows indicate the directions in which the basic flows will change. Note that the signs alternate in the cells of the cycle. Figure 6.6 shows these steps for the first iteration of the algorithm.

The cells that have negative signs determine the maximum flow change in the cycle and hence the variable that will leave the basis. In the example, cells $(2, 3)$ and $(3, 5)$ undergo decreases in flow. The minimum flow in these two cells determines the flow change,

	D1		D2		D3		D4		D5		Supply	u_i
S1	15		15		16	−3	11	−4	11	−5		
		5		**10**							15	8
S2	13	2	11		15		9	−2	6	−6		
				0	→−	**15**			↓+		15	4
S3	8	1	12	**5**	11		7		8			
					↑+	**0**		**5**	←−	**10**	15	0
Demand		**5**		**10**		**15**		**5**		**10**		
v_j		7		7		11		7		8		

Figure 6.6 First iteration for example; x_{25} enters the basis and x_{35} leaves, with $\delta = 10$

$\delta = \min\{15, 10\} = 10$. The basis change will have cell $(2, 5)$ entering the basis and cell $(3, 5)$ leaving. Figure 6.7 shows the tableau for the second iteration.

The second iteration selects cell $(1, 4)$ to enter the basis. The cycle formed is shown by the arrows, with the directions of flow change indicated by the adjacent plus and minus signs. The minimum flow in the cells containing negative signs is the value of the flow change, $\delta = \min\{10, 5, 5\} = 5$. Both cells $(2, 3)$ and $(3, 4)$ have this value, and we select $(2, 3)$ arbitrarily. Because of the tie for the leaving variable, the flow x_{34} will also be zero in the next tableau, indicating a degenerate solution.

The remaining iterations are shown in Figures 6.8 through 6.10. Note that at the end of iteration 3, the flow change is $\delta = 0$, so the entering variable has $x_{31} = 0$. This corresponds to a degenerate pivot and can lead to cycling. Fortunately, the algorithm continues

	D1		D2		D3		D4		D5		Supply	u_i
S1	15		15		16	−3	11	−4	11	1		
		5	→−	10			↓+				15	4
S2	13	2	11		15		9	−2	6			
			↑+	0	←−	5				10	15	0
S3	8	1	12	5	11		7		8	6		
					↑+	10	←−	5			15	−4
Demand	5		10		15		5		10			
v_j	11		11		15		11		6			

Figure 6.7 Second iteration for example; x_{14} enters the basis and x_{23} leaves, with $\delta = 5$

	D1		D2		D3		D4		D5		Supply	u_i
S1	15		15		16	1	11		11	1		
	→−	5		5			↓+	5			15	0
S2	13	2	11		15	4	9	2	6			
				5						10	15	−4
S3	8	−3	12	1	11		7		8	2		
	↑+					15	←−	0			15	−4
Demand	5		10		15		5		10			
v_j	15		15		15		11		10			

Figure 6.8 Third iteration; x_{31} enters the basis and x_{34} leaves, with $\delta = 0$

	D1	D2	D3	D4	D5	Supply	u_i
S1	15 →- 5	15 5	16 -2 ↓+	11 5	11 1	15	0
S2	13 2	11 5	15 1	9 2	6 10	15	-4
S3	8 ↑+ 0	12 4	11 ←- 15	7 3	8 5	15	-7
Demand	5	10	15	5	10		
v_j	15	15	18	11	10		

Figure 6.9 Fourth iteration; x_{13} enters the basis and x_{11} leaves, with $\delta = 5$

	D1	D2	D3	D4	D5	Supply	u_i
S1	15 2	15 5	16 5	11 5	11 1	15	0
S2	13 4	11 5	15 3	9 2	6 10	15	-4
S3	8 5	12 2	11 10	7 1	8 3	15	-5
Demand	5	10	15	5	10		
v_j	13	15	16	11	10		

Figure 6.10 Optimal solution

to make progress at iteration 4. In the final tableau shown in Figure 6.10, all the values of w_{ij} are nonnegative, indicating that the optimal solution has been found.

Finding an Initial Basic Feasible Solution

The simplex algorithm requires a basic feasible solution to start. We now present a general algorithm that can be used to find such a solution followed by three different implementation schemes. The algorithm is based on the idea of crossing out rows and columns until all the supply has been exhausted and all the demand has been met. Once again, we assume that the feasibility property holds.

Initialization Step   Construct a tableau showing all problem parameters. None of the rows or columns is crossed out.

Iterative Step   According to some criterion, select a cell that has not been crossed out. Let that cell be basic. Assign a value to the basic variable for the cell that will use up either the remaining supply in the row or the remaining demand in the column (i.e., pick the smaller of the two). Reduce the remaining supply for the row and the remaining demand for the column by the amount assigned. Cross out the row or column whose supply or demand goes to zero. If both go to zero, cross out only one. This situation indicates a degenerate BFS.

Stopping Rule   If only one row or column remains, let all the cells not yet crossed out be basic and assign them all the remaining supply or demand. Stop with a BFS defined by the selected cells. If more than one row and more than one column remain, go to the iterative step.

There are many possibilities for the selection rule in the iterative step. They differ with respect to the work required to carry them out and the quality of the initial solutions they generate. Generally, methods requiring greater effort yield better solutions. For hand calculations, the effort may be justified. For computer implementations, it is often true that the simpler methods are more effective because the iterative step of the transportation simplex algorithm is so efficient. Several alternatives will now be described.

Northwest Corner Rule

The upper left-hand corner of the tableau is called the northwest corner. By the northwest corner rule, which is simplest to apply, the uncrossed cell closest to the northwest corner is selected. The initial solution in Figure 6.3 was obtained by using this rule. Note that on selecting cell (1, 2) at the second iteration, both the supply in row 1 and the demand in column 2 are exhausted, so $s_1 = 0$ and $d_2 = 0$. Either the row or column could have been crossed out. For our example, we crossed out row 1, which left (2, 2) as the upper left-most uncrossed cell. This cell is thus selected and becomes basic with $x_{22} = 0$, indicating a degenerate solution. Column 2 is crossed out, and the algorithm continues.

Vogel's Method

This procedure looks ahead one step and constructs a penalty for not being able to assign a flow to the remaining cells in a row or column whose cost is smallest, but instead having to pick the cell with the second-smallest cost. In economic terms, the idea is to determine the opportunity cost associated with each possible assignment and then pick the cell with the largest opportunity cost. To begin, for each uncrossed row, compute the difference between the smallest cost in the row and the second-smallest cost. Do the same for all uncrossed columns. Select the row or column with the largest difference. The rule is to choose the uncrossed cell in the selected row or column with the smallest cost.

When done by hand, Vogel's method is usually performed directly on the tableau. The selection of basic variables in each iteration is accompanied by the crossing out of a row or column and the recomputation of supplies, demands, and corresponding differences. The tableau becomes so messy, however, that we suggest using the format shown in Table 6.1 together with the transportation matrix. Each row of the table lists the row and column differences, the basic variable assigned, the adjustment of demands and supplies, and the row or column crossed out. When Vogel's method is applied to the example, it yields the optimal solution shown in Figure 6.10 without additional iterations.

Table 6.1 Format for Vogel's Computations

Iteration number	Row differences			Column differences					Basic variable	Action
	1	2	3	1	2	3	4	5		
1	0	3	1	5	1	4	2	2	$x_{31} = 5$	$d_1 = 0, s_3 = 10$
										Cross out column 1
2	0	3	1	—	1	4	2	2	$x_{33} = 10$	$d_3 = 5, s_3 = 0$
										Cross out row 3
3	0	3	—	—	4	1	2	5	$x_{25} = 10$	$d_5 = 0, s_2 = 5$
										Cross out column 5
4	4	2	—	—	4	1	2	—	$x_{14} = 5$	$d_4 = 0, s_1 = 10$
										Cross out column 4
5	1	4	—	—	4	1	—	—	$x_{22} = 5$	$d_2 = 5, s_2 = 0$
										Cross out row 2
6		Only row 1 remains							$x_{12} = 5$	$d_2 = 0, s_1 = 5$
7		Only row 1 remains							$x_{13} = 5$	$d_3 = 0, s_1 = 0$

Russell's Method

This method approximates the value of the reduced cost associated with each cell not yet assigned and then selects for entry into the basis the cell using the steepest descent rule. At each iteration for each uncrossed row and column, we first find the following.

$$\overline{u}_i = \text{value of the largest uncrossed cost, } c_{ij}, \text{ in row } i$$
$$\overline{v}_j = \text{value of the largest uncrossed cost, } c_{ij}, \text{ in column } j$$

Then, for each uncrossed cell we compute

$$\Delta_{ij} = c_{ij} - \overline{u}_i - \overline{v}_j$$

The rule is to choose the uncrossed cell with the most negative value of Δ_{ij} (smallest value).

Illustrating Russell's method for our example, we obtain the results in Table 6.2. As can be seen, the basic feasible solution obtained is also optimal.

Degeneracy and Cycling

For network-type problems, it is common to encounter a degenerate basic feasible solution while executing the simplex method. Therefore, some degenerate pivots may be encountered in the process of solving the TP. This does not present any major difficulties per se, but there is the possibility of cycling. When this phenomenon occurs, the algorithm cycles through a series of degenerate basic solutions with no actual flow change and never terminates. Although this does not seem to be a problem in practical applications, we describe a perturbation method for resolving this issue. The theoretical foundations are discussed by Murty [1992], among others.

Referring to Equations (1) and (2), assume that s_i and d_j are strictly positive and the feasibility property holds—i.e., $\Sigma_i s_i = \Sigma_j d_j$. We now perturb s_i to $\hat{s}_i = s_i + \varepsilon$ for $i = 1, \ldots, m$. We leave $d_1, \ldots, d_{n-1}$ unchanged but perturb d_n to $\hat{d}_n = d_n + m\varepsilon$. Here, $\varepsilon > 0$ is an arbitrarily small number. If s_i and d_j are positive integers, ε can be taken to be any positive number less than $1/2m$. However, it is not necessary to assign a specific value to ε; it can be

Table 6.2   Computations for Russell's Method

Iteration number	Rows ($\bar{u}_i$)			Columns ($\bar{v}_j$)					Basic variable	Action
	1	2	3	1	2	3	4	5		
1	16	15	12	15	15	16	11	11	$x_{25} = 10$	$d_5 = 0, s_2 = 5$
										Cross out column 5
2	16	15	12	15	15	16	11	—	$x_{22} = 5$	$d_2 = 5, s_2 = 0$
										Cross out row 2
3	16	—	12	15	15	16	11	—	$x_{31} = 5$	$d_1 = 0, s_3 = 10$
										Cross out column 1
4	16	—	12	—	15	16	11	—	$x_{33} = 10$	$d_3 = 5, s_3 = 0$
										Cross out row 3
5		Only row 1 remains							$x_{12} = 5$	$d_2 = 0, s_1 = 10$
6		Only row 1 remains							$x_{13} = 5$	$d_3 = 0, s_1 = 5$
7		Only row 1 remains							$x_{14} = 5$	$d_4 = 0, s_1 = 0$

left as a parameter that is treated as a positive number smaller than any positive number with which it is compared in the computations.

Making the substitutions discussed above, we solve the perturbed problem with the transportation simplex algorithm. All basic feasible solutions will be nondegenerate and of the form $\bar{x}_{ij} + \varepsilon \hat{x}_{ij}$, with $\bar{x}_{ij}$ being the solution for the original right-hand side constants s_i, d_j, and $\hat{x}_{ij}$ being the basic solution for the right-hand-side constants equal to the coefficients of ε in the perturbed problem. For implementation purposes, it is most efficient to solve the perturbed problem by maintaining and updating $\bar{x}_{ij}$ and $\hat{x}_{ij}$ separately, and always remembering that the value of x_{ij} in the solution is $\bar{x}_{ij} + \varepsilon \hat{x}_{ij}$.

Assignment Problem

A special case of the TP arises when all the sources have one unit of supply and all the destinations have one unit of demand. The resultant model is known as the assignment problem (AP) and can be interpreted in the balanced case (i.e., the feasibility property holds) of shipping one unit of commodity from each source to each destination to minimize total cost. For the model to be balanced, the number of sources must equal the number of destinations. Call this number n. Common applications include assigning n workers to n tasks or assigning n jobs to n machines but do not have an explicit shipment or flow interpretation.

The LP model can be stated as follows.

$$\text{Minimize } \sum_{i=1}^{n} \sum_{j=1}^{n} c_{ij} x_{ij}$$

$$\text{subject to } \sum_{j=1}^{n} x_{ij} = 1, \quad i = 1, \ldots, n$$

$$\sum_{i=1}^{n} x_{ij} = 1, \quad j = 1, \ldots, n$$

$$x_{ij} \geq 0, \quad i = 1, \ldots, n \text{ and } j = 1, \ldots, n$$

The same procedures used to solve the transportation problem can be used to solve the assignment problem. Based on its special structure, however, several highly efficient algorithms have been devised for the AP. These algorithms are discussed in several of the network texts listed at the end of this chapter. In practice, large-scale transportation and assignment problems are generally solved with network simplex codes.

6.2 SHORTEST PATH PROBLEM

The problem investigated in this section is one of finding a collection of arcs that constitutes the shortest path on a directed network of m nodes and n arcs from a specified node s, called the *source*, to a second specified node t, called the *destination* or *sink*. Figure 6.11 shows a typical network for which $s = 1$ and $t = 10$. The parameters on the arcs enclosed in parentheses are the arc lengths. Each arc is numbered sequentially and is directed. Note that it is always possible to convert a (partially) undirected network into a directed one by replacing each undirected edge $e(i, j)$ of length $L(i, j)$ with two directed arcs (i, j) and (j, i), each of length $L(i, j)$.

A closely related problem that we also consider is to find the set of shortest paths from the source node to all other nodes in the network. This *shortest path tree problem* is solved with very little additional difficulty. Figure 6.12 shows the shortest path tree with the source

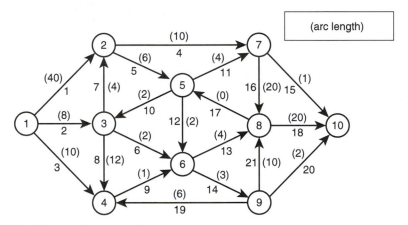

Figure 6.11 Network showing arc lengths

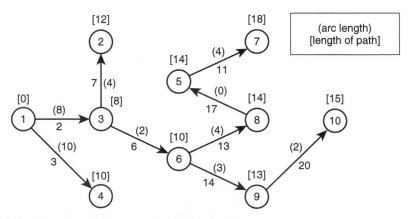

Figure 6.12 Shortest path tree rooted at node 1

at node 1. The network in the figure is an example of a spanning tree, which, by definition, is a subset of arcs that connects all nodes but contains no cycles.

We now present two algorithms for solving the shortest path problem. The first is a greedy approach that was developed by Dijkstra and yields the optimal solution when all arc lengths are nonnegative. If some lengths are negative but no negative cycles exist, a primal simplex solution algorithm may be used. This is the second algorithm that we will discuss.

Dijkstra's Algorithm

When all arc lengths are nonnegative, as in Figure 6.11, an efficient greedy algorithm can solve the shortest path (tree) problem. Starting with the source node alone, the algorithm finds the shortest path from the source node to one additional node in each subsequent iteration. This procedure requires $m - 1$ iterations to find the shortest path tree, where m is the number of nodes in the network.

Dijkstra's algorithm uses a set S, called the "set of solved nodes." This set includes the nodes for which the shortest path has already been determined at the current point in the algorithm. The nodes in the unsolved set $\overline{S}$ are those that are not in S. While iterating, the algorithm assigns the numbers π_i to each node in the network, where π_i is the length of the shortest path to node i from the source node s through the members of S. Note that the π values are equivalent to the dual variables associated with the LP formulation of the problem. At the end of the algorithm, π_i is the length of the shortest path to node i. In the following discussion, A is the set of all arcs and c_k is the length of arc $k \in A$.

Algorithm

Initially, let $S = \{s\}$, $\pi_s = 0$.

Repeat until S is the set of all nodes:

Find an arc $k(i, j)$ that passes from a solved node to an unsolved node such that:

$$k(i, j) = \operatorname{argmin}\{\pi_{i'} + c_{k'} : k'(i', j') \in A, i' \in S, j' \in \overline{S}\}$$

Add node j and arc $k(i, j)$ to the tree. Add node j to the solved set S. Let $\pi_j = \pi_i + c_k$, where $k \equiv k(i, j)$.

At each iteration, the algorithm computes the length of the path to every unsolved node through solved nodes only. The unsolved node with the shortest path length is added to the solved set S. The process terminates when a spanning tree is obtained. Because a node is added to the tree at each step, the algorithm requires $m - 1$ iterations. The algorithm works because of the assumption of nonnegative arc lengths.

This algorithm is easily adapted to graphs with undirected edges, where each edge has nonnegative length. In the iterative step, we replace the arc set A with the edge set E and replace the word "arc" with the work "edge." The result of the algorithm is an undirected spanning tree from which one can determine the shortest path from node s to each of the other nodes.

Example

For the network in Figure 6.11, we wish to find the shortest path tree rooted at node 1. For hand computations, the algorithm is easily accomplished on the figure representing the problem. Figure 6.13 shows the intermediate situation with five nodes assigned to the tree, $S = \{1, 3, 4, 6, 2\}$ with arcs $\{2, 3, 6, 7\}$ (note that the order of the items in the sets shows the order in which they were added by the algorithm). The boldface numbers in brackets indicate the values of π associated with the nodes in S. For example, the length of the shortest path to node 6 is 10. The numbers in square brackets adjacent to the nodes in $\overline{S}$ show the lengths of the shortest paths to unsolved nodes passing through only nodes in S. For example, the length of the shortest path passing through S to node 8 is 14. The algorithm now selects the arc and node associated with the smallest π_i value for $i \in \overline{S}$. The choice is min{18, 22, 14, 13} = 13, so node 9 and arc 14 join the tree.

Tabular Implementation

An alternative statement of the algorithm is represented by a table with seven columns.

Initialization   Let $h = 1$ and let s be the originating node. Let

$$S = \{s\}, \pi_s = 0$$

Iterative step   Repeat until all nodes are in the set S.

    a. Develop the following table.

       Column 1: Value of h. At any step, h is the number of nodes in S.

       Column 2: Members of S (the solved set) that have at least one arc connected to an unsolved node.

       Column 3: For each node listed in column 2, find the closest unsolved node and list it in column 3.

       Column 4: Let i be the index of a node listed in column 2, let j be the index of the corresponding node listed in column 3, and let k be the index of the arc connecting nodes i and j. For each case, compute

$$\pi_j' = \pi_i + c_k$$

       Show these numbers in column 4.

       Column 5: Select the smallest number in column 4, breaking ties arbitrarily. Let i be the node in column 2 and let j be the node in column 3 from which the smallest number was computed. Show node j in column 5.

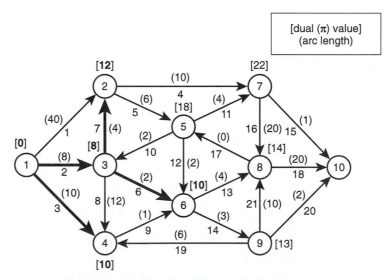

Figure 6.13   Intermediate point in Dijkstra's algorithm with $S = \{1, 3, 4, 6, 2\}$

Column 6: Show the length of the shortest path to the node added. This is the minimum obtained from column 4.

Column 7: Show the arc $k(i, j)$. Add node j and arc k to the shortest path tree.

b. Add node j to S. Let $\pi_j = \pi_j'$.

The steps of the algorithm for the example depicted in Figure 6.11 are shown in Table 6.3. The optimal tree is displayed in Figure 6.12.

Primal Simplex Algorithm

The primal simplex algorithm can also be used to solve the shortest path tree problem but is not limited to the case of all nonnegative arc lengths. It requires, however, that there exist no directed cycle with negative total length. If such a cycle is present, the algorithm terminates with the selected arcs identifying the cycle. This means that the associated LP has an unbounded solution.

As with all simplex-based procedures, the optimal solution to the shortest path problem is basic. The algorithm for the shortest path problem makes use of the critical property that any collection of arcs that forms a spanning tree is a basic solution. As we shall see, it is a variant on the pure minimum-cost flow programming algorithm discussed in Section 6.4.

Table 6.3   Shortest Path Computations for Network in Figure 6.11

h	Solved nodes	Closest unsolved node	Length of path to unsolved node	Node added to solved set	Length of shortest path	Arc added to tree
1	1	3	8	3	8	2
2	1	4	10			
	3	6	10	4	10	3
3	1	2	40			
	3	6	10			
	4	6	11	6	10	6
4	1	2	40			
	3	2	12			
	6	9	13	2	12	7
5	2	5	18			
	6	9	13	9	13	14
6	2	5	18			
	6	8	14			
	9	10	15	8	14	13
7	2	5	18			
	8	5	14			
	9	10	15	5	14	17
8	2	7	22			
	5	7	18			
	8	10	34			
	9	10	15	10	15	20
9	2	7	22			
	5	7	18	7	18	11

Algorithm

Step 1: Start with a basis described by the arcs of a spanning tree. Assign the dual value $\pi_s = 0$ for the source node s.

Step 2: Repeat until all reduced costs are nonnegative.

a. Compute the dual values for all nodes except the source node s such that if arc $k(i, j)$ is in the basis,

$$\pi_j = \pi_i + c_k$$

b. Compute the reduced costs, d_k, for each nonbasic arc $k(i, j)$, where

$$d_k = c_k + \pi_i - \pi_j$$

c. Select some nonbasic arc for which $d_k < 0$ and call it the entering arc. The leaving arc is the one in the tree that currently enters node j.

d. Change the basis by removing the leaving arc from the tree and adding the entering arc. If a cycle is formed by this operation, stop, because the network contains a negative cycle. Otherwise, repeat Step 2.

Example

We use the example problem in Figure 6.11 to illustrate the computations but assume that arc 4(2, 7) has length -10 rather than 10. The arcs shown in Figure 6.12 form the initial basis. The node labels in that figure are equivalent to the values of π computed in Step 2a of the algorithm. With $c_4 = -10$, we compute for arc 4

$$d_4 = c_4 + \pi_2 - \pi_7 = -10 + 12 - 18 = -16$$

Because d_4 is negative, we select arc 4 to enter the basis. According to Step 2c, the arc that currently enters node 7, arc 11, must leave the basis. We change the basis and obtain the new spanning tree shown in Figure 6.14. The dual values are shown adjacent to the nodes in brackets.

At the next iteration, we find that arc 15 is a candidate to enter the basis, so the solution in Figure 6.14 is not optimal. If we allow arc 15 to enter the basis, arc 20 must leave. This leads to the solution in Figure 6.15, which is optimal.

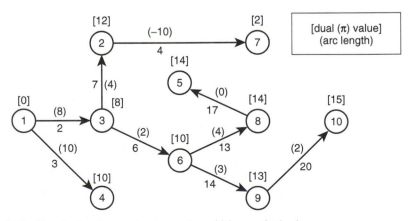

Figure 6.14 New basis after arc 4 enters and arc 11 leaves the basis

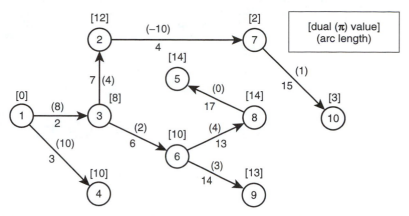

Figure 6.15 Optimal basis after arc 15 enters and arc 20 leaves the basis

6.3 MAXIMUM FLOW PROBLEM

In this section, we consider a directed network with m nodes and n arcs in which the only relevant parameter is the upper bound on arc flow, called *arc capacity*. The problem is to find the maximum flow that can be sent through the network from some specified node s, called the source, to a second specified node t, called the sink. Applications include finding the maximum flow of orders through a job shop, the maximum flow of water through a storm sewer system, and the maximum flow of a product through a distribution system, among others.

A specific instance of the problem with source node $s = A$ and sink node $t = F$ is shown in Figure 6.16 with its solution. In particular, the solution is the assignment of flows to arcs. For feasibility, flow balance or conservation of flow is required at each node (flow in = flow out) except at the source and sink, and each arc flow must be less than or equal to its capacity. The solution in the figure indicates that the flow from A to F is 15, which turns out to be the maximum.

A cut is a set of arcs whose removal will interrupt all paths from the source to the sink. The capacity of a cut is the sum of the arc capacities of the set. The *minimal cut* is the cut with the smallest capacity. Given a solution to the maximum flow problem, one can always find at least one minimal cut, as illustrated in Figure 6.17. The minimal cut is a set of arcs that limit the value of the maximum flow. Not coincidentally, the example shows that the total capacity of the arcs in the minimal cut equals the value of the maximum flow (this

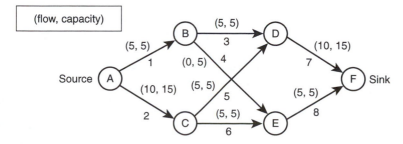

Figure 6.16 Example maximum flow problem with solution

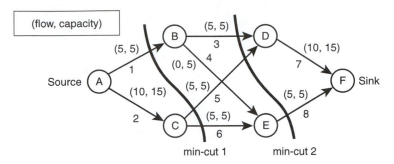

Figure 6.17   Minimal cuts determined by the maximum flow

result is called the *max-flow min-cut theorem*). The algorithm described in this section solves both the maximum flow and minimal cut problems.

Flow Augmenting Algorithm

The traditional way to solve the maximum flow problem is with the flow augmenting algorithm developed by Ford and Fulkerson [1962]. The algorithm begins with a feasible set of arc flows obtaining some value v_0 for the flow out of the source and into the sink. A search is then made in the network for a set of connected arcs from source to sink whose remaining capacity is positive. This is called a flow augmenting path. Flow is increased along that path as much as possible. The process continues until no such path can be found, at which time the algorithm terminates.

Before stating the procedure formally, we illustrate the general ideas with the network in Figure 6.16. In the presentation, let x_k be the flow on arc k. For an initial solution, we assign zero flows to all arcs, yielding $v_0 = 0$, as shown in Figure 6.18.

Flow may be increased in an arc when the current flow is less than capacity ($x_k < u_k$). Flow can be decreased in an arc when the current flow is greater than zero ($x_k > 0$). As we will see, a flow augmenting path traverses an arc in the forward direction when the arc flow is to be increased and in the reverse direction when the arc flow is to be decreased. For the initial solution, there are several flow augmenting paths. We choose the path $P_1 = (1, 4, 8)$, identified by the list of arcs in the path, and note that the flow may be increased by 5 to obtain the solution in Figure 6.19.

By observation we discover another flow augmenting path in Figure 6.19, $P_2 = (2, 5, 7)$. Augmenting the flow on this path by 5, we obtain the solution in Figure 6.20.

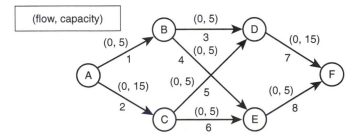

Figure 6.18   Initial flow with $v_0 = 0$

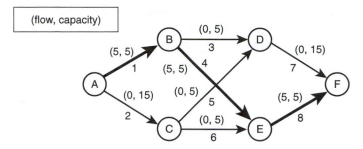

Figure 6.19 Augmenting flow on path (1, 4, 8), yielding $v_1 = 5$

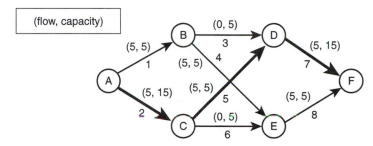

Figure 6.20 Augmenting flow on path (2, 5, 7), yielding $v_2 = 10$

From Figure 6.20 we discover a flow augmenting path of a different nature than the two previously described. Searching from A, we find that flow can be increased in arcs 2 and 6, but that further progress is blocked in arc 8. We can, however, increase the flow from node E to node B by decreasing the flow in arc 4, since its flow is greater than zero. Finally, flow can be increased in arcs 3 and 7 because the flows in these arcs are less than the arc capacities. The path describing this sequence of flow changes is $P_3 = (2, 6, -4, 3, 7)$, where the positive numbers show arcs in which flow is to be increased and the negative numbers show arcs in which flow is to be decreased. The flow can be increased by 5 along this path to obtain the results shown in Figure 6.21.

An examination of Figure 6.21 reveals no additional flow augmenting paths, so the maximum flow has been obtained. To find the minimal cut, we identify a set of nodes, call it **S**, such that one or more paths exist from the source node to the nodes in **S** on which additional flow can be delivered. The source node is necessarily in **S**. The set of arcs leaving **S** comprises the minimal cut. For example, starting at node A, we can reach only node C through

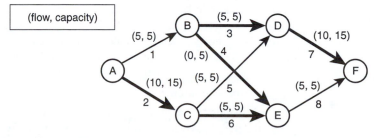

Figure 6.21 Augmenting flow on path (2, 6, -4, 3, 7), yielding $v_3 = 15$

an arc that has additional capacity. Thus, we have the node set $S = \{A, C\}$ to which additional flow could be advanced. The remaining nodes define the set $\overline{S} = \{B, D, E, F\}$. The minimal cut consists of all the arcs that pass from S to $\overline{S}$, arcs in the set $C = \{1, 5, 6\}$ in the example. The flows in the arcs on the minimal cut are at their respective arc capacities, because the cut is the bottleneck for the total flow. Logically, the flows in arcs that pass from $\overline{S}$ to S are at zero.

Formal Algorithm

We first identify the source as node s and the sink as node t. An initial feasible flow solution is assumed to be available for the algorithm with flow into node t (and out of node s) equal to v_0.

Do the following until no flow augmenting paths can be found.

a. Find a flow augmenting path defined by the sequence of arcs $P = (k_1, k_2, \ldots, k_p)$, where p is the number of arcs in the path. (An arc will have a positive sign if it is traversed in the forward direction and a negative sign if it is traversed in the reverse direction.)

b. Determine the maximum flow increase along the path.

$$\delta = \min \begin{cases} \min(u_k - x_k : k > 0, k \in P) \\ \min(x_{-k} : k < 0, k \in P) \end{cases}$$

c. Change the flow in the arcs on the path. Let x'_k be the new value of flow. For every arc k on the path P,

$$x'_k = x_k + \delta \text{ if } k > 0 \text{ and } x'_k = x_{-k} - \delta \text{ if } k < 0$$

This algorithm does not derive from the simplex method and does not maintain a basic solution. It is a representative of a large class of nonsimplex algorithms for network flow problems.

Finding Flow Augmenting Paths

In the example problem, flow augmenting paths were discovered by observation. For larger networks and for computer implementation, a more formal procedure is required. The algorithm described below is a search procedure that begins at s and labels all nodes to which a flow augmenting path from s can be found. When node t is labeled, we have a *breakthrough*, and the required flow augmenting path to t has been discovered. When the algorithm begins, all nodes are unlabeled except s. We "check" a labeled node after all avenues for finding paths from the node have been explored. Initially, all nodes are unchecked.

Labeling Algorithm   Do the following until node t is labeled or there exists no labeled unchecked node.

a. Select a node i that is labeled but unchecked. For each arc $k(i, j)$ originating at node i such that node j is unlabeled and $x_k < u_k$, label node j with the index of the arc k. For each arc $k(j, i)$ terminating at node i such that node j is unlabeled and $x_k > 0$, label node j with the negative index $-k$.

b. Check node i.

If the algorithm terminates with t labeled, the flow augmenting path is found by tracing the path backward from t and constructing P from the labels encountered. If the algorithm terminates with t unlabeled, no flow augmenting path exists. We illustrate the former

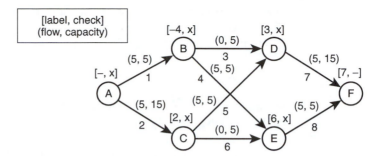

Figure 6.22 Illustration of labeling algorithm

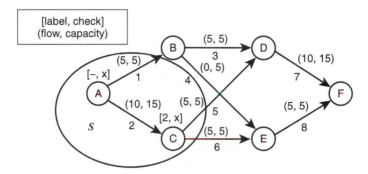

Figure 6.23 The minimal cut discovered by the labeling algorithm

case with the situation depicted in Figure 6.20. Node C is the first to be labeled and checked, followed by nodes E, B, D, and F. The resultant node labels and checks are shown adjacent to the nodes in Figure 6.22. An "x" indicates that a node is checked.

When node F is labeled, the flow augmenting path is complete. The label on node F indicates that the last arc on the path is 7. Node D is the origin of this arc, so we find the next arc on the path from the label on node D. The process continues until the path $P = (2, 6, -4, 3, 7)$ is discovered.

The sink node is not labeled when we apply the algorithm to the network in Figure 6.21. The labels and checks for this case are shown adjacent to the nodes in Figure 6.23. Although no flow augmenting path is found, we do identify the minimal cut. Let S be the set of nodes labeled in this final attempt. The arcs in the minimal cut are those that leave S. In this case, the cut is $C = \{1, 5, 6\}$. Adding the capacities of these arcs, we obtain 15, the same value as the maximum flow from s to t.

6.4 PURE MINIMUM-COST FLOW PROBLEM

Networks are especially convenient for modeling because of their simple nonmathematical structure that can be easily depicted with a graph. This simplicity also reaps benefits with regard to algorithmic efficiency. Although the simplex method is one of the primary solution techniques, we have already seen that its implementation for network problems allows many procedural simplifications. Very large instances can be solved much faster than with standard simplex codes. This section provides a solution algorithm for the pure minimum-cost flow problem or, equivalently, the pure NFP problem, on a directed graph. The procedure for solving a generalized model that includes nonunity arc gains is similar but more complicated.

Problem Statement

A pure NFP problem is defined by a given set of arcs and a given set of nodes, where each arc has a known capacity and unit cost and each node has a fixed external flow. The optimization problem is to determine the minimum-cost plan for sending flow through the network to satisfy supply and demand requirements. The arc flows must be nonnegative and no greater than the arc capacities, and they must satisfy conservation of flow at the nodes. In this section, it is assumed that all arc lower bounds on flow are zero and all arc gains are unity. Although it is possible to deal with a relaxation of these conditions, the accompanying discussion would be beyond the scope of this book.

Formulation

We now formulate the problem as a linear program for a directed, connected graph with m nodes and n arcs. It is assumed that there is at least one supply node and one demand node. Regarding notation, recall that in the formulation of the transportation problem in Section 6.1, the arc that connects nodes i and j was denoted by (i, j) and the decision variable associated with that arc was x_{ij}. Here, we adopt the notation $k(i, j)$, or simply k, to denote the arc between nodes i and j. This is done primarily to simplify the algorithmic presentation and the labeling of arcs in the figures to follow. Accordingly, the decision variables are

$$x_k = \text{flow through arc } k(i, j) \text{ from node } i \text{ to node } j$$

and the given data are

$$c_k = \text{unit cost of flow through arc } k$$
$$b_i = \text{net supply (arc flow out – arc flow in) at node } i$$
$$u_k = \text{capacity of arc } k(i, j)$$

The value of b_i is determined by the nature of node i. In particular,

$$b_i > 0 \text{ if } i \text{ is a supply node}$$
$$b_i < 0 \text{ if } i \text{ is a demand node}$$
$$b_i = 0 \text{ if } i \text{ is a transshipment node}$$

Also, let

$$K_{O(i)} = \text{set of arcs leaving node } i$$
$$K_{T(i)} = \text{set of arcs terminating at node } i$$

The mathematical model is as follows.

$$\text{Minimize} \quad \sum_{k=1}^{n} c_k x_k \tag{7}$$

$$\text{subject to} \quad \sum_{k \in K_{O(i)}} x_k - \sum_{k \in K_{T(i)}} x_k = b_i \quad i = 1, \ldots, m \tag{8}$$

$$0 \le x_k \le u_k, \quad k = 1, \ldots, n \tag{9}$$

The objective function, Equation (7), sums the arc costs over all arcs in the network. Equation (8) defines the flow balance or conservation-of-flow constraints. The first summation represents the flow out of node i, and the second represents the flow into node i. The difference between the two represents the net flow supplied at node i. The right-hand-side values b_i are positive if node i supplies flow, negative if it consumes flow, and zero otherwise. In some applications, the lower bound in Equation (9) is not zero but some value l_k. As we saw in the Chapter 5, it is always possible to transform such a lower bound to zero by introducing the variable

$$\hat{x}_k = x_k - l_k$$

and substituting $\hat{x}_k + l_k$ for x_k throughout the model. The upper bound must also be changed to

$$\hat{u}_k = u_k - l_k$$

For the models in this chapter, all arcs were required to have both origin and terminal nodes. Because this is somewhat restrictive for modeling purposes, we will now allow for arcs that are incident to a single node only, as shown in Figure 6.24. Arcs 6, 7, and 8 represent variable external flows at nodes 3, 2, and 1, respectively. For theoretical and algorithmic purposes, we introduce an additional node into the network called the *slack node*. Arcs representing the variable external flows originate or terminate at the slack node, as shown in Figure 6.25. Now all arcs again have both origin and terminal nodes. Generally, the index m is assigned to the slack node.

For a pure NFP problem, the gains on the arcs are 1, and so for a feasible solution, the external flows at the nodes must sum to zero. Note the following property.

Feasibility property: A necessary condition for a pure NFP problem to have a feasible solution is

$$\sum_{i=1}^{m} b_i = 0$$

In other words, the total flow being supplied at the nodes must equal the total demand being absorbed by the nodes in the network. The feasibility property will hold if we designate the external flow of the slack node to be

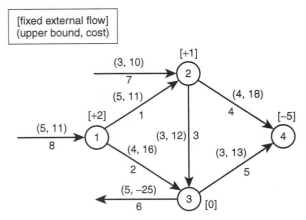

Figure 6.24   Model with variable external flows

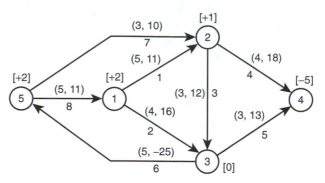

Figure 6.25 Model with a slack node (node 5)

$$b_m = -\sum_{i=1}^{m-1} b_i$$

This is shown in Figure 6.25, where $m = 5$.

Because of the feasibility property, when we add the m conservation-of-flow constraints in Equation (8), we obtain 0 on both sides of the equal sign. This means that one of the equations is redundant. In fact, there are exactly $m - 1$ linearly independent equations, so one can be discarded arbitrarily without changing the solution.

The feasibility property is necessary but not sufficient for a problem to be well posed. Whether or not a particular instance of the model represented by Equations (7) to (9) has a feasible solution depends on the network structure, arc capacities, and supply and demand values.

Integrality Property

In many applications, the flows on the arcs must take integer values. This is implicitly the case for the assignment problem, in which, for example, one worker must be assigned to a single job. In all the network examples worked out earlier in this chapter, we saw that the solutions were always integral. This was not a coincidence but a direct consequence of the structure of the **A** matrix in Equation (8). In more general terms, let the feasible region be given by the set $S = \{x \in \Re^n : Ax = b, 0 \le x \le u\}$, where **A** and **b** are assumed to be integer matrices. As discussed in the Chapter 3, let us partition **A** as (**B**, **N**), such that **B** is non-singular, and write the constraints in S as $Bx_B + Nx_N = b$. Recall that a basic solution is $x_B = B^{-1}b$, $x_N = 0$. Thus, a sufficient condition for x_B to be integer valued is for B^{-1} to be an integer matrix. To derive testable conditions, we introduce the notion of unimodularity.

Definition 1: A square integer matrix **B** is called *unimodular* if the absolute value of its determinant equals 1—i.e., if $|\det B| = 1$. An integer $m \times n$ matrix **A** is *totally unimodular* if every square submatrix has determinant +1, −1, 0 (or, equivalently, if every nonsingular square submatrix is unimodular).

From linear algebra we know that if **B** is nonsingular, then $B^{-1} = B^+/\det B$, where B^+ is the adjoint of **B** and is similarly an integer matrix. This implies the following theorem.

Theorem 1: If **A** is totally unimodular, every basic solution $(x_B, x_N) = (B^{-1}b, 0)$ associated with the feasible region S is integer valued.

As a direct consequence, we have the following property.

Integrality property: For the pure NFP problem [Equations (7) to (9)], when all supply and demand values b_i and all upper bounds on arc flows u_k are integer valued, all basic solutions are integral.

The algebraic implication of this statement is that every basis inverse matrix has components with values of 0, 1, or −1. If the standard simplex algorithm were applied to the problems represented by Equations (7) to (9), we would see that every pivot element had a value of ±1 and so fractions would never appear in the tableau. In our adaptation of the simplex method for solving this problem, when the integrality property holds, the decision variables similarly never become fractional.

Vector Notation

It is convenient to identify the vectors that represent the parameters and variables of a problem. These vectors are $\mathbf{x}$ for arc flows, $\mathbf{u}$ for arc flow capacities or upper bounds, $\mathbf{c}$ for arc unit costs, and $\mathbf{b}$ for node external flow. For the network in Figure 6.25, upper bounds and costs are given alongside the arcs and the external flows are given in square brackets adjacent to the nodes. In vector notation, we have the following.

$$\text{Arc flows:} \quad \mathbf{x} = (x_1, x_2, x_3, x_4, x_5, x_6, x_7, x_8)$$
$$\text{Upper bounds:} \; \mathbf{u} = (5, 4, 3, 4, 3, 5, 3, 5)$$
$$\text{Unit costs:} \quad \mathbf{c} = (11, 16, 12, 18, 13, -25, 10, 11)$$
$$\text{External flows:} \; \mathbf{b} = (2, 1, 0, -5, 2)$$

Basic Solutions

Because an optimal solution must be among the finite set of bases, the simplex algorithm examines only basic solutions as it iterates. When applied to the NFP problem, the simplex algorithm maintains the needed information more efficiently and hence is able to access it more quickly than when applied to a general LP problem. This section describes basic solutions for the NFP problem and provides procedures for computing the primal and dual solutions associated with a given basis.

Basis Tree

We know that if an optimal solution to the LP, and to the NFP problem by extension, exists, then there exists a basic solution that is optimal. The basis is defined by a selection of independent variables equal in number to the number of linearly independent constraints. Because the network model contains $m - 1$ independent conservation-of-flow constraints and the variables are the arc flows, a basis is determined by selecting $m - 1$ independent arcs. These arcs are identified by the arc indices of the basic variables. Call the corresponding set $\mathbf{n}_B$. For the pure NFP problem, we have the interesting characteristic that every basis defines a spanning tree subnetwork. To illustrate, let $\mathbf{n}_B = \{2, 4, 7, 8\}$ for the network shown in Figure 6.25. Drawing only the selected arcs forms the subnetwork shown in Figure 6.26. Node 5 is defined as the root node of the tree.

Nonbasic Arcs

The arcs not selected as basic are, by definition, the nonbasic arcs. In the *bounded variable simplex method* of general LP, each nonbasic variable takes the value 0 or the value

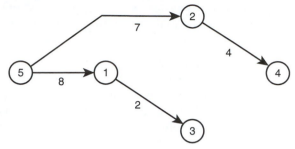

Figure 6.26   Basis tree for $\mathbf{n}_B = \{2, 4, 7, 8\}$

of its upper bound in the definition of a basic solution. The same is true when solving the NFP problem.

In a basic solution, each nonbasic arc k has its flow at 0 or u_k, the upper bound. We let $\mathbf{n}_0$ denote the set of nonbasic arcs with 0 flow, and we let $\mathbf{n}_1$ denote the set of arcs with upper bound flow. To represent a specific case graphically, we add the members of $\mathbf{n}_1$ to the basis tree as dotted lines. To illustrate, Figure 6.27 shows the basis tree representing $\mathbf{n}_B = \{2, 4, 7, 8\}$, $\mathbf{n}_1 = \{5\}$, and $\mathbf{n}_0 = \{1, 3, 6\}$.

Primal Basic Solution

Given a selection of basic arcs and an assignment of nonbasic arcs to either $\mathbf{n}_0$ or $\mathbf{n}_1$, there is a unique assignment of flows to the basic arcs that satisfies the conservation-of-flow requirement at the nodes. We let $\mathbf{x}_B$ be the vector of flows on the basic arcs, and we let $\mathbf{x}_1$ be the vector of flows on the arcs in $\mathbf{n}_1$. The flows on the arcs in $\mathbf{n}_0$ are zero.

To solve for the basic arc flows, we must first adjust the external flows in the network to account for nonbasic arcs at their upper bounds. One way to do this is to take each member of $\mathbf{n}_1$ in turn. Say arc $k(i, j)$ is at its upper bound. To accommodate the flow on arc $k(i, j)$ for the following calculations, we reduce the external flow at its originating node i by u_k and increase the external flow at its terminal node j by u_k. For the network in Figure 6.25, the external flow vector is

$$\mathbf{b} = (2, 1, 0, -5, 2)^T$$

Arc 5 is in the set $\mathbf{n}_1$, so we adjust the external flows at both ends of the arc: $b_3' = b_3 - u_5$, $b_4' = b_4 + u_5$. Since $u_5 = 3$, the adjusted external flows are

$$\mathbf{b}' = (2, 1, -3, -2, 2)^T$$

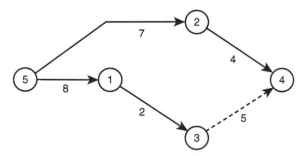

Figure 6.27   Basis tree with a nonbasic arc with flow at its bound

Equation (10) provides a general expression for the adjusted external flow b_i at node i—that is, the original external flow reduced by the flow of the upper bound arcs leaving the node and increased by the flow on the upper bound arcs entering the node. Once again, $K_{O(i)}$ is the set of arcs leaving node i and $K_{T(i)}$ is the set of arcs terminating at node i.

$$b_i' = b_i - \sum_{k \in (K_{O(i)} \cap \mathbf{n}_1)} u_k + \sum_{k \in (K_{T(i)} \cap \mathbf{n}_1)} u_k \tag{10}$$

Given the adjusted external flows, there is a unique assignment of flows to the basic arcs that satisfies conservation of flow at the nodes. The solution for the basis in Figure 6.27 is shown in Figure 6.28. Comparison of the arc flows to the upper and lower bounds for the basic arcs will show that the flows fall within these limits. This is a primal basic feasible solution (BFS).

$$\mathbf{b} = (2, 1, 0, -5, 2)^\mathrm{T}$$
$$\mathbf{b}' = (2, 1, -3, -2, 2)^\mathrm{T}$$
$$\mathbf{n}_\mathrm{B} = \{2, 4, 7, 8\}$$
$$\mathbf{n}_1 = \{5\}$$
$$\mathbf{n}_0 = \{1, 3, 6\}$$
$$\mathbf{x}_\mathrm{B} = (3, 2, 1, 1)$$
$$\mathbf{x}_1 = (3)$$
$$\mathbf{x}_0 = (0, 0, 0)$$

Dual Basic Solution

Dual variables, or, alternatively, the dual values, are used explicitly in the solution algorithm for network flow problems. There is a dual variable for each node i, denoted by π_i and interpreted as the cost of bringing one unit of flow to node i from the slack node. Given a basis tree, the dual values are assigned using the requirement of complementary slackness. For every basic arc $k(i, j)$,

$$c_k + \pi_i - \pi_j = 0 \tag{11}$$

Because one of the conservation-of-flow constraints in the LP model is redundant, one of the dual variables can have an arbitrary value. We choose to set the dual value for the slack variable, node 5 in the example, equal to zero. The other dual variables are set equal to values

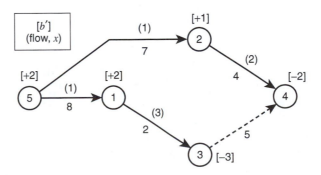

Figure 6.28   Solution for the basic flows

that satisfy the complementary slackness conditions. The values are computed in the following order.

$$\pi_5 = 0, \; \pi_1 = \pi_5 + 11 = 11, \; \pi_3 = \pi_1 + 16 = 27,$$
$$\pi_2 = \pi_5 + 10 = 10, \; \pi_4 = \pi_2 + 18 = 27$$

There are a several different orders that can be followed in this computation, but they all must start at the slack node and work outward. Figure 6.29 shows the basis tree previously considered in Figure 6.27, with costs shown on the arcs and dual values shown adjacent to the nodes. The assignment of nonbasic arcs to $\mathbf{n}_0$ and $\mathbf{n}_1$ does not affect the value of π.

$$\mathbf{n}_B = \{2, 4, 7, 8\}$$
$$\pi = (\pi_1, \pi_2, \pi_3, \pi_4, \pi_5)$$
$$= (11, 10, 27, 28, 0)$$

Solution Characteristics

Every selection of arcs $\mathbf{n}_B$ that forms a spanning tree together with a specification of $\mathbf{n}_1$ determines primal and dual basic solutions for the network flow problem. The spanning tree determines a unique assignment of dual variables (π) that satisfies complementary slackness. The spanning tree and $\mathbf{n}_1$ determine a unique primal solution ($\mathbf{x}$) that satisfies conservation of flow at each node—that is,

$$\sum_{k \in \mathbf{K}_{O(i)}} x_k - \sum_{k \in \mathbf{K}_{T(i)}} x_k = b_i, \quad i = 1, \dots, m$$

The process of assigning arc flows, however, does not ensure feasibility with respect to bounds. If the requirement that $\mathbf{0} \le \mathbf{x} \le \mathbf{u}$ is not satisfied, then the basic solution is infeasible. A solution for which either $x_k = 0$ or $x_k = u_k$ for some basic arcs is called a *primal degenerate* solution. The solution in Figure 6.28 is feasible because all the arc flows are nonnegative and no greater than the arc capacities. Also, all arc flows are strictly between their bounds, so the solution is not primal degenerate.

As in LP, there is a reduced cost, call it d_k, that can be computed for each arc $k(i, j)$ in the network using the formula

$$d_k = c_k + \pi_i - \pi_j$$

We will use the reduced costs for determining the optimality of a basic solution.

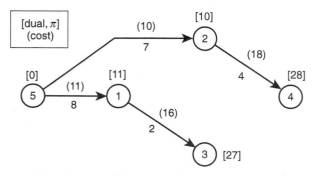

Figure 6.29 Computation of dual variables

The condition of complementary slackness ensures that

$$d_k = 0 \text{ for all } k(i, j) \in \mathbf{n_B}$$

A solution for which $d_k = 0$ for some nonbasic arcs is called a *dual degenerate* solution. For Figure 6.29, the nonbasic arcs are 1, 3, 4, and 6. Computing the values of d_k for these arcs, we find $d_1 = 12$, $d_3 = -5$, $d_5 = 12$, and $d_6 = 2$, so the solution is not dual degenerate.

Simplified Computation of Primal Solutions

The tree structure of the basis subnetwork makes possible simplified procedures for finding primal and dual basic solutions. Observe again the example basis in Figure 6.30, and note that there is a directed path from node 5 to every other node in the tree. This is called a *directed spanning tree rooted at node 5*.

The basic flows are easily determined by assigning flows to the basic arcs in a sequential manner. At any time in the sequence, there is enough information to assign one or more basic flows. The process starts at nodes at the extremes of the tree (the nodes incident to only one arc are called the *leaves* of the tree). We assign flows to the arcs incident to the leaves and work backward through the tree toward the root. The order in which arc flows are assigned is not unique, but for a given basis tree and set $\mathbf{n_1}$, there is a unique solution for the basic flows. For our example, we find two leaves of the tree in Figure 6.30—nodes 3 and 4. We assign the flows to arcs 2 and 4 such that

$$x_2 = -b_3' = -(-3) = 3, \, x_4 = -b_4' = -(-2) = 2$$

Now the flows for arcs 7 and 8 can be assigned.

$$x_7 = -b_2' + x_4 = -(1) + 2 = 1, \, x_8 = -b_1' + x_2 = -(2) + 3 = 1$$

This completes the determination of the basic flows.

The arcs selected for the basis in Figure 6.30 naturally form a directed spanning tree. Other selections may not. For instance, consider the basis consisting of arcs 2, 3, 5, and 6 as shown in Figure 6.31 where

$$\mathbf{n_B} = \{2, 3, 5, 6\}$$
$$\mathbf{n_1} = \{4, 7\}$$
$$\mathbf{n_0} = \{1, 8\}$$

Although the subnetwork defines a tree, it does not form a directed spanning tree. To put the network into the desired form, the direction of some of the arcs must be reversed, as shown in Figure 6.32.

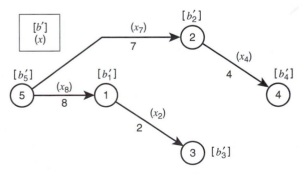

Figure 6.30   A directed spanning tree used to compute the primal solution

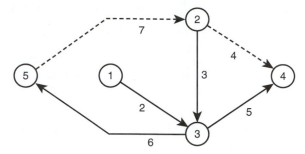

Figure 6.31   Basis that is not a directed spanning tree

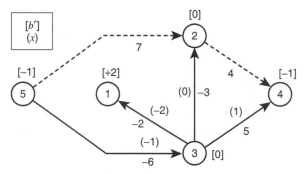

Figure 6.32   Solution for the basic flows

The reversed arcs, called *mirror arcs*, play a major role in the minimum-cost flow algorithm. We call an arc with a positive index (k) a *forward arc*, and one with a negative index ($-k$) a *mirror arc*. A forward arc has the direction and parameters given in the problem statement. The parameters of the mirror arc are derived from those of the forward arc as follows:

$$l_{-k} = -u_k, \quad u_{-k} = -l_k, \quad c_{-k} = -c_k$$

Similarly, for the flow variable, we set $x_{-k} = -x_k$.

It is always possible to construct a directed spanning tree by replacing some forward arcs with mirror arcs. The basic flows for the tree in Figure 6.32 are easily obtained by following the tree backward from its leaves. The flows in the forward arcs are found by negating the flows in the mirror arcs. For the original basis $\mathbf{n}_B = \{2, 3, 5, 6\}$, the directed tree uses $\mathbf{n}_B = \{-2, -3, 5, -6\}$. Now, working backward from the leaves yields $\mathbf{x}_B = (-2, 0, 1, -1)$. To find the equivalent flow in the forward (original) arcs, we negate the flow in the mirror arcs, yielding $\mathbf{x}_B = (2, 0, 1, 1)$, which, along with $\mathbf{x}_1 = (4, 3)$, satisfies the conservation-of-flow constraints for the network in Figure 6.31. The data associated with the solution in Figure 6.32 are as follows

$$\mathbf{b} = (2, 1, 0, -5, 2)^T$$
$$\mathbf{b}' = (2, 0, 0, -1, -1)^T$$
$$\mathbf{n}_B = \{-2, -3, 5, -6\}$$
$$\mathbf{n}_1 = \{4, 7\}$$
$$\mathbf{n}_0 = \{1, 8\}$$
$$\mathbf{x}_B = (-2, 0, 1, -1)$$
$$\mathbf{x}_1 = (4, 3)$$
$$\mathbf{x}_0 = (0, 0)$$

Simplified Computation of Dual Solutions

The directed spanning tree can also be used to compute the dual variables, but here we start at the root and work outward toward the leaves. The basis in Figure 6.33 will be used to illustrate the approach with the parameters as before.

We begin by setting $\pi_5 = 0$. When the dual variable for a node on one end of a basic arc is known, the value for the node on the other end can be computed using the complementary slackness condition defined by Equation (11). Accordingly, we compute the values of π_1 and π_2 as follows.

$$\pi_1 = \pi_5 + c_8 = 0 + 11 = 11, \quad \pi_2 = \pi_7 + c_8 = 0 + 10 = 10$$

Now the values of π_3 and π_4 can be calculated.

$$\pi_3 = \pi_1 + c_2 = 11 + 16 = 27, \quad \pi_4 = \pi_2 + c_4 = 10 + 18 = 28$$

The solution is $\boldsymbol{\pi} = (11, 10, 27, 28, 0)$.

The process of assigning the dual variables begins at the root node of the tree and progresses out toward the branches. At each step, one more dual variable is assigned. The direction of the assignment process is the reverse of the procedure for assigning the primal variables. The nature of the spanning tree ensures that the values of all the dual variables can be computed in this manner.

When the basic arcs do not naturally define a spanning tree, mirror arcs are used to construct one. For $\mathbf{n}_B = \{2, 3, 5, 6\}$, we replace arcs 2, 3, and 6 with mirror arcs. The relation $c_{-k} = -c_k$ provides the unit costs for the mirror arcs. The spanning tree and the computed dual variables are shown in Figure 6.34.

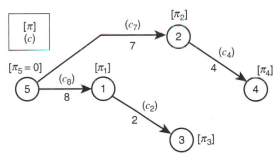

Figure 6.33   Directed spanning tree used to compute the dual solution

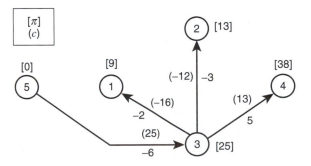

Figure 6.34   Solution for the dual variables for $\mathbf{n}_B = \{-2, -3, 5, -6\}$

$$\mathbf{n}_B = \{-2, -3, 5, -6\}$$
$$\pi = (\pi_1, \pi_2, \pi_3, \pi_4, \pi_5)$$
$$= (9, 13, 25, 38, 0)$$

Optimality Conditions

The solution to the minimum-cost flow problem is achieved by finding values of $\mathbf{x}$ and π for the primal and dual problems, respectively, that satisfy the following optimality conditions.

1. *Primal feasibility*
   a. The vector $\mathbf{x}$ is basic and satisfies conservation of flow at all nodes except the slack node. (Note that for pure problems, conservation of flow at the slack node is automatic.)
   b. $0 \le x_k \le u_k$ for all arcs k

2. *Complementary slackness*

   Given x_k for each arc k and corresponding reduced cost $d_k = c_k + \pi_i - \pi_j$
   a. if $0 < x_k < u_k$, then $d_k = 0$
   b. if $x_k = 0$, then $d_k \ge 0$
   c. if $x_k = u_k$, then $d_k \le 0$

Recall that a primal solution is degenerate when a basic flow x_k is either zero or at the upper bound u_k. Degeneracy may affect the progress of the solution algorithm as it moves toward the optimal solutions.

Condition 2a is possible only for basic variables; the method that we use for computing the values of π ensures that it is satisfied. When one of the last two conditions, 2b or 2c, holds for every nonbasic arc, both the primal and dual solutions are optimal. We describe presently an algorithm that constructively finds values of $\mathbf{x}$ and π that satisfy these optimality conditions.

Examples

For the primal and dual solutions illustrated in Figures 6.28 and 6.29, respectively, we see that $\mathbf{x}$ satisfies both parts of condition 1. Condition 2a is satisfied for the basic variables $\mathbf{x}_B = (x_2, x_4, x_7, x_8) = (3, 2, 1, 1)$, so the solution is optimal if condition 2b or condition 2c is satisfied for each nonbasic arc. For the nonbasic arcs with flows at the lower bound, $\mathbf{n}_0 = \{1, 3, 6\}$ and dual solution $\pi = [\pi_1, \pi_2, \pi_3, \pi_4, \pi_5] = (11, 10, 27, 28, 0)$, we get the following reduced costs.

$$d_1 = c_1 + \pi_1 - \pi_2 = 11 + 11 - 10 = 12: \text{satisfies condition 2b}$$
$$d_3 = c_3 + \pi_2 - \pi_3 = 12 + 10 - 27 = -5: \text{violates condition 2b}$$
$$d_6 = c_6 + \pi_3 - \pi_5 = -25 + 27 - 0 = 2: \text{satisfies condition 2b}$$

For the nonbasic arcs with flows at their upper bound, $\mathbf{n}_1 = \{5\}$, the reduced cost is

$$d_5 = c_5 + \pi_3 - \pi_4 = 13 + 27 - 28 = 12: \text{violates condition 2c}$$

Thus, the solution is not optimal, because arcs 3 and 5 violate the optimality conditions. In fact, these two arcs are candidates to enter the basis.

For the basic feasible solution illustrated in Figures 6.32 and 6.34, we have $\mathbf{n}_B = \{2, 3, 5, 6\}$, $\mathbf{x}_B = (x_2, x_3, x_5, x_6) = (2, 0, 1, 1)$, and $\pi = (\pi_1, \pi_2, \pi_3, \pi_4, \pi_5) = (9, 13, 25, 38, 0)$. Examining complementary slackness, for the nonbasic arcs with flows at the lower bound, $\mathbf{n}_0 = \{1, 8\}$, we have

$$d_1 = c_1 + \pi_1 - \pi_2 = 11 + 9 - 13 = 7: \text{satisfies condition 2b}$$
$$d_8 = c_8 + \pi_5 - \pi_1 = 11 + 0 - 9 = 2: \text{satisfies condition 2b}$$

For the nonbasic arcs with flows at the upper bound, $\mathbf{n}_1 = \{4, 7\}$ we have

$$d_4 = c_4 + \pi_2 - \pi_4 = 18 + 13 - 38 = -7: \text{satisfies condition 2c}$$
$$d_7 = c_7 + \pi_5 - \pi_2 = 10 + 0 - 13 = -3: \text{satisfies condition 2c}$$

Since primal feasibility and complementary slackness are satisfied, this is the optimal solution.

Summary

This section has illustrated how the primal and dual basic solutions are computed directly from the graphical structure associated with the basis spanning tree. Because it is not necessary to store a representation of the basis inverse matrix as it is in general LP, a significant savings in both time and memory requirements is achieved. This is the source of the major computational advantage attending network flow programming algorithms. For any variation of the pure network flow problem, when the parameters are integer valued we also have the advantage of the integrality property. Thus, all computations can be carried out in integer arithmetic, resulting in numerical stability and computational efficiency. An algorithm that implements these methods to find an optimal solution to the pure NFP problem is included in the supplement to this chapter on the accompanying CD.

EXERCISES

Exercises 1 through 7 relate to a company with two warehouses and four customers. The costs of transportation between warehouses and customers are shown in the table. In each case, set up and solve the problem using the transportation algorithm.

		Customer			
		1	2	3	4
Warehouse	A	10	15	8	13
	B	3	5	7	10

1. Each warehouse has a supply of 30 units, and each customer has a demand of 15 units. The objective is to minimize total shipping cost. Demand must be satisfied.

2. All demands are 15 units and all supplies are 40 units. All demands must be met, but not all supplies must be shipped. The objective is to minimize total shipping cost.

3. All demands are 20 units and all supplies are 30 units. Not all demands need to be met, but the company wants to ship as many units as possible. The objective is to minimize total shipping cost.

4. All demands are 20 units and all supplies are 30 units. Not all demands need to be met, but each customer must receive at least five units. The company wants to ship as many units as possible. The objective is to minimize total shipping cost.

5. The maximum shipment to each customer is 15 units and the maximum shipment from each warehouse is 30 units; however, it is not necessary to meet these maximums amounts. The product is made

at the warehouses; production costs are $8 and $10 per unit at warehouses A and B, respectively. Revenues are $14, $17, $20, and $23 for customers 1, 2, 3, and 4, respectively. The objective is to maximize total profit.

6. The company needs to establish a shipping schedule for the next 2 months. The demands for each customer are 15 units in the first month and 20 units in the second month. These demands must be met. Assume that the warehouses are also manufacturing plants where the products are made. Plant A has a manufacturing capacity of 30 units per month, whereas plant B has a capacity of 50 units per month. In the first month, the manufacturing costs are $8 per unit at plant A and $10 per unit at plant B. In the second month, manufacturing costs are $10 per unit at both plants. Products can be stored at the customer sites from one month to the next at a cost of $1 per unit. Products cannot be stored at the plants. Shipping costs are as given in the preceding table except that the shipping company is giving a discount of $1 per unit on all routes during the first month. The goal is to minimize total production, shipping, and inventory costs over the 2 months. Note that not all production capacity will be utilized by the solution.

7. Modify the model developed in Exercise 6 to allow shortages in the first month to be filled by production in the second month. The back-order cost is $2 per unit.

8. Consider the data for a transportation problem in the matrix.

Source	Destination				Supplies
	D1	D2	D3	D4	
S1	10	10	6	15	10
S2	5	15	10	12	15
S3	11	8	7	21	8
Demands	5	3	8	17	

   (a) Find the optimal distribution from sources to destinations.

   (b) Add another source to the problem with a supply of five units and shipping costs to the four destinations of $4, $9, $7, and $13, respectively. Find the new optimal solution. Not all supplies need to be used.

9. Data for an assignment problem are displayed in the matrix. The columns represent jobs, and the rows represent workers. The numbers are the costs of making assignments. When *M* appears in a cell, that assignment is not possible. Solve the problem by hand using the transportation algorithm.

	J1	J2	J3	J4	J5
W 1	*M*	8	6	12	1
W 2	15	12	7	*M*	10
W 3	10	*M*	5	14	*M*
W 4	12	*M*	12	16	15
W 5	18	17	14	*M*	13

10. A company has three workers. On a particular day, six jobs are scheduled to be completed. A cost is estimated for each worker-job combination and is shown in the table. Set up and solve the transportation model for each of the following situations.

    (a) Find the minimum cost assignment when each worker can do two jobs.

    (b) Find an assignment that completes as many jobs as possible at minimum cost when each worker can do only one job.

    (c) Find a minimum-cost assignment when each worker can do any number of jobs.

				Job		
Worker	1	2	3	4	5	6
A	3	2	2	6	4	6
B	4	3	7	5	3	3
C	9	9	7	9	7	6

Shortest Path Problems

Use Dijkstra's algorithm for Exercises 11 through 16. Solve manually.

11. In the directed network shown in the figure, find the shortest paths from the root node (node 1) to all other nodes.

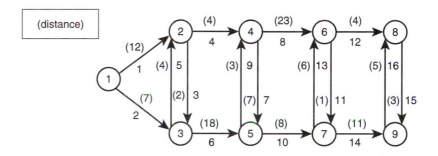

12. The matrix given in the table shows the distances between pairs of nodes in a network. An "X" in a cell indicates that there is no connection between the corresponding pair of nodes. Find the shortest path tree rooted at node 1 by hand.

				To		
From	1	2	3	4	5	6
1	X	10	3	1	X	X
2	X	X	0	1	2	X
3	X	X	X	0	2	4
4	0	X	X	X	2	9
5	1	6	X	X	X	6
6	2	3	4	X	X	X

13. The table shows the distances between all pairs of six points. Find the shortest path from point D to point E.

	To					
From	A	B	C	D	E	F
A	—	13	12	15	22	10
B	2	—	9	8	26	13
C	14	8	—	12	7	9
D	7	4	22	—	30	12
E	17	8	14	10	—	3
F	13	8	10	15	15	—

14. Find the shortest path tree for the network shown in the figure when the origin is node L. All links are two-way links.

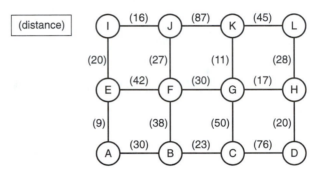

15. The table shows the arc lengths between pairs of nodes in a network. Find the shortest path tree that connects node A to all the other nodes.

	To					
From	A	B	C	D	E	F
A	—	27	43	16	30	26
B	7	—	16	1	30	25
C	20	13	—	35	5	0
D	21	16	25	—	18	18
E	12	46	27	48	—	5
F	23	5	5	9	5	—

16. Solve the following problems for the network shown in the figure. All edges may be traveled in either direction.

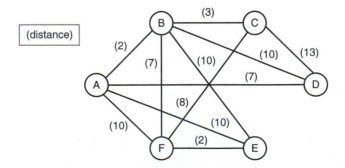

(a) Find the shortest path tree with origin at node A.

(b) Find the shortest path tree with origin at node C.

17. Write an LP formulation for the shortest path problem based on the minimum-cost flow model described by Equations (7) to (9).

Maximum Flow Problems

For Exercises 18 to 20, use the flow augmenting algorithm both to solve and to identify a minimal cut.

18. We are seeking the maximum flow from node A to node G.

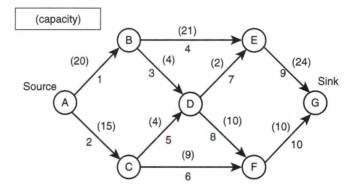

19. We are seeking the maximum flow from node 1 to node 14.

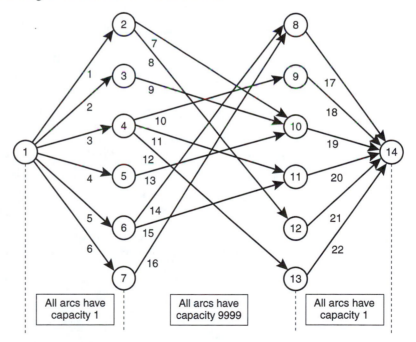

20. The source is node 1 and the sink is node 6. Find the maximum flow.

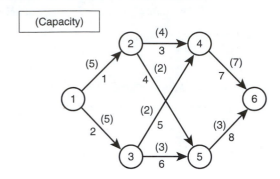

21. Write an LP formulation for the maximum flow problem based on the NFP model described by Equations (7) to (9).

Minimum Cost Flow Problems

22. Consider the following network flow diagram.

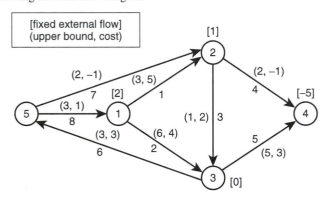

Using this diagram, determine the primal and dual basic solutions when

(a) $\mathbf{n}_B = \{8, 7, 2, 5\}$, $\mathbf{n}_0 = \{1, 3, 6\}$, $\mathbf{n}_1 = \{4\}$.

(b) $\mathbf{n}_B = \{8, 1, 2, 4\}$, $\mathbf{n}_0 = \{6, 7\}$, $\mathbf{n}_1 = \{3, 5\}$.

(c) $\mathbf{n}_B = \{-2, 1, -6, 4\}$, $\mathbf{n}_0 = \{3, 5, 7, 8\}$.

Classify each as feasible, infeasible, or no solution. Classify feasible solutions as optimal or non-optimal. In each case, indicate the reason for your classification.

23. For the network shown in the diagram, determine if the indicated flows are optimal. The basic arcs are 1, 4, 7, and 9. If the solution is not optimal, which arcs violate the optimality conditions?

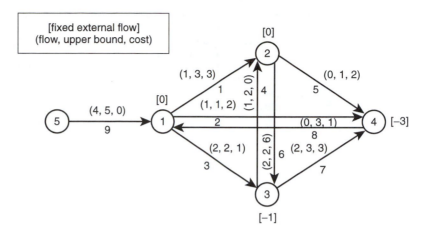

24. For the network shown in the diagram, compute the primal and dual solutions for the basis consisting of arcs 1, 2, 4, 6, 8, 10, 12, and 14. Test the solution for optimality. If it is not optimal, which arcs violate the optimality conditions?

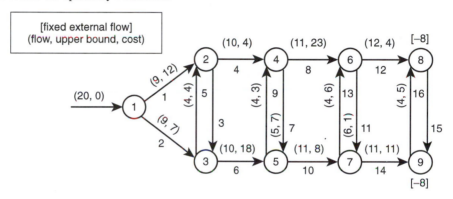

BIBLIOGRAPHY

Ahuja, R.K., T.L. Magnanti, and J.B. Orlin, "Some Recent Advances in Network Flows," *SIAM Review*, Vol. 33, pp. 175–219, 1991.

Ahuja, R.K., T.L. Magnanti, and J.B. Orlin, *Network Flows: Theory, Algorithms, and Applications*, Prentice–Hall, Engelwood Cliffs, NJ, 1993.

Aronson, J.E., "A Survey of Dynamic Network Flows," *Annals of Operations Research*, Vol. 20, pp. 1–66, 1989.

Bazaraa, M.S., J.J. Jarvis, and H.D. Sherali, *Linear Programming and Network Flows*, Second Edition, Wiley, New York, 1990.

Bertsekas, D.P., *Linear Network Optimization*, MIT Press, Cambridge, MA, 1991.

Bertsekas, D.P. and P. Tseng, "RELAX: A Computer Code for Minimum Cost Network Flow Problems," *Annals of Operations Research*, Vol. 13, pp. 127–190, 1988.

Evans, J.R. and E. Minieka, *Optimization Algorithms for Networks and Graphs*, Second Edition, Dekker, New York, 1992.

Florian, M.S., S. Nguyen, and S. Pallottino, "A Dual Simplex Algorithm for Finding all Shortest Paths," *Networks*, Vol. 11, pp. 367–378, 1981.

Ford, L.R. and D.R. Fulkerson, *Flows in Networks*, Princeton Univ. Press, Princeton, NJ, 1962.

Fulkerson, D.R., "An Out-of-Kilter Method for Minimal Cost Flow Problems," *SIAM Journal on Applied Mathematics*, Vol. 9, pp. 18–27, 1963.

Glover, F., D. Karney, and D. Klingman, "Implementation and Computational Comparisons of Primal, Dual and Primal-Dual Computer Codes for Minimum Cost Network Flow Problem," *Networks*, Vol. 4, pp. 191–212, 1974.

Glover, F., D. Klingman, and N. Philips, *Network Models and Their Applications in Practice*, Wiley, New York, 1992.

Grigoriadis, M.D., "An Efficient Implementation of the Network Simplex Method," *Mathematical Programming Study*, Vol. 26, pp. 83–111, 1986.

Jensen, P.A. and J.W. Barnes, *Network Flow Programming*, Wiley, New York, 1980.

Klingman, D., A. Napier, and J. Stutz, "NETGEN: A Program for Generating Large Scale Capacitated Assignment, Transportation, and Minimum Cost Flow Network Problems," *Management Science*, Vol. 20, pp. 814–821, 1974.

Lawler, E.L., *Combinatorial Optimization: Networks and Matroids*, Holt, Rinehart & Winston, New York, 1976.

Murty, K.G., *Network Programming*, Prentice–Hall, Engelwood Cliffs, NJ, 1992.

Chapter 7

Integer Programming Models

Integer programming (IP) is concerned with optimization problems in which some of the variables are required to take on discrete values. Rather than allow a variable to assume all real values in a given range, only predetermined discrete values within the range are permitted. In most cases, these values are integers, giving rise to the name of this class of models.

The integrality requirement underlies a wide variety of applications. As we have seen in the previous chapters, there are many situations, such as distributing goods from warehouses to factories or finding the shortest path through a network, in which the flow variables are logically required to be integer valued. In manufacturing, products are often indivisible, so a production plan that calls for fractional output is not acceptable. There are also many situations that require logical decisions of the form yes/no, go/no go, assign/don't assign. Clearly, these are discrete decisions that when quantified allow only two values. They can be modeled with binary variables that assume values of 0 or 1. Designers faced with selecting from a finite set of alternatives, schedulers seeking the optimal sequence of activities, or transportation planners searching for minimum cost vehicle routes all face discrete decision problems.

When optimization models contain both integer and continuous variables, they are referred to as *mixed-integer programs*, and may or may not be linear. The power and usefulness of these models in representing real-world situations cannot be overstated, but along with modeling convenience comes substantial computational difficulty. Only relatively small problems that contain integer variables can be solved to optimality in most cases. At first glance, this might seem counterintuitive given our ability to solve huge linear programs (LPs). As we will see in the next chapter, however, the discrete nature of the variables gives rise to a combinatorial explosion of possible solutions. In the worst case, a majority of these solutions must be enumerated before optimality can be confirmed. Consequently, as the number of integer variables in a problem grows large, heuristic methods that do not guarantee optimality must be used to find solutions.

7.1 SITE SELECTION EXAMPLE

A manufacturer is planning to construct new buildings at four local sites designated 1, 2, 3, and 4. At each site, there are three possible building designs labeled A, B, and C. There is also the option of not using a site. The problem is to select the optimal combination of building sites and building designs. Preliminary studies have determined the required investment and net annual income for each of the 12 options. This information is shown in Table 7.1, with A1, for example, denoting design A at site 1. The company has an investment budget of \$100 million (\$100M). The goal is to maximize total annual income without exceeding the investment budget. As the OR analyst, you are given the job of finding the optimal plan.

This example introduces one of the major differences between LP and IP—i.e., the indivisibility of decisions. It is an obvious requirement here that only whole buildings may be built and only whole designs may be selected. To begin creating a model, variables must be defined to represent each decision. Let $I = \{A, B, C\}$ be the set of design options, and let $J = \{1, 2, 3, 4\}$ be the set of site options.

$$\text{Let } y_{ij} = \begin{cases} 1 & \text{if design } i \text{ is used at site } j \\ 0 & \text{otherwise} \end{cases} \quad \text{for } i \in I \text{ and } j \in J$$

Also, denote by p_{ij} the annual net income and by a_{ij} the investment required for the design/site combination i, j. As a first try, you propose the following model for finding the maximum net annual income.

$$\text{Maximize } z = \sum_{i \in I} \sum_{j \in J} p_{ij} y_{ij}$$

$$\text{subject to} \qquad \sum_{i \in I} \sum_{j \in J} a_{ij} y_{ij} \leq 100$$

$$y_{ij} \in \{0,1\}, i \in I, j \in J$$

This is a typical IP formulation. It resembles an LP model in that the objective function and constraints are linear expressions, but the variables are restricted to integer values. Solving the model with an appropriate algorithm for the parameter values given in Table 7.1, the optimal solution is

$$y_{A1} = y_{A3} = y_{B3} = y_{B4} = y_{C1} = 1$$

with all other values of y_{ij} equal to zero and $z = 40$. Of the available budget, \$99M is used.

Your supervisor reviews the solution and questions your basic reasoning. You seem to have omitted some of the logic of the problem, because two designs are built on the same site—that is, A1 and C1, and also A3 and B3, are all in the solution. In addition, your supervisor now realizes that you were not alerted to several other logical restrictions imposed by the owners and architects—i.e., site 2 must have a building, design A can be used at sites 1, 2, and 3 only if it is also selected for site 4, and at most two of the designs may be included in the plans.

Your solution violates all of these restrictions and must be discarded. The following additional constraints are needed to guarantee a feasible solution.

Site 2 must have a building: $\sum_{i \in I} y_{i2} = 1$

There can be at most one building at each of the other sites: $\sum_{i \in I} y_{ij} \leq 1$ for $j = 1, 3, 4$

Table 7.1   Data for Site Selection Example

Option	A1	A2	A3	A4	B1	B2	B3	B4	C1	C2	C3	C4
Net income (\$M)	6	7	9	11	12	15	5	8	12	16	19	20
Investment (\$M)	13	20	24	30	39	45	12	20	30	44	48	55

Design A can be used at sites 1, 2, and 3 only if it is also selected for site 4:

$$y_{A1} + y_{A2} + y_{A3} \le 3y_{A4}$$

To formulate the constraints associated with design selection, three new binary variables are introduced. Let

$$w_i = \begin{cases} 1 & \text{if design } i \text{ is used} \\ 0 & \text{otherwise} \end{cases} \quad \text{for } i = \text{A, B, C}$$

At most two designs may be used: $w_A + w_B + w_C \le 2$

Finally, the y_{ij} and w_i variables must be tied together:

$$\sum_{j=1}^{4} y_{ij} \le 4w_i \quad \text{for } i = \text{A, B, C}$$

The new model has 15 variables and 10 constraints not including the integrality requirement. Solving, you find that the optimal solution is $y_{A1} = y_{A4} = y_{B2} = y_{B3} = w_A = w_B = 1$ with all other variables equal to zero and $z = 37$. All the budget is spent, but the profit has decreased.

Your supervisor is disappointed with the reduced profit and asks you which constraints are most restrictive. Unfortunately, sensitivity analysis is not available for IP as it is for LP. You can observe that the budget constraint and the constraint limiting the number of designs are both tight, and the equality constraints are obviously satisfied because of the requirement for feasibility. However, the optimal solution provides no numerical information regarding the effects of changing the problem parameters. The concept of a dual variable does not apply to IP models. To gain insight on the robustness of the solution, the best that you can do is change the right-hand sides of the constraints and re-solve the problem.

Logical Constraints

The logical constraints used to enforce restrictions on the investment options can be generalized to accommodate many situations involving two or more interrelated decisions. Let us define a series of n binary variables

$$y_j = 0 \text{ or } 1 \text{ for } j = 1, \ldots, n$$

that represents a subset of all the variables in a problem. Also, let k be an integer constant. Common relationships and their mathematical programming equivalents are defined by the following expressions.

Bound Constraints

- The n decisions are mutually exclusive: $y_1 + y_2 + \cdots + y_n \le 1$
- At most k in the subset may be chosen: $y_1 + y_2 + \cdots + y_n \le k$
- At least k in the subset must be chosen: $y_1 + y_2 + \cdots + y_n \ge k$
- Exactly k must be chosen: $y_1 + y_2 + \cdots + y_n = k$

Implication Constraints

In the following situations, let w be a binary variable that corresponds to a decision implied by one or more related decisions.

- Decision w is implied if any one of the other n decision variables has a value of 1:

$$y_1 + y_2 + \cdots + y_n \leq nw$$

- Decision w is implied if all of the other n decision variables are 1:

$$y_1 + y_2 + \cdots + y_n \leq n - 1 + w$$

- Decision w is implied if at least k of the other n decision variables are 1:

$$y_1 + y_2 + \cdots + y_n \leq (k - 1) + [n - (k - 1)]w$$

Binary Variable Implied by a Real Variable

Consider the binary variable y representing the decision whether or not to build a facility and the real variable x representing the number of products produced by the facility. A logical constraint that restricts the variable x to be 0 unless y is 1 is

$$x \leq uy$$

where u is an upper bound on x.

7.2 GENERAL CONSIDERATIONS

Integer programming (IP) models have at least some decision variables that must take on discrete rather than continuous values. We use the term "integer" because most often these values are the integers $0, 1, 2, \ldots$; however, any specific set of discrete values can be accommodated. Most often, IP models are limited to the linear form, but one can certainly conceive of situations that call for nonlinear terms to be included in the objective function and constraints. We call the more general class nonlinear integer programming models. In this section, we introduce several transformations that extend our capabilities to represent various problems. We also discuss a variety of computational issues that the practitioner should be aware of before beginning an analysis. In subsequent sections, we describe specific applications and the accompanying models.

Model Structure

The following model will serve as a prototype for this discussion.

$$\text{Maximize } z = \sum_{j=1}^{n} c_j x_j$$

$$\text{subject to } \sum_{j=1}^{p} a_{ij} x_j + \sum_{j=p+1}^{n} a_{ij} x_j \begin{Bmatrix} \geq \\ \leq \\ = \end{Bmatrix} b_i, \quad i = 1, \ldots, m$$

$$x_j \geq 0 \text{ and integer, } j = 1, \ldots, p, \text{ and } x_j \geq 0, j = p + 1, \ldots, n$$

The first p variables are restricted to integer values, whereas the remaining $n - p$ variables can assume any nonnegative real values. When $p = n$, we have a pure integer linear program (ILP). When $p = 0$, we have a linear program (LP). When $0 < p < n$, we have a mixed-integer linear program (MILP).

An important case occurs when all the variables are constrained to the values 0 and 1. This restriction allows us to represent binary decisions such as yes/no and gives rise to what is called a binary or 0-1 programming problem. There are several other special classes that are identified by the characteristics of the objective function or constraints. Some of these classes will be described as the chapter progresses. For now, we restrict our attention to situations where all functions are linear. Most solution algorithms require this form.

Simple Transformations

Converting a General IP to a 0-1 Model

Although it is somewhat inefficient from a computational point of view, it is possible to use only binary variables in IP models. A simple transformation replaces a bounded general integer variable with the weighted sum of several binary variables. Let x be an integer variable bounded by zero from below and by u from above. Let t be the smallest integer such that $2^{t+1} > u$. The transformation that can be used to remove x from the problem is

$$x = y_1 + 2y_2 + 4y_3 + \cdots + 2^t y_{t+1} = \sum_{j=0}^{t} 2^j y_{j+1} \leq u \qquad (1)$$

$$y_j = 0 \text{ or } 1, \quad j = 1, \ldots, t + 1$$

where $2^t \leq u < 2^{t+1}$.

For example, if x is bounded by 15, then $t = 3$ (because $2^4 > 15 \geq 2^3$). The transformation is $x = y_1 + 2y_2 + 4y_3 + 8y_4$. Every integer value from 0 to 15 is generated by some selection of the binary variables. Thus we have obtained a binary representation but at a cost of more variables. If x had a simple upper bound less than 15 but greater than 7 (because otherwise only three binary variables would be needed), an additional constraint enforcing the bound must be included.

Discrete Decision Values and Functions

When a decision variable can take on discrete values that are not necessarily consecutive or even integer values, it is possible to replace the variable with a weighted sum of binary terms. Suppose that x can take on only one of a finite set of values—i.e., $x \in D = \{d_1, d_2, \ldots, d_r\}$. We can model this situation as

$$x = d_1 y_1 + d_2 y_2 + \cdots + d_r y_r \qquad (2)$$
$$y_1 + y_2 + \cdots + y_r = 1$$
$$y_j = 0 \text{ or } 1, \quad j = 1, \ldots, r$$

These expressions allow only one of the y_j variables to be positive, which in turn forces x to take the value of the corresponding element in D. Similarly, when a function $g(x_1, \ldots, x_n)$ of the decision variables is restricted to one of a finite set of values, we can use the same approach by simply substituting $g(x_1, \ldots, x_n)$ for x in Equation (2).

When x is an integer and lies in the interval $0 \leq x \leq u$, the transformation given as Equation (2) can be applied by defining $D = \{0, 1, \ldots, u\}$; however, in this simple case the transformation presented as Equation (1) requires fewer variables and hence is preferred.

Complemented Variables

For some algorithms, it is necessary to have all positive signs in the objective function. When only binary variables appear in the model, replacing the corresponding variable with its complement can reverse a negative coefficient. The complement of binary variable x_j is $1 - \bar{x}_j$, where $\bar{x}_j$ is also binary.

Computational Complexity

The class of IP problems includes both the easiest and hardest types of mathematical programs. We have already identified as an easy class the pure network flow problems with integer parameters. Some special cases are the transportation problem, the assignment problem, the shortest path problem, and the maximum flow problem, all of which exhibit the integrality property. This means that LP or special-purpose algorithms much more efficient than the standard simplex method can be used to find solutions. Also easy are those models that can be solved with a greedy algorithm, such as the minimal spanning tree problem and certain sequencing problems.

The general class of integer programs, represented by arbitrary values of the parameters a_{ij}, b_i, and c_j, present difficulties even for the most powerful computers. What makes them hard is the fact that the effort required to find an optimal solution increases exponentially with problem size. In the worst case, it may be necessary to examine all combinations of integer values before arriving at the optimal solution. For example, suppose that the time required to solve a problem is proportional to 2^n, where in the case of 0-1 programming, say, n is the number of integer variables. A problem with $n = 10$, for which the difficulty is proportional to 2^{10} ($= 1024$ or approximately 10^3), may be easy to solve. With $n = 100$, the difficulty is proportional to about 10^{30}. Thus, a 10-fold increase in size makes the problem about 10^{27} times more difficult to solve. To give further perspective to these numbers, we note that the number of atoms in the known universe is less than 10^{100}.

No algorithm has yet been discovered for the general ILP whose running time does not have this exponential characteristic. Nevertheless, it would be wrong simply to dismiss IP models as unsolvable. Many real-world applications have been successfully tackled with exponential-time algorithms. For the general case, problem instances with up to 100 or 200 integer variables can usually be solved on today's computers. Tractability, however, is very problem dependent, and advanced computational techniques can bring much larger instances into the solvable range. Still, this is a far cry from the practical range of LP algorithms that can handle problems with tens of thousands or even hundreds of thousands of variables with surprising speed. For IP models, it is important to remember that solution techniques that guarantee optimality are limited. The analyst should not get carried away in developing models that require excessive numbers of integer variables.

Heuristic Methods

For those IPs that are difficult to solve, it is natural to seek methods that run quickly and yield good, although not necessarily optimal, solutions. If a computational procedure can reliably find solutions in a few seconds that are within a percent or two of the optimal solution, it may be roundly preferred to an exact procedure that takes hours to reach the optimal solution. Such procedures are called heuristic methods or *heuristics*. Because they have been applied successfully to problems that are not solvable with exact optimization algorithms, they have gained considerable standing among researchers and practitioners alike. Over the last two decades many different classes of heuristics have been developed, including tabu search, GRASP, simulated annealing, and genetic algorithms, each of which has its own cadre of supporters.

7.3   SYSTEM DESIGN WITH FIXED CHARGES

Consider a telecommunications network in which traffic flows from various origins to various destinations. Figure 7.1 illustrates a case with three source nodes $S = \{1, 3, 7\}$, four destination nodes $D = \{2, 4, 5, 8\}$, and one transshipment node $T = \{6\}$. The arcs drawn as solid lines represent existing communication links, whereas those drawn as dashed lines are proposed links. The complete set of arcs is denoted by $A = \{1, \ldots, 17\}$.

Associated with each link k is an upper bound on flow (u_k) and a unit flow cost (c_k). If one of the proposed links is constructed, it will have the capacity and cost indicated on the arc. Let $A' = \{1, \ldots, 5\}$ be the set of proposed links with corresponding construction costs $f_1 = 8, f_2 = 6, f_3 = 9, f_4 = 7$, and $f_5 = 7$. The problem is to determine which links to build, if any, and how much traffic to put on each link in the network so that the sum of the flow costs and construction costs is minimized. For a solution to be feasible, it must satisfy flow balance at each node and stay within the arc capacities.

Fixed Charge Model

In many situations, undertaking an activity means that a fixed charge or setup cost is incurred in addition to the variable cost associated with the level of the activity. In the telecommunications network design problem, there is a tradeoff between construction costs and operating costs for a given demand. In a manufacturing problem, there is usually a fixed charge for setting up a machine and a variable cost for each item produced. The same is true for building and operating most facilities.

To construct a model for the network design problem, let x_k be the flow on proposed link $k \in A'$. The total cost for link k can be modeled by the following nonlinear concave function.

$$h_k(x_k) = \begin{cases} f_k + c_k x_k & \text{when } x_k > 0 \\ 0 & \text{when } x_k = 0 \end{cases} \tag{3}$$

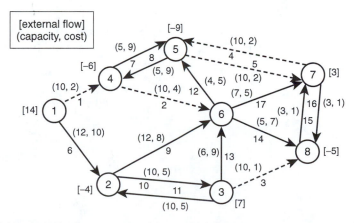

Figure 7.1   Telecommunications network with proposed links

where f_k is the fixed cost coefficient and c_k is the variable cost coefficient. A graph of $h_k(x_k)$ is sketched in Figure 7.2 and can be seen to be linear except for a jump at the origin.

LP formulations cannot handle this kind of function directly. To get around the discontinuity at the origin, it is necessary to introduce a binary variable that represents the fixed charge portion of the cost.

$$\text{Let } y_k = \begin{cases} 1 & \text{if link } k \text{ is built} \\ 0 & \text{otherwise} \end{cases} \quad \text{for all } k \in A'$$

When $h_k(x_k)$ is to be minimized and $f_k > 0$, the following transformation allows us to formulate the objective as a linear function.

$$h_k(x_k) = f_k y_k + c_k x_k$$
$$x_k \geq 0, \, y_k = 0 \text{ or } 1$$

We can now write the network design problem as an MILP. Using the notation in the network programming chapters, where $K_{O(i)}$ is the set of arcs originating at node i and $K_{T(i)}$ is the set of arcs terminating at node i, we have

$$\text{Minimize } z = \sum_{k \in A'} f_k y_k + \sum_{k \in A} c_k x_k$$

C1:   subject to $$\sum_{k \in K_{O(i)}} x_k - \sum_{k \in K_{T(i)}} x_k = b_i, \quad i \in S \cup D \cup T$$

C2: $x_k - u_k y_k \leq 0, \qquad k \in A'$

C3: $0 \leq x_k \leq u_k, \qquad k \in A$

C4: $y_k = 0 \text{ or } 1, \qquad k \in A'$

All of the data for the model are included in Figure 7.1. Constraint C1 represents conservation of flow. The RHS constants b_i are positive for a source node, negative for a destination node, and zero for a transshipment node. Constraint C2 represents the implication that if x_k is greater than zero, y_k must be 1. The implication constraint forces the fixed charge to be incurred if arc k is used in the solution. Constraint C3 places upper and lower bounds on the flows. Constraint C4 specifies the binary nature of the fixed charge variables.

When the problem is solved as an MILP, we obtain the flow solution shown in Figure 7.3. It is optimal to build only arcs 1, 3, and 4.

Time Value of Money

The model just considered assumes that the construction costs f_k and the flow costs c_k are expressed in terms of the same time unit. In fact, the construction costs are usually speci-

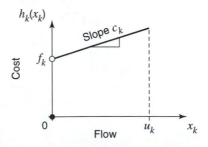

Figure 7.2   Cost of new link

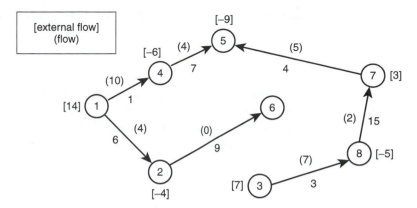

Figure 7.3   Optimal solution for network design problem

fied as a single dollar value without a time unit, whereas the flow costs are specified as the amounts transferred over an interval of time. For example, say the flows x_k are specified in terms of a monthly rate (product units per month). Because flow costs c_k are measured in dollars per product unit, the continuous component of the objective function is in units of dollars per month.

To be consistent, it is necessary to specify construction costs in dollars per month. This can be done with the capital recovery factor commonly used in engineering economics. This factor, denoted by $(A/P, i, n)$, converts a single payment, call it P, incurred at the time of construction to uniform periodic equivalents, call them A, over the life of the project, say n months. The interest rate or discount rate is denoted by i and can have various interpretations depending on whether P is borrowed from a bank or is viewed as an opportunity cost.

When f_k is the fixed cost associated with link k (or, more generally, with facility k), the per-month equivalent is

$$\overline{f}_k = f_k(A/P, i, n)$$

where i is the interest or discount rate per month determined through economic considerations, n is the life of the project (or facility) in months, and $(A/P, i, n)$ is the capital recovery factor for n periods and interest rate i. Assuming that the time unit for flow is months, we now replace f_k in the objective function of the network design problem with $\overline{f}_k$ and solve to get the minimum monthly cost.

Note that the general formula for the capital recovery factor is

$$(A/P, i, n) = \frac{i(1+i)^n}{(1+i)^n - 1}$$

For large n, a reasonable approximation is $(A/P, i, n) = (A/P, i, \infty) = i$. If one assumes an interest rate of 0, the capital recovery factor is $1/n$.

As an alternative to this approach, we could convert all costs to an equivalent lump sum at time zero. This lump sum would be the present worth of the lifetime costs of the system and would be derived by finding the present value of the monthly operating costs associated with each link k (facility k). The first step is to find the equivalent of the monthly unit cost c_k in terms of a single unit cost at time zero. The appropriate formula is

$$\overline{c}_k = c_k(P/A, i, n)$$

where the factor $(P/A, i, n) = 1/(A/P, i, n)$. The adjusted cost $\bar{c}_k$ then replaces c_k in the objective function.

Nonlinear Variable Costs

When the unit costs associated with the variable flow or production in a fixed charge problem are not constant but decrease as the activity level increases, they are said to exhibit *economies of scale*. One way to represent this type of nonlinearity is with the following polynomial function

$$c_k(x_k) = a(x_k)^b$$

where a and b are constants such that $a > 0$ and $0 < b \leq 1$. When $b = 1$, the linear model results. Because $b \leq 1$, $c_k(x_k)$ is a concave function of x_k and exhibits economies of scale since its derivative is decreasing with x_k. The marginal cost of link k (facility k) decreases with the activity level. In Section 7.8, we describe a procedure for replacing nonlinear functions with piecewise linear approximations. By doing so, we can find approximate solutions to nonlinear integer programs by solving the MILP equivalents.

7.4   FACILITY LOCATION PROBLEM

A logistics company wants to set up a distribution network in a new region of the country. There are five possible locations for warehouses and five customer locations that use the commodities supplied by the warehouses. Table 7.2 displays the data defining the problem, including unit shipping costs between potential warehouse sites and customers, fixed and variable costs for constructing the warehouses, customer demands, and warehouse capacities.

The data concerning the warehouse sites are shown in the last three columns in Table 7.2. The maximum capacity is given in terms of the number of units shipped per week. This is the largest facility that can be built at the location. The fixed and variable costs associated with construction are given as amortized values over the life of the facility. Both are expressed in weekly equivalents. A fixed cost is incurred if the facility is built, and this cost is independent of the size of the facility; the variable cost is the cost of adding one unit of size to the facility. The cost function is of the type depicted in Figure 7.2.

The goal is to select warehouse sites and sizes and to establish a shipping pattern between warehouses and customers that minimizes total shipping costs as well as the amortized cost

Table 7.2   Data for Facility Location Problem

| Warehouse | Unit shipping cost for customer | | | | | Maximum capacity | Fixed cost | Variable cost |
	1	2	3	4	5			
1	8	21	42	12	37	80	1000	20
2	21	10	31	24	40	80	1500	17
3	42	31	4	14	32	80	1700	13
4	12	24	14	7	12	80	1400	25
5	37	40	32	12	10	80	1200	33
Demand	30	40	50	35	40			

of construction. All demands must be met and the capacity at each facility must not be exceeded.

The classic transportation problem has m sources (warehouses) with known supplies and n destinations (customers) with known demands. The cost of shipping from each source to each destination is specified. The problem is to determine the minimum cost shipping plan that satisfies all demands. The facility location problem addresses the additional question of where the warehouses should be opened. The former problem is concerned with an operating question as to how existing facilities should be used most efficiently. The latter problem is concerned with a design question as to how the system should be configured.

We now construct a general model for the facility location problem that is easily adapted to related situations. Let us say that m potential sites for warehouses have been identified and that the locations and demands of the n customers are known. Let d_j be the demand for customer j. The shipping cost between each potential warehouse site i and each customer j has been estimated as c_{ij}. The cost of establishing a warehouse at location i consists of a fixed cost f_i and a variable cost v_i per unit of warehouse capacity. The various costs must be in comparable units. The maximum capacity at warehouse site i is u_i.

To develop a linear model, we define the following variables.

$$y_i = \begin{cases} 1 & \text{if a warehouse is located at site } i \\ 0 & \text{if a warehouse is not located at site } i \end{cases}$$

z_i = size of warehouse at location i

x_{ij} = amount of product shipped from warehouse i to customer j

The objective is to minimize total cost subject to the requirements that all demands be met and warehouse capacities not be exceeded. The mathematical programming model is as follows.

$$\text{Minimize } z = \sum_{i=1}^{m} f_i y_i + \sum_{i=1}^{m} v_i z_i + \sum_{i=1}^{m}\sum_{j=1}^{n} c_{ij} x_{ij}$$

All demands must be met:
$$\sum_{i=1}^{m} x_{ij} = d_j, \quad j = 1, \ldots, n$$

Supply must not be exceeded:
$$\sum_{j=1}^{n} x_{ij} \leq z_i, \quad i = 1, \ldots, m$$

Shipping from a location implies that the warehouse has been built:
$$z_i \leq u_i y_i, \qquad i = 1, \ldots, m$$

Nonnegative shipments: $\quad x_{ij} \geq 0, \qquad i = 1, \ldots, m, \; j = 1, \ldots, n$

Nonnegative size: $\qquad z_i \geq 0, \qquad i = 1, \ldots, m$

Integrality: $\qquad\qquad y_i = 0 \text{ or } 1, \quad i = 1, \ldots, m$

This formulation can be solved directly as an MILP. It can also be viewed as a network model with additional integer variables, side constraints, and a revised objective function.[1] Figure 7.4 partially depicts the network model. Note that customer demands are given as negative quantities for conformance with standard network notation. The optimal solution calls for warehouses to be built at locations 1, 3, and 4, with the flows as shown in Figure 7.5.

---

[1] This problem has been solved both as a network model with integer variables and side constraints and as a transportation model with integer variables and side constraints. See CD for solution.

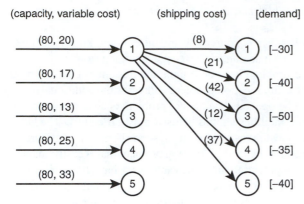

Figure 7.4   Network model of facility location problem

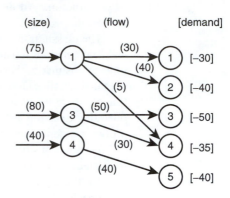

Figure 7.5   Solution to facility location problem

Uncapacitated Facility Location Problem

A variant on the preceding problem, called the *uncapacitated facility location problem*, arises when it is assumed that there is no limit on the size of the warehouses. Although the same mathematical programming model applies with u_i set at an arbitrarily large value, we present a slightly different formulation that allows for a more efficient solution technique. This requires a different definition of the decision variables. Let

$$y_i = \begin{cases} 1 & \text{if warehouse } i \text{ is built} \\ 0 & \text{otherwise} \end{cases}$$

x_{ij} = proportion of demand j satisfied by warehouse i

Because capacity is unlimited, it can be shown that it is optimal to meet the demand of each customer from a single warehouse. As such, the unit transportation cost and the variable facility cost can be combined with the demand to obtain the cost coefficient associated with the new variable x_{ij}. This is the cost of supplying the entire demand of customer j from warehouse i.

$$\bar{c}_{ij} = (v_i + c_{ij})d_j$$

The new model is as follows.

Minimize $\qquad\qquad\qquad z = \sum_{i=1}^{m} f_i y_i + \sum_{i=1}^{m}\sum_{j=1}^{n} \bar{c}_{ij} x_{ij}$

All demands must be met: $\qquad \sum_{i=1}^{m} x_{ij} = 1, \quad j = 1, \ldots, n$

Shipping from a location implies that the warehouse has been built:

$$\sum_{j=1}^{n} x_{ij} \le n y_i, \qquad i = 1, \ldots, m$$

Simple bounds: $\qquad\quad 0 \le x_{ij} \le 1, \qquad i = 1, \ldots, m; j = 1, \ldots, n$

Integrality: $\qquad\qquad y_i = 0 \text{ or } 1, \qquad i = 1, \ldots, m$

For the second constraint, it is necessary to multiply the y_i variable on the RHS by n to allow for the extreme case in which all customers are serviced by warehouse i. In an alternative formulation, these m implication constraints representing the potential warehouse sites are replaced by mn implication constraints each representing the relation between an individual transportation link and a site.

$$x_{ij} \le y_i, \ i = 1, \ldots, m; \ j = 1, \ldots, n$$

Although this is inefficient from a modeling point of view because we have increased the number of constraints by a factor of n, IP algorithms may now be able to find solutions more quickly. This increase in efficiency results from the fact that if we relax the integrality requirement on all y_i, the feasible region associated with the expanded model is much tighter than that of the original. As a consequence, the LP solution is likely to be closer to the IP solution. This is particularly important when a branch and bound method, as described in the next chapter, are used to solve the original IP model.

7.5   COVERING AND PARTITIONING PROBLEMS

There are many situations in which a decision maker is faced with a large but finite number of options from which to formulate a plan that meets a given set of demands. Examples include scheduling nurses over a 24-hour period so that adequate coverage is provided in a hospital and routing trash collectors over a 5-day period so that all residences in a city are visited. What each of these problems has in common is a set of demands (the number of nurses needed per hour in the first case and the requirement that each customer be visited once per week in the second) that can be met in various ways. If each nurse is required to work an 8-hour shift, the options are shift starting times and the number of nurses assigned to each shift; in the trash collection example, the options are the individual pickup points assigned to each route during each of the 5 days.

To state this type of problem in general terms, let S be a set of items and let S_j be a subset of S that includes one or more of the items. Assume that there are n subsets, each with cost c_j. The covering problem is to choose a finite number of subsets so that the entire set S is included in their union and the overall cost is minimized. Each item in S must be included at least once but perhaps more than once. The partitioning problem is the same except that each item must be included exactly once in the selected subsets. The pure forms of these problems arise when the subsets and items are unique. The trash collection problem meets these criteria, but the nurse scheduling problem does not, because any number of nurses can work the same shift and because demand per period need not be unity.

For $N = \{1, \ldots, n\}$ and $S = \{1, \ldots, m\}$, the set covering problem can be written as

$$\underset{T \subseteq N}{\text{Minimize}} \left\{ \sum_{j \in T} c_j \ : \ \cup_{j \in T} S_j = S \right\}$$

We now show how this problem can be converted into an IP model.

Covering Problem

Consider a microelectronics company that would like to manufacture six new products. Initial cost estimates indicate that the equipment needed to make any of the products is very expensive, and that to make each product individually would probably require too much investment. It is possible, however, to produce composite devices that through different interconnections can perform the functions of two or more of the products. In fact, a very complex device can be constructed that would have the same functionality as all six products, but it would require the use of unproven technology and so has been ruled out. After studying the design and manufacturing issues, the company's engineers have come up with 14 options, each performing a subset of functions, for management to consider. The first six options are the products themselves.

To define the problem mathematically, let c_j be the equipment cost for device j and let the column vector $\mathbf{A}_j$ represent the set of functions that the device performs. For instance, the vector $\mathbf{A}_{14} = (1, 0, 0, 1, 1, 0)^{\mathrm{T}}$ indicates that device 14 can be used for products 1, 4, and 5. The problem is to find the set of devices with the minimum total equipment cost that can perform the functions of all six products.

For convenience, Figure 7.6 shows the 14 vectors $\mathbf{A}_1, \mathbf{A}_2, \ldots, \mathbf{A}_{14}$ arrayed in a 6×14 matrix in which each row corresponds to a particular function. We call this the **A** matrix. The cost vector **c** is also shown in Figure 7.6.

Covering Model

To develop an IP formulation, assume that there are, in general, n devices or options, and let

$$x_j = \begin{cases} 1 & \text{if device } j \text{ is selected for manufacture} \\ 0 & \text{otherwise} \end{cases}$$

For $\mathbf{x} \in \Re^n$, the model is

$$\begin{array}{ll} \text{Minimize} & \mathbf{cx} \\ \text{subject to} & \mathbf{Ax} \geq \mathbf{e} \\ & x_j = 0 \text{ or } 1, \ \ j = 1, \ldots, n \end{array}$$

$$\begin{array}{rcccccccccccccc} \text{Column number} & 1 & 2 & 3 & 4 & 5 & 6 & 7 & 8 & 9 & 10 & 11 & 12 & 13 & 14 \end{array}$$

$$\mathbf{A} = \begin{bmatrix} 1 & 0 & 0 & 0 & 0 & 0 & 1 & 1 & 0 & 0 & 0 & 1 & 0 & 1 \\ 0 & 1 & 0 & 0 & 0 & 0 & 1 & 0 & 1 & 1 & 0 & 1 & 1 & 0 \\ 0 & 0 & 1 & 0 & 0 & 0 & 1 & 1 & 1 & 1 & 1 & 0 & 0 & 0 \\ 0 & 0 & 0 & 1 & 0 & 0 & 0 & 1 & 1 & 1 & 1 & 0 & 1 & 1 \\ 0 & 0 & 0 & 0 & 1 & 0 & 0 & 0 & 1 & 0 & 1 & 1 & 0 & 1 \\ 0 & 0 & 0 & 0 & 0 & 1 & 0 & 0 & 1 & 1 & 0 & 1 & 1 & 0 \end{bmatrix}$$

$$\mathbf{c} = (12 \ \ 17 \ \ 13 \ \ 10 \ \ 13 \ \ 17 \ \ 24 \ \ 24 \ \ 60 \ \ 38 \ \ 27 \ \ 45 \ \ 25 \ \ 35)$$

Figure 7.6 Technology matrix and cost vector for manufacturing example

where $\mathbf{e} = (1, \ldots, 1)^T$ is an n-dimensional vector of 1's. We must select devices to cover all functions. The ith row of $\mathbf{A}$ identifies the set of devices that includes the ith function. The corresponding constraint ensures that at least one of the devices selected will perform the function.

In the optimal solution for this example, $x_5 = x_7 = x_{13} = 1$, and all other variables are zero, with cost $z = 62$. Each function is covered by this solution, and function 2 is covered twice.

Partitioning Model

If we now add the restriction that each function must be included once and only once in the selected devices, the partitioning problem results. The general formulation is as follows.

$$\text{Minimize} \quad \mathbf{cx}$$
$$\text{subject to} \quad \mathbf{Ax} = \mathbf{e}$$
$$x_j = 0 \text{ or } 1, \quad j = 1, \ldots, n$$

For this problem, the subsets cannot overlap, and thus the solution obtained for the covering model is not feasible because function 2 is covered twice. Solving, we find that $x_1 = x_3 = x_5 = x_{13} = 1$ and that all other variables are zero, with cost $z = 63$.

Computations

A typical set covering or partitioning problem might have hundreds or thousands of rows. The general problem statement assumes that the feasible subsets S_j are given and that there is one binary variable for each subset. This means that there may be an exponential number of decision variables, since the number of nonempty subsets of m items is $2^m - 1$. An important component of any algorithm for solving large covering-type problems is a procedure for generating attractive subsets to include in the formulation. It is usually impossible or unwise to include all feasible subsets. For the airline crew scheduling problem, for example, the columns of the $\mathbf{A}$ matrix are called pairings and represent sets of flights that crew members may fly. Pairings begin and end at the same base, must satisfy all legal restrictions related to duty time and rest, and generally span 2 to 4 weeks (see Andersson *et al.* [1998] for more discussion).

A second point about set covering and partitioning problems is that they have a special structure in that the $\mathbf{A}$ matrix is composed entirely of 1's and 0's and the RHS vector is composed entirely of 1's. As is true for many types of IP problems with special structures, researchers have developed algorithms uniquely designed to handle these problems. Empirically speaking, the corresponding algorithms are much more efficient than those designed for solving general IP problems, but their running times are still exponential functions of problem size in the worst case. These issues and the related literature are further discussed in Chapter 8 of Garfinkel and Nemhauser [1972] and in Chapter 13 of Salkin and Mathur [1989].

7.6 DISTANCE PROBLEMS

There are a variety of problems involving the selection of routes through networks that can be formulated as IP problems but are not necessarily total unimodular. In this section, we consider several of these problems in which distance minimization is the objective. As an example, suppose there are six locations or nodes identified by the numbers 1 through 6. The matrix shown in Figure 7.7 gives the distance between each pair of nodes. The dashes along the diagonal rule out self-loops. Also, the data imply an asymmetric problem structure, because the distance from node i to node j is generally not equal to the distance from node j to node i.

Traveling Salesman Problem

The distance matrix gives rise to a directed network with six nodes and 30 arcs. Starting at an arbitrary location, we would like to find a tour that visits each node exactly once and returns to the original location. The goal is to minimize the sum of the lengths of the selected arcs.

This is the classical asymmetric traveling salesman problem (TSP). One possible tour is shown in Figure 7.8, and is indicated by a series of node pairs: $(1, 2) \rightarrow (2, 3) \rightarrow (3, 4) \rightarrow (4, 5) \rightarrow (5, 6) \rightarrow (6, 1)$. The total length of this tour is $z = 124$.

To write a general IP model for this problem, let n be the number of nodes in the network and let c_{ij} be the length of the arc passing from i to j. When it is infeasible to go from i to j, either the coefficient c_{ij} is set to an arbitrarily large number or the corresponding variable is omitted from the formulation. We specify the set of nodes in the problem as $N = \{1, \ldots, n\}$ and a proper subset as $S \subset N$. A solution is feasible if each of the n nodes is visited exactly once and the corresponding tour forms a closed path returning to the starting node. A subtour is a sequence of unique nodes that starts and ends at the same location but visits only a subset of the n nodes. A solution containing subtours is not feasible.

We define the following decision variables

$$x_{ij} = \begin{cases} 1 & \text{if the arc from } i \text{ to } j \text{ is selected} \\ 0 & \text{otherwise} \end{cases} \quad \text{for all } i \neq j = 1, \ldots, n$$

	1	2	3	4	5	6
1	—	27	43	16	30	26
2	7	—	16	1	30	25
3	20	13	—	35	5	0
4	21	16	25	—	18	18
5	12	46	27	48	—	5
6	23	5	5	9	5	—

Figure 7.7   Distance between pairs of nodes

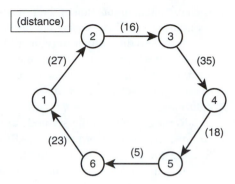

Figure 7.8   Traveling salesman tour

When $i = j$, x_{ij} does not exist so it is not included in the model. We now give the mathematical programming formulation of the asymmetric TSP.

Tour length: Minimize $z = \sum_{i=1}^{n} \sum_{j=1}^{n} c_{ij} x_{ij}$

C1:   Exactly one successor for each node: $\sum_{j=1}^{n} x_{ij} = 1, \quad i = 1, \ldots, n$

C2:   Exactly one predecessor for each node: $\sum_{i=1}^{n} x_{ij} = 1, \quad j = 1, \ldots, n$

C3:   Subtour elimination: $\sum_{i \in S} \sum_{j \in S} x_{ij} \leq |S| - 1, \quad S \subset N, 2 \leq |S| \leq {}^{n}\!/_{2}$

C4:   Integrality: $x_{ij} = 0$ or $1, i \neq j = 1, \ldots, n$

Constraint C1 guarantees that each node has a single successor, and Constraint C2 guarantees that each node has a single predecessor. The number of arcs entering and leaving each node must be 1. In network terminology, these constraints ensure continuity of flow and are the same as those found in the assignment problem.

Constraint C3 prevents formation of subtours or cycles of size less than n. It is based on the observation that any subtour constructed from the nodes in a subset S must have exactly $|S|$ arcs. In words, it says that for any subset of nodes S whose cardinality is less than n, the number of arcs connecting the nodes in that subset must be at least 1 less than the number of nodes in S. This is true whether or not the arcs form a subtour. For $|S| = 2$, for example, we have $\binom{n}{2}$ constraints of the form

$$x_{12} + x_{21} \leq 1, \quad x_{13} + x_{31} \leq 1, \quad x_{14} + x_{41} \leq 1, \ldots$$

We only consider subsets with no more than $n/2$ nodes because eliminating subtours of length k automatically eliminates subtours of length $n - k$. A major difficulty in solving a TSP is dealing with the subtour elimination constraints, of which there are $2^{n-1} - n - 1$. This number is an exponential function of the problem size and it is not practical to write out the corresponding inequalities for any but the smallest values of n.

A second difficulty in solving a TSP is the loss of the integrality property commonly associated with network flow problems. The subtour elimination constraints do not admit a total unimodular structure, and so solving the relaxed problem obtained by replacing the integrality constraint with $0 \leq x_{ij} \leq 1$ for all i and j will not, in general, yield integer solutions.

Returning now to the six-node example, if one relaxes the subtour elimination constraints, the assignment problem (AP) results, which automatically has integer solutions. Solving the AP relaxation, we obtain the solution {(1, 4), (4, 2), (2, 1), (3, 5), (5, 6), (6, 3)}. The corresponding objective function value $z = 54$ provides a lower bound on the length of the optimal TSP tour. This solution is not feasible, because it contains two subtours: $(1, 4) \rightarrow (4, 2) \rightarrow (2, 1)$ and $(3, 5) \rightarrow (5, 6) \rightarrow (6, 3)$.

To eliminate the first subtour associated with $S = \{1, 2, 4\}$, we add the constraint $x_{12} + x_{21} + x_{14} + x_{41} + x_{24} + x_{42} \leq 2$ to the relaxed formulation. This constraint indirectly eliminates the second subtour associated with the subset $S = \{3, 5, 6\}$ as well as the subtour in the opposite direction: $(1, 2) \rightarrow (2, 4) \rightarrow (4, 1)$. Solving the new problem, we obtain the solution {(1, 4), (4, 3), (3, 5), (5, 6), (6, 2), (2, 1)}, with $z = 63$. This solution contains no subtours and so is optimal (see Figure 7.9).

The idea of solving a relaxed version of the problem and then adding constraints sequentially to remove subtours is the strategy followed by the most sophisticated algorithms. The basic approach is simply to try to solve the full model using implicit enumeration (*branch and bound*). This is taken up in the next chapter. The combination of implicit enumeration and the sequential addition of constraints is called *branch and cut*.

Directed Minimal Spanning Tree Problem

Given a directed graph $G = (N, A)$ with node set N and arc set A, a tree is a subgraph $G' = (N, A')$ that has no cycles and is connected (contains a directed path from the *root* node to all other nodes in N). The directed minimal spanning tree (MST) problem is to find a tree rooted at, say, node 1 with a directed path to every other node. The goal is to minimize the sum of the arc lengths used in the tree. Figure 7.10 depicts a feasible solution with a total arc length of 100 for the six-node example.

The model for this problem is similar to that of the TSP except now one or more arcs may leave node 1; however, there is no requirement that any arcs leave nodes 2, . . . , n. The mathematical programming formulation is

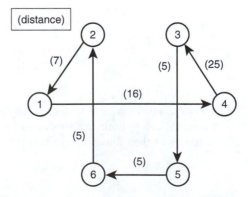

Figure 7.9 Optimal TSP tour

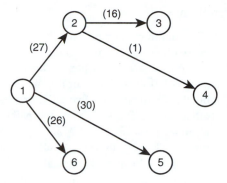

Figure 7.10 A directed spanning tree

Tree length:	Minimize $z = \sum\limits_{i=1}^{n} \sum\limits_{j=1}^{n} c_{ij} x_{ij}$
At least one arc must leave root node 1:	$\sum\limits_{j=1}^{n} x_{1j} \geq 1$
One predecessor for all other nodes:	$\sum\limits_{j=1}^{n} x_{ij} = 1, \quad i = 2, \ldots, n$
TSP subtour elimination constraints:	$\sum\limits_{j \in S} \sum\limits_{j \in S} x_{ij} \leq |S| - 1, S \subset N, 2 \leq |S| \leq n/2$
Integrality:	$x_{ij} = 0$ or $1, \quad i \neq j = 1, \ldots, n$

Again we solve the relaxation of the problem obtained by removing the subtour elimination constraints. The solution to this problem is $\{(1, 4), (3, 6), (3, 5), (6, 2), (6, 3)\}$, with $z = 31$. It contains the cycle $(6, 3) \rightarrow (3, 6)$, so we add the constraint $x_{63} + x_{36} \leq 1$ and resolve. At the next two iterations, we add $x_{56} + x_{65} \leq 1$ and $x_{35} + x_{53} + x_{56} + x_{65} + x_{63} + x_{36} \leq 2$, respectively, to eliminate the corresponding cycles. The solution at this step is $\{(1, 6), (6, 2), (6, 3), (6, 5), (2, 4)\}$, with $z = 42$, and is optimal (see Figure 7.11).

We have used a procedure similar to that used for the TSP, but more efficient polynomial algorithms have been discovered that solve both the directed and undirected versions of the minimal spanning tree problem. In the next chapter, we present a greedy algorithm for the undirected case.

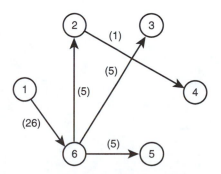

Figure 7.11 Optimal solution to minimal spanning tree problem

Shortest Path Tree Problem

Given a directed graph, we wish to find a tree rooted at, say, node 1 with a path leading to every other node. The goal is to minimize the sum of the individual path lengths.

As we saw in Chapter 5, this problem can be cast as a network flow program. A feasible solution is shown in Figure 7.12 for the six-node example. The numbers in square brackets indicate that five units of flow enter at node 1, and one unit leaves each of the other nodes. The lengths of the paths to nodes 2, . . . , 6 for this solution are $P_2 = 27$, $P_3 = 43$, $P_4 = 28$, $P_5 = 57$, and $P_6 = 46$. The flows along the arcs indicate the number of times each arc is used in the solution. The sum of the path lengths is $z = 231$.

To develop a mathematical programming formulation for the problem, we must define the variables in a way that allows the arcs to be used multiple times. Accordingly, let

$$x_{ij} = \text{number of paths using the arc from } i \text{ to } j$$

The model for the shortest path tree problem is as follows.

$$\text{Length of the } n - 1 \text{ paths:} \quad \text{Minimize } z = \sum_{i=1}^{n} \sum_{j=1}^{n} c_{ij} x_{ij}$$

C1:   At node 1, supply $= n - 1$: $\sum_{j=2}^{n} x_{1j} = n - 1$

C2:   Conservation of flow: $\sum_{j=1}^{n} x_{ij} - \sum_{j=1}^{n} x_{ji} = -1, \quad i = 2, \ldots, n$

C3:   Nonnegativity: $x_{ij} \geq 0, i \neq j = 1, \ldots, n$

Constraint C1 requires that exactly $n - 1$ units of flow leave the root node 1, thus establishing $n - 1$ paths. For every other node, the amount of flow entering that node must be one unit greater than the amount leaving. This is guaranteed by Constraint C2. Nevertheless, as is true for all pure network problems, there is one redundant conservation of flow constraint. The flow out of node 1 will surely be 5 if the flow out of the others totals 5, so Constraint C1 can be removed. Upper bounds on the variables could be included in the formulation, but they are implied by the other constraints and so would be redundant.

Note that we have not required the variables to have integer values in Constraint C3. Since this is a pure network flow problem, the constraint matrix is totally unimodular. The solution will automatically be integer valued.

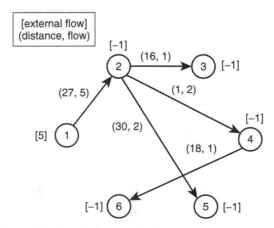

Figure 7.12   A directed tree rooted at node 1

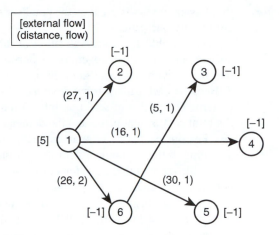

Figure 7.13   Optimal solution to shortest path tree problem

The optimal solution to the shortest path tree problem is shown in Figure 7.13. The minimal path lengths to each node are $P_2 = 27$, $P_3 = 31$, $P_4 = 16$, $P_5 = 30$, and $P_6 = 26$, with objective $z = 130$. The optimal tree identifies the shortest paths from the root node (node 1), to each of the other five nodes. The corresponding spanning tree has a size of 104. This is certainly different from the minimal directed spanning tree in Figure 7.11, which has a size of 42.

7.7   EXAMPLES

Pattern Selection: The Cutting Stock Problem

A paper company sells rolls of paper of fixed length in five standard widths: 5, 8, 12, 15, and 17 feet. Its manufacturing process produces 25-foot-wide rolls only, so all orders must be cut from stock of this size. The demands for the 5-, 8-, 12-, 15-, and 17-foot rolls are 40, 35, 30, 25, and 20, respectively. The problem is to cut the manufactured rolls in a fashion that minimizes the total number required. The manufactured rolls can be cut in 11 different patterns, as shown in Figure 7.14. Some patterns result in excess paper (shown in black) that must be discarded. To simplify the situation somewhat, no pattern is included

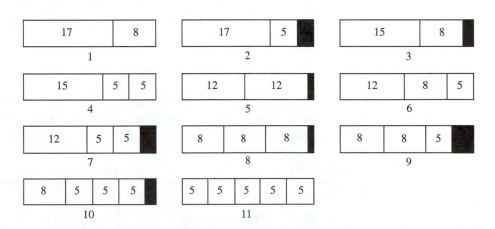

Figure 7.14   Permissible patterns for the cutting stock problem

that has an excess as great as the smallest standard width (5 feet). Also, each roll can be cut in only one pattern.

This problem is a generalization of the covering problem in that a set of patterns is defined, each of which covers some portion of the demand for some item. However, neither the individual demands nor the amounts covered by the various patterns are limited to 1. Because each pattern is cut from a 25-foot-wide roll, we can assign a unit cost to each pattern. The goal is to select an integral number of rolls to be cut in each pattern so that the total cost is minimized and the demand satisfied.

For our example, the mathematical programming model has five constraints corresponding to the requirements for the five standard widths. We associate with each pattern a variable representing the number of manufactured rolls cut in that pattern. The structural coefficient a_{ij} indicates the number of rolls of standard width i that can be cut in pattern j, for $i = 1, \ldots, 5$ and $j = 1, \ldots, 11$. For example, for pattern 4 ($j = 4$), we can cut two rolls 5 feet wide ($i = 1$) and one roll 15 feet wide ($i = 4$), so $a_{14} = 2$ and $a_{44} = 1$. To formulate the model, let

$$x_j = \text{number of rolls cut in pattern } j$$

The following integer program minimizes material usage while meeting all demand.

Minimize
$$z = x_1 + x_2 + x_3 + x_4 + x_5 + x_6 + x_7 + x_8 + x_9 + x_{10} + x_{11}$$
subject to
$$x_2 \quad + 2x_4 + \quad x_6 + 2x_7 \quad + x_9 + 4x_{10} + 5x_{11} \geq 40$$
$$\text{(Demand for 5-foot rolls)}$$
$$x_1 \quad + x_3 \quad + x_6 \quad + 3x_8 + 2x_9 + x_{10} \quad \geq 35$$
$$\text{(Demand for 8-foot rolls)}$$
$$2x_5 + x_6 + x_7 \quad \geq 30$$
$$\text{(Demand for 12-foot rolls)}$$
$$x_3 + x_4 \quad \geq 25$$
$$\text{(Demand for 15-foot rolls)}$$
$$x_1 + x_2 \quad \geq 20$$
$$\text{(Demand for 17-foot rolls)}$$
$$x_j \geq 0 \text{ and integer}, \quad j = 1, \ldots, 11$$

Three optimal solutions are shown in Table 7.3, with the total number of rolls cut being equal to 64 in each case. Only the second solution satisfies the demand exactly. The first solution produces three extra rolls of the 5-foot width, and the third produces two extra 8-foot rolls. There may be other optimal solutions.

Production Scheduling

You are given the demands for two products A and B over a 5-day period. Both products are manufactured on the same machine, but on any given day the machine can be used for only one product because of the extensive changeover time. Your job is to establish a pro-

Table 7.3 Several Optimal Solutions for Cutting Stock Problem

Variable	x_1	x_2	x_3	x_4	x_5	x_6	x_7	x_8	x_9	x_{10}	x_{11}
Solution 1 (number of rolls)	19	1	9	16	14	2	0	1	0	2	0
Solution 2 (number of rolls)	16	4	7	18	15	0	0	4	0	0	0
Solution 3 (number of rolls)	20	0	5	20	15	0	0	4	0	0	0

duction schedule for the machine that shows which product, and the amount of that product, that should be manufactured on each day. The schedule must ensure that all demands are met without shortages.

The machine is set up for production each morning. Let d_{At} and d_{Bt} be the respective product demands, with t ranging from 1 to 5. The corresponding setup costs are $100 for product A and $50 for product B. These are the only relevant manufacturing costs. For each of the 5 days, you are to determine if product A, product B, or neither is to be manufactured.

When product A is being manufactured, the machine can produce at most 10 units per day. When product B is being manufactured, the machine can produce at most 20 units per day. On any given day, you can produce more than is needed and place the excess in inventory. The cost of storing product A is $2 per unit per day and the cost of storing product B is $3 per unit per day. Initially, there are 10 units of each product in inventory. At the end of the 5-day planning period, inventories are to be zero. For convenience, assume that all additions to and withdrawals from inventory occur at the beginning of each day.

This problem can be formulated as an MILP. In the model, the production quantities and inventory levels are treated as real variables, and the decisions regarding which product to manufacture on a specific day are treated as binary variables.

Definition of Variables

Real variables:

$$A_t = \text{quantity of product A produced on day } t$$
$$B_t = \text{quantity of product B produced on day } t$$
$$I_t = \text{inventory of product A at the end of day } t$$
$$J_t = \text{inventory of product B at the end of day } t$$

Binary variables:

$$x_t = \text{decision to produce product A on day } t$$
$$y_t = \text{decision to produce product B on day } t$$

Objective Function

$$\text{Minimize } z = \sum_{t=1}^{5}(100x_t + 50y_t + 2I_t + 3J_t)$$

Constraints

Conservation of inventory ($t = 1, \ldots, 5$):

C1:
$$A_t - I_t + I_{t-1} = d_{At}$$
$$B_t - J_t + J_{t-1} = d_{Bt}$$

Initial and final inventories:

C2:
$$I_0 = 10, \quad J_0 = 10, \quad I_5 = 0, \quad J_5 = 0$$

Capacity implications ($t = 1, \ldots, 5$):

C3:
$$A_t \le 10x_t$$
$$B_t \le 20y_t$$

At most one product on any day ($t = 1, \ldots, 5$):

C4:
$$x_t + y_t \le 1$$

Integrality and nonnegativity ($t = 1, \ldots, 5$):

C5:
$$x_t, y_t = 0 \text{ or } 1; \quad A_t, B_t, I_t, J_t \ge 0$$

Except for the discrete decisions, this is a typical production scheduling problem involving two products. Constraint C1 requires that inventory is conserved from one day to the next and suggests that the model has a network structure underpinning it. These constraints are better viewed by moving the inventory on day t to the RHS of the equation and the demand to the left. Doing this for product A, for example, yields $A_t - d_{At} + I_{t-1} = I_t$, which says that the inventory at the end of day t equals the inventory at the end of day $t-1$ plus the amount produced on day t minus the demand on day t. Constraint C3 illustrates the logical relation between the amount produced and the decision variable representing the setup operation. The logical condition that at most one product can be manufactured on any day is enforced by Constraint C4. Except for x_t and y_t, the model is a pure network model. Enforcing integrality for x_t and y_t assures that the remainder of the variables will be integer.

Production Scheduling with Continued Setup

We now modify the preceding example so that the setup for a particular product can be carried over from one day to the next. In this situation, if the machine is set up for product A on day t at a cost of $100, it is available to manufacture product A on any number of consecutive days with no additional setup cost.

This feature can be easily accommodated in the existing model by noting that the expression

$$x_t(1 - x_{t-1})$$

has the value 1 only if $x_t = 1$ and $x_{t-1} = 0$. This will occur whenever production of product A starts on day t. The model can be modified by inserting this factor (and a similar one for product B) into the objective function, yielding

$$\text{Minimize } z = \sum_{t=1}^{5}\left[100x_t(1 - x_{t-1}) + 50y_t(1 - y_{t-1}) + 2I_t + 3J_t\right]$$

Unfortunately, the objective now has nonlinear terms that are simple in appearance but are not allowed in linear-integer programming models. An acceptable alternative is to introduce two binary variables u_t and v_t to replace the multiplicative terms in the objective function. Let

$$u_t = \begin{cases} 1 & \text{if production begins on product A on day } t \\ 0 & \text{otherwise} \end{cases}$$

$$v_t = \begin{cases} 1 & \text{if production begins on product B on day } t \\ 0 & \text{otherwise} \end{cases}$$

We must now add constraints that determine the values of u_t and v_t in terms of x_t and y_t. Note that the event of production beginning on product A is the intersection of the event that product A is not produced on day $t-1$ and the event that it is produced on day t. The logical intersection of these two events can be modeled by Constraint C6. The other constraints remain the same.

Constraints defining production initiation ($t = 1, \ldots, 5$):

C6:
$$x_t + (1 - x_{t-1}) \le 1 + u_t$$
$$y_t + (1 - y_{t-1}) \le 1 + v_t$$

The new objective function is

$$\text{Minimize } z = \sum_{t=1}^{5}\left[100u_t + 50v_t + 2I_t + 3J_t\right]$$

which now has a linear form. The primary reason why this transformation works is that we are minimizing cost and so there is no incentive for u_t or v_t to be positive unless there is a desire to begin manufacturing the respective products on day t.

Days-Off Scheduling

Short-term workforce planning gives rise to three types of problems. The first problem involves the determination of how many different shifts are needed per day to meet labor requirements. The second problem involves sizing and then scheduling the workforce to minimize cost under the constraint that each employee receive a fixed number of days off per week. The third problem is called tour scheduling and combines the results of the first two into weekly assignments for individual workers.

In this section, we present an ILP model for the days-off scheduling problem, which must be solved routinely by businesses that operate 6 or 7 days a week. Examples include hospitals, airlines, municipal transportation companies, and the postal service. In general terms, we wish to investigate the (k, m)-cyclic staffing problem, whose objective is to minimize the cost of assigning workers to an m-period cyclic schedule so that (1) sufficient workers are available during time period i to meet requirements r_i and (2) each person works k consecutive time periods and is idle for the remaining $m - k$ periods. Note that periods m and 1 are considered to be consecutive to account for the cyclic nature of the problem.

The most common example is the $(5, 7)$-cyclic staffing problem, in which each employee works 5 days per week and is given two consecutive days off. To formulate the model for this problem, let

$$x_j = \text{number of employees assigned to days-off pattern } j$$
$$c_j = \text{weekly cost of days-off pattern } j \text{ per employee}$$
$$r_i = \text{number of employees required on day } i$$

In the objective function, the coefficient c_j can account for such factors as premium pay for weekend assignments or different pay rates for different labor categories, assuming a more elaborate model that included, say, full-time and part-time employees.

The ILP formulation is

$$\text{Minimize } z = \sum_{j=1}^{7} c_j x_j$$

subject to
$$\left(\sum_{j=1}^{7} x_j \right) - x_i - x_{i-1} \ge r_i, \quad i = 1, \ldots$$

$$x_j \ge 0 \text{ and integer, } j = 1, \ldots, 7 \text{ where } x_0 = x_7$$

or

$$\text{Minimize } z = \mathbf{cx}$$

subject to

$$\begin{bmatrix} 0 & 1 & 1 & 1 & 1 & 1 & 0 \\ 0 & 0 & 1 & 1 & 1 & 1 & 1 \\ 1 & 0 & 0 & 1 & 1 & 1 & 1 \\ 1 & 1 & 0 & 0 & 1 & 1 & 1 \\ 1 & 1 & 1 & 0 & 0 & 1 & 1 \\ 1 & 1 & 1 & 1 & 0 & 0 & 1 \\ 1 & 1 & 1 & 1 & 1 & 0 & 0 \end{bmatrix} \mathbf{x} \ge \mathbf{r}$$

$$\mathbf{x} \ge \mathbf{0} \text{ and integer}$$

Each column of the **A** matrix represents a feasible days-off pattern. It is easy to see, therefore, how the formulation can be generalized to handle any 7-day pattern. After solving the model to get the optimal solution $\mathbf{x}^*$, the minimum-cost workforce has $W = \sum_{j=1}^{7} x_j^*$ employees.

Although the **A** matrix is not totally unimodular, it does have a special structure called the *circular property*. In particular, a 0-1 vector is said to be circular if its 1's occur consecutively, where the first and last entries are considered to be adjacent. Veinott and Wagner [1962] recognized that a related ILP with consecutive 1's could be transformed into network flow problems. Building on that work, Bartholdi, Orlin, and Ratliff [1980] transformed the (k, m)-cyclic staffing ILP into a bounded series of network flow problems with a nonsingular transformation of the decision variables. This was possible because the **A** matrix is both column and row circular. In addition, they provided an alternative solution technique to the original problem based on LP rounding. The steps of this technique are as follows.

1. Ignoring the integrality restrictions in the days-off ILP, solve the LP relaxation to obtain the solution $\bar{x}_1, \ldots, \bar{x}_7$. If these values are all integer, this is the optimal solution to the ILP; if not, go to Step 2.

2. Form two linear programs LP1 and LP2 from the relaxation in Step 1 by adding, respectively, the constraints $x_1 + x_2 + \cdots + x_7 = \lceil \bar{x}_1 + \cdots + \bar{x}_7 \rceil$ and $x_1 + x_2 + \cdots + x_7 = \lfloor \bar{x}_1 + \cdots + \bar{x}_7 \rfloor$, where $\lceil y \rceil$ is the smallest integer greater than or equal to y and $\lfloor y \rfloor$ is the largest integer less than or equal to y. LP1 is always feasible, but it is possible that LP2 is infeasible. In either case, it can be shown that if an optimal solution exists, then an integral optimal solution exists, so the better of the two solves the days-off ILP.

This procedure illustrates the important point that many IP models can be solved efficiently by taking advantage of something unique in their formulation. In this case, we are able to solve the IP by solving just two LPs. If it were desirable to include part-time employees who worked less than 5 days a week or perhaps less than a full shift each day, the appropriate columns would have to be added to the model. The resultant model would be a standard ILP without any special structure.

Assembly Line Balancing

A wide range of products are assembled on fixed-paced flow lines consisting of a collection of workstations. One or more operations are performed at each station, with some restrictions on the order; these are called *precedence relations*. There is also a limit on the amount time a product can spend at any particular workstation. This value is known as the *cycle time* and is assumed to be fixed.

Consider an example of a product whose manufacture requires five operations. The decisions involve allocating each operation to a particular workstation so that the number of stations is minimized. Table 7.4 gives the precedence relations and the time needed to complete each operation.

Logically, operation i is either performed or not performed at station j. This is an either/or situation that can be readily modeled with 0-1 variables.

$$\text{Let } x_{ij} = \begin{cases} 1 & \text{if operation } i \text{ is done at station } j \\ 0 & \text{otherwise} \end{cases}$$

Suppose that the maximum time available at each workstation is 12 minutes. Along with the data in Table 7.4, this implies that at most four stations are needed. Hence, we have the following time constraints.

Table 7.4   Data for Assembly Line Balancing

Operation i	Processing time p_i (minutes)	Precedence
1	6	—
2	5	—
3	7	—
4	6	3
5	5	2, 4

$$\sum_{i=1}^{5} p_i x_{ij} \leq 12, \quad j = 1, \ldots, 4$$

(i.e., the total time taken for all operations assigned to station j must be no more than 12 minutes). Expanding these inequalities yields

$$6x_{11} + 5x_{21} + 7x_{31} + 6x_{41} + 5x_{51} \leq 12$$
$$6x_{12} + 5x_{22} + 7x_{32} + 6x_{42} + 5x_{52} \leq 12$$
$$6x_{13} + 5x_{23} + 7x_{33} + 6x_{43} + 5x_{53} \leq 12$$
$$6x_{14} + 5x_{24} + 7x_{34} + 6x_{44} + 5x_{54} \leq 12$$

Next, we address the precedence relations among the operations. By saying that operation 3 must be done before operation 4, we mean that the former must be performed either at the same workstation as the latter or at a prior workstation. In mathematical terms, operation i is done at or before station k if $\sum_{j=1}^{k} x_{ij} = 1$ and is done after station k if $\sum_{j=1}^{k} x_{ij} = 0$. If $x_{4k} \leq \sum_{j=1}^{k} x_{3j}$, then operation 4 cannot be done at station k unless operation 3 has been done, because $x_{4k} = 1$ only if $\sum_{j=1}^{k} x_{3j} = 1$. For the precedence relations to be satisfied, this must hold at all stations, so

$$x_{4k} \leq \sum_{j=1}^{k} x_{3j}, \quad k = 1, \ldots, 4$$

If neither operation is done at or prior to station k, this expression holds trivially (i.e., $0 \leq 0$); it also holds if both operations are done at station k (i.e., $1 \leq 1$).

One set of constraints is required for each precedence relation. For job 5, the constraints are

$$\left. \begin{array}{l} x_{5k} \leq \sum_{j=1}^{k} x_{2j} \\[2ex] x_{5k} \leq \sum_{j=1}^{k} x_{4j} \end{array} \right\} \quad k = 1, \ldots, 4$$

To ensure that each operation is performed once and only once, we require

$$\sum_{j=1}^{4} x_{ij} = 1, \quad i = 1, \ldots, 5$$

The objective is to find the minimum number of workstations needed to assemble the product. This can be achieved by assigning a smaller "cost" to operation i performed at station 1 than for operation i performed at station 2, and so on. By minimizing these costs, we force each operation to be allocated to the earliest possible workstation. One way to specify the objective function coefficients to obtain the desired result is to let $c_{ij} = j$ for all i.

This means that the cost of performing operation i at station j is simply j. Consequently, we have

$$\text{Minimize } z = \sum_{i=1}^{5} x_{i1} + 2\sum_{i=1}^{5} x_{i2} + 3\sum_{i=1}^{5} x_{i3} + 4\sum_{i=1}^{5} x_{i4}$$

The preceding equations together with the nonnegativity and integrality conditions on the variables

$$x_{ij} \geq 0 \text{ and integer for all } i \text{ and } j$$

make up the integer program for this problem. Note that we do not need to impose an upper bound of 1 on each x_{ij}, because the last constraint does this implicitly.

7.8 NONLINEAR OBJECTIVE FUNCTION

In the material supplementary to Chapter 2 on the CD, we describe how a maximization problem with a separable concave objective function can be modeled with arbitrary accuracy using a piecewise linear approximation. Linear programming can then be applied to obtain a near-optimal solution to the original nonlinear formulation. When an objective function includes cost terms that are concave or revenue terms that are convex, this method does not work because the pieces of the approximation enter the solution in the wrong order. These cases arise quite often, as we have seen in Section 7.3, where we introduced problems with fixed charges and economies of scale.

In this section, we present an MILP model for the more general case involving separable nonlinear terms in the objective function. This model can easily handle maximization problems with convex terms in the objective as well as minimization problems with concave terms in the objective. We begin with the specific case involving a convex function of a single variable.

Piecewise Linear Approximation of Convex Terms

The mathematical programming problem we wish to consider has linear constraints but a nonlinear objective function that is separable in the variables x_j.

$$\text{Maximize } z = \sum_{j=1}^{n} f_j(x_j)$$

$$\text{subject to} \quad \sum_{j=1}^{n} a_{ij}x_j \leq b_i, \quad i = 1, \ldots, m$$

$$x_j \geq 0, \quad j = 1, \ldots, n$$

Assume that each term $f_j(x_j)$ is convex. An example is given in Figure 7.15.

A piecewise linear approximation identifies r points along the x_j axis: $d_1, d_2, \ldots, d_r$, and r points along the f_j axis: $c_1, c_2, \ldots, c_r$. Now we have r pieces representing the objective with the first starting at the origin. If $f_j(0)$ does not equal 0, the objective term can be replaced by $f_j(x_j) - f_j(0)$ without affecting the optimal solution, so define $d_0 = c_0 = 0$.

Procedure for Changing the Model

The following steps can be used to obtain piecewise linear approximations of nonlinear convex terms in problems with maximization objectives or of nonlinear concave terms in problems with minimization objectives.

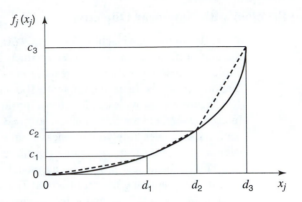

Figure 7.15   Piecewise linear approximation of a convex function

Define Variables for the Pieces

Select the values of the r_j breakpoints for x_j. Compute the corresponding objective function values $c_k = f_j(d_k)$ for $k = 1, \ldots, r_j$. Define $x_{j1}, x_{j2}, \ldots, x_{jr_j}$, where x_{jk} is the amount of the kth piece used in the solution.

Compute the Coefficient (Slope) of Each Piece

$$s_{jk} = (c_k - c_{k-1})/(d_k - d_{k-1})$$

(Note that the coefficients c_k and d_k really depend on j as well.)

Replace the Objective Function Terms with Linear Approximation

$$f_j(x_j) = \sum_{k=1}^{r_j} s_{jk} x_{jk} \tag{4}$$

Add the Linking Constraint Relating New Variables to Old

$$x_j = \sum_{k=1}^{r_j} x_{jk} \tag{5}$$

Define the Binary Variables

$$\text{Let } y_{jk} = \begin{cases} 1 & \text{if piece } k \text{ is included} \\ 0 & \text{if piece } k \text{ is not included} \end{cases} \quad \text{for } k = 2, \ldots, r_j \tag{6}$$

Add the Bound Constraints

$$d_1 y_{j2} \leq x_{j1} \leq d_1$$
$$(d_k - d_{k-1})y_{j,k+1} \leq x_{jk} \leq (d_k - d_{k-1})y_{jk}, \qquad k = 2, \ldots, r_j - 1$$
$$0 \leq x_{jr_j} \leq (d_{r_j} - d_{r_j - 1})y_{jr_j}$$

When a convex objective function term is present and the problem is one of minimization, or if we have a concave term and the problem is maximization, it is not necessary to introduce the binary variables in Equation (6). In this case, the bound constraints reduce to $0 \leq x_{jk} \leq d_k - d_{k-1}$, $k = 1, \ldots, r_j$.

Manufacturing Problem with Nonlinear Objective

We return to the four-machine production example in Section 2.1, but now we assume that the unit profit for each of the three products is not constant but varies with sales volume. Table 2.1 gives the machine usage data. Table 7.5 shows revenues as functions of sales for the three products. In each case, up to 100 units can be sold.

Figure 7.16 provides a plot of total profit as a function of sales for the three products. In each case, there are three ranges, so $r_j = 3$ for $j = 1, 2,$ and 3. For product P, unit profit declines with sales, implying a concave curve. The profit function for product Q is convex, because unit profit increases with sales. Product R has an S-shaped profit function.

The mathematical programming model of the problem is as follows, with nonlinear functions representing the profit terms. The variables P, Q, and R represent the numbers of units of products P, Q, and R that are manufactured.

$$\text{Maximize } Z = f_1(P) + f_2(Q) + f_3(R)$$

$$\begin{array}{lll}
\text{subject to} & 20P + 10Q + 10R \leq 2400 & \text{(Machine A)} \\
& 12P + 28Q + 16R \leq 2400 & \text{(Machine B)} \\
& 15P + 6Q + 16R \leq 2400 & \text{(Machine C)} \\
& 10P + 15Q + 0R \leq 2400 & \text{(Machine D)} \\
& P \leq 100, Q \leq 100, R \leq 100 & \text{(Marketing)} \\
& P \geq 0, Q \geq 0, R \geq 0 & \text{(Nonnegativity)}
\end{array}$$

To linearize this model, we identify three market segments for each product. From Figure 7.16, we have $d_1 = 30$, $d_2 = 60$, and $d_3 = 100$ in each case. Now define

$P_1, Q_1,$ and R_1 as the numbers of sales in the first price range

$P_2, Q_2,$ and R_2 as the numbers of sales in the second price range

$P_3, Q_3,$ and R_3 as the numbers of sales in the third price range

Reading the values of c_k for each product from Figure 7.16 and using Equation (4), the objective function becomes

$$\text{Maximize } Z = 60P_1 + 45P_2 + 35P_3 + 40Q_1 + 60Q_2 + 65Q_3$$
$$+ 20R_1 + 70R_2 + 20R_3 \tag{Obj}$$

Additional constraints in the form of Equation (5) must be added to link the segments to the total sales:

$$P_1 + P_2 + P_3 = P, \quad Q_1 + Q_2 + Q_3 = Q, \quad R_1 + R_2 + R_3 = R \tag{Link}$$

It is possible to replace P, Q, and R with their equivalent summations in all the constraints, but it is more convenient to leave them in the formulation.

Table 7.5 Profit Data for Products

	Profit/unit, $		
Number of sales	P	Q	R
0 to 30	60	40	20
30 to 60	45	60	70
60 to 100	35	65	20

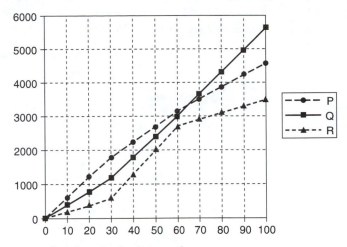

Figure 7.16   Revenue functions for the three products

For the model to represent the nonlinear functions correctly, the segments must be used in the proper order. Sales in the first segment must be at their maximum before sales in the second segment take place, and both the first and second segments must be exhausted before the third segment becomes available. In the case of product P, this happens automatically, because the unit profits are decreasing. This is a situation in which we are maximizing a concave function, so it is not necessary to add integer variables of the type given in Equation (6) to ensure compliance. Only constraints that limit sales for each segment are necessary.

$$0 \leq P_1 \leq 30, \quad 0 \leq P_2 \leq 30, \quad 0 \leq P_3 \leq 40 \qquad \text{(P bounds)}$$

For product Q, the unit profits are increasing with sales. If the model treated Q in the same manner in which it treats P, the most profitable sales would be made first. To prevent this, we introduce integer variables of the type specified in Equation (6) to control the order of sales. Two variables are necessary.

$$y_{Q2} = \begin{cases} 1 & \text{if sales in segment 2 are made} \\ 0 & \text{otherwise} \end{cases}$$

$$y_{Q3} = \begin{cases} 1 & \text{if sales in segment 3 are made} \\ 0 & \text{otherwise} \end{cases}$$

To establish bounds on the pieces and ensure that they enter in the proper order, we add the bound constraints.

$$30 y_{Q2} \leq Q_1 \leq 30, \quad 30 y_{Q3} \leq Q_2 \leq 30 y_{Q2}, \quad 0 \leq Q_3 \leq 40 y_{Q3} \qquad \text{(Q bounds)}$$

An analysis of these constraints tells us the following.

If $(y_{Q2}, y_{Q3}) = (0, 0)$, only Q_1 can be positive
If $(y_{Q2}, y_{Q3}) = (1, 0)$, $Q_1 = 30$, $0 \leq Q_2 \leq 30$, and $Q_3 = 0$
If $(y_{Q2}, y_{Q3}) = (1, 1)$, $Q_1 = 30$, $Q_2 = 30$, and $0 \leq Q_3 \leq 40$
$(y_{Q2}, y_{Q3}) = (0, 1)$ is impossible

For product R, the first and second pieces define a convex curve. This means that we must control only the entry of the first piece. One integer variable is necessary for this purpose.

Segments 2 and 3 form a concave function over the corresponding range, so sales in segment 2 will always be preferred to sales in segment 3. Let

$$y_{R2} = \begin{cases} 1 & \text{if sales in segment 2 or 3 are made} \\ 0 & \text{otherwise} \end{cases}$$

The bound constraints are

$$30y_{R2} \le R_1 \le 30, \quad 0 \le R_2 \le 30y_{R2}, \quad 0 \le R_3 \le 40y_{R2} \qquad \text{(R bounds)}$$

The complete model uses the objective presented as Equation (Obj), the original machine constraints, the linking constraints (Link), and the three sets of bound constraints (P bounds), (Q bounds), and (R bounds). The original market constraints are redundant, since they are implied by the bound constraints.

As illustrated, integer variables are necessary when objective function terms are convex in a maximization (or concave in a minimization). Although we can handle such cases with this approach, we pay a price in terms of additional integer variables. As the number of additional integer variables grows, the problem becomes increasingly more difficult to solve.

General Piecewise Linear Approximations

In the manufacturing example, product R had an S-shaped profit function of which the first part was convex and the second part was concave. The appropriate transformation was then applied to each part. The more general case occurs when $f_j(x_j)$ is any continuous nonlinear function appearing in either the objective function or the constraints, and is much more complex. We now consider the problem of approximating a nonlinear function $f_j(x_j)$ for $0 \le x_j \le u_j$ over a grid of size r. An example is depicted in Figure 7.17 for $r = 5$. The curve has been approximated by four line segments, producing a piecewise linearization. Once the grid $D_j = \{0 = d_{1j}, d_{2j}, \ldots, d_{rj} = u_j\}$ has been chosen, the function $g_j(x_j)$, given by the set of line segments, can be represented as follows.

$$x_j = \sum_{i=1}^{r} \alpha_i d_{ij}$$

and

$$g_j(x_j) = \sum_{i=1}^{r} \alpha_i f_{ij} \qquad (7)$$

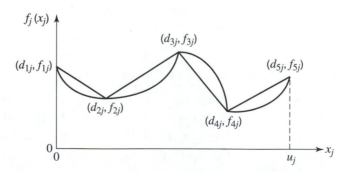

Figure 7.17   Nonlinear function approximated by line segments

where

$$\sum_{i=1}^{r} \alpha_i = 1, \quad \alpha_i \geq 0, \quad i = 1, \ldots, r$$

with the additional restriction that at most two adjacent α_i terms are nonzero.

For a given $\hat{x}_j$, $g_j(\hat{x}_j)$ as determined by Equation (7) is a point on one of the line segments if no more than two of the α_i terms are positive and adjacent—i.e., of the form α_i, α_{i+1}. This condition can be achieved with the additional constraints

$$\alpha_1 \leq y_1$$
$$\alpha_i \leq y_{i-1} + y_i, \quad i = 2, \ldots, r-1$$
$$\alpha_r \leq y_{r-1}$$
$$\sum_{i=1}^{r-1} y_i = 1$$
$$y_i = 0 \text{ or } 1, \quad i = 1, \ldots, r-1$$

From these relationships, it follows that for some q, $1 \leq q \leq r-1$, $y_q = 1$, and $y_i = 0$, $i \neq q$. Thus, $\alpha_q \leq 1$, $\alpha_{q+1} \leq 1$, and $\alpha_i = 0$ for all $i \neq q$ and $i \neq q+1$.

EXERCISES

1. Consider the IP model

$$\begin{aligned}
\text{Maximize } & 2x_1 + 5x_2 \\
\text{subject to } & x_1 + x_2 \leq 15 \\
& -x_1 + x_2 \leq 2 \\
& x_1 - x_2 \geq 2 \\
& x_1 + x_2 \geq 2 \\
& x_1, x_2 \geq 0 \text{ and integer}
\end{aligned}$$

The following problems are cumulative in that each part is based on the answer(s) to the previous part(s).

(a) Rewrite the model using only binary variables.

(b) Rewrite the model as a minimization problem with all "less than or equal to" constraints.

(c) Rewrite the model as a minimization problem with all positive objective function coefficients.

2. A company is considering three major research projects labeled A, B, and C. Each of the projects can be selected for any of the next 3 years or may be omitted from the portfolio entirely. The total return for each project based on the year it is selected is given in the table. This return captures all relevant cash flows, including investments. It also includes the effects of the time value of money. The investment required for each project occurs entirely within the year for which it is selected. The goal is to maximize the total return. This problem has the following constraints. No more than two projects can be selected in any year. Total investment in any year cannot exceed 9. Project B must be selected after project A. Project A and B cannot be selected in the same year. A project can be selected at most once. Formulate and solve the problem as an IP. Define all notation.

Year	Total return for project		
	A	B	C
1	7	6	4
2	5	4	4
3	8	7	5
Investment	5	3	5

3. A computer service company needs to establish communications among five cities. An analysis of various media has determined that the monthly cost of connecting a pair of cities i and j with a link is c_{ij}, as shown in the matrix. The connection allows communications in both directions. The cost of establishing interconnection facilities at each city depends on the number of links incident to the city. Note that these are node costs rather than arc costs.

City	1	2	3	4	5
1	—	15	13	19	21
2	15	—	10	24	14
3	13	10	—	14	17
4	19	24	14	—	12
5	21	14	17	12	—

- If one link touches a city, the cost is d_1
- The second link touching the city adds the cost d_2
- The third link touching the city adds the cost d_3
- These costs are related as follows: $d_1 > d_2 > d_3$

Formulate and solve a 0-1 ILP model that incorporates the following information.

- The objective is to minimize monthly cost
- Each city must be touched by at least one connection
- The links selected must form a tree
- No more than three links can touch any one city
- $d_1 = 5$, $d_2 = 3$, $d_3 = 1$

4. The figure shows a road network between two cities A and B located in different states. The federal government wants to place inspection stations on the roads so that all traffic moving between the cities must pass through at least one station. The cost of establishing a station on road k is c_k, as indicated by the numbers in parentheses in the figure.

(a) Show that the problem of selecting the minimum cost locations of stations can be modeled as a set covering problem. (*Hint*: the rows of the **A** matrix will represent paths between the two cities, and the columns will represent individual roads.)

(b) Describe a more efficient way to solve this problem using one of the standard network models.

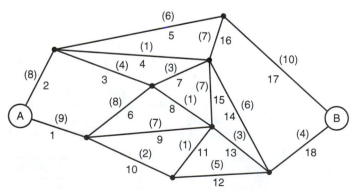

Road network between cities A and B

5. (*Symmetric TSP*) Consider an undirected graph with m edges and n nodes. Develop an ILP model for the symmetric traveling salesman problem. In the model, let x_e be a binary variable equal to 1 if edge e is used, and 0 otherwise. Also, let S be a proper subset of the node set N, let $E(S)$ be the set

of edges whose two endpoints are contained in S, and let $\delta(j)$ be the set of edges incident to node j. The cost of traversing edge e is c_e.

6. (*Prize collecting TSP*) A variant of the traveling salesman problem occurs when the salesman receives a profit of f_j for visiting city $j \in N$. His tour must start at city 1 and include at least two other cities. A cost of c_e is incurred if he traverses edge e. Unlike the traditional TSP, however, the salesman does not have to visit all n cities. The goal is to find a tour that maximizes the difference between profits and travel costs subject to these restrictions. Formulate an ILP that can be used to solve this problem. Use the notation introduced in the preceding exercise as well as any new notation deemed necessary.

7. It is possible to replace the subtour elimination constraint C3 in the text with the following constraint:

$$\sum_{i \in S} \sum_{j \notin S} x_{ij} \geq 1, \quad S \subset N, S \neq \varnothing \qquad \text{(SEC)}$$

To see why this is true, show that C3 is a linear combination of the assignment constraints C1 and C2, and (SEC). C1, C2, and C3 are from the TSP discussion in Section 7.6.

8. Give a general IP formulation for the cutting stock problem described in Section 7.7.

9. For the cutting stock example in Section 7.7, write the objective function for each of the following two cases. Solve each case.

   (a) The goal is to find the number of each pattern so that the excess that must be discarded is minimized

   (b) In practice, each pattern requires a certain number of cuts. For instance, pattern 1 requires only one cut whereas pattern 2 requires two cuts. The goal is to find the selection of patterns that minimizes the total number of cuts

10. The figure shows well targets identified by geologists in an offshore oil field. The wells are to be drilled from platforms that are very expensive to erect. To reduce the cost, several wells may be drilled from the same platform. The cost of a platform designed to drill k wells is c_k. To drill a set of wells, the platform is located at the centroid of the wells and directional drilling is used to reach each target. The operational cost of drilling a well is $a(h + d)^2$, where h is the horizontal distance from the platform location to the target, d is the depth, and a is a constant. The depth is assumed to be constant over the field.

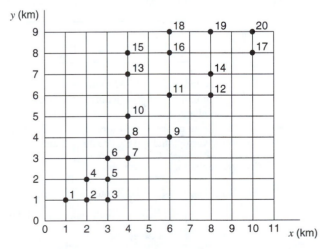

Well targets for offshore drilling

Eighteen platform designs are proposed, as shown in the following table. When a platform is constructed, assume that all wells associated with it are drilled. The problem is to find the selection of platforms that permits the drilling of all 20 wells at minimum total cost.

Platform	Set	Platform	Set
I	1, 2, 3	X	14, 15, 18
II	4, 5, 6	XI	12, 13, 14, 15
III	7, 8, 9	XII	18, 19, 20
IV	6, 8, 9	XIII	16, 17, 18, 19, 20
V	7, 8, 9, 10, 11	XIV	10, 11, 16, 17
VI	10, 11	XV	1, 2, 3, 4, 5, 6, 7, 8, 9
VII	16, 17	XVI	1, 2, 3, 4, 5
VIII	12, 13	XVII	19, 20
IX	14, 15	XVIII	all

Solve the problem for $a = 1$, $d = 0.5$, and c_k as given in the following table. Specific values of h must be computed.

Number of wells (k)	1	2	3	4	5	6	7	8	9	20
Platform cost (c_k)	10	19	27	34	40	45	49	52	54	90

To illustrate the computation of the cost of placing a platform, consider platform II. The coordinates of the centroid are the average of the coordinates of the wells in the set. Platform II covers wells 4, 5, and 6 with coordinates (2, 2), (3, 2), and (3, 3), respectively. The location of the centroid is (x_c, y_c) = (8/3, 7/3). To compute the distance between well 4 and the centroid, we use the theory of right triangles:

$$h_4 = \sqrt{(x_4 - x_c)^2 + (y_4 - y_c)^2} = \sqrt{(2/3)^2 + (1/3)^2} \cong 0.745$$

Similarly, $h_5 = 0.471$ and $h_6 = 0.745$. The cost of building the platform and drilling the wells is as follows.

$$c_3 + a(h_4 + d)^2 + a(h_5 + d)^2 + a(h_6 + d)^2 = 31.044$$

This problem requires several calculations, so set up a spreadsheet program to automate the calculations.

11. The demand for a company's product over the next 12 months is given in the table.

Month	1	2	3	4	5	6	7	8	9	10	11	12
Demand	5	10	7	12	13	3	10	12	6	11	12	13

There is a fixed charge of $50 to make a production run. The cost of holding the product in inventory for 1 month is $2 for each unit remaining in inventory at the end of the month. There is no lead time for production, initial inventory is zero, and no inventory cost is charged for items held less than 1 month. It can be shown that it is never optimal to produce when product remains in inventory, and that it is never optimal to produce more than enough to cover an integral number of future months.

(a) Define the following variable:

$$x_{ij} = \begin{cases} 1 & \text{if the product is produced in month } i \text{ and} \\ & \text{covers the demand through months } j \\ 0 & \text{otherwise} \end{cases} \quad \begin{array}{l} i = 1, \ldots, 12 \\ j = 1, \ldots, 12 - i + 1 \end{array}$$

Formulate the minimum cost production plan as a set partitioning problem.

(b) Define the following variables:

x_j = amount of production in period j

y_j = 1 if there is nonzero production in period j; 0 otherwise

z_j = inventory remaining at the end of period j

Give an MILP formulation for the problem.

(c) Develop a network model for the problem such that the shortest path through the network yields the optimal production plan. (*Hint*: Let the nodes correspond to months and let the costs of the arcs be the costs of producing and perhaps holding items in inventory for the corresponding numbers of months.)

12. An electronic system has n components operating in series. The reliability of a component is the probability that it will not fail while it is in use. The reliability of component i is given as r_i. The reliability of the system is the probability that none of the components will fail—that is, the product of the component reliabilities. To increase the reliability of the system, extra components may be included for backup. These are called redundant components, since they are not required unless the original components fail. Thus, the reliability of a component for which there are x redundant components is

$$R_i = 1 - (1 - r_i)^{(1+x)}$$

The reliability R of the system is the product of the component reliabilities:

$$R = \prod_{i=1}^{n} R_i$$

The data in the table describe the components of a system in which $n = 4$. Find the optimal number of redundant components of each type subject to constraints on total cost and weight. For redundant components, the maximum total cost must be less than or equal to \$1000, and the maximum total weight must be less than or equal to 80 pounds. In addition, no more than five redundant components of each type may be installed.

Component	1	2	3	4
Reliability	0.9	0.8	0.95	0.75
Cost (\$)	100	50	40	200
Weight (lb)	8	12	7	5

13. The figure represents a grid on which electric power is generated and distributed. Each node is a demand point and also a potential location for a power generator. The numbers in parentheses indicate power demands in megawatts (MW). Assume that the grid is in the xy-plane, where $x, y = 0, 1, 2, 3$.

Power will be supplied by generators constructed at the nodes. The maximum size of a plant in terms of power output is 100 MW. The cost of a plant in scaled terms consists of a fixed charge of \$100 plus a variable cost of $2 per megawatt. In the model, all costs are expressed as lifetime figures, so no adjustment need be made to account for the time value of money or the differences between operating and investment costs. Power is transmitted along the lines that connect the nodes (all parallel to the grid axes). The cost of transmission is $1.5 per megawatt per mile. Note that the distance between adjacent nodes is 1 mile. Power may flow in either direction along a line and is delivered to the customers at the nodes. Excess power arriving at a node may be transshipped along the lines to other nodes. The problem is to select the nodes at which generators should be built as well as to determine the sizes of the generators. The solution must satisfy the following environmental restrictions.

- No two generators may be less than 1.5 miles apart
- No more that two generators may be built on any coordinate line. For instance, of the four locations on the vertical axis $x = 1$, at most two can be used
- If a generator is built at location (3, 3), one must also be built at location (0, 0)

Set up and solve a mathematical programming model for this problem

14. The figure shows several towns along a river that generate pollution in the amounts indicated in square brackets adjacent to their locations. The unit of pollution is pounds per hour. Without control measures, the pollutants enter the river directly and do not dissipate, so any quantity discharged at one point remains in the river at every point downstream. In other words, assuming steady-state flows, if 10 lb/hr is discharged at point A, then 10 lb/hr will flow past every point downstream of point A. Pollutant quantities are additive, so the amount flowing past any point is the sum of the contributions of all points upstream.

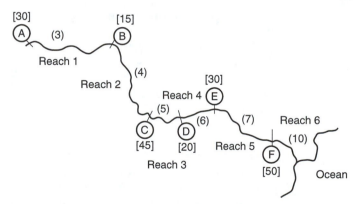

Pollution control in a river basin

For analytic purposes, the river has been divided into segments called reaches. The quantity in parentheses adjacent to each reach is the flow of water in that segment measured in acre-feet per hour. An acre-foot is a unit volume of water. The concentration in a reach is the ratio of pollutant flow to water flow and is measured in pounds per acre-foot. The river basin authority requires that the concentration be no more that 10 pounds per acre-foot in a reach. To satisfy this requirement, control facilities must be constructed in each city. Two kinds of facilities are available: one removes 90% of the pollutants and the other removes 95%. Construction costs are in terms of volume of pollutant input to the facility (pounds per hour) and can be approximated by fixed charge functions. The functions are

$$C_{0.9} = 1{,}000{,}000 + 100{,}000x \quad (\text{\$/hr})$$

$$C_{0.95} = 5{,}000{,}000 + 300{,}000x \quad (\text{\$/hr})$$

where x is the number of pounds of pollutants per hour. Of course, the costs are zero if x is zero. Operating costs for processing the pollutants are as follows.

$$c_{0.9} = 0.10 \quad (\$/lb)$$
$$c_{0.95} = 0.20 \quad (\$/lb)$$

The facilities have a life of 20 years; the interest rate (discount rate) to be used in the economic cal-culations is 6% per year. Although a treatment facility may be built at a city, it is not necessary that all pollution at the city be treated. Set up a network model for this problem. For each city, the model should include a node corresponding to the pollutant source and a node corresponding to the river entry point, yielding 12 nodes in all. A terminal node should also be included. Note that once the pol-lutants enter the river they cannot be treated. To determine which arcs to include, consider the paths that pollutants can take from the source to the river. Define the necessary variables, write out all the constraints and the objective function, and find the solution that minimizes total costs.

15. A company must determine on which of four possible sites to build warehouses. The amortized monthly cost at each site has two components. The first is a construction cost of $1000 and the second is an operating cost of $5 per unit shipped. There are five customers with monthly demands and transportation costs ($ per unit), as shown in the accompanying table.

Warehouse	Transportation cost for customer				
	1	2	3	4	5
1	8	12	6	15	9
2	11	17	3	4	7
3	9	17	10	2	6
4	15	20	5	12	3
Demand	20	30	40	50	20

(a) Assume that warehouse capacities are limited to shipments of 70 units per month. Set up the asso-ciated linear model and solve as both an LP and an MILP. Use the software accompanying the text. Of course, you must neglect the integrality constraints for the LP. Comment on the results and indicate the deficiency of the LP solution.

(b) Assume that there is no limit on warehouse capacities. Set up three separate models using the three formulations in Section 7.4. Recall that in the first formulation, x_{ij} was defined as the num-ber of units shipped from i to j; in the second (and the third), the definition of x_{ij} was the pro-portion of units shipped from i to j; and in the third, implication constraints of the form $x_{ij} \leq y_i$ were introduced. Solve each formulation as both an LP and an MILP. Provide the same types of comments as those in part (a).

16. (*Diet problem*) The space agency is preparing the menu for a lunar mission. The astronauts have pro-vided utility measures for single units of eight food types, which are labeled A through H in the table. Nutritional contents of each food type are also given.

Food type	A	B	C	D	E	F	G	H
Utility	9	12	6	8	10	3	6	7
Vitamins	10	8	13	12	4	6	9	6
Fat	8	9	10	6	18	9	10	5
Sugar	7	7	2	14	13	9	8	4

(a) Find the menu that will maximize the total utility subject to the following nutritional require-ments. The diet must have a minimum of 40 vitamin units, a maximum of 35 fat units, and a maximum of 30 sugar units. In addition, no more than two units of any one food type are allowed, and only whole units may be selected.

(b) Compare the answer found in part (a) with the solution of the LP model obtained when the inte-grality conditions are dropped.

(c) An alternative goal is to maximize vitamin intake subject to the other restrictions. To reach this goal, what should the new diet be? How does it affect the menu utility as compared with the answer found in part (a)?

17. A convenience store chain is planning to enter a growing market and must determine where to open several new stores. The map shows the major streets in the area being considered.

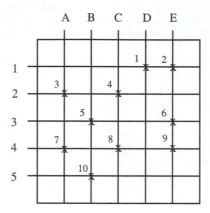

Proposed locations of convenience stores

Adjacent streets are 1 mile apart. The symbol X indicates possible store locations. All travel must follow the street network, so distance is determined with a rectilinear metric. For instance, the distance between corners A1 and C2 is 3 miles. Set up an IP that can be used to find the optimal store locations. Use the following data and restrictions.

- The costs of constructing stores at the various locations are as follows.

Location	1	2	3	4	5	6	7	8	9	10
Cost	100	80	90	50	80	90	100	70	90	80

- No two stores selected can be on the same street (either north–south or east–west)
- Stores must be at least 3 miles apart. Stores exactly 3 miles apart are acceptable
- Every grid point (A1, B2, etc.) must be no more than 3 miles from a store

18. When solving an IP problem, it is very helpful to be able to compute efficiently the bounds on the optimal objective function value. Consider the days-off scheduling problem in Section 7.7 with coefficient $c_j = 1$ for $j = 1, \ldots, 7$. With this data, the goal is to minimize the size of the workforce (Burns and Carter [1985]).

(a) Derive two simple lower bounds on W, the minimum number of workers needed given daily requirements r_i, $i = 1, \ldots, 7$. The first should be based on the daily demand, and the second on the total weekly demand.

(b) Assume that another constraint has been added to the problem—namely, that each employee must be given A out of every B weekends off where $B > A$. For example, employees might be given three out of every five weekends off. Let r_1 be the demand on Sunday and r_7 the demand on Saturday, with $r = \max\{r_1, r_7\}$. Derive a third lower bound based on this constraint. *Hint*: in B weeks, each employee is available for $(B - A)$ weekends. (Somewhat surprisingly, the largest of these lower bounds is the optimal value of W when the requirement that each employee work 5 consecutive days is changed to the requirement that each employee work 5 out of 7 days, but not more than 6 consecutive days.)

19. For the days-off scheduling ILP in Section 7.7, assume that $c_j = 1$ for $j = 1, \ldots, 7$, and write its dual problem. Find two feasible solutions to the dual problem that yield objective function values equal to the two bounds found in Exercise 18a.

20. Consider the following nonlinear program.

$$
\begin{aligned}
\text{Maximize } & f_1(x_1) + f_2(x_2) + f_3(x_3) \\
\text{subject to } & x_1 + 2x_2 + 3x_3 \leq 13 \\
& 2x_1 \qquad\quad - x_3 \geq 0 \\
& 0 \leq x_1 \leq 5 \text{ and integer, } 0 \leq x_2 \leq 9, 0 \leq x_3 \leq 5
\end{aligned}
$$

Note that x_2 and x_3 are not required to be integer valued. The values of $f_1(x_1)$ are given in the table, and the functions f_2 and f_3 are displayed in the figure. Set up the MILP model that properly represents the nonlinear terms in the objective function.

x_1	0	1	2	3	4	5
$f_1(x_1)$	10	20	-5	15	0	0

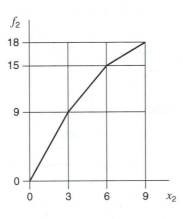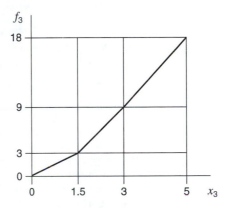

21. The figure shows an oil field with eight wells spaced uniformly around the perimeter of a circle. The length of a radial line is 1000 feet. Oil is to be collected at the wells and delivered to the tanks that are at the center of the circle by a pipeline system. The lines shown in the figure represent potential pipeline segments. The cost of building a segment is given by the formula

$$
f_k(x_k) = a_k(x_k)^b
$$

where a_k is the length of the segment, x_k is the flow of oil through the segment, and b is an economy of scale factor. Each well produces one unit of oil in the time period of interest. The goal is to find the pipeline design that minimizes total construction cost.

Collection network for wells

(a) Explain why the problem can be solved as a shortest path problem when $b = 1$, and find the corresponding solution.

(b) Explain why the problem can be solved as a minimum spanning tree problem when $b = 0$, and find the corresponding solution.

(c) Set up and solve the mathematical programming model for this problem when $b = 1/2$ for the radial lines and $b = 1$ for the chordal lines. Use three segments in the piecewise linear approximation for the radial lines with breakpoints at 2, 4, and 8.

22. (*Car pooling*) A total of m persons have filled out a questionnaire for the purpose of joining a car pool. Based on this information, analysts have determined a compatibility index between person i and person j, denoted by c_{ij}, for all pairs of persons. A small value for this index implies incompatibility. Let U and L represent the maximum and minimum number of people in a car pool, respectively. Also, let

$$x_{ik} = \begin{cases} 1 & \text{if person } i \text{ is assigned to car pool } k \\ 0 & \text{otherwise} \end{cases}$$

The objective is to construct n car pools such that the associated total compatibility index is maximized and each person is assigned to one of the vehicles.

(a) Using the aforementioned notation, write out the objective function that is to be maximized.

(b) Write all the constraints (using linear functions only).

(c) Write a linear constraint that would ensure that persons 1 and 2 are not assigned to the same vehicle.

(d) Write a linear constraint that ensures that persons 3 and 4 are assigned to the same vehicle.

(Define any additional notation used.)

23. (*Vehicle routing*) A trucking company has a fleet of m identical vehicles, each with a capacity of Q. The company must visit n customers each day starting from its depot. Customer i requires that load d_i be picked up, where d_i is measured in the same units as Q. The cost of traveling from customer i to customer j is c_{ij}. After a route has been completed, the truck must return to the depot with its load. Denote the depot by index 0. Each truck is assigned at most one route per day and starts with an empty cargo bay. Consider the case where all demand $d_i = 1$. Develop an IP model that can be used to find the minimum cost of visiting all the customers exactly once. (This problem is an extension of the traveling salesman problem. Here, any subset of n customers forming a route that does not contain the depot is infeasible.)

(a) Let x_{ij} be a binary decision variable equal to 1 if customer i immediately precedes customer j on any route, and equal to 0 otherwise. Write out your model, and explain the objective function and the meanings of all the constraints.

(b) How would the model change if the objective were to minimize the number of vehicles used to visit all the customers? Write the new objective function.

(c) If the integrality requirement on the decision variables x_{ij} is relaxed so that now we require only that $0 \le x_{ij} \le 1$ for all i and j, and the corresponding LP is solved, will the x_{ij}'s still turn out to be 0 or 1? Explain.

BIBLIOGRAPHY

Andersson, E., E. Housos, N. Kohl, and D. Wedelin, "Crew Pairing Optimization," in G. Yu (editor), *Operations Research in the Airline Industry*, pp. 228–258, Kluwer Academic, Boston, 1998.

Askin, R.G. and C.R. Standridge, *Modeling and Analysis of Manufacturing Systems*, John Wiley & Sons, New York, 1993.

Bartholdi, J.J., J.B. Orlin, and H.D. Ratliff, "Cyclic Scheduling via Integer Programs with Circular Ones," *Operations Research*, Vol. 28, No. 5, pp. 1074-1085, 1980.

Burns, R.N. and M.W. Carter, "Work Force Size and Single Shift Schedules with Variable Demands," *Management Science*, Vol. 31, No. 5, pp. 599-607, 1985.

Daskin, M.S., *Network and Discrete Location*, John Wiley & Sons, New York, 1995.

Desrosiers, J., Y. Dumas, M.M. Solomon, and F. Soumis, "Time Constrained Routing and Scheduling," in M.O. Ball, T.L. Magnanti, C.L. Monma, and G.L. Nemhauser (editors), *Handbook in Operations Research and Management Science*, Vol. 8: *Network Routing*, Elsevier Science Publishers, North-Holland, Amsterdam, pp. 35–139, 1995.

Feo, T.A. and J.F. Bard, "Flight Scheduling and Maintenance Base Planning," *Management Science*, Vol. 35, No. 12, pp. 1415–1432, 1989.

Feo, T.A. and M.G.C. Resende, "Greedy Randomized Adaptive Search Procedures," *Journal of Global Optimization*, Vol. 6, pp. 109–133, 1995.

Feo, T.A., J.F. Bard, and K. Venkatraman, "A GRASP for a Difficult Single Machine Scheduling Problem," *Computers & Operations Research*, Vol. 18, No. 8, pp. 635–643, 1991.

Fisher, M., "Vehicle Routing," in M.O. Ball, T.L. Magnanti, C.L. Monma, and G.L. Nemhauser (editors), *Handbook in Operations Research and Management Science*, Vol. 8: *Network Routing*, Elsevier Science Publishers, North-Holland, Amsterdam, pp. 1–33, 1995.

Garfinkel, R.S. and G.L. Nemhauser, *Integer Programming*, Wiley, New York, 1972.

Glover, F., "Tabu Search: A Tutorial," *Interfaces*, Vol. 20, No. 4, pp. 74–94, 1990.

Glover, F. and M. Laguna, *Tabu Search*, Kluwer Academic, Boston, 1997.

Glover, F., D. Klingman, and N. Philips, *Network Models and Their Applications in Practice*, Wiley, New York, 1992.

Goldberg, D.E., *Genetic Algorithms in Search, Optimization & Machine Learning*, Addison-Wesley, Reading, MA, 1989.

Jarrah, A.I.Z., J.F. Bard, and A.H. deSilva, "Solving Large-Scale Tour Scheduling Problems," *Management Science*, Vol. 40, No. 9, pp. 1124–1144, 1994.

Kirkpatrick, S., C.D. Gelatt, and M. P. Vecchi, "Optimization by Simulated Annealing," *Science*, Vol. 220, No. 4598, pp. 671–680, 1983.

Nanda, R. and J. Browne, *Introduction to Employee Scheduling*, Van Nostrand–Reinhold, New York, 1992.

Papadimitrou, C.H., *Computational Complexity*, Addison-Wesley, Reading, MA, 1994.

Salkin, H.M and K. Mathur, *Foundations of Integer Programming*, North-Holland, New York, 1989.

Veinott, A.F. and H.M. Wagner, "Optimal Capacity Scheduling—I and II," *Operations Research*, Vol. 10, pp. 518–546, 1962.

G. Yu (editor), *Industrial Applications of Combinatorial Optimization*, Kluwer Academic, Boston, 1998.

G. Yu (editor), *Operations Research in the Airline Industry*, Kluwer Academic, Boston, 1998.

Chapter 8

Integer Programming Methods

Integer programming (IP) problems can be characterized as being either very easy or very hard. This is true even when all the functions are linear and all the data are known. The easy problems include those that exhibit total unimodularity and hence can be solved directly as linear programs (LPs) without recourse to more advanced techniques. Most models that exhibit this property fit the definition of a pure network. Their special structure has led to the development of many uniquely tailored algorithms that have proven to be much faster than either the general simplex method or the most popular interior point approaches.

A second class of easy IP problems are those that can be solved with *greedy* algorithms. The minimal spanning tree problem, the equally weighted knapsack problem, and several single machine sequencing problems fall into this class. The computational effort required to solve each of them grows only linearly with problem size—as opposed to the general case, in which the computational effort grows exponentially. Several such problems are considered in Section 8.1.

The *hard* IP problems form a class of their own and are among the most difficult optimization problems to solve. For these IPs, the number of steps required to find optimal solutions in the worst case increases exponentially with problem size. To appreciate such growth, assume that n is the problem size and that a solution algorithm for a particular class of problems requires on the order of n^3 steps [written $O(n^3)$]. Doubling n thus increases the effort by approximately 2^3 or 8. If the solution algorithm takes $O(3^n)$ steps, however, doubling n increases the effort by a factor of 3^n, which can be enormous even for relatively small values of n. The implication is that, although we may be able to solve a problem of a given size, there is some problem perhaps a bit larger for which the $O(3^n)$ algorithm will not be practical. The majority of this chapter deals with these hard problems. Two general types of methods are described: enumerative approaches and cutting plane techniques. These are the most common and form the basis of more advanced methods.

8.1 GREEDY ALGORITHMS

A greedy algorithm for solving an optimization problem is one that iteratively constructs a solution such that at each step the current partial solution is augmented by an attempt to realize maximum immediate gain. Two additional characteristics of such algorithms are (1) that the selection at each step is not altered in subsequent steps and (2) that the number of steps is a polynomial function of problem size. Greedy algorithms are shortsighted in the sense that they do not look ahead. As a consequence, they usually incur high opportunity costs for making good decisions at the early stages, which in turn limit the choices at later stages to the best of a bad set of options. The nearest neighbor algorithm for constructing a solution to the traveling salesman problem is one such example. In this algorithm, the node nearest to the home base is selected as the first node visited on the tour. At the second iteration, the node nearest to the one just selected is placed second on the tour. This process continues until all nodes have been sequenced, with the last node on the tour being linked to the home base. What happens in practice is that the tour is often nearly optimal until the final steps. At that point, sequencing of the last few nodes causes the tour length to increase dramatically because the only choices available have long distances associated with them. The tour derived from this procedure may be very far from the optimal solution. On an incomplete directed graph, it may not even find a feasible solution.

In this section, we consider several IP models in which the application of a greedy algorithm does indeed lead to the optimal solution. For these cases, being shortsighted is the best policy.

Minimal Spanning Tree Problem

One of the best known greedy algorithms in operations research was designed to solve the minimal spanning tree problem on a graph with m nodes and n edges. Consider the undirected network shown in Figure 8.1. The graph consists of 11 nodes and 20 edges, where each edge has an associated length. The problem is to select a set of edges such that there is a path between each two nodes. The sum of the edge lengths is to be minimized.

When the edge lengths are all nonnegative, as assumed here, the optimal selection of edges forms a spanning tree. This characteristic gives rise to the name minimal spanning tree, or MST. The problem can be solved with a greedy algorithm attributable to R. Prim, as described below. A survey of greedy algorithms for this problem can be found in Cheriton and Tarjan [1976].

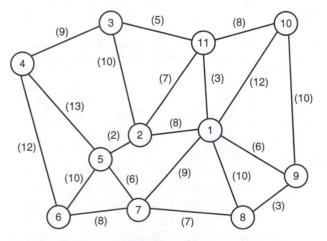

Figure 8.1   Example of a minimum spanning tree problem

Greedy Algorithm

Step 1: (*Initialization*) Let m be the number of nodes in the graph, and let S_1 and S_2 be two disjoint sets. Arbitrarily select any node i and place it in S_1. Place the remaining $m - 1$ nodes in S_2.

Step 2: (*Selection*) From all the edges with one end in S_1 and the other end in S_2, select the edge with the smallest length. Call this edge (i, j), where $i \in S_1$ and $j \in S_2$.

Step 3: (*Construction*) Add edge (i, j) to the spanning tree. Remove node j from set S_2 and place it in set S_1. If set $S_2 = \varnothing$, stop with the MST; otherwise, repeat Step 2.

The greedy choice is made at the *Selection* step by identifying the shortest edge among all remaining candidates and then adding it to the tree. It is a myopic (shortsighted) policy in that it selects a component of the solution without regard to which choices will be precluded at future steps. In this case, the greedy policy is optimal, because the resultant tree spans all nodes and has minimum length. Note that it would never be optimal to have a cycle in the solution. If a cycle were present, it would always be possible to remove one of its edges and still satisfy the connectivity requirement.

Applying the greedy algorithm to the graph in Figure 8.1, we obtain the sequence of iterations given in Table 8.1. The columns of the table show the set S_1 with the node numbers listed in the order in which they are added to the set, the edge selected at Step 2, the length of that edge, and the cumulative length of the tree. Node 2 was arbitrarily selected as the starting point. The length of the MST is the sum of the edge lengths, or 57 for our example. This number is obtained regardless of the starting node and regardless of the selection in the case of ties. Alternative optimal trees exist when there are ties at Step 2.

The algorithm just described is one of the simplest in optimization theory. A crude implementation has complexity $O(m^2)$. The optimal solution is found after executing the selection step $m - 1$ times, where at most $m - 1$ edges must be checked at each iteration, illustrating that discrete optimization problems are sometimes the easiest to solve. Nevertheless, if we add a simple constraint requiring, for example, that each node have no more than, say, four edges incident to it, the problem becomes as difficult as in any general IP. The MST problem on a directed network also can be solved in polynomial time, but the solution is not obtained with a greedy algorithm (Lawler, [1976]).

Table 8.1 Minimal Spanning Tree for Figure 8.1

S_1	Selected edge	Length	Total
{2}	(2, 5)	2	—
{2, 5}	(5, 7)	6	2
{2, 5, 7}	(7, 8)	7	8
{2, 5, 7, 8}	(8, 9)	3	15
{2, 5, 7, 8, 9}	(9, 1)	6	18
{2, 5, 7, 8, 9, 1}	(1, 11)	3	24
{2, 5, 7, 8, 9, 1, 11}	(11, 3)	5	27
{2, 5, 7, 8, 9, 1, 11, 3}	(7, 6)	8	32
{2, 5, 7, 8, 9, 1, 11, 3, 6}	(11, 10)	8	40
{2, 5, 7, 8, 9, 1, 11, 3, 6, 10}	(3, 4)	9	48
{2, 5, 7, 8, 9, 1, 11, 3, 6, 10, 4}	—	—	57

A Machine Sequencing Problem

Another problem for which a greedy algorithm yields an optimal solution is the machine sequencing problem with no constraints and a linear penalty cost objective. In this case, we have n jobs that we wish to sequence through a single machine. The time required to perform job i once it is on the machine is $p(i)$. At completion, a penalty cost is incurred equal to $c(i)T(i)$, where $c(i)$ is a positive quantity and $T(i)$ is the time at which job i is finished. Also, it is assumed that the time associated with setting up the machine for each job is negligible.

Table 8.2 lists the data for a 10-job example. The last column shows the ratio of the penalty cost to the processing time. This value is used in the solution algorithm.

A sequence of jobs is described by the vector $(J_1, J_2, \ldots, J_n)$, where $J_k = j$ indicates that job j is the kth job in the sequence. One possible sequence $(4, 1, 10, 5, 7, 2, 8, 3, 6, 9)$ is shown in Table 8.3 along with the completion time and penalty cost of each job. In this example, job 4 is done first, job 1 second, and so on. The job completion times and the total penalty cost are determined by the following equations.

Completion time through ith job: $\quad T(i) = \sum_{k=1}^{i} p(J_k) \quad \text{for } i = 1, \ldots, n$

Total penalty cost: $\quad z = \sum_{k=1}^{n} c(J_k)T(J_k)$

The problem is to determine the sequence that minimizes the total penalty cost z. In all, there are $n!$ feasible solutions. It happens that this seemingly complex problem can be solved with a greedy algorithm. Intuitively, one reasons that the jobs with the greatest per unit penalty costs should be scheduled earlier in the sequence because they will then be completed earlier. One would also like to schedule the job with the shortest processing time

Table 8.2   Linear Sequencing Example

Job, (i)	Processing time, $p(i)$	Penalty cost, $c(i)$	Penalty/time, $c(i)/p(i)$
1	5	13	2.60
2	3	10	3.33
3	10	2	0.20
4	6	12	2.00
5	12	12	1.00
6	2	3	1.50
7	15	5	0.33
8	8	20	2.50
9	4	8	2.00
10	9	6	0.67

Table 8.3   Results for Arbitrary Sequence

Sequence	4	1	10	5	7	2	8	3	6	9
Completion time	6	11	20	32	47	50	58	68	70	74
Penalty cost[a]	72	143	125	384	235	500	1160	136	210	592

[a] Total penalty cost = 3557.

first, since the completion times of all subsequent jobs will be affected by this time. The optimal policy is a combination of these two ideas. It first computes the $c(i)/p(i)$ ratios for all i and then sequences the jobs in decreasing order of this ratio. Ties may be broken arbitrarily. For the data in Table 8.2, we get

Optimal sequence: (2, 1, 8, 4, 9, 6, 5, 10, 7, 3)

The corresponding total penalty cost is 2252, which is the minimum. Because sorting of n items can be done in $O(n \log n)$ time and computing the ratios can be done while sorting, this is the complexity of the algorithm.

In general, sequencing problems are quite difficult to solve because all feasible job permutations may have to be examined before the optimal sequence is confirmed. The number of permutations becomes prohibitively large for n greater than 25 or 30. The problem discussed here, however, can be solved with a greedy algorithm, and so it is among the simplest of all optimization problems.

Heuristic Methods

Many heuristic procedures that yield good, but not necessarily optimal, solutions use greedy criteria when selecting the values of the decision variables. For example, consider the following one-constraint 0-1 IP problem in which the coefficients c_j and a_j are positive for all j.

$$
\left.
\begin{aligned}
\text{Maximize } z &= \sum_{j=1}^{n} c_j x_j \\[2mm]
\text{subject to} \quad & \sum_{j=1}^{n} a_j x_j \leq b \\[2mm]
& x_j = 0 \text{ or } 1, \ j = 1, \ldots, n
\end{aligned}
\right\} \tag{1}
$$

This is called the *knapsack problem*, because it describes the dilemma of a camper selecting items for a knapsack. The variable x_j represents the decision whether or not to include item j. The quantities c_j and a_j are the benefit and weight of item j. The maximum weight the camper can carry is b. The problem is to select the set of items that provides the maximum total benefit without exceeding the weight limitation.

A heuristic for finding solutions uses the benefit/weight ratio for each item. The item with the largest ratio appears to be the best from a greedy, myopic point of view, so it is chosen first. Subsequent items are considered in order of decreasing ratio. If the item will fit within the remaining weight limitation, it is included in the knapsack; if it will not, it is excluded. This greedy method does not always find the optimal solution, but it is very easy to execute and often yields a good solution. Methods, such as dynamic programming, that guarantee the optimal solution require substantially more computational effort.

8.2   SOLUTION BY ENUMERATION

If a problem does not fall into the easy class of IP problems, its solution becomes much more difficult to find. In the remainder of this chapter, we identify several techniques for solving *hard* problems. A common and easily understood way to begin when discrete variables are included in the formulation is to enumerate and evaluate all possible solutions. Say we have a mixed-integer linear program (MILP) with three 0-1 variables and all others real. One can set the integer variables to the eight possible combinations and then solve the resulting LPs. For a maximization problem, the combination that yields the largest value of the objective is the optimal solution.

This approach is called *exhaustive enumeration* and is practical for problems with only a few discrete variables. No new algorithms are required for the MILP, since linear programming is used to evaluate the alternatives. We now show that even if we take an enumerative approach, it is often possible to reduce the effort substantially without overlooking the optimal solution.

Example 1

Consider the following integer linear program (ILP).

$$\text{Maximize } z = -2x_1 + x_2$$
$$\text{subject to} \quad 9x_1 - 3x_2 \geq 11$$
$$x_1 + 2x_2 \leq 10$$
$$2x_1 - x_2 \leq 7$$
$$x_1, \quad x_2 \geq 0 \text{ and integer}$$

The feasible region of the corresponding LP is displayed in Figure 8.2. The parallel dashed lines represent isovalue contours of the objective function z, and the black dots represent feasible points. From this figure it is clear that there are 10 feasible solutions and that the optimal solution is $x_1 = 2$, $x_2 = 2$, with $z = -2$.

Even without examining Figure 8.2, it is possible to get an upper bound on the number of feasible points. From the nonnegativity conditions coupled with the second constraint, we see that $0 \leq x_1 \leq 10$ and $0 \leq x_2 \leq 5$. When $x_2 = 0$, the first constraint tells us that $x_1 \geq 2$; and when $x_2 = 5$, the third constraint implies that $x_1 \leq 6$. Thus, the constraints $2 \leq x_1 \leq 6$ and $0 \leq x_2 \leq 5$ along with the integrality requirements limit the number of feasible points to no more than 30. For such a small problem, these 30 points could be completely enumerated to find that 10 are feasible, 20 are infeasible, and $\mathbf{x} = (2, 2)$ is optimal.

Using a bit more imagination, we could have added the second and third constraints to arrive at $3x_1 + x_2 \leq 17$. When $x_2 = 0$, $3x_1 \leq 17$ and integrality imply that $x_1 \leq 5$. This reduces the upper bound on the number of feasible points from 30 to 24. Also, multiplying

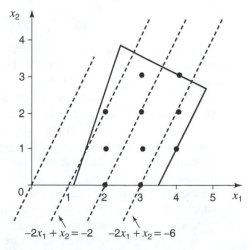

Figure 8.2   Feasible region and isovalue contours for Example 1

the first constraint by -1 and the second by 4 and adding yields $9x_2 \leq 35$ or $x_2 \leq 3$. This further reduces the number of feasible points to no more than 16.

Next we can pick a feasible solution such as $x_1 = 2$, $x_2 = 0$ and evaluate the objective function, yielding $z(2, 0) = -4$. Thus, every optimal solution to Example 1 must satisfy $-2x_1 + x_2 \geq -4$. Now, $x_2 \leq 3$ implies $2x_1 \leq 7$ or $x_1 \leq 3$. This means that the candidates for optimality have been reduced to the set

$$F = \{(x_1, x_2): x_1 = 2, 3; x_2 = 0, 1, 2, 3\}$$

Of these eight points, $(3, 1)$ and $(3, 0)$ yield objective values smaller than -4. The optimal value of the corresponding LP is $z_{\text{LP}} = -1\frac{4}{21}$ at $x_{\text{LP}} = \left(2\frac{10}{21}, 3\frac{16}{21}\right)$. Since the objective value at optimality for Example 1 must be integral, it follows that $-2x_1 + x_2 \leq -2$. This constraint is not satisfied at $(2, 3)$, so the candidates for optimality have been reduced to the set

$$F = \{(2, 0), (2, 1), (2, 2), (3, 2), (3, 3)\}$$

The intent of this simple analysis has been to show how enumeration techniques can be systematically applied to reduce the size of the feasible region, perhaps to the point where it is possible to identify the optimal solution by inspection. In the analysis, we made use of implied bounds on variables, rounding, linear combinations of constraints, and bounds on the objective function value. In subsequent sections, we will elaborate on these ideas when we describe several implicit enumeration methods. In the next section, we introduce notation and several useful procedures for enumerating all discrete values. As mentioned, this approach is practical only when a problem contains no more than a handful of discrete variables.

Exhaustive Enumeration

The pure 0-1 integer program is written, in general, as

$$\left. \begin{aligned} \text{Maximize } z &= \sum_{j=1}^{n} c_j x_j \\ \text{subject to} \quad & \sum_{j=1}^{n} a_{ij} x_j \leq b_i, \quad i = 1, \ldots, m \\ & x_j = 0 \text{ or } 1, \quad j = 1, \ldots, n \end{aligned} \right\} \tag{2}$$

With respect to chapter notation, when we are dealing with a pure IP, the decision variables will be x_j or variants thereof. For MILPs, the designation x_j is used for the continuous variables and the designation y_j for the integer variables.

We now demonstrate how to solve problem (2) by enumerating all possible solutions. Such a brute force approach can be carried out, at least in theory, because every instance of (2) has a finite number of solutions. Each could be generated and tested for feasibility, and if it is feasible, the corresponding objective function could be compared with those of the other candidates. The one with the largest z is the optimal solution. Unfortunately, the number of possible solutions is 2^n, where n is the number of variables. For $n = 20$, there are more than 1,000,000 candidates; for $n = 30$, the number is greater than 1,000,000,000, which is large enough to stymie even today's computers.

In any case, exhaustive enumeration requires a formal way of generating all possible solutions. This is done with a search tree or enumeration tree, as shown in Figure 8.3 for

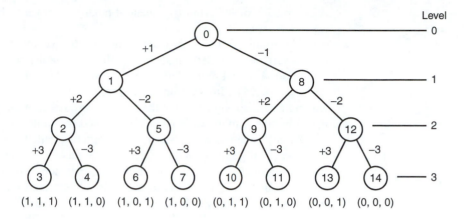

Figure 8. 3 Exhaustive search tree

three binary variables. As is customary in discussions such as this, we draw the tree with the *root* at the top and the *leaves* at the bottom. The circles are called *nodes*, and the lines are called *branches*. At the very top of the tree, we have node 0 or the root. As we descend the tree, decisions are made as indicated by the numbers on the branches. A negative number, $-j$, implies that the variable x_j has been set equal to 0, whereas a positive number, $+j$, implies that x_j has been set equal to 1.

The nodes are numbered sequentially as the variables are fixed to either 0 or 1. The sequence will vary depending on the enumeration scheme. Each node k inherits all the restrictions defined by the branches on the path joining it to the root. This path is given the designation P_k. For example, at node 1 the decision +1 is indicated by the branch joining node 0 to node 1. This means we have set variable x_1 equal to 1. At node 5, the decision −2 is indicated by the branch joining nodes 1 and 5, so we have the additional restriction $x_2 = 0$. The leaves at the bottom of the tree signal that all variables have been fixed. Each of these eight nodes in Figure 8.3 represents a complete solution that can be identified by tracing the path from the leaf node to the root and noting the decisions associated with the branches traversed along the way. Thus, node 6 represents the solution $\mathbf{x} = (1, 0, 1)$, whereas node 10 represents $\mathbf{x} = (0, 1, 1)$.

Each node of the tree resides at a particular level that indicates the number of decisions that have been made to reach that point. Node 0 is at level 0, for example, indicating that no decisions have been made; node 14 at level 3 indicates that three decisions have been made. A complete search tree will have $2^{n+1} - 1$ nodes.

For each node k, there is a path P_k leading to it from node 0, which corresponds to an assignment of binary values to a subset of the variables. Such an assignment is called a *partial solution*. We denote the index set of assigned variables by $W_k \subseteq N = \{1, \ldots, n\}$ and let

$$S_k^+ = \left\{ j : j \in W_k \text{ and } x_j = 1 \right\}$$

$$S_k^- = \left\{ j : j \in W_k \text{ and } x_j = 0 \right\}$$

$$S_k^0 = \left\{ j : j \notin W_k \right\}$$

A completion of W_k is an assignment of binary variables to the *free* variables specified by the index set S_k^0.

The solution set at a particular node includes vectors with all possible combinations of free variables. The set at node 0 with all free variables consists of all eight solutions, the set at node 1 has all four solutions with $x_1 = 1$, and the set at node 3 has only one solution with $\mathbf{x} = (1, 1, 1)$. In constructing a binary search tree, it is common for the left branch leaving a node to correspond to the variable, say x_j, being assigned the value 1 ($x_j = 1$), and the right branch to the variable being assigned the value 0 ($x_j = 0$).

At any node that is not a leaf, some variable, called the *separation variable*, is fixed at its two possible values at the next level. The set of solutions at the current node is divided into two mutually exclusive subsets by this operation. For example, the set of eight solutions at node 0 in Figure 8.3 is separated into two sets of four each at node 1 (at which all the solutions have $x_1 = 1$) and node 8 (at which all the solutions have $x_1 = 0$). Choosing a separation variable and moving to the next level is called *branching*, because this is how the branches of the tree are formed.

In the enumeration process, the most common option is to pursue a "depth-first" search strategy. That is, we first create a direct path from the root node to some leaf of the tree and then *backtrack* to explore other paths diverging from this first path. The node numbers assigned in Figure 8.3 indicate the order in which the nodes are enumerated under this strategy. Reviewing the figure, we note that nodes 0, 1, 2, and 3 are first created to reach the solution $\mathbf{x} = (1, 1, 1)$. At node 3, no separation variables remain (since $S_3^0 = \varnothing$), so we backtrack on the path to the highest level where there exists a right branch that has yet to be explored. This happens to be level 2, where we create node 4 at level 3. The solution $(1, 1, 0)$ is generated at node 4, and since no variables remain free (i.e., $S_4^0 = \varnothing$), we must again backtrack. Now, the alternative at level 2 has already been explored, so we backtrack to level 1 and branch from node 1 to create node 5 by fixing $x_2 = 0$. The process continues in this fashion, guided by the following rules.

- If a free variable remains at node k ($S_k^0 \neq \varnothing$), choose a separation variable ($j \in S_k^0$) and branch to the next level to create node $k + 1$.

- If no free variables remain ($S_k^0 = \varnothing$), evaluate the solution and backtrack to the highest level that contains a node whose right branch has not been explored; generate the node along this branch one level down.

Depth-First Search—Branching to the Left

To implement this process, we need a data structure that gives the status of the tree at any point. The vector P_k will be used for this purpose. For node k at level l of the tree, P_k is defined as follows.

- The length of the vector is l and is written $P_k = (j_1, j_2, \ldots, j_l)$.

- The absolute magnitude of j_i is the separation variable at level i.

- The sign of j_i indicates the value of the separation variable on the current path. A negative sign indicates that the variable is set equal to zero, and a positive sign (or no sign) indicates that it is set equal to 1.

- The component j_i can be underlined or not. If it is underlined, the alternative node at level i has already been explored. If it is not underlined, the alternative has yet to be explored.

- The variables not mentioned in P_k are the free variables.

Table 8.4 lists the P_k vectors for the nodes in Figure 8.3 in the order in which they were generated.

When branching to $x_s = 1$ from node k, we simply changed P_k to (P_k, s) to get P_{k+1}. In backtracking, we underline the rightmost nonunderlined entry in P_k and erase all entries to its right. The rightmost remaining entry is underlined and its sign is changed. (In a computer implementation, of course, it is not possible to "underline" indices, so one way to account for this condition is to introduce a companion vector U_k containing a +1 or −1 in each component depending on whether or not it is supposed to be underlined.) To illustrate how to work with the path vector, assume that we are at node 7 in Figure 8.3, where $P_7 = (1, -\underline{2}, -\underline{3})$. To backtrack, we erase the second and third components, underline the first, and then change the sign of the first. This puts us at node 8, where $P_8 = (-\underline{1})$. To branch to node 9 with the additional constraint $x_2 = 1$, we set $P_9 = (P_8, 2) = (-\underline{1}, 2)$. The enumeration is complete when all the entries in the path vector P have been underlined. When we use the left-first branching rule, all the entries will also be negative at termination.

Depth-First Search—Arbitrary Branching

For the left-first branching rule, the path vector P_k has more information than is really needed. Because an underlined component $\underline{s}$ means that we have already considered $x_s = 1$, and a nonunderlined component s means that we haven't considered $x_s = 0$, using + and − signs is redundant. A more general rule would allow us to create either branch first—i.e., $x_s = 1$ or $x_s = 0$. This is what the original definition of the path vector permits. To summarize, if $j \in W_k$, we let it appear in P_k as

$$\begin{cases} j & \text{if } j \in S_k^+ \text{ and } x_j = 0 \text{ has not been considered} \\ \underline{j} & \text{if } j \in S_k^+ \text{ and } x_j = 0 \text{ has been considered} \\ -j & \text{if } j \in S_k^- \text{ and } x_j = 1 \text{ has not been considered} \\ -\underline{j} & \text{if } j \in S_k^- \text{ and } x_j = 1 \text{ has been considered} \end{cases}$$

For example, if the order of nodes considered in Figure 8.4 had been 1, 3, 2, 4, the sequence of P_k vectors would have been (3), (3, −2), (3, $\underline{2}$), (−$\underline{3}$). Branching can be further generalized by choosing to fix more than one variable at a time. Instead of branching to a successor of node k, one can branch to any node on the path out of k. As long as backtracking

Table 8.4   Notation for Search Tree in Figure 8.3

Node, k	Level, l	P_k	Node, k	Level, l	P_k
0	0	∅			
1	1	(1)	8	1	(−$\underline{1}$)
2	2	(1, 2)	9	2	(−$\underline{1}$, 2)
3	3	(1, 2, 3)	10	3	(−$\underline{1}$, 2, 3)
4	3	(1, 2, −$\underline{3}$)	11	3	(−$\underline{1}$, 2, −$\underline{3}$)
5	2	(1, −$\underline{2}$)	12	2	(−$\underline{1}$, −$\underline{2}$)
6	3	(1, −$\underline{2}$, 3)	13	3	(−$\underline{1}$, −$\underline{2}$, 3)
7	3	(1, −$\underline{2}$, −$\underline{3}$)	14	3	(−$\underline{1}$, −$\underline{2}$, −$\underline{3}$)

is done as previously specified, there is no danger of omitting a portion of the search tree. For example, in the tree shown in Figure 8.5, one could branch directly from 1 to 6. The vector sequence would be (3), (3, −2, −4). This option is especially valuable for starting the enumeration. It allows us to begin at a node other than node 0. In particular, if a good feasible solution $\mathbf{x}^0$ is known at the outset, we can start at the corresponding node.

Algorithm for Exhaustive Enumeration

For explicit enumeration of all solutions, we select the separation variables in their natural order and use the left-first branching rule. The branch with the separation variable set to 1 is explored first. Of course, for exhaustive enumeration the order is immaterial since all solutions are to be generated. In practice, more flexibility is desired, so every free variable would be a candidate for separation.

The following notation is used in the description of the algorithm. Let z_B be the best objective function value in problem (2) found so far with corresponding solution $\mathbf{x}_B$. Let k be the node index, let l be the level index, and let j_l be the current variable index at level l.

Initialize: Create node 0 at level 0. Let $k = 0$, $l = 0$, and $j_0 = 0$.
          Set $z_B = -M$ (a large number in absolute terms).
          Let $P_0 = \varnothing$.

Branch: Branch by performing the following:
          $l \leftarrow l + 1, k \leftarrow k + 1$.
          Set $j_l = j_{l-1} + 1$.
          Create node k with decision j_l.
          Add j_l to current path vector to get P_k.
          Go to *Update*.

Update: If $l < n$, go to *Branch*.
          Otherwise, $l = n$ so a complete solution has been generated.
          Let $\mathbf{x}^k$ be the solution vector defined by P_k.
          If $\mathbf{x}^k$ is feasible, compute the objective value z^k.

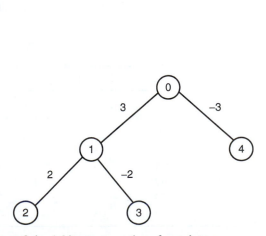

Figure 8.4 Arbitrary generation of search tree

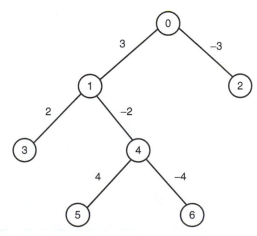

Figure 8.5 Arbitrary starting point

If $\mathbf{x}^k$ is feasible and $z^k > z_B$, put $\mathbf{x}_B \leftarrow \mathbf{x}^k$ and $z_B \leftarrow z^k$.

Whether or not $\mathbf{x}^k$ is feasible, go to *Backtrack*.

Backtrack: In the vector P_k, find the element farthest to the right (at the highest level) that is not underlined.

If all elements are underlined, stop: $\mathbf{x}_B$ is the optimal solution.

Otherwise, let this element be j_i. Backtrack by doing the following:

Let the level be $l = i$.

Delete all elements of P_k to the right of j_i.

Put $j_i \leftarrow -j_i$ (i.e., change the sign of j_i).

Now underline j_i to get P_{k+1}.

Put $k \leftarrow k + 1$ and create node k.

Go to *Update*.

To illustrate this procedure, consider the following knapsack problem.

$$\text{Maximize } z = 4x_1 + 9x_2 + 6x_3$$
$$\text{subject to} \quad 5x_1 + 8x_2 + 6x_3 \leq 12$$
$$x_j = 0 \text{ or } 1, \quad j = 1, 2, 3$$

The nodes of the search tree, along with the path vectors and the partial solutions determined in the course of the iterations, appear in Table 8.5. The notation j_k is used here to identify the separation variable at node k.

A tree identical to the one shown in Figure 8.3 is generated during the iterative process. Note that all eight possible solutions are enumerated, although three of them are infeasible. In this example, the optimal solution is found, at node 6, to be $\mathbf{x}_B = (1, 0, 1)$ with $z_B = 10$.

Table 8.5 Exhaustive Enumeration for Knapsack Example

Node, k	Level, l	P_k	z^k	$\mathbf{x}^k$	j_k	Action
0	0	$\varnothing$	—	—	1	Separate on x_1
1	1	(1)	—	—	2	Separate on x_2
2	2	(1, 2)	—	—	3	Separate on x_3
3	3	(1, 2, 3)	$-M$	(1, 1, 1)	—	Infeasible solution, backtrack
4	3	(1, 2, $-\underline{3}$)	$-M$	(1, 1, 0)	—	Infeasible solution, backtrack
5	2	(1, $-\underline{2}$)	—	—	3	Separate on x_3
6	3	(1, $-\underline{2}$, 3)	10	(1, 0, 1)	—	Feasible solution, update z_B, backtrack
7	3	(1, $-\underline{2}$, $-\underline{3}$)	4	(1, 0, 0)	—	Feasible solution, backtrack
8	1	($-\underline{1}$)	—	—	2	Separate on x_2
9	2	($-\underline{1}$, 2)	—	—	3	Separate on x_3
10	3	($-\underline{1}$, 2, 3)	$-M$	(0, 1, 1)	—	Infeasible solution, backtrack
11	3	($-\underline{1}$, 2, $-\underline{3}$)	9	(0, 1, 0)	—	Feasible solution, backtrack
12	2	($-\underline{1}$, $-\underline{2}$)	—	—	3	Separate on x_3
13	3	($-\underline{1}$, $-\underline{2}$, 3)	6	(0, 0, 1)	—	Feasible solution, backtrack
14	3	($-\underline{1}$, $-\underline{2}$, $-\underline{3}$)	0	(0, 0, 0)	—	Feasible solution, stop

8.3 BRANCH AND BOUND

It should be clear by now that enumerating all potential solutions of an integer program is not, in general, a practical analytic approach. This idea, however, is at the center of what is known as *branch and bound* (B&B), a group of procedures that attempts to perform the enumeration intelligently so that not all combinations of variables need be examined. Depending on the implementation, the terms *implicit enumeration*, *tree search*, and *strategic partitioning* are sometimes used. Regardless of the name, B&B has two appealing qualities. First, it can be applied to the mixed-integer problem and to the pure integer problem in essentially the same way, so a single method works for both problems. Second, it typically yields a succession of feasible integer solutions, so if the computations have to be terminated as a result of time restrictions, the current best solution can be accepted as an approximate solution. We will see that the traditional cutting plane approaches discussed in Section 8.4 do not obtain feasibility until the problem is solved.

Branch and bound methods are often tailored to exploit special problem structures, thereby allowing these structures to be handled with greater efficiency and reduced computer memory (a critical feature when solving large-scale problems). In fact, except in very general terms, there is no one method but rather a whole collection of methods that share several common characteristics. The tailoring aspect further permits B&B to be applied directly to many kinds of combinatorial problems without first going through an intermediate stage of introducing specific integer variables and linear constraints to yield an IP formulation. The direct application of methods to problems is invaluable when the use of an intervening IP formulation may drastically increase the model size or otherwise obscure exploitable problem features.

For both pure LPs and MILPs, many of the most effective B&B procedures are based on the use of the simplex method. These procedures will be examined first. For the moment, let us assume that we are dealing with a 0-1 IP, although the ideas are the same for the more general case. For different models, the primary changes are in bookkeeping methods and notation.

Simplex-based methods begin by solving the original problem as an LP. The appropriate LP is obtained by replacing the 0-1 constraint $x_j \in \{0, 1\}$, $j = 1, \ldots, n$, with the relaxed constraint $0 \leq x_j \leq 1$, for all j. If an integer solution is not realized, one of the integer variables that is fractional in the LP solution, denoted by, say, x_s, where s is the variable index, is selected and two descendants of the original problem are created—one in which $x_s = 0$ and the other in which $x_s = 1$. As we have seen in the preceding section, this operation is called *branching*. Because x_s is a 0-1 variable, and the descendant problems are exactly the same as the original problem except for the assignment of a specific value to x_s, the solution to one of the two descendants must in fact be the solution to the original problem. Thus, the latter has been replaced with two IPs that now must be solved. The expectation is that these modified IPs will be easier to solve than their parent because each is more restricted than the parent, having fewer variables that must be assigned integer values.

The process is then repeated, selecting one of the problems that remains to be solved as the current IP and treating it exactly the same as the original. This in turn may create two new problems to replace the one currently under consideration unless, for example, the current problem is infeasible. Repeated iterations eventually produce an integer solution (if one exists) for one of the current IPs that becomes a candidate for the optimal solution to the original problem. Keeping track of the best of these candidates and its attendant objective function value provides an additional way to weed out descendants of the original IP (perhaps several generations removed) that would be unprofitable to explore because they could not possibly yield the overall optimal solution. The current best candidate is called the *incumbent*, and the weeding out process is called *fathoming*. When a depth-first strategy is used, the branching operations are represented in a rooted binary search tree in which each node has one predecessor and at most two successors. A node that is not fathomed and is not a leaf of the tree is said to be *live*. Branching means choosing a live node to consider next for fathoming or separation. Note that other search strategies are possible, such as breadth-first, in which more than one separation variable is chosen at a node for branching, but they will not be pursued here.

General Ideas

To formalize B&B concepts, let z_B again denote the objective function value of the incumbent and let z^k represent the objective function value of the corresponding LP relaxation at some node k. Then, whenever an IP maximization problem is solved as an LP, one of the following four alternatives arises.

1. The LP has no feasible solution (in which case the current IP also has no feasible solution).

2. The LP has an optimal solution $z^k \leq z_B$ (in which case the current IP optimal solution $z_{IP}^k \leq z_B$ and so cannot provide an improvement over the incumbent).

3. The optimal solution to the LP is integer valued and feasible, and yields $z^k > z_B$ (in which case the solution is optimal for the current IP and provides an improved incumbent for the original IP, and thus z_B is reset to z^k).

4. None of the foregoing occurs—i.e., the optimal LP solution satisfies $z^k > z_B$, but is not integer valued.

In each of the first three cases, the IP at node k is disposed of simply by solving the LP. That is, the IP is fathomed. A problem that is fathomed as a result of case 3 yields particularly useful information because it allows us to update the incumbent. If the problem is not fathomed and hence winds up in case 4, further exploration or branching is required.

Once again we note that the relaxed problem associated with each node does not have to be an LP. A second choice could be an IP that is easier to solve than the original. Typical relaxations of the traveling salesman problem, for instance, are the assignment problem and the MST problem.

B&B Subroutines

We now elaborate on the aforementioned ideas and present the basic steps that are needed for solving a 0-1 integer program using B&B. Although most of the steps are general in that they are appropriate for a variety of problem classes, several computational procedures are problem dependent. We begin with a description of the operations that each procedure is designed to perform and discuss implementation issues as they arise. Although a maximization objective is assumed, if the goal is to minimize, the problem can be solved with the same algorithm after making a few modifications, or directly by converting it to a maximization problem. The five routines below are used to guide the search for the optimal solution and to extract information that can be used to reduce the size of the B&B tree.

Bound: This procedure examines the relaxed problem at a particular node and tries to establish a bound on the optimal solution. It has two possible outcomes:

1. An indication that there is no feasible solution in the set of integer solutions represented by the node

2. A value z_{UB}—an upper bound on the objective function for all solutions at the node and its descendent nodes

Approximate: This procedure attempts to find a *feasible* integer solution from the solution of the relaxed problem. If one is found, it will have an objective value, call it z_{LB}, that is a lower bound on the optimal solution for a maximization problem.

Variable Fixing: This procedure performs logical tests on the solution found at a node. The goal is to determine if any of the free binary variables are necessarily 0 or 1 in an optimal integer solution at the current node or at its descendents, or whether they must be set to 0 or 1 to ensure feasibility as the computations progress.

Branch: A procedure aimed at selecting one of the free variables for separation. Also decided is the first direction (0 or 1) to explore.

Backtrack: This is primarily a bookkeeping procedure that determines which node to explore next when the current node is fathomed. It is designed to enumerate systematically all remaining live nodes of the B&B tree while ensuring that the optimal solution to the original IP is not overlooked.

The implementation of these routines in a complete algorithm requires many details, most of which are likely to be problem specific, to be filled in. Before providing a step-by-step description of a B&B algorithm for solving a 0-1 IP, we illustrate some of the concepts using the knapsack problem introduced in the preceding section. It is repeated below for easy reference.

$$\text{Maximize } z = 4x_1 + 9x_2 + 6x_3$$
$$\text{subject to} \quad 5x_1 + 8x_2 + 6x_3 \le 12$$
$$x_j = 0 \text{ or } 1, \ j = 1, 2, 3$$

When solved with our branch and bound algorithm, the search tree depicted in Figure 8.6 results. Further explanation will be given presently, but for now we note that the variables are considered in what appears to be an arbitrary order. In fact, the order is determined by a specific set of rules designed to provide tight upper bounds and to increase the likelihood of fathoming. This will speed convergence.

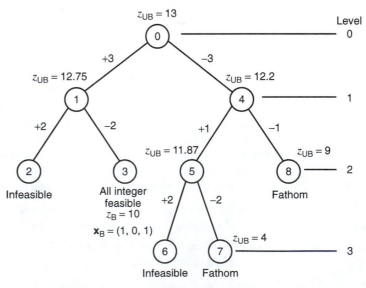

Figure 8.6   Search tree for the knapsack problem

Fathoming by Bounds

Assume that the enumeration process has arrived at some node in the search tree and that, in some manner, a feasible solution $\mathbf{x}_B$ with objective value z_B has been obtained. As we have mentioned, this solution is called the incumbent. Because $\mathbf{x}_B$ is feasible for the original IP, z_B is a lower bound on the value of the optimal solution z_{IP}—that is, $z_B \leq z_{IP}$.

The current node represents some set of feasible solutions that could be enumerated and identified by setting the free variables equal to every possible combination. By solving a relaxed problem (or solving a problem by some other method), we obtain an upper bound z_{UB} on the objective values for all solutions in this set (but not on the solution to the original IP unless we are at node 0). If it happens that

$$z_{UB} \leq z_B \tag{3}$$

then all the solutions in the set can be judged nonoptimal or at least no better than the solution currently in hand. All solutions in the tree that are descendants of the node under consideration can be fathomed and the procedure can backtrack to another part of the tree. On the other hand, if

$$z_{UB} > z_B$$

the node cannot be fathomed, and so we must continue to branch.

Returning now to the knapsack problem, an upper bound can be obtained by relaxing the integrality restrictions on the variables. The resulting problem is

$$\text{Maximize } z = 4x_1 + 9x_2 + 6x_3$$
$$\text{subject to} \quad 5x_1 + 8x_2 + 6x_3 \leq 12$$
$$0 \leq x_j \leq 1, \quad j = 1, 2, 3$$

which is an LP with a single constraint. Computing the benefit/cost ratio for each variable and then using a greedy algorithm will easily solve this problem. Here we are interpreting the problem as having the goal of maximizing the benefit subject to a budget constraint. The benefit/cost ratio for variable j is its objective coefficient c_j divided by its constraint coefficient a_j. For our example, the data are presented in Table 8.6.

The solution is obtained by considering the variables in decreasing order of the benefit/cost ratio, which is (2, 3, 1). Each variable is set equal to 1 until the constraint is violated. The variable for which this occurs is then reduced to the fractional value that will just use up the budget. For the example problem, $x_2 = 1$, $x_3 = 0.667$, $x_1 = 0$, with $z_{UB} = 13$.

The optimal objective value for the relaxed problem is the desired upper bound. No solution to the integer problem can have an objective greater than z_{UB}. Since this is a pure integer problem and the objective coefficients are integer valued, we can round this down to the nearest integer. The rounded value is indicated as follows.

$$z_{UB} \leftarrow \lfloor z_{UB} \rfloor \tag{4}$$

For this example, the solution value 13 is already an integer, so we do not round down.

The principle that we have illustrated is that an upper bound on a maximization problem can be obtained by relaxing one or more constraints. In so doing, we obtain a feasible

Table 8.6   Ratio Data for Knapsack Example

Variable, j	1	2	3
Benefit/cost, c_j/a_j	0.8	1.25	1.0

region that is larger than that of the original problem. At any node in the search tree, then, the optimal objective value of the relaxed problem is always at least as great as the optimal objective value of the IP at that node. For the knapsack example, the feasible region is the set of all 0–1 integer points satisfying the single constraint. When the integrality restriction is relaxed, all fractional solutions are added, thus expanding the feasible region.

The value $z_{UB} = 13$ previously obtained is an upper bound on all solutions of the knapsack problem, because it was found at node 0 of the search tree. At some other node k, the value of z_{UB}^k is an upper bound on all solutions represented by that node. As the iterations progress, some variables will be fixed at 0 or 1, so the relaxed problem will change. We use the same notation introduced in the preceding section to represent the search. For example, at node 4 in Figure 8.6, we have $S_4^+ = \varnothing$, $S_4^- = \{3\}$, $S_4^0 = \{1, 2\}$, and $z_B = 10$. The relaxation is

$$z_{UB}^4 = \text{Maximize } 4x_1 + 9x_2$$
$$\text{subject to } 5x_1 + 8x_2 \leq 12$$
$$0 \leq x_j \leq 1, \ j = 1, 2$$

which has the solution $x_1 = 0.8$, $x_2 = 1$, with $z_{UB}^4 = 12.2$. If we use Equation (4), we can lower the upper bound to the next smaller integer: $z_{UB}^4 \leftarrow \lfloor z_{UB}^4 \rfloor = \lfloor 12.2 \rfloor = 12$. Since $z_{UB}^4 > z_B$, the node cannot be fathomed.

Now, examining node 8, we have $S_8^+ = \varnothing$, $S_8^- = \{1, 3\}$, and $S_8^0 = \{2\}$. The corresponding relaxation is

$$z_{UB}^8 = \text{Maximize } 9x_2$$
$$\text{subject to } 8x_2 \leq 12$$
$$0 \leq x_2 \leq 1$$

with solution $x_2 = 1$ and $z_{UB}^8 = 9$. Because this upper bound is less than the current best solution $z_B = 10$, we can fathom the node. No solution represented by this node [i.e., (0, 0, 1) and (0, 0, 0)] can have an objective value greater than 10, so there is no need to examine them explicitly.

Branch and bound can be applied to a wide variety of discrete optimization problems, but, for the general case, the simple idea of using the benefit/cost ratio with a greedy algorithm to solve the relaxed problem will not work. Part of the art of IP is the ability to discover a good relaxation for the problem at hand. LP is usually considered first, but there are often several alternatives. Virtually every optimization method considered in other parts of this book has been used for a relaxation of a discrete programming problem. In addition, techniques based on surrogate constraints and Lagrangian duality have proven quite successful. These are discussed in several of the references.

With respect to elimination by bounds, the efficiency of implicit enumeration depends on several criteria, including the following.

1. The quality of the upper bound (for a maximization problem, the smaller the bound the better)
2. The computational difficulty in obtaining the upper bound (the easier the better)
3. The quality of the lower bound (the larger the better)
4. The computational difficulty in obtaining the lower bound (the easier the better)
5. The sensitivity of the objective function value to the decision variables in **x** (the more sensitive the better)

The first two criteria are usually in conflict. The quality of the upper bound is most likely to be directly related to the computational difficulty in obtaining it. For many problems, a poor upper bound can be found very easily, whereas a good upper bound can be found only with considerable effort. Similarly, the third and fourth criteria are also contradictory. Lower bounds are often provided by heuristics that construct feasible solutions from fractional solutions obtained during the enumerative procedure. Rounding a fractional solution may lead to a feasible point, but it is not likely to be close to the optimal solution. The analyst must trade off the strength of a bound and the computational effort required to obtain it.

The sensitivity of the objective function value to changes in the solution vector is a characteristic of the problem being solved. The bounding test will be more effective for problems for which the optimal value of z_{IP} is significantly greater than values for other decisions. If there are many near-optimal solutions, the bounding test will not be as effective. By "near-optimal solutions" we mean feasible solutions whose objective values are within a percent or two of z_{IP}.

The size of the search tree is directly related to the effectiveness of the fathoming tests. With 20 binary variables, the full tree has more than 2×10^6 nodes. If the lower bound test is 90% effective, 2×10^5 nodes still need to be evaluated. This illustrates why the computational effort mentioned in criteria 2 and 4 is so important. Since any procedure used to compute bounds will be executed at each node of the tree, efficiency is paramount.

Implicit enumeration procedures, often coupled with advanced decomposition or cutting plane techniques, have been used to solve many real-world problems. Nevertheless, when more than a few hundred integer variables are present in a model, algorithmic convergence cannot be ensured. To speed convergence, a compromise is needed between the quality of the solution and the finiteness of the algorithm.

One way to achieve this is to strengthen the elimination test by using a parameter ρ that specifies an allowed percentage deviation from the optimal solution. To implement this idea, we replace the bounding test given by Condition (3) with

$$z_{UB} \leq z_B + \rho \left(\frac{z_B}{100} \right)$$

Then, if at some node k, z_{UB}^k satisfies this inequality, the node is fathomed. Although this test may preclude the optimal solution from being found, we are assured that the final solution is always within ρ% of the optimal. In many situations, this may be a useful compromise for obtaining a good, but not necessarily optimal, solution within an acceptable amount of time. Many commercial codes set the default value of ρ to some value in the neighborhood of 0.5%.

Fathoming by Infeasibility

Because a relaxed problem has a larger feasible region than that of the original problem, when the relaxation has no feasible solution, the original integer program must also be infeasible. This is the second test that is used to fathom nodes.

Determining a Feasible Solution

The importance of Equation (3), in which z_{UB} is compared with z_B in an attempt to fathom the current node, cannot be overstated. The effectiveness of this test increases as the quality of $\mathbf{x}_B$ increases or, in other words, as z_B gets closer to the optimal IP solution z_{IP}. Accordingly, a valuable component of any B&B scheme is a heuristic procedure for obtaining good feasible solutions. This is often done using the relaxed solution as a starting point but may involve more intelligent techniques such as tabu search, simulated annealing, and GRASP. We call such a procedure *Approximate*.

Actually, implicit enumeration works without an *Approximate* procedure, since feasible solutions are automatically determined at the leaves of the tree. Virtually all problems, however, can be solved more quickly if good feasible solutions are available early on. In most implementations, an approximation procedure is applied at least at the root node to provide a value of z_B for fathoming tests before a leaf node is reached. For problems with more than a handful of variables, however, experience suggests that if nodes are not fathomed before the leaves are reached, it is unlikely that convergence will be achieved in a reasonable amount of time.

Returning now to the knapsack example, recall that the *Bound* procedure yields a solution with some variables set to 1, some variables set to 0, and at most one fractional variable. The simplest way to obtain a feasible solution is to truncate the fractional variable to 0. This solution is obviously feasible, so its objective provides an acceptable lower bound on the optimal solution. The solutions obtained by the *Bound* and *Approximate* procedures for three of the nodes in Figure 8.6 are given in Table 8.7. Here, $\mathbf{x}_{UB}$ represents the relaxed solution with objective z_{UB}, and $\mathbf{x}_F$ is the rounded solution with objective value z_F.

The design of *Approximate* procedures is part of the art of IP and must balance the quality of the bound against the difficulty of finding it. The better the solution, and hence the larger the lower bound, the more difficult it is to obtain.

When successful, an *Approximate* procedure returns two results: a feasible solution $\mathbf{x}_F$ and the corresponding objective value z_F. This information can be used to update the incumbent and to provide an improved bound for the fathoming. If $z_F > z_B$, we can update the best feasible solution as follows.

$$z_B \leftarrow z_F \text{ and } \mathbf{x}_B \leftarrow \mathbf{x}_F$$

We now have a stronger bound for Condition (3), so any live node k such that $z_{UB}^k \leq z_F$ can be fathomed.

Fixing Variables

Some algorithms incorporate a variety of logic tests to determine if any of the free variables must be set to 1 or 0 in order to ensure a feasible solution at a particular node. For example, in the knapsack problem under consideration, when x_3 is set to 1, it is logically clear that x_1 must be set to 0 to ensure feasibility. This step is optional and is not included in all algorithms. It plays a particularly important role, however, in the additive algorithm for pure 0-1 problems that we describe in a supplement on the CD.

Table 8.7 Some Upper and Lower Bounds for the Knapsack Example

Node, k	S_k^0	S_k^+	S_k^-	$\mathbf{x}_{UB}$	z_{UB}	$\mathbf{x}_F$	z_F
0	{1, 2, 3}	∅	∅	(0, 1, 0.667)	$\lfloor 13 \rfloor = 13$	(0, 1, 0)	9
1	{1, 2}	{3}	∅	(0, 0.75, 1)	$\lfloor 12.75 \rfloor = 12$	(0, 0, 1)	6
4	{1, 3}	∅	{3}	(0.8, 1, 0)	$\lfloor 12.2 \rfloor = 12$	(0, 1, 0)	9

When the relaxed problem is an LP, information from the solution can be used to fix variables. Given an optimal LP solution, the reduced costs $\bar{c}_j$ are nonnegative for all non-basic variables x_j at their lower bounds (typically zero), and nonpositive for all nonbasic variables at their upper bounds. This leads to the following result that is valid for *any* MILP.

Proposition 1: Let $\mathbf{x} \in \Re^n$ be the decision variables in an ILP such that $x_j \geq 0$ for all j, and let z_{UB} and z_{B} be the objective values of the LP relaxation and incumbent, respectively. If x_j is nonbasic at its lower bound (zero) in the solution to the LP relaxation, and $z_{\text{UB}} - \bar{c}_j \leq z_{\text{B}}$, there exists an optimal solution to the integer program with x_j at its lower bound. Similarly, if x_j is nonbasic at its upper bound in the solution to the LP relaxation, and $z_{\text{UB}} + \bar{c}_j \leq z_{\text{B}}$, there exists an optimal solution to the integer program with x_j at its upper bound.

Just about all commercial B&B codes use LP to solve the relaxed problem, and most include the tests specified by Proposition 1 as a subroutine. The more advanced codes also include several logical and implication tests at each node of the search tree (e.g., see Savelsbergh [1994] and Suhl and Szymanski [1994]).

Finding a Separation Variable

Whenever a node is not eliminated by a fathoming test, a separation variable must be chosen for branching to the next level. The direction of exploration (either 0 or 1) must also be specified. The particular choice governs the sequence of solutions obtained by the enumeration process and may play an important role in the efficiency of the algorithm. The only requirement for the separation variable is that it be a member of S_k^0, the set of free variables at node k.

There are several possible strategies for selecting the separation variables, including the two that follow.

1. Select the variable and direction such that the node created will be most likely to contain the optimal solution.

2. Select the variable and direction that will most reduce the infeasibility of the relaxed solution.

The motivation for the first rule is that once we have found an optimal solution, even if we are unable to prove it immediately, we will have the largest possible value of z_{B}. This is extremely important for subsequent fathoming. When nodes are fathomed at the lower levels, large numbers of solutions are eliminated. A node fathomed at level l eliminates 2^{n-l} solutions. Moreover, if the computations are halted before convergence, the incumbent is optimal even if it has not been verified. For the knapsack problem, this rule would imply that the variables are selected in the ranked order of their benefit/cost ratios. Thus x_2 would be chosen first and set equal to 1.

The second rule often produces a feasible solution more rapidly than the first. The relaxation of the knapsack problem at node 0, for example, results in the solution $x_1 = 0.25$, $x_2 = 1$, $x_3 = 1$. The infeasibility in this solution is caused by the nonintegrality of x_1. The rule would choose x_1 as the separation variable at node 0. Exploring the 0 direction first, one would obtain the optimal solution as the relaxed solution at node 1. The search would then backtrack to set x_1 equal to 1.

There are a variety of ways to select the separation variable and direction of exploration. Because there is no theoretically correct approach, the best strategy depends on the problem under investigation and can be justified only on the basis of empirical computational comparisons. Most commercial codes allow the user to specify the branching rules.

Generic B&B Algorithm for 0-1 Integer Programs

We now extend the exhaustive enumeration algorithm presented in Section 8.2 to incorporate fathoming, variable fixing, and the use of heuristics to find feasible solutions. In the algorithm, k is a counter used to identify nodes, l is the level in the tree, and n is the number of integer variables. The l-dimensional vector P_k identifies the path from the root to node k, and S_k^+, S_k^-, and S_k^0 indicate the current status of the variables. Vectors $\mathbf{x}^k$ and $\mathbf{x}_B$ refer to a feasible solution at node k and the incumbent, respectively.

Initialize: Create node 0 at level 0. Let $k = 0$, $l = 0$, and $P_0 = \varnothing$.

Perform the *Approximate* procedure in an attempt to find a feasible solution.

 If a feasible solution is found,

  let $\mathbf{x}_B$ be the vector of integer variables with objective value z_B.

 Otherwise, let $z_B = -M$ (a large number in absolute terms).

Update: If $l = n$, an integer solution has been generated.

 Let $\mathbf{x}^k$ be the solution vector defined by P_k.

 If $\mathbf{x}^k$ is feasible,

  compare the objective value z^k with z_B.

  If $z^k > z_B$, put $\mathbf{x}_B \leftarrow \mathbf{x}^k$ and $z_B \leftarrow z^k$.

 Whether or not $\mathbf{x}^k$ is feasible, go to *Backtrack*.

If $l < n$, continue with *Variable Fixing*.

Variable Fixing: Use logical and other tests to determine if a free variable x_j should be set to 1 or 0, where $j \in S_k^0$.

 Say variable x_s should be set to 1. Perform the following:

  $l \leftarrow l + 1$, $k \leftarrow k + 1$.

  Add $+s$ to the current path vector to get P_k.

  Create node k with decision $+s$.

 Alternatively, say variable x_s should be set to 0. Perform the following:

  $l \leftarrow l + 1$, $k \leftarrow k + 1$.

  Add $-s$ to the current path vector to get P_k.

  Create node k with decision $-s$.

If any changes are made, repeat this step;

otherwise, continue with *Bound*.

Bound: Solve the relaxed problem at node k.

If the result shows no feasible solution, fathom the node and go to *Backtrack*.

If the procedure returns z_{UB}^k, put $z_{UB}^k \leftarrow \lfloor z_{UB}^k \rfloor$ and compare this value with the incumbent.

 If $z_{UB}^k \leq z_B$, fathom the node and go to *Backtrack*;

 otherwise, continue with *Approximate*.

Approximate: Attempt to find a feasible solution in the set of solutions for node *k*. If a feasible solution is found, call it $\mathbf{x}^k$ and let the objective value be z^k.

If $z^k > z_B$, put $\mathbf{x}_B \leftarrow \mathbf{x}^k$ and $z_B \leftarrow z^k$.
If $z_{UB}^k = z_B$, fathom the node and go to *Backtrack*;
otherwise, continue with *Branch*.

Branch: Choose a separation variable x_s, such that $s \in S_k^0$, and a direction of exploration.

Put $k \leftarrow k + 1$, $l \leftarrow l + 1$, and create node *k* at level *l* of the tree.
If x_s is to be set to 1, add $+s$ to the current path vector to get P_k.
If x_s is to be set to 0, add $-s$ to the current path vector to get P_k.
Go to *Update*.

Backtrack: In the vector P_k, find the element farthest to the right that is not underlined.

If all elements are underlined, stop—$\mathbf{x}_B$ is the optimum.
Otherwise, let this element be j_i. Backtrack by doing the following:

Let the level be $l = i$.
Delete all elements of P_k to the right of j_i.
Put $j_i \leftarrow -j_i$ (i.e., change the sign of j_i).
Now underline j_i to get P_{k+1}.
Put $k \leftarrow k + 1$ and create node *k*.
Go to *Update*.

Table 8.8 shows the results when the algorithm is applied to the knapsack example. No *Variable Fixing* or *Approximate* procedures were used. During the *Bound* step, the LP relaxation was solved using the simple ranking scheme previously described. Bounds were improved using Equation (4). At each node, the free variable, say x_s, with a fractional value was chosen for separation. If its value was greater than or equal to 0.5, the branch $x_s = 1$

Table 8.8   B&B Results for the Knapsack Example

Node, k	Level, l	P_k	z_{UB}	z_B	$\mathbf{x}_B$	s	Action
0	0	$\varnothing$	13	$-M$	—	3	Set $x_3 = 1$
1	1	(+3)	12	$-M$	—	2	Set $x_2 = 1$
2	2	(+3, +2)	Infeasible	$-M$	—	—	Backtrack
3	2	(+3, $-\underline{2}$)	10	10	(1, 0, 1)	—	Feasible, backtrack
4	1	($-\underline{3}$)	12	10	(1, 0, 1)	1	Set $x_1 = 1$
5	2	($-\underline{3}$, +1)	11	10	(1, 0, 1)	2	Set $x_2 = 1$
6	3	($-\underline{3}$, +1, +2)	Infeasible	10	(1, 0, 1)	—	Backtrack
7	3	($-\underline{3}$, +1, $-\underline{2}$)	4	10	(1, 0, 1)	—	Fathom, backtrack
8	2	($-\underline{3}$, $-\underline{1}$)	9	10	(1, 0, 1)	—	Fathom, backtrack, stop

was selected first; otherwise, the branch $x_s = 0$ was selected first. Because the linear program had only one constraint, there was at most one fractional variable in each solution.

The search tree corresponding to data in Table 8.8 is depicted in Figure 8.6. As we can see, the optimal solution is uncovered at node 3 and is the first feasible point found. Five more nodes had to be examined before convergence occurred. This type of performance is typical of B&B algorithms. The optimal (or a near-optimal solution) is often found in the early enumerative stages, especially when good feasibility heuristics are used, but are not confirmed until later in the search.

This was a very simple implementation of B&B. Other improvements are possible, and many more details have to be considered than we can provide here. Good design calls for a preprocessing step to eliminate redundant constraints and to fix variables, a modular structure to accommodate various relaxation and approximation schemes, and problem-dependent branching and backtracking rules.

Branch and Bound with General Integer Variables

When a depth-first search strategy is used, it is possible to generalize the B&B approach taken for 0-1 IPs to include variables that can take on any nonnegative integer value, rather than just 0 and 1.

Example 2

To illustrate, consider the six-variable problem shown in detached coefficient form in Figure 8.7. The constraint coefficients are randomly generated integer values. The objective coefficient for a variable is the sum of the constraint coefficients. This results in a random problem that is not very easy to solve because each variable has a similar effect on the constraints and the objective function.

	x_1	x_2	x_3	x_4	x_5	x_6		
Max z	7	18	23	21	16	12		
C1	1	5	9	8	2	6	≤	17
C2	1	8	10	2	10	2	≤	18
C3	2	2	2	7	3	3	≤	10
C4	3	3	2	4	1	1	≤	7

$$x_j \geq 0 \text{ and integer, } j = 1, \ldots, 6$$

Figure 8.7   Coefficients for Example 2

The decision tree constructed by a branch and bound procedure is shown in Figure 8.8 A single line under a node indicates fathoming by bounds, $z_{UB} \le z_B$. Two lines under a node indicate fathoming as a result of infeasibility.

For the relaxation, we drop the integrality requirements and solve the resultant LP. For node 0, there are no restrictions on the variables other than those defined by the original constraints, and the solution is

$$\mathbf{x}^0 = (0.4341, 0, 1.0341, 0.765, 0.5695, 0), \text{ with } z_{UB}^0 = 52$$

The separation variable is selected as the one with the largest fractional part, where the fractional part is the distance to the nearest whole number. For example, the fractional part of x_1 is 0.4341, the interval from 0 to x_1, whereas the fractional part of x_5 is 0.4305, the interval from x_5 to 1. The fractional part of x_2 is 0. We branch in the direction of the closest integer. For this example, the first branch will set $x_1 \le 0$.

Note that the solution set is divided by two inequalities. Node 0 represents all solutions, node 1 represents all solutions with $x_1 \le 0$, and node 14 represents all solutions with $x_1 \ge 1$. In general, if a separation variable j is chosen that is noninteger and has the numerical value $\bar{x}_j$, we have the two restrictions

$$x_j \le \lfloor \bar{x}_j \rfloor \quad \text{and} \quad x_j \ge \lfloor \bar{x}_j \rfloor + 1$$

The brackets select the largest integer less than $\bar{x}_j$. No integer solutions are lost by this separation, since only fractional solutions are in the open interval

$$\lfloor \bar{x}_j \rfloor < x_j < \lfloor \bar{x}_j \rfloor + 1$$

Returning to the example, the feasible integer solution

$$\mathbf{x}_B = (0, 2, 0, 0, 0, 1), \text{ with } z_B = 48$$

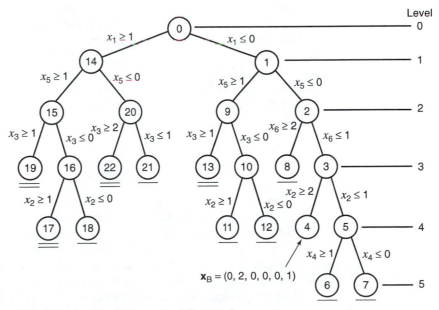

Figure 8.8　Decision tree for the general IP example

is found at node 4. It is interesting to note that x_2 and x_6 were the only two zero variables in the LP solution at node 0. The complete solution has 22 nodes with no better integer solution found. The last 18 nodes in the tree were necessary to prove that the solution found at node 4 was optimal.

Data Structure for the General Case

Because each branch adds a constraint, rather than a value as in the 0-1 case, a different data structure is used to keep track of the tree. For each branch, we must know the level at which the constraint was imposed, the right-hand-side (RHS) value of the constraint, the direction of the inequality, and an indication of whether or not the opposite branch has been explored. We use two vectors to store this information.

P: Path vector whose length is the depth of the search tree (number of additional bound constraints imposed). Each element corresponds to a variable being restricted. Again, we will underline an element when its opposite member has been explored. We use the sign of the element to indicate if the restriction is an upper or a lower bound.

Q: Value vector associated with P. Each element gives the RHS value of the bound in the inequality.

Consider node 12 in Figure 8.8. The two-part notation describing the node is

$$P_{12} = (-1, +\underline{5}, -3, -\underline{2}), \quad Q_{12} = (0, 1, 0, 0)$$

This information allows us to construct the constraints that have been imposed on the problem up to this point. They are

$$x_1 \le 0, \quad x_5 \ge 1, \quad x_3 \le 0, \quad x_2 \le 0$$

Since the variables are constrained to be nonnegative, the ≤ 0 constraints have the effect of setting the variable equal to zero.

Solving the relaxation with these constraints yields $z_{UB}^{12} = 45.67$, a value below the best solution $z_B = 48$. The appropriate action is to fathom the node and backtrack. This is accomplished by finding the rightmost component of P_{12} that is not underlined (-3), underlining it, and eliminating all components ($-\underline{2}$) to its right. The corresponding components in Q_{12} are also deleted. The final step is to adjust the rightmost component of Q_{12} to reflect the new constraint implied by the rightmost component of P_{12}. For the example, we backtrack to node 13, which is defined by the vectors

$$P_{12} = (-1, +\underline{5}, +\underline{3}), \quad Q_{12} = (0, 1, 1)$$

The corresponding constraints are

$$x_1 \le 0, \quad x_5 \ge 1, \quad x_3 \ge 1$$

With this data structure, the same variable can appear in P more than once. Because a particular variable is not restricted to a unique value at a node, it may be constrained several times along a path. This does not occur in the example.

Table 8.9 provides the search information for the first 11 nodes. We leave the construction of the remainder of the table as an exercise for the reader. In solving this problem, we did not use Equation (4) to provide a tighter upper bound. This would have been appropriate, because the problem has all integer variables and all integer objective coefficients.

Table 8.9   First 11 Nodes for Example 2

Node, k	Level, l	P_k	Q_k	z_{UB}	z_B	s	Action
0	0	$\varnothing$	$\varnothing$	52	$-M$	1	$x_1 \le 0$
1	1	(-1)	(0)	52	$-M$	5	$x_5 \le 0$
2	2	$(-1, -5)$	$(0, 0)$	49.5	$-M$	6	$x_6 \le 1$
3	3	$(-1, -5, -6)$	$(0, 0, 1)$	49.5	$-M$	2	$x_2 \ge 2$
4	4	$(-1, -5, -6, +2)$	$(0, 0, 1, 2)$	48	48	—	Feasible, integer; set z_B, backtrack
5	4	$(-1, -5, -6, -\underline{2})$	$(0, 0, 1, 1)$	49	48	4	$x_4 \ge 1$
6	5	$(-1, -5, -6, -\underline{2}, +4)$	$(0, 0, 1, 1, 1)$	42	48	—	Backtrack
7	5	$(-1, -5, -6, -\underline{2}, -4)$	$(0, 0, 1, 1, 0)$	46.3	48	—	Backtrack
8	3	$(-1, -5, +\underline{6})$	$(0, 0, 2)$	42	48	—	Backtrack
9	2	$(-1, +\underline{5})$	$(0, 1)$	49.3	48	3	$x_3 \le 0$
10	3	$(-1, +\underline{5}, -3)$	$(0, 1, 0)$	49	48	2	$x_2 \ge 1$
11	4	$(-1, +\underline{5}, -3, +2)$	$(0, 1, 0, 1)$	34	48	—	Backtrack

8.4   CUTTING PLANE METHODS

A second approach to solving IPs is also based on the idea of relaxing the integrality requirements and solving the resultant LP. But rather than iteratively imposing restrictions on the fractional variables, as is done in branch and bound, we now generate a series of constraints or "cuts," add them to the formulation, and re-solve. If care is taken in generating the constraints so that no feasible solutions are eliminated, we will see that eventually the LP solution will be integral and optimal to the original IP. Of course, if the LP solution is integral before any constraints are added, it is optimal to the original IP, but this is not likely to be the case unless the model is totally unimodular.

Cutting Planes

To illustrate the cutting plane approach, we return to the IP in Example 1. One way to ensure that the LP solution is integral is to reduce the relaxed feasible region to the point where all its vertices are integral. Figure 8.9 shows two constraints or cutting planes that partially accomplish this goal. When the first ($x_1 - x_2 \ge 0$) is added and the LP is re-solved, the new solution occurs as the intersection of cutting plane 1 and constraint $4x_1 - x_2 = 5$; i.e., at $\mathbf{x}^1 = (5/3, 5/3)$. After the second cutting plane ($x_1 \ge 2$) is added, the IP and LP solutions coincide at $\mathbf{x} = (2, 2)$. Notice that it was not necessary for all vertices of the feasible region to be integral in order for the LP solution to be integral.

To formalize the use of cutting planes, let $S = \{\mathbf{x} \in Z_+^n : \mathbf{Ax} = \mathbf{b}, \mathbf{x} \ge 0\}$ be the set of feasible solutions to an IP and assume that S is bounded. We use the notation Z_+^n to indicate the n-dimensional space of nonnegative integers. Let $\hat{n}$ be the number of points in S and define its *convex hull* as

$$\text{conv}(S) = \left\{ \mathbf{y} : \mathbf{y} = \sum_{k=1}^{\hat{n}} \alpha_k \mathbf{x}^k, \alpha_k \ge 0, \sum_{k=1}^{\hat{n}} \alpha_k = 1, \mathbf{x}^k \in S \right\}$$

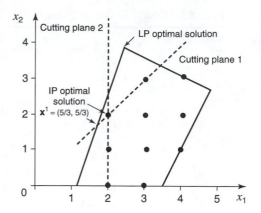

Figure 8.9   Use of cutting planes to reduce feasible region of IP in Example 1

Note that

$$S \subseteq \text{conv}(S) \subseteq \bar{S} = \{\mathbf{x} : \mathbf{Ax} = \mathbf{b}, \mathbf{x} \geq \mathbf{0}\}$$

where $\bar{S}$ is the set of feasible solutions to the corresponding LP. For Example 1, in order to construct conv(S), it would be necessary to introduce three additional cutting planes: $x_2 \leq 3$, $x_1 \leq 4$, and $x_1 - x_2 \leq 3$. In general, it is very difficult to identify all the cutting planes (actually called *facets*) needed to construct the convex hull of integer solutions to an IP. For special cases such as the traveling salesman problem and the knapsack problem, research over the last 20 years has produced a significant number of these facets (e.g., see Grötschel and Padberg [1985] and Nemhauser and Wolsey [1988]). However, as this example demonstrates, it is rarely necessary to construct the entire convex hull in order for the IP optimal solution and the corresponding LP optimal solution to coincide.

When working with cutting planes, we will add one or more constraints at each iteration, each having the following properties.

1. The current LP solution is not feasible for the constraint.
2. No integer solution that is feasible for the original constraints is made infeasible by the new constraint.

The new constraints remove some of the noninteger solutions from the relaxed feasible region, but none that are integer. After they are added, the model is again solved as an LP. The process continues until an integer solution is obtained. This solution must be optimal for the original IP.

The ultimate convergence of the procedure depends on the type(s) of cutting plane(s) added at each iteration. There are several options, not all of which guarantee finite convergence. In any case, the identification of a constraint that has the aforementioned properties usually depends on the specific type of problem being solved. In this section, we are concerned with pure IPs and discuss two types of cutting planes.

Dantzig Cuts

Consider the integer linear program

$$\text{Maximize } \mathbf{cx}$$
$$\text{subject to } \mathbf{Ax} = \mathbf{b}$$
$$\mathbf{x} \geq \mathbf{0} \text{ and integer}$$

and assume that all components of the $m \times (n + 1)$ matrix $(\mathbf{A}, \mathbf{b})$ are integer. If this is not the case, the procedures of this section are not valid. As noted, an LP relaxation of this problem simply drops the integrality requirement on the variables.

The simplest kind of cut is called a **Dantzig cut**. From LP theory, we know that the basic variables are functions of the problem parameters and the values of the nonbasic variables—i.e.,

$$\mathbf{x}_B = \mathbf{B}^{-1}\mathbf{b} - \mathbf{B}^{-1}\mathbf{N}\mathbf{x}_N$$

where $\mathbf{A} = (\mathbf{B}, \mathbf{N})$, $\mathbf{x} = (\mathbf{x}_B, \mathbf{x}_N)^T$. A basic solution has $\mathbf{x}_N$ equal to zero, but every other feasible solution of the LP, including integer solutions, can be obtained by setting some of the components of $\mathbf{x}_N$ equal to positive values. In particular, consider the optimal solution of the LP relaxation given by

$$\mathbf{x}_B = \mathbf{B}^{-1}\mathbf{b}$$

When this vector is integer, it must be optimal for the IP problem. When it is not integer, the optimal integer solution must have some nonbasic variables greater than zero. Let Q be the set of nonbasic variables. Because the smallest positive integer is 1, the following constraint must hold for every integer solution.

$$\sum_{j \in Q} x_j \geq 1 \tag{5}$$

It should be evident that the current LP solution does not satisfy Constraint (5) but that every integer solution (including the optimal solution) does, so this cut may be added to the relaxation.

Example 3

Consider the following pure IP.

$$
\begin{array}{llrcrcrcrcl}
& \text{Maximize} & z = & x_1 & + & x_2 & & & & \\
\text{C1:} & \text{subject to} & & 5x_1 & - & 3x_2 & + & x_{s1} & & & = 5 \\
\text{C2:} & & & -3x_1 & + & 5x_2 & & & + & x_{s2} & = 5 \\
& & & \multicolumn{9}{l}{x_1 \geq 0,\ x_2 \geq 0 \text{ and integer}} \\
& & & \multicolumn{9}{l}{x_{s1} \geq 0,\ x_{s2} \geq 0}
\end{array}
$$

The slack variables x_{s1} and x_{s2} will automatically be integer in an optimal solution, because all problem parameters are integer, and thus it is not necessary to include this restriction in the formulation. The graph of the feasible region in the (x_1, x_2)-plane is shown in Figure 8.10a. The crosses indicate the integer points. One can identify five feasible points in the figure. The optimal solution is $\mathbf{x}^* = (2, 2)$. The shaded area depicts the feasible region of the LP obtained by dropping the integrality requirements.

Solving the LP relaxation, we obtain the fractional solution

$$(x_1, x_2) = (2.5, 2.5), \text{ with } z = 5$$

The nonbasic variables are the slacks for Constraints C1 and C2, so the Dantzig cut is

$$x_{s1} + x_{s2} \geq 1$$

To show this in the figure, we note that the slack variables can be expressed as linear functions of the structural variables using the original equations

$$x_{s1} = 5 - 5x_1 + 3x_2 \text{ and } x_{s2} = 5 + 3x_1 - 5x_2$$

Substituting these values into the Dantzig cut yields the equivalent inequality

C3: $$2x_1 + 2x_2 \leq 9$$

Let x_{s3} be the slack variable for this constraint. Adding Constraint C3 to the LP and re-solving yields the solution $(x_1, x_2) = (37/16, 35/16)$, with $z = 9/2$, as shown in Figure 8.10b. Two more iterations are depicted in Figures 8.10c and 8.10d. As we can see, the cuts continue to tighten the feasible region and reduce the objective function value.

The example illustrates that with more cuts we obtain a better estimate of the convex hull of feasible integer points. The shaded region in Figure 8.11 represents conv(*S*) and includes all integer points feasible for the original IP. The constraints defining the convex hull for the example are the nonnegativity conditions and

$$2x_1 - x_2 \leq 2$$
$$-x_1 + 2x_2 \leq 3$$

Solving the problem as an LP with these constraints will yield the optimal integer solution. When 20 more Dantzig cuts were added to the LP relaxation, however, neither of these two constraints was generated. As a consequence, the procedure failed to converge to the optimal solution.

The goal of cutting plane approaches is to approximate the convex hull, at least in the region of the optimal solution, with as few cuts as possible. Unfortunately, when Dantzig cuts are used, there is no guarantee that the LP solution will converge to the optimal integer solution in a finite number of iterations. The most we can say is that if $\mathbf{x}^*$ is an optimal solution to an IP, a necessary (but not sufficient) condition for an algorithm based on Constraint (5) to converge to $\mathbf{x}^*$ is that $\mathbf{x}^*$ be on an edge (a line joining two adjacent extreme points) of the constraint set of the LP relaxation $\overline{S} = \{\mathbf{x} : \mathbf{Ax} = \mathbf{b}, \mathbf{x} \geq \mathbf{0}\}$.

Improved versions of Constraint (5) can be obtained by considering the ith basic variable $x_{B(i)}$ in the relaxed LP solution. Assume that $x_{B(i)} = \overline{b}_i \neq$ integer. Writing out the ith equation yields

$$x_{B(i)} = \overline{b}_i - \sum_{j \in Q} \overline{a}_{ij} x_j$$

or

$$x_{B(i)} = \lfloor \overline{b}_i \rfloor + f_i - \sum_{j \in Q} \overline{a}_{ij} x_j$$

where $\lfloor b \rfloor$ denotes the greatest integer less than or equal to b and

$$f_i = \overline{b}_i - \lfloor \overline{b}_i \rfloor$$

—i.e., the fractional component of $\overline{b}_i$. In order for $x_{B(i)}$ to be integer, the following constraint must hold.

$$\sum_{j \in Q_i} x_j \geq 1, \ Q_i = \left\{ j : j \in Q, \ \overline{a}_{ij} \neq 0 \right\} \tag{6}$$

The validity of Constraint (6) is based on the observation that the current basic solution (all nonbasic variables at 0) is noninteger. Therefore, at least one of the current nonbasic variables whose coefficient is nonzero must be positive. The improvement removes some of the slack variables from the Dantzig constraints, increasing their effectiveness.

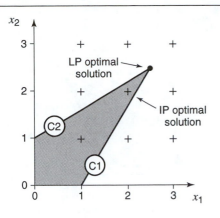

a. Original problem

LP optimal solution:
$(x_1, x_2) = (2.5, 2.5), z = 5$
New cut C3:
$x_{s1} + x_{s2} \geq 1$ or $2x_1 + 2x_2 \leq 9$

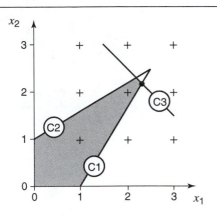

b. After first cut

LP optimal solution:
$(x_1, x_2) = (37/16, 35/16), z = 9/2$
New cut C4:
$x_{s1} + x_{s3} \geq 1$, or $7x_1 - x_2 \leq 13$

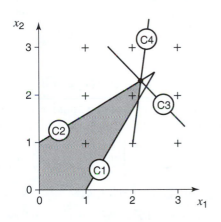

c. After second cut

LP optimal solution:
$(x_1, x_2) = (35/16, 37/16), z = 9/2$
New cut C5:
$x_{s3} + x_{s4} \geq 1$, or $9x_1 + x_2 \leq 21$

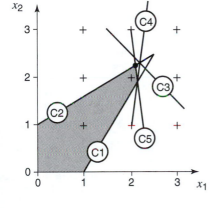

d. After third cut

LP optimal solution:
$(x_1, x_2) = (25/12, 27/12), z = 13/3$
New cut C6:
$x_{s2} + x_{s5} \geq 1$, or $6x_1 + 6x_2 \leq 25$

Figure 8.10 Dantzig cuts applied to Example 3

These improved Dantzig cuts [Constraint (6)] will yield a finite algorithm for solving an IP. The details are unimportant, however, because it is unlikely that an algorithm based on Constraint (6) is going do better than an algorithm based on the cuts derived next.

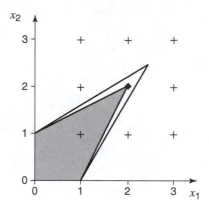

Figure 8.11   Convex hull of feasible integer points

Gomory Cuts

A second type of cut that uses more information from the LP solution than Dantzig cuts use and generally removes more of the feasible region is attributed to Gomory. The corresponding approach is known as the *method of integer forms*. Consider again the matrix equations describing the solution space in terms of a given basis where it is again assumed that the components of the original $(\mathbf{A}, \mathbf{b})$ matrix are all integer.

$$\mathbf{x}_B = \mathbf{B}^{-1}\mathbf{b} - \mathbf{B}^{-1}\mathbf{N}\mathbf{x}_N$$

or

$$\mathbf{x}_B + \mathbf{B}^{-1}\mathbf{N}\mathbf{x}_N = \mathbf{B}^{-1}\mathbf{b} \tag{7}$$

Referring to Example 3, we note that the optimal basis for the original LP relaxation consists of the first two columns of the $\mathbf{A}$ matrix. The basis and its inverse are

$$\mathbf{B} = \begin{bmatrix} 5 & -3 \\ -3 & 5 \end{bmatrix} \text{ and } \mathbf{B}^{-1} = \begin{bmatrix} 5/16 & 3/16 \\ 3/16 & 5/16 \end{bmatrix}$$

Using these data, we can write Equation (7) as

$$\mathbf{x}_B + \begin{bmatrix} 5/16 & 3/16 \\ 3/16 & 5/16 \end{bmatrix}\begin{bmatrix} 1 & 0 \\ 0 & 1 \end{bmatrix}\begin{pmatrix} x_{s1} \\ x_{s2} \end{pmatrix} = \begin{bmatrix} 5/16 & 3/16 \\ 3/16 & 5/16 \end{bmatrix}\begin{pmatrix} 5 \\ 5 \end{pmatrix}$$

which reduces to

$$x_1 + \frac{5}{16}x_{s1} + \frac{3}{16}x_{s2} = 2.5$$

$$x_2 + \frac{3}{16}x_{s1} + \frac{5}{16}x_{s2} = 2.5$$

From the discussion of Dantzig cuts, we know that at least one nonbasic variable must be positive to attain integrality, can we make a stronger statement using the additional information in these equations? Gomory cuts are based on the idea that the values of the fractional parts of the terms involving the nonbasic variables must balance out the fractional parts of the constant terms on the right-hand side in order for the basic variables, x_1 and x_2 in this case, to be integer. The operation "modulo 1," abbreviated (mod 1), takes only the fractional part of a number. In general,

$$a(\text{modulo } 1) = a - \lfloor a \rfloor$$

For instance, $3(\text{mod } 1) = 0$, $2.3(\text{mod } 1) = 0.3$, and $-2.3(\text{mod } 1) = 0.7$.

Using only the fractional part of the equation defining x_1 for the example and recalling that x_1 must be an integer, we write

$$x_1(\text{mod } 1) + \frac{5}{16}x_{s1}(\text{mod } 1) + \frac{3}{16}x_{s2}(\text{mod } 1) = 2.5(\text{mod } 1)$$

$$0 + \frac{5}{16}x_{s1}(\text{mod } 1) + \frac{3}{16}x_{s2}(\text{mod } 1) = 0.5$$

Substitution shows that $(x_{s1}, x_{s2}) = (1, 1)$, $(3, 3)$, $(8, 0)$, and $(0, 8)$ are all solutions to this equation. For example, when $(x_{s1}, x_{s2}) = (1, 1)$, we get

$$0 + \frac{5}{16}(\text{mod } 1) + \frac{3}{16}(\text{mod } 1) = \frac{5}{16} + \frac{3}{16} = 0.5$$

Nevertheless, an equation involving modulo 1 is nonlinear and cannot be added as a cut. Rather, we note that any values of the nonbasic variables that satisfy the modulo 1 equation must also satisfy the less restrictive linear requirement

$$\frac{5}{16}x_{s1} + \frac{3}{16}x_{s2} \geq 0.5$$

This is a Gomory fractional cut for the first constraint. Its formal derivation is given below, but for the moment let us assume that it is valid. Now, replacing x_{s1} and x_{s2} with their equivalents in terms of x_1 and x_2 yields the constraint

C3: $$x_1 \leq 2$$

In a similar manner, we can derive another Gomory cut from the second constraint.

$$x_2 + \frac{3}{16}x_{s1} + \frac{5}{16}x_{s2} = 2.5$$

implying $$\frac{3}{16}x_{s1} + \frac{5}{16}x_{s2} \geq 0.5 \quad \text{or}$$

C4: $$x_2 \leq 2$$

Adding Constraints C3 and C4 to the original LP relaxation and solving, we obtain the optimal integer solution $\mathbf{x}^* = (2, 2)$, as shown in Figure 8.12.

The fact that the use of Gomory cuts yielded the optimal solution on the first iteration is not an indication that this will always occur. If an integer solution is not forthcoming, additional cuts must be derived from the new LP solutions. Each solution yields as many Gomory cuts as there are noninteger basic variables. When the objective function has integer coefficients, it can also be used to derive a cut. In general, the value of the objective function is

$$z = \pi\mathbf{b} - (\pi\mathbf{N} - \mathbf{c}_N)\mathbf{x}_N$$

or

$$z + (\pi\mathbf{N} - \mathbf{c}_N)\mathbf{x}_N = \pi\mathbf{b}$$

where $\mathbf{c} = (\mathbf{c}_B, \mathbf{c}_N)$ are the original objective coefficients and $\pi = \mathbf{c}_B\mathbf{B}^{-1}$ is the vector of dual variables. When $\pi\mathbf{b}$ is noninteger, this equation can be used in the same manner as a constraint equation to obtain a Gomory cut

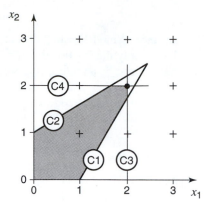

Figure 8.12 Optimal solution for Example 3 obtained with Gomory cuts

$$\sum_{j \in Q} \bar{c}_j \bmod(1) x_j \geq \bar{z} \bmod(1)$$

where $\bar{c}_j$ is the jth reduced cost (recall that $\bar{\mathbf{c}} = \boldsymbol{\pi}\mathbf{N} - \mathbf{c}_N$) and $\bar{z}$ is the LP objective function value. For our example, the first LP solution (see Figure 8.10a) gives $z = \bar{z} = 5$, so no cut is available.

Derivation of Gomory Cut

Consider an IP in equality form with all-integer data, and note that any set of values of the decision variables satisfying the constraints must also satisfy any relations derived by row operations on the individual constraints. For example, we could add all the constraints to obtain

$$\sum_{i=1}^{m} \sum_{j=1}^{n} a_{ij} x_j = \sum_{i=1}^{m} b_i$$

Now suppose that we perform some arithmetic to obtain a new constraint of the form

$$\sum_{j=1}^{n} a_j x_j = b \tag{8}$$

Using the notation $a_j = \lfloor a_j \rfloor + f_j$, $b = \lfloor b \rfloor + f$, we can rewrite Equation (8) as

$$\sum_{j=1}^{n} \left(\lfloor a_j \rfloor + f_j \right) x_j = \lfloor b \rfloor + f \tag{9}$$

Dropping the fractions from the left-hand side of (9), we obtain the inequality

$$\sum_{j=1}^{n} \lfloor a_j \rfloor x_j \leq \lfloor b \rfloor + f \tag{10}$$

Since the left-hand side of Equation (10) must be integer, we can drop f from the right

$$\sum_{j=1}^{n} \lfloor a_j \rfloor x_j \leq \lfloor b \rfloor \tag{11}$$

Subtracting Equation (11) from Equation (9) and inserting a slack variable yields

$$\sum_{j=1}^{n} f_j x_j - x_s = f \tag{12}$$

This relationship defines a Gomory fractional cut. For $x_{B(i)}$ a basic variable with a noninteger value, the actual cutting plane is

$$\sum_{j \in Q} f_j x_j - x_s = f_i \tag{13}$$

where Q is the set of nonbasic variables. Although a more general version of Equation (13) can be derived, it yields no additional insights and so is left as an exercise.

Notice that x_s must be negative in Equation (13) at the current LP solution so that it is not feasible. Moreover, if the cut is added to an optimal tableau, the current solution will still satisfy the optimality conditions, implying that the dual simplex algorithm should be used to reoptimize. It can be shown that Equation (13) will be satisfied by all integer-valued feasible solutions, and so it will never eliminate the optimal solution. This leads to the following rudimentary algorithm.

Algorithm for the Method of Integer Forms

Step 1: (*Initialization*) Solve the LP relaxation of the IP and go to Step 2.

Step 2: (*Optimality test*) If the solution to the LP relaxation is all-integer, it is optimal for the original IP; if not, go to Step 3.

Step 3: (*Cutting and pivoting*) Choose a row r with $f_r > 0$ and add Equation (14) to the bottom of the tableau. Reoptimize using the dual simplex method. (This may take more than one iteration.) Go to Step 2.

At Step 3, there may be many rows that could serve as the source for the cut. To guarantee finite convergence, we perform the following.

1. Delete the row corresponding to a slack variable from a cut if such a variable becomes basic during the restoration of primal feasibility.

2. Choose the source row to be the topmost row with $f_r > 0$.

In addition, theory requires that we maintain the nonnegativity of the first component of each nonbasic column. If there is no dual degeneracy, this will always be the case, because the reduced costs will always be positive at optimality. Otherwise, it may be necessary to rearrange the rows or add a redundant constraint of the form $\sum_{j \in Q} x_j \leq M$ (big number). The details are not important and so are omitted.

Example 4

Consider the problem

$$
\begin{aligned}
\text{Maximize } z = \ & 4x_1 + 5x_2 + x_3 \\
\text{subject to } \quad & 3x_1 + 2x_2 \qquad\quad + x_4 \qquad\qquad\qquad = 10 \\
& x_1 + 4x_2 \qquad\qquad\quad + x_5 \qquad\quad = 11 \\
& 3x_1 + 3x_2 + x_3 \qquad\qquad\quad + x_6 = 13 \\
& x_j \geq 0 \text{ and integer}, \ j = 1, \ldots, 6
\end{aligned}
$$

Dropping the integrality requirement and solving the relaxed LP leads to the solution

$$\mathbf{x}_B = (x_1, x_2, x_3) = (1.8, 2.3, 0.7), \text{ with } z = 19.4$$

which is noninteger. Therefore, we must find a constraint to add to the formulation that cuts off this fractional solution but no integer solutions. Let us find the Gomory cuts that are available for this purpose.

The basis and basis inverse along with the primal and dual formulations of the LP optimal solution are

$$\mathbf{B} = \begin{bmatrix} 3 & 2 & 0 \\ 1 & 4 & 0 \\ 3 & 3 & 1 \end{bmatrix} \text{ and } \mathbf{B}^{-1} = \begin{bmatrix} 0.4 & -0.2 & 0 \\ -0.1 & 0.3 & 0 \\ -0.9 & -0.3 & 1 \end{bmatrix}$$

$$\mathbf{x}_B = \begin{pmatrix} x_1 \\ x_2 \\ x_3 \end{pmatrix} = \begin{pmatrix} 1.8 \\ 2.3 \\ 0.7 \end{pmatrix} - \begin{bmatrix} 0.4 & -0.2 & 0 \\ -0.1 & 0.3 & 0 \\ -0.9 & -0.3 & 1 \end{bmatrix} \begin{pmatrix} x_4 \\ x_5 \\ x_6 \end{pmatrix}$$

$$\boldsymbol{\pi} = \mathbf{c}_B \mathbf{B}^{-1} = (0.2, 0.4, 1) \text{ and } \bar{\mathbf{c}} = (0, 0, 0, 0.2, 0.4, 1)$$

The four Gomory cuts obtained from the three equations and the objective function are as follows.

Constraint 1: $x_1 + 0.4x_4 - 0.2x_5 = 1.8$ yields
Cut 1: $0.4x_4 + 0.8x_5 \geq 0.8$
Constraint 2: $x_2 - 0.1x_4 + 0.3x_5 = 2.3$ yields
Cut 2: $0.9x_4 + 0.3x_5 \geq 0.3$
Constraint 3: $x_3 - 0.9x_4 - 0.3x_5 + x_6 = 0.7$ yields
Cut 3: $0.1x_4 + 0.7x_5 \geq 0.7$
Objective: $z + 0.2x_4 + 0.4x_5 + x_6 = 19.4$ yields
Cut 0: $0.2x_4 + 0.4x_5 \geq 0.4$

Although all these cuts are valid, the best strategy might be to add only a subset of them to the formulation. This cautionary note is based on the fact that the computational effort increases with the number of constraints, and that some cuts or linear combinations of cuts may dominate others. Dominated cuts are redundant and so retard rather than speed up convergence. A heuristic for selecting a *best* cut is to use the one with the largest RHS value, reasoning that a large RHS will provide a deep cut. Although this rule may violate convergence condition 2, it rarely causes any difficulties in obtaining the optimal solution. For this example, we added cut 1 through cut 3 at the same time and obtained the solution $\mathbf{x}^* = (2, 2, 1)$.

Example 5

Let us now try to solve the following two-variable IP.

$$\begin{aligned} \text{Maximize } z = -x_1 - &\ x_2 \\ \text{subject to} \quad 4x_1 + 10x_2 &\geq 12 \\ 10x_1 + 4x_2 &\geq 12 \\ x_1 \geq 0, x_2 \geq 0 \text{ and} &\text{ integer} \end{aligned}$$

By relaxing the integrality requirement and solving the resulting LP, we obtain the tableau presented in Table 8.10.

The feasible region and the optimal solution are graphically displayed in Figure 8.13. The feasible region lies above the two constraint lines, and the optimal solution lies at their intersection. Clearly, the solution $\mathbf{x}_{LP} = (6/7, 6/7)$ is noninteger.

Starting from the tableau in Table 8.10, we generate one Gomory cut per iteration and add it to the tableau. For the first cut, the row associated with the basic variable x_2 is used. This yields the inequality $(37/42)x_{s1} + (1/21)x_{s2} \geq 6/7$. Substituting for $x_{s1} = 4x_1 + 10x_2 - 12$ and $x_{s2} = 10x_1 + 4x_2 - 12$ yields C1 below. Cuts C2 through C4 are generated at the next three iterations (they are not generated from the tableau in Table 8.10).

C1:	$4x_1 + 9x_2 \geq 12$	Iteration 1
C2:	$4x_1 + 8x_2 \geq 12$	Iteration 2
C3:	$9x_1 + 4x_2 \geq 12$	Iteration 3
C4:	$8x_1 + 4x_2 \geq 12$	Iteration 4

After each cut is added, the dual simplex method is used to reoptimize. The corresponding solutions are shown in Table 8.11. The variables x_{s3} through x_{s6} are the slacks for the new cuts.

Table 8.10   Relaxed LP Solution to Example 5

Row number	Basic variables	Coefficients					RHS
		z	x_1	x_2	x_{s1}	x_{s2}	
0	z	1	0	0	3/42	3/42	−12/7
1	x_2	0	0	1	−5/42	2/42	6/7
2	x_1	0	1	0	2/42	−5/42	6/7

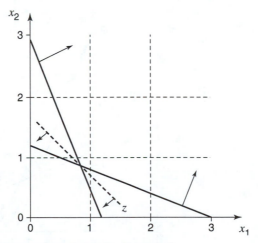

Figure 8.13   Linear programming solution for Example 5

Table 8.11 Optimal Solutions for Sequence of Gomory Cuts

Iteration	Cut	z	x_1	x_2	x_{s1}	x_{s2}	x_{s3}	x_{s4}	x_{s5}	x_{s6}
1	C1	−1.78	0.81	0.97	0.97	0	0	—	—	—
2	C2	−1.88	0.75	1.13	2.25	0	1.13	0	—	—
3	C3	−1.93	0.86	1.07	2.14	0.86	1.07	0	0	—
4	C4	−2	1	1	2	2	1	0	1	0

Four iterations are required before an all-integer solution is found and the computations can be halted. A graphical depiction of the sequence of cuts appears in Figure 8.14. The behavior exhibited in the figure is typical of all traditional cutting plan methods. The cuts remove only small portions of the feasible region, so the algorithm may take a long time to converge, especially if the optimal IP solution is far from the initial LP solution.

It is interesting to observe that when the cuts are written in terms of the original problem variables, as they are in cuts C1 to C4, the coefficients and RHS constants are all integer valued. This will always be the case when the original constraint data [**A**, **b**] are integer valued. For empirical confirmation, try solving a few problems with the Teach IP Excel Add-in.

All cutting plane methods can be applied to MILPs regardless of the mix of variables. Gomory's method of integer forms can be viewed as a dual approach because the current solution is always dual feasible. It will not produce an integer-valued solution, however, if the algorithm is terminated prior to convergence, perhaps because of excessive run time. Primal methods, designed to maintain a primal feasible tableau, are available to counter this disadvantage, but their run times are usually inferior.

The real difficulty with traditional cutting plane methods is that the cuts they produce are not very deep. Ideally, we would like to be able to generate facets of the convex hull of integer feasible points. This is just about impossible for general IPs and MILPs. However, when we are dealing with combinatorial optimization problems (such as the TSP, set

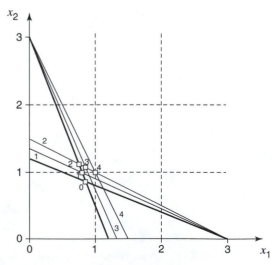

Figure 8.14 Sequence of cuts added for Example 5

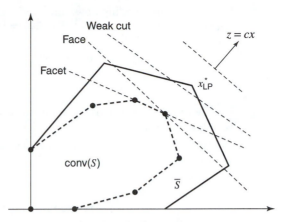

Figure 8.15   Different types of cutting plans for integer programs

covering problems, node packing problems, vehicle routing problems, and machine scheduling problems, to name a few), the results are much more promising. These problems are characterized, for the most part, by the use of binary decision variables and constraints that exhibit a special structure, such as the requirement that the sum of a subset T of the variables equals 1 ($\Sigma_{j \in T} x_j = 1$) or that all the elements of the $\mathbf{A}$ matrix are 0 or 1 ($a_{ij} = 0$ or 1). When dealing with combinatorial optimization problems, it is often possible to use a branch of mathematics called *polyhedral theory* to develop facets of the underlying convex hull. As shown in Figure 8.15, facets provide the deepest cuts without removing any feasible integer solutions. The line labeled "weak cut" typically results from Dantzig's or Gomory's method. For more discussion of these issues, see Grötschel and Padberg [1985] and Wolsey [1998], among others.

8.5 ADDITIONAL CUTS

The central concept behind cutting planes is the concept of a valid inequality. We have already been using this idea informally throughout this chapter, but a definition is in order.

Definition 1: Let $S \subseteq \Re^n$ be a feasible region for an optimization problem. A constraint $\mathbf{a}\mathbf{x} \le b$ is a *valid inequality* for S if $\mathbf{a}\mathbf{x} \le b$ for all $\mathbf{x} \in S$.

If $S = \{\mathbf{x} \in Z_+^n : \mathbf{A}\mathbf{x} \le \mathbf{b}\}$ and conv$(S) = \{\mathbf{x} \in \Re^n : \tilde{\mathbf{A}}\mathbf{x} \le \tilde{\mathbf{b}}\}$, the constraints $\mathbf{a}^i\mathbf{x} \le b_i$ and $\tilde{\mathbf{a}}^i\mathbf{x} \le \tilde{b}_i$ for all i are easily seen to be valid inequalities for S. For cutting plane algorithms, valid inequalities are most effective when they are facets of conv(S). Although it is generally not easy to find facets, there are several simple valid inequalities that can be identified from the constraints defining the feasible region of an IP. When the current relaxed solution violates one of them, they can be added to the formulation much like Dantzig or Gomory cuts. Several are presented below.

Example 6 (Integer Rounding)

Consider the feasible region $S = \{\mathbf{x} \in Z_+^4 : 9x_1 + 4x_2 + 13x_3 + 19x_4 \le 23\}$. Dividing by 9 yields a valid inequality for S.

$$x_1 + \tfrac{4}{9}x_2 + \tfrac{13}{9}x_3 + \tfrac{19}{9}x_4 \le \tfrac{23}{9} = 2\tfrac{5}{9}$$

Because $\mathbf{x} \geq \mathbf{0}$, rounding down the coefficients on the left-hand side (LHS) to the nearest integer yields $x_1 + x_3 + 2x_4 \leq x_1 + \frac{4}{9}x_2 + \frac{13}{9}x_3 + \frac{19}{9}x_4 \leq 2\frac{5}{9}$ which leads to the weaker constraint

$$x_1 + x_3 + 2x_4 \leq 2\tfrac{5}{9}$$

The fact that x_j must be integer valued means that the LHS of this constraint must also be integer valued. An integer that is less than or equal to $2\frac{5}{9}$ is no greater than 2, so we can round the RHS constant down to the nearest integer to arrive at a new valid inequality for S

$$x_1 + x_3 + 2x_4 \leq 2$$

which cuts off part of the relaxed feasible region. Different valid inequalities can be generated in the same manner by dividing the original constraint by any nonnegative integer.

Example 7 (Logical Implications)

Let B^n be the n-dimensional space of binary variables and consider the 0-1 knapsack set

$$S = \{\mathbf{x} \in B^5 : 2x_1 + x_2 - 3x_3 - 6x_4 + 4x_5 \leq -2\}$$

If $x_3 = x_4 = 0$, the LHS of the constraint $2x_1 + x_2 + 4x_5 \geq 0$. But the RHS equals –2, which is impossible, so to ensure feasibility we must have $x_3 + x_4 \geq 1$.

Also, if $x_5 = 1$ and $x_4 = 0$, the LHS equals $2x_1 + x_2 - 3x_3 + 4$. The smallest the LHS can be, then, is 1 when $x_3 = 1$. But the RHS is –2, so the situation leads to an infeasible solution. Consequently, we must have $x_5 \leq x_4$, which is also a valid inequality.

Example 8 (Mixed 0-1 Set)

The approaches discussed in Section 8.4 for generating Gomory cuts can be extended to the MILP with a bit more analysis. For the moment, consider the simple set

$$S = \{(x, y) : x \leq 9999y, \; 0 \leq x \leq 8, \; y \in B^1\}$$

which represents a portion of the feasible region for a fixed charge problem. It is easy to check that the constraint

$$x \leq 8y$$

is a valid inequality for S. In fact, $S = \{(0, 0), (x, 1) : 0 \leq x \leq 8\}$, and, by adding the constraint $x \leq 8y$, we can check graphically that we get the convex hull of S, where conv$(S) = \{(x, y) : 0 \leq x \leq 8, \; x \leq 8y, \; 0 \leq y \leq 1\}$.

Example 9 (Mixed-Integer Set)

Consider the set

$$S = \{(x, y) : x \leq 12y, \; 0 \leq x \leq 15, \; y \in Z_+^1\}$$

It can be verified that $x \leq 9 + 3y$ [or, written another way, $x \leq 15 - 3(2 - y)$] is a valid inequality for S. Figure 8.16 depicts this set graphically. The bold lines represent the feasible points. When $x \leq 9 + 3y$ is added, we get conv(S).

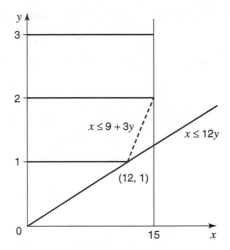

Figure 8.16   Mixed-integer valid inequality

For the general case in which

$$S = \{(x, y): x \le \alpha y, 0 \le x \le \beta, y \in Z_+^1\}$$

and α does not divide evenly into β, we obtain the valid inequality $x \le \beta - \gamma(d - y)$, where

$$d = \left\lceil \frac{\beta}{\alpha} \right\rceil \text{ and } \gamma = \beta\left(\left\lceil \frac{\beta}{\alpha} \right\rceil - 1\right)\alpha.$$

Example 10 (Cover Inequalities for *0–1* Knapsack Constraints)

Consider the set

$$S = \left\{ \mathbf{x} \in B^n: \sum_{j=1}^{n} a_j x_j \le b \right\}$$

which represents a knapsack constraint. Assume that $a_j > 0$ for all j and $b > 0$. For a problem with $a_j < 0$ it is always possible to use a complementary variable $\hat{x}_j = 1 - x_j$ to achieve the desired form. Let $N = \{1, \ldots, n\}$, $C \subseteq N$, and denote by $\mathbf{x}^C$ the n-dimensional vector such that $x_j = 1$ if $j \in C$, 0 otherwise.

Definition 2: A set $C \subseteq N$ is a *cover* if $\Sigma_{j \in C}\, a_j > b$. A cover is *minimal* if $C\backslash\{j\}$ is not a cover for any $j \in C$.

From this definition, we see that C is a cover if and only if its associated incidence vector $\mathbf{x}^C$ is infeasible for S. This gives rise to the following result, which says that at least one of the variables in a cover must be zero in a feasible solution.

Proposition 2: If $C \subseteq N$ is a cover, the cover inequality

$$\sum_{j \in C} x_j \le |C| - 1 \tag{14}$$

is valid for S.

To see how we can make use of Equation (14), consider the feasible region

$$S = \{\mathbf{x} \in B^6 : 13x_1 + 8x_2 + 5x_3 + 5x_4 + 4x_5 + x_6 \leq 21\}$$

The minimal cover inequalities for S are

$$
\begin{aligned}
x_1 + x_2 + x_3 &\leq 2 \\
x_1 + x_2 + x_4 &\leq 2 \\
x_1 + x_2 + x_5 &\leq 2 \\
x_1 + x_2 + x_6 &\leq 2 \\
x_1 + x_3 + x_5 &\leq 2 \\
x_1 + x_3 + x_4 &\leq 2 \\
x_1 + x_4 + x_5 &\leq 2 \\
x_2 + x_3 + x_4 + x_5 &\leq 3
\end{aligned}
$$

Each of these constraints can be added to the formulation to obtain a tighter relaxed feasible region.

Example 11 (Enhanced Cover Inequalities)

In general, Equation (14) is not a facet of the knapsack polyhedron, and hence can be strengthened to provide a better cut. The simple way to do this is given in the following proposition.

Proposition 3: If C is a cover of S, the *enhanced cover inequality*

$$\sum_{j \in E(C)} x_j \leq |C| - 1$$

is valid for S, where $E(C) = C \cup \{j : a_j > a_i \text{ for all } i \in C\}$.

Continuing with Example 10, the only cover that can be enhanced is $C = \{2, 3, 4, 5\}$, the one associated with the last inequality. Proposition 3 says that we can add all variables that have coefficients at least as large as the largest a_i for $i \in C$. The only candidate is $a_1 = 12$, so adding x_1 to the cover inequality gives the enhanced version $x_1 + x_2 + x_3 + x_4 + x_5 \leq 3$. Observe, however, that this new constraint is dominated by the valid inequality

$$2x_1 + x_2 + x_3 + x_4 + x_5 \leq 3 \tag{15}$$

which implies that if $x_1 = 1$, only one of the remaining four variables can be 1. The validity of Equation (16) can be seen by substituting $x_1 = 1$ into the knapsack constraint to get $8x_2 + 5x_3 + 5x_4 + 4x_5 \leq 21 - 13 = 8$. The general procedure that was used to derive Equation (15) is called *lifting* and can be applied to all cover-type inequalities. Lifting and other advanced techniques for strengthening inequalities fall under the heading of polyhedral theory.

EXERCISES

1. Find the minimal spanning tree for the network in the figure starting from node 1 and then from node 5. Are your answers the same?

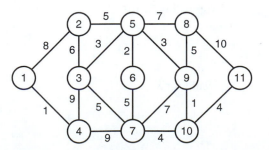

2. Solve the linear sequencing problem for the set of jobs shown in the table.

Job	1	2	3	4	5	6	7	8	9	10
Job time	10	12	2	8	3	15	9	2	19	6
Penalty	5	10	12	2	5	4	14	9	10	2

3. The problem of finding the sequence of jobs that has the minimum average completion time can also be solved with a greedy algorithm. Note that penalties do not play a part in the criterion. Propose a greedy algorithm to solve this problem and use the data in Exercise 2 to illustrate the computations.

4. Redefine the table in Exercise 2 to represent data for a knapsack problem. Let the term "Job" become "Item," let "Job time" become "Benefit," and let "Penalty" become "Weight." Use a greedy algorithm to find a selection of items for a knapsack with a weight capacity of 40 units. Also solve the problem with an LP algorithm and a 0-1 IP algorithm. Comment on the different solutions obtained.

5. State the conditions for which your greedy algorithm developed in Exercise 4 will find the optimal solution for the knapsack problem. For the given data, indicate for what size knapsacks the algorithm will provide the optimal solution.

6. The table shows the cost of building a road between each pair of towns in a remote Alaskan area. Roads can be traveled in both directions. The goal is to select the roads to be built such that total cost is minimized. There must be a path between every pair of towns in the final network. Model the problem as an IP, and solve it. Also solve the problem as a MST.

Town pair	A–B	A–C	A–D	A–E	B–C	B–D	B–E	C–D	C–E	D–E
Cost	100	150	250	90	140	230	40	110	130	70

7. Use exhaustive enumeration to solve the 0-1 IP. Draw the corresponding search tree and provide a table similar to Table 8.5 listing the results obtained at each iteration.

$$\text{Minimize } 3x_1 + 7x_2 + 4x_3 + 5x_4$$
$$\text{subject to } 2x_1 + x_2 + 3x_3 + 4x_4 \geq 6$$
$$x_1 + 2x_2 + 4x_3 + 2x_4 \geq 6$$
$$3x_1 + 4x_2 + x_3 + x_4 \geq 6$$
$$x_j = 0 \text{ or } 1, \ j = 1, 2, 3, 4$$

8. The tableau gives the unit costs of transporting a commodity from each of three proposed warehouses to each of six customers. The demands of the six customers for the commodity are also listed.

Warehouse	Customer					
	1	2	3	4	5	6
1	9	13	2	12	3	3
2	7	2	13	4	12	10
3	1	12	14	6	7	13
Demand	50	60	70	20	30	40

We must decide which of the three locations should have a warehouse. The capacity of a warehouse, if built, is 200, and the construction costs at locations 1, 2, and 3 are 500, 600, and 700, respectively. Find the optimal locations of the warehouses using exhaustive enumeration. For the locations selected, solve the transportation problem with software accompanying the text for each complete solution obtained from the enumeration. Also show the search tree generated by the procedure.

9. The following path vector was generated by a depth-first search, branch and bound algorithm for a 0-1 integer program containing 10 decision variables.

$$P_k = (-6, +\underline{5}, +\underline{9}, -4, +\underline{3})$$

(a) Identify each problem variable as fixed to 1, fixed to 0, or free.

(b) Sketch the current path in the search tree, showing which alternative branches have already been explored and which have yet to be explored.

(c) Starting from P_k, separate on x_8 in the direction $x_8 = 0$. Show the next path vector P_{k+1}.

(d) Again starting from P_k, backtrack and show the new P_{k+1}.

10. Consider the problem

$$\text{Maximize } z = 3x_1 + 5x_2 + 8x_3 + x_4 + 3x_5 + 10x_6$$
$$\text{subject to} \quad 2x_1 + 2x_2 + 5x_3 + 3x_4 + x_5 + 8x_6 \leq 12$$
$$x_j = 0 \text{ or } 1, \ j = 1, \ldots, 6$$

A feasible solution is $\mathbf{x}_B = (1, 1, 0, 0, 0, 1)$, with $z_B = 18$. Determine the relaxed solution for each of the following cases. Apply the fathoming rules described in Section 8.3 whenever possible. Solve the LP relaxations by hand.

(a) All variables are free.

(b) $S^- = \{2\}$, $S^+ = \{5\}$

(c) $S^- = \{1, 6\}$, $S^+ = \varnothing$

(d) $S^- = \varnothing$, $S^+ = \{3, 6\}$

(e) $S^- = \{4, 6\}$, $S^+ = \{3\}$

11. Solve the knapsack problem in Exercise 10 using branch and bound coupled with LP relaxation. At each iteration, branch on the fractional variable in the relaxed solution and use a rounding technique to find feasible solutions. Show your results in a table and draw the corresponding search tree.

12. Solve the following 0-1 IP using branch and bound. Draw the search tree and at each node provide the relaxed LP solution and any other relevant information.

$$\text{Maximize } z = 10x_1 + 30x_2 + 20x_3 + 20x_4 + 10x_5$$
$$\text{subject to} \quad 8x_1 + 12x_2 + x_3 + 8x_4 + 2x_5 \leq 15$$
$$9x_1 + 7x_2 + 4x_3 + 10x_4 + 5x_5 \leq 20$$
$$x_1 + x_2 + 8x_3 + 3x_4 + 7x_5 \leq 11$$
$$x_j = 0 \text{ or } 1, \ j = 1, \ldots, 5$$

13. You wish to solve the following IP with a cutting plane technique.

$$\text{Maximize} \quad 4x_1 + 2x_2 + x_3$$
$$\text{subject to} \quad 14x_1 + 10x_2 + 11x_3 \leq 32$$
$$-10x_1 + 8x_2 + 9x_3 \geq 0$$
$$x_1, x_2, x_3 \geq 0 \text{ and integer}$$

After relaxing the integrality requirements and solving the resulting LP, you obtain the solution given in the table with objective value = 7.849.

Variable number	Name	Value	Status
1	x_1	1.208	BASIC–1
2	x_2	1.509	BASIC–2
3	x_3	0	ZERO
4	SLK–1(x_4)	0	ZERO
5	SLK–2(x_5)	0	ZERO

The basis and basis inverse for this solution are

$$\mathbf{B} = \begin{bmatrix} 14 & 10 \\ -10 & 8 \end{bmatrix} \text{ and } \mathbf{B}^{-1} = \frac{1}{212} \begin{bmatrix} 8 & -10 \\ 10 & 14 \end{bmatrix}$$

(a) What Gomory cuts can be derived from the current information? Express the cuts in terms of the nonbasic variables in the solution given in the table.

(b) Write the cuts found in part (a) in terms of the original structural variables.

(c) What Dantzig cut should be added to continue the cutting plane procedure?

(d) Write the Dantzig cut in terms of the original structural variables.

(e) Add the Gomory cut with the largest RHS value to the original LP and use an LP code to find the solution. Report the values of the variables in the optimal solution. How has the objective value changed with the addition of the cut?

14. The tableau gives the LP solution to the relaxation of an IP problem with maximization objective. Assume all a_{ij} and b_i coefficients in the original problem are integer valued.

Row number	Basic variables	Coefficients							RHS
		z	x_1	x_2	x_3	x_4	x_5	x_6	
0	z	1	0.1	0.3	0	0.2	0	0	23.1
1	x_6	0	1.3	−0.3	0	−1.0	0	1	5.3
2	x_3	0	0	1.1	1	0.4	0	0	1.6
3	x_5	0	−0.8	−0.2	0	−0.5	1	0	3.7

(a) Write out all Gomory cuts that can be derived from the tableau.

(b) Write out the Dantzig cut that can be derived from the tableau.

(c) Add the Gomory cut with the largest RHS value to the tableau and use the dual simplex method to find the new solution. Perform these computations by hand.

15. Find the relaxed LP solution to the following IP.

$$\text{Maximize } 2x_1 + 5x_2$$
$$\text{subject to } \quad x_1 + x_2 \leq 5$$
$$-x_1 + x_2 \leq 2$$
$$x_1 - x_2 \leq 2$$
$$x_1 + x_2 \geq 3$$
$$x_1, x_2 \geq 0 \text{ and integer}$$

Now identify which of the following cuts are valid for use in a cutting plane algorithm. The constraints are not cumulative, so each part should be analyzed separately. Justify your conclusion in each case.

(a) $x_1 \leq 3$

(b) $x_2 \leq 4$

(c) $x_1 + 3x_2 \leq 10$

(d) $1.75x_3 - 0.75x_4 \geq 0.25$ (Note that x_3 and x_4 are, respectively, the slack variables for the first two constraints.)

16. An oil production company has eight wells that have failed and are not producing oil. Engineers have estimated the time it takes for each well to be "worked over" in order that production may resume. Also estimated is the amount of lost production per day for each well. A single "workover crew" is to repair all the wells.

(a) Write a mathematical programming model that can be used to find a well repair schedule that minimizes the total oil lost. Define all terms.

(b) Using the data from the table in conjunction with a greedy algorithm, determine the schedule of well repair that will minimize the total oil lost.

Well number	Workover time (days)	Production lost (barrels per day)
1	10	30
2	5	10
3	20	15
4	8	12
5	4	14
6	15	40
7	2	5
8	10	18

A note not important to the solution: Even though all oil is eventually produced from a well, the amount not produced while the well is down is delayed until the end of the well's life. Because of the time value of money, the income from the oil may be reduced significantly in value, so we assume for this problem that it is lost.)

17. Propose greedy heuristics for the problems below taken from Chapter 7. For each problem find an answer using your heuristic and compare it to the optimal solution. Discuss why the heuristic might fail to find the optimum for all instances of the problem class.

   (a) Solve the telecommunications network design problem described in Section 7.3 with data given in Figure 7.1.
   (b) Solve the facility location problem described in Section 7.4 with the data given in Table 7.2.
   (c) Solve the covering problem described in Section 7.5 with the data given in Figure 7.6.
   (d) Solve the traveling salesman problem described in Section 7.6 with the data given in Figure 7.7. Use the nearest neighbor heuristic.
   (e) Solve the directed minimal spanning tree problem described in Section 7.6 with the data given in Figure 7.7. Use the nearest neighbor heuristic.
   (f) Solve the cutting stock problem described in Section 7.7 with data given in Figure 7.14.

18. (*Lagrangian relaxation*) Consider the following IP in which the m constraints have been partitioned into two groups of sizes m_1 and m_2, respectively, such that $m = m_1 + m_2$, and the original matrix $\mathbf{A} = (\mathbf{A}_1, \mathbf{A}_2)^\mathsf{T}$ and the original RHS vector $\mathbf{b} = (\mathbf{b}_1, \mathbf{b}_2)^\mathsf{T}$.

$$
\begin{aligned}
z_{IP} = \text{Maximize } & \mathbf{cx} \\
\text{subject to } & \mathbf{A}_1\mathbf{x} \le \mathbf{b}_1 \\
& \mathbf{A}_2\mathbf{x} \le \mathbf{b}_2 \\
& \mathbf{x} \in Z_+^n
\end{aligned}
$$

A relaxation of this problem can be created by multiplying each constraint, say in the first group, by a nonnegative number (a weight) and adding the result to the objective function. Let λ be an m_1-dimensional, nonnegative row vector. The relaxed problem as a function of λ is

$$
\begin{aligned}
z_{LR}(\lambda) = \text{Maximize } & \mathbf{cx} + \lambda(\mathbf{b}_1 - \mathbf{A}_1\mathbf{x}) \\
\text{subject to } & \mathbf{A}_2\mathbf{x} \le \mathbf{b}_2 \\
& \mathbf{x} \in Z_+^n
\end{aligned}
$$

   (a) Why is this problem considered a relaxation of the original integer program?
   (b) Show that when the relaxed IP is solved for any $\lambda \ge \mathbf{0}$, we have $z_{LR}(\lambda) \ge z_{IP}$.
   (c) To find the "best" value of λ, it is necessary to solve the *Lagrangian dual*:

$$
z_{LD} = \text{Minimize } \{z_{LR}(\lambda): \lambda \ge \mathbf{0}\}
$$

   Explain why $z_{LD} \ge z_{IP}$. Under what circumstances might $z_{LD} = z_{IP}$?

19. In solving the directed traveling salesman problem (TSP), it is possible to use the assignment problem (AP) as a relaxation. This approach treats the intercity distance matrix of the TSP as the cost matrix of the AP. Solving the AP for this matrix yields a set of cells such that no more than one cell is selected from each row and column.

    (a) Explain why this approach is considered a relaxation of the TSP.

    (b) Interpret the AP solution in terms of a TSP.

20. (*Branch and cut*) How could a cutting plane technique be incorporated into an implicit enumeration technique for solving the pure IP problem?

    (a) Write out the steps.

    (b) Under what circumstances could the cuts be used at each node of the search tree?

21. (*Generalized Gomory cut*) Suppose we wish to use a cutting plane technique to solve a pure IP problem. Assume that the LP relaxation has been solved and that the ith constraint in the simplex tableau is

$$x_{B(i)} + \sum_{j \in Q} \bar{a}_{ij} x_j = \bar{b}_i$$

    where $x_{B(i)}$ is the ith basic variable and Q is the set of nonbasic variables. Multiply this equation by any nonzero rational number h. Now, using the same approach outlined in Section 8.4 to derive the basic Gomory fractional cut given in Equation (14), derive from the preceding equation a more generalized version of Equation (14).

22. (*Mixed-integer cuts*) From the discussion in the text on rounding, we know that when $y \leq b$, $y \in Z^1$, the inequality $y \leq \lfloor b \rfloor$ is valid for any IP. This idea is now extended for mixed-integer programs (Wolsey [1998]).

    (a) Let $X^{\geq} = \{(x, y) \in \mathfrak{R}_+^1 \times Z^1 : x + y \geq b\}$ and $f = b - \lfloor b \rfloor > 0$. Prove that the inequality

$$x \geq f(\lceil b \rceil - y) \text{ or } \frac{x}{f} + y \geq \lceil b \rceil$$

    is valid for $X^{\geq}$. (Note that this inequality holds when $x \leftarrow \sum_j a_j x_j$ for a_j, $x_j \geq 0$, and when $y \leftarrow \sum_k d_k y_k$ for d_k, $y_k \geq 0$ and integer.)

    (b) Let $X^{\leq} = \{(x, y) \in \mathfrak{R}_+^1 \times Z^1 : -x + y \leq b\}$ and $f = b - \lfloor b \rfloor > 0$. Prove that the inequality

$$-\frac{x}{1-f} + y \leq \lfloor b \rfloor$$

    is valid for $X^{\leq}$. [See "Note" in part (a).]

    (c) Consider the set $X^{\text{MIR}} = \{(x, y) \in \mathfrak{R}_+^1 \times Z_+^2 : -x + a_1 y_1 + a_2 y_2 \leq b\}$ where a_1, a_2, and b are scalars with $b \notin Z^1$. Let $f = b - \lfloor b \rfloor$, $f_i = a_i \rfloor - \lfloor a_i$ for $i = 1, 2$ and suppose that $f_1 \leq f \leq f_2$. Prove that

$$-\frac{x}{1-f} + \lfloor a_1 \rfloor y_1 + \left(\lfloor a_2 \rfloor + \frac{f_2 - f}{1 - f} \right) y_2 \leq \lfloor b \rfloor$$

    is valid for X^{MIR}. *Hint:* make use of the result in part (b) and the relation $a_2 = \lceil a_2 \rceil - (1 - f_2)$. (Note that this inequality holds when $x \leftarrow \sum_j a_j x_j$ for a_j, $x_j \geq 0$, and when $y \leftarrow \sum_k d_k y_k$ for $y_k \geq 0$ and integer and $d_k \geq 0$ and any two d_k fractional with the rest integer.)

23. Prove the validity of the greedy algorithm for the minimal spanning tree problem by

    (a) induction (prove that it is valid for two nodes, assume that it is valid for n nodes, and then prove that it is valid for $n + 1$ nodes)

    (b) contradiction (assume that a better solution exists that does not satisfy the conditions of the algorithm and then show this to be impossible)

24. Consider the following four models. For every pair (there are six), determine whether or not one is a relaxation of the other. Also show the relation ($\leq$, $=$, or $\geq$) between the optimal objective values of each pair. If no relation can be established, so indicate.

(a) $z_a^* =$ Maximize $4x_1 + 3x_2 + x_3$
       subject to $2x_1 - x_2 + 4x_3 \leq 5$
                  $3x_1 + 2x_2 - x_3 \leq 3$
                  $x_j = 0$ or 1, $j = 1, 2, 3$

(b) $z_b^* =$ Maximize $4x_1 + 3x_2 + x_3$
       subject to $2x_1 - x_2 + 4x_3 \leq 5$
                  $3x_1 + 2x_2 - x_3 \leq 3$
                  $0 \leq x_j \leq 1$, $j = 1, 2, 3$

(c) $z_c^* =$ Maximize $4x_1 + 3x_2 + x_3$
       subject to $5x_1 + x_2 + 3x_3 \leq 8$
                  $x_j = 0$ or 1, $j = 1, 2, 3$

(d) $z_d^* =$ Maximize $4x_1 + 3x_2 + x_3$
       subject to $5x_1 + x_2 + 3x_3 \leq 8$
                  $0 \leq x_j \leq 1$, $j = 1, 2, 3$

25. Consider the following 0-1 knapsack problem.

$$\text{Maximize } 14x_1 + 10x_2 + 8x_3$$
$$\text{subject to } \quad 8x_1 + 6x_2 + 5x_3 \leq 10$$
$$x_j = 0 \text{ or } 1, \ j = 1, 2, 3$$

Solve this problem using branch and bound with LP relaxations to provide bounds.

26. Consider the following IP model and the optimal tableau of its LP relaxation.

$$\text{Maximize } z = 2x_1 + x_2 - 3x_3 + 5x_4$$
$$\text{subject to} \quad 3x_1 - x_2 + x_3 + 2x_4 \leq 8$$
$$x_1 + x_2 + 4x_3 - x_4 \leq 6$$
$$2x_1 + 3x_2 - x_3 + x_4 \leq 10$$
$$x_1 \quad\quad + x_3 + x_4 \leq 7$$
$$x_j \geq 0 \text{ and integer, } j = 1, \ldots, 4$$

Row number	Basic variables	z	x_1	x_2	x_3	x_4	x_{s1}	x_{s2}	x_{s3}	x_{s4}	RHS
						Coefficients					
0	z	1	6	0	4	0	2	0	1	0	26
1	x_4	0	11/7	0	2/7	1	3/7	0	1/7	0	34/7
2	x_{s2}	0	17/7	0	33/7	0	4/7	1	−1/7	0	64/7
3	x_2	0	1/7	1	−3/7	0	−1/7	0	2/7	0	12/7
4	x_{s4}	0	−4/7	0	5/7	0	−3/7	0	−1/7	1	15/7

(a) Write the Dantzig cut that can be derived from the tableau.

(b) From the set of possible Gomory cuts, write the one with the greatest RHS value.

(c) Write the Gomory cut that can be derived from row 0.

(d) Write the Gomory cut that can be obtained from row 4. Add it to the tableau and reoptimize using the dual simplex method.

(e) Solve the problem with the cutting plane algorithm embodied in the Teach IP Excel Add-in.

27. Consider the IP

$$\text{Maximize} \quad x_1 + x_2$$
$$\text{subject to} \quad -2x_1 + 5x_2 \le 5$$
$$4x_1 - 3x_2 \le 4$$
$$x_1, x_2 \ge 0 \text{ and integer}$$

When this problem is solved as an LP, both constraints are tight at optimality.

(a) Identify the basic variables, and show the basis, basis inverse, and solution for the optimal LP.

(b) Add a Gomory cut and use the dual simplex method to reoptimize.

28. Consider the IP

$$\text{Maximize } z = \quad x_1 + x_2$$
$$\text{subject to} \quad -11x_1 + 4x_2 - x_3 \qquad\qquad = 4$$
$$2x_1 + x_2 \qquad + x_4 \qquad = 3$$
$$-6x_1 + 4x_2 \qquad\qquad + x_5 = 7$$
$$x_j \ge 0 \text{ and integer}, \quad j = 1, \ldots, 5$$

After solving the LP relaxation, we obtain the following tableau.

Basic variables	Basic variables	Coefficients						RHS
		z	x_1	x_2	x_3	x_4	x_5	
0	z	1	0	0	0	0.714	0.071	2.64
1	x_2	0	0	1	0	0.428	0.142	2.28
2	x_1	0	1	0	0	0.285	−0.071	0.36
3	x_3	0	0	0	1	−1.42	1.35	1.21

(a) Show that $3x_4 + x_5 \ge 2$ is a valid cut that can be derived from this tableau.

(b) What other cuts can be derived from this tableau?

29. Solve the following problem below with branch and bound. Perform several iterations by hand and then use the accompanying Excel add-in.

$$\text{Maximize } 2x_1 + 3x_2 + 4x_3 + 7x_4$$
$$\text{subject to } 4x_1 + 6x_2 - 2x_3 + 8x_4 = 20$$
$$x_1 + 2x_2 - 6x_3 + 7x_4 = 10$$
$$x_j \ge 0 \text{ and integer}, \quad j = 1, 2, 3, 4$$

BIBLIOGRAPHY

Ahuja, R.K., T.L. Magnanti, and J.B. Orlin, *Network Flows: Theory, Algorithms, and Applications*, Prentice–Hall, Englewood Cliffs, NJ, 1993.

Balas, E., "An Additive Algorithm for Solving Linear Programs with Zero-One Variables," *Operations Research*, Vol. 13, pp. 517–546, 1965.

Balas, E. and P. Toth, "Branch and Bound Methods," in E.L. Lawler, J.K. Lenstra, A.H.G. Rinnooy Kan, and D.B. Shmoys (editors), *The Traveling Salesman Problem: A Guided Tour of Combinatorial Optimization*, Wiley, New York, pp. 361–401, 1985.

Balas, E., S. Ceria, G. Cornuéjols, and N. Natraj, "Gomory Cuts Revisited," *Operations Research Letters*, Vol. 19, pp. 1–9, 1996.

Bard, J.F. and J.T. Moore, "An Algorithm for the Discrete Bilevel Programming Problem," *Naval Research Logistics*, Vol. 39, pp. 419–435, 1992.

Bard, J.F., G. Kontoravdis, and G. Yu, "A Branch-and-Cut Procedure for the Vehicle Routing Problem with Time Windows," *Transportation Science*, Vol. 36, No. 2, pp. 250–269, 2002.

Barnhart, C., E.L. Johnson, G.L. Nemhauser, M.W.P. Savelsbergh, and P.H. Vance, "Branch and Price: Column Generation for Solving Huge Integer Programs," *Operations Research*, Vol. 46, No. 3, pp. 316–329, 1998.

Cheriton, D. and R. E. Tarjan, "Finding Minimum Spanning Trees," *SIAM Journal on Computing*, Vol. 5, pp. 724–742, 1976.

Crowder, H., E.L. Johnson, and M.W. Padberg, "Solving Large-Scale Zero-One Linear Programming Problems," *Operations Research*, Vol. 31, No. 5, pp. 803–834, 1983.

Fisher, M.L., "An Applications Oriented Guide to Lagrangian Relaxation," *Interfaces*, Vol. 15, No. 2, pp. 10–21, 1985.

Garfinkel, R.S. and G.L. Nemhauser, *Integer Programming*, Wiley, New York, 1972.

Geoffrion, A.M., "Lagrangian Relaxation for Integer Programming," *Mathematical Programming Study*, Vol. 2, pp. 82–114, 1974.

Glover, F. "Tabu Search—Part I," *ORSA Journal on Computing*, Vol. 1, No. 3, pp. 190–206, 1989.

Grötschel, M. and M.W. Padberg, "Polyhedral Theory," in E.L. Lawler, J.K. Lenstra, A.H.G. Rinnooy Kan, and D.B. Shmoys (editors), *The Traveling Salesman Problem: A Guided Tour of Combinatorial Optimization*, Chapter 8, pp. 251–305, Wiley, New York, 1985.

Lawler, E.L., *Combinatorial Optimization: Networks and Matroids*, Saunders College Publishing, 1976.

Lawler, E.L., J.K. Lenstra, A.H.G. Rinnooy Kan, and D.B. Shmoys, "Sequencing and Scheduling: Algorithms and Complexity," in S.S. Graves, A.H.G. Rinnooy Kan, and P. Zipkin (editors), *Handbook in Operations Research and Management Science*, Vol. 4: *Logistics of Production and Inventory*, Elsevier Science Amsterdam, and North-Holland, Amsterdam, pp. 445–522, 1993.

Murty, K.G., *Network Programming*, Prentice–Hall, Englewood Cliffs, NJ, 1992.

Nemhauser, G.L. and L.A. Wolsey, *Integer and Combinatorial Optimization*, Wiley, New York, 1988.

Papadimitrou, C.H. and K. Stieglitz, *Combinatorial Optimization: Algorithms and Complexity*, Prentice–Hall, Englewood Cliffs, NJ, 1982.

Parker, G. and R. Rardin, *Discrete Optimization*, Academic Press, New York, 1988.

Pinedo, M., *Scheduling: Theory, Algorithms, and Systems*, Prentice-Hall, Englewood Cliffs, NJ, 1995.

Savelsbergh, M.W.P., "Preprocessing and Probing Techniques for Mixed Integer Programming Problems," *ORSA Journal on Computing*, Vol. 6, No. 4, pp. 445–454, 1994.

Suhl, U.H. and R. Szymanski, "Supernode Processing of Mixed-Integer Models," Working paper, Institut für Wirtschaftsinformatik und Operations Research, Free University of Berlin, Berlin, 1994.

Taillard, E., "Some Efficient Heuristic Methods for the Flow Shop Sequencing Problem," *European Journal of Operational Research*, Vol. 47, No. 1, pp. 65–74, 1990.

Wolsey, L.A., *Integer Programming*, Wiley, New York, 1998.

Chapter 9

Nonlinear Programming Models[1]

The principal abstraction of the linear programming (LP) model is that all functions are linear when the constant terms are ignored. This leads to several powerful results that greatly facilitate our ability to find solutions. The first is that all local optima are global optima; the second is that if the optimal value of the objective function is finite, at least one of the extreme points in the set of feasible solutions will be an optimal solution. Furthermore, starting at any extreme point in the feasible region, it is possible to reach an optimal extreme point in a finite number of steps by moving only to an adjacent extreme point in an improving direction. The simplex method embodies these ideas and has proven to be extremely efficient.

Nevertheless, much of the world is nonlinear, and so it is natural to ask if it is possible to achieve the same efficiency with nonlinear models. In many contexts, the elements of a linear model are really approximations of more complex relationships. Economies of scale in manufacturing, for example, lead to decreasing costs, while biological systems commonly exhibit exponential growth. In the design of a simple hatch cover, the shearing stress, bending stress, and degree of deflection are polynomial functions of flange thickness and beam height. Similar relationships abound in engineering design, economics, and distribution systems, to name a few.

The appeal of nonlinear programming (NLP) is strong because of the modeling richness it affords. In this chapter, we examine a variety of applications and discuss the difficulties arising in the solution process. Unfortunately, NLP solvers have not yet achieved the same level of performance and reliability associated with LP solvers. For all but the most structured problems, the solution obtained from an NLP solver may not be globally optimal. This argues for caution. Before taking any action, the decision maker should have a full understanding of the nonlinearities governing the system under study.

9.1 MANUFACTURING EXAMPLE

To show how nonlinear relationships can make a mathematical programming model more realistic, we return to the manufacturing example presented in Chapter 2, where the concepts of linear programming were introduced. The only difference in the two models is that now the amount of raw material used is treated as a variable. Tables 9.1 to 9.3 provide revenue and cost data, as well as information relating production to machine capacity and raw

---

[1] J. Wesley Barnes contributed the initial draft of this chapter.

Table 9.1 Market Data

Product	P	Q	R
Revenue	$90	100	70
Maximum sales	100	40	60

Table 9.2 Machine Data

| Machine | Processing time (min/unit) | | | Availability (min) |
	P	Q	R	
A	20	10	10	2400
B	12	28	16	2400
C	15	6	16	2400
D	10	15	0	2400

Table 9.3 Raw Material Data

| Material (parts) | Requirements (parts/unit) | | | Cost ($/unit) |
	P	Q	R	
M1	1	0	0	$20
M2	1	1	0	20
M3	0	1	1	20
M4	1	0	0	5

material usage. The LP model will now be repeated. After examining its solution, we will introduce several nonlinearities and discuss their implications.

Linear Programming Model

Variables:

P number of units of product P to produce during the week

Q number of units of product Q to produce during the week

R number of units of product R to produce during the week

M_j number of units of raw material j to purchase, $j = 1, \ldots, 4$

Objective:

Maximize profit
$$Z = 90P + 100Q + 70R - 20M_1 - 20M_2 - 20M_3 - 5M_4$$

Machine Constraints:

A: $20P + 10Q + 10R \leq 2400$

B: $12P + 28Q + 16R \leq 2400$

C: $15P + 6Q + 16R \leq 2400$

D: $10P + 15Q + \leq 2400$

Raw Material Constraints:

M1: $P - M_1 = 0$

M2: $P + Q - M_2 = 0$

M3: $Q + R - M_3 = 0$

M4: $P - M_4 = 0$

Nonnegativity and Upper Bounds:

$0 \leq P \leq 100, \ 0 \leq Q \leq 40, \ 0 \leq R \leq 60, \ M_j \geq 0$ for $j = 1, \ldots, 4$

Solution: Profit = 7663.6

	P	Q	R
Product sales ($)	81.8	16.4	60

	M1	M2	M3	M4
Materials purchased	81.8	98.2	76.4	81.8

	A	B	C	D
Machine usage (min)	2400	2400	2285	1064

The accompanying solution agrees with that found in Chapter 2. Note that if we were to solve for each M_j in the raw material constraints and substitute the results into the objective function, we would get the same coefficients as those in the original model. Also, because this is an LP problem, the solution occurs at a vertex of the polyhedral feasible region.

For this type of manufacturing problem, there are many nonlinearities that might arise in practice. Some possibilities follow.

- Revenue is a nonlinear function of sales. The marginal revenue may increase or decrease with sales.
- Material cost is a nonlinear function of amount purchased. The marginal cost may increase or decrease with the amount purchased.
- Machine usage is a nonlinear function of production volume. This might result from congestion on the shop floor.

Focusing on the objective function, we now examine several nonlinear relationships that serve to illustrate the important difference between local and global optima. The feasible region will remain the same.

Product Revenue as a Concave Function of Sales

Linearity requires that the revenue for each unit of product sold remains constant over the range from zero to the value given in Table 9.1 for maximum sales. In many situations, however, companies may reduce the unit prices of their products to boost sales. When such a policy is employed, a product's marginal revenue decreases with the amount sold.

In economic terms, the marginal revenue is the derivative of the total revenue function with respect to sales volume. Call the marginal revenues for the three products r_P, r_Q, and r_R, and assume that they are linear functions of P, Q, and R, respectively. In the following modification, the relevant objective function coefficients are defined such that the marginal revenue reduces to half of the original value at maximum sales. For product P, the original value of revenue is $90 per unit and maximum sales are 100. We model the marginal revenue as $r_P = 90 - (45/100)P$. The first sale provides a revenue of $90 while the last sale provides a revenue of $45. Similarly, the other marginal revenues are

$$r_Q = 100 - (50/40)Q \text{ and } r_R = 70 - (35/60)R$$

The total revenue associated with the products is the integral of the marginal values. In particular,

$$f_P(P) = 90P - 0.225P^2, \quad f_Q(Q) = 100Q - 0.625Q^2, \quad \text{and} \quad f_R(R) = 70R - 0.292R^2$$

Replacing the original revenue terms for P, Q, and R generates the following nonlinear objective function.

$$Z = 90P - 0.225P^2 + 100Q - 0.625Q^2 + 70R - 0.292R^2$$
$$- 20M_1 - 20M_2 - 20M_3 - 5M_4$$

This is called a *quadratic, separable function* because the highest order of the nonlinear terms is 2, and each term is a function of only one variable. By comparing the new solution given below with the one obtained for the linear model, we see that the profit is reduced, production of product Q is increased, and production of products P and R is reduced. The bottleneck is machine B, which is used for the entire 2400 minutes it is available.

Solution: Profit = 5004.4

	P	Q	R	
Product sales ($)	70.8	23.5	55.7	

	M1	M2	M3	M4
Materials purchased	70.8	94.4	79.2	70.8

	A	B	C	D
Machine usage (min)	2209	2400	2095	1061

In contrast to linear programming, this solution is not an extreme point even though the feasible region is polynomial. This observation follows from the fact that all three product variables, all four raw material variables, and three of the four machine usage variables—10 variables in all—are not at extreme values. Because there are eight structural constraints, an extreme point solution will have at most eight variables falling strictly within their lower and upper bounds.

As discussed later in this chapter, we note that the objective function Z is a concave function of the decision variables P, Q, and R. For a maximization problem, when the objective function is concave and the constraints form a convex feasible region, we can be sure that the solution obtained is a global maximum (i.e., no other feasible solution provides a greater objective value). This is discussed in the supplementary material on linear programming models on the accompanying CD, where it is shown how to approximate a nonlinear concave function with a piecewise linear one. We now illustrate what may happen when the objective function is convex rather than concave. Various difficulties arise that are not readily apparent.

Revenue as a Convex Function of Sales

A slight change in the assumptions alters the structure of the problem significantly and makes it much harder to solve. To see this, suppose the products have increasing rather than decreasing marginal revenues. For our example, assume the following relationships.

$$r_P = 45 + (45/100)P, \quad r_Q = 50 + (50/40)Q, \quad r_R = 35 + (35/60)R$$

This is just the opposite of the previous situation: now the marginal revenue of P starts at 45 and ends at 90. Similar relations hold for Q and R. Although this may seem odd for most products, increasing marginal revenues might be appropriate when goods are rationed. In any case, these functional forms serve to illustrate what happens when revenue is a convex function of the decision variables. The new objective function is

$$Z = 45P + 0.225P^2 + 50Q + 0.625Q^2 + 35R + 0.292R^2$$
$$- 20M_1 - 20M_2 - 20M_3 - 5M_4$$

The resultant problem was solved with the NLP code available in the Excel Solver. The algorithm embedded in this code, as well as in all nonlinear solvers, is an iterative procedure that starts with an initial solution and searches over the decision space until a feasible solution is reached such that no improvement is possible within its neighborhood. Informally, this means that if we move in any direction from the current point, the objective value will either remain constant or degrade. Such a solution is called a *local maximum* because it has the largest objective value in a local region. In contrast, a *global maximum* is a solution that provides the largest objective value among all feasible solutions.

For this example, we tried five different starting points for the variables P, Q, and R. The initial values of all M_j were then chosen such that the raw material constraints were satisfied. The results are shown in Table 9.4 in the rows labeled "Final." In the second run, for example, the initial values were $P_0 = 50$, $Q_0 = 50$, $R_0 = 50$; the algorithm converged to $P^* = 26.7$, $Q^* = 40$, $R^* = 60$, yielding $Z^* = 3510$. In the third run, the algorithm made no progress. It terminated at the same point at which it started, implying that $P_0 = 100$, $Q_0 = 40$, $R_0 = 0$ is a local optimum.

The results in Table 9.4 indicate the difficulty encountered in maximizing a convex function (or minimizing a concave function), regardless of the feasible region. To a large extent, the path taken by a solution algorithm and the point to which it converges depend on the initial point chosen. All the solutions identified in Table 9.4 are local optima, but since we don't know if all local optima have been identified, we may not have found the global optimum. Unfortunately, this predicament is all too common in attempts to optimize nonlinear functions and becomes even more disconcerting when there are nonlinearities in the constraints. As the number of nonlinear terms in a problem increases, the amount of work required to find solutions typically goes up at an exponential rate.

The extension of the original linear model to include nonlinear relationships adds more realism to the analysis, but one factor is still missing: we have not included any requirement

Table 9.4   Illustration of Local Optima

	Solution				
	1	2	3	4	5
Initial P	0	50	100	90	50
Initial Q	0	50	40	0	20
Initial R	0	50	0	50	60
Final P	0	26.7	100	90	81.8
Final Q	40	40	40	0	16.4
Final R	60	60	0	60	60
Final A	1000	1533	2400	2400	2400
Final B	2080	2400	2320	2040	2400
Final C	1200	1600	1740	2310	2285
Final D	600	867	1600	900	1063
Final Z	3350	3510	3650	3772	3787

Machine usage (brace spanning Final A–Final D)

for integrality of production and sales. In fact, integer NLP problems are much more difficult to solve than their integer LP counterparts and are not considered in this book. Most codes for such problems use ad hoc procedures that combine branch and bound with standard NLP solvers.

Before presenting additional examples, we will investigate the issue of local versus global optimality and the role convexity plays in finding solutions. Understanding these concepts is critical to the development and use of NLP techniques.

9.2 GENERAL CONSIDERATIONS

The vocabulary of nonlinear programming is primarily concerned with describing the characteristics of the mathematical expressions that define the objective function and constraints. These characteristics determine the kind of algorithm that will be used to solve the problem and indicate how the results are to be interpreted. We now introduce the general NLP model, identify the principal mathematical forms that its components can take, and provide several examples. As is common in the field, the basic model will be stated in terms of minimizing rather than maximizing an objective function $f(\mathbf{x})$.

The Nonlinear Programming Problem

The optimization problem under consideration is

$$\text{Minimize } f(\mathbf{x}) \tag{1}$$

$$\text{subject to } g_i(\mathbf{x}) \le b_i, \quad i = 1, \ldots, m \tag{2}$$

where $\mathbf{x} \in \Re^n$ is the n-dimensional vector of decision variables, $f(\mathbf{x})$ is the objective function, and $g_1(\mathbf{x}), \ldots, g_m(\mathbf{x})$ are the constraint functions. All functions are assumed to be continuous but of arbitrary form, and all RHS values b_i are assumed to be known constants. Simple upper and lower bounds for the variables are included implicitly in the constraints defined by Equation (2).

The "$\le$" form of the constraints is for convenience only and does not restrict the generality of the developments in any way. A constraint that has the "$\ge$" relation can be changed to "$\le$" by negating both sides, whereas an equality constraint can be replaced by two equivalent inequalities. In a similar vein, a minimization objective can be changed to a maximization objective by negation. A final assumption we make, mainly to conform to the assumptions underlying most NLP solvers, is that all problem functions are twice differentiable. If nondifferentiable functions were present, it would be necessary to rely on a higher level of mathematics than is suitable for an introductory text.

For the LP problem, recall that at least one optimal solution must occur at an extreme point of the feasible region, and that any extreme point that yields an objective function value that is at least as good as its neighbors (i.e., adjacent extreme points) is guaranteed to be the optimal solution of the problem. Unfortunately, neither of these properties is necessarily true when nonlinear functions are involved. To proceed, several new definitions are required.

Minima and Maxima

Let $S \subseteq \Re^n$ be the set of feasible points associated with the constraints defined by Equation (2).

Definition 1: A *global minimum* is any $\mathbf{x}^0 \in S$ such that

$$f(\mathbf{x}^0) \le f(\mathbf{x})$$

for all feasible $\mathbf{x}$ not equal to $\mathbf{x}^0$. A *unique global minimum* implies that the weak inequality ($\le$) is replaced by the strong inequality ($<$).

Definition 2: A *weak local minimum* is any $\mathbf{x}' \in S$ such that

$$f(\mathbf{x}') \le f(\mathbf{x})$$

for all feasible $\mathbf{x}$ not equal to $\mathbf{x}'$ in the neighborhood of $\mathbf{x}'$. A *strong local minimum* has $<$ in place of $\le$ in this inequality.

The terms *global maximum* and *local maximum* are defined in a similar manner by changing the sense of the inequalities in the preceding statements.

Convex Functions and Convex Sets

The importance of *convex functions* and *convex sets* stems from the following fact. In minimizing a convex function over a feasible region defined by a convex set, any local minimum is also a global minimum. A problem that has both of these characteristics is called a *convex program*. If the objective function is strictly convex, any local minimum is certain to be the unique global minimum and the problem is a *strict convex program*.

A convex function has the property that all points on a line connecting any two points $\mathbf{x}_1$ and $\mathbf{x}_2$ on the graph of the function lie on or above the function. Figure 9.1 illustrates this situation in one dimension.

Definition 3: A function $f(\mathbf{x})$ is convex if and only if

$$f(\lambda \mathbf{x}_1 + (1 - \lambda)\mathbf{x}_2) \le \lambda f(\mathbf{x}_1) + (1 - \lambda)f(\mathbf{x}_2)$$

for all $0 < \lambda < 1$. It is *strictly convex* if the inequality sign $\le$ is replaced with the sign $<$.

For a strictly convex function, the line lies completely above the graph of the function except at the endpoints of the line (where $\lambda = 0$ and $\lambda = 1$). Furthermore, a nonnegative linear combination of convex (strictly convex) functions is also a convex (strictly convex) function. Stated formally, we have the following lemma.

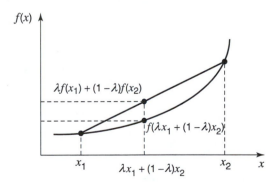

Figure 9.1   A strictly convex function in one dimension

Lemma 1: Let $f_i(\mathbf{x})$, $i = 1, \ldots, s$, be (strictly) convex functions defined on a convex set $S \subseteq \mathfrak{R}^n$. Then $f(\mathbf{x}) = \sum_{i=1}^{s} \alpha_i f_i(\mathbf{x})$ is (strictly) convex on S for all $\alpha_i \geq 0$.

A function is *concave* when the line joining any two points on its graph lies on or below the function.

Definition 4: A function $f(\mathbf{x})$ is concave if and only if

$$f(\lambda \mathbf{x}_1 + (1 - \lambda)\mathbf{x}_2) \geq \lambda f(\mathbf{x}_1) + (1 - \lambda)f(\mathbf{x}_2)$$

for all $0 < \lambda < 1$. If the $\geq$ sign is replaced by the strict inequality sign $>$, the function is *strictly concave*.

These definitions imply that the negative of a (strictly) convex function is a (strictly) concave function. It follows that in maximizing a concave function over a convex region we are certain that any local maximum is a global maximum. Indeed, if such a problem has a strictly concave objective function, any local maximum will be a unique global maximum. From a theoretical point of view, these facts allow us to focus exclusively on convex functions, because any concave function can be converted to a convex function by negation.

A second implication that can be drawn from these definitions is that linear functions are both convex and concave, and that they are the only functions that have this property. A linear function, however, cannot be strictly convex or strictly concave, as evidenced by the fact that LP solutions need not be unique.

Definition 5: A set $S \subseteq \mathfrak{R}^n$ is convex if any point on the line segment connecting any two points $\mathbf{x}_1, \mathbf{x}_2 \in S$ is also in S. Mathematically, this is equivalent to

$$\bar{\mathbf{x}} = \lambda \mathbf{x}_1 + (1 - \lambda)\mathbf{x}_2 \in S \text{ for all } \lambda \text{ such that } 0 \leq \lambda \leq 1$$

Figure 9.2 depicts several examples of convex and nonconvex sets. Note that we do not speak of "concave sets."

The feasible region of interest is denoted by the set

$$S = \{\mathbf{x} : g_i(\mathbf{x}) \leq b_i, i = 1, \ldots, m\}$$

The first question we ask is: What conditions must the constraint functions $g_i(\mathbf{x})$ satisfy in order to ensure that S is convex? The simplest answer is that when $g_i(\mathbf{x})$ is convex, the region defined by the inequality $g_i(\mathbf{x}) \leq b_i$ is a convex set. This result follows directly from Definition 5 and is illustrated in Figure 9.3 for a one-dimensional example in which $g(x)$ is con-

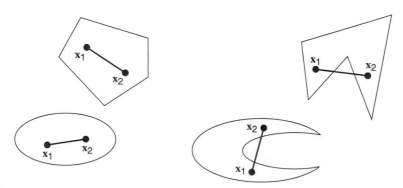

Figure 9.2 Examples of convex sets (left) and nonconvex sets (right)

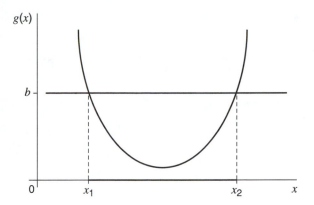

Figure 9.3   Convex constraint function defining a convex feasible region

vex. Clearly, the set of all x satisfying the constraint $g(x) \leq b$ is represented by the line join-ing x_1 and x_2. This is a convex region. The points satisfying the constraint $g(x) \geq b$ fall either below x_1 or above x_2, and do not define a convex region.

Lemma 2: Let $S_1, \ldots, S_m \subseteq \mathfrak{R}^n$ be convex sets. Then $S = S_1 \cap \cdots \cap S_m$ is a convex set—that is, the intersection of convex sets is also convex.

Because the feasible region of a mathematical program is the intersection of the fea-sible regions defined by the constraints, when all $g_i(\mathbf{x})$ are convex, S is convex. This argu-ment implies that the feasible region of a linear program is a convex set. Alternatively, it is easy to show that if $g_i(\mathbf{x})$ is strictly concave, the region defined by the constraint $g_i(\mathbf{x}) \leq b_i$ is not convex. A problem that includes a constraint defining a nonconvex region will have a nonconvex feasible region unless the constraint is redundant in the sense that it can be eliminated without changing the feasible region.

The word "convex" has two different meanings as used in this chapter. When applied to a function, it describes the shape of the function. A function may be convex, concave, or neither; linear functions are both convex and concave. When applied to a constraint, "con-vex" refers to the associated feasible region. When a set of points does not have the con-vex property, it is said to be nonconvex. In the remainder of this section, we will present several examples to illustrate these concepts.

Examples

Consider the unconstrained problem shown in Figure 9.4—the minimization of a function in two variables, $f(\mathbf{x}) = (x_1 - 1)^2 + (x_2 + 3)^2$. Such problems are often found in basic cal-culus. The two-dimensional plot on the left in the figure depicts isovalue contours of $f(\mathbf{x})$, whereas the three-dimensional plot on the right is the graph of $f(\mathbf{x})$. As we can see, $f(\mathbf{x})$ is strictly convex, and so it has a unique global minimum that occurs at $x_1 = 1$, $x_2 = -3$, yield-ing $f(1, -3) = 0$. If we add the nonnegativity constraint $x_2 \geq 0$ to the problem, the unique constrained global minimum occurs at $x_1 = 1$, $x_2 = 0$, where $f(1, 0) = 9$.

A constrained optimization problem is depicted in Figure 9.5. The objective function, $f(x) = \sin(y)$, is neither convex nor concave, whereas the feasible region $S = \{x: 0 \leq x \leq 5\pi\}$ is convex. Two global minima exist at $x = 3\pi/2$ and $x = 7\pi/2$, and two local minima exist at the constraint boundaries $x = 0$ and $x = 5\pi$. Three global maxima are also present.

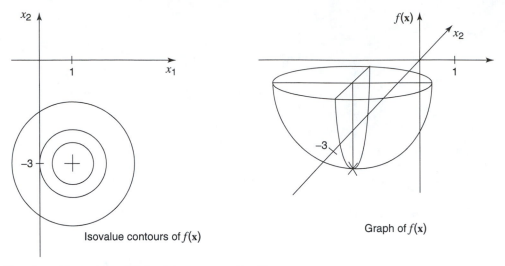

Figure 9.4   Function with a unique global minimum at $\mathbf{x} = (1, -3)$

The problem of maximizing a damped sinusoidal objective function is shown in Figure 9.6. There is a unique global maximum at the boundary $x = 0$ with $f(0) = 3$ and a unique global minimum at $x = 1.498$ radians with $f(\pi/2) = -0.6517$. In addition, there are five strong local maxima and five strong local minima exclusive of the global optima. Again, this objective function is neither convex nor concave.

Figure 9.7 depicts the isovalue contours of a parabola to be maximized over the shaded polyhedral constraint region. The objective function $f(\mathbf{x}) = (x_1 - 2)^2 + (x_2 - 2)^2$ is convex, and the constraint set is polyhedral. As we can see, two global maxima are present, one at $\mathbf{x} = (2, 5)$ and the other at $\mathbf{x} = (5, 2)$, with $f(5, 2) = f(2, 5) = 9$. Moreover, two strong local maxima exist at $\mathbf{x} = (2, 0)$ with $f(2, 0) = 4$, and $\mathbf{x} = (0, 3)$ with $f(0, 3) = 5$. These observations can be formalized in theoretical terms.

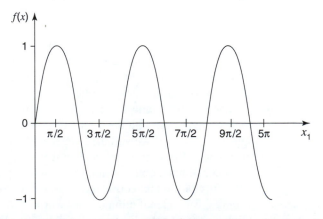

Figure 9.5   Constrained function with two global minima and three global maxima

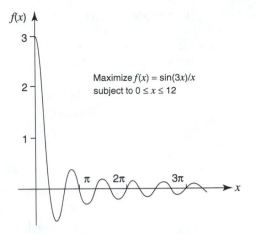

Figure 9.6   Constrained function with a unique global maximum, a unique global minimum, and strong local minima and maxima

Theorem 1: Let $\mathbf{x} \in \Re^n$ and let $f(\mathbf{x})$ be a convex function defined over a polyhedral constraint set S. If there is a finite solution to the problem

$$\text{Maximize}\{f(\mathbf{x}): \mathbf{x} \in S\} \tag{3}$$

then a global optimum exists at a vertex of S.

Corollary 1: Let $f(\mathbf{x})$ be a concave function defined over a polyhedral constraint set S. If there is a finite solution to the problem

$$\text{Minimize}\{f(\mathbf{x}): \mathbf{x} \in S\}$$

then a global optimum exists at a vertex of S.

Corollary 2: If the conditions of Theorem 1 hold and $f(\mathbf{x})$ is strictly convex, then all global optima are strong.

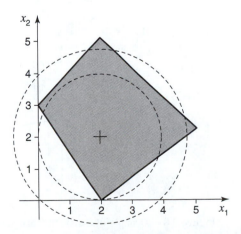

Figure 9.7   Linearly constrained function with two global maxima and two strong local maxima

The importance of Theorem 1 and its corollaries comes into play in the design of algorithms. Because a solution to Problem (3) occurs at a vertex of the feasible region, it is basic. This means that we can use simplex-type methods in the search for an optimal solution; however, we should not forget that local solutions that are not global may also exist at vertices.

For the problem

$$\text{Maximize } f(\mathbf{x}) = (x_1 - 2)^2 + (x_2 - 2)^2$$
$$\text{subject to } -3x_1 - 2x_2 \leq -6$$
$$-x_1 + x_2 \leq 3$$
$$x_1 + x_2 \leq 7$$
$$2x_1 - 3x_2 \leq 4$$

in Figure 9.7, the unique global minimum lies at $\mathbf{x}^* = (2, 2)$ and is interior to the feasible region. This leads to our second principal result.

Theorem 2: Let $\mathbf{x} \in \mathfrak{R}^n$ and let $f(\mathbf{x})$ be a convex function defined over a convex constraint set S. If there is a finite solution to the problem

$$\text{Minimize}\{f(\mathbf{x}): \mathbf{x} \in S\} \qquad (4)$$

then all local optima are global optima. If $f(\mathbf{x})$ is strictly convex, the optimal solution is unique.

As stated previously, Problem (4) is called a convex program and has been studied extensively. Notice that the feasible region is not restricted to be a polyhedron, but can be any convex set. When $f(\mathbf{x})$ is concave and the objective is to maximize, the same result holds. Figure 9.8 illustrates a nonconvex feasible region S formed by two nonlinear constraints.

$$S = \{(x_1, x_2): (0.5x_1 - 0.6)x_2 \leq 12x_1^2 + 3x_2^2 \geq 27; \ x_1, x_2 \geq 0\}$$

This region is said to be disjoint because the constraints partition the decision space into two regions that are not connected. Clearly, a disjoint feasible region cannot be convex.

The problem shown in Figure 9.9 provides an example of a nonlinear objective function and a nonconvex feasible region (model is below figure). Here we have a strong local minimum, this time owing its existence primarily to the nonlinear constraints rather than the mathematical form of the objective function. The strong local minimum lies at $\mathbf{x} = (0, 30)$ with $f(0, 30) = -810$ and is obviously inferior to the unique global minimum at $\mathbf{x} = (7, 23)$ with $f(7, 23) = -1162.1$.

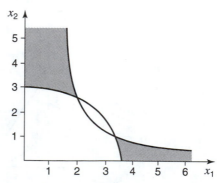

Figure 9.8   Example of a disjoint feasible region

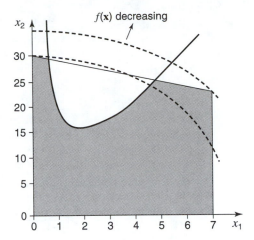

Figure 9.9   Nonconvex feasible region (shaded) with a strong local minimum

$$\text{Minimize}\quad f(\mathbf{x}) = -14x_1^2 - 0.9x_2^2$$

$$\text{subject to}\quad 3x_1 - \frac{13}{x_1} + 0.8x_2 \le 1.7$$

$$x_1 + x_2 \le 30$$

$$x_1 \le 7$$

$$x_1, x_2 \ge 0$$

The next example introduces a nonsmooth function in the objective and an equality constraint. When the feasible region is in part defined by a nonlinear equality constraint, it is never convex unless there is only one point that is feasible (degenerate case). To see this, consider the following problem, which is illustrated in Figure 9.10.

$$\text{Minimize}\quad f(x_1, x_2) = |x_1 - 2| + |x_2 - 2|$$

$$\text{subject to}\quad g_1(x_1, x_2) = x_1^2 + x_2^2 = 1$$

$$g_2(x_1, x_2) = x_1 - x_2^2 \ge 0$$

The dashed lines represent isovalue contours of the objective function. The feasible region is the arc of the circle lying within the parabola. The unique solution is seen by inspection to be $\mathbf{x}^* = \left(\sqrt{2}/2, \ \sqrt{2}/2 \right)$, yielding $f(\mathbf{x}^*) = 2.586$. If the equality constraint is removed, the solution becomes $\mathbf{x}^* = \left(2, \ \sqrt{2}/2 \right)$ with $f(\mathbf{x}^*) = 2.585$. If both constraints are removed, the solution is at $(2, 2)$, which is termed an unconstrained minimum.

To conclude this section, we note that all the functions in the preceding examples are differentiable with the exception of the objective function in the last example. Usually, we require that each of the $m + 1$ functions in the NLP, the objective and m constraints, be continuous. However, much of NLP theory concerns the case in which the functions are at least once, and often twice, continuously differentiable. In these instances, it is possible to prove theorems that characterize solutions to the NLP and in turn influence the development of algorithms.

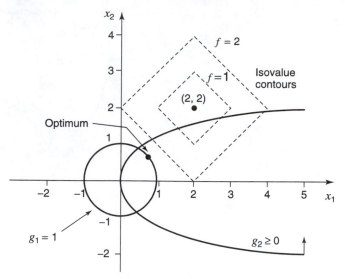

Figure 9.10   Geometry of example with nonsmooth objective function

9.3   DETERMINING CONVEXITY

Graphs and diagrams provide insights into the nature of convexity but they are not very convenient for dealing with functions of more than one or two variables. In this section, we use more sophisticated methods for ascertaining the contours of a given function. In the discussion, it will be assumed that all functions are at least twice continuously differentiable and are defined over convex sets. In general, a function f that can be differentiated at least k times is written mathematically as $f \in C^k$, where C^k is the set of continuous functions that have k or more derivatives.

Single Dimension

The concavity or convexity of functions of a single variable is easy to determine using the second derivative.

- A function $f(x) \in C^2$ is convex if and only if its second derivative is nonnegative. If the inequality is strict (>), the function is strictly convex.

$$f(x) \text{ convex if } \frac{d^2 f(x)}{dx^2} \geq 0 \text{ for all } x$$

- A function $f(x) \in C^2$ is concave if and only if its second derivative is nonpositive. If the inequality is strict (<), the function is strictly concave.

$$f(x) \text{ concave if } \frac{d^2 f(x)}{dx^2} \leq 0 \text{ for all } x$$

Multiple Dimensions

To determine convexity in multiple dimensions, we define the following.

- The *gradient* of $f(\mathbf{x})$, written $\nabla f(\mathbf{x})$, is the n-dimensional column vector formed by the first partial derivatives of $f(\mathbf{x})$ with respect to each component of $\mathbf{x}$.

$$\nabla f(\mathbf{x}) = \left(\frac{\partial f(\mathbf{x})}{\partial x_j} \right)$$

- The *Hessian matrix* $\mathbf{H}(\mathbf{x})$ associated with $f(\mathbf{x})$ is the $n \times n$ symmetric matrix of second partial derivatives of $f(\mathbf{x})$ with respect to the components of $\mathbf{x}$.

$$\mathbf{H}(\mathbf{x}) = \left(\frac{\partial^2 f(\mathbf{x})}{\partial x_i x_j} \right)$$

For example, for $f(\mathbf{x}) = 3x_1^2 + 4x_2^3 - 5x_1 x_2 + 4x_1$ we have

$$\nabla f(\mathbf{x}) = \begin{pmatrix} 6x_1 - 5x_2 + 4 \\ 12x_2^2 - 5x_1 \end{pmatrix} \text{ and } \mathbf{H}(\mathbf{x}) = \begin{bmatrix} 6 & -5 \\ -5 & 24x_2 \end{bmatrix}$$

When $f(\mathbf{x})$ is quadratic, $\mathbf{H}(\mathbf{x})$ has only constant terms; when $f(\mathbf{x})$ is linear, $\mathbf{H}(\mathbf{x})$ does not exist.

We now define several useful properties of the Hessian matrix.

- $\mathbf{H}(\mathbf{x})$ is *positive semidefinite* if and only if $\mathbf{x}^T \mathbf{H} \mathbf{x} \geq 0$ for all $\mathbf{x}$ and there exists an $\mathbf{x} \neq \mathbf{0}$ such that $\mathbf{x}^T \mathbf{H} \mathbf{x} = 0$.
- $\mathbf{H}(\mathbf{x})$ is *positive definite* if and only if $\mathbf{x}^T \mathbf{H} \mathbf{x} > 0$ for every $\mathbf{x} \neq \mathbf{0}$.
- $\mathbf{H}(\mathbf{x})$ is *indefinite* if and only if $\mathbf{x}^T \mathbf{H} \mathbf{x} > 0$ for some $\mathbf{x}$, and $\hat{\mathbf{x}}^T \mathbf{H} \hat{\mathbf{x}} < 0$ for some other $\hat{\mathbf{x}}$.

From Definition 3, Taylor's Theorem, and limit concepts used in calculus, it is straightforward to show the following.

- $f(\mathbf{x})$ is convex if and only if $f(\mathbf{x}_1) \geq f(\mathbf{x}_2) + \nabla^T f(\mathbf{x}_2)(\mathbf{x}_1 - \mathbf{x}_2)$ for all $\mathbf{x}_1$ and $\mathbf{x}_2$.
- $f(\mathbf{x})$ is concave if and only if $f(\mathbf{x}_1) \leq f(\mathbf{x}_2) + \nabla^T f(\mathbf{x}_2)(\mathbf{x}_1 - \mathbf{x}_2)$ for all $\mathbf{x}_1$ and $\mathbf{x}_2$.
- $f(\mathbf{x})$ is convex (strictly convex) if its associated Hessian matrix $\mathbf{H}(\mathbf{x})$ is positive semidefinite (definite) for all $\mathbf{x}$.
- $f(\mathbf{x})$ is neither convex nor concave if its associated Hessian matrix $\mathbf{H}(\mathbf{x})$ is indefinite.

The terms *negative definite* and *negative semidefinite* are also appropriate for the Hessian matrix and provide symmetric results for concave functions. Recall that a function $f(\mathbf{x})$ is concave if $-f(\mathbf{x})$ is convex. Consequently, it follows that a function is concave (strictly concave) if the Hessian matrix is everywhere negative semidefinite (definite).

Determining the "Definiteness" Characteristic

Because convexity plays such an important role in nonlinear programming and because the convexity of a function is determined by the definiteness of its Hessian matrix, it is important to develop a test for assessing this characteristic. For example, consider the second-order polynomial

$$f(\mathbf{x}) = 2x_1^2 + x_2^2 + 3x_3^2 + x_1 x_2 + 2x_1 x_3 + 3x_2 x_3, \text{ with } \mathbf{H} = \begin{bmatrix} 4 & 1 & 2 \\ 1 & 2 & 3 \\ 2 & 3 & 6 \end{bmatrix}$$

We would like to determine whether $f(\mathbf{x})$ is convex. Because the Hessian matrix is independent of $\mathbf{x}$ when f is quadratic, we can conclude that a quadratic is convex if its Hessian is positive definite or positive semidefinite.

One way to determine if $\mathbf{x}^T \mathbf{H} \mathbf{x} \geq 0$ for all $\mathbf{x}$ is to diagonalize $\mathbf{H}$ through a linear transformation. Let $\mathbf{S}$ be a nonsingular $n \times n$ matrix, and define the vector $\mathbf{z} = \mathbf{S}\mathbf{x}$. We wish to find an $\mathbf{S}$ such that

$$\mathbf{x}^T \mathbf{H} \mathbf{x} = (\mathbf{S}\mathbf{x})^T \mathbf{D}(\mathbf{S}\mathbf{x}) = \mathbf{z}^T \mathbf{D} \mathbf{z}$$

where $\mathbf{D}$ is a diagonal matrix. Once this result is obtained, it is easy to determine the character of the new form

$$\mathbf{z}^T \mathbf{D} \mathbf{z} = d_{11} z_1^2 + d_{22} z_2^2 + \cdots + d_{nn} z_n^2$$

and hence of $\mathbf{x}^T \mathbf{H} \mathbf{x}$. If all the coefficients d_{jj} are positive, the matrix $\mathbf{D}$ is positive definite as is $\mathbf{H}$. If all are nonnegative, but some are zero the matrix is positive semidefinite. If the signs are mixed, the matrix is indefinite. Similarly, we can conclude that the matrix is negative definite if all the coefficients are negative, and negative semidefinite if all are nonpositive but some are zero.

A popular method of diagonalizing a square symmetric matrix is based on first finding its eigenvalues and eigenvectors. As an alternative, we describe a more efficient method given by Beightler *et al.* [1979] that is similar to Gaussian reduction and equivalent to the "completing the square" technique of linear algebra (see Hadley [1973]). The steps will be presented and then demonstrated on the preceding Hessian matrix with $n = 3$. To begin, we set

$$\mathbf{S}_0 = \mathbf{H} = \begin{bmatrix} 4 & 1 & 2 \\ 1 & 2 & 3 \\ 2 & 3 & 6 \end{bmatrix}$$

The method requires n iterations to transform the matrix $\mathbf{S}_0$ into $\mathbf{S}_n$, the desired transformation $\mathbf{S}$ that will be an upper diagonal matrix with 1's on the main diagonal. Let s_{ij} be a component of $\mathbf{S}_k$. The Gaussian pivot element at iteration k is s_{kk} and corresponds to diagonal element d_{kk} of $\mathbf{D}$. A general statement of the algorithm follows.

Iteration k to Determine $\mathbf{S}_k$ from $\mathbf{S}_{k-1}$

1. If the kth diagonal element of $\mathbf{S}_{k-1}$ is nonzero, record $d_{kk} = s_{kk}$ as the diagonal element of row k. Divide row k by d_{kk} and perform the necessary row operations to make all elements of the rows $i = k + 1, \ldots, n$ in column k equal to zero. The matrix obtained is $\mathbf{S}_k$.

2. If the kth diagonal element of $\mathbf{S}_{k-1}$ is zero but some other diagonal element in row $i > k$ is not zero, interchange row i and column i with row k and column k. Element s_{kk} is now nonzero; go to Step 1.

3. If the kth diagonal element of $\mathbf{S}_{k-1}$ is zero and no other diagonal element for row $i > k$ is nonzero, stop. The following two cases are possible at this point.

   a. If some elements in rows k through n are nonzero, $\mathbf{H}$ is indefinite.

   b. If all the elements in rows k through n are zero, $\mathbf{S} = \mathbf{S}_{k-1}$ and the remaining diagonal elements of $\mathbf{D}$ are zero.

Applying the algorithm to the example, we get the results shown in Table 9.5. The final matrix in the series is the transformation **S**, and the matrix **D** is formed from the pivot elements.

$$\mathbf{S} = \begin{bmatrix} 1 & 1/4 & 1/2 \\ 0 & 1 & 10/7 \\ 0 & 0 & 1 \end{bmatrix} \text{ and } \mathbf{D} = \begin{bmatrix} 4 & 0 & 0 \\ 0 & 7/4 & 0 \\ 0 & 0 & 10/7 \end{bmatrix}$$

Given that the elements of **D** are all positive, **D** and also **H** are positive definite. The reader can verify that

$$\mathbf{z}^T \mathbf{D} \mathbf{z} = d_{11} z_1^2 + d_{22} z_2^2 + d_{33} z_3^2 = 4 z_1^2 + \frac{7}{4} z_2^2 + \frac{10}{7} z_3^2$$

$$= 4 \left(x_1 + \frac{1}{4} x_2 + \frac{1}{2} x_3 \right)^2 + \frac{7}{4} \left(x_2 + \frac{10}{7} x_3 \right)^2 + \frac{10}{7} (x_3)^2$$

is the same as our original quadratic function.

Step 1 of the algorithm performs Gaussian elimination to zero out the lower rows in column k. Step 2 has the effect of interchanging x_i and x_k when a zero diagonal element is encountered. Step 3 terminates the algorithm when there are no remaining nonzero diagonal elements. If some other nonzero elements remain in rows k through n, the matrix is indefinite; otherwise, the matrix is at best semidefinite.

Testing for Definiteness with Determinants

A second but generally more cumbersome approach that can be used to determine definiteness is based on computing determinants of submatrices of the Hessian. Let the ijth element of **H** be h_{ij}. The ith leading principal submatrix of **H** is the matrix formed by taking the intersection of its first i rows and i columns. Let H_i be the value of this determinant. Then,

$$H_1 = h_{11}$$

$$H_2 = \begin{vmatrix} h_{11} & h_{12} \\ h_{21} & h_{22} \end{vmatrix}$$

and so on until H_n is obtained.

- **H** is positive definite if and only if the determinants of all the leading principal submatrices are positive—that is, $H_i > 0$ for $i = 1, \ldots, n$.

- **H** is negative definite if and only if $H_1 < 0$ and the remaining leading principal determinants alternate in sign as the index of the determinant increases:

$$H_2 > 0, \ H_3 < 0, \ H_4 > 0, \ldots$$

Table 9.5   Diagonalization Algorithm Applied to Example Function

k	0	1	2	3
S_k	$\begin{bmatrix} 4 & 1 & 2 \\ 1 & 2 & 3 \\ 2 & 3 & 6 \end{bmatrix}$	$\begin{bmatrix} 1 & 1/4 & 1/2 \\ 0 & 7/4 & 5/2 \\ 0 & 5/2 & 5 \end{bmatrix}$	$\begin{bmatrix} 1 & 1/4 & 1/2 \\ 0 & 1 & 10/7 \\ 0 & 0 & 10/7 \end{bmatrix}$	$\begin{bmatrix} 1 & 1/4 & 1/2 \\ 0 & 1 & 10/7 \\ 0 & 0 & 1 \end{bmatrix}$
d_{kk}		4	7/4	10/7

- **H** is indefinite if the leading principal determinants are all nonzero and neither of the foregoing conditions is true.

- **H** is positive semidefinite if and only if **all** principal submatrices have nonnegative determinants.

- **H** is negative semidefinite if and only if all principal submatrices of odd order have nonpositive determinants and all principal submatrices of even order have nonnegative determinants—that is,

$$H_i \leq 0 \text{ for } i \text{ odd and } H_i \geq 0 \text{ for } i \text{ even}$$

If **H** is positive or negative definite, or if it is indefinite, the determinant test can establish the characteristic with n determinant evaluations. If some of the determinants are zero, the test involves *all* principal submatrices. The first-order submatrices are the numbers along the main diagonal of **H**. Of course, these are evaluated by observation. An mth-order submatrix is formed by deleting the rows and columns for all but m variables. When the number of variables exceeds 3 or 4, the number of principal submatrices becomes rather large, making it difficult to determine that **H** is semidefinite.

Quadratic Forms

The general statement of a quadratic function in n variables is

$$f(\mathbf{x}) = a + \mathbf{c}^T \mathbf{x} + \tfrac{1}{2} \mathbf{x}^T \mathbf{Q} \mathbf{x} \tag{5}$$

where a is a constant, $\mathbf{c} \in \Re^n$ is a vector of coefficients associated with the linear terms, and **Q** is a symmetric $n \times n$ matrix describing the coefficients of the quadratic terms. Differentiating $f(\mathbf{x})$ with respect to the components of **x** gives the gradient and Hessian matrices.

$$\nabla f(\mathbf{x}) = \mathbf{c} + \mathbf{Q}\mathbf{x} \text{ and } \mathbf{H} = \mathbf{Q}$$

As an example, consider the ellipsoid

$$f(\mathbf{x}) = 3x_1 x_2 + x_1^2 + 3x_2^2$$

which yields

$$\nabla f(\mathbf{x}) = (3x_2 + 2x_1, 3x_1 + 6x_2)^T \text{ and } \mathbf{Q} = \mathbf{H} = \begin{bmatrix} 2 & 3 \\ 3 & 6 \end{bmatrix}$$

(When written in matrix form, the element q_{jj} is twice the value of the coefficient of x_j^2 to account for the multiplier $\tfrac{1}{2}$ in Equation (5); the elements q_{ij} and q_{ji} are both equal to the coefficient of $x_i x_j$.) Using the diagonalization algorithm, we get

$$\mathbf{S} = \begin{bmatrix} 1 & 3/2 \\ 0 & 1 \end{bmatrix} \text{ and } \mathbf{D} = \begin{bmatrix} 2 & 0 \\ 0 & 3/2 \end{bmatrix}$$

Thus, the Hessian is positive definite, so the function is strictly convex. We also note that the determinants of the leading principal submatrices of **H** are

$$H_1 = 2 \text{ and } H_2 = 3$$

which is sufficient to conclude that **H** is positive definite.

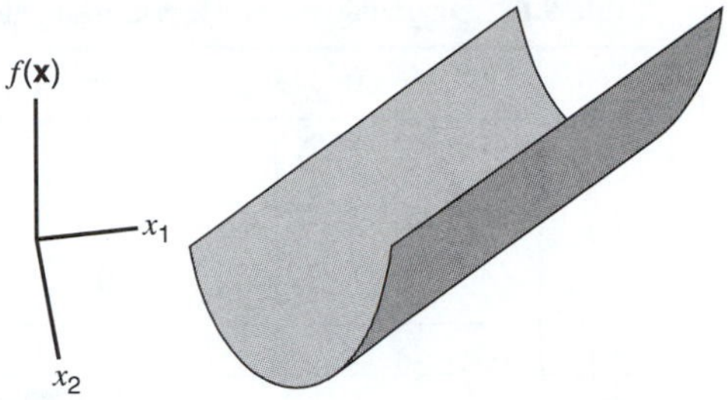

Figure 9.11   A parabolic contour

Now consider the function

$$f(\mathbf{x}) = 24x_1x_2 + 9x_1^2 + 16x_2^2$$

which is pictured in Figure 9.11. For this case,

$$\mathbf{H} = \begin{bmatrix} 18 & 24 \\ 24 & 32 \end{bmatrix}, \ \mathbf{S} = \begin{bmatrix} 1 & 4/3 \\ 0 & 0 \end{bmatrix}, \text{ and } \mathbf{D} = \begin{bmatrix} 18 & 0 \\ 0 & 0 \end{bmatrix}$$

The desired equivalence is

$$\mathbf{x}^{\mathrm{T}}\mathbf{H}\mathbf{x} = 18z_1^2, \text{ with } z_1 = x_1 + \tfrac{4}{3}x_2$$

Evidently, by virtue of the absence of a z_2^2 term, it is possible for $\mathbf{x}^{\mathrm{T}}\mathbf{H}\mathbf{x}$ to be equal to 0 even when $\mathbf{x} \neq \mathbf{0}$. Thus, $\mathbf{H}$ cannot be positive definite but is, in fact, positive semidefinite, so the function is convex (but not *strictly* convex). The determinant test tells us this because $H_1 > 0$, $H_2 = 0$, and the determinants of all other principal submatrices are zero. As may be observed, $f(\mathbf{x})$ has an infinite number of global minima, all resident on the locus of the valley of the parabolic contour.

Finally, consider a quadratic in three variables.

$$f(\mathbf{x}) = x_1x_2 + x_1x_3 + x_2x_3 + \tfrac{1}{2}x_1^2 + \tfrac{1}{2}x_2^2$$

Applying the diagonalization algorithm with $\mathbf{S}_0 = \mathbf{H}$, we obtain the results shown in Table 9.6. After we derive $\mathbf{S}_1$, all the elements in row 2 are zero; however, the diagonal element $s_{33} \neq 0$, so we interchange rows 2 and 3 and columns 2 and 3 to get the matrix shown in the column in which the index k first equals 2. Dividing row 2 by $d_{22} = -1$ gives the matrix $\mathbf{S}_2$ shown in the second column where $k = 2$. The algorithm terminates at this point with

$$\mathbf{S} = \begin{bmatrix} 1 & 1 & 1 \\ 0 & 1 & 0 \\ 0 & 0 & 0 \end{bmatrix} \text{ and } \mathbf{D} = \begin{bmatrix} 1 & 0 & 0 \\ 0 & -1 & 0 \\ 0 & 0 & 0 \end{bmatrix}$$

The linear transform is $\mathbf{z} = \mathbf{S}\mathbf{x} = (x_1 + x_2 + x_3, x_2, 0)^{\mathrm{T}}$, so

Table 9.6 Diagonalization Algorithm Applied to Quadratic Function

k	0	1	2	2
$\mathbf{S}_k$	$\begin{bmatrix} 1 & 1 & 1 \\ 1 & 1 & 1 \\ 1 & 1 & 0 \end{bmatrix}$	$\begin{bmatrix} 1 & 1 & 1 \\ 0 & 0 & 0 \\ 0 & 0 & -1 \end{bmatrix}$	$\begin{bmatrix} 1 & 1 & 1 \\ 0 & -1 & 0 \\ 0 & 0 & 0 \end{bmatrix}$	$\begin{bmatrix} 1 & 1 & 1 \\ 0 & 1 & 0 \\ 0 & 0 & 0 \end{bmatrix}$
d_{kk}		1		−1
Note		Interchange 2 and 3		

$$\mathbf{X}^T\mathbf{H}\mathbf{x} = (\mathbf{S}\mathbf{x})^T\mathbf{D}(\mathbf{S}\mathbf{x}) = \mathbf{z}^T\mathbf{D}\mathbf{z}$$

$$= (x_1 + x_2 + x_3, x_2, 0) \begin{bmatrix} 1 & 0 & 0 \\ 0 & -1 & 0 \\ 0 & 0 & 0 \end{bmatrix} \begin{pmatrix} x_1 + x_2 + x_3 \\ x_2 \\ 0 \end{pmatrix}$$

$$= (x_1 + x_2 + x_3)^2 - (x_2)^2$$

which can take on both positive and negative values. Consequently, we conclude that the Hessian matrix is indefinite, so the function is neither convex nor concave. The determinants of the principal leading matrices are $H_1 > 1$, $H_2 = 0$, and $H_3 = 0$, but the principal submatrix formed by striking out the first row and column of $\mathbf{H}$ has determinant −1. This indicates that $f(\mathbf{x})$ is indefinite.

For n more than 3 or 4, the diagonalization algorithm is much more computationally efficient than the determinant test for determining the definiteness of a matrix. This is especially true when it is necessary to examine all principal submatrices. Therefore, the determinant test is mostly of theoretical interest.

Nonquadratic Forms

Now let us consider a Hessian matrix derived from a *nonquadratic* function $f(\mathbf{x})$. In this case, some or all of the elements of $\mathbf{H}$ will depend on $\mathbf{x}$. Suppose that

$$f(\mathbf{x}) = (x_2 - x_1^2)^2 + (1 - x_1)^2$$

so

$$\nabla f(\mathbf{x}) = \begin{pmatrix} -4x_1(x_2 - x_1^2) - 2(1 - x_1) \\ 2(x_2 - x_1^2) \end{pmatrix}$$

and

$$\mathbf{H}(\mathbf{x}) = \begin{bmatrix} -4x_2 + 12x_1^2 + 2 & -4x_1 \\ -4x_1 & 2 \end{bmatrix}$$

The Hessian depends on the point under consideration, as does the function $\mathbf{x}^T\mathbf{H}\mathbf{x}$, which can be both positive and negative for some values of $\mathbf{x}$. For instance, at $\mathbf{x} = (1, 1)$ we have

$$\mathbf{H} = \begin{bmatrix} 10 & -4 \\ -4 & 2 \end{bmatrix}, \ \mathbf{S} = \begin{bmatrix} 1 & -0.4 \\ 0 & 1 \end{bmatrix}, \text{ and } \mathbf{D} = \begin{bmatrix} 10 & 0 \\ 0 & 0.4 \end{bmatrix}$$

which indicates that **H**(1, 1) is positive definite. This allows us to conclude that $f(\mathbf{x})$ is strictly convex in the "neighborhood" of $\mathbf{x} = (1, 1)$. At $\mathbf{x} = (0, 1)$, however, $\mathbf{H}(0, 1) = \begin{bmatrix} -2 & 0 \\ 0 & 2 \end{bmatrix}$, which is indefinite. Therefore, $f(\mathbf{x})$ is *not* strictly convex for all $\mathbf{x}$.

9.4   APPLICATIONS

We now present a number of situations that require the use of nonlinear functions in the model. For the most part, we concentrate on examples in which the constraints are linear and the objective functions are separable in the decision variables, thus making the determination of convexity or concavity a simple matter of differentiation.

Distance Problem

The distance between two points depends on the metric used to measure it, but invariably involves nonlinearities. Problems with the rectilinear distance metric have linear programming equivalents, as described in the supplementary material to Chapter 2 on the CD. Consider three customers located in a plane, as shown in Figure 9.12. Each customer i has a demand w_i for some service. We want to locate a depot (a fourth point) to provide the service such that the total weighted distance between the customers and the depot is minimized.

In general, for a problem with m customers located at the coordinates (a_i, b_i), $i = 1$, ..., m, let the decision variables be x and y, the coordinates of the depot. Using the Euclidean norm as the distance metric, the objective is to

$$\text{Minimize } f(x, y) = \sum_{i=1}^{m} w_i \sqrt{(x - a_i)^2 + (y - b_i)^2}$$

This is an unconstrained NLP model in two variables. The function $f(x, y)$ is convex, so it has a unique local minimum that is the global minimum. In fact, it is strictly convex unless all the points are on the same line (Francis *et al.* [1992]).

The solution for the special case of three customers with demand 1 (i.e., $w_i = 1$ for all i) is shown in Figure 9.13. The lines connecting the customers to the depot form 120° angles. This will always be true for three customers where the triangle formed by connecting the customers has internal angles less than 120°. When the triangle formed by the customers includes an angle greater than or equal to 120°, the depot is placed at the customer forming the angle. One instance of this case is shown in Figure 9.14.

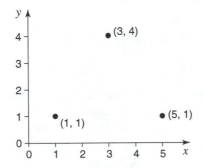

Figure 9.12   Distance problem

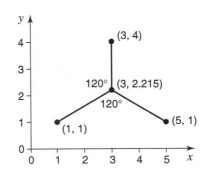

Figure 9.13   Solution to distance problem

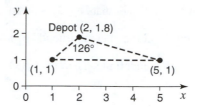

Figure 9.14   Solution to a problem with an interior angle greater than 120°

When more than three customers are involved and the demands are not equal to 1, generalizations are difficult to discover, but the problem remains a convex program.

Dimensions of a Geometric Solid

Consider the problem of finding the dimensions of a box that must hold a given volume, say $V = 1000$ cubic centimeters, such that its surface area is minimized. Because the areas and volumes of solids are nonlinear functions of their dimensions, regardless of the coordinate system used, optimization problems involving physical shapes give rise to NLP models. Using x, y, and z as the box's length, width, and height, respectively, the minimum surface area can be found by solving the following NLP.

$$\text{Minimize } f(x, y, z) = 2xy + 2xz + 2yz$$
$$\text{subject to} \qquad xyz = V$$
$$x \geq 0, y \geq 0, z \geq 0$$

One way to approach this problem is to ignore the nonnegativity constraints and use the classical Lagrange multiplier method of calculus. If the solution has all variables positive, a stationary point has been found. It would then be necessary to determine whether the stationary point was a maximum, a minimum, or a point of inflection. It must be a global minimum if the objective function $f(x, y, z)$ is convex and the feasible region is a convex set.

For this problem, the Lagrangian is

$$\mathcal{L}(x, y, z, \lambda) = f(x, y, z) + \lambda(b - g(x, y, z))$$
$$= 2xy + 2xz + 2yz + \lambda(V - xyz)$$

where λ is the unrestricted Lagrange multiplier. Taking partial derivatives with respect to each variable and setting the results equal to zero leads to four equations in four unknowns. Solving yields $x = y = z = 10$ centimeters per side, $\lambda = 0.4$, and $f = 600$.

If we add the information that the bottom and top of the box cost twice as much as the sides, the objective is

$$\text{Minimize } f(x, y, z) = 4xy + 2xz + 2yz$$

Which has the solution $x = y = 7.94$, $z = 15.88$, with $f = 756$. That is, the length and the width are one-half the height.

To determine convexity, we note that the objective function is a convex function of x, y, and z, but we have a nonlinear equality constraint. An equality constraint can never define a convex region unless it is linear. Since the feasible region is not convex, we cannot conclude that the result is a global minimum.

The problem can be changed into an unconstrained minimum if we use the equality to eliminate the variable z and ignore the nonnegativity conditions for the moment. The new objective is

$$\text{Minimize } f(x, y) = 2xy + 2V/y + 2V/x$$

which suggests that a solution must have all the variables positive. The Hessian of $f(x, y)$ is

$$\mathbf{H}(x, y) = \begin{bmatrix} \dfrac{4V}{x^3} & 2 \\ 2 & \dfrac{4V}{y^3} \end{bmatrix}$$

The first-order principal minors have $H_1 > 0$ for positive values of the variables. The second-order principal minor is $\mathbf{H}$, and we have $H_2 > 0$ for $x^3y^3 < 4V^2$. When $x = 10$ and $y = 10$, $\mathbf{H}(10, 10)$ is positive definite, indicating that $f(x, y)$ is a strictly convex function in the current neighborhood. Thus, $(x, y) = (10, 10)$ is a strong local minimum. Because no other points satisfy the Lagrangian equations, we can conclude that it is the global minimum.

As a final word, we note that the original problem has no finite maximum surface area. This is suggested by the fact that when $x^3y^3 > 4V^2$, the Hessian $\mathbf{H}(x, y)$ is negative definite. Thus, $f(x, y)$ is strictly concave in the region satisfying this inequality. When the product $xy \to \infty$, the surface area goes to ∞ and $z = V/xy \to 0$, ensuring that the volume remains V.

Inventory Problem

A company keeps m items in inventory. Each item i has a unit cost c_i, a demand rate d_i, and a reorder cost K_i. The reorder cost is incurred each time a replenishment order is placed. The items come from different suppliers, so reordering is done independently. After several reviews, the company believes that its investment in inventory is too high and so has commissioned a study to determine an optimal control policy.

The underlying assumptions of the optimization problem are partly illustrated in Figure 9.15, which plots the inventory level for one item as a function of time. The inventory level ranges from 0 to the amount Q. The fact that it never goes below 0 indicates that no shortages or stockouts are allowed. Periodically, an order of size Q is placed at a cost of K. The arrival of the order is assumed to occur instantaneously, so there is no need for replenishment before the stock goes to 0. Between orders, the inventory decreases at a constant

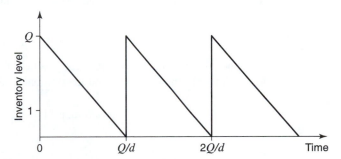

Figure 9.15   Lot-size model with no shortages

rate d. While an item remains in inventory it incurs a holding cost at a rate h (dollars per unit time). For a single item, the total system cost expressed per unit time is

$$\text{Cost/unit time} = \text{setup cost} + \text{product cost} + \text{holding cost}$$
$$= dK/Q + dc + hQ/2$$

The problem is to determine the value of Q, called the *economic order quantity*, that minimizes this cost subject to a limit on investment. For multiple products, we assign indices to the variables and parameters and sum over the m items. Disregarding the product cost, which is independent of the decision variables, the objective is to minimize

$$\text{Inventory system cost: } z(\mathbf{Q}) = \sum_{i=1}^{m}(d_i K_i/Q_i + h_i Q_i/2)$$

where $\mathbf{Q} = (Q_1, \ldots, Q_m)$. The average inventory for an item is half of the lot size. With $I_{\max}$ specified as the maximum average value of the inventory, the constraint on investment is

$$\text{Investment limit: } \sum_{i=1}^{m} c_i Q_i/2 \le I_{\max}$$

The model has a separable, nonlinear objective with a single linear constraint. Differentiation of the objective function shows that each term is strictly convex (that is, the Hessian is a diagonal matrix and each element $h_{ii} > 0$ for $Q_i > 0$), so the sum denoted by $z(\mathbf{Q})$ is strictly convex. Because the feasible region is also convex, there is a unique local minimum, which is the global minimum.

Optimal Redundancy Problem

Consider an electronic system with n components in series. The reliability of a component, r_i, is the probability that it will not fail during its operation. The reliability of the system is the probability that none of its components will fail. To increase the reliability, extra components may be included as backups for the originals. These are called redundant components since they are not required unless the originals fail.

Assume that each component costs c_i and weighs w_i. The problem is to find the optimal number of redundant components of each type, subject to constraints on maximum total cost C and weight W.

Although the decision variables are really discrete, the relaxation of the integrality restrictions results in a standard NLP. Because joint probabilities are products of terms, this kind of problem will be nonlinear. In particular, the reliability of component type i with x_i backup units is

$$R_i = 1 - (1 - r_i)^{1+x_i}$$

The reliability of the system is the product of the component type reliabilities. The following model can be used to find the optimal levels of redundancy.

$$\text{Maximize } \prod_{i=1}^{n}(1-(1-r_i)^{1+x_i})$$

$$\text{subject to } \sum_{i=1}^{n} c_i x_i \le C$$

$$\sum_{i=1}^{n} w_i x_i \le W$$

$$x_i \ge 0, \quad i = 1, \ldots, n$$

It is typical in a problem with a product forming the objective function to take the natural log of the function to create a summation of terms. The solution of the transformed problem also optimizes the original. For our example, we have

$$\ln(\prod_{i=1}^{n} R_i) = \sum_{i=1}^{n} \ln\left(1 - (1 - r_i)^{1+x_i}\right)$$

Because each term in the sum is negative, we now minimize rather than maximize the transformed objective function.

$$\text{Minimize } \sum_{i=1}^{n} -\ln\left(1 - (1 - r_i)^{1+x_i}\right)$$

With some algebraic gymnastics, the new objective function can be shown to be convex. Thus, we are minimizing a convex function over a polyhedral feasible region, so all local minima are global minima. Alternatively, if the goal were to minimize cost subject to attaining some nominal level of reliability, the model would now have a linear objective and a nonlinear constraint.

$$\prod_{i=1}^{n} R_i \geq R_{\min} \rightarrow \sum_{i=1}^{n} -\ln(R_i) \leq -\ln(R_{\min})$$

We would still have a convex program, however, because the set $\{\mathbf{x} \in \Re^n : g(\mathbf{x}) \leq b\}$ is convex when $g(\mathbf{x})$ is convex. Therefore, a reliable NLP code would return a global minimum.

Routing Through a Queuing Network

Products are typically manufactured or assembled in flow shops. Figure 9.16 illustrates a situation in which a single product must pass through a series of workstations during its processing. Given that several different paths exist, the problem is to determine the optimal routing as a function of the arrival rates and services rates at each station when the system is in steady state. The product arrives at station A at a rate of $\lambda = 1.5$ units per minute and leaves station F at the same rate. The time between arrivals is a random variable governed by an exponential distribution.

Each station is assigned a single worker who performs his or her tasks at a rate μ. The data in Figure 9.16 indicate that stations B and E have a service rate of $\mu = 3/\text{min}$ whereas stations A, C, D, and F have a service rate of $\mu = 2/\text{min}$. The service times are also exponentially distributed random variables. Because of the fluctuations in the system, queues form at each station. We wish to determine how to direct the flow out of these stations so that either the total waiting time or the total number of units in the facility is minimized. A system of interconnected queuing stations in which all interarrival times and all service times are exponentially distributed is called a *Markov queuing network*.

A queuing station can be described by an arrival rate λ, a service rate per channel μ, and the number of servers s. If one assumes that the time between arrivals and the time for

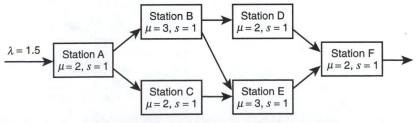

Figure 9.16   Markov queuing network with alternative routings

service have exponential distributions, the average number of units at a station either waiting (L_q) or waiting plus being served (L) is a complicated function of the parameters. For a single server facility in steady state, however, the total number at the station on average is

$$L = \lambda/(\mu - \lambda) \text{ or } L = \rho/(1 - \rho), \text{ where } \rho = \lambda/\mu \tag{6}$$

and their total waiting plus service time on average is

$$W = 1/\mu(1 - \rho)$$

These formulas come from the study of queuing theory, which is discussed in Chapter 16 along with the more complicated formulas that result when $s > 1$. Figure 9.17 shows that L is a convex function of λ for $\mu = 2$.

The optimization problem addressed here involves routing a product through a network of queues. When the arrival and service processes follow exponential distributions, Equation (6) gives the total number of units at each station i. The arrival rates λ_i are determined by the routing and are the decision variables of the problem. An objective or constraint involving the total number or total waiting time in the system results in a convex nonlinear program.

The easiest way to determine the optimal routing for, say, minimizing the total number of units in the system on average is to solve an equivalent nonlinear network flow problem. In the network diagram depicted in Figure 9.18, all the arcs except the one between nodes 2 and 3 represent stations. The arc from station 2 to station 3 allows flow from station B to station E. A nonlinear cost function of the type given in Equation (6) is associated with each station arc. For example, the flow through the arc from station 1 to station 2 is the flow rate entering station A, f_A. The cost measure is the average number of units at that station, $f_A/(2 - f_A)$.

Because each arc cost function is convex, the total cost, which represents the total number of units in the system, is also convex. The fact that the conservation-of-flow constraints are linear implies that all local solutions are global. The optimal flow for the example is given in parentheses along the arcs in Figure 9.18. From Little's law described in Section 16.1, we know that the solution also minimizes the average number in the system also minimizes the average flow time through the system.

Many variations of this problem can be solved with the same approach. It is relatively easy to include such factors as multiple servers and losses in product owing to defects, among others.

Curve Fitting and Regression

Regression analysis is commonly used in statistics to establish a functional relationship been a dependent variable y and a set of m independent variables $x_1, \ldots, x_m$. In the simplest case, the relationship is hypothesized to be linear.

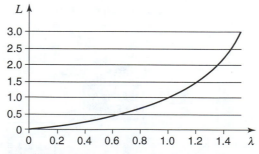

Figure 9.17   Average number at station for $\mu = 2$

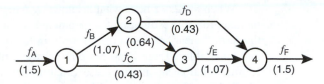

Figure 9.18   Queuing network solved as a nonlinear network flow problem

$$y = b_0 + b_1 x_1 + b_2 x_2 + \cdots + b_m x_m \tag{7}$$

The goal is to estimate the values of the parameters $b_0, b_1, \ldots, b_m$ such that some measure of deviation between the actual data and the function in Equation (7) is minimized.

Assume that we have n sample observations from a population $(y_i, x_{1i}, \ldots, x_{mi})$, $i = 1, \ldots, n$. The ith deviation can then be written as

$$\alpha_i = y_i - (b_0 + b_1 x_{1i} + b_2 x_{2i} + \cdots + b_m x_{mi}) \tag{8}$$

where α_i may be positive or negative. The case in which there is only one independent variable, call it x, is illustrated in Figure 9.19. The graph is referred to as a *scatter diagram*.

To find the "best" values of the parameters, we solve an unconstrained nonlinear optimization problem of the general form

$$\text{Minimize } z = \sum_{i=1}^{n} \left| y_i - (b_0 + b_1 x_{1i} + b_2 x_{2i} + \cdots + b_m x_{mi}) \right|^p \tag{9}$$

where $p > 0$ and integer. Note that $(y_i, x_{1i}, \ldots, x_{mi})$ are data and $b_0, b_1, \ldots, b_m$ are the decision variables for this problem. The most common values of p are 1, 2, and ∞. When $p = 1$, the objective is to minimize the sum of the absolute values of the deviations. This problem can be converted to a linear program using the ideas in the supplementary material for Chapter 2 on the CD. One way to proceed is to observe that $|\alpha| = \max\{\alpha, -\alpha\}$, so Problem (9) is equivalent to

$$\text{Minimize } \left\{ \sum_{i=1}^{n} |\alpha_i| : \text{subject to (8)} \right\} = \text{Minimize } \left\{ \sum_{i=1}^{n} \max\{\alpha_i, -\alpha_i\} : \text{subject to (8)} \right\}$$

When $p = \infty$, the objective is to minimize the maximum absolute deviation, sometimes referred to as the Chebyshev criterion—i.e., minimize $\{\max(|\alpha_i| : i = 1, \ldots, n)\}$. This problem can also be converted to a linear program. In both cases, the details are left as an exercise.

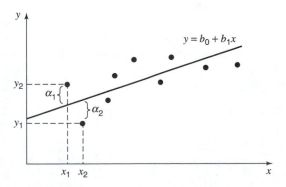

Figure 9.19   Typical scatter diagram

When $p = 2$, we have what is known as *least squares regression*. This is the most common approach to estimating the parameters of the linear model given by Equation (7). Here the term "linear" refers to the linearity in the parameters and not to the linearity in the x and y variables. There is no reason why we could not hypothesize a polynomial or logarithmic relationship among these variables instead of a straight-line relationship (really a hyperplane).

We now focus on the bivariate case characterized by a single independent variable x and perform the analysis associated with minimizing the sum of the squared deviations between the estimated line and the data. The variables in this problem are the parameters b_0 and b_1, the intercept and slope of the line in Equation (7). The sum of the squared deviations is

$$z(b_0, b_1) = \sum_{i=1}^{n} (y_i - (b_0 + b_1 x_i))^2$$

The Hessian matrix of the objective function $z(b_0, b_1)$ with respect to the variables b_0 and b_1 is

$$\mathbf{H}(b_0, b_1) = 2 \begin{bmatrix} n & \sum_{i=1}^{n} x_i \\ \sum_{i=1}^{n} x_i & \sum_{i=1}^{n} (x_i)^2 \end{bmatrix}$$

This matrix can be shown to be positive definite, so the least squares deviation function is convex. To find the optimal solution, we set the first partials of $z(b_0, b_1)$ equal to zero.

$$\frac{\partial z}{\partial b_0} = -2 \sum_{i=1}^{n} (y_i - (b_0 + b_1 x_i)) = 0$$

and

$$\frac{\partial z}{\partial b_1} = -2 \sum_{i=1}^{n} x_i (y_i - (b_0 + b_1 x_i)) = 0$$

Solving, we get

$$b_1 = \frac{n \sum_{i=1}^{n} x_i y_i - \sum_{i=1}^{n} x_i \sum_{i=1}^{n} y_i}{n \sum_{i=1}^{n} (x_i)^2 - \left(\sum_{i=1}^{n} x_i \right)^2} \quad \text{and} \quad b_0 = \frac{\sum_{i=1}^{n} y_i - b_1 \sum_{i=1}^{n} x_i}{n}$$

Similar results are obtained when the multiple regression model [Problem (9)] with m independent variables is solved with $p = 2$. In that case, the objective function is also convex, so a global optimum is ensured. The question of the quality of the results, however, still remains. In most situations, the "goodness of fit" is measured by the coefficient of determination R^2, which is computed from the formula

$$R^2 = \frac{\sum_{i=1}^{n} (\hat{y}_i - \bar{y})^2}{\sum_{i=1}^{n} (y_i - \bar{y})^2}$$

where, in general, $\bar{y} = \dfrac{1}{n} \sum_{i=1}^{n} y_i$ and $\hat{y}_i = b_0 + b_1 x_{1i} + b_2 x_{2i} + \cdots + b_m x_{mi}$. The term $\hat{y}_i$ is the value of the dependent variable y predicated by the regression line when the data $x_{1i}, x_{2i}, \ldots, x_{mi}$ are inserted for the independent variables.

It can be shown that $0 \leq R^2 \leq 1$. When $R^2 = 1$, we have a perfect fit and all the points on the scatter diagram line on the regression line. Alternatively, when $R^2 = 0$, we have $\hat{y}_i = \bar{y}$, implying that the model has no predictive power because each value of $\hat{y}_i$ equals the mean $\bar{y}$ regardless of the value of x_i. When $m > 1$, most analysts prefer to use a normalized version of R^2 known as *adjusted* R^2, which is given by

$$R_a^2 = 1 - (1 - R^2)\left(\frac{n-1}{n-m-1}\right)$$

where n is the total number of observations and $m + 1$ is the number of coefficients to be estimated. By working with R_a^2, it is possible to compare different regression models used to estimate the same dependent variable with different numbers of independent variables.

Example 1 (*Linear regression model*)

A high-tech firm decides to use a regression model to estimate the time required to develop a new e-commerce website. The candidate list of independent variables is as follows.

x_1 = number of requests for customer information in the program

x_2 = average number of lines of code associated with each request

x_3 = number of modules or subprograms

Table 9.7 summarizes the data collected on 10 similar development efforts. The time required in person-months, denoted by y, is the dependent variable (the duration is given by the number of person-months divided by the number of programmers assigned to the project). Using the Excel Solver to minimize the sum of the errors squared for the data yields the equation

$$y = -0.76 + 0.13x_1 + 0.045x_2$$

with $R^2 = 0.972$ and $R_a^2 = 0.964$. The value of R_a^2 is always smaller than R^2.

Table 9.7   Data for Linear Regression Model

Development effort, i	Time required, y	Number of requests, x_1	Number of lines, x_2	Number of modules, x_3
1	7.9	50	100	4
2	6.8	30	60	2
3	16.9	90	120	7
4	26.1	110	280	9
5	14.4	65	140	8
6	17.5	70	170	7
7	7.8	40	60	2
8	19.3	80	195	7
9	21.3	100	180	6
10	14.3	75	120	3

When the third candidate x_3 is introduced into the regression model, the value of R_a^2 is reduced to 0.963; consequently, it is best to use only the independent variables x_1 and x_2 as predictors, although the difference is minimal.

If a new website similar to the previous 10 is to be developed, and it contains $x_1 = 45$ information requests with an average of $x_2 = 170$ lines of code for each request, the estimated development time is

$$y = -0.76 + (0.13)(45) + (0.045)(170) = 12.7 \text{ person-months}$$

Example 2 (*Nonlinear regression model*)

There are many situations in which the parameters to be estimated are required to satisfy certain constraints such as nonnegativity. When a convex feasible region results, the least squares problem of finding the optimal fit remains a convex program. We now go one step further and look at an example in which the regression model is nonlinear. The particular model arises in an economic context and is known as *Mitcherlisch's law of diminishing returns*.

$$y = b_0 + b_1 e^{b_2 x}$$

Here, b_0, b_1, and b_2 are the parameters to be determined subject to the single constraint that $b_2 \leq 0$. Data for six observations are given in Table 9.8. The optimization problem to be solved using the least squares criterion is

$$\text{Minimize } z = \sum_{i=1}^{6} \left[y_i - (b_0 + b_1 e^{b_2 x_i}) \right]^2$$

$$\text{subject to} \quad b_2 \leq 0$$

This is a constrained convex programming model with the solution

$$b_0 = 523.2, \ b_1 = -156.9, \text{ and } b_2 = -0.1997$$

Physical Network Flow Problems

Most engineering disciplines face problems that involve the flow of some physical quantity through a network. In the distribution of electricity, for example, the engineer is concerned with the flow of current through wires. In the distribution of water, the concern is with the flow of fluids through pipes. In a heating system, the thermodynamicist is concerned with the flow of heat through conduction paths.

Table 9.8 Data for Nonlinear Regression Model

Dependent variable, y	Independent variable, x
127	−5
151	−3
379	−1
421	5
460	3
426	1

Figure 9.20 shows a simple direct current network of the type we consider here. The problem is to determine the current flows $I_1, I_2, \ldots, I_7$ such that the total *content* is minimized. This measure, along with the symbols used in the figure, are described in Table 9.9.

Kirchhoff's laws can be used to find the currents and voltages in a circuit whose power source is given. Other disciplines have similar means of characterizing flows and finding solutions. The methods of operations research as applied to networks are quite different. Whereas these methods attempt to optimize some economic criterion by prescribing flow, the methods used to analyze and design physical systems attempt to simulate the distribution of some quantity whose flow is governed by natural laws.

There is, however, an equivalence between the solution techniques for physical flow problems and those for network flow programming problems. Each of the former can be stated and solved as a network flow program with convex arc cost functions.

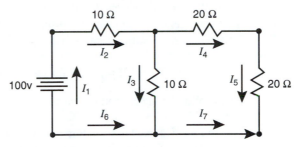

Figure 9.20   Direct current electrical network

Table 9.9   Voltage and Content Functions for Electrical Components

Component	Symbol	Voltage, $v(I)$	Content function, $G(I)$
Linear resistor	R I	IR	$I^2R/2$
Diode	R_1, R_2 I	$I \geq 0,\ IR_1$ $I < 0,\ IR_2$ $R_1 \ll R_2$	$I \geq 0,\ I^2R_1/2$ $I < 0,\ I^2R_2/2$
Battery	E I	$-E$	$-EI$
Nonlinear resistor	R, k I	$I \geq 0,\ RI^k$ $I < 0,\ -R(-I)^k$	$I \geq 0,\ \dfrac{RI^{k+1}}{k+1}$ $I < 0,\ \dfrac{R(-I)^{k+1}}{k+1}$

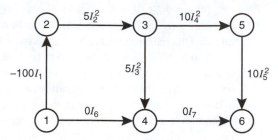

Figure 9.21   Network model of the DC circuit problem

Table 9.9 lists the types of components that might be used in a direct current (DC) network, the symbols assigned to those components, the voltages $v(I)$ across the terminals of the components as a function of current, and the content functions $G(I)$ of the components. This function is defined separately for positive and negative current directions, as follows.

$$G(I) = \int_0^I v(i)\,di \text{ for } I \geq 0 \text{ and } G(I) = \int_I^0 v(i)\,di \text{ for } I < 0$$

Performing the integration for each component generates the functions in the last column of the table.

In electrical networks, nature sets the currents such that the total content is minimized. This provides the NLP equivalent of the physical flow problem. Figure 9.21 depicts a nonlinear network flow model whose solution provides the answer directly. The values along the arcs represent the contents of the corresponding components.

A mathematical programming statement of the problem is as follows.

$$\text{Minimize } Z = -100I_1 + 5I_2^2 + 5I_3^2 + 10I_4^2 + 10I_5^2$$

$$\text{subject to} \quad I_1 - I_2 = 0, \, I_2 - I_3 - I_4 = 0, \, I_5 - I_6 = 0,$$

$$I_5 + I_7 = 0, \, I_3 + I_6 - I_7 = 0, \, -I_1 - I_6 = 0$$

The objective function is convex, and the constraints, representing conservation of flow at each node, are linear, so the solution should be a global minimum. Solving the problem yields the solution

$$I_1 = I_2 = 50/9, \, I_3 = 40/9, \, I_4 = I_5 = 10/9, \, I_6 = -50/9, \, I_7 = -10/9$$

where the flow is in amperes. Those familiar with Kirchhoff's laws can easily verify the accuracy of these results.

Although this example is a simple one, any complex network with linear and nonlinear resistors, diodes, and batteries can be solved as a convex programming problem with equality (conservation-of-flow) constraints. A related problem in water distribution involves the determination of fluid flows in pipes. For such a system, the equivalents of currents are fluid flows, the equivalents of resistors are pipe segments, and the equivalents of batteries are pressure pumps. Because the head loss in a pipe is generally a nonlinear function of flow, this would be analogous to including nonlinear resistors in the circuit.

9.5   PROBLEM CLASSES

Within the past three decades, a wide variety of mathematical programming problems have been tackled successfully. As might be expected, the linear model has received the most attention, with great strides being made for large, sparse problems and network formulations. Impressive results have also been obtained for the quadratic programming (QP) problem, in which $f(\mathbf{x})$ assumes a positive semidefinite quadratic form and the constraints are linear. Most commercial interior point codes for solving LPs contain a QP solver as well.

Several unique algorithms have been developed for the problem in which $f(\mathbf{x})$ is a convex, separable function and the constraints are linear. Special-purpose algorithms also exist for the slightly more general case in which $f(\mathbf{x})$ is convex and twice differentiable and the constraints are linear. The smoothness of the functions makes the problem well-behaved, and the convexity assumption ensures that the feasible region is convex and, most importantly, that any local solution is a global solution. When the functions assume an arbitrary form, this is no longer true and the problem remains rather intractable.

The approach to solving a nonlinear program begins by attempting to assign the problem to one of the classes listed in Figure 9.22. For our purposes here, it will be assumed that all the functions $f(\mathbf{x})$ and $g_i(\mathbf{x})$, $i = 1, \ldots, m$, are at least once continuously differentiable. Identification of a class directs the analyst to the proper solution algorithm and helps to categorize the results at termination.

Unconstrained Problems

Convex Objective

When the objective function is strictly convex and no constraints exist on the set of feasible solutions, the minimum will be unique if it is finite. Thus, any method that can discover a local minimum will also discover the global minimum. Classical optimization theory or direct search methods are often useful for uncovering the optimal solution in this case. When the objective function decreases indefinitely in some direction, the problem has no minimum. Functions that are convex but not strictly convex may have multiple (infinite) minima.

General Objective

When the objective function is not convex, there may be more than one local minimum. In this case, all local minima must be uncovered and the objective function evaluated at each. The local minimum with the smallest objective value will be the global minimum if the objective function does not decrease indefinitely in some direction. The possible existence of multiple local minima greatly complicates the development of solution algorithms.

Constrained Problems

Convex Programming Problem

When the objective function and all constraint functions are convex, a local minimum is also a global minimum. The convex constraint functions, when written in $\leq$ form, ensure that the feasible region is convex. When the objective function is strictly convex, there will be no more than one global minimum lying either interior to or on the boundary of the feasible region.

Quadratic Programming Problem

A problem with a convex quadratic objective function and linear constraints is a quadratic program. Since this is a special case of a convex program, a local minimum is also a global minimum. The structure of this problem allows for solutions with a modified LP algorithm.

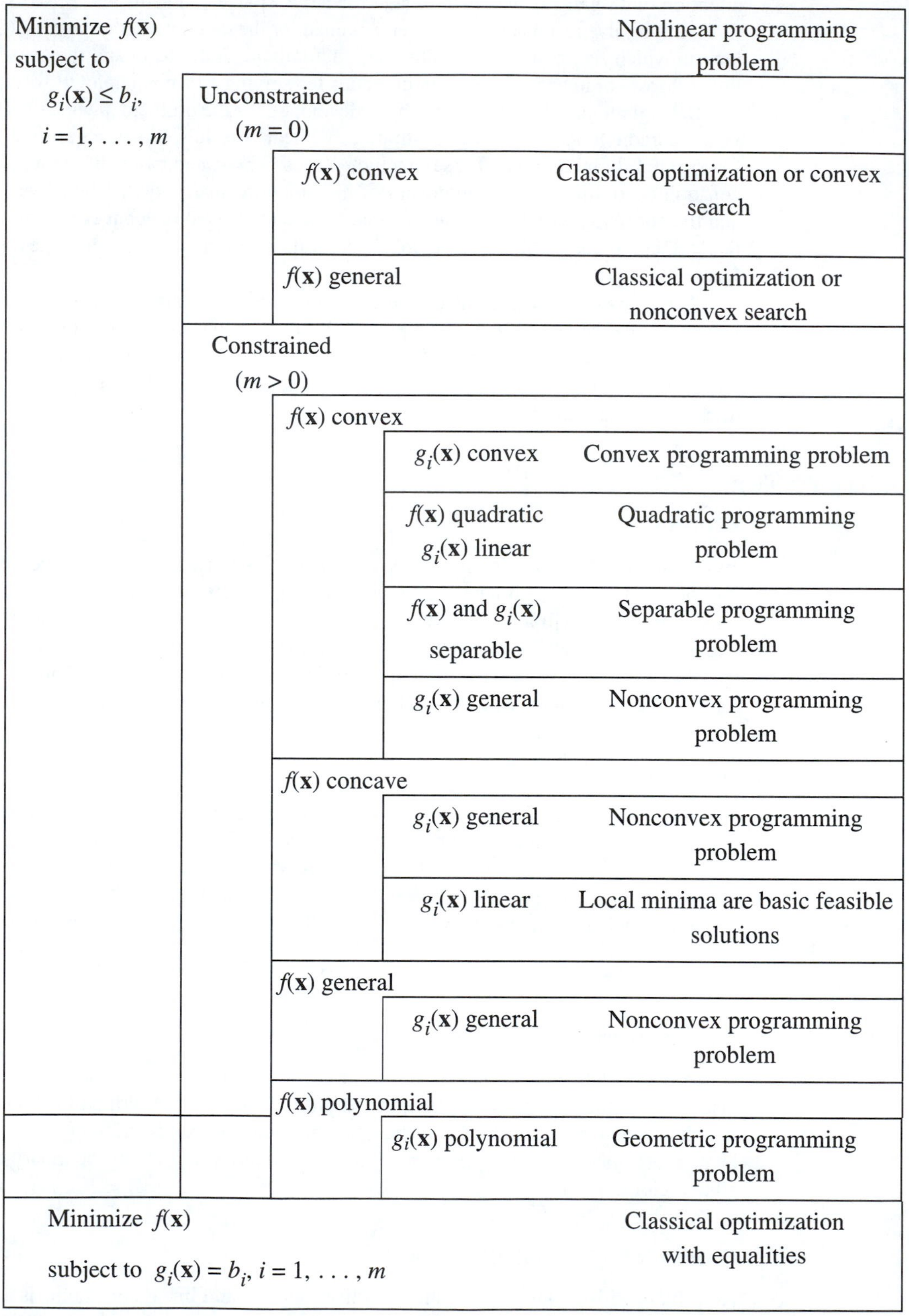

Minimize $f(\mathbf{x})$ subject to $g_i(\mathbf{x}) \leq b_i$, $i = 1, \ldots, m$				Nonlinear programming problem
	Unconstrained $(m = 0)$			
		$f(\mathbf{x})$ convex		Classical optimization or convex search
		$f(\mathbf{x})$ general		Classical optimization or nonconvex search
	Constrained $(m > 0)$			
		$f(\mathbf{x})$ convex		
			$g_i(\mathbf{x})$ convex	Convex programming problem
			$f(\mathbf{x})$ quadratic $g_i(\mathbf{x})$ linear	Quadratic programming problem
			$f(\mathbf{x})$ and $g_i(\mathbf{x})$ separable	Separable programming problem
			$g_i(\mathbf{x})$ general	Nonconvex programming problem
		$f(\mathbf{x})$ concave		
			$g_i(\mathbf{x})$ general	Nonconvex programming problem
			$g_i(\mathbf{x})$ linear	Local minima are basic feasible solutions
		$f(\mathbf{x})$ general		
			$g_i(\mathbf{x})$ general	Nonconvex programming problem
		$f(\mathbf{x})$ polynomial		
			$g_i(\mathbf{x})$ polynomial	Geometric programming problem
Minimize $f(\mathbf{x})$ subject to $g_i(\mathbf{x}) = b_i$, $i = 1, \ldots, m$				Classical optimization with equalities

Figure 9.22 Categorization of nonlinear programs

Separable Programming Problem

A separable function $f(\mathbf{x})$ can be expressed as the sum of terms, each of which is a function $f_j(x_j)$ of a single decision variable $x_j, j = 1, \ldots, n$—that is, $f(\mathbf{x}) = \Sigma_j f_j(x_j)$. A separable program results when the objective function and constraints are separable functions. A separable function is convex if and only if each term is convex, so the convexity of the objective function or a constraint is easily determined by analyzing each term.

Nonconvex Programming Problem

This is a broad class that includes all problems that are not convex; for example, $f(\mathbf{x})$ may be concave or any of the $g_i(\mathbf{x})$ may be concave. For these problems, a local minimum is not necessarily a global minimum. In fact, there may be many local minima, each of a different value. Algorithms generally find just one local minimum and so cannot guarantee that a particular solution is the global minimum.

Concave Objective and Linear Constraints

This problem class exhibits the interesting characteristic that if a minimizing solution exists, there will be at least one basic (extreme point) solution, as in linear programming. Usually there are several basic solutions that are local minima among which the global minimum will lie. Uncovering the global minimum is difficult because of the large number of basic solutions for most problems.

Geometric Programming Problem

For this class, both objective and constraint functions take the form

$$g(\mathbf{x}) = \sum_{t=1}^{T} c_t P_t(\mathbf{x})$$

where c_t are given coefficients and $P_t(\mathbf{x}) = (x_1^{a_{t1}})(x_2^{a_{t2}}) \ldots (x_n^{a_{tm}})$ for $t = 1, \ldots, T$. Special procedures are available to solve this type of problem, especially when the coefficients c_t are all positive. In this case, the preceding expression is called a *posinomial*. The resulting mathematical program can be written as

$$\text{Minimize} f(\mathbf{x}) = \sum_{t=1}^{T} c_{0t} \prod_{j=1}^{n} x_j^{a_{tj}}$$

$$\text{subject to} \quad \sum_{t=1}^{T} c_{it} \prod_{j=1}^{n} x_j^{d_{itj}} \le b_i, \quad i = 1, \ldots, m$$

$$x_j > 0, \quad j = 1, \ldots, n$$

where c_{it}, a_{tj}, d_{itj}, and b_i are model parameters. The requirement that all the variables be positive indicates that the model is appropriate only in a limited number of situations.

Equality Constrained Problems

Classical optimization procedures based on the Lagrangian function introduced in calculus are available for dealing with problems having only equality constraints. These procedures provide the basis for a variety of algorithms designed to solve inequality constrained problems but cannot be used directly on them. Inequalities, even simple nonnegative conditions, greatly complicate solution algorithms.

The feasible region for a problem with one or more nonlinear equality constraints will be nonconvex. Only linear equality constraints yield a convex feasible region. Consequently, most NLP algorithms will yield only local solutions when nonlinear equalities are present.

Characteristics of Solutions

Solutions to convex programming problems are generally very different from those obtained when minimizing concave or linear cost functions. As an illustration, consider the power distribution problem shown in Figure 9.23. Power sources are represented by square nodes with the positive number in brackets adjacent to each node indicating the amount of power available. Customers are shown as circles with the adjacent negative number in square brackets indicating the amount demanded. The length of the arcs are roughly proportional to the cost of distribution. A conservation-of-flow constraint must be written for each node.

The numbers in parentheses along the emboldened arcs in Figure 9.23 constitute the LP solution. The fact that the optimal solution is basic leads to a tree structure for the nonzero flows. The same kind of result would be obtained if the flow costs on the arcs were concave.

When the costs are proportional to the square of the flow, a convex quadratic objective function results. The solution shown in Figure 9.24 was obtained for such a function. The fact that all arcs have flow implies that the nonnegativity constraints are not active. Many customers now receive power from more than one source, so if there is an outage at a particular plant, those customers will not be completely without power. For many planning problems, a solution similar to the one obtained for the convex case is more satisfactory than the solution obtained for the linear case because of the inherent uncertainty in planning that argues against an all-or-nothing strategy.

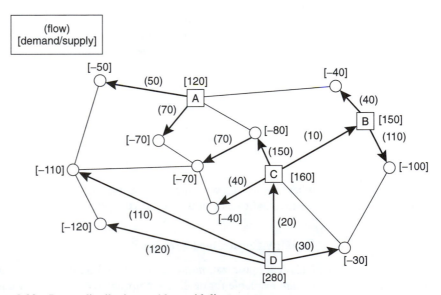

Figure 9.23   Power distribution problem with linear costs

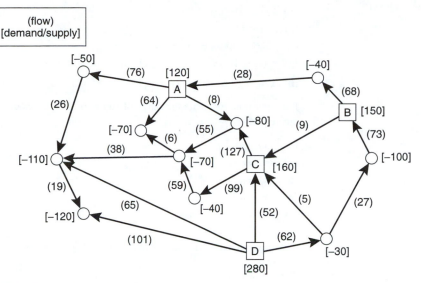

Figure 9.24   Power distribution problem with convex costs

EXERCISES

1. Plot the following functions and indicate whether each is convex, concave, or neither. Provide a mathematical explanation for your conclusions.

   (a) $f(x) = 4x^2 - 20x + 10$

   (b) $f(x) = -4x^2 - 20x$

   (c) $f(x) = 8x^3 + 15x^2 + 9x + 6$

2. Plot several isovalue contours for each of the following functions on the (x_1, x_2)-plane and determine if each is convex, concave, or neither. Provide a mathematical explanation for your conclusions.

   (a) $f(\mathbf{x}) = 2x_1^2 + 4x_2^2$

   (b) $f(\mathbf{x}) = -2x_1^2 + 4x_1x_2 - 4x_2^2 + 4x_1 + 4x_2 + 10$

   (c) $f(\mathbf{x}) = -0.5x_1^2 - 2x_1x_2 - 2x_2^2$

3. Determine whether the following functions are convex, concave, or neither.

   (a) $f(\mathbf{x}) = x_1^2 + x_1x_2 + {}_2^2$

   (b) $f(\mathbf{x}) = x_1^2 + 2x_1x_2 + x_2^2$

   (c) $f(\mathbf{x}) = x_1^2 + 4x_1x_2 + x_2^2$

   (d) $f(\mathbf{x}) = x_1^2 - 6x_1x_2 + 2x_2^2$

   (e) $f(\mathbf{x}) = 2x_1^2 + 3x_1x_3 + x_2^2 + 3x_2x_3 + 4x_3^2 + x_1 + 2x_2 + 3x_3$

   (f) $f(\mathbf{x}) = 2x_1^2 + 2x_1x_3 + x_2^2 - 2x_2x_3 + \frac{1}{2}x_3^2 + x_1 + 2x_2 + 3x_3$

4. Determine whether or not the following constraint sets define convex regions.

   (a) $0 \leq x \leq 12$

   (b) $|x| \geq 6$

   (c) $|x| \leq 6$

(d) $4x^2 - 20x \geq 0$

(e) $\ln x \geq 1$

(f) $-2x_1^2 + 4x_1x_2 - 4x_2^2 + 4x_1 + 4x_2 \leq 0$

(g) $x_1 + 2x_2 + x_3 \leq 10;\ x_1 - 2x_3 \leq 9;\ x_1 \geq 2,\ x_2 \geq 0,\ x_3 \geq 0$

(h) $x_1^2 + x_2^2 + x_3^2 \leq 9;\ 1.5x_1 + x_3 \geq 4$

5. For each function, state the range of values of the parameters a and b for which the function $c(x)$ is convex when $0 < x \leq 1$.

   (a) $c(x) = ax + b$

   (b) $c(x) = ax^b$

   (c) $c(x) = a^{bx}$

   (d) $c(x) = a \ln(bx)$

   (e) $c(x) = 1 + ae^{-x} + bx^2$

6. Determine whether the quadratic $f(\mathbf{x}) = \frac{1}{2}\mathbf{x}^T\mathbf{Q}\mathbf{x}$ is concave, convex, or neither for the following $\mathbf{Q}$ matrices.

   (a) $\begin{bmatrix} 1 & -1 & -1 \\ -1 & 2 & 4 \\ -1 & 4 & 6 \end{bmatrix}$

   (b) $\begin{bmatrix} 4 & 2 & -2 \\ 2 & 4 & 2 \\ -2 & 2 & 5 \end{bmatrix}$

7. Use the diagonalization algorithm and the determinant test to ascertain the definiteness of the following matrices.

   (a) $\begin{bmatrix} 1 & 2 & 1 \\ 2 & 1 & -2 \\ 2 & -2 & 1 \end{bmatrix}$

   (b) $\begin{bmatrix} 1 & 2 & 3 \\ 2 & 0 & -1 \\ 3 & -1 & 1 \end{bmatrix}$

   (c) $\begin{bmatrix} 0 & 1 & -1 & 2 \\ 1 & 1 & 0 & -1 \\ -1 & 0 & -1 & 1 \\ 2 & -1 & 1 & 0 \end{bmatrix}$

   (d) $\begin{bmatrix} 0 & 1 & 2 & 3 \\ 1 & 0 & 1 & 2 \\ 2 & 1 & 0 & 1 \\ 3 & 2 & 1 & 0 \end{bmatrix}$

8. Consider the functions.

$$a(\mathbf{x}) = 2x_1^2 - 30x_1 + 10 \qquad d(\mathbf{x}) = 5x_1^2 + 3x_2^2$$

$$b(\mathbf{x}) = -4x_2^2 + 20x_2 \qquad e(\mathbf{x}) = -2x_1^2 + x_1x_2 - 4x_2^2$$

$$c(\mathbf{x}) = \ln(x_1 + 1) + \ln(x_2 + 1) \qquad f(\mathbf{x}) = x_1 + x_2$$

Classify the following mathematical programs based on the types of problems described in Section 9.5.

(a) Minimize $\{a(\mathbf{x}) - b(\mathbf{x}): f(\mathbf{x}) \leq 10\}$

(b) Maximize $\{b(\mathbf{x}) + c(\mathbf{x}): d(\mathbf{x}) \leq 10, \mathbf{x} \geq \mathbf{0}\}$

(c) Maximize $\{d(\mathbf{x}): c(\mathbf{x}) \geq 4, e(\mathbf{x}) \geq -5, f(\mathbf{x}) \leq 10, \mathbf{x} \geq \mathbf{0}\}$

(d) Minimize $\{d(\mathbf{x}): c(\mathbf{x}) \geq 4, e(\mathbf{x}) \geq -5, f(\mathbf{x}) \geq 2, a(\mathbf{x}) \geq 8, \mathbf{x} \geq \mathbf{0}\}$

(e) Maximize $\{d(\mathbf{x})/f(\mathbf{x}): f(\mathbf{x}) \geq 4\}$

9. Decide whether each of the following objective functions could be part of a convex separable programming model. In all cases, the variables are nonnegative.

(a) Maximize $f(\mathbf{x}) = -2x_1^2 + 4x_1 x_2 - 4x_2^2 + 4x_1 + 4x_2$

(b) Maximize $f(\mathbf{x}) = \ln(x_1 + 1) + \ln(x_2 + 1)$

(c) Maximize $f(\mathbf{x}) = x_1^2 + 4x_2^2 - 3x_1 x_2 + 10x_1 + 20x_2$

(d) Minimize $f(\mathbf{x}) = x_1^2 + 4x_2^2 - 10x_1 + 20x_2$

(e) Minimize cost:

$$f(\mathbf{x}) = \sum_{j=1}^{n} f_j(x_j), \text{ where } f_j(x_j) = a_j x_j^{-b} \text{ with } a_j > 0 \text{ and } 0 < b < 1$$

(f) Maximize profit = revenue − cost:

$$f(\mathbf{x}) = \sum_{j=1}^{n} r_j(x_j) - \sum_{j=1}^{n} c(x_j), \text{ where } r_j(x_j) = a_j(1 - e^{-b_j x_j}) \text{ with } a_j > 0, b_j > 0,$$
$$\text{and } c_j(x_j) = d_j x_j^h \text{ with } d_j > 0 \text{ and } h > 1$$

10. Determine whether the objective function is convex, concave, or neither.

$$\begin{aligned} \text{Maximize } &-2x_1^2 - 4x_2^2 - 3x_3^2 + 4x_1 x_2 + 2x_1 x_3 - 2x_2 x_3 + 10x_1 + 15x_2 + 20x_3 \\ \text{subject to } \quad x_1 + \;\; x_2 + \;\; x_3 &\leq 15 \\ 2x_1 + \;\; 4x_2 \quad\quad\;\; &\leq 26 \\ x_2 + \;\; 3x_3 &\leq 20 \\ x_j \geq 0, \quad j = 1, 2, 3 \end{aligned}$$

Solve this problem using the software that accompanies this book.

11. Use the definition of convexity to prove Lemma 1.

12. Prove Theorem 1. Make use of the fact that Definition 3 can be extended to any finite number of points, say s, so that $f(\mathbf{x})$ is convex if and only if

$$f(\overline{\mathbf{x}}) \leq \sum_{i=1}^{s} \lambda_i f(\mathbf{x}_i), \text{ where } \overline{\mathbf{x}} = \sum_{i=1}^{s} \lambda_i \mathbf{x}_i, \;\; \sum_{i=1}^{s} \lambda_i = 1, \;\; \lambda_i \geq 0, \;\; i = 1, \ldots, s$$

Also make use of the fact that any point $\overline{\mathbf{x}}$ in a polyhedron S can be represented by a convex combination of its vertices. Note that the preceding equation for $\overline{\mathbf{x}}$ defines what is meant by a *convex combination* of points.

13. Use the definition of convexity to prove Theorem 2. Let $\mathbf{x}_1$ be a local minimum and let $\mathbf{x}_2$ be a global minimum such that $f(\mathbf{x}_2) < f(\mathbf{x}_1)$. Show that this leads to a contradiction.

14. Following the suggestions in the text, prove that $f(\mathbf{x})$ is convex if

$$f(\mathbf{x}_1) \geq f(\mathbf{x}_2) + \nabla^{\mathrm{T}} f(\mathbf{x}_2)(\mathbf{x}_1 - \mathbf{x}_2) \text{ for all } \mathbf{x}_1, \mathbf{x}_2$$

15. Five oil wells located in the (x, y)-plane have the coordinates (a_i, b_i) given by

$(a_1, b_1) = (5, 0)$ $(a_2, b_2) = (0, 10)$ $(a_3, b_3) = (10, 10)$
$(a_4, b_4) = (50, 50)$ $(a_5, b_5) = (-10, 50)$

It is desired to connect all the wells to a single collection point using the minimum total amount of pipe. Each pipe segment will be placed in a straight line from the well to the collection point.

(a) Formulate an NLP model that can be used to determine where the collection point should be located in the plane.

(b) Classify this problem and find the optimal solution for the data provided.

16. You are in charge of providing labor for a manufacturing shop for the next 6 months. Currently, there are 20 employees in the shop. Each worker costs $1000 per month and can manufacture five units of product during the month. The cost of hiring and training a new worker is $500. The cost of laying off a worker is $1000. At the end of the 6-month period you want 20 persons working in the shop. All products must be sold in the month they are produced.

Demand in each month is a random variable uniformly distributed over the ranges specified in the table. If demand is lower than production capacity, some workers will be idle. If demand is higher than capacity, sales are lost with a penalty of $200 per item not sold. To compute the expected cost of lost sales, consider month i with minimum demand a_i, and maximum demand b_i. Say x_i is the capacity for production in month i. The expected cost of lost sales is

$$C_i(x_i) = \begin{cases} 100(b_i - a_i) + 200(a_i - x_i) \text{ for } 0 \le x_i \le a_i \\ \dfrac{100}{b_i - a_i}(b_i - x_i)^2 \text{ for } a_i \le x_i \le b_i \\ 0 \text{ for } b_i \le x_i \end{cases} \qquad \text{for } i = 1, \ldots 6$$

Set up and solve the NLP to determine how many workers to hire or lay off at the beginning of each month to minimize the expected cost of lost sales plus the cost of providing the workforce. Workers may also be laid off at the end of month 6.

Month	Minimum demand	Maximum demand
1	50	150
2	100	200
3	75	175
4	50	150
5	200	250
6	100	200

17. The figure represents a reservoir system. The amounts of water entering at nodes 1, 2, and 3 are indicated in square brackets. All the flow is to leave the network at node 4. There are six possible pipelines (arcs) labeled 1 through 6 that can carry water. The problem is to assign the flow to arcs such that the total cost is minimized. The flows are directed and must be conserved at each node.

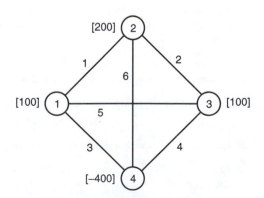

The arc costs are $a_i(f_i)^{b_i}$, where f_i is the absolute value of the flow on arc i, and a_i and b_i are constants for $i = 1, \ldots, 6$. Classify the problem for the various arc cost functions in the tables. Solve the problems for the given data. Hint: Define a variable for each direction along an arc.

(a)

Arc, i	1	2	3	4	5	6
a_i	1	2	1	2	1.4	2.8
b_i	0.9	0.7	0.9	0.7	0.9	0.7

This kind of cost function is often used to represent economies of scale with respect to size, so this is a reasonable model for systems design.

(b)

Arc, i	1	2	3	4	5	6
a_i	1	2	1	2	1.4	2.8
b_i	2	2	2	2	2	2

This kind of cost function is often used to represent the energy used for transmitting flow through an arc, so this is a reasonable model for choosing a distribution scheme.

18. A company wants an aggregate production plan for the next 6 months. Projected sales for its product are listed in the table.

Month	Sales goals (units)	Production cost ($/unit)
1	1300	100
2	1400	105
3	1000	110
4	800	115
5	1700	110
6	1900	110

Production that exceeds sales in each month may be put in inventory and sold in some future month. Because of seasonal factors, the production costs vary from month to month. An additional charge of $4 per unit per month is incurred for units in inventory.

The major costs center on production, but the company also wishes to account for failures to meet monthly sales, production, and inventory goals by penalizing deviations from either predetermined targets or policies. The inventory goal is to have 100 units in stock at the end of each month. The production goal is to minimize labor and raw material fluctuations from one month to the next. To account for these goals, squared penalty terms of the following form are added to the objective function.

$$w_S(\text{actual sales} - \text{sales goal})^2 + w_I(\text{actual inventory} - \text{inventory goal})^2$$
$$+ w_P(\text{production} - \text{production in previous month})^2$$

where w_S, w_I, and w_P are predetermined weights. Assume in your model that the initial and final inventories are 100, and assume the production prior to month 1 is 1000.

Set up the mathematical programming model that minimizes production, inventory, and goal penalty costs. Classify the model and solve for the case in which each penalty weight is equal to 20.

19. (*Economic Order Quantity with Shortages Back-ordered*) A variation of the inventory problem is to allow shortages, as shown in the figure. The shortage, equal to $Q - S$, is back-ordered at a cost of p per unit and is satisfied when the replenishment is received.

Justify the following formula for cost per unit time and characterize the shape of the function. The constants d, K, h, and c are defined in the text. Find the optimal solution in algebraic form using calculus.

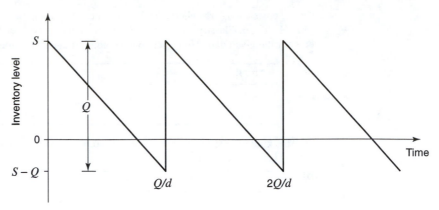

Lot-size model with shortages

$$\text{Cost/time} = \text{setup cost} + \text{product cost} + \text{holding cost} + \text{back-order cost}$$

$$f(Q, S) = dK/Q + dc + hS^2/2Q + p(Q - S)^2/2Q$$

20. Write the nonlinear program model for the Markov queuing network associated with the product routing problem represented by Figure 9.18. Let the objective be to minimize the average time in the system for a unit of product. Assume that each station has a single server. Find the solution when the arrival rate at station A is 30 units per hour and the service rates for the six stations are

$$\mu_A = 40, \ \mu_B = 20, \ \mu_C = 30, \ \mu_D = 10, \ \mu_E = 30, \ \mu_F = 40$$

21. The figure depicts the geographical area (2 miles wide and 4 miles long) for which emergency service is to be provided. The area has been divided into eight square cells (1 mile by 1 mile), each of which is identified by a number in the upper left-hand corner. The average number of calls per hour is given in the center of each cell. One service facility is to be located in cell 2 and another in cell 8. For cell 2, the service rate is 30 clients per hour, whereas the rate for cell 8 is 60 clients per hour.

1	2	3	4
8	10	15	7
5	6	7	8
12	8	2	10

Geographic grid: (2 miles) × (4 miles)

For modeling purposes, it is assumed that travel from a cell to a service facility originates and terminates at the center of the respective cell and follows a rectilinear path (movement is either horizontal or vertical). Travel time is 10 minutes per mile. The problem is to assign the demand to the service facilities such that the total time of travel plus service is minimized.

(a) Model the facilities as single-channel queues with exponential distributions for interarrival times and service times. Write the corresponding mathematical program that describes the full situation.

(b) Solve the optimization problem.

22. The regression problem when $p = 1$ is

$$\text{Minimize } \sum_{i=1}^{n} | y_i - (b_0 + b_1 x_{1i} + b_2 x_{2i} + \cdots + b_m x_{mi}) |$$

Write this unconstrained nonlinear program as an LP.

23. The regression problem when $p = \infty$ is

$$\text{Minimize } \left\{ \text{Maximum}_{i=1, \ldots, n} \left(| y_i - (b_0 + b_1 x_{1i} + b_2 x_{2i} + \cdots + b_m x_{mi}) | \right) \right\}$$

Write this unconstrained nonlinear program as an LP.

24. The data in the table represent 20 observations of cotton quality (y) as a function of amount of fertilizer applied per acre (x_1), average daily temperature (x_2), and amount of insecticide applied per acre (x_3).

Data for Regression Exercise

i	y_i	x_1	x_2	x_3
1	99	85	76	44
2	93	82	78	42
3	99	75	73	42
4	97	74	72	44
5	90	76	73	43
6	96	74	69	46
7	93	73	69	46
8	130	96	80	36
9	118	93	78	36
10	88	70	73	37
11	89	82	71	46
12	93	80	72	45
13	94	77	76	42
14	75	67	76	50
15	84	82	70	48
16	91	76	76	41
17	100	74	78	31
18	98	71	80	29
19	101	70	83	39
20	80	64	79	38

(a) Use a spreadsheet application such as Excel and its built-in functions to solve the bivariate least squares regression model $y = b_0 + b_1 x_1$. What values of b_0, b_1, and R^2 do you find? Repeat the analysis using an NLP solver.

(b) Solve the full least squares regression model

$$y = b_0 + b_1 x_1 + b_2 x_2 + b_3 x_3$$

What are the values of b_0, b_1, b_2, b_3, R^2, and R_a^2?

(c) Let $p = 1$ and solve the corresponding LP developed in Exercise 22 to find the regression coefficients and R^2 values.

(d) Let $p = \infty$ and solve the corresponding LP developed in Exercise 23 to find the regression coefficients and R^2 values.

(e) Repeat part (b) for the situation in which each deviation $\alpha_i = y_i - (b_0 + b_1 x_{1i} + b_2 x_{2i} + b_3 x_{3i})$ is weighted by $1/y_i$. The corresponding problem is to

$$\text{Minimize } \sum_{i=1}^{20} \left| \frac{\alpha_i}{y_i} \right|^2$$

BIBLIOGRAPHY

Beightler, C.S. and D.T. Phillips, *Applied Geometric Programming*, Wiley, New York, 1976.

Beightler, C.S., D.T. Phillips, and D.J. Wilde, *Foundations of Optimization*, Second Edition, Prentice-Hall, Englewood Cliffs, NJ, 1979.

Bracken, J. and G.O. McCormick, *Selected Applications of Nonlinear Programming*, Wiley, New York, 1968.

Draper, N.R. and H. Smith, *Applied Regression Analysis*, Third Edition, Wiley, New York, 1998.

Francis, R.L., L. McGinnis, and J. White, *Facility Layout and Location: An Analytical Approach*, Prentice-Hall, Englewood Cliffs, NJ, 1992.

Hadley, G., *Linear Algebra*, Addison-Wesley, Reading, MA, 1973.

Lay, D.C., *Linear Algebra and Its Applications*, Second Edition, Addison-Wesley, Reading, MA, 2000.

Liebman, J.S., L. Lasdon, L. Shrage, and A. Waren, *Modeling and Optimization with GINO*, Boyd and Fraser, Danvers, MA, 1986.

Nash, S.G., "Software Survey: NLP" *OR/MS Today*, Vol. 22, No. 2, pp. 60–71, 1995.

Chapter 10

Nonlinear Programming Methods[1]

Solution techniques for nonlinear programming (NLP) are much more complex and much less effective than those for linear programming (LP). The greater complexity is partly attributable to the large variety of functional forms that may appear in a problem. Although the simplex method and its interior point counterparts provide acceptable results for almost all LP models, many procedures have been invented to solve nonlinear problems. The success of a particular procedure depends a great deal on the type of model involved. Linear programming codes will provide optimal solutions for problems with hundreds of thousands of variables, but there is a reasonable chance that an NLP code will fail on a problem containing only a handful of variables. To sharpen this contrast, recall that all interior point methods for solving LP problems include ideas originally developed to solve NLP problems.

Notwithstanding the difficulties, NLP is a very important subject, because many problems simply cannot be modeled otherwise. The field is continually evolving, and good computer codes are available for numerous problem classes. Commercial products that permit algebraic input are relatively easy to use, but if a code is going to be embedded in a decision support system, the analyst should generally have a detailed knowledge of how it works. Useful solutions are rarely obtained on the first run; parameters must be tuned to the problem instance and convergence criteria must be set for termination. Even so, a code may take a long time to converge or may converge to a point that is not the optimal solution. There is a risk that the inexperienced user will unconditionally accept a termination point as the global optimal solution when it may not even be a local optimum. The user of an NLP code should be well acquainted with the mathematical characteristics of the model and should be very careful about interpreting the output.

The optimization problem that we wish to solve is

$$\left. \begin{array}{l} \text{Minimize} \ \ f(\mathbf{x}) \\ \text{subject to} \ \ g_i(\mathbf{x}) \le b_i, \ \ i = 1, \ \ldots, m \end{array} \right\} \tag{1}$$

where $\mathbf{x} \in \Re^n$ is the n-dimensional vector of decision variables, $f(\mathbf{x})$ is the objective function, and $g_1(\mathbf{x}), \ldots, g_m(\mathbf{x})$ are the constraint functions, which implicitly include variable bounds. All functions are assumed to be twice continuously differentiable,

---

[1] J. Wesley Barnes contributed to the initial draft of this chapter.

and all right-hand-side (RHS) values b_i are assumed to be known constants. Although the model as written includes only one type of constraint, it is completely general. All equality (=) constraints and "greater than or equal to" (≥) inequalities can be put in the form of Problem (1) with the help of simple transformations.

The way most continuous mathematical programs are solved is by applying an algorithm that searches for a point in the feasible region that satisfies a set of first-order optimality conditions—i.e., conditions based on first derivatives. Depending on the formulation and the properties of the underlying functions, it may be possible only to verify that a trial point satisfies first-order conditions that are necessary for a local (relative) extremum. In other cases, usually when the functions meet certain convexity requirements, it may be possible to establish sufficiency. As such, our initial aim in this chapter is to outline the optimality conditions associated with various classes of nonlinear programs. All the results are stated for minimization objectives, but comparable results hold for maximization as well.

We begin with the unconstrained case in which $m = 0$ and then look at a simple extension in which the variables are required to be nonnegative. Next, we examine the classical case that contains only equalities. The most general version of the problem in which the feasible region is defined by a combination of equality and inequality constraints is then discussed. Special attention is given to the separable programming problem and the quadratic programming problem. The latter part of the chapter is devoted to search methods. We start with the problem of finding the minimum of a one-dimensional unconstrained function and then move on to multidimensional functions. The supplement to this chapter highlights several techniques for solving the general constrained NLP given in Problem (1).

10.1 CLASSICAL OPTIMIZATION

The simplest situation that we address concerns the minimization of a function f in the absence of any constraints. This problem can be written as

$$\text{Minimize } \{f(\mathbf{x}): \mathbf{x} \in \mathfrak{R}^n\}$$

where $f \in C^2$ (twice continuously differentiable). Without additional assumptions on the nature of f, we will most likely have to be content with finding a point that is a local minimum. Elementary calculus provides a necessary condition that must be true for an optimal solution of a nonlinear function with continuous first and second derivatives. Specifically, the gradient is zero at every stationary point that is a candidate for a maximum or minimum. Sufficient conditions derived from convexity properties are also available in many cases.

The ideas presented in Chapter 9 can be used to develop the theoretical foundations that form the basis for modern NLP techniques. By initially limiting our attention to the optimization of nonlinear functions with no constraints on the decision variables we can simplify the presentation of the basic concepts and create a natural bridge to the study of constrained optimization problems.

Unconstrained Optimization

The first-order *necessary condition* that any point $\mathbf{x}^*$ must satisfy to be a minimum of f is that the gradient must vanish.

$$\nabla f(\mathbf{x}^*) = \mathbf{0} \tag{2}$$

This property is most easily illustrated for a univariate objective function in which the gradient is simply the derivative or the slope of $f(x)$. Consider, for example, Figure 10.1. The function in part (a) has a unique global minimum x^* at which the slope is zero. Any movement from that point yields a greater, and therefore less favorable, value. The graph in part (b) exhibits a range of contiguous global minima where the necessary condition holds; however, we should note that the corresponding $f(x)$ is not twice continuously differentiable at all points.

Figure 10.2 shows why Equation (2) is only a necessary condition and not a *sufficient* condition. In all three parts of the figure there are points at which the slope of $f(x)$ is zero but the global minima are not attained. Figure 10.2a illustrates a strong local maximum at x_1^* and a strong local minimum at x_2^*. Figure 10.2b shows a point of inflection at x_1^* that is a one-dimensional *saddle point*. Finally, Figure 10.2c presents the case of a unique global maximum at x_1^*.

The ideas embodied in Figures 10.1 and 10.2 can be easily generalized to functions in a higher-dimensional space at both the conceptual and mathematical levels. Because the necessary condition that the gradient be zero ensures only a *stationary point*—i.e., a local minimum, a local maximum, or a saddle point at $\mathbf{x}^*$—let us consider *sufficient conditions* for $\mathbf{x}^*$ to be either a local or a global minimum.

- If $f(\mathbf{x})$ is strictly convex in the neighborhood of $\mathbf{x}^*$, then $\mathbf{x}^*$ is a strong local minimum.
- If $f(\mathbf{x})$ is convex for all $\mathbf{x}$, then $\mathbf{x}^*$ is a global minimum.
- If $f(\mathbf{x})$ is strictly convex for all $\mathbf{x}$, then $\mathbf{x}^*$ is a unique global minimum.

To be precise, a neighborhood of $\mathbf{x}$ is an open sphere centered at $\mathbf{x}$ with arbitrarily small radius $\varepsilon > 0$. It is denoted by $N_\varepsilon(\mathbf{x})$, where $N_\varepsilon(\mathbf{x}) = \{\mathbf{y} : \|(\mathbf{y} - \mathbf{x})\| < \varepsilon\}$. Recall from Section

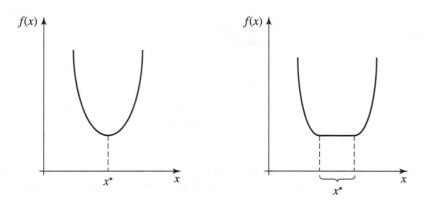

a. Function with a unique minimum   b. Function with a range of minima

Figure 10.1   Univariate functions in which the gradient is zero at a minimum

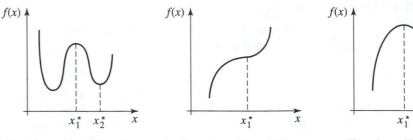

a. Function with two local minima and a local maximum

b. Function with a saddle point

c. Function with a unique global maximum

Figure 10.2 Univariate functions for which a zero gradient does not indicate a global minimum

9.3 that $f(\mathbf{x})$ is strictly convex if its Hessian matrix $\mathbf{H}(\mathbf{x})$ is positive definite for all $\mathbf{x}$. In this case, a stationary point must be a unique global minimum. $f(\mathbf{x})$ is convex if its Hessian matrix $\mathbf{H}(\mathbf{x})$ is positive semidefinite for all $\mathbf{x}$. For this case a stationary point will be a global (but perhaps not unique) minimum. If we do not know the Hessian for all $\mathbf{x}$, but we evaluate $\mathbf{H}(\mathbf{x}^*)$ at a stationary point $\mathbf{x}^*$ and find it to be positive definite, the stationary point is a strong local minimum.

Functions of a Single Variable

Let $f(x)$ be a convex function of $x \in \Re^1$. A necessary and sufficient condition for x^* to be a global minimum is that the first derivative of $f(x)$ be zero at that point.

$$\left.\frac{df(x)}{dx}\right|_{x=x^*} = 0$$

This is also a necessary and sufficient condition for the maximum of a concave function. The optimal solution is determined by setting the derivative equal to zero and solving the corresponding equation for x. If no solution exists, there is no finite optimal solution.

A sufficient condition for a local minimum (maximum) point of an arbitrary function is that the first derivative of the function be zero and the second derivative be positive (negative) at the point.

Example 1

Let us find the minimum of $f(x) = 4x^2 - 20x + 10$. The first step is to take the derivative of $f(x)$ and set it equal to zero.

$$df(x)/dx = 8x - 20 = 0$$

Solving this equation yields $x^* = 2.5$, which is a candidate solution. We now examine the curvature of $f(x)$ to determine if it is convex. Looking at the second derivative, we see

$$d^2f(x)/dx^2 = 8 > 0 \text{ for all } x$$

so f is strictly convex. Therefore, x^* is a global minimum.

Example 2

As a variation of Example 1, let us find the minimum of $f(x) = -4x^2 - 20x$. Taking the first derivative and setting it equal to zero yields $df(x)/dx = -8x - 20 = 0$, so $x^* = -2.5$. The second derivative is $d^2f(x)/dx^2 = -8 < 0$ for all x, so f is strictly concave. This means that x^* is a global maximum. There is no minimum solution because $f(x)$ is unbounded from below.

Example 3

Now let us minimize the cubic function $f(x) = 8x^3 + 15x^2 + 9x + 6$. Taking the first derivative and setting it equal to zero yields $df(x)/dx = 24x^2 + 30x + 9 = (6x + 3)(4x + 3) = 0$. The roots of this quadratic are at $x = -0.5$ and $x = -0.75$, so we have two candidates. Checking the second derivative

$$d^2f(x)/dx^2 = 48x + 30$$

we see that it is > 0 for $x > 0$ and < 0 for $x < 0$. Therefore, $f(x)$ is neither convex nor concave. At $x = -0.5$, $d^2f(-0.5)/dx^2 = 6$, so we have a local minimum. At $x = -0.75$, $d^2f(-0.75)/dx^2 = -6$, which indicates a local maximum. These points are not global optima, because the function is actually unbounded from both above and below.

Functions of Several Variables

Optimality conditions for the general case are given in the following theorems.

Theorem 1: Let $f(\mathbf{x})$ be twice continuously differentiable throughout a neighborhood of $\mathbf{x}^*$. Necessary conditions for $\mathbf{x}^*$ to be a local minimum of f are

    a. $\nabla f(\mathbf{x}^*) = \mathbf{0}$

    b. $\mathbf{H}(\mathbf{x}^*)$ is positive semidefinite.

Theorem 2: Let $f(\mathbf{x})$ be twice continuously differentiable throughout a neighborhood of $\mathbf{x}^*$. Then a sufficient condition for $f(\mathbf{x})$ to have a strong local minimum at $\mathbf{x}^*$, where Equation (2) holds, is that $\mathbf{H}(\mathbf{x}^*)$ be positive definite.

Quadratic Forms

A common and useful nonlinear function is the *quadratic* function

$$f(\mathbf{x}) = a + \mathbf{c}\mathbf{x} + \tfrac{1}{2}\mathbf{x}^{\mathrm{T}}\mathbf{Q}\mathbf{x}$$

that has coefficients $a \in \mathfrak{R}^1$, $\mathbf{c} \in \mathfrak{R}^n$, and $\mathbf{Q} \in \mathfrak{R}^{n \times n}$. $\mathbf{Q}$ is the Hessian matrix of $f(\mathbf{x})$. Setting the gradient

$$\nabla f(\mathbf{x}) = \mathbf{c}^{\mathrm{T}} + \mathbf{Q}\mathbf{x}$$

to zero results in a set of n linear equations in n variables. A solution will exist whenever $\mathbf{Q}$ is nonsingular. In such instances, the stationary point is

$$\mathbf{x}^* = -\mathbf{Q}^{-1}\mathbf{c}^{\mathrm{T}}$$

For a two-dimensional problem, the quadratic function is

$$f(\mathbf{x}) = a + c_1 x_1 + c_2 x_2 + \tfrac{1}{2} q_{11} x_1^2 + \tfrac{1}{2} q_{22} x_2^2 + q_{12} x_1 x_2$$

For this function, setting the partial derivatives with respect to x_1 and x_2 equal to zero results in the following linear system.

$$c_1 + q_{11} x_1 + q_{12} x_2 = 0, \quad c_2 + q_{12} x_1 + q_{22} x_2 = 0$$

These equations can be solved using Cramer's rule from linear algebra. The first step is to find the determinant of the $\mathbf{Q}$ matrix. Let

$$\det \mathbf{Q} = \begin{vmatrix} q_{11} & q_{12} \\ q_{12} & q_{22} \end{vmatrix} = q_{11} q_{22} - (q_{12})^2$$

The appropriate substitutions yield

$$x_1^* = \frac{-c_1 q_{22} + c_2 q_{12}}{\det \mathbf{Q}} \text{ and } x_2^* = \frac{-c_2 q_{11} + c_1 q_{12}}{\det \mathbf{Q}}$$

which is the desired stationary point.

When the objective function is a quadratic, the determination of definiteness is greatly facilitated because the Hessian matrix is constant. For more general forms, it may not be possible to determine conclusively whether the function is positive definite, negative definite, or indefinite. In such cases, we can only make statements about local optimality. In the following examples, we use $\mathbf{H}$ to identify the Hessian. For quadratic functions, $\mathbf{Q}$ and $\mathbf{H}$ are the same.

Example 4

Find the local extreme values of $f(\mathbf{x}) = 25x_1^2 + 4x_2^2 - 20x_1 + 4x_2 + 5$.

Solution: Using Equation (2) yields

$$50x_1 - 20 = 0 \text{ and } 8x_2 + 4 = 0$$

The corresponding stationary point is $\mathbf{x}^* = (2/5, -1/2)$. Because $f(\mathbf{x})$ is a quadratic, its Hessian matrix is constant.

$$\mathbf{H} = \begin{bmatrix} 50 & 0 \\ 0 & 8 \end{bmatrix}$$

The determinants of the leading submatrices of $\mathbf{H}$ are $H_1 = 50$ and $H_2 = 400$, so $f(\mathbf{x})$ is strictly convex, implying that $\mathbf{x}^*$ is the global minimum.

Example 5

Find the local extreme values of the nonquadratic function

$$f(\mathbf{x}) = 3x_1^3 + x_2^2 - 9x_1 + 4x_2$$

Solution: Using Equation (2) yields

$$\nabla f(\mathbf{x}) = (9x_1^2 - 9, 2x_2 + 4)^{\mathrm{T}} = (0, 0)^{\mathrm{T}}$$

so $x_1 = \pm 1$ and $x_2 = -2$. Checking $\mathbf{x} = (1, -2)$, we have

$$\mathbf{H}(1, -2) = \begin{bmatrix} 18 & 0 \\ 0 & 2 \end{bmatrix}$$

which is positive definite since $\mathbf{v}^T\mathbf{H}(1, -2)\mathbf{v} = 18v_1^2 + 2v_2^2 > 0$ when $\mathbf{v} \neq \mathbf{0}$. Thus $(1, -2)$ yields a strong local minimum. Next, consider $\mathbf{x} = (-1, -2)$ with Hessian matrix

$$\mathbf{H}(-1, -2) = \begin{bmatrix} -18 & 0 \\ 0 & 2 \end{bmatrix}$$

Now we have $\mathbf{v}^T\mathbf{H}(-1, -2)\mathbf{v} = -18v_1^2 + 2v_2^2$, which may be less than or equal to 0 when $\mathbf{v} \neq \mathbf{0}$. Thus, the sufficient condition for $(-1, -2)$ to be either a local minimum or a local maximum is not satisfied. Actually, the second necessary condition (b) in Theorem 1 for either a local minimum or a local maximum is not satisfied. Therefore, $\mathbf{x} = (1, -2)$ yields the only local extreme value of f.

Example 6

Find the extreme values of $f(\mathbf{x}) = -2x_1^2 + 4x_1x_2 - 4x_2^2 + 4x_1 + 4x_2 + 10$.

Solution: Setting the partial derivatives equal to zero leads to the linear system

$$-4x_1 + 4x_2 + 4 = 0 \text{ and } 4x_1 - 8x_2 + 4 = 0$$

which yields $\mathbf{x}^* = (3, 2)$. The Hessian matrix is

$$\mathbf{H} = \begin{bmatrix} -4 & 4 \\ 4 & -8 \end{bmatrix}$$

Evaluating the principal determinants of $\mathbf{H}$, we find $H_1 = -4$ and $H_2 = 16$. Thus, $f(\mathbf{x})$ is strictly concave and $\mathbf{x}^*$ is a global maximum.

Example 7

Find the extreme values of $f(\mathbf{x}) = -0.5x_1^2 - 2x_1x_2 - 2x_2^2$.

Solution: Equation (2) yields $\nabla f(\mathbf{x}) = (-x_1 - 2x_2, -4x_2 - 2x_1)^T = (0, 0)^T$. The only stationary point in this system is $\mathbf{x}^* = (0, 0)$. Using the diagonalization algorithm, we discover

$$\mathbf{x} = \begin{bmatrix} -1 & 0 \\ 0 & 0 \end{bmatrix}$$

which is a negative semidefinite matrix. Therefore, $f(\mathbf{x})$ is concave and the stationary point is a global maximum.

Nonquadratic Forms

When the objective function is not quadratic (or linear), the Hessian matrix will depend on the values of the decision variables $\mathbf{x}$. This was demonstrated in Example 5. We now consider two more examples. Suppose

$$f(\mathbf{x}) = (x_2 - x_1^2)^2 + (1 - x_1)^2$$

The gradient of this function is

$$\nabla f(\mathbf{x}) = \begin{pmatrix} -4x_1(x_2 - x_1^2) - 2(1 - x_1) \\ 2(x_2 - x_1^2) \end{pmatrix}$$

For the second component of the gradient to be zero, we must have $x_2 = x_1^2$. Taking this into account, the first component is zero only when $x_1 = 1$, so $\mathbf{x}^* = (1, 1)$ is the sole stationary point. It was previously shown (in Section 9.3) that the Hessian matrix $\mathbf{H}(\mathbf{x})$ at this point is positive definite, indicating that it is a local minimum. Because we have not shown that the function is everywhere convex, further arguments are necessary to characterize the point as a global minimum. Logically, $f(\mathbf{x}) \geq 0$ because each of its two component terms is squared. The fact that $f(1, 1) = 0$ implies that $(1, 1)$ is a global minimum.

As a further example, consider

$$f(\mathbf{x}) = (x_1 - 2x_2^2)(x_1 - 3x_2^2)$$

where

$$\nabla f(\mathbf{x}) = \begin{pmatrix} 2x_1 - 5x_2^2 \\ -10x_1 x_2 + 24x_2^3 \end{pmatrix} \text{ and } \mathbf{H}(\mathbf{x}) = \begin{bmatrix} 2 & -10x_2 \\ -10x_2 & 72x_2^2 - 10x_1 \end{bmatrix}$$

A stationary point exists at $\mathbf{x}^* = (0, 0)$. Also, $H_1 = 2$ and $H_2 = 44x_2^2 - 20x_1$, implying that $\mathbf{H}(\mathbf{x})$ is indefinite. Although $\mathbf{H}(\mathbf{x})$ is positive semidefinite at $(0, 0)$ this does not allow us to conclude that $\mathbf{x}^*$ is a local minimum. Notice that $f(\mathbf{x})$ can be made arbitrarily small or large with the appropriate choices of $\mathbf{x}$.

These last two examples suggest that for nonquadratic functions of several variables, the determination of the character of a stationary point can be difficult even when the Hessian matrix is semidefinite. Indeed, a much more complex mathematical theory is required for the general case.

Summary for Unconstrained Optimization

Table 10.1 summarizes the relationship between the optimality of a stationary point $\mathbf{x}^*$ and the character of the Hessian evaluated at $\mathbf{x}^*$. It is assumed that $f(\mathbf{x})$ is twice differentiable and $\nabla f(\mathbf{x}^*) = \mathbf{0}$.

If $\mathbf{H}(\mathbf{x})$ exhibits either of the first two definiteness properties for all $\mathbf{x}$, then "local" can be replaced with "global" in the associated characterizations. Furthermore, if $f(\mathbf{x})$ is quadratic, a positive semidefinite Hessian matrix implies a nonunique global minimum at $\mathbf{x}^*$.

Notice that although convexity in the neighborhood of $\mathbf{x}^*$ *is* sufficient to conclude that $\mathbf{x}^*$ is a weak local minimum, the fact that $\mathbf{H}(\mathbf{x}^*)$ is positive semidefinite *is not* sufficient, in general, to conclude that $f(\mathbf{x})$ is convex in the neighborhood of $\mathbf{x}^*$. When $\mathbf{H}(\mathbf{x}^*)$ is positive

Table 10.1   Relation between Hessian Matrix and Stationary Point

$\mathbf{H}(\mathbf{x}^*)$	$\mathbf{x}^*$
Positive definite	Strong local minimum
Negative definite	Strong local maximum
Indefinite	Saddle point
Positive or negative semidefinite	No conclusion: resort to higher-order analysis

semidefinite, it is possible that points in a small neighborhood of $\mathbf{x}^*$ can exist such that $f(\mathbf{x})$ evaluated at those points will produce smaller values than $f(\mathbf{x}^*)$. This would invalidate the conclusion of convexity in the neighborhood of $\mathbf{x}^*$.

As a final example in this section, consider

$$f(x) = 2x_1^3 + 4x_1^2 x_2^2 - 2x_1 x_2^3 - 5x_1 x_3^3 + x_2^2 x_3 + 3x_3^3$$

for which

$$\nabla f(\mathbf{x}) = \begin{pmatrix} 6x_1^2 + 8x_1 x_2^2 - 2x_2^3 - 5x_3^3 \\ 8x_1^2 - 6x_1 x_2^2 + 2x_2 x_3 \\ -15x_1 x_3^2 + x_2^2 + 9x_3^2 \end{pmatrix}$$

and

$$H(\mathbf{x}) = \begin{bmatrix} 12x_1 + 8x_2^2 & 16x_1 x_2 - 6x_2^2 & -15x_3^2 \\ 16x_1 x_2 - 6x_2^2 & 8x_1^2 - 12x_1 x_2 + 2x_3 & 2x_2 \\ -15x_3^2 & 2x_2 & -30x_1 x_3 + 18x_3 \end{bmatrix}$$

Looking at the Hessian matrix, it is virtually impossible to make any statements about the convexity of $f(\mathbf{x})$. This gives us a glimpse of the difficulties that can arise when one attempts to solve unconstrained nonlinear optimization problems by directly applying the classical theory. In fact, the real value of the theory is that it offers insights into the development of more practical solution approaches. Moreover, once we have a stationary point $\mathbf{x}^*$ obtained from one of those approaches, it is relatively easy to check the properties of $\mathbf{H}(\mathbf{x}^*)$, because only numerical evaluations are required.

Nonnegative Variables

A simple extension of the unconstrained optimization problem involves the addition of non-negativity restrictions on the variables.

$$\text{Minimize } \{f(\mathbf{x}) : \mathbf{x} \geq \mathbf{0}\} \tag{3}$$

Suppose that f has a local minimum at $\mathbf{x}^*$, where $\mathbf{x}^* \geq \mathbf{0}$. Then there exists a neighborhood $N_\varepsilon(\mathbf{x}^*)$ of $\mathbf{x}^*$ such that whenever $\mathbf{x} \in N_\varepsilon(\mathbf{x}^*)$ and $\mathbf{x} \geq \mathbf{0}$, we have $f(\mathbf{x}) \geq f(\mathbf{x}^*)$. Now write $\mathbf{x} = \mathbf{x}^* + t\mathbf{d}$, where $\mathbf{d}$ is a direction vector and $t > 0$. Assuming that f is twice continuously differentiable throughout $N_\varepsilon(\mathbf{x}^*)$, a second-order Taylor series expansion of $f(\mathbf{x}^* + t\mathbf{d})$ around $\mathbf{x}^*$ yields

$$f(\mathbf{x}^*) \leq f(\mathbf{x}) = f(\mathbf{x}^* + t\mathbf{d}) = f(\mathbf{x}^*) + \nabla f(\mathbf{x}^*) t\mathbf{d} + \frac{t}{2} d^T \nabla^2 f(\mathbf{x}^* + \alpha t\mathbf{d}) t\mathbf{d}$$

where $\alpha \in [0, 1]$. Canceling terms and dividing through by t yields

$$0 \leq \nabla f(\mathbf{x}^*)\mathbf{d} + \frac{t}{2} v^T \nabla^2 f(\mathbf{x}^* + \alpha t\mathbf{d})\mathbf{d}$$

As $t \to 0$, the inequality becomes $0 \leq \nabla f(\mathbf{x}^*)\mathbf{d}$, which says that f must be nondecreasing in any feasible direction $\mathbf{d}$. If $\mathbf{x}^* > \mathbf{0}$, we know that $\nabla f(\mathbf{x}^*) = \mathbf{0}$. But what can we say about the individual pairs x_j and $\partial f(x_j)/\partial x_j$ at optimality? With a bit more analysis, it can be shown that the following conditions are necessary for $\mathbf{x}^*$ to be a local minimum of $f(\mathbf{x})$.

$$\frac{\partial f(\mathbf{x}^*)}{\partial x_j} = 0, \text{ if } x_j^* > 0$$

$$\frac{\partial f(\mathbf{x}^*)}{\partial x_j} \geq 0, \text{ if } x_j^* = 0$$

These results are summarized as follows.

Theorem 3: Necessary conditions for a local minimum of f in Problem (3) to occur at $\mathbf{x}^*$ include

$$\nabla f(\mathbf{x}^*) \geq \mathbf{0}, \quad \nabla f(\mathbf{x}^*)\mathbf{x}^* = 0, \quad \mathbf{x}^* \geq \mathbf{0} \tag{4}$$

where f is twice continuously differentiable throughout a neighborhood of $\mathbf{x}^*$.

Example 8

$$\text{Minimize } f(\mathbf{x}) = 3x_1^2 + x_2^2 + x_3^2 - 2x_1x_2 - 2x_1x_3 - 2x_1$$
$$\text{subject to } x_1 \geq 0, \; x_2 \geq 0, \; x_3 \geq 0$$

Solution: From Conditions (4), we have the following necessary conditions for a local minimum.

a. $\quad 0 \leq \dfrac{\partial f}{\partial x_1} = 6x_1 - 2x_2 - 2x_3 - 2$

b. $\quad 0 = x_1 \dfrac{\partial f}{\partial x_1} = x_1(6x_1 - 2x_2 - 2x_3 - 2)$

c. $\quad 0 \leq \dfrac{\partial f}{\partial x_2} = 2x_2 - 2x_1$

d. $\quad 0 = x_2 \dfrac{\partial f}{\partial x_2} = x_2(2x_2 - 2x_1)$

e. $\quad 0 \leq \dfrac{\partial f}{\partial x_3} = 2x_3 - 2x_1$

f. $\quad 0 = x_3 \dfrac{\partial f}{\partial x_3} = x_3(2x_3 - 2x_1)$

g. $\quad x_1 \geq 0, x_2 \geq 0, x_3 \geq 0$

From condition (d), we see that either $x_2 = 0$ or $x_1 = x_2$. When $x_2 = 0$, conditions (c) and (g) imply that $x_1 = 0$. From condition (f) then, $x_3 = 0$. But this contradicts condition (a), so $x_2 \neq 0$ and $x_1 = x_2$.

Condition (f) implies that either $x_3 = 0$ or $x_1 = x_3$. If $x_3 = 0$, then conditions (d), (e), and (g) imply that $x_1 = x_2 = x_3 = 0$. But this situation has been ruled out. Thus, $x_1 = x_2 = x_3$, and from condition (b) we get $x_1 = 0$ or $x_1 = 1$. Since $x_1 \neq 0$, the only possible relative minimum of f occurs when $x_1 = x_2 = x_3 = 1$. To characterize the solution at $\mathbf{x}^* = (1, 1, 1)$ we evaluate the Hessian matrix.

$$\mathbf{H} = \begin{bmatrix} 6 & -2 & -2 \\ -2 & 2 & 0 \\ -2 & 0 & 2 \end{bmatrix}$$

which is easily shown to be positive definite. Thus, f is strictly convex and has a strong local minimum at $\mathbf{x}^*$. It follows from Theorem 2 in Chapter 9 that $f(\mathbf{x}^*) = 1$ is a global minimum.

10.2   EQUALITY CONSTRAINTS

Classical optimization also deals with problems that contain equality constraints. Much theory is available, but applying it to all but the simplest of instances is very difficult. As in the unconstrained case, the most that we can expect from an algorithm is a stationary point that may be a local extremum, a global extremum, or neither, depending on convexity. Although second-order necessary and sufficient conditions exist for constrained problems, their implicit nature makes them virtually impossible to check. Therefore, they are mostly of theoretical interest and are not presented here.

From a practical perspective, if any of the equality constraints can be solved explicitly for one of the decision variables, doing so allows problem size to be reduced by substituting out that variable. The corresponding constraint is also eliminated. Once the smaller problem has been solved, the values of the removed variables can be found from the relationships used in the substitution. For a problem with n variables and m constraints, at most m variables can be eliminated, where it is assumed that $n > m$. Of course, this procedure works only if the variables are unconstrained. If x_j is required to be nonnegative, for example, and the first constraint is used in the substitution—i.e., $x_j = g_1(x_1, \ldots, x_{j-1}, x_{j+1}, \ldots, x_n)$—there is no guarantee that the solution to the reduced problem without x_j and g_1 will return values of the remaining decision variables such that when substituted back into g_1 we will have $x_j \geq 0$.

From a modeling point of view, redundant variables and equations are often included in a formulation to represent the problem more clearly. Although it may be possible to eliminate some of them at the solution stage, this may not be a good idea for a number of reasons. First, it may be awkward to do so; second, the model may be more difficult to debug; and third, the results might not lend themselves as readily to interpretation.

Necessary Conditions for Optimality

Limiting our attention to equality constraints leads to the following mathematical statement of the problem to be solved.

$$\text{Minimize } f(\mathbf{x}) \text{ subject to } g_i(\mathbf{x}) = 0, \ i = 1, \ldots, m \tag{5}$$

Once again, the objective and constraint functions are assumed to be at least twice continuously differentiable. Furthermore, each of the $g_i(\mathbf{x})$ subsumes the constant term b_i.

To provide intuitive justification for the general results, consider the special case of Problem (5) with two decision variables and one constraint—i.e.,

$$\text{Minimize } f(x_1, x_2)$$
$$\text{subject to } g(x_1, x_2) = 0$$

To formulate the first-order necessary conditions, we construct the *Lagrangian*

$$\mathscr{L}(x_1, x_2, \lambda) = f(x_1, x_2) + \lambda g(x_1, x_2)$$

where λ is an unconstrained variable called the Lagrange multiplier. The constraint has been removed from the problem and placed in the objective function as a penalty term. Our goal now is to minimize the unconstrained function $\mathscr{L}(x_1, x_2, \lambda)$. As in Section 10.1, we construct the gradient of the Lagrangian with respect to its decision variables x_1 and x_2 and the multiplier λ. Setting the gradient equal to zero, we obtain

$$\nabla \mathcal{L}(x_1, x_2, \lambda) = \begin{pmatrix} \dfrac{\partial f(x_1, x_2)}{\partial x_1} + \lambda \dfrac{\partial g(x_1, x_2)}{\partial x_1} \\ \dfrac{\partial f(x_1, x_2)}{\partial x_2} + \lambda \dfrac{\partial g(x_1, x_2)}{\partial x_2} \\ g(x_1, x_2) \end{pmatrix} = \begin{pmatrix} 0 \\ 0 \\ 0 \end{pmatrix} \tag{6}$$

which represents three equations in three unknowns. Using the first two equations to eliminate λ, we have

$$\frac{\partial f}{\partial x_1} \frac{\partial g}{\partial x_2} - \frac{\partial f}{\partial x_2} \frac{\partial g}{\partial x_1} = 0, \quad g(x_1, x_2) = 0$$

which yields a stationary point $\mathbf{x}^* = (x_1^*, x_2^*)$ and λ^* when solved. From Equation (6), we see that $\nabla f(x_1, x_2)$ and $\nabla g(x_1, x_2)$ are coplanar at this solution.

It is a simple matter to extend these results to the general case. The Lagrangian is

$$\mathcal{L}(\mathbf{x}, \lambda) = f(\mathbf{x}) + \sum_{i=1}^{m} \lambda_i g_i(\mathbf{x})$$

where $\lambda = (\lambda_1, \ldots, \lambda_m)$ is an m-dimensional row vector. Here, every constraint i has an associated unconstrained multiplier λ_i. Setting the partial derivatives of the Lagrangian with respect to each decision variable and each multiplier equal to zero yields the following system of $n + m$ equations. These equations represent the first-order necessary conditions for an optimum to exist at $\mathbf{x}^*$.

$$\frac{\partial \mathcal{L}}{\partial x_j} = \frac{\partial f(\mathbf{x})}{\partial x_j} + \sum_{i=1}^{m} \lambda_i \frac{\partial g_i(\mathbf{x})}{\partial x_j} = 0, \quad j = 1, \ldots, n \tag{7a}$$

$$\frac{\partial \mathcal{L}}{\partial \lambda_i} = g_i(\mathbf{x}) = 0, \quad i = 1, \ldots, m \tag{7b}$$

A solution to Equations (7a) and (7b) yields a stationary point $(\mathbf{x}^*, \lambda^*)$; however, an additional qualification must be placed on the constraints in Equation (7b) if these conditions are to be valid. This is illustrated in Example 10.

Because Equations (7a) and (7b) are identical regardless of whether a minimum or maximum is sought, additional work is required to distinguish between the two. Indeed, it may be that some selection of the decision variables and multipliers that satisfies these conditions determines a saddle point of $f(\mathbf{x})$ a rather than a minimum or maximum.

To specify which kind of extremum resides at a stationary point, we must look at second-order conditions. Several approaches are available in this regard. The first uses *constrained derivatives*. Once the constrained gradient and the constrained Hessian matrix have been constructed, the same techniques described for unconstrained problems may be applied. A second approach is to investigate the convexity of the *bordered* Hessian matrix. Although informative, these ideas will not be pursued here because it is rare that problems arising in operations research have equality constraints only. Such problems are simply a special case of the more general model. Therefore, we defer the characterization of an extremum to the next section where methods for handling problems with inequality constraints and nonnegative restrictions on the variables are discussed.

Example 9

We wish to solve the following problem.

$$\text{Minimize } f(\mathbf{x}) = 3x_1^2 + x_2^2$$
$$\text{subject to } x_1 + x_2 = 6$$

Solution: The Lagrangian is $\mathscr{L}(\mathbf{x}, \lambda) = 3x_1^2 + x_2^2 + \lambda(x_1 + x_2 - 6)$. Setting the partial derivatives equal to zero yields three equations in three unknowns:

$$\frac{\partial \mathscr{L}}{\partial x_1} = 6x_1 + \lambda = 0, \quad \frac{\partial \mathscr{L}}{\partial x_2} = 2x_2 + \lambda = 0, \quad \frac{\partial \mathscr{L}}{\partial \lambda} = x_1 + x_2 - 6 = 0$$

The solution to this linear system is

$$(x_1, x_2, \lambda) = (1.5, 4.5, -9)$$

Because the objective function is convex, the constraint is linear, and both decision variables are positive, these values constitute a global minimum over the set of feasible points.

Alternatively, we could have solved for x_2 in the single constraint and substituted the result

$$x_2 = 6 - x_1$$

into the objective function, yielding $f(x_1) = 4x_1^2 - 12x_1 + 36$. The minimum is at $x_1 = 1.5$. Thus, the same solution is found by either approach.

Example 10

$$\text{Minimize } f(\mathbf{x}) = (x_1 + x_2)^2$$
$$\text{subject to } -(x_1 - 3)^3 + x_2^2 = 0$$

The Lagrangian is $\mathscr{L}(\mathbf{x}, \lambda) = (x_1 + x_2)^2 + \lambda[-(x_1 - 3)^3 + x_2^2]$. Now, setting partial derivatives equal to zero gives three highly nonlinear equations in three unknowns:

$$\frac{\partial \mathscr{L}}{\partial x_1} = 2(x_1 + x_2) - 3\lambda(x_1 - 3)^2 = 0, \quad \frac{\partial \mathscr{L}}{\partial x_2} = 2(x_1 + x_2) + 2\lambda x_2 = 0$$

$$\frac{\partial \mathscr{L}}{\partial \lambda} = -(x_1 - 3)^3 + x_2^2 = 0$$

The feasible region is illustrated in Figure 10.3. Notice that the two parts of the constraint corresponding to the positive and negative values of x_2 form a *cusp*. At the endpoint $(3, 0)$, the second derivatives are not continuous, foreshadowing trouble. In fact, $\mathbf{x} = (3, 0)$ is the constrained global minimum, but on substitution of this point into the necessary conditions, we find that the first two equations are not satisfied. Further analysis reveals that no values of x_1, x_2, and λ will satisfy all three equations.

The difficulty is that the constraint surface is not smooth, implying that the second derivatives are not everywhere continuous. Depending on the objective function, when such a situation arises the first-order necessary conditions [Equations (7a) and (7b)] may not yield

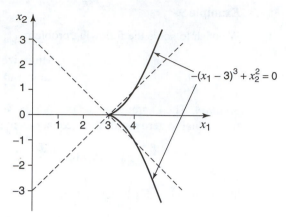

Figure 10.3 Feasible region with cusp

a stationary point. These conditions are valid only when $g_i(\mathbf{x}) = 0$, $i = 1, \ldots, m$, satisfy one of several constraint qualifications. The most common qualification is that the gradients of the binding constraints are linearly independent at a solution. For the preceding problem, we see that $\partial g(3, 0)/\partial x_1 = 3(x_1 - 3)^2 = 0$ and $\partial g(3, 0)/\partial x_2 = 2x_2 = 0$, which violates the linear independence condition. An in-depth treatment of this subject can be found in most NLP textbooks.

10.3   INEQUALITY CONSTRAINTS

Realistically speaking, we wish to solve mathematical programs that include both equality and inequality constraints. The theory used to derive optimality conditions for the inequality is an extension of the case in which only equalities are present in the model. The major difference is that the Lagrange multipliers associated with the inequality constraints are now restricted to be nonnegative. In fact, these multipliers are nothing more than the equivalents of the dual variables in linear programming. Similarly, we must also consider the equivalents of the primal–dual complementarity conditions.

Problem Statement

The most general NLP model that we investigate is

$$\left.\begin{aligned} \text{Minimize}\ \ & f(\mathbf{x}) \\ \text{subject to}\ \ & h_i(\mathbf{x}) = 0,\ \ i = 1,\ \ldots, p \\ & g_i(\mathbf{x}) \le 0,\ \ i = 1,\ \ldots, m \end{aligned}\right\} \tag{8}$$

where an explicit distinction is now made between the equality and inequality constraints. In the model, all functions are assumed to be twice continuously differentiable, and any RHS constants are subsumed in the corresponding functions $h_i(\mathbf{x})$ or $g_i(\mathbf{x})$. Problems with a maximization objective or $\ge$ constraints can easily be converted into the form of Problem (8). Although it is possible and sometimes convenient to treat variable bounds explicitly, we assume that they are included as a subset of the m inequalities.

Karush–Kuhn–Tucker Necessary Conditions

To derive first- and second-order optimality conditions for Problem (8), it is necessary to suppose that the constraints satisfy certain regularity conditions or constraint qualifications, as mentioned previously. The accompanying results are important from a theoretical point of view but less so for the purposes of designing algorithms. Consequently, we take a practical approach and simply generalize the methodology used in the developments associated with the equality constrained Problem (5).

In what follows, let $\mathbf{h}(\mathbf{x}) = (h_1(\mathbf{x}), \ldots, h_p(\mathbf{x}))^{\mathrm{T}}$ and $\mathbf{g}(\mathbf{x}) = (g_1(\mathbf{x}), \ldots, g_m(\mathbf{x}))^{\mathrm{T}}$. For each equality constraint we define an unrestricted multiplier, λ_i, $i = 1, \ldots, p$, and for each inequality constraint we define a nonnegative multiplier, μ_i, $i = 1, \ldots, m$. Let $\lambda \in \Re^p$ and $\mu \in \Re^m$ be the corresponding row vectors. This leads to the Lagrangian for Problem (8).

$$\mathscr{L}(\mathbf{x}, \lambda, \mu) = f(\mathbf{x}) + \sum_{i=1}^{p} \lambda_i h_i(\mathbf{x}) + \sum_{i=1}^{m} \mu_i g_i(\mathbf{x})$$

Definition 1: Let $\mathbf{x}^*$ be a point satisfying the constraints $\mathbf{h}(\mathbf{x}^*) = \mathbf{0}$, $\mathbf{g}(\mathbf{x}^*) \leq \mathbf{0}$ and let K be the set of indices k for which $g_k(\mathbf{x}^*) = 0$. Then $\mathbf{x}^*$ is said to be a *regular point* of these constraints if the gradient vectors $\nabla h_i(\mathbf{x}^*)$ $(1 \leq i \leq p)$, $\nabla g_k(\mathbf{x}^*)$ $(k \in K)$ are linearly independent.

This definition says that $\mathbf{x}^*$ is a regular point if the gradients of the binding or active constraints are linearly independent. It rules out certain pathological situations in which, for instance, the solutions occur at a cusp of the feasible region, as discussed in Section 10.2, and it leads to perhaps the most important result in differential optimization.

Theorem 4 (*Karush–Kuhn–Tucker Necessary Conditions*): Let $\mathbf{x}^*$ be a local minimum for Problem (8) and suppose that $\mathbf{x}^*$ is regular point for the constraints. Then there exists a vector $\lambda^* \in \Re^p$ and a vector $\mu^* \in \Re^m$ such that

$$\frac{\partial \mathscr{L}}{\partial x_j} = \frac{\partial f(\mathbf{x}^*)}{\partial x_j} + \sum_{i=1}^{p} \lambda_i^* \frac{\partial h_i(\mathbf{x}^*)}{\partial x_j} + \sum_{i=1}^{m} \mu_i^* \frac{\partial g_i(\mathbf{x}^*)}{\partial x_j} = 0, \quad j = 1, \ldots, n \tag{9a}$$

$$\frac{\partial \mathscr{L}}{\partial \lambda_i} = h_i(\mathbf{x}^*) = 0, \quad i = 1, \ldots, p \tag{9b}$$

$$\frac{\partial \mathscr{L}}{\partial \mu_i} = g_i(\mathbf{x}^*) \leq 0, \quad i = 1, \ldots, m \tag{9c}$$

$$\mu_i^* g_i(\mathbf{x}^*) = 0, \quad\quad i = 1, \ldots, m \tag{9d}$$

$$\mu_i^* \geq 0, \quad\quad\quad i = 1, \ldots, m \tag{9e}$$

Constraints (9a) to (9e) were derived in the early 1950s and are known as the *Karush–Kuhn–Tucker* (KKT) conditions in honor of their developers. They are first-order necessary conditions and postdate Lagrange's work on the equality constrained Problem (5) by 200 years. The first set of equations [Constraint (9a)] is referred to as the *stationarity conditions* and is equivalent to dual feasibility in linear programming. Constraints (9b) and (9c) represent primal feasibility, and Constraint (9d) represents complementary slackness. Nonnegativity of the "dual" variables appears explicitly in Constraint (9e). In vector form, the system can be written as

$$\nabla f(\mathbf{x}^*) + \lambda^* \nabla \mathbf{h}(\mathbf{x}^*) + \mu^* \nabla \mathbf{g}(\mathbf{x}^*) = \mathbf{0}$$

$$\mathbf{h}(\mathbf{x}^*) = \mathbf{0}, \ \mathbf{g}(\mathbf{x}^*) \leq \mathbf{0}$$

$$\mu^* \mathbf{g}(\mathbf{x}^*) = 0$$

$$\mu^* \geq \mathbf{0}$$

The KKT conditions are highly nonlinear and present a formidable computational challenge even when only a handful of variables and constraints are involved. When all the functions are linear, we have, of course, a linear program. Recall that the simplex method and its variants do not try to solve Constraints (9a) to (9e) directly but maintain two of the following three conditions—primal feasibility, dual feasibility, and complementary slackness—and then iterate toward the third. Interior point methods tackle Constraints (9a) to (9e) head on.

For the linear program, the KKT conditions are necessary and sufficient for global optimality. This is a result of the convexity of the problem and suggests the following, more general result.

Theorem 5 (*Karush–Kuhn–Tucker Sufficient Conditions*): For Problem (8), let $f(\mathbf{x})$ and $g_i(\mathbf{x})$ be convex, $i = 1, \ldots, m$, and let $h_i(\mathbf{x})$ be linear, $i = 1, \ldots, p$. Suppose that $\mathbf{x}^*$ is a regular point for the constraints and that there exist a $\lambda^* \in \mathfrak{R}^p$ and a $\mu^* \in \mathfrak{R}^m$ such that $(\mathbf{x}^*, \lambda^*, \mu^*)$ satisfies Constraints (9a) to (9e). Then $\mathbf{x}^*$ is a global optimal solution to Problem (8). If the convexity assumptions on the objective and constraint functions are restricted to a neighborhood $N_\varepsilon(\mathbf{x}^*)$ for some $\varepsilon > 0$, then $\mathbf{x}^*$ is a local minimum of Problem (8).

This theorem states that if the objective function and feasible region are convex, a solution to the KKT conditions provides a global optimal solution as long as $\mathbf{x}^*$ is a regular point. The only equality constraints that ensure convexity are of the form $\mathbf{h}(\mathbf{x}) = \mathbf{Ax} - \mathbf{b} = \mathbf{0}$. A local optimal solution is ensured if the problem is locally convex. For the more general case, second-order necessary (and sufficient) conditions are more complex, involving the Hessian matrix of the Lagrangian and the tangent space of the binding constraints. Because they are very difficult to assess for real problems and do not provide insight into algorithmic development, we will bypass them. For our purposes here, we simply state that if $(\mathbf{x}^*, \lambda^*, \mu^*)$ satisfies the conditions of Theorem 4, then a necessary condition for $\mathbf{x}^*$ to be a local solution of Problem (8) is that the Hessian matrix of the Lagrangian is positive semidefinite at $\mathbf{x}^*$.

Sufficient Conditions

The foregoing discussion has shown that under certain convexity assumptions and a suitable constraint qualification, the first-order KKT conditions are necessary and sufficient for at least local optimality. Actually, the KKT conditions are sufficient to determine if a particular solution is a global minimum if it can be shown that the solution $(\mathbf{x}^*, \lambda^*, \mu^*)$ is a saddle point of the Lagrangian function.

Definition 2: The triplet $(\mathbf{x}^*, \lambda^*, \mu^*)$ is called a *saddle point* of the Lagrangian function if $\mu^* \geq \mathbf{0}$ and

$$\mathcal{L}(\mathbf{x}^*, \lambda, \mu) \leq \mathcal{L}(\mathbf{x}^*, \lambda^*, \mu^*) \leq \mathcal{L}(\mathbf{x}, \lambda^*, \mu^*)$$

for all $\mathbf{x}$ and λ, and $\mu \geq \mathbf{0}$.

Hence, $\mathbf{x}^*$ minimizes $\mathcal{L}$ over $\mathbf{x} \in \mathfrak{R}^n$ when (λ, μ) is fixed at (λ^*, μ^*), and (λ^*, μ^*) maximizes $\mathcal{L}$ over $(\lambda, \mu) \in \mathfrak{R}^{p \times n}$ with $\mu \geq \mathbf{0}$ when $\mathbf{x}$ is fixed at $\mathbf{x}^*$. This leads to the definition of the *dual problem* in nonlinear programming.

Lagrangian Dual: Maximize $\{\psi(\lambda, \mu) : \lambda \text{ free}, \mu \geq 0\}$ (10)

where $\psi(\lambda, \mu) = \text{Min}_x\{f(\mathbf{x}) + \lambda\mathbf{h}(\mathbf{x}) + \mu\mathbf{g}(\mathbf{x})\}$. When all the functions in Problem (8) are linear, Problem (10) reduces to the familiar LP dual. In general, $\psi(\lambda, \mu)$ is a concave function; for the LP it is piecewise linear as well as concave.

Theorem 6 (*Saddle Point Conditions for Global Minimum*): A solution $(\mathbf{x}^*, \lambda^*, \mu^*)$ with $\mu^* \geq \mathbf{0}$ is a saddle point of the Lagrangian function $\mathcal{L}(\mathbf{x}, \lambda, \mu) = f(\mathbf{x}) + \lambda\mathbf{h}(\mathbf{x}) + \mu\mathbf{g}(\mathbf{x})$ if and only if

    a. $\mathbf{x}^*$ minimizes $\mathcal{L}(\mathbf{x}, \lambda^*, \mu^*)$

    b. $\mathbf{g}(\mathbf{x}^*) \leq \mathbf{0}, \mathbf{h}(\mathbf{x}^*) = \mathbf{0}$

    c. $\mu^*\mathbf{g}(\mathbf{x}^*) = 0$

Moreover, $(\mathbf{x}^*, \lambda^*, \mu^*)$ is a saddle point if and only if $\mathbf{x}^*$ solves Problem (8) and (λ^*, μ^*) solves the *dual* Problem (10) with no *duality gap*—that is, $f(\mathbf{x}^*) = \psi(\lambda^*, \mu^*)$.

It should be underscored that Theorem 6 offers indirectly a *sufficient* condition for $\mathbf{x}^*$ to solve Problem (8) but not a *necessary* condition. In particular, part (a) may not be satisfied at an optimal solution. Minimizing the Lagrangian for the multiplier values fixed at (λ^*, μ^*) is an unconstrained problem and is not the same as solving Problem (8). This means that a global minimum of the original problem may exist in the absence of the associated Lagrangian that has a saddle point. For example, consider the following problem in a single variable x.

$$\text{Minimize } \{f(x) = -x^2 + 4x + 5 : 0 \leq x \leq 5\}$$

The global optimal solution is $x^* = 5$ with $f(5) = 0$. The Lagrangian is $\mathcal{L}(x, \mu_1, \mu_2) = -x^2 + 4x + 5 + \mu_1(x - 5) + \mu_2(-x)$, which is unbounded from below for all finite nonnegative values of μ_1 and μ_2. This indicates that at an optimal solution, part (a) of Theorem 6 may not hold. If $\mu_1 \to \infty$ as $x \to -\infty$, part (c) would be violated. Consequently, there is no saddle point. The fact that we are minimizing a concave function foreshadows difficulties with the theorem.

On the positive side, the saddle point condition does not require differentiability of any of the problem functions, a constraint qualification, or any stipulation of convexity or concavity. In other words, if one can show that any $(\bar{\mathbf{x}}, \bar{\lambda}, \bar{\mu})$ is a saddle point of the Lagrangian, it can be safely concluded that $\bar{\mathbf{x}}$ is a global minimum.

Under the convexity assumptions in Theorem 4, the KKT conditions are sufficient for optimality. Under weaker assumptions such as nondifferentiability of the objective function, however, they are not applicable. Table 10.2 summarizes the various cases that can arise and the conclusions that can be drawn from each.

A qualifying comment should be made in regard to the entries in Table 10.2. In the first two cases, it is assumed that the optimal solution occurs at a regular point. This rules

Table 10.2   Applicability of Necessary and Sufficient Conditions for Optimality

All functions differentiable	Convex objective and feasible region	Karush–Kuhn–Tucker conditions	Saddle point conditions
Yes	Yes	Necessary and sufficient	Necessary and sufficient
Yes	No	Necessary	Sufficient
No	Yes	Not applicable	Sufficient
No	No	Not applicable	Sufficient

out constraints that give rise to unusual forms such as cusps on the boundary of the feasibility region. In the third and fourth cases, the KKT conditions are not applicable because they depend on the differentiability of all problem functions. To address these cases, it would be necessary to introduce a more sophisticated definition of a gradient based on limit concepts.

Example 11

Use the KKT conditions to solve the following problem.

$$\text{Minimize } f(\mathbf{x}) = 2(x_1 + 1)^2 + 3(x_2 - 4)^2$$
$$\text{subject to } x_1^2 + x_2^2 \leq 9, \quad x_1 + x_2 \geq 2$$

Solution: It is straightforward to show that both the objective function and the feasible region are convex. Therefore, we are assured that a global minimum exists and that any point $\mathbf{x}^*$ that satisfies the KKT conditions will be a global minimum. Figure 10.4 illustrates the constraint region and the isovalue contour $f(\mathbf{x}) = 2$.

The partial derivatives required for the analysis are

$$\frac{\partial f}{\partial x_1} = 4(x_1 + 1), \quad \frac{\partial g_1}{\partial x_1} = 2x_1, \quad \frac{\partial g_2}{\partial x_1} = -1$$

$$\frac{\partial f}{\partial x_2} = 6(x_2 - 4), \quad \frac{\partial g_1}{\partial x_2} = 2x_2, \quad \frac{\partial g_2}{\partial x_2} = -1$$

Note that we have rewritten the second contraint as a $\leq$ constraint prior to evaluating the partial derivatives. Based on this information, the KKT conditions are as follows.

a. $4(x_1 + 1) + \mu_1(2x_1) - \mu_2 = 0, \ 6(x_2 - 4) + \mu_1(2x_2) - \mu_2 = 0$

b. $x_1^2 + x_2^2 - 9 \leq 0, \ -x_1 - x_2 + 2 \leq 0$

c. $\mu_1(x_1^2 + x_2^2 - 9) = 0, \quad \mu_2(-x_1 - x_2 + 2) = 0$

d. $\mu_1 \geq 0, \mu_2 \geq 0$

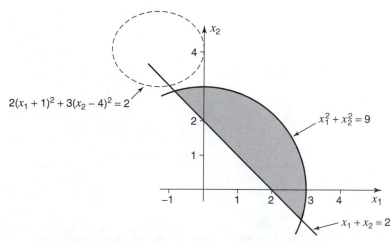

Figure 10.4   Constrained convex programming example

From Figure 10.4 it appears that the minimum may lie at the intersection of the two constraint boundaries where $x_1 = -0.87$ and $x_2 = 2.87$. Because the constraints are binding at this point, conditions (b) and (c) are satisfied. Also, ∇g_1 and ∇g_2 are linearly independent at $(-0.87, 2.87)$ so it is a regular point.

Using the equations in condition (a) to solve for μ at $\mathbf{x} = (-0.87, 2.87)$, we get

$$\mu_1 = 0.976 \text{ and } \mu_2 = -1.178$$

which, unfortunately, violates the nonnegativity restriction [condition (d)]. Thus, the KKT conditions are not satisfied at this point, so it cannot be a minimum.

Other candidates must be investigated. The only way to do this is through trial and error. Essentially, we have to "guess" which constraints are binding at the optimal solution and use them in conjunction with the stationarity conditions [condition (a)] to solve for $\mathbf{x}$ and μ. If there are m inequality constraints, there are 2^m possibilities. This implies that the computational effort required to solve a nonlinear program increases exponentially with the problem size, at least in the worst case. This puts an NLP in the same category as a general integer program (IP).

Returning to the example, consider that the unconstrained minimum of $f(\mathbf{x})$ lies at $x_1 = -1$ and $x_2 = 4$, and at this point only the first constraint is violated. From these observations we might conclude that the constrained minimum lies on the surface $x_1^2 + x_2^2 = 9$ whereas the second constraint is loose. If this is the case, then μ_2 must be 0 in order to satisfy condition (c). Consequently, we obtain the following updated conditions.

    a. $4(x_1 + 1) + \mu_1(2x_1) = 0$, $6(x_2 - 4) + \mu_1(2x_2) = 0$

    b. $x_1^2 + x_2^2 - 9 = 0$, $-x_1 - x_2 + 2 \le 0$

    c. $\mu_1 \ge 0$

which must be solved for x_1, x_2, and μ_1. This is not a trivial exercise if one attempts to attack it with a brute force approach involving algebraic manipulation. However, limiting the search to points on the circle $x_1^2 + x_2^2 = 9$, we may solve by trial and error for the values of x_1 and x_2 that minimize $f(\mathbf{x})$. From Figure 10.4 we see that it is sufficient to examine the points that lie on the arc between the intersection with the line $x_1 + x_2 = 2$ and the x_2 axis. This procedure leads to

$$x_1 = -0.646 \text{ and } x_2 = 2.9297$$

Subsequent substitution into either of the two relations in condition (a) yields $\mu_1 = 1.097$. The solution $\mu^* = (1.097, 0)$, $\mathbf{x}^* = (-0.656, 2.9297)$ satisfies the KKT conditions, and because the objective function and constraints are convex, we are assured that it is a global minimum.

This example raises two issues that merit further discussion. The first is that finding a solution to the KKT conditions is rarely a simple matter, even for small problems. Just finding a feasible solution to the constraints can be a challenge (many algorithms require a phase 1 procedure as in linear programming). In general, an iterative approach is needed that goes beyond solving a system of nonlinear equations. Dealing with the nonnegativity restrictions on the variables is another major difficulty that must be overcome.

The second point concerns the relationship between nonlinear programming and integer programming. It is always possible to convert a 0-1 IP into an NLP by rewriting the integrality requirement as $x_j(1 - x_j) = 0$ for all j. These constraints are analogous to the complementarity conditions [Constraint (9d)]; however, they are very difficult to work with numerically. On the other hand, it is always possible to rewrite Constraint (9d) as ordinary mixed-integer constraints of the form $g_i(\mathbf{x}) \ge -Mz_i$ and $\mu_i \le M(1 - z_i)$, where z_i is a binary variable and M is a sufficiently large constant. Coupled with Constraints (9c) and (9e), when $z_i = 0$, $g_i(\mathbf{x}) = 0$, and $\mu_i \ge 0$; when $z_i = 1$, $g_i(\mathbf{x}) \le 0$ and $\mu_i = 0$. In either case, these transformations have limited utility and should be considered only in special circumstances.

Explicit Consideration of Nonnegativity Restrictions

Nonnegativity is often required of the decision variables. When this is the case, the first-order necessary conditions listed as Constraints (9a) to (9e) can be specialized in a way that gives a slightly different perspective. Omitting explicit treatment of the equality constraints, the problem is now

$$\text{Minimize } \{f(\mathbf{x}): g_i(\mathbf{x}) \le 0, \, i = 1, \ldots, m; \, \mathbf{x} \ge \mathbf{0}\}$$

The Karush–Kuhn–Tucker conditions for a local minimum are as follows.

$$\frac{\partial \mathscr{L}}{\partial x_j} = \frac{\partial f(\mathbf{x}^*)}{\partial x_j} + \sum_{i=1}^{m} \mu_i^* \frac{\partial g_i(\mathbf{x}^*)}{\partial x_j} \ge 0, \quad j = 1, \ldots, n \tag{11a}$$

$$\frac{\partial \mathscr{L}}{\partial \mu_i} = g_i(\mathbf{x}^*) \le 0, \quad i = 1, \ldots, m \tag{11b}$$

$$x_j \frac{\partial \mathscr{L}}{\partial x_i} = 0, \quad j = 1, \ldots, n \tag{11c}$$

$$\mu_i^* g_i(\mathbf{x}^*) = 0, \quad i = 1, \ldots, m \tag{11d}$$

$$x_j^* \ge 0, \quad j = 1, \ldots, n; \quad \mu_i^* \ge 0, \quad i = 1, \ldots, m \tag{11e}$$

Once again, the Lagrange multipliers $\boldsymbol{\mu}$ can be interpreted as the dual variables of the problem. Analogous to linear programming, Constraints (11a) and (11b) correspond to dual and primal feasibility, respectively; Constraints (11c) and (11d) represent complementary slackness; and Constraint (11e) requires both sets of variables to be nonnegative.

Example 12

Find a point that satisfies the first-order necessary conditions for the following problem.

$$\text{Minimize } f(\mathbf{x}) = x_1^2 + 4x_2^2 - 8x_1 - 16x_2 + 32$$
$$\text{subject to } x_1 + x_2 \le 5, \, x_1 \ge 0, \, x_2 \ge 0$$

Solution: We first write out the Lagrangian function excluding the nonnegative conditions.

$$\mathscr{L}(\mathbf{x}, \mu) = x_1^2 + 4x_2^2 - 8x_1 - 16x_2 + 32 + \mu(x_1 + x_2 - 5)$$

The specialized KKT conditions [Constraints (11a) to (11e)] are

    a. $2x_1 - 8 + \mu \ge 0$, $8x_2 - 16 + \mu \ge 0$

    b. $x_1 + x_2 - 5 \le 0$

    c. $x_1(2x_1 - 8 + \mu) = 0$, $x_2(8x_2 - 16 + \mu) = 0$

    d. $\mu(x_1 + x_2 - 5) = 0$

    e. $x_1 \ge 0$, $x_2 \ge 0$, $\mu \ge 0$

Let us begin by examining the unconstrained optimal solution $\mathbf{x} = (4, 2)$. Because both primal variables are nonzero at this point, condition (c) requires that $\mu = 0$. This solution satisfies all the constraints except condition (b), primal feasibility, suggesting that the inequality $x_1 + x_2 \le 5$ is binding at the optimal solution. Let us further suppose that $\mathbf{x} > \mathbf{0}$ at the optimal solution. Condition (c) then requires $2x_1 - 8 + \mu = 0$ and $8x_2 - 16 + \mu = 0$. Coupled with $x_1 + x_2 = 5$, we have three equations in three unknowns. Their solution is $\mathbf{x} = (3.2, 1.8)$

and $\mu = 1.6$, which satisfies Constraints (11a) to (11e) and is a regular point. Given that the objective function is convex and the constraints are linear, these conditions are also sufficient. Therefore, $\mathbf{x}^* = (3.2, 1.8)$ is the global minimum.

10.4   SEPARABLE PROGRAMMING

Separable programming is important because it allows a convex nonlinear program to be approximated with arbitrary accuracy by a linear programming model. The idea is to replace each nonlinear function with a piecewise linear approximation. Global solutions can then be obtained with any number of efficient LP codes. For nonconvex problems, the approach is still valid, but an LP no longer results. Either a mixed-integer linear programming (MILP) problem must be solved, as discussed in Section 7.8, or a modified version of the simplex algorithm with a restricted basis entry rule can be applied to the model directly. The candidates for the entering variable must be restricted to maintain the validity of the LP approximation. In this case, a local optimal solution is obtained, but it is possible to find the global optimal solution with the help of branch and bound.

Problem Statement

Consider again the general NLP problem

$$\text{Minimize } \{f(\mathbf{x}) : g_i(\mathbf{x}) \leq b_i, i = 1, \ldots, m\}$$

with two additional provisions: (1) the objective function and all constraints are *separable*, and (2) each decision variable x_j is bounded below by 0 and above by a known constant u_j, $j = 1, \ldots, n$. Recall that a function $f(\mathbf{x})$ is separable if it can be expressed as the sum of functions of the individual decision variables.

$$f(\mathbf{x}) = \sum_{j=1}^{n} f_j(x_j)$$

The separable NLP has the following structure.

$$\text{Minimize } \sum_{j=1}^{n} f_j(x_j)$$

$$\text{subject to } \sum_{j=1}^{n} g_{ij}(x_j) \leq b_i, \quad i = 1, \ldots, m$$

$$0 \leq x_j \leq u_j, \quad j = 1, \ldots, n$$

The key advantage of this formulation is that the nonlinearities are mathematically independent. This property, in conjunction with the finite bounds on the decision variables, permits the development of a piecewise linear approximation for each function in the problem.

Linearization

Consider the general nonlinear function $f(x)$ depicted in Figure 10.5. To form a piecewise linear approximation using, say, r line segments, we must first select $r + 1$ values of the scalar x within its range $0 \leq x \leq u$ (call them $\bar{x}_0, \bar{x}_1, \ldots, \bar{x}_r$) and let $f_k = f(\bar{x}_k)$ for $k = 0, 1, \ldots, r$. At the boundaries we have $\bar{x}_0 = 0$ and $\bar{x}_r = u$. Notice that the values of $\bar{x}_k$ do not have to be evenly spaced.

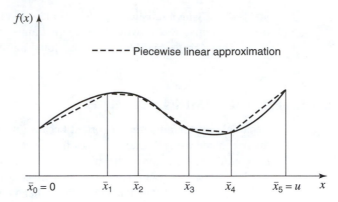

Figure 10.5 Piecewise linear approximation of a nonlinear function

Recall that any value of x lying between the two endpoints of the kth line segment may be expressed as

$$x = \alpha \bar{x}_{k+1} + (1 - \alpha)\bar{x}_k \text{ or } x - \bar{x}_k = \alpha(\bar{x}_{k+1} - \bar{x}_k) \text{ for } 0 \le \alpha \le 1$$

where $\bar{x}_k$ $(k = 0, 1, \ldots, r)$ are data and α is the decision variable. This relationship leads directly to an expression for the kth line segment.

$$\hat{f}(x) = f_k + \frac{f_{k+1} - f_k}{\bar{x}_{k+1} + \bar{x}_k}(x - \bar{x}_k) = \alpha f_{k+1} + (1 - \alpha)f_k \text{ for } 0 \le \alpha \le 1$$

The approximation $\hat{f}(x)$ becomes increasingly more accurate as r gets larger. Unfortunately, there is a corresponding growth in the size of the resultant problem.

For the kth segment, let $\alpha = \alpha_{k+1}$ and let $(1 - \alpha) = \alpha_k$. As such, for $\bar{x}_k \le x \le \bar{x}_{k+1}$, the expression for x becomes

$$x = \alpha_{k+1}\bar{x}_{k+1} + \alpha_k \bar{x}_k \text{ and } \hat{f}(x) = \alpha_{k+1}f_{k+1} + \alpha_k f_k$$

where $\alpha_k + \alpha_{k+1} = 1$ and $\alpha_k \ge 0$, $\alpha_{k+1} \ge 0$. Generalizing this procedure to cover the entire range over which x is defined yields

$$x = \sum_{k=0}^{r} \alpha_k \bar{x}_k, \quad \hat{f}(x) = \sum_{k=0}^{r} \alpha_k f_k, \quad \sum_{k=0}^{r} \alpha_k = 1, \quad \alpha_k \ge 0, \quad k = 0, \ldots, r$$

such that at least one and no more than two α_k can be greater than zero. Furthermore, we require that if two α_k are greater than zero, their indices must differ by exactly 1. In other words, if α_s is greater than zero, then only one of either α_{s+1} or α_{s-1} can be greater than zero. If this last condition, known as the *adjacency criterion*, is not satisfied, the approximation to $f(x)$ will not lie on $\hat{f}(x)$.

To apply the preceding transformations, a grid of $r_j + 1$ points must be defined for each variable x_j over its range. This requires the use of an additional index for each variable and function. For the jth variable, for example, $r_j + 1$ data points result: $\bar{x}_{j0}, \bar{x}_{j1}, \ldots, \bar{x}_{jr_j}$. With this notation in mind, the separable programming problem in $\mathbf{x}$ becomes the following "almost" linear program in $\boldsymbol{\alpha}$.

$$\text{Minimize } f(\boldsymbol{\alpha}) = \sum_{j=1}^{n} \sum_{k=0}^{r_j} \alpha_{jk} f_{jk}(\overline{x}_{jk})$$

$$\text{subject to } g_i(\boldsymbol{\alpha}) = \sum_{j=1}^{n} \sum_{k=0}^{r_j} \alpha_{jk} g_{ijk}(\overline{x}_{jk}) \le b_i, \quad i = 1, \ldots, m$$

$$\sum_{k=0}^{r_j} \alpha_{jk} = 1, \quad j = 1, \ldots, n$$

$$\alpha_{jk} \ge 0, \; j = 1, \ldots, n, \; k = 0, \ldots, r_j$$

The reason that this is an "almost" linear programming problem is that the adjacency criterion must be imposed on the new decision variables α_{jk} when any of the functions are nonconvex. This can be accomplished with a restricted basis entry rule. When all the functions are convex, the adjacency criterion will be satisfied automatically, and so no modifications of the simplex algorithm are necessary. Note that the approximate problem has $m + n$ constraints and $\sum_j r_j + n$ variables.

From a practical point of view, one might start off with a rather large grid and find the optimal solution to the corresponding approximate problem. This should be easy to do, but the results may not be very accurate. To improve on the solution, we could then introduce a smaller grid in the neighborhood of the optimal solution and solve the new problem. This idea is further discussed by Bard *et al.* [2000].

Example 13

Consider the following problem, whose feasible region is shown graphically in Figure 10.6. All the functions are convex, but the second constraint is $g_2(\mathbf{x}) \ge 10$. Because $g_2(\mathbf{x})$ is not linear, this implies that the feasible region is not convex, and so the solution to the approximate problem may not be a global optimal solution.

$$\text{Minimize } f(\mathbf{x}) = 2x_1^2 - 3x_1 + 2x_2$$

$$\text{subject to } g_1(\mathbf{x}) = 3x_1^2 + 4x_2^2 \le 8$$

$$g_2(\mathbf{x}) = 3(x_1 - 2)^2 + 5(x_2 - 2)^2 \ge 10$$

$$g_3(\mathbf{x}) = 3(x_1 - 2)^2 + 5(x_2 - 2)^2 \le 21$$

$$0 \le x_1 \le 1.75, \; 0 \le x_2 \le 1.5$$

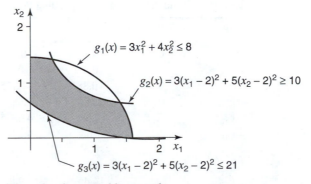

Figure 10.6   Feasible region for separable example

The upper bounds on the variables have been selected to be redundant. The objective function and constraints are separable, with the individual terms being identified in Table 10.3.

To develop the piecewise linear approximations, we select six grid points for each variable and evaluate the functions at each point. The results are given in Table 10.4. For this example, $n = 2$, $m = 3$, $r_1 = 5$, and $r_2 = 5$. As an illustration, the piecewise linear approximations of $f_1(x_1)$ and $g_{12}(x_2)$, along with the original graphs, are depicted in Figure 10.7. The full model has five constraints and 12 variables. The coefficient matrix is given in Table 10.5 where the last two rows correspond to the summation constraints on the two sets of α variables.

The problem will be solved with a linear programming code modified to enforce the adjacency criterion. In particular, for the jth variable we do not allow an α_{jk} variable to enter the basis unless $\alpha_{j,k-1}$ or $\alpha_{j,k+1}$ is already in the basis, or no α_{jk} ($k = 0, 1, \ldots, 5$) is currently basic. The following slack and artificial variables are used to put the problem into standard simplex form.

s_1 = slack for constraint 1, g_1

s_2 = surplus for constraint 2, g_2

a_2 = artificial for constraint 2, g_2

s_3 = slack for constraint 3, g_3

a_4 = artificial for constraint 4, $\Sigma_k \alpha_{1k}$

a_5 = artificial for constraint 5, $\Sigma_k \alpha_{2k}$

The initial basic solution is

$$\mathbf{x}_B = (s_1, a_2, s_3, a_4, a_5) = (8, 10, 21, 1, 1)$$

Table 10.3 Separable Functions for Example

j	1	2
$f_j(x_j)$	$2x_1^2 - 3x_1$	$2x_2$
$g_{1j}(x_j)$	$3x_1^2$	$4x_2^2$
$g_{2j}(x_j)$	$3(x_1 - 2)^2$	$5(x_2 - 2)^2$
$g_{3j}(x_j)$	$3(x_1 - 2)^2$	$5(x_2 - 2)^2$

Table 10.4 Grid Points and Corresponding Function Values

k	0	1	2	3	4	5
x_{1k}	0	0.4	0.75	1.0	1.25	1.75
x_{2k}	0	0.3	0.6	0.9	1.2	1.5
f_{1k}	0	−0.88	−1.125	−1	−0.625	0.875
g_{11k}	0	0.48	1.6875	3	4.6875	9.1875
g_{21k}	12	7.68	4.6875	3	1.6875	0.1875
g_{31k}	12	7.68	4.6875	3	1.6875	0.1875
f_{2k}	0	0.6	1.2	1.8	2.4	3
g_{12k}	0	0.36	1.44	3.24	5.76	9
g_{22k}	20	14.45	9.8	6.05	3.2	1.25
g_{32k}	20	14.45	9.8	6.05	3.2	1.25

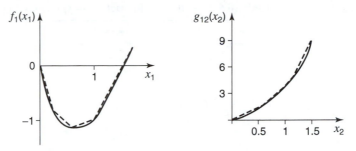

Figure 10.7   Piecewise linear approximations of $f_1(x_1)$ and $g_{12}(x_2)$

Table 10.5   Coefficients of the Linear Programming Model

	α_{10}	α_{11}	α_{12}	α_{13}	α_{14}	α_{15}	α_{20}	α_{21}	α_{22}	α_{23}	α_{24}	α_{25}
f	0	0.88	1.125	1	0.625	−0.875	0	−0.6	−1.2	−1.8	−2.4	−3
g_1	0	0.48	1.6875	3	4.6875	9.1875	0	0.36	1.44	3.24	5.76	9
g_2	12	7.68	4.6875	3	1.6875	0.1875	20	14.45	9.8	6.05	3.2	1.25
g_3	12	7.68	4.6875	3	1.6875	0.1875	20	14.45	9.8	6.05	3.2	1.25
x_1	1	1	1	1	1	1	0	0	0	0	0	0
x_2	0	0	0	0	0	0	1	1	1	1	1	1

which can be seen to contain three artificial variables. The phase 1 procedure required five iterations to drive the objective function to zero. Phase 2 also required five iterations. The corresponding data are presented in Table 10.6. The most negative rule was used to select the entering variable, but if that variable led to a violation of the adjacency criterion, the next one on the list was selected. For example, at iteration 3.1, α_{12} was selected to enter the basis, but since it was not adjacent to α_{15}, it was not permitted to do so. The nonbasic variable with the next smallest reduced cost was α_{11}, but it, too, did not satisfy the adjacency criterion. Finally, at iteration 3.4 the variable with the smallest reduced cost not yet examined was α_{14}, which met the criterion and so was allowed to enter the basis. A pivot was executed with a_4 the leaving variable.

At the end of iteration 9, the algorithm terminates with the optimal solution

$$s_1 = 4.87, \; s_2 = 11, \; \alpha_{13} = 1, \; \alpha_{20} = 0.6396, \; \alpha_{21} = 0.3604, \text{ and } \hat{f} = -0.7838$$

In terms of the original problem variables, this corresponds to

$$x_1 = \alpha_{13}\bar{x}_{31} = 1 \text{ and } x_2 = \alpha_{20}\bar{x}_{02} + \alpha_{21}\bar{x}_{12} = 0.1081$$

When comparing these values with the true optimal solution $x_1^* = 0.9227$, $x_2^* = 0.1282$, and $f(\mathbf{x}^*) = -0.8089$, we can observe the effect of the linearization process on the accuracy of the solution. With respect to the objective function, the six-point approximation is off by 3.2%. For this particular problem, however, we can get as close as desired by making the grid size arbitrarily small in the neighborhood of the true optimal solution.

It is interesting to note that although the basic solutions given in Table 10.6 lie at vertices in the α space, they do not necessarily have any relation to the vertices in the original $\mathbf{x}$ space. This is illustrated in Figure 10.8, which plots the feasible region and several isovalue contours of $f(\mathbf{x})$. Also shown are the six basic solutions that comprise phase 2. The first basic solution corresponds to iteration 5.3 in Table 10.6. As can be seen, none of the

Table 10.6   Iterations of Restricted Entry Simplex Algorithm

Iteration	Entering	Leaving	Basic variables
0			s_1, a_2, s_3, a_4, a_5
1	α_{20}	a_2	$s_1, \alpha_{20}, s_3, a_4, a_5$
2	α_{15}	s_1	$\alpha_{15}, \alpha_{20}, s_3, a_4, a_5$
3.1	α_{12}	—	Not adjacent to α_{15}
3.2	α_{11}	—	Not adjacent to α_{15}
3.3	α_{13}	—	Not adjacent to α_{15}
3.4	α_{14}	a_4	$\alpha_{15}, \alpha_{20}, s_3, \alpha_{14}, a_5$
4.1	α_{25}	—	Not adjacent to α_{20}
4.2	α_{24}	—	Not adjacent to α_{20}
4.3	α_{23}	—	Not adjacent to α_{20}
4.4	α_{22}	α_{20}	$\alpha_{15}, \alpha_{22}, s_3, \alpha_{14}, a_5$
5.1	α_{25}	—	Not adjacent to α_{22}
5.2	α_{24}	—	Not adjacent to α_{22}
5.3	α_{23}	a_5	$\alpha_{15}, \alpha_{22}, s_3, \alpha_{14}, \alpha_{23}$
		Begin phase 2	
5.3			$\alpha_{15}, \alpha_{22}, s_3, \alpha_{14}, \alpha_{23}$
5.4	α_{20}	—	Not adjacent to α_{22}
5.5	α_{24}	—	Not adjacent to α_{22}
5.6	s_1	α_{15}	$\alpha_{14}, \alpha_{23}, s_3, s_1, \alpha_{22}$
6	α_{13}	α_{14}	$\alpha_{13}, \alpha_{23}, s_3, s_1, \alpha_{22}$
7	s_2	α_{23}	$\alpha_{13}, s_2, s_3, s_1, \alpha_{22}$
8.1	α_{20}	—	Not adjacent to α_{22}
8.2	α_{21}	α_{22}	$\alpha_{13}, s_2, s_3, s_1, \alpha_{21}$
9	α_{20}	s_3	$\alpha_{13}, s_2, \alpha_{20}, s_1, \alpha_{21}$

six solutions in the figure lies at a vertex of the feasible region. At the optimal solution only g_3 is binding.

If the objective in the example is changed from minimization to maximization and the same approximation is used, the algorithm converges to $\mathbf{x} = (1.458, 0.6)$. This point is a local maximum approximating the rightmost intersection of the boundaries of constraints

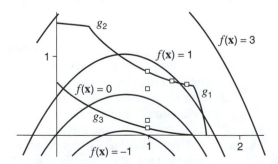

Figure 10.8   Sequence of six basic solutions during phase 2

g_1 and g_2 located at $\mathbf{x} = (1.449, 0.652)$. The global maximum resides at $\mathbf{x}^* = (0, 1.414)$. The fact that it was not found can be attributed to the nonconvexity of the constraint region. We are assured that a local maximum is a global maximum if and only if a concave function is being maximized over a convex constraint region. Although the algorithm found the global minimum of the example, this was only because we were minimizing a convex objective function and the global optimal solution happened to reside on the boundary of a constraint, g_3, whose feasible region was convex.

Convex Programming Problems

These observations stem directly from the fact that the separable programming method guarantees an approximate global optimal solution to the original problem only when one is minimizing a convex function (maximizing a concave function) over a convex set. When these conditions hold, the accuracy of the approach is limited only by the coarseness of the grid used to obtain the piecewise linear approximations. Furthermore, when solving a convex programming problem, we may solve the approximate problem as an ordinary linear program without enforcing the adjacency restriction.

Nonconvex Programming

If the conditions that define a convex program are not present, several outcomes may occur.

1. An approximate global optimal solution is found (as in the preceding minimization example).
2. An approximate local optimal solution is found that is not the global optimal solution.
3. The solution to the approximate problem may be infeasible with respect to the original problem or be nowhere near a corresponding local or global optimal solution. These outcomes result when an insufficient number of line segments are chosen for the approximation. In many cases, however, infeasible solutions will be only *slightly* infeasible, and thus will present no practical difficulties.

Notwithstanding the possibility of obtaining erroneous results, separable programming methods have proven to be very useful in a variety of practical applications. In addition, it is possible to modify the basic transformations by introducing integer variables and related constraints so that approximate global optima are always obtained, regardless of the convexity of the original problem. An experimental code called MOGG was developed by Falk and Soland [1969] along these lines. Testing has shown, however, that real applications are often impossible to solve within acceptable time limits because of the excessive size of the expanded model.

10.5   QUADRATIC PROGRAMMING

A linearly constrained optimization problem with a quadratic objective function is called a quadratic program (QP). Because of its many applications, quadratic programming is often viewed as a discipline in and of itself. More importantly, however, it forms the basis for several general NLP algorithms. We begin this section by examining the Karush–Kuhn–Tucker conditions for the QP and discovering that they turn out to be a set of linear equalities and complementarity constraints. Much as for the separable programming problem, a modified version of the simplex algorithm can be used to find solutions.

Problem Statement

The general quadratic program can be written as

$$\text{Minimize } f(\mathbf{x}) = \mathbf{c}\mathbf{x} + \tfrac{1}{2}\mathbf{x}^T\mathbf{Q}\,\mathbf{x}$$

$$\text{subject to } \mathbf{A}\mathbf{x} \le \mathbf{b} \text{ and } \mathbf{x} \ge \mathbf{0}$$

where $\mathbf{c}$ is an n-dimensional row vector describing the coefficients of the linear terms in the objective function and $\mathbf{Q}$ is an $(n \times n)$ symmetric matrix describing the coefficients of the quadratic terms. If a constant term exists, it is dropped from the model. As in linear programming, the decision variables are denoted by the n-dimensional column vector $\mathbf{x}$, and the constraints are defined by an $(m \times n)$ $\mathbf{A}$ matrix and an m-dimensional column vector $\mathbf{b}$ of RHS coefficients. We assume that a feasible solution exists and that the constraint region is bounded.

When the objective function $f(\mathbf{x})$ is strictly convex for all feasible points, the problem has a unique local minimum which is also the global minimum. A sufficient condition to guarantee strict convexity is for $\mathbf{Q}$ to be positive definite.

Karush–Kuhn–Tucker Conditions

We now adapt the first-order necessary conditions given in Section 10.3 to the quadratic program. These conditions are sufficient for a global minimum when $\mathbf{Q}$ is positive definite; otherwise, the most we can say is that they are necessary.

Excluding the nonnegativity conditions, the Lagrangian function for the quadratic program is

$$\mathscr{L}(\mathbf{x},\boldsymbol{\mu}) = \mathbf{c}\mathbf{x} + \tfrac{1}{2}\mathbf{x}^T\mathbf{Q}\mathbf{x} + \boldsymbol{\mu}(\mathbf{A}\mathbf{x} - \boldsymbol{\beta})$$

where $\boldsymbol{\mu}$ is an m-dimensional row vector. The KKT conditions for a local minimum are as follows.

$$\frac{\partial\mathscr{L}}{\partial x_j} \ge 0, \quad j = 1,\ \ldots,\ n \qquad \mathbf{c} + \mathbf{x}^T\mathbf{Q} + \boldsymbol{\mu}\mathbf{A} \ge 0 \tag{12a}$$

$$\frac{\partial\mathscr{L}}{\partial \mu_i} \le 0, \quad i = 1,\ \ldots,\ m \qquad \mathbf{A}\mathbf{x} - \mathbf{b} \le \mathbf{0} \tag{12b}$$

$$x_j\frac{\partial\mathscr{L}}{\partial x_j} = 0, \quad j = 1,\ \ldots,\ n \qquad \mathbf{x}^T(\mathbf{c}^T + \mathbf{Q}\mathbf{x} + \mathbf{A}^T\boldsymbol{\mu}) = 0 \tag{12c}$$

$$\mu_i g_i(\mathbf{x}) = 0,\ i = 1, \ldots, m \qquad \boldsymbol{\mu}(\mathbf{A}\mathbf{x} - \mathbf{b}) = 0 \tag{12d}$$

$$x_j \ge 0, j = 1, \ldots, n \qquad\qquad \mathbf{x} \ge \mathbf{0} \tag{12e}$$

$$\mu_i \ge 0,\ i = 1, \ldots, m \qquad\quad\ \boldsymbol{\mu} \ge \mathbf{0} \tag{12f}$$

To put Conditions (12a) to (12f) into a more manageable form, we introduce nonnegative surplus variables $\mathbf{y} \in \mathfrak{R}^n$ to the inequalities in Condition (12a) and nonnegative slack variables $\mathbf{v} \in \mathfrak{R}^m$ to the inequalities in Condition (12b) to obtain the equations

$$\mathbf{c}^T + \mathbf{Q}\mathbf{x} + \mathbf{A}^T\boldsymbol{\mu}^T - \mathbf{y} = \mathbf{0} \text{ and } \mathbf{A}\mathbf{x} - \mathbf{b} + \mathbf{v} = \mathbf{0}$$

The KKT conditions can now be written with the constants moved to the right-hand-side.

$$\mathbf{Qx} + \mathbf{A}^T\mathbf{\mu}^T - \mathbf{y} = -\mathbf{c}^T \qquad \text{(13a)}$$

$$\mathbf{Ax} + \mathbf{v} = \mathbf{b} \qquad \text{(13b)}$$

$$\mathbf{x} \geq \mathbf{0},\ \mathbf{\mu} \geq \mathbf{0},\ \mathbf{y} \geq \mathbf{0},\ \mathbf{v} \geq \mathbf{0} \qquad \text{(13c)}$$

$$\mathbf{y}^T\mathbf{x} = 0,\ \mathbf{\mu v} = 0 \qquad \text{(13d)}$$

The first two expressions are linear equalities, the third restricts all the variables to be non-negative, and the fourth prescribes complementary slackness.

Solving for the Optimal Solution

The simplex algorithm can be used to solve Equations (13a) to (13d) by treating the complementary slackness conditions [Equation (13d)] implicitly with a restricted basis entry rule. The procedure for setting up the LP model follows.

- Let the structural constraints be Equations (13a) and (13b) defined by the KKT conditions.
- If any of the RHS values are negative, multiply the corresponding equation by −1.
- Add an artificial variable to each equation.
- Let the objective function be the sum of the artificial variables.
- Convert the resultant problem into simplex form.

The goal is to find the solution to the linear program that minimizes the sum of the artificial variables with the additional requirement that the complementarity slackness conditions be satisfied at each iteration. If the objective value is zero, the solution will satisfy Equations (13a) to (13d). To accommodate Equation (13d), the rule for selecting the entering variable must be modified with the following relationships in mind.

x_j and y_j are complementary for $j = 1, \ldots, n$

μ_i and v_i are complementary for $i = 1, \ldots, m$

The entering variable will be the one whose reduced cost is most negative provided that its complementary variable is not in the basis or would leave the basis on the same iteration. At the conclusion of the algorithm, the vector $\mathbf{x}$ defines the optimal solution and the vector $\mathbf{\mu}$ defines the optimal dual variables.

This approach has been shown to work well when the objective function is positive definite, and requires computational effort comparable to an LP problem with $m + n$ constraints, where m is the number of constraints and n is the number of variables in the QP. Positive semidefinite forms of the objective function, however, can present computational difficulties. Van De Panne [1975] presents an extensive discussion of the conditions that will yield a global optimal solution even when $f(\mathbf{x})$ is not positive definite. The simplest practical approach to overcoming any difficulties caused by semidefiniteness is to add a small constant to each of the diagonal elements of $\mathbf{Q}$ in such a way that the modified $\mathbf{Q}$ matrix becomes positive definite. Although the resultant solution will not be exact, the difference will be insignificant if the alterations are kept small.

Example 14

Solve the following problem.

$$\text{Minimize } f(\mathbf{x}) = -8x_1 - 16x_2 + x_1^2 + 4x_2^2$$

$$\text{subject to } x_1 + x_2 \leq 5,\ x_1 \leq 3,\ x_1 \geq 0,\ x_2 \geq 0$$

Table 10.7   Simplex Iterations for QP Example

Iteration	Basic variables	Solution	Objective value	Entering variable	Leaving variable
1	(a_1, a_2, a_3, a_4)	(8, 16, 5, 3)	32	x_2	a_2
2	(a_1, x_2, a_3, a_4)	(8, 2, 3, 3)	14	x_1	a_3
3	(a_1, x_2, x_1, a_4)	(2, 2, 3, 0)	2	μ_1	a_4
4	(a_1, x_2, x_1, μ_1)	(2, 2, 3, 0)	2	μ_1	a_1
5	(μ_2, x_2, x_1, μ_1)	(2, 2, 3, 0)	0	—	—

Solution: The data and variable definitions are given below. As we can see, the **Q** matrix is positive definite, so the KKT conditions are necessary and sufficient for a global optimal solution.

$$\mathbf{x}^{\mathrm{T}} = \begin{pmatrix} -8 \\ -16 \end{pmatrix}, \quad \mathbf{Q} = \begin{bmatrix} 2 & 0 \\ 0 & 8 \end{bmatrix}, \quad \mathbf{A} = \begin{bmatrix} 1 & 1 \\ 1 & 0 \end{bmatrix}, \quad \mathbf{b} = \begin{pmatrix} 5 \\ 3 \end{pmatrix}$$

$$\mathbf{x}^{\mathrm{T}} = (x_1, x_2), \quad \mathbf{y}^{\mathrm{T}} = (y_1, y_2), \quad \boldsymbol{\mu} = (\mu_1, \mu_2), \quad \mathbf{v}^{\mathrm{T}} = (v_1, v_2)$$

The linear constraints [Equations (13a) and (13b)] take the following form.

$$
\begin{aligned}
2x_1 && + \mu_1 + \mu_2 - y_1 && && = 8 \\
& 8x_2 + \mu_1 && - y_2 && && = 16 \\
x_1 + & x_2 && && + v_1 && = 5 \\
x_1 && && && + v_2 & = 3
\end{aligned}
$$

To create the appropriate linear program, we add artificial variables to each constraint and minimize their sum.

Minimize $a_1 + a_2 + a_3 + a_4$

$$
\begin{aligned}
\text{subject to} \quad 2x_1 && + \mu_1 + \mu_2 - y_1 && && + a_1 && && = 8 \\
& 8x_2 + \mu_1 && - y_2 && && + a_2 && = 16 \\
x_1 + & x_2 && && + v_1 && && + a_3 && = 5 \\
x_1 && && && + v_2 && && + a_4 & = 3
\end{aligned}
$$

All variables ≥ 0 and subject to complementarity conditions

Applying the modified simplex technique to this example yields the sequence of iterations given in Table 10.7. The optimal solution to the original problem is $(x_1^*, x_2^*) = (3, 2)$, $(\mu_1^*, \mu_2^*) = (0, 2)$, and all other variables 0.

10.6   ONE-DIMENSIONAL SEARCH METHODS

The basic approach to solving almost all mathematical programs in continuous variables is to select an initial point $\mathbf{x}^0$ and a direction $\mathbf{d}^0$ in which the objective function is improving, and then move in that direction until either an extremum is reached or a constraint is violated. In either case, a new direction is computed and the process is repeated. A check for convergence is made at the end of each iteration. At the heart of this approach is a one-dimensional search by which the length of the move, called the *step size*, is determined. That is, given a point $\mathbf{x}^k$ and a direction $\mathbf{d}^k$ at iteration k, the aim is to find an optimal step size t_k that moves us to the next point $\mathbf{x}^{k+1} = \mathbf{x}^k + t_k \mathbf{d}^k$.

The simplex method for linear programming is essentially a very clever implementation of this idea. The initial point is the initial basic solution, the direction of improvement is determined by the entering variable, and the step size is determined by the ratio test. Convergence occurs when there is no improving direction—that is, when all the reduced costs are nonnegative. A closer look at the simplex method reveals that what it is trying to do is to satisfy the KKT conditions [Equations (9a) to (9e)], which turn out to be sufficient for global optimality. For the general NLP problem, the most that we can expect a standard algorithm to provide is a solution to these first-order necessary conditions. We rely on the convexity properties of the functions to tell us more about the solution. For the nonconvex case, more advanced techniques are required to obtain a global optimal solution.

Up to this point in our study of nonlinear programming, we have only tried to solve the KKT conditions by hand or have used the simplex algorithm to solve separable and quadratic problems. We will now take a step back and develop several techniques for solving the simplest case—minimizing a continuous unconstrained function of a single variable. In fact, a critical component of every NLP algorithm is a numerical procedure for finding the extremum of a one-dimensional function. Of course, if the function has a known first derivative, the problem reduces to finding the point that satisfies $\nabla f(x) = 0$—but even in this case, a numerical method is usually required for the calculations. Therefore, we proceed from the point of view that either the function is so complex that evaluations of its first and second derivatives can be made only at a prohibitive cost, or the function and its derivatives are not known explicitly and only evaluations at specific points are available. If this is not the case, different methods of approaching the optimization problem may be called for.

Unimodal Functions

Out of practical considerations, we define an interval of uncertainty $[a, b]$ in which the minimum of $f(x)$ must lie. This leads to the one-dimensional problem

$$\text{Minimize } \{f(x) : x \in [a, b]\} \tag{14}$$

For simplicity, it will also be assumed that f is continuous and *unimodal* in the interval $[a, b]$, implying that f has a single minimum x^*—that is, for $x \in [a, b]$ and $f(x) \neq f(x^*)$, f is strictly decreasing when $x < x^*$ and strictly increasing when $x > x^*$. In the case of a minimization problem, the stronger property of strict convexity implies unimodality, but unimodality does not imply convexity. This fact is illustrated by the unimodal functions shown in Figure 10.9. Each function is both concave and convex in subregions but exhibits only one relative minimum in the entire range.

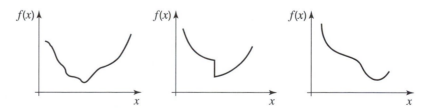

Figure 10.9   Three unimodal functions

During a search procedure, if we could exclude portions of $[a, b]$ that did not contain the minimum, then the interval of uncertainty would be reduced. The following theorem shows that it is possible to obtain a reduction by evaluating two points within the interval.

Theorem 7: Let f be a continuous, unimodal function of a single variable defined over the interval $[a, b]$. Let $x_1, x_2 \in [a, b]$ such that $x_1 < x_2$. If $f(x_1) \geq f(x_2)$, then $f(x) \geq f(x_2)$ for all $x \in [a, x_1]$. If $f(x_1) \leq f(x_2)$, then $f(x) \geq f(x_1)$ for all $x \in [x_2, b]$.

The two cases are illustrated in Figure 10.10. Of course, if $f(x_1) = f(x_2)$, then $x^* \in [x_1, x_2]$, a fact that should be taken into account in any implementation. In particular, a line search algorithm is one that systematically evaluates f, eliminating subintervals of $[a, b]$ until the location of the minimum point x^* is determined to a desired level of accuracy. We now present a few of the most popular approaches to solving Problem (14).

Dichotomous Search Method

Under the restriction that we may evaluate $f(x)$ only at selected points, our goal is to find a technique that will provide either the minimum or a specified *interval of uncertainty* after a certain number n of evaluations of the function. The simplest method of doing this is known as the *dichotomous search method*. Without loss of generality, we restrict our attention to Problem (14). Let the unknown location of the minimum value be denoted by x^*.

The dichotomous search method requires a specification of the minimal distance $\varepsilon > 0$ between two points x_1 and x_2 such that one can still be distinguished from the other. The first two measurements are made at $\varepsilon/2$ on either side of the center of the interval $[a, b]$, as shown in Figure 10.11.

$$x_1 = 0.5(a + b - \varepsilon) \text{ and } x_2 = 0.5(a + b + \varepsilon)$$

On evaluating the function at these points, Theorem 7 allows us to draw one of three conclusions.

- If $f(x_1) < f(x_2)$, x^* must be located between a and x_2. This indicates that the value of b should be updated by setting b to x_2.

- If $f(x_2) < f(x_1)$, x^* must be located between x_1 and b. This indicates that the value of a should be updated by setting a to x_1.

- If $f(x_2) = f(x_1)$, x^* must be located between x_1 and x_2. This indicates that both endpoints should be updated by setting a to x_1 and b to x_2.

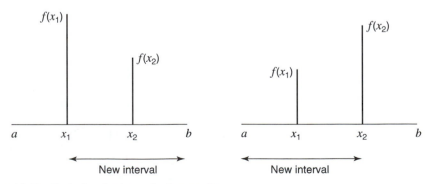

Figure 10.10 Reducing the interval of uncertainty

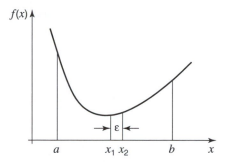

Figure 10.11   Placement of x_1 and x_2 for the dichotomous search method

These results are based on the assumption of unimodality and the information provided by two measurements; a single evaluation of $f(x)$ provides no useful information. Viewed another way, the measurements at x_1 and x_2 provide an estimate of the slope at the center point. When the slope of $f(x)$ is positive, we are assured that x^* lies to the left of x_2. When the slope is negative, we are assured that x^* lies to the right of x_1. When the slope is zero, x^* must lie between x_1 and x_2, although this case is unlikely for all but contrived problems.

Following the updating of a or b, the process is simply repeated until the predetermined number of points have been evaluated or the required interval of uncertainty has been attained. It is easy to see that after one iteration the interval of uncertainty is $0.5(b - a + \varepsilon)$. It may be shown that, in general, after n iterations the interval of uncertainty, d_n, is

$$d_n = \frac{b-a}{2^n} + \left(1 - \frac{1}{2^n}\right)\varepsilon \tag{15}$$

Thus, by using this method it is possible to compute precisely what the final interval of uncertainty will be for a given n, or we could specify d_n and compute n. With each iteration, the interval is nearly halved. In fact, if ε were zero, the reduction at each iteration would be exactly one-half. From this observation, we may surmise that placing the two points at each iteration in the manner prescribed by the dichotomous search algorithm is the best possible strategy; no other placement would give as much reduction in the interval of uncertainty.

Example 15

To illustrate the dichotomous search procedure, consider the function

$$f(x) = 3x + \frac{13}{x} + 1.7$$

as shown in Figure 10.12. We will search for the minimum in the range [1, 3] using six pairs of observations ($n = 6$). The computations are presented in Table 10.8 for $\varepsilon = 0.01$. The last column, labeled "Decision," indicates which endpoint is updated at each iteration.

From these results, we can see that the minimal point found is $x = 2.088$ with $f = 14.1901$. However, owing to the nature of the algorithm, all that we can conclude is that the global minimum x^* lies in the interval [2.057, 2.098], which indicates an interval of uncertainty of 0.041 unit. Similarly, using Equation (15), we get $d_6 = 0.041$, which is about 2% of the original interval [1, 3]. Note that the true minimum occurs at $x^* = 2.08167$ with $f = 14.19$.

With this kind of performance using only two points at each iteration, it is natural to consider whether three or more points at each iteration would provide even better results.

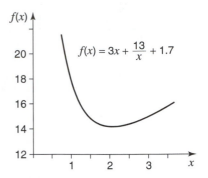

Figure 10.12   Function used to illustrate the dichotomous search method

If one considers the amount of additional reduction in the interval of uncertainty relative to the additional computational effort, the answer to this question is no.

Golden Section Search Method

In the preceding approach, all new evaluations were used at each iteration. Suppose instead that at each iteration after the first we use a combination of one new evaluation and one old evaluation. This should result in a significant reduction of computational effort if comparable results can be achieved. One method of implementing this approach was inspired by a number commonly observed in nature. In the architecture of ancient Greece, for example, a method of dividing a distance measured from point a to point b at a point c was called a *golden section* if

$$\frac{c-a}{b-a} = \frac{b-c}{c-a} = \frac{(b-a)-(c-a)}{c-a}$$

Dividing the numerators and denominators of each term by $b - a$ and letting $\gamma = (c - a)/(b - a)$ yields

$$\gamma = \frac{1-\gamma}{\gamma}$$

where γ is known as the *golden section ratio*. Solving for γ is equivalent to solving the quadratic equation $\gamma^2 + \gamma - 1 = 0$, whose positive root is $\gamma = (\sqrt{5} - 1)/2 \cong 0.618$. The negative root would imply a negative ratio, which has no meaning from a geometric point of view.

Table 10.8   Iterations for the Dichotomous Search Method

Iteration	a	b	x_1	$f(x_1)$	x_2	$f(x_2)$	Decision
1	1	3	1.995	14.2012	2.005	14.1987	$a = 1.995$
2	1.995	3	2.493	14.3931	2.503	14.4023	$b = 2.503$
3	1.995	2.503	2.244	14.2251	2.254	14.2294	$b = 2.254$
4	1.995	2.254	2.119	14.1920	2.129	14.1932	$b = 2.129$
5	1.995	2.129	2.057	14.1908	2.067	14.1903	$a = 2.057$
6	2.057	2.129	2.088	14.1901	2.098	14.1904	Stop

We now use the concept of the golden section to develop what is called the *golden section search method*. This method requires that the ratio of the new interval of uncertainty to the preceding one always be the same. This can be achieved only if the constant of proportionality is the golden section ratio γ.

To implement the algorithm, we begin with the initial interval $[a, b]$ and place the first two search points symmetrically at

$$x_1 = a + (1 - \gamma)(b - a) = b - \gamma(b - a) \text{ and } x_2 = a + \gamma(b - a) \tag{16}$$

as illustrated in Figure 10.13. By construction, we have $x_1 - a = b - x_2$, which is maintained throughout the computations.

For successive iterations, we determine the interval containing the minimal value of x, just as we did in the dichotomous search method. The next step of the golden section method, however, requires only one new evaluation of $f(x)$ with x located at the new *golden section point* of the new interval of uncertainty. At the end of each iteration, one of the following two cases arises (see Figure 10.13).

- Case 1: If $f(x_1) > f(x_2)$, the left endpoint a is updated by setting a to x_1 and the new x_1 is set equal to the old x_2. A new x_2 is computed from Equation (16).

- Case 2: If $f(x_1) \leq f(x_2)$, the right endpoint b is updated by setting b to x_2 and the new x_2 is set equal to the old x_1. A new x_1 is computed from Equation (16).

We stop when $b - a < \varepsilon$, an arbitrarily small number. At termination, one point remains in the final interval, either x_1 or x_2. The solution is taken as that point.

It can be shown that after k evaluations, the interval of uncertainty, call it d_k, has width

$$d_k = \gamma^{k-1} d_1 \tag{17}$$

where $d_1 = b - a$ (initial width). From this it follows that

$$\frac{d_{k+1}}{d_k} = \lambda \cong 0.618 \tag{18}$$

Table 10.9 provides the results for the same example used to illustrate the dichotomous search method. From the table we see that after 12 function evaluations (11 iterations) the minimum point found is $x_2 = 2.082$ with $f = 14.189996$. The true optimal solution is guaranteed to lie in the range $[2.0782, 2.0882]$. The width of this interval is 0.01 unit, which is less than one-fourth of the interval yielded by the dichotomous search method with the same

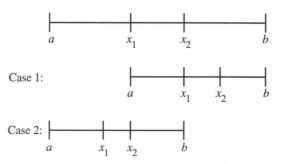

Figure 10.13   Point placement for the golden section search method

Table 10.9 Iterations for the Golden Section Search Method

Iteration	a	b	x_1	x_2	$f(x_1)$	$f(x_2)$	Decision
1	1	3	1.7639	2.2361	14.3617	14.2220	$a = 1.7639$
2	1.7639	3	2.2361	2.5279	14.2220	14.4263	$b = 2.5279$
3	1.7639	2.5279	2.0557	2.2361	14.1910	14.2220	$b = 2.2361$
4	1.7639	2.2361	1.9443	2.0557	14.2191	14.1910	$a = 1.9443$
5	1.9443	2.2361	2.0557	2.1246	14.1910	14.1926	$b = 2.1246$
6	1.9443	2.1246	2.0132	2.0557	14.1970	14.1910	$a = 2.0132$
7	2.0132	2.1246	2.0557	2.0820	14.1910	14.1900	$a = 2.0557$
8	2.0557	2.1246	2.0820	2.0983	14.1900	14.1904	$b = 2.0983$
9	2.0557	2.0983	2.0720	2.0820	14.1901	14.1900	$a = 2.0720$
10	2.0720	2.0983	2.0820	2.0882	14.189996	14.190058	$b = 2.0882$
11	2.0720	2.0882	2.0782	2.0820	14.190013	14.189996	Stop

number of evaluations. Equation (17) indicates that the interval of uncertainty after 12 evaluations is similarly 0.01 unit. The reader can verify that successive ratios are all (approximately) equal to γ, as specified by Equation (18). For example, for $k = 7$ we have at the completion of iteration 6 the ratio $d_6/d_5 = (2.1246 - 1.9443)/(2.2361 - 1.9443) = 0.61789 \cong \gamma$, with the error attributable to rounding.

The golden section search method is an outgrowth of the well-known Fibonacci search method, which is designed to minimize the final interval of uncertainty d_n, where n is the total number of function evaluations and is fixed at the outset. The solution to this optimization problem is based on the Fibonacci sequence generated by the recurrence relation

$$F_\nu = F_{\nu-1} + F_{\nu-2}, \text{ where } F_0 = F_1 = 1 \text{ and } \nu = 2, 3, \ldots$$

The resulting sequence is 1, 1, 2, 3, 5, 8, 13, . . .

The procedure for reducing the interval of uncertainty to d_n is almost identical to that of the golden section method. At iteration k, suppose that the interval of uncertainty is $[a_k, b_k]$. Consider the following two symmetric placements.

$$x_{1k} = a_k + \frac{F_{n-k-1}}{F_{n-k+1}}(b_k - a_k), k = 1, \ldots, n-1$$

$$x_{2k} = a_k + \frac{F_{n-k}}{F_{n-k+1}}(b_k - a_k), k = 1, \ldots, n-1$$

By Theorem 7, the new interval of uncertainty, $[a_{k+1}, b_{k+1}]$, is given by $[x_{1k}, b_k]$ if $f(x_{1k}) > f(x_{2k})$ and by $[a_k, x_{2k}]$ if $f(x_{1k}) \leq f(x_{2k})$.

The first two placements are made symmetrically at a distance of $(F_{n-1}/F_n)d_1$ from the ends of the initial interval, now $[a_1, b_1]$. As stated, a portion of the interval is discarded and the process is repeated. The width of the interval of uncertainty after the kth measurement is $d_k = (F_{n-k+1}/F_n)d_1$. As $n \to \infty$, $F_{n-1}/F_n \to \gamma$, so the Fibonacci search method converges to the golden section method.

Newton's Method

When more information than just the value of the function can be computed at each iteration, convergence is likely to be accelerated. Suppose that $f(x)$ is unimodal and twice continuously differentiable. In approaching Problem (14), also suppose that at a point x_k where a measurement is made, it is possible to determine the following three values: $f(x_k), f'(x_k)$, and $f''(x_k)$. This means that it is possible to construct a quadratic function $q(x)$ that agrees with $f(x)$ up to second derivatives at x_k. Let

$$q(x) = f(x_k) + f'(x_k)(x - x_k) + \frac{1}{2} f''(x_k)(x - x_k)^2$$

As shown in Figure 10.14a, we may then calculate an estimate x_{k+1} of the minimum point of f by finding the point at which the derivative of q vanishes. Thus, setting

$$0 = q'(x_{k+1}) = f'(x_k) + f''(x_k)(x_{k+1} - x_k)$$

we find

$$x_{k+1} = x_k - \frac{f'(x_k)}{f''(x_k)} \tag{19}$$

which, incidentally, does not depend on $f(x_k)$. This process can then be repeated until some convergence criterion is met, typically $|x_{k+1} - x_k| < \varepsilon$ or $|f'(x_k)| < \varepsilon$, where ε is some small number.

Newton's method can more simply be viewed as a technique for iteratively solving equations of the form $\phi(x) = 0$, where $\phi(x) \equiv f'(x)$ when applied to the line search problem. In this notation, we have $x_{k+1} = x_k - \phi(x_k)/\phi'(x_k)$. Figure 10.14b geometrically depicts how the new point is found. The following theorem gives sufficient conditions under which the method will converge to a stationary point.

Theorem 8: Consider the function $f(x)$ with continuous first and second derivatives $f'(x)$ and $f''(x)$. Define $\phi(x) = f'(x)$ and $\phi'(x) = f''(x)$ and let x^* satisfy $\phi(x^*) = 0$, $\phi'(x^*) \neq 0$. Then, if x_1 is sufficiently close to x^*, the sequence $\{x_k\}_{k=1}^{\infty}$ generated by Newton's method [Equation (19)] converges to x^* with an order of convergence of at least 2.

The phrase "convergence of order ρ" will be defined presently, but for now it means that when the iterate x_k is in the neighborhood of x^*, the distance from x^* at the next iteration is reduced by the ρth power. Mathematically, this can be stated as $||x_{k+1} - x^*|| \leq \beta ||x_k - x^*||^\rho$, where $\beta < \infty$ is some constant. The larger the order ρ, the faster the convergence.

When second derivative information is not available, it is possible to use first-order information to estimate $f''(x_k)$ in the quadratic $q(x)$. By letting $f''(x_k) \cong (f'(x_{k-1}) - f'(x_k)) / (x_{k-1} - x_k)$, the equivalent of Equation (19) is

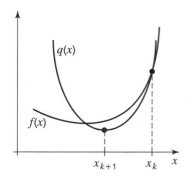

a. Minimizing the quadratic

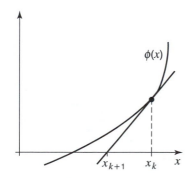

b. Finding the zero of the equation

Figure 10.14   Geometric view of Newton's method

$$x_{k+1} = x_k - f'(x_k) \left(\frac{x_{k-1} - x_k}{f'(x_{k-1}) - f'(x_k)} \right)$$

which gives rise to what is called the *method of false position*. Comparing this formula with that of Newton's method [Equation (19)], we see again that the value $f(x_k)$ does not enter.

A final word about one-dimensional search methods concerns the assumption of unimodality. If the function $f(x)$ is not unimodal, several local solutions will exist, so a more elaborate strategy is needed to identify the global minimum. One way to proceed is to overlay a grid on the original interval of uncertainty and apply to each subinterval one of the search methods discussed in this section. Refinements are possible, but there is no way of ensuring that any procedure will uncover the global minimum without an exhaustive and costly search.

10.7   MULTIDIMENSIONAL SEARCH METHODS

We now turn our attention to methods for the unconstrained minimization of functions defined on $\Re^n$. When $f(\mathbf{x})$ is linear and no constraints exist, the problem has no bounded solution. Unconstrained nonlinear functions, however, may have many local extrema. As we have stated, a necessary condition for a differentiable function f to have a minimum at $\mathbf{x}^* \in \Re^n$ is that $\nabla f(\mathbf{x}^*) = \mathbf{0}$. Recall that such an $\mathbf{x}^*$ is called a stationary point. Thus, we are faced with the problem of finding a solution to the system of nonlinear equations $\nabla f(\mathbf{x}) = \mathbf{0}$. A principal aspect of any procedure for solving these equations involves executing a sequence of line searches of the type just discussed; however, we note that a stationary point may be a local minimum, a local maximum, or a saddle point. For f twice continuously differentiable, sufficient conditions for $\mathbf{x}^*$ to be a local minimum are that the gradient $\nabla f(\mathbf{x}^*) = \mathbf{0}$ and the Hessian matrix $\nabla^2 f(\mathbf{x}^*)$ is positive definite.

General Descent Algorithm

The general descent algorithm starts at an arbitrary point, $\mathbf{x}^0$ and proceeds for some distance in a direction that improves (decreases) the objective function. Arriving at a point that has a smaller objective value than $\mathbf{x}^0$, the process finds a new improving direction and moves in that direction to a new point with a still smaller objective. In theory, the process could continue until there are no improving directions, at which point the algorithm would report a local minimum. In practice, the process stops when one or more numerical convergence criteria are satisfied. The algorithm is stated more formely below.

1. Start with an initial point $\mathbf{x}^0$. Set the iteration counter k to 0.

2. Choose a descent direction $\mathbf{d}^k$.

3. Perform a line search to choose a step size t_k such that
$$\omega_k(t_k) \equiv f(\mathbf{x}^k + t_k \mathbf{d}^k) < \omega_k(t_{k-1})$$

4. Set $\mathbf{x}^{k+1} = \mathbf{x}^k + t_k \mathbf{d}^k$.

5. Evaluate convergence criteria. If satisfied, stop; otherwise, increase k by 1 and go to Step 2.

An exact line search is one that chooses t_k as the first local minimum of $\omega_k(t_k)$ at Step 3—i.e., the one with the smallest t value. Finding this minimum to high accuracy is overly time consuming, so modern NLP codes use a variety of inexact line search techniques often involving polynomial fits, as in the method of false position. With regard to termination, stationarity is commonly used as one of several criteria at Step 5. In this case, termination occurs when the gradient is small. However, if f is scaled by multiplying it by a positive scale factor c, then $\nabla(cf(\mathbf{x})) = c\nabla f(\mathbf{x})$, so what constitutes "small" may be arbitrary. As a consequence, it is necessary to add other criteria, such as

$$\frac{|f(\mathbf{x}^{k+1}) - f(\mathbf{x}^k)|}{1 + |f(\mathbf{x}^k)|} < \varepsilon, \text{ say, for } s \text{ consecutive values of } k$$

In this case, the computations stop whenever the fractional change in the objective function is less than a user-defined tolerance $\varepsilon > 0$ for s consecutive iterations. The NLP code GRG2 (Lasdon *et al.* [1996]), for example, uses $\varepsilon = 10^{-4}$ and $s = 3$ as default values.

The remaining issue concerning the general descent algorithm centers on the determination of the improving direction $\mathbf{d}^k$ at Step 2. Several procedures will now be discussed.

Steepest Descent

In the absence of superior information, the best direction in which to proceed from a given point $\mathbf{x}^0$ is the direction that yields the most favorable improvement in $f(\mathbf{x})$ in the neighborhood of $\mathbf{x}^0$. For an LP problem, the nonbasic variable with the most negative reduced cost provides this direction. In general, it is well known that the gradient $\nabla f(\mathbf{x})$ is the direction of maximal immediate increase of $f(\mathbf{x})$, so the negative gradient is the direction of maximal immediate decrease. The steepest descent method moves in the direction of the gradient to maximize and in the opposite direction of the gradient to minimize.

The gradient, however, measures the rate of change of the objective function only at the point for which it is calculated. This means that if you move more than a short distance away from the current point you may no longer be moving in a favorable direction. It is this latter fact that raises the question of how far to proceed before seeking a new direction. Moving only a very small distance is unacceptable from a practical viewpoint, because not much progress will be made. A line search is needed to answer the question.

When minimizing is performed by means of the steepest descent method, the $k+1$st point of the search is obtained from the kth point using the expression

$$\mathbf{x}^{k+1} = \mathbf{x}^k - t_k \nabla f(\mathbf{x}^k) \tag{20}$$

where $\nabla f(\mathbf{x}^k)$ is a column vector evaluated at $\mathbf{x}^k$. When maximizing, we replace the minus sign in Equation (20) with a plus sign. The optimal step size t_k is found by solving the following one-dimensional optimization problem.

$$f(\mathbf{x}^k - t_k \nabla f(\mathbf{x}^k)) = \min_{t \geq 0} f(\mathbf{x}^k - t\nabla f(\mathbf{x}^k)) \tag{21}$$

assuming that an exact line search is performed. Any of the procedures discussed in Section 10.6 can be used.

For purposes of discussion, we will terminate the algorithm when the norm of the gradient (i.e., $\|\nabla f(\mathbf{x}^k)\|$ is less than some specified small number ε, or when the search step is infinite). The latter occurrence indicates an unbounded solution.

Algorithm

The goal is to find a point $\mathbf{x}^*$ that minimizes the function $f(\mathbf{x})$. When $f(\mathbf{x})$ is convex, the steepest descent method will converge to a global minimum; otherwise, it will converge to a local minimum. If a finite solution exists on termination, the current trial point $\mathbf{x}^k$ is an estimate of the optimum $\mathbf{x}^*$.

Initialization step: Select an initial trial point $\mathbf{x}^0$ and a small number $\varepsilon > 0$ for the termination test. Let $k = 0$ and compute the gradient $\nabla f(\mathbf{x}^0)$. Stop if $\| \nabla f(\mathbf{x}^0) \| < \varepsilon$.

Iterative step: Construct the line that passes through the trial solution and goes in the direction of the negative gradient.

$$\mathbf{x}(t) = \mathbf{x}^k - t\nabla f(\mathbf{x}^k)$$

Find the value of t that minimizes $f(\mathbf{x}(t))$ along this line—that is, solve Equation (21) to obtain t_k. If no solution is found, stop: the problem is unbounded. Otherwise, let the new trial solution be

$$\mathbf{x}^{k+1} = \mathbf{x}^k - t_k\nabla f(\mathbf{x}^k)$$

Termination step: Compute the gradient at the trial point $\mathbf{x}^{k+1}$. If $\|\nabla f(\mathbf{x}^{k+1})\| < \varepsilon$, the termination criterion is satisfied, so stop; otherwise, increase k by 1 and repeat the iterative step.

Formally, we have the following result.

Theorem 9: Suppose that f has continuous partial derivatives and that $\{\mathbf{x} : f(\mathbf{x}) \leq f(\mathbf{x}^0)\}$ is closed and bounded. Then any limit point of the sequence generated by the steepest descent algorithm is a stationary point.

Application to a Quadratic in Two Dimensions

For purposes of illustration, let us consider the problem of minimizing a two-dimensional quadratic function.

$$f(\mathbf{x}) = \mathbf{c}\mathbf{x} + \tfrac{1}{2}\mathbf{x}^T\mathbf{Q}\mathbf{x}$$

$$= c_1 x_1 + c_2 x_2 + \tfrac{1}{2}(q_{11}x_1^2 + q_{22}x_2^2 + 2q_{12}x_1 x_2)$$

The gradient of $f(\mathbf{x})$ is

$$\begin{aligned}
\nabla f(\mathbf{x}) &= \mathbf{c} + \mathbf{Q}\mathbf{x} \\
&= ((c_1 + q_{11}x_1 + q_{12}x_2), (c_2 + q_{12}x_1 + q_{22}x_2))^T \\
&= (\nabla_1 f, \nabla_2 f)^T
\end{aligned}$$

Thus, starting from the initial point $\mathbf{x}^0$, we must solve Problem (21) over the line

$$\mathbf{x}(t) = \mathbf{x}^0 - t\nabla f(\mathbf{x}) = \begin{pmatrix} x_1^0 \\ x_2^0 \end{pmatrix} - t\begin{pmatrix} \nabla_1 f^0 \\ \nabla_2 f^0 \end{pmatrix}$$

to find the new point. The optimal step size, call it t^*, can be determined by substituting the right-hand side of the expression above into $f(\mathbf{x})$ and finding the value of t that minimizes $f(\mathbf{x}(t))$. For this simple case, it can be shown with some algebra that

$$t^* = \frac{\left(\nabla_1 f^0\right)^2 + \left(\nabla_2 f^0\right)^2}{q_{11}\left(\nabla_1 f^0\right)^2 + q_{22}\left(\nabla_2 f^0\right)^2 + 2q_{12}\nabla_1 f^0\nabla_2 f^0}$$

Table 10.10 gives the sequence of points generated during the first 10 iterations of the steepest descent method applied to the problem

$$\text{Minimize } f(\mathbf{x}) = 32 - 8x_1 - 16x_2 + x_1^2 + 4x_2^2$$

The termination parameter ε was set equal to 0.01. Figure 10.15 plots the trajectory of the trial solutions, which make rapid progress toward the optimal solution during the first few iterations. Subsequently, progress stalls even though $f(\mathbf{x})$ is convex. The difficulty is that when the isovalue contours of the function to be minimized differ significantly from concentric circles, steepest descent is accompanied by excessive *zigzagging*. This is highly inefficient. In mathematical terms, if t_k minimizes $\omega_k(t_k)$, then $\omega_k'(t_k) = \nabla f(\mathbf{x}^k - t_k\mathbf{d}^k)\mathbf{d}^k = 0$. If $\mathbf{d}^k = -\nabla f(\mathbf{x}^k)$, then the latter equation indicates that successive search directions are orthogonal, and hence the zigzagging for eccentric isovalue contours.

Gradient Search on a More General Function

To illustrate a more general case, consider the location problem described in Chapter 9. It is desired to connect five oil wells to a single collection point using the minimum amount of pipe. The five wells have coordinates (a_i, b_i) in the (x, y)-plane given by

Table 10.10   Steepest Descent Method Applied to Quadratic Function

k	x_1	x_2	$f(x_1, x_2)$	$\nabla_1 f$	$\nabla_2 f$	t_k
0	0.00	0.00	32.000	−8.000	−16.000	0.147
1	1.18	2.35	8.471	−5.647	2.824	0.313
2	2.94	1.47	2.242	−2.118	−4.235	0.147
3	3.25	2.09	0.594	−1.495	0.747	0.313
4	3.72	1.86	0.157	−0.561	−1.121	0.147
5	3.80	2.02	0.042	−0.396	0.198	0.313
6	3.93	1.96	0.011	−0.148	−0.297	0.147
7	3.95	2.01	0.003	−0.105	0.052	0.313
8	3.98	1.99	0.001	−0.039	−0.079	0.147
9	3.99	2.00	0.000	−0.028	0.014	0.313
10	3.99	2.00	0.000	−0.010	−0.021	0.147

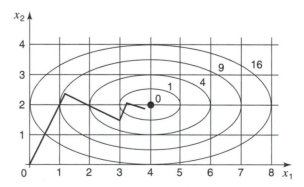

Figure 10.15   Zigzagging nature of steepest descent method

$$(a_1, b_1) = (5, 0), \quad (a_2, b_2) = (0, 10), \quad (a_3, b_3) = (10, 10),$$
$$(a_4, b_4) = (50, 50), \quad (a_5, b_5) = (-10, 50)$$

The problem is to determine the (x, y) coordinates of the optimal collection point. For m wells, we wish to

$$\text{Minimize } f(x, y) = \sum_{i=1}^{m} D_i(x, y)$$

where $D_i(x, y)$ is the distance from the ith well to the collection point given by the Euclidean norm.

$$D_i(x, y) = \sqrt{(x - a_i)^2 + (y - b_i)^2}$$

It can be shown that $D_i(x, y)$ and hence $f(x, y)$ are strictly convex, which suggests that we should be able to find a global optimal solution. The gradient of f with respect to the decision variables is

$$\frac{\partial f}{\partial x} = \sum_{i=1}^{m} \frac{a_i - x}{D_i} \text{ and } \frac{\partial f}{\partial y} = \sum_{i=1}^{m} \frac{b_i - y}{D_i}$$

One way to proceed is to set these expressions equal to zero and try to solve for x and y. There is no closed-form solution, however, so a numerical method would have to be applied to find an approximate solution. Alternatively, we will use these expressions as part of the steepest descent algorithm to determine an improving direction. The step size subproblem will be solved with the golden section method. As opposed to the case in which we were minimizing a two-dimensional quadratic function, however, no closed-form solution exists either for t^*. The final issue concerns the fact that the partial derivatives are not defined for solutions at any of the well locations where the corresponding distance D_i is zero. This difficulty can be handled by perturbing each distance by a small amount δ so that we now have

$$D_i(x, y) = \sqrt{(x - a_i)^2 + (y - b_i)^2 + \delta}$$

The results for seven iterations are given in Table 10.11, which shows convergence to two decimal places. Accordingly, the optimal solution is to place the collection point at $(x^*, y^*) = (7.34, 11.32)$.

Table 10.11 Steepest Descent Results for Location Problem

k	Trial point, (x^k, y^k)	Value, f^k	Gradient, ∇f^k	Step size, t_k
0	(20.00, 20.00)	146.36	(2.2015, 0.5401)	6.560
1	(5.56, 16.46)	125.48	(−0.2579, 1.0716)	4.776
2	(6.79, 11.34)	122.01	(−0.1328, 0.0179)	2.870
3	(7.17, 11.39)	121.97	(−0.0649, 0.0304)	0.667
4	(7.21, 11.37)	121.97	(−0.0589, 0.0258)	0.651
5	(7.25, 11.35)	121.97	(−0.0535, 0.0223)	1.638
6	(7.34, 11.32)	121.96	(−0.0411, 0.0152)	0.036
7	(7.34, 11.32)	121.96	—	—

Numerical Evaluation of the Gradient

Directly determining the values of the elements of the gradient is feasible when $f(\mathbf{x})$ is known explicitly and is not too complex. When it is impractical or impossible to represent the gradient analytically, an approximate value of each partial derivative can be computed from the following finite difference equations.

$$\frac{\partial f(\mathbf{x})}{\partial x_i} \cong \frac{f(\mathbf{x}_j^+) - f(\mathbf{x}_j^-)}{2\Delta}, \quad j = 1, \ldots, n$$

Here, Δ is some small number, $\mathbf{x}_j^+ = (x_1, \ldots, x_{j-1}, x_j + \Delta, x_{j+1}, \ldots, x_n)$, and $\mathbf{x}_j^- = (x_1, \ldots, x_{j-1}, x_j - \Delta, x_{j+1}, \ldots, x_n)$. That is, all the components of $\mathbf{x}^+$ and $\mathbf{x}^-$ remain the same except x_j, which is perturbed by Δ in the forward and backward directions, respectively. The approximations introduced by numerical estimates of the gradient must be properly accounted for in the implementation of the algorithm.

Rate of Convergence

The steepest descent algorithm converges in a linear fashion when optimal values of t are used and $f(\mathbf{x})$ is a strictly convex quadratic. In this context, convergence is defined to mean that as the number of iterations increases without bound, the gradient approaches the zero vector. *Linear convergence*, or convergence of order 1, is defined to mean that the value of $f(\mathbf{x})$ decreases by a constant proportion at each iteration

$$\lim_{k \to \infty} \frac{\left\| \nabla f\left(\mathbf{x}^{k+1}\right) \right\|}{\left\| \nabla f\left(\mathbf{x}^k\right) \right\|} = \beta$$

where $0 \leq \beta < 1$. When $\beta = 0$, we have superlinear convergence. For $f(\mathbf{x})$ quadratic and strictly convex,

$$\beta \leq \left(\frac{A - a}{A + a}\right)^2 = \left(\frac{r - 1}{r + 1}\right)^2$$

where A and a are the largest and smallest eigenvalues of $\mathbf{Q}$, respectively, and $r = A/a$ is known as the *condition number*. For strictly convex functions $f(\mathbf{x})$, the eigenvalues of the Hessian matrix $\mathbf{H}$ are always positive. As r gets larger, the contours of f become more eccentric and convergence slows. When f is strictly convex in the neighborhood of the optimal solution $\mathbf{x}^*$, but not quadratic, the same results hold; however, r is now determined from the eigenvalues of $\mathbf{H}(\mathbf{x}^*)$. This is a theoretical statement because $\mathbf{H}(\mathbf{x}^*)$ is not known before convergence.

For the simplest case, in which $f(\mathbf{x})$ is an n-dimensional sphere, steepest descent will converge to the global optimal solution in one iteration. It is important to realize, however, that linear convergence does not imply that the algorithm will find the optimal solution in a finite number of iterations for the more general case.

Method of Parallel Tangents

The method of parallel tangents or *gradient PARTAN search* is an easy way to avoid, or at least reduce, zigzagging while using the same information and techniques that have been developed for steepest descent. If you were to take a ruler and draw a straight line through $\mathbf{x}^0$ and $\mathbf{x}^2$ in Figure 10.15, you would see that the line would pass precisely through the optimal solution, $\mathbf{x}^* = (4, 2)$. This is illustrated in Figure 10.16.

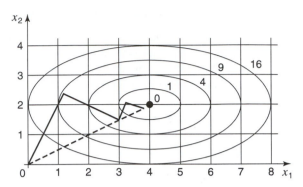

Figure 10.16   PARTAN search for the quadratic objective function

Performing a one-dimensional search over the line $\mathbf{x}(t) = \mathbf{x}^0 + t(\mathbf{x}^2 - \mathbf{x}^0)$ leads immediately to the optimal solution. Indeed, this same idea can be applied to any two points two iterations apart when using the steepest descent method in conjunction with an optimal line search for t at each iteration. To see this, draw a straight line through $\mathbf{x}^1$ and $\mathbf{x}^3$ and note that it, too, passes through $\mathbf{x}^*$. The search along the line formed by connecting the two points $\mathbf{x}^k$ and $\mathbf{x}^{k+2}$ is called an *acceleration step*.

The preceding observation is the basis of a somewhat more involved algorithm for objective functions in n dimensions. If $f(\mathbf{x})$ has contours formed by n-dimensional ellipsoids—convex quadratic functions—it may be shown that the PARTAN method will find the minimum point in n gradient based steps plus $n - 1$ acceleration steps. The original method attempted to extend the acceleration idea using all previous steps and was based largely on a special geometric property of the tangents to the contours of a quadratic function. Current implementations do not make use of this property.

The algorithm will be defined with reference to Figure 10.17. The boldface lines indicate the path taken. Starting at an arbitrary point $\mathbf{x}^0$, the point $\mathbf{x}^1$ is found with a standard steepest descent step. After that, from the point $\mathbf{x}^k$ a corresponding $\mathbf{y}^k$ is found, also with a standard steepest descent step. A one-dimensional search is then conducted over the line joining $\mathbf{y}^k$ and $\mathbf{x}^{k-1}$ to determine the point $\mathbf{x}^{k+1}$ (see Figure 10.17a). The process is repeated until $\mathbf{x}^n$ is determined. When $f(\mathbf{x})$ is a strictly convex quadratic function, $\mathbf{x}^* = \mathbf{x}^n$. In other cases, we typically replace $\mathbf{x}^0$ with $\mathbf{x}^n$ and restart the algorithm with a standard steepest descent step. The computations continue until some convergence criterion is satisfied. Except for the first iteration, the values of $\mathbf{x}^k$ are found by the acceleration step and the values of $\mathbf{y}^k$ are found by the steepest descent step.

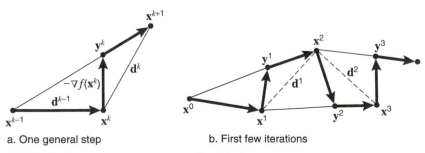

a. One general step   b. First few iterations

Figure 10.17   Illustration of the PARTAN search method

Figure 10.18 illustrates the application of the PARTAN method to a three-dimensional function. Basically, PARTAN first attacks the problem in a single plane, as if only two variables were present. In so doing, the algorithm finds that $\mathbf{x}^2$ is the best value of $f(\mathbf{x})$ lying in that plane. This is accomplished in the expected three steps—two gradient based steps to determine $\mathbf{x}^1$ and $\mathbf{y}^1$, and one acceleration step to determine $\mathbf{x}^2$. [In Figure 10.18, for simplicity, the first plane is shown to be the (x_1, x_2)-space.] Following the initial optimization, the algorithm proceeds to work in a plane perpendicular to the first. This is achieved by moving in the direction of the gradient, departing from the point $\mathbf{x}^2$. The new gradient direction is necessarily orthogonal to the first plane. The endpoint $\mathbf{y}^2$ of the optimal step into the new plane, combined with $\mathbf{x}^1$ and $\mathbf{x}^2$, defines the second search plane. Now we need only perform an optimal search along the direction that passes through $\mathbf{x}^1$ and $\mathbf{y}^2$. Just as in the two-dimensional case, once the second plane has been identified, three search points are sufficient to identify the *best* point in the *second* plane. For a convex quadratic objective function in three dimensions, the process terminates with the optimal solution at $\mathbf{x}^3$.

PARTAN is generally viewed as a particular implementation of the method of *conjugate gradients*. One of its attractive features is its simplicity and ease of implementation. Also, it has strong global convergence characteristics. Each step of the process is at least as good as steepest descent, because going from $\mathbf{x}^k$ to $\mathbf{y}^k$ is exactly steepest descent whereas the additional move to $\mathbf{x}^{k+1}$ provides a further decrease in the objective function. An undesirable feature of the PARTAN algorithm is that two line searches are required at each step except the first.

Newton's Method

The steepest descent algorithm and its variants rely on first-order approximations to the function f at the points $\mathbf{x}^k$. For a highly nonlinear function with continuous second partial derivatives, it is preferable to use second-order approximations. In addition to tracking more closely the curvature of f, the second-order approximations have better convergence properties. This is the idea behind Newton's method for functions of more than one variable. Using a second-order Taylor series approximation

$$f(\mathbf{x}) \cong q(\mathbf{x} = f(\mathbf{x}^k) + \nabla f(\mathbf{x}^k)(\mathbf{x} - \mathbf{x}^k) + \tfrac{1}{2}(\mathbf{x} - \mathbf{x}^k)^{\mathrm{T}} \mathbf{F}(\mathbf{x}^k)(\mathbf{x} - \mathbf{x}^k)$$

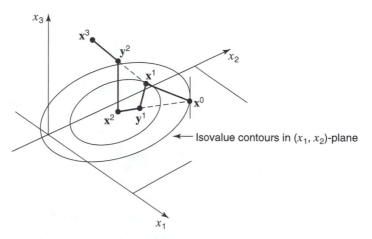

Figure 10.18   PARTAN method in three dimensions

where $\mathbf{F}(\mathbf{x}^k) \equiv \mathbf{H}(\mathbf{x}^k)$ is the $n \times n$ Hessian matrix, the point $\mathbf{x}^{k+1}$ is chosen to minimize the approximation. These points must satisfy $\nabla q(\mathbf{x}^{k+1}) = \mathbf{0}$ or $\nabla f(\mathbf{x}^k) + (\mathbf{x}^{k+1} - \mathbf{x}^k)^{\mathrm{T}} \mathbf{F}(\mathbf{x}^k) = \mathbf{0}$. This is a linear system with zero, one, or an infinite number of solutions.

The drawback of Newton's method is that it cannot be expected to converge to a stationary point. It is intended for use primarily in regions of $\mathfrak{R}^n$ close to a strict local minimum of f. In such a region, it can be shown that the function f is strictly convex, implying that the Hessian matrix is positive definite and invertible. The descent algorithm is given by

$$\mathbf{x}^{k+1} = \mathbf{x}^k - \nabla f(\mathbf{x}^k)\mathbf{F}(\mathbf{x}^k)^{-1} \tag{22}$$

Comparing the steepest descent algorithm [Equation (20)] with Equation (22), we see that they are both special cases of the general algorithm

$$\mathbf{x}^{k+1} = \mathbf{x}^k - t_k \nabla f(\mathbf{x}^k)\mathbf{M}_k$$

for suitable matrices $\mathbf{M}_k$. In regions distant from a local minimum, the step can be chosen by selecting t_k as we did for the steepest descent algorithm where $\mathbf{M}_k = \mathbf{I}$. Near a local minimum, we take $t_k = 1$ and $\mathbf{M}_k = \mathbf{F}(\mathbf{x}^k)^{-1}$ to obtain Equation (22).

An inherent difficulty with the aforementioned second-order methods is the necessity of calculating and inverting the Hessian matrix $\mathbf{F}$ at each step. Procedures have been devised for approximating the Hessian matrix and its inverse, resulting in faster iterative schemes for finding a minimum of f. These procedures are known as *conjugate direction methods* and *quasi-Newton methods*.

Nonconvex Functions

The foregoing gradient based approaches will converge to a global optimal solution when $f(\mathbf{x})$ is convex. For nonconvex functions, they, as well as any NLP search technique, can be expected at most to converge to a local minimum that is not necessarily a global minimum. Their performance will be highly dependent on the "goodness" of the starting point chosen to initiate the algorithm. Often, some direct search over a range or lattice of possible starting points is warranted to improve the results. It should be noted that although proximity to the optimal point is often the largest contributor to the goodness of a starting point, there may be other attributes that must be considered. This is particularly true when $f(\mathbf{x})$ is highly nonconvex. Particular points may doom the search at the outset simply because of the nature of the function in the immediate neighborhood of the point.

Our attention to the quadratic form in this section is motivated by the fact that no matter what form the objective function takes, a second-order Taylor series expansion will provide a good approximation near the optimal solution. Thus, a procedure that is effective for quadratic functions will also be effective for nonquadratic functions, at least near the optimal solution. On the other hand, a procedure that is not guaranteed to find the minimum of a positive semidefinite quadratic is not likely to be useful in optimizing any continuous function.

As we mentioned earlier in this chapter, the addition of constraints to the problem of minimizing a nonlinear function can vastly increase the complexity of the optimality conditions and hence make the analysis much more difficult. Nevertheless, problems with nonlinear objective functions and constraints must be addressed, because they are common in practice. In the supplementary material on the CD, we describe several procedures that use the gradient based search methods developed here to optimize constrained problems.

EXERCISES   *For Exercises 1 to 5, find the stationary points of f(**x**) by setting the first partial derivatives equal to zero and solving the resulting equations. If possible, determine whether the points obtained are local or global maxima or minima. Justify your conclusions.*

1. $f(\mathbf{x}) = -2x^3 + 3x^2 + 72x + 30$

2. $f(\mathbf{x}) = 3x_1^2 + 4x_2^2 - x_1 x_2 - 5x_1 - 8x_2$

3. $f(\mathbf{x}) = -2x_1^2 - x_2^2 - 4x_3^2 - 3x_1 x_3 + 3x_2 x_3 - x_1 - 2x_2 - 3x_3$

4. $f(\mathbf{x}) = -2x_1^2 - x_2^2 - 4x_3^2 - 8x_1 x_3 + 4x_2 x_3 - x_1 - 2x_2 - 3x_3$

5. $f(\mathbf{x}) = 100x_1^4 + 100x_2^2 - 200x_1^2 x_2 + x_1^2 - 2x_1$

6. Solve the following problem using the Lagrangian technique.

$$\text{Maximize } \left\{ x_1 + 2x_2 + 3x_3 \ : \ x_1^2 + x_2^2 + x_3^2 = 14 \right\}$$

7. Find the stationary points of the function $f = xyz$ subject to the condition that $x^2 + y^2 + z^2 = 1$. Indicate which points are maxima, minima, and saddle points.

8. Find the point in the set

$$\{(x, y, z) : x^2 - xy + y^2 - z^2 = 1, \ x^2 + y^2 = 1\}$$

that is nearest the origin in the (x, y, z)-space.

9. Consider the problem of maximizing $f(\mathbf{x})$ subject to $\mathbf{x} \in S$, where

$$S = \{(x_1, x_2) : x_1 + x_2 \le 4, \ -x_1 + x_2 \le 1, \ x_1 \ge 0, \ x_2 \ge 0\}$$

For each of the following cases, determine whether the KKT conditions are satisfied at the points $(1.5, 2.5)$, $(4, 0)$, and $(3, 0.5)$. Indicate which of these points, if any, are local solutions. Find the global optimal solutions.

   (a) $f(\mathbf{x}) = (x_1 - 3)^2 + (x_2 - 5)^2$
   (b) $f(\mathbf{x}) = (x_1 - 3)^2 + (x_2 - 2.5)^2$
   (c) $f(\mathbf{x}) = (x_1 - 3)^2 + (x_2 - 0.5)^2$

10. Consider the following QP problem.

$$\begin{aligned}
\text{Minimize } \ & 2x_1^2 + 4x_2^2 + 3x_3^2 - 4x_1 x_2 - 2x_1 x_3 + 2x_2 x_3 - 10x_1 - 15x_2 - 20x_3 \\
\text{subject to } \ & x_1 + x_2 + x_3 \le 15 \\
& 2x_1 + 4x_2 \le 26 \\
& x_2 + 3x_3 \le 20 \\
& x_j \ge 0, \ j = 1, 2, 3
\end{aligned}$$

   (a) Write out the KKT conditions and solve by hand.
   (b) Alternatively, set up the LP model that can be used to find the solution and solve with the Teach LP add-in accompanying this book (you must control the entering variables to ensure that complementarity is satisfied at each iteration).

11. Set up the LP model that can be used to solve the following QP problem. Use the simplex algorithm with a restricted basis entry rule to find the solution. The Teach LP add-in allows control of the entering basic variable.

$$\text{Minimize } \ 2x_1^2 + 2x_2^2 + 3x_3^2 + x_1 - 3x_2 - 5x_3 + 2x_1 x_2 + 2x_2 x_3$$
$$\text{subject to } \ x_1 + x_2 + x_3 \ge 1, \ 3x_1 + 2x_2 + x_3 \le 6, \ x_1 \ge 0, \ x_2 \ge 0, \ x_3 \ge 0$$

Also solve with a general NLP code and compare the results.

12. Consider the following problem.

$$\text{Minimize } (x_1 - x_5)^2 - x_2^2 - x_3^2$$
$$\text{subject to } x_1^2 + x_2^2 + x_3^2 \leq 9$$
$$(2 - \sqrt{5})x_1 + 2x_3 \geq 4$$
$$x_1 \geq 0, x_2 \geq 0, x_3 \geq 0$$

Test the KKT conditions for each of the following points and determine which is the global minimum.

(a) $\mathbf{x} = (0, \sqrt{5}, 2)$

(b) $\mathbf{x} = (0, 0, 2)$

(c) $\mathbf{x} = (2, 0, \sqrt{5})$

(d) $\mathbf{x} = (0, 3, 0)$

(e) $\mathbf{x} = (3, 0, 0)$

13. Consider the following problem.

$$\text{Minimize } (x_1 - 4)^2 + (x_2 - 4)^2 + (x_3 - 4)^2$$
$$\text{subject to } x_1 + x_2 + x_3 \geq 1$$
$$x_1 + x_2 + x_3 \leq 6$$
$$x_1 \geq 0, x_2 \geq 0, x_3 \geq 0$$

(a) Using the KKT conditions, investigate the points (1, 0, 0), (1, 1, 1), (0, 0, 6), and (2, 2, 2), and determine which is the optimal solution.

(b) Using the restricted basis entry rule, set up and solve the associated linear program.

14. Solve the problem in Exercise 13 using piecewise linear approximations to the nonlinear separable functions. Let $x_j \leq u_j = 6$ for $j = 1, 2, 3$, and use a grid size of 0.5.

15. Use the separable programming approach to solve the following problems. Pick a suitable grid for the variables.

(a) Minimize $\dfrac{1}{x_1 + 1} + x_2^3$

subject to $x_1^2 - 2x_2^3 \leq 5$

$x_1 \geq 8, x_2 \geq 8$

(b) Minimize $e^{x_1} + x_1^2 + 3x_1 + 2x_2^2 - 5x_2 + 2x_3$

subject to $-2x_1^2 + e^{x_2} - 6x_3 \leq 12$

$x_1^4 - 3x_2 + 5x_3 \leq 23$

$0 \leq x_1 \leq 4, 0 \leq x_2 \leq 2, 0 \leq x_3$

16. Identify the computational steps associated with the one-dimensional search methods listed below. Include an *initialization step* and a *main step* with appropriate subcomponents. Be sure to specify input parameters and termination conditions. The main step should be written in terms of iteration k.

(a) Dichotomous search method

(b) Golden section search method

(c) Newton's method

(d) Fibonacci search method

(e) Method of false position

17. Write a computer code to implement the methods listed in Exercise 16.

18. For the following problems of a single variable, use the dichotomous search method to find the solution for $\varepsilon = 0.01$.

 (a) Minimize $\{f(x) = 105x + 8/x - 45 : 0 \le x \le 1\}$
 (b) Minimize $\{f(x) = 3x^2 - 5x - 4 : 0 \le x \le 2\}$
 (c) Minimize $\{f(x) = 2x^4 - 2x - 1 : 0 \le x \le 2\}$
 (d) Maximize $\{f(x) = 4x^3 - 7x^2 + 14x + 6 : 0 \le x \le 1\}$
 (e) Minimize $\{f(x) = 3x^2 - 2x^{3/2} + 1 : 0 \le x \le 1\}$
 (f) Minimize $\{f(x) = e^x - x : 0 \le x \le 1\}$

19. Use the golden section method to solve the problems in Exercise 18 with stopping parameter $\varepsilon = 0.01$.

20. Use Newton's method to solve the problems in Exercise 18 with $\varepsilon = 0.01$. Start at the midpoint of the range.

21. Use the steepest descent method to find the minimum of the function

$$f(\mathbf{x}) = 3x_1^2 + 4x_2^2 - x_1 x_2 - 5x_1 - 8x_2$$

 Start at the point $\mathbf{x}^0 = (0, 0)$ and do four iterations. Determine the value of the step size t at each iteration with the special procedure described in the text for quadratic functions.

22. Repeat Exercise 21 using
 (a) The PARTAN method
 (b) Newton's method

23. Use the steepest descent procedure to find the minimum of the function

$$f(\mathbf{x}) = 2x_1^2 + x_2^2 + 4x_3^2 + 3x_1 x_3 - 3x_2 x_3 + x_1 + 2x_2 + 3x_3$$

 Start at the point $\mathbf{x}^0 = (1, 1, 1)$ and do four iterations. Find the closed-form solution for the step size t that minimizes the function in the search direction and use it in the computations.

24. Repeat Exercise 23, but instead of using the closed-form solution for t, use a one-dimensional search procedure to find the optimal step size. Assume that the initial range for t is $[0, 1]$ and that $\varepsilon = 0.1$.

25. Repeat Exercise 23 using
 (a) The PARTAN method
 (b) Newton's method

26. Use Newton's method to minimize $f(\mathbf{x}) = x_1^2 - 4x_1 + 2x_2^2 - 2x_2 + (x_1 x_2)^{3/2}$.

27. Use the Teach NLP add-in and steepest descent to solve the following problem. Start the search at the origin.

$$\text{Maximize } f(\mathbf{x}) = -100\left(x_3 - [0.5(x_1 + x_2)]^2\right)^2 - 25(x_1 - x_2^2 + x_3^2)^2$$
$$- (10 - x_1 + x_2 - x_3)^2 - (x_1 x_2 x_3)^2$$

28. Perform three iterations of steepest descent (ascent) for each of the following problems. Plot isovalue contours of $f(\mathbf{x})$ and indicate how the method progresses from one iteration to the next.

 (a) Minimize $f(\mathbf{x}) = 2x_1^2 + x_2^2 - x_1 x_2 - 7x_1$. Start at $\mathbf{x}^0 = (1, 0)$.
 (b) Minimize $f(\mathbf{x}) = x_1^3 - 9x_1^2 + 5x_2^2 - 20x_2$. Start at $\mathbf{x}^0 = (1, 1)$.
 (c) Repeat part (b) starting at $\mathbf{x}^0 = (-1, 1)$.
 (d) Maximize $f(\mathbf{x}) = -x_1^4 + 8x_1^3 - 10x_1^2 - x_2^2 + 2x_2$. Start at $\mathbf{x}^0 = (0, 0)$. Note that there are local optima at $(0, 1)$ and $(5, 1)$.

29. Perform three iterations of the PARTAN method for the following problems. In each case, let $\mathbf{x}^0 = (0, 0)$.

(a) Minimize $f(\mathbf{x}) = x_1^2 + x_2^2 - 4x_1x_2 - 16x_1 - 20x_2$

(b) Minimize $f(\mathbf{x}) = 2x_1^2 + (3 - x_2)^2$

(c) Minimize $f(\mathbf{x}) = x_1^2 + x_2^2 - 6x_1 - 8x_2$

(d) Minimize $f(\mathbf{x}) = (1 - 2x_1)^2 + (2 - 3x_2)^2 + x_1x_2$

30. Use Newton's method to solve the problems in Exercise 29 with the same starting points.

31. Find the minima of the functions in Exercises 2 to 5 using a multidimensional search method. You may use the Teach NLP add-in.

32. For each of the following objective functions, solve the problem

$$\text{Minimize } \{f(\mathbf{x}): \mathbf{x} \geq \mathbf{0}\}$$

using the conditions given in Theorem 3.

(a) $f(\mathbf{x}) = (x_1 - 5)^2 + (x_2 - 3)^2 - 4x_1x_2$

(b) $f(\mathbf{x}) = (x_1 + 5)^2 + (x_2 - 3)^2$

(c) $f(\mathbf{x}) = (x_1 - 5)^2 + (x_2 + 3)^2 + 2x_1x_2 + 4x_1 - 3x_2$

(d) $f(\mathbf{x}) = (x_1 + 5)^2 + (x_2 + 3)^2 - 6x_1x_2$

BIBLIOGRAPHY

Bard, J.F., *Practical Bilevel Optimization: Algorithms and Applications*, Kluwer Academic, Boston, 1998.

Bard, J.F., J. Plummer, and J.C. Sourie, "A Bilevel Programming Approach to Determining Tax Credits for Bio-fuel Production," *European Journal of Operational Research*, Vol. 120, No. 1, pp. 30–46, 2000.

Bazaraa, M.S., H.D. Sherali, and C.M. Shetty, *Nonlinear Programming: Theory and Algorithms*, Second Edition, Wiley, New York, 1993.

Falk, J.E. and R.M. Soland, "An Algorithm for Separable Nonconvex Programming Problems," *Management Science*, Vol. 15, pp. 550–569, 1969.

Fan, Y., S. Sarkar, and L. Lasdon, "Experiments with Successive Quadratic Programming Algorithms," *Journal of Optimization Theory and Applications*, Vol. 56, No. 3, pp. 359–383, 1988.

Fiacco, A.V. and G.P. McCormick, *Nonlinear Programming: Sequential Unconstrained Minimization Techniques*, Wiley, New York, 1968.

Fletcher, R., *Practical Methods of Optimization*, Second Edition, Wiley, New York, 1987.

Gill, P.E., W. Murray, and M.H. Wright, *Practical Optimization*, Academic Press, New York, 1981.

Horst, R. and H. Tuy, *Global Optimization: Deterministic Approaches*, Third Edition, Springer-Verlag, Berlin/New York, 1995.

Lasdon, L., J. Plummer, and A. Warren, "Nonlinear Programming," in M. Avriel and B. Golany (eds.), *Mathematical Programming for Industrial Engineers*, Chapter 6, pp. 385–485, Dekker, New York, 1996.

Liebman, J.S., L. Lasdon, L. Shrage, and A. Waren, *Modeling and Optimization with GINO*, Boyd and Fraser, Danvers, MA, 1986.

Luenberger, D.G., *Linear and Nonlinear Programming*, Second Edition, Addison-Wesley, Reading, MA, 1984.

Nash, S.G. and A. Sofer, "A Barrier Method for Large-Scale Constrained Optimization," *ORSA Journal on Computing*, Vol. 5, No. 1, pp. 40–53, 1993.

Nash, S.G. and A. Sofer, *Linear and Nonlinear Programming*, McGraw-Hill, New York, 1996.

Van de Panne, C., *Methods for Linear and Quadratic Programming*, North-Holland, Amsterdam, 1975.

Zoutendijk, G., *Methods of Feasible Directions*, Elsevier, Amsterdam, 1960.

Chapter 11

Models for Stochastic Processes

In many common situations, the attributes of a system randomly change over time. Examples include the number of customers in a checkout line, congestion on a highway, the number of items in a warehouse, and the price of a financial security, to name a few. In certain instances, it is possible to describe an underlying process that explains how the variability occurs. When aspects of the process are governed by probability theory, we have a stochastic process. This chapter introduces the general components of stochastic processes that are central to many different kinds of operations research studies. Subsequent chapters provide analytic results for determining the statistical characteristics of the random variables associated with the underlying activities.

The first step in modeling a dynamic process is to define the set of states over which it can range and to characterize the mechanisms that govern its transitions. A state is like a snapshot of the system at a point in time. It is an abstraction of reality that describes the attributes of the system of interest. Time is the linear measure through which the system moves, and can be thought of as a parameter. Because of time there is a past, a present, and a future. We usually know the trajectory a system has followed to arrive at its present state. Using this information, our goal is to predict the future behavior of the system in terms of a basic set of attributes. As we shall see, a variety of analytic techniques are available for this purpose.

Although the terms "stochastic process models" and "stochastic models" are sometimes used interchangeably, an important distinction needs to be made. The latter constitute a larger set that may include a decision-making component. The former represent phenomena governed by random variables and are descriptive only; they are not prescriptive. The words "state" and "time" are used in the chapters on dynamic programming[*] to describe models that, in fact, can be stochastic. The difference is that an elementary dynamic program is a representation of a decision process defined over a set of state and control variables. The model is used to determine optimal policies. In contrast, a stochastic process is aimed at predicting the behavior of a system rather than making optimal decisions. A stochastic dynamic program is the integration of a dynamic decision process governed by probabilistic state transitions. It is sometimes referred to as a Markov decision process when the transition probabilities satisfy the "memoryless" property discussed in Section 11.1.

From a modeling point of view, state and time can be treated as either continuous or discrete. Theoretical and computational considerations, however, argue in favor of the

---

[*]Supplements on the CD accompanying this text include chapters on dynamic programming models and methods.

discrete state case, so this will be our focus for the remainder of the book. The model of a stochastic process describes activities that culminate in events. The events cause a transition from one state to another. When activity durations are assumed to be continuous random variables, events occur in the continuum of time. Section 11.1 provides the vocabulary used in conjunction with a particular type of stochastic process known as a continuous-time Markov chain. Several examples are given to highlight the usefulness of the model. Later in the chapter we introduce discrete-time Markov chains, for which time is considered to be a discrete parameter.

11.1   CONTINUOUS-TIME MARKOV CHAINS

To illustrate the components of a stochastic process, we use the example of a single automated teller machine (ATM) located in the foyer of a bank. The ATM performs banking functions for arriving customers. Only one person at a time can use the machine, and that person is said to be *in service*. Other customers arriving when the machine is busy must wait in a single queue; these persons are said to be *in the queue*. Following the rule of *first-come-first-served* and assuming that the average arrival rate is less than the average service rate, a person in the queue will eventually step up to the ATM, execute a series of transactions, and then leave the system. The number of persons in the system at any point in time is determined, in general, by summing the number in service and the number in the queue.

Time

Many situations are conveniently described by a sequence of events that occur over time. For example, the number of persons in front of the ATM depends on the particular sequence of arrivals and departures.

In Figure 11.1, time is treated as a continuous parameter, but events occur at specific instances. We use t to represent time and subscript notation, such as t_0, t_1, and t_2, to identify the specific instances.

State

The state describes the attributes of a system at some point in time. For the ATM, the state indicates the number of persons in the system (either in service or in the queue). It may be limited to a maximum value determined by the physical size of the bank foyer, or it may be unlimited for practical purposes if a line is allowed to form outside the building. The minimum number, of course, is zero, which occurs when there are no customers in service. In this case, the machine is said to be *idle*.

Depending on the situation, the state may be very complex or very simple. To provide for this variety, we use a v-component vector, $\mathbf{s} = (s_1, s_2, \ldots, s_v)$, to describe the state. The ATM example allows a simple state definition with only one dimension. Here

$$\mathbf{s} = (k)$$

where k is the number in the system. To speak of the state at time t_i, we attach the subscript i to the state vector to get $\mathbf{s}_i$.

Figure 11.1   Events occurring in continuous time

It is often convenient to assign a unique nonnegative integer index to each possible value of the state vector. We call this index X and require that for each **s** there is a unique value of X. For the ATM example, an obvious assignment is $X = k$. For more complex state definitions, there may be several possible assignments describing different characteristics of the system. We use the notation X_t to represent the value of the index variable at time t. Because of the random nature of a stochastic process, the future realization of X_t is usually not predictable with certainty. Rather, X_t is a random variable. To understand the behavior of the system under study, it is necessary to describe the statistical properties of this random variable.

The set of all possible states is called the *state space* and is denoted by S. In general, a state space may be either continuous or discrete, but since the vast majority of applications in operations research can be modeled with a discrete state space, we limit our attention to that case only. For the ATM example, the set of all nonnegative integers can be used to describe the number of persons in the system at any instance, so

$$S = \{0, 1, 2, \ldots \}$$

This is an infinite state space and assumes that there is no limit to the number of persons that can be in the system. If the physical dimensions of the bank foyer restrict the number of customers to a maximum of K, the state space would be finite, so

$$S = \{0, 1, 2, \ldots, K\}$$

Activities and Events

The dynamic nature of a system is evidenced by its *activities*. Each activity has some time interval between its initiation and its completion, called the *duration*. An activity culminates in an *event*. In the ATM example, there are two activities, one associated with service and the other with arrivals. The service activity begins when a customer first appears in front of the ATM and ends when all the customer's banking transactions are completed and the customer departs. This event is typically referred to as a *service completion*. The duration of the activity that leads to this event is the *time for service*.

The arrival activity is the process that generates customers for the ATM and is a bit more abstract. It begins immediately after a customer arrives and ends with the arrival of the next customer. The event is an *arrival*, and the duration of the accompanying activity is the *time between arrivals*.

Symbolically, we use letters to designate activities and events. Because each activity leads to a unique event, we use the same letter for both the activity and the subsequent event. For the current example, a is used to represent the arrival activity as well as the event of an arrival, and d is used to represent the service activity as well as the event of service completion (and departure of the customer).

The duration of each activity is a random variable whose probability distribution may or may not be known. This is the probabilistic element of the stochastic process. The time for service, t_d, and the time between arrivals, t_a, may be described by their respective cumulative distribution functions should they be known.

$$F_d(t) \equiv \Pr\{t_d \le t\} \text{ and } F_a(t) \equiv \Pr\{t_a \le t\}$$

Here, $\Pr\{\tau \le t\}$ denotes the probability that the random variable τ is less than or equal to some number t.

Generally speaking, the set of possible activities, and thus the set of possible events, depends on the state of the system. For example, when there are no customers in the system, the service activity is not active so no departure can occur. If there is some upper limit on the number of customers who can wait in the queue, no arrival can occur when the queue is full. We use the notation $Y(\mathbf{s})$ to describe the set of possible activities (events) for state $\mathbf{s}$. For the ATM example, we have

$$Y(0) = \{a\}; \quad Y(k) = \{a, d\} \text{ for } 0 < k < K; \quad Y(K) = \{d\}$$

The cardinality of $Y(\mathbf{s})$ may be infinite or finite, depending on the situation. In what follows, the term *calendar* is used in reference to $Y(\mathbf{s})$. The set of activities that may occur while the system is in state $\mathbf{s}$ determines the calendar for the state.

Given the current state $\mathbf{s}$, one of the events in $Y(\mathbf{s})$ will occur next. This *next event* and the time at which it takes place have important effects on the state of the system. Let x be the next event and let t_x be the time at which it occurs. When the activity durations are random variables, x and t_x are random variables. By definition,

$$t_x = \text{Minimum}\{t_e : e \in Y(\mathbf{s})\}$$

where x is the event for which the minimum is obtained.

Transition

A transition is caused by an event and results in movement from one state to another. Thus, it always relates to two states. For convenience, we will say that a transition describes the movement from the *current* state to the *next* state. There are cases, however, in which the current state is allowed to be the same as the next state.

A state-transition network is used to describe the components of the stochastic model. In the diagram of the model, each state is represented by a circle or node and the possible transitions by directed arcs. Each arc originates at one state and terminates at another. The event leading to the transition is shown adjacent to the arc. Figure 11.2 depicts the network for the ATM example. An arrival causes a transition to the next higher state, and a departure causes a transition to the next lower state.

When the stochastic process is relatively simple, the transitions can be represented graphically as in Figure 11.2. For complex situations, we use a *transition function*. This function gives the value of the next state in terms of the current state and the next event. Let the current state be $\mathbf{s}$ and let the next event be x. Let $\mathbf{s}'$ be the next state determined by the transition function $T(\mathbf{s}, x)$, such that

$$\mathbf{s}' = T(\mathbf{s}, x)$$

When the state vector has more than one component, the transition function is multidimensional. It describes how each component of the state vector changes when the event occurs.

For the ATM example, the transition equation is

$$s' = s + 1 \text{ if } x = a$$
$$s' = s - 1 \text{ if } x = d$$

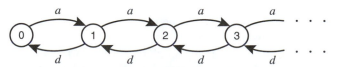

Figure 11.2 State-transition network for ATM example

The Process

The state, the activity and event definitions, the probability distributions for activity durations, and the state-transition network are sufficient to describe a stochastic process. In more formal terms, we have the following.

Definition 1: A *stochastic process* is a collection of random variables $\{X_t\}$, where t is a time index that takes values from a given set T.

Both X_t and T may be continuous or discrete, giving rise to four different categories of models (as mentioned, however, we do not discuss the case in which X_t is continuous). Later chapters provide solution methods for determining estimates of the statistical properties of $\{X_t\}$ when S is discrete. For the most part, we focus on processes that have the Markovian property, which is defined as follows.

Definition 2 (*Markovian property*)**:** Given that the current state is known, the conditional probability of the next state is independent of the states prior to the current state.

A system that has this property is said to be memoryless because the future realization depends only on the current state and in no way on the past. For a discrete-state and discrete-time stochastic process, the conditional probability of the next state ($X_{t+1} = j$) given the current state ($X_t = i$) and given all states prior to the current state (that is, given $X_0 = k_0$, $X_1 = k_1, \ldots, X_{t-1} = k_{t-1}$) is identical to the conditional probability of a specific next state given the present state. This can be expressed as

$$\Pr\{X_{t+1} = j \mid X_0 = k_0, X_1 = k_1, X_1 \neq k_1, \ldots, X_t = i\} = \Pr\{X_{t+1} = j \mid X_t = i\} \qquad (1)$$

for $t = 0, 1, \ldots$, and all possible sequences $i, j, k_0, k_1, \ldots, k_{t-1}$. Note that the uppercase letters represent the random variables while the lower case letters represent specific values. A similar equation can be written for the continuous version of the process.

Chapters 12 and 13 discuss discrete-time Markov chains, and Chapters 14 and 15 discuss continuous-time Markov chains, both of which are examples of discrete state-space stochastic processes. In the continuous-time case, we will see that the Markovian property is rather restrictive because all activity durations must follow the exponential probability distribution. The Markovian assumption is useful, however, in that it allows one to obtain analytic solutions for the statistical properties associated with the process. In Chapter 16, we present several types of queuing models that satisfy this assumption and hence are relatively easy to analyze.

11.2   REALIZATION OF THE PROCESS

Let us suppose that for a particular system the initial state is given and that its evolution is governed by known probability distributions. By recording the state of the system from one transition to the next as a function of time, one obtains a *realization* or *sample path* of the stochastic process. As mentioned, we represent the state by the random variable X_t, which is some measure of interest. In this section, we illustrate the transitional flow from one state to the next by first describing a deterministic process and then a related stochastic process.

A Deterministic Process

Consider a manufacturing system in which individual parts arrive at a robotic assembly station at 1-minute intervals. The robot requires 0.75 minute to service the part and does so whenever one is present. The first part arrives at time zero. For purposes of analysis, we record the number of parts either in service or in the queue. The network diagram in Figure 11.2 is applicable to this process, because the state definition, the types of activities, and the transition function are the same as they were for the ATM example.

The absence of randomness in the system implies that the assembly process is deterministic. The probability distributions for the activity times are presented in Table 11.1. They are necessarily degenerate owing to the deterministic nature of the process.

With the first part arriving at time 0, there is only one possible realization of this process, as shown in Figure 11.3. The graph plots the number in the system, n, as a function of time t. As one might imagine, the process admits a regular pattern.

Consider again the deterministic assembly system, but now assume that two parts are waiting for service at time 0. The response is shown in Figure 11.4. The system oscillates a bit before reaching the steady pattern that emerges after $t = 4$. The first portion of the realization depends on the initial condition and is termed the *transient response*. The pattern that ultimately emerges is the *steady-state response*. Similar results are often observed when randomness is introduced.

It is interesting to note the spike at $t = 3$ in Figure 11.4. When the system starts, there are two parts in the queue. At time 3, the third part arrives (or the fifth, considering the initial number). At exactly that time the fourth part is finished and ready to go ($4 \times 0.75 = 3$), so an arrival and a service completion occur simultaneously at $t = 3$. For continuous-time systems with randomness, the probability of two events occurring at exactly the same time is virtually zero, so such spikes would not be seen in the corresponding graphs.

A Stochastic Process

Continuing with the assembly example, suppose that the robot is successful in performing its task on the first attempt in only six out of 10 tries. When unsuccessful, it must perform the task again. The probability of success remains 0.6 for the second try. After two failures, the part is discarded. Parts arriving when the robot is busy must wait. Now the process is stochastic, because the service time is a random variable governed by the probability distribution presented in Table 11.2. The part is surely finished after 1.5 minutes, but the assembly operation may or may not be successfully completed within this period of time.

Table 11.1   Examples of Degenerate Probability Distributions

Time between arrivals	$\Pr\{t_a \leq t\} = \begin{cases} 0 \text{ for } t < 1\,\text{min} \\ 1 \text{ for } t \geq 1\,\text{min} \end{cases}$	Arrivals occur every minute.
Time for servicing part	$\Pr\{t_s \leq t\} = \begin{cases} 0 \text{ for } t < 0.75\,\text{min} \\ 1 \text{ for } t \geq 0.75\,\text{min} \end{cases}$	Processing takes exactly 0.75 min

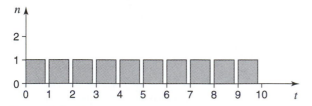

Figure 11.3   Realization of a deterministic process

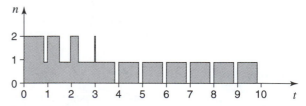

Figure 11.4   A deterministic process with a transient response

Table 11.2　Example of Discrete Probability Distribution

Time for servicing part	$\Pr\{t_s \leq t\} = \begin{cases} 0 & \text{for } t < 0.75\,\text{min} \\ 0.6 & \text{for } 0.75 \leq t < 1.5\,\text{min} \\ 1 & \text{for } t \geq 1.5\,\text{min} \end{cases}$

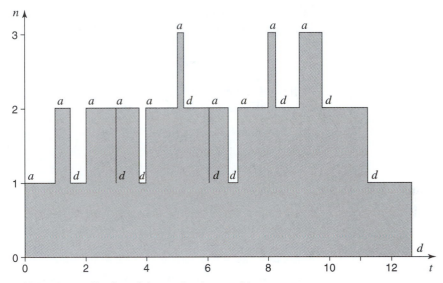

Figure 11.5　One realization of the stochastic assembly process

The state of the system is plotted in Figure 11.5 for one realization of the process. The particular sequence and timing of events that generated this plot were determined with a discrete event simulation model. If the simulation were run again, a different realization would most likely result.

For processes governed by probabilistic activities, it is difficult, if not impossible, to identify exactly the transient and steady-state portions of the realization. In fact, the entire realization depends to some extent on the initial state. For some situations, however, the statistical properties of the realization become less and less affected by the initial state as one considers intervals well beyond time 0. When the time of observation is sufficiently removed from the starting time so that the statistical properties are not affected by the initial state, we are observing the steady-state behavior of the system. When the process is observed near the starting time, we are observing the transient behavior. Most of the results in subsequent chapters are associated with the steady-state properties of the system.

11.3　DISCRETE-TIME MARKOV CHAINS

For some systems, events occur naturally in steps or at specific intervals of time. For example, a game might involve a series of moves followed by an assessment of the state after each move. In the game of craps, the steps of the process are the sequential throws of the dice; a change of state is induced as a result of each throw. Other examples involve time; at the beginning or end of an interval, a count is made of some relevant variable. Time is often divided into operational or accounting periods such as days, weeks, or months. Many events might occur during an interval, causing the state to change more than once; however, our only concern for the moment is with the states at the end of the interval. In such instances, we keep track not of which event occurs next but of which transition occurs during a single step or time period.

The model just outlined contains the elements of a discrete-time Markov chain and has desirable computational properties as well as a host of real-world applications. To make the model complete, we need to introduce the one-step transition probabilities.

Definition 3: Let $S = \{0, 1, \ldots\}$ be the state space for a discrete-time Markov chain and let p_{ij} be the probability of going from state i to state j in one transition. Then the *state-transition matrix* for the Markov chain is $\mathbf{P} = (p_{ij})$, where

$$\sum_{j \in S} p_{ij} = 1 \text{ for all } i \in S \text{ and } 0 \le p_{ij} \le 1 \text{ for all } i, j \in S \tag{2}$$

Definition 4: A discrete-time Markov chain (or Markov chain, for short) is a stochastic process with the following characteristics.

1. A discrete state space
2. Markovian property
3. The one-step transition probabilities, p_{ij}, from time n to time $n + 1$ remain constant over time (termed *stationary* transition probabilities)

If the state space is finite, then we have a *finite-state Markov chain*. In any case, a Markov chain is completely specified by the state definition, the transition matrix, and the set of unconditional probabilities for initial states. This knowledge allows us to determine the probability of being in any particular state at any future point in time. To distinguish a discrete-time Markov chain from a more general stochastic process, we use the following notation.

$$\{X_n\} \text{ for } n \in N = \{0, 1, 2, \ldots\}$$

Here, n rather than t denotes the time index.

In general, the state-transition network for a Markov chain has an arc passing from state i to state j when the associated transition probability is positive—i.e., $p_{ij} > 0$. This is illustrated for a three-state example in Figure 11.6. Note that if the process is in state 2, it always goes to state 0 at the next step.

$$\mathbf{P} = \begin{array}{c} \\ 0 \\ 1 \\ 2 \end{array} \begin{array}{ccc} 0 & 1 & 2 \\ \begin{bmatrix} 0.6 & 0.3 & 0.1 \\ 0.8 & 0.2 & 0 \\ 1 & 0 & 0 \end{bmatrix} \end{array}$$

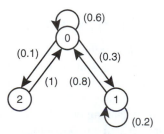

Figure 11.6   State-transition diagram for a simple Markov chain

Income Tax Audit

A variety of circumstances can trigger an audit of a tax return by the Internal Revenue Service (IRS). Table 11.3 shows the 10-year audit history of a particular taxpayer. Given these data and the assumption that the probability of an audit in year $n + 1$ is dependent only on whether or not there was an audit in year n, what is the Markov chain model that reflects IRS actions in this case?

In mathematical terms, the event of an audit can be represented by the discrete-state random variable X_n, which takes on one of two values. Let $X_n = 0$ if there is no audit in year n and let $X_n = 1$ if there is an audit. The collection of random variables $\{X_n\}$ then is a discrete-state stochastic process for all $n \in N = \{1, 2, \ldots\}$ with state space $S = \{0, 1\}$. For a 10-year period, $\{X_n\} = \{X_1, X_2, \ldots, X_{10}\}$ is a general representation of this stochastic process and $\{0, 0, 0, 1, 0, 0, 1, 1, 1, 0\}$ is a possible realization.

In light of the Markovian property holding, the conditional probability of an audit in year 11 can be stated according to Equation (1):

$$\Pr\{X_{11} = 1 \mid X_1 = 0, X_2 = 0, \ldots, X_{10} = 0\} = \Pr\{X_{11} = 1 \mid X_{10} = 0\}$$

Suppose that the transition matrix between year 10 and year 11 is given by

$$\mathbf{P} = \begin{bmatrix} p_{00} & p_{01} \\ p_{10} & p_{11} \end{bmatrix} = \begin{bmatrix} 0.6 & 0.4 \\ 0.5 & 0.5 \end{bmatrix}$$

In this case, row 1 ($i = 0$) represents a state of no audit in year 10 (i.e., $X_{10} = 0$) and row 2 ($i = 1$) represents an audit in year 10 ($X_{10} = 1$); column 1 ($j = 0$) represents no audit in year 11 ($X_{11} = 0$) and column 2 ($j = 1$) represents an audit in year 11 ($X_{11} = 1$).

The transition probability p_{00} indicates that the probability of not having an audit in year 11 ($j = 0$) given that no audit was performed in year 10 ($i = 0$) is 0.6. Mathematically,

$$p_{00} = \Pr\{X_{11} = 0 \mid X_{10} = 0\} = 0.6$$

Similarly,

$$p_{01} = \Pr\{X_{11} = 1 \mid X_{10} = 0\} = 0.4$$

Note that both properties of $\mathbf{P}$ specified in Equation (2) are satisfied by this matrix.

The Game of Craps

An occasional gambler loves to go to Las Vegas and play dice or craps, as it is known in the casinos. Not being naïve, she is aware that the house must have an advantage but would nevertheless like to know her probability of winning and losing.

In this game, the player rolls a pair of dice and sums the numbers showing. A total of 7 or 11 on the first roll wins for the player, whereas a total of 2, 3, or 12 loses. Any other number is called the point. The player then rolls the dice again. If she rolls the point number, she wins. If she throws a 7, she loses. Any other number requires another roll. The process continues until either a 7 or the point is thrown. The game is pure chance, devoid of

Table 11.3   Audit History for Income Tax Example

Year, n	1	2	3	4	5	6	7	8	9	10
Audit	No	No	No	Yes	No	No	Yes	Yes	Yes	No

strategy, so it should be straightforward to calculate the probability of winning and losing on each turn. Given that the payoff is one for one, the closer the probability of winning is to 0.5, the "fairer" the game is said to be.

In constructing a model, the natural interval is a roll of the dice. The state-transition network for the problem is shown in Figure 11.7. An activity is the roll of the dice culminating in an event represented by arcs in the diagram. The gambler begins at the node labeled "Start." The first roll leads to "Win," "Lose," or one of the states representing a point (P4,

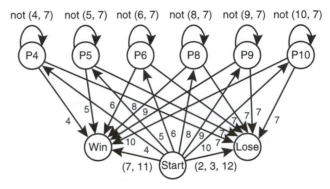

Figure 11.7 States and events for craps

	Start	Win	Lose	P4	P5	P6	P8	P9	P10
Start	0	0.222	0.111	0.083	0.111	0.139	0.139	0.111	0.083
Win	0	1	0	0	0	0	0	0	0
Lose	0	0	1	0	0	0	0	0	0
P4	0	0.083	0.167	0.75	0	0	0	0	0
P = P5	0	0.111	0.167	0	0.722	0	0	0	0
P6	0	0.139	0.167	0	0	0.694	0	0	0
P8	0	0.139	0.167	0	0	0	0.694	0	0
P9	0	0.111	0.167	0	0	0	0	0.722	0
P10	0	0.083	0.167	0	0	0	0	0	0.75

Figure 11.8 Transition matrix for the game of craps

P5, P6, P8, P9, P10). The process stops when the system enters either state Win or Lose. If the game were to begin again after reaching one of these terminal states, transitions back to the Start node would be included.

Using simple probability analysis concerning the throw of a pair of dice, we can determine the probabilities of the events shown in Figure 11.7. These are the transition probabilities, p_{ij}, of going from state i to state j, and are constant for each throw. For example, there are six ways of obtaining 7 and two ways of obtaining 11 out of 36 total possibilities (six for each die). Therefore, the probability of winning on the first roll is $8/36 = 2/9 \cong 0.222$. Note that the transition from Start to Win is accomplished by either of two mutually exclusive events, throw a 7 or throw an 11. The probability of the transition is the sum of the two event probabilities. The collection of the transition probabilities forms the transition matrix **P** shown in Figure 11.8, which, along with the state definitions, completely describes the Markov chain. We consider this example and justify these numbers in Chapter 12.

11.4   EXAMPLES OF STOCHASTIC PROCESSES

A surprising variety of dynamic systems in which uncertainty is present can be modeled as stochastic processes. In this section, we describe several generic situations and the accompanying state-transition diagrams. The examples are based primarily on situations that can be modeled as continuous-time stochastic processes, although we will see that these situations can be approximated numerically by discrete-time processes. Moreover, every continuous-time process has a closely related embedded discrete-time process. The primary objective in this section is to illustrate the definition of the state vector and the construction of the state-transition diagrams, which are central to the modeling process. Whenever the process has the Markovian property, the models can be analyzed using the methods described in Chapters 12 through 16.

A Single-Stage Process with a Single Worker

Consider an appliance technician who does in-home repairs. At each home the service time is a random variable governed by some probability distribution with a mean value of m_s. When finished, the technician must travel to the next job, where the travel time is governed by a second probability distribution with mean m_a. There are two activities: the service activity results in the departure event (d), and the travel activity results in the arrival event (a).

In this situation, there are only two states, the technician is traveling (state 0) or working (state 1). The arrival event triggers the service activity, and the service completion event triggers the travel activity. The state diagram appears in Figure 11.9. Only two states are necessary, because a queue is not permitted.

A Multistage Process with a Single Worker

Now we assume that the service activity of the appliance technician consists of a series of phases. Perhaps the technician must first gain entrance to the home, then locate the cause

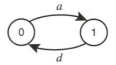

Figure 11.9   Single server, service triggers arrival, and arrival triggers service

of the problem, then find the necessary parts, and so on through a total of five phases. The process begins with the arrival and goes through each phase sequentially. When all five phases have been completed, the technician travels to the next home.

The state-transition diagram for this example is shown in Figure 11.10. The state is s, where $s = 0$ when the technician is traveling and $s = k$ when the technician is performing phase k, for $k = 1, \ldots, 5$. The activities are the traveling activity, which concludes with the arrival event a, and the k working phases. The completion of phase k results in the event d_k. An arrival causes a transition from state 0 to state 1. The completion of each phase moves the system to the next phase until the completion of phase 5 causes the technician to travel to the next job.

A Multistage Process with a Single Worker and a Queue

In the last two examples, we assumed that that the next customer was always available when the technician finished a job. We now assume that customers call into a dispatcher. We identify a new activity associated with the time between arrivals of calls. That activity culminates with the arrival of a call, a. When a call arrives while the technician is busy, the customer waits in a queue. To simplify the situation, we reduce the number of phases in the process to three. We include the travel time to the next job as the third phase of the job. Thus, d_3 now represents the event of arriving at the next job.

To incorporate this change in the model, we must associate with each state the number of customers in the queue plus the number receiving service. The following pair of state variables describes each state.

$$\mathbf{s} = (s_1, s_2), \text{ where } \begin{cases} s_1 = \text{number of customers in system} \\ s_2 = \text{current phase being performed} \end{cases}$$

When $\mathbf{s} = (0, 0)$, the technician is idle and waiting for the next job.

The state-transition network is depicted in Figure 11.11. The state space is infinite in this case, because there is no upper limit on the first state variable.

Queuing Model with Two Different Servers

Suppose that two machines are available to perform the same manufacturing operation. Although both can do the job equally well, machine 1 is more efficient than machine 2 and works faster. Machine 2 was put into service because of a sudden increase in demand. When an arrival occurs and both machines are idle, the job is assigned to machine 1. When machine 1 is busy and machine 2 is idle, the work is assigned to machine 2. When both machines are busy, the arrival waits in a first-come-first-served queue. Once a machine has started a job, the job remains on that machine until it is finished—that is, there is no preemption. The situation is depicted in Figure 11.12.

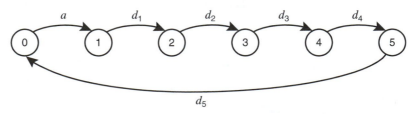

Figure 11.10   Multistage process in which the service completion triggers arrival

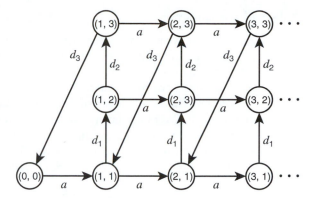

Figure 11.11   Multistage process with infinite queue

Three state variables are needed to model the process—two for the activities of the servers and one for the number in the queue. Let $\mathbf{s} = (s_1, s_2, s_3)$, where

$$s_i = \begin{cases} 0 & \text{if server } i \text{ is idle} \\ 1 & \text{if server } i \text{ is busy} \end{cases} \quad \text{for } i = 1, 2$$

and

$$s_3 = \text{number in queue}$$

We define event a as an arrival and events d_1 and d_2 as service completions for machines 1 and 2, respectively. The state-transition network for the situation is depicted in Figure 11.13. Once again, nodes correspond to states whereas arcs correspond to possible transitions. For convenience, the nodes are labeled sequentially and their unique state vector is placed adjacent to them.

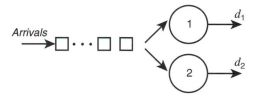

Figure 11.12   Queuing system with two different servers

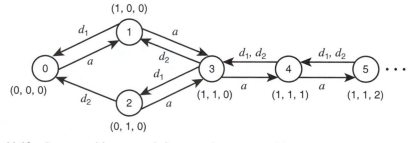

Figure 11.13   State-transition network for a queuing system with two servers

Matching Two Classes of Customers

Assume that two parts are assembled at a robotic workstation within a manufacturing facility. The parts arrive independently, but whenever one part of each type is available the assembly operation can take place. When only one of the pair is present or the workstation is busy, arriving parts must wait in a queue.

Here we have two classes of customers and a single server. Whenever there is at least one of each class of customers available, the server performs some function and releases the pair. For simplicity, assume that arrivals of a particular class are blocked or turned away whenever there are three of that class already waiting for service.

For this process we define two state variables:

$$\mathbf{s} = (s_1, s_2), \text{ where } s_i = \text{number in the queue of class } i$$

When we restrict the maximum of s_i to 3, the state space is

$$S = \{(s_1, s_2) : 0 \le s_1 \le 3, 0 \le s_2 \le 3\}$$

The events associated with the process are defined as follows.

a_i = arrival of a class i customer, $i = 1, 2$

d = completion of a service activity (assembly operation)

The network in Figure 11.14 shows the calendar and possible transitions for each state. Once again, the nodes are labeled sequentially and the state vector $\mathbf{s}_i$ for $i = 0, 1, \ldots$ is adjacent to node i. If the system is at node 11, where $\mathbf{s}_{11} = (3, 2)$, for example, a service completion triggers a transition to node 6, where $\mathbf{s} = (2, 1)$. Regarding blocking, if the system is at node 13, where $\mathbf{s}_{13} = (1, 3)$, for example, all class 2 arrivals are turned away.

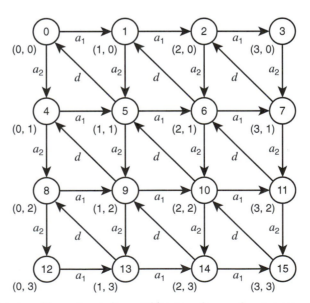

Figure 11.14   State-transition network for matching two classes of customers

11.5   ASSESSMENT OF STOCHASTIC MODELS

Waiting lines, Markov chains, and related stochastic models are everyday occurrences. We experience them in our normal routines, from standing in traffic and checking out at the supermarket to playing games of chance. They also occur extensively within an economic, industrial, or social context, sharing common features of people or objectives arriving at a service facility, enduring delays, and transiting through some sort of organizational structure.

In contrast to the deterministic models that comprise a large portion of operations research, models of stochastic processes are not directly amenable to optimization. For example, queuing theory is chiefly concerned with determining, for a given service facility and proposed mode of operation, certain performance characteristics, such as average waiting time, average queue length, or average idle time for servers. These characteristics provide input to the decision-making process about the design of the facility—a process that often reduces to the economic evaluation of a small number of possible alternatives, with the "best" solution being found by enumeration.

Most real-life applications involving random systems that evolve over time are highly complex and may defy any formal analytic treatment. When this is the case, simulation is the only approach that offers any hope of capturing their essential features. In other cases, the methodologies available for analyzing stochastic processes may only serve as a first approximation, providing some quantitative information about the behavior of the more complex system. To some extent, then, the models and methods presented in the upcoming chapters are primarily intended to provide qualitative insights into the behavior of stochastic processes rather than to give formulas for predicting the future with 100% accuracy.

In Chapter 18 we discuss simulation, which is one of the most flexible, comprehensive, and frequently used OR tools. To simulate is to duplicate the dynamic behavior of some aspect of a system, real or proposed, by substituting the properties of another system for the critical properties of the system being studied. In operations research, we usually build a descriptive mathematical model to represent the properties of the system being studied. This model is then used to trace step by step how the system responds to various inputs. Through simulation, the analyst has at his or her disposal a laboratory for observation and experimentation, which has long been part of the scientific method in the physical, biological, and medical sciences. With the help of high-speed computers, the viability of proposed policies or new designs can be explored and compared with relative ease, whereas practical evaluations that would take years to accomplish by real-life observation would be either very costly or impossible to carry out. Simulation often provides the only vehicle for analyzing and designing stochastic systems.

EXERCISES

1. A military air base has K airplanes. The planes fail independently, and each has its own repair crew. Denote the event of a failure by a and the event of a repair by d. Construct the state-transition network for the corresponding stochastic process. Define all terms.

2. A truck has four rolling tires and two spares. The tires in use are identical, and each may fail while the truck is in motion. The event of a single failure is denoted by a. When a tire fails, it is replaced with a spare, provided one is available. The truck cannot operate when less than four good tires are available. Draw the separate state-transition networks under the following operating rules. Denote the repair operation by d.

   (a) Whenever a tire fails, it is sent to the shop (assume that can be done instantaneously). The truck continues to operate as long as it has four good tires. Repaired tires are returned to the truck one at a time.

(b) Failed tires are sent to the shop when no spares remain. The truck continues to operate as long as four good tires are available. Repaired tires are returned to the truck one at a time.

(c) No repair is possible. The truck continues to operate until there are fewer than four good tires available.

3. An apartment service has hired a specialist to match students as roommates. Individuals arrive independently, fill out the necessary forms, and wait for a match. The arrival event is *a* and the matching event is *d*. A matched pair departs from the system. Describe the stochastic process that determines the number of individuals awaiting service and draw the corresponding state-transition network.

4. A queuing system has two identical servers. Separate queues form in front of each server, with arriving customers joining the shortest line. Customers do not switch lines once they are in the system (i.e., there is no *jockeying*). The completion of a service is called *d*. The arrival of a customer is called *a*. Describe the stochastic process that governs the numbers of customers in the two lines. Draw the corresponding state-transition network.

5. A production system has three workstations in series. Each station is in one of three states: idle with no product (*i*), idle holding a product (*h*), and working on a product (*w*). There is sufficient backlog so that whenever station 1 is idle with no product it is immediately provided with one to work on. When station 3 completes its operations, the product leaves the system. Stations 1 and 2 cannot pass a product on to the next station unless the station is idle and not holding. This restriction represents a version of blocking. Construct the state-transition network that describes the overall process.

6. A self-paced course consists of six units, numbered 1 through 6, that must be taken in sequence. The time it takes to study a unit is a random variable. After completing a unit, the student takes a test and has a probability *p* of passing. If the student does not pass, he must repeat the unit and then retake the test. Passing the test allows him to go on to the next unit. When all six units have been passed, the course is complete. Describe the stochastic process under the following conditions and draw the corresponding state-transition network. Treat each part separately.

(a) There is no limit on the number of tests the student may take.

(b) The student fails the course when he fails any three tests (i.e., fails three different units, fails the same unit three times, or any combination).

(c) Whenever the student fails more than two tests, he must start the course over again.

(d) The student must take a test each week. If he fails a test he must take another test on the same unit in the following week. To pass the course, the student must finish in 12 weeks; otherwise, he fails.

7. In each case, show the state-transition network that describes the process.

(a) A single machine makes a product. To make one unit of the product, an operator must load the machine with the raw material and then process it. Raw material is continuously available. The machine can be in one of two states: loading (0) and processing (1). There are two activities— the loading activity (*a*) and the processing activity (*b*).

(b) Expand the situation in part (a) in the following way. When a product has been completed, the operator performs a test. The test time is included in the operating time. The product will pass with probability *p*. When the product fails the test, the operator must spend additional time repairing the failure. Let *c* represent the repair activity which is always successful. When repair is complete, the operator loads the raw material for another unit. Now the operator has three states: loading (0), operating (1), and repairing (2).

(c) Change the situation in part (b). Now the repair activity is not always successful. The probability of success is *q*. When it is successful, the operator begins loading raw material. When the repair activity is not successful, it is repeated. The repair continues in this fashion until it is successful.

(**d**) Change the situation in part (c) such that the repair activity is repeated only two times. The product is discarded after the second unsuccessful attempt.

8. Answer the following questions with respect to the IRS auditing problem described in Section 11.3.

   (**a**) What are the implications and plausibility of each of the three assumptions (properties) needed to treat this problem as a Markov chain?

   (**b**) In estimating **P**, would you prefer more than 10 observations? Explain.

   (**c**) Suppose that a year has passed. If an audit were performed in year 11, how would you update **P**? Would you say that the history of the stochastic process is relevant when **P** is not known a priori? Explain. If so, does collecting more data necessarily violate the Markovian property? Explain.

   (**d**) Let f_{ij} represent the number of observations in cell (i, j) of a contingency table (see Tables 12.4 and 12.5) for **P**. Write a general expression for calculating all p_{ij}.

9. The number of students arriving at a university's financial aid office was recorded over a period of 100 hours. The table indicates the number that arrived each hour.

Students per hour	Number of times recorded
0	10
1	20
2	30
3	15
4	15
5	10
	Total: 100

   (**a**) Compute the average arrival rate per hour.

   (**b**) Compute the average interarrival time in minutes.

   (**c**) Construct a frequency diagram of arrivals. Try to fit the resultant curve to some well-known distribution. Use statistical methods to determined the goodness of fit (see Section 18.7).

10. The table provides data associated with a repair operation for 550 customers.

Hours per repair	Number of times recorded
1	110
2	165
3	165
4	85
5	25
	Total: 550

   (**a**) Compute the average repair time in hours.

   (**b**) Compute the average number of repairs per day (24 hours).

   (**c**) Construct a frequency diagram of service times. Try to fit the resultant curve to some well-known distribution. Use statistical methods to determined the goodness of fit (see Section 18.7).

11. A manufacturing facility uses two identical machines that are attended by a single operator. Each machine requires the operator's attention at random points in time. The probability that the machine requires service in a period of 5 minutes is $p = 0.4$. The operator is able to service a machine in 5 minutes. Let us approximate this situation by assuming that when service is required, it is always initiated at the beginning of a 5-minute period. Construct a state-transition network for this problem and find the transition matrix associated with it. Define all terms.

12. Discrete-time Markov chains are frequently used as models for infection. A nonfatal tropical disease takes three weeks to run its course. The chance of a healthy person contracting the disease in any week is 0.1. Two drugs are available to shorten the length of the disease. The first drug can be used only during week 1, and it cures 50 percent of the patients immediately. The second drug must be used in week 2, and it also has an immediate cure rate of 50 percent.

(a) Define the states of the system and write the transition matrix when no drugs are taken.

(b) Write the transition matrix when the drug options are included. If both drugs are used when appropriate, what fraction of the infected population will experience a reduction in the length of the disease?

13. Gwynne Hodler bought several thousand shares of a high-flying internet stock at $38, and has given orders to her broker to sell the stock as soon as its price rises to $40 or more or falls to $37 or less. From observations about this stock over the last few weeks, Gwynne estimates that the probability of a price increase of $1 is 0.5 and that the probability of a price decline of $1 is 0.2 per day. Find the transition matrix for this process. Which states are transient states, and which states are absorbing (see Section 13.2 for definitions)?

BIBLIOGRAPHY

Brigandi, A.J., D.R. Dargon, M.J. Sheehan, and T. Spencer III, "AT&T's Call Processing Simulator (CAPS) Operational Design for Inbound Call Centers," *Interfaces*, Vol. 24, No. 1, pp. 6–28, 1994.

Çinlar, E., *Introduction to Stochastic Processes*, Prentice-Hall, Englewood Cliffs, NJ, 1975.

Feldman, R.M. and C. Valdez-Flores, *Applied Probability & Stochastic Processes*, PWS, Boston, 1996.

Golabi, K. and R. Shepard, "Pontis: A System for Maintenance Optimization and Improvement of US Bridge Networks," *Interfaces*, Vol. 27, No. 1, pp. 71–88, 1997.

Gross, D. and C.M. Harris, *Fundamentals of Queuing Theory*, Third Edition, Wiley, New York, 1998.

Heyman, D.P. and M.J. Sobel, "Stochastic Models," *Handbook in Operations Research*, Vol. 2, North-Holland, Amsterdam, 1990.

Mehdi, J., *Stochastic Models in Queuing Theory*, Academic Press, Boston, 1991.

Puterman, M.L., *Markov Decision Processes*, Wiley, New York, 1991.

Ross, S.M., *Introduction to Probability Models*, Fifth Edition, Academic Press, San Diego, 1993.

Ross, S.M., *Introduction to Probability and Statistics for Engineers and Scientists*, Second Edition, Academic Press, San Diego, 1999.

Wolff, R.W., *Stochastic Modeling and the Theory of Queues*, Prentice-Hall, Englewood Cliffs, NJ, 1989.

Chapter 12

Discrete-Time Markov Chains

In this chapter, we investigate a finite-state stochastic process in which the defining random variables are observed at discrete points in time. Here, time is viewed as a parameter. As we saw in Chapter 11, when the future probabilistic behavior of the process depends only on the present state regardless of when the present state is measured, the resultant model is called a *discrete-time Markov chain*, or simply *Markov chain*, for short. Unlike most stochastic processes, Markov chains have very agreeable properties that allow for easy study. They are often used to approximate quite complex physical systems, even when it is clear that the actual behavior of the system being analyzed may depend on more than just the present state, or when the number of states is not really finite.

The name of this field derives from the probabilist A. A. Markov, who published a series of papers starting in 1907 that laid the theoretical foundations for finite-state problems. The foundations for infinite-state problems were developed by A. N. Kolmogorov in the mid-1930s. An interesting example predating their work was investigated by British scientist F. Galton and British mathematician H. W. Watson. The question they addressed centered on when and with what probability a given family name would become extinct. In the nineteenth century, the propagation or extinction of aristocratic family names was important because land and titles stayed with the name.

The Galton–Watson process can be described as follows. Generation 0 starts with a single ancestor. Generation 1 consists of all the sons of the ancestor (daughters were not included in the model because only the sons carried the family name). The next generation consists of all the sons of each male offspring from the first generation (i.e., grandsons of the ancestor), generations continuing ad infinitum or until extinction. The assumption is that for each male in a generation, the probability of having zero, one, two, or more sons is given by some specified (and unchanging) probability mass function, and that the mass function is identical for all individuals of any generation. Although Galton and Watson did not use Markov chains in their analysis, they were able to determine the likelihood of extinction and how many generations it would take.

The characteristic that makes the Galton–Watson process a Markov chain is the fact that in any generation, the number of males in the next generation is completely independent of the numbers of males in previous generations, as long as the number of males in the current generation is known. It is processes such as this (in which the future is independent of the past, given the present) that we now study.

To develop a model of a Markov chain, we need to define the system states S and specify the one-step transition matrix $\mathbf{P}$. We concentrate in this chapter on the modeling process. Our ultimate goal is to answer questions related to the steady-state behavior of the Markov chain, the probability of residing in a particular state after a fixed number of transitions, and the expected cost of operating the system being modeled. Using formulas derived in Chapter 13, we can also compute the following information: the n-step transition matrix, transient and steady-state probability vectors, absorbing state probabilities, and first passage probability distributions. Integrating these results with economic data leads directly to a framework for systems design and optimal decision making under conditions of uncertainty.

12.1 TRANSITION MATRIX

The first applications of Markov models were developed in the physical sciences. Of particular note were studies on the behavior of gas particles in confined spaces, the growth of biological populations, and the forecasting of weather patterns. In more recent years, the focus has shifted to managerial applications, including the analysis of inventory and queuing systems, the development of replacement and maintenance policies for machines, brand loyalty in marketing, time series of economic data such as stock prices, hospital systems including the movement of coronary patients, and expected payout of life insurance policies, to name a few. In this section, we highlight several prominent applications to illustrate the computational requirements and versatility of the model. In each case, a finite, discrete-time Markov chain is assumed. Continuous-time processes are investigated in Chapters 14 and 15.

A (discrete-time) Markov chain is a special case of a stochastic process whose state space is referenced by the index set

$$S = \{0, 1, \ldots, m - 1\}$$

This set identifies all possible states in which the system may be observed. The notation X_n is used to represent the state of the system at time n. X_n is a random variable restricted to the discrete values in the set S. The subscript n is the time or step index and takes the values in the set $N = \{0, 1, 2 \ldots \}$, which can be measured in any appropriate unit such as seconds, minutes, or weeks (or, say, moves in a game). The general time interval is called the *period*.

The system begins in state X_0, and over successive time periods jumps to the states X_1, X_2, and so on. These jumps are governed entirely by transition probabilities that are assumed to be time invariant. We denote the transition probability of going from i to j as p_{ij}, and assume that the process has the Markovian property. This means that each p_{ij} depends only on the current state i and not on the particular path the process takes to reach state i.

Under these restrictions, the stochastic process is entirely defined by the $m \times m$ transition matrix denoted by $\mathbf{P} = (p_{ij})$. The rows and columns of $\mathbf{P}$ are labeled 0 through $m - 1$. An element of the matrix, p_{ij}, is the probability that, given that the system is in state i at some time n, it will be in state j at time $n + 1$.

$$\mathbf{P} = \begin{bmatrix} p_{00} & p_{01} & p_{02} & \cdots & p_{0,m-1} \\ p_{10} & p_{11} & p_{12} & \cdots & p_{1,m-1} \\ \vdots & \vdots & \vdots & \cdots & \vdots \\ p_{m-1,0} & p_{m-1,1} & p_{m-1,2} & \cdots & p_{m-1,m-1} \end{bmatrix}$$

Since the system must always be in one of the m states, every row of the transition matrix must sum to 1.

$$\sum_{j=0}^{m-1} p_{ij} = 1, \quad i = 0, 1, \ldots, m-1 \tag{1}$$

Also, because the elements of **P** are probabilities, we must have

$$0 \le p_{ij} \le 1, \quad i, j = 0, 1, \ldots, m-1 \tag{2}$$

To model a problem as a Markov chain, one must first describe the states of the system and then define the transition matrix **P**. A variety of examples are given throughout this chapter. Section 13.1 describes the mathematical properties of **P** in more detail.

Computer Repair

A real estate office that relies on two aging computers for word processing is suffering high costs and great inconvenience as a result of chronic machine failures. It has been observed that when both computers are working in the morning, there is a 30% chance that one will fail by evening and a 10% chance that both will fail. If it happens that only one computer is working at the beginning of the day, there is a 20% chance that it will fail by the close of business. If neither computer is working in the morning, the office sends all work to a typing service. In this case, of course, no machines will fail during the day.

Service orders are placed with a local repair shop. The computers are picked up during the day and returned the next morning in operating condition. The one-day delay occurs when either one or both machines are under repair.

Time

For a Markov chain, the system is observed at discrete points in time that are indexed with the nonnegative integers. The initial point of observation is denoted by $n = 0$. It is important to identify carefully the exact moment when the system is to be observed in relation to the events described by the problem statement. For our example, the system is to be observed in the morning after any repaired computers have been returned and before any failures have occurred during the current day.

State

The state describes the situation at a point in time. Because the states are required to be discrete, they can be identified with nonnegative integers 0, 1, 2, 3, and so on. There may be a finite or an infinite number of states. For this introductory discussion, we concentrate on the finite case and use $m - 1$ as the maximum state index[1]. The sequence of random variables $X_0, X_1, X_2, \ldots$ is the stochastic process that describes the system as it evolves over time. Each X_n can take one of m values.

Depending on the situation, the state may be very complex or very simple. We use a v-dimensional vector of state variables to define the state.

$$\mathbf{s} = (s_1, s_2, \ldots, s_v)$$

---

[1] For some applications it is more natural to number the states starting at 1 rather than 0. The state indices will then be 1, 2, ..., m. All definitions can be easily modified to accommodate this numbering scheme.

In constructing a model, it must be made clear what the one-to-one relationship is between the possible state vectors and the nonnegative integers used to identify a state in a Markov chain. We denote the state associated with index i as s_i. Depending on the context, i typically ranges from 0 to $m - 1$ or from 1 to m. The state definition must encompass all possible states; however, the system can reside in only one state at any given time.

The computer repair example allows a simple state definition with only one component ($v = 1$).

> $s = (k)$, where k is the number of computers that have failed when the system is observed in the morning. Note that the system is observed after the repaired units have been delivered but before any failures have occurred.

The list of possible states for the example appears in Table 12.1. The value of m is 3. In this case, the state index is conveniently identical to the variable defining the state.

Events

To understand the behavior of a Markov chain, it is necessary to identify the events that might occur during a single time period and to describe their probabilities of occurrence. Generally, the set of possible events and their probabilities depend on the state s.

Given some current state, s, at the beginning of a period and one or more events occurring during the period, the system will be in some new (*next*) state, s', at the beginning of the next period. This occurrence is called a transition. Recall that for the model of a continuous-time stochastic process, an event causes a change in the state. Here, one or more events may occur within the period, and by observing them we must identify the resulting new state at the beginning of the next period.

For the computer example, we list in Table 12.2 the current states together with the set of possible events that might occur during the day. Given the current state and the problem description, one must be able to determine the probability of every possible transition for the upcoming period.

State-Transition Matrix

The following transition matrix for the computer example is easily constructed from the data in Table 12.2. The state indices are shown above and to the right of the matrix.

Table 12.1   States for the Computer Repair Example

Index	s	State definitions
0	$s = (0)$	No computers have failed. The office starts the day with both computers functioning properly.
1	$s = (1)$	One computer has failed. The office starts the day with one computer working and the other in the shop until the next morning.
2	$s = (2)$	Both computers have failed. All work must be sent out for the day.

Table 12.2   Events and Probabilities for the Computer Repair Example

Index	Current state	Events	Probability	Next state
0	$\mathbf{s} = (0)$	Neither computer fails.	0.6	$\mathbf{s}' = (0)$
		One computer fails.	0.3	$\mathbf{s}' = (1)$
		Both computers fail.	0.1	$\mathbf{s}' = (2)$
1	$\mathbf{s} = (1)$	The remaining computer does not fail and the other is returned.	0.8	$\mathbf{s}' = (0)$
		The remaining computer fails and the other is returned.	0.2	$\mathbf{s}' = (1)$
2	$\mathbf{s} = (2)$	Both computers are returned.	1.0	$\mathbf{s}' = (0)$

$$\mathbf{P} = \begin{array}{c} \\ \begin{array}{ccc} 0 & 1 & 2 \end{array} \\ \left[\begin{array}{ccc} 0.6 & 0.3 & 0.1 \\ 0.8 & 0.2 & 0 \\ 1 & 0 & 0 \end{array}\right] \begin{array}{c} 0 \\ 1 \\ 2 \end{array} \end{array}$$

We note the following, which must be true for all transition matrices.

a. All rows sum to 1. This follows from the logical requirement that the states define every possible condition for the system.

b. Every element has a value between 0 and 1. This follows from the definition of probability.

State-Transition Network

The information in the transition matrix can also be displayed in a directed network that has a node for each state and an arc passing from state i to state j if p_{ij} is nonzero. Figure 12.1 depicts the network for the computer repair example. Transition probabilities are shown adjacent to the arcs. A requirement is that the sum of all probabilities leaving a node must be 1.

Complete Model

A Markov chain model requires specification of the following.

a. The times when the system is to be observed.

b. The discrete states in which the system may be found. The list of states must be exhaustive. In addition, a one-to-one correspondence must be prescribed between the states with the nonnegative integers.

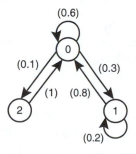

Figure 12.1   State-transition network for computer repair example

c. The state-transition matrix showing the transition probabilities in one time interval. Transition probabilities must satisfy the Markovian property and be time invariant.

Although the model structure is easily stated, it is not always easy to realize. For example, one might propose time and state definitions [parts (a) and (b) above] for which the Markovian property is not satisfied. This may sometimes be remedied by a more complex state definition.

Because a Markov chain model is very general, it can be used to describe a variety of stochastic systems. In many cases, however, the number of states required for adequate definition of the model is very large. As with dynamic programming, this "curse of dimensionality" frequently arises when we try to identify all possible states.

Variations on the Computer Repair Problem

Simple variations in the situation require that the model be changed in some fashion. The easiest changes involve the transition probabilities. If one finds that for a given state definition it is impossible to determine logically the transition probabilities, it is usually necessary to enlarge the state space. The requirement in every case is that every transition probability p_{ij} be entirely determined by information available in the definitions of states i and j.

Technician Works on One Machine at a Time

Now consider the case in which there is only a single technician and so only one machine can be repaired per day (assume that it takes a full day to diagnose, fix, and test a unit). If two machines are at the shop, the second must wait until the first is repaired.

With zero or one computer in the shop, this variation does not affect the situation, so the analyses of states 0 and 1 are the same. With two computers in the shop (state 2), the technician can fix only one, so the system will surely return to state 1 at the next time period. The transition matrix has been changed to accommodate this scenario.

$$\mathbf{P} = \begin{matrix} & \begin{matrix} 0 & \;\; 1 & \;\; 2 \end{matrix} \\ \begin{bmatrix} 0.6 & 0.3 & 0.1 \\ 0.8 & 0.2 & 0 \\ 0 & 1 & 0 \end{bmatrix} & \begin{matrix} 0 \\ 1 \\ 2 \end{matrix} \end{matrix}$$

Repair Operation Takes Two Days

Assume again that there is one technician, but now she requires two days to perform all the steps needed for the repair. In this situation, it is no longer possible to allow the states to be identified entirely by the number of failed machines. We must also know how many days a failed machine has been in the shop. For purposes of state definition, we order the machines on the basis of when they entered the service facility.

Let $s_1 =$ the number of days the first machine has been in the shop.

Let $s_2 =$ the number of days the second machine has been in the shop.

For s_1, we assign a value of 0 if the machine has not failed, 1 if it is in the first day of repair, and 2 if it is in the second day of repair. The complete set of definitions is given in Table 12.3. As we can see, each state represents a relatively complex situation with more information than simply the number of failed machines. The information provided is necessary to determine the transition matrix.

The easiest way to construct this matrix is to consider each state in turn and ask what can possibly happen in the next time period. If the current state is $\mathbf{s} = (0, 0)$, only failures can occur. One failure leads to $(1, 0)$ and two failures lead to $(2, 0)$ with probabilities $p_{01} = 0.3$ and $p_{03} = 0.1$, respectively. If the current state is $\mathbf{s} = (1, 0)$, the next state will be $(2, 0)$ if the second computer does not fail and $(2, 1)$ if it does. From $\mathbf{s} = (1, 1)$, the state will surely progress to $(2, 1)$ the next day, so $p_{34} = 1$. The remaining probabilities can be analyzed in a similar fashion. The full transition matrix is as follows.

$$\mathbf{P} = \begin{array}{c} \\ \\ \\ \\ \\ \\ \end{array} \begin{array}{ccccc} 0 & 1 & 2 & 3 & 4 \\ \begin{bmatrix} 0.6 & 0.3 & 0.0 & 0.1 & 0.0 \\ 0.0 & 0.0 & 0.8 & 0.0 & 0.2 \\ 0.8 & 0.2 & 0.0 & 0.0 & 0.0 \\ 0.0 & 0.0 & 0.0 & 0.0 & 1.0 \\ 0.0 & 1.0 & 0.0 & 0.0 & 0.0 \end{bmatrix} & \begin{array}{c} 0 \\ 1 \\ 2 \\ 3 \\ 4 \end{array} \end{array}$$

Repair Operation Takes Two Days but Computers Fail Independently

We continue with the computer repair scenario, but now we assume that failures are independent events. The probability that either machine will fail during the day (assuming that it is working at the beginning of the day) is 0.2. The probability that it does not fail is 0.8.

As such, if both computers are working at the beginning of the day, the number that fail, call it N, is a random variable with a binomial distribution. For parameter $p \in (0, 1)$ and positive integer n, the probability mass function of a binomial distribution is given by

$$\Pr\{N = k\} = P(k) = \frac{n!}{k!(n-k)!}(p)^k(1-p)^{n-k} \text{ for } k = 0, 1, \ldots, n$$

In this example, we have $n = 2$ and $p = 0.2$. Substituting yields

$$P(k) = \frac{2!}{k!(2-k)!}(0.2)^k(0.8)^{2-k} \text{ for } k = 0, 1, 2$$

Table 12.3 State Definitions for Two-Day Repair Times

Index	s	State definitions
0	$\mathbf{s} = (0, 0)$	No machines have failed.
1	$\mathbf{s} = (1, 0)$	One machine has failed and is in the first day of repair.
2	$\mathbf{s} = (2, 0)$	One machine has failed and is in the second day of repair.
3	$\mathbf{s} = (1, 1)$	Both machines have failed and both are in the first day of repair.
4	$\mathbf{s} = (2, 1)$	Two machines have failed and one is in the second day of repair.

so $P(0) = 0.64$, $P(1) = 0.32$, and $P(2) = 0.04$. These probabilities replace the three nonzero elements for state 0; the remaining values in the preceding **P** matrix remain the same, yielding

$$
\mathbf{P} =
\begin{array}{c}
\begin{array}{ccccc}
0 & 1 & 2 & 3 & 4
\end{array} \\
\left[
\begin{array}{ccccc}
0.64 & 0.32 & 0.0 & 0.04 & 0.0 \\
0.0 & 0.0 & 0.8 & 0.0 & 0.2 \\
0.8 & 0.2 & 0.0 & 0.0 & 0.0 \\
0.0 & 0.0 & 0.0 & 0.0 & 1.0 \\
0.0 & 1.0 & 0.0 & 0.0 & 0.0
\end{array}
\right]
\begin{array}{c}
0 \\ 1 \\ 2 \\ 3 \\ 4
\end{array}
\end{array}
$$

This scenario illustrates that individual elements of the state-transition matrix may be determined by a special discrete probability distribution.

Brand Switching

Markov chains for brand switching behavior have been used for many years as diagnostic tools for devising marketing strategies. To illustrate a specific formulation, consider the brand switching behavior depicted in Table 12.4 for a group of 500 consumers. According to the first row, of the 100 consumers who purchased brand 1 in week 26, 90 repurchased brand 1 in week 27, seven switched to brand 2, and three switched to brand 3. Note, however, that five customers switched from brand 2 to brand 1 and 30 switched from brand 3 to brand 1 (according to the first column); hence, for brand 1, the loss of 10 customers $(7 + 3)$ was more than compensated for by the gain of 35 customers $(5 + 30)$, yielding a net gain of 25 customers from one week to the next (125 versus 100). The market share for brand 1 increased from 0.2 (100/500) to 0.25 (125/500).

Contingency tables of this type are useful because they not only show the net changes and market shares but also identify the sources of change. For example, brand 1 showed a net loss of two customers $(5 - 7)$ to brand 2 and a net gain of 27 customers $(30 - 3)$ from brand 3. Moreover, such tables directly yield the current one-step matrix of transition probabilities, as shown in Table 12.5. Note that in this case **P** represents a sample estimate of the underlying or true transition matrix. Further note that p_{ii} is a reflection of the "holding power" of brand i because it represents the probability that a consumer will purchase brand i given that the preceding purchase was brand i. Similarly, p_{ij} reflects the "attraction power" of brand j in that it is an estimate of the probability that brand j will be purchased next given that brand i was the preceding purchase.

If we define the state variable X_n as the brand purchased in week n, then $\{X_n\}$ represents a discrete state and a discrete parameter stochastic process, where $S = \{1, 2, 3\}$ and $N = \{0, 1, 2, \ldots\}$. If we can establish that $\{X_n\}$ exhibits the Markovian property and that

Table 12.4 Number of Consumers Switching from Brand i in Week 26 to Brand j in Week 27

Brand (j)	1	2	3	Total
(i)				
1	90	7	3	100
2	5	205	40	250
3	30	18	102	150
Total	125	230	145	500

Table 12.5   Empirical Transition Probabilities (p_{ij})

Brand (j)	1	2	3
(i)			
1	$\dfrac{90}{100} = 0.90$	$\dfrac{7}{100} = 0.07$	$\dfrac{3}{100} = 0.03$
2	$\dfrac{5}{250} = 0.02$	$\dfrac{205}{250} = 0.82$	$\dfrac{40}{250} = 0.16$
3	$\dfrac{30}{150} = 0.20$	$\dfrac{18}{150} = 0.12$	$\dfrac{102}{150} = 0.68$

P is stationary, then a Markov chain should be a reasonable representation of aggregate consumer brand switching behavior.

In this context, Markov analysis can be useful in the following types of studies, to name a few.

1. Predict market shares at specific future points in time
2. Assess rates of change in market shares over time
3. Predict market share equilibria (if they exist)
4. Assess the specific effects of marketing strategies when market shares are unfavorable
5. Evaluate the process for introducing new products

Exercises at the end of the chapter ask the reader to address several of these issues

Income Tax Audit

Let us return to the IRS example presented in Section 11.3. Table 12.6 repeats the 10-year audit history of a particular taxpayer. If we assume that the probability of an audit in year $n + 1$ depends only on whether or not there was an audit in year n, we have a Markov chain. As previously defined, the discrete state variable X_n for the Markov chain takes on one of the two values; $X_n = 0$ if there is no audit in year n and $X_n = 1$ if there is an audit, $n = 1$, 2, . . .

If a finite-state Markov chain is to be assumed, the transition probabilities must remain constant over time. Typically, the "true" transition matrix is unknown to the analyst. The usual procedure for specifying **P** first assumes that the observed (historical) stochastic process is a random sample from the true process. If this is the case, then there is theoretical justification for basing point estimates of transition probabilities on the empirical probabilities obtained from a contingency table. To illustrate, the historical stochastic process given by

$$\{X_1, X_2, \ldots, X_{10}\} = \{0, 0, 0, 1, 0, 0, 1, 1, 1, 0\}$$

Table 12.6   Audit History of a Taxpayer

Year, n	1	2	3	4	5	6	7	8	9	10
Audit	No	No	No	Yes	No	No	Yes	Yes	Yes	No

Table 12.7 Contingency Table for Historical Audit Data

From a specific state in a given year	To a specific state in the following year		Row sum
	No audit	Audit	
No audit	3	2	5
Audit	2	2	4

can be described by such a contingency table. Three times, for example, the state of the system went from "no audit" to "no audit" in successive years (that is, from $X_1 = 0$ to $X_2 = 0$, from $X_2 = 0$ to $X_3 = 0$, and from $X_5 = 0$ to $X_6 = 0$). Hence, a 3 is placed in cell (1, 1) in Table 12.7. Other entries are found in a similar fashion.

By definition, p_{00} is the probability of "no audit" in a succeeding year given that there is no audit in the current year. From the contingency table, out of five "no audit" current years (the sum of row 1), three succeeding "no audit" years followed. Empirically, then $p_{00} = 3/5$. Similarly, $p_{01} = 2/5$, which means that two "audit" years followed "no audit" years out of a possible five. For the entire matrix, we have

$$\mathbf{P} = \begin{bmatrix} p_{00} = 3/5 & p_{01} = 2/5 \\ p_{10} = 2/4 & p_{11} = 2/4 \end{bmatrix}$$

12.2 MULTISTEP TRANSITIONS

The matrix $\mathbf{P}$ provides direct information about one-step transition probabilities but is much more versatile and can also be used to calculate the probabilities for transitions involving more than one step. The formulas used in this section are derived in Section 13.3.

Soliciting Potential Clients

A young stockbroker working for a Wall Street firm is expected to generate his own client base. Each week he receives a list of prospects that he uses to solicit business. The sales department prepares three different types of lists—call them a, b, and c—each representing a different group of customers served by the firm. The use of a new list every week can be modeled as a Markov chain with state space $S = \{a, b, c\}$ where the random variable X_n corresponds to the working list in week n. The stockbroker's history over the last few months suggests that the transition matrix can be approximated as follows.

$$\mathbf{P} = \begin{array}{c} \\ a \\ b \\ c \end{array} \begin{array}{c} \begin{array}{ccc} a & b & c \end{array} \\ \begin{bmatrix} 0 & 0.50 & 0.50 \\ 0.75 & 0 & 0.25 \\ 0.75 & 0.25 & 0 \end{bmatrix} \end{array}$$

The zero entries along the diagonal indicate that the broker receives a different list every week. If he is working with list b in the current week, the probability of receiving list a after one step (the next week) is 0.75. That is, the probability of going from state b to state a in one step is $p_{ba} = 0.75$. What is the probability that the broker will be in state a after two weeks? Figure 12.2 illustrates the paths that go from b to a in two steps (some of the paths shown have zero probability).

To compute the probability, we need to sum over all possible routes. In other words, we must perform the following calculations.

$$
\begin{aligned}
\Pr\{X_2 = a \mid X_0 = b\} &= \Pr\{X_1 = a \mid X_0 = b\} \times \Pr\{X_2 = a \mid X_1 = a\} \\
&\quad + \Pr\{X_1 = b \mid X_0 = b\} \times \Pr\{X_2 = a \mid X_1 = b\} \\
&\quad + \Pr\{X_1 = c \mid X_0 = b\} \times \Pr\{X_2 = a \mid X_1 = c\} \\
&= p_{ba}p_{aa} + p_{bb}p_{ba} + p_{bc}p_{ca}
\end{aligned}
$$

The final equation identifies the terms required to compute element (b, a) for $\mathbf{P}^{(2)}$, where $\mathbf{P}^{(2)} = \mathbf{PP}$. In particular,

$$
\Pr\{X_2 = a \mid X_0 = b\} = p_{ba}^{(2)}
$$

In general, define the n-step transition matrix as

$$
\mathbf{P}^{(n)} = \mathbf{P}^{(n-1)}\mathbf{P}
$$

Then the n-step transition probability $p_{ij}^{(n)}$ is element (i, j) of $\mathbf{P}^{(n)}$.

Let $\{X_n : n = 0, 1, \ldots\}$ be a Markov chain with state space S and state-transition matrix $\mathbf{P}$. Then, for i and $j \in S$ and $n = 1, 2, \ldots,$

$$
\Pr\{X_n = j \mid X_0 = i\} = p_{ij}^{(n)}
$$

where the right-hand side represents the ijth element of the matrix $\mathbf{P}^{(n)}$. This definition is Property 1 from Section 13.3.

It is important to remember that the notation $p_{ij}^{(n)}$ does not represent the element p_{ij} raised to the nth power. Rather, it is the ijth element of the transition matrix raised to the nth power. For our example, the squared matrix is

$$
\mathbf{P}^{(2)} = \begin{array}{c} \\ \\ \\ \\ \end{array}
\begin{array}{ccc}
a & b & c \\
\end{array}
\left[
\begin{array}{ccc}
0.75 & 0.125 & 0.125 \\
0.1875 & 0.4375 & 0.375 \\
0.1875 & 0.375 & 0.4375
\end{array}
\right]
\begin{array}{c}
a \\
b \\
c
\end{array}
$$

Element (b, a) shows that there is a 0.1875 probability that the broker will receive list a two weeks after receiving list b.

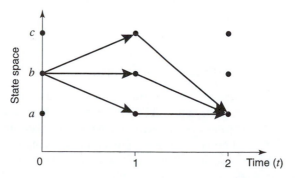

Figure 12.2   Possible paths of a two-step transition for a three-state Markov chain

IRS Revisited

For another illustration, we continue the IRS auditing example from Section 12.1 with the transition matrix given by

$$\mathbf{P} = \begin{bmatrix} 0.6 & 0.4 \\ 0.5 & 0.5 \end{bmatrix}$$

and $S = \{0, 1\}$. The first row represents the conditional probability of possible states in the *next* period given that state 0 (no audit) is observed in the *current* period. Thus, given "no audit" in the current period, $p_{00} = 0.6$ and $p_{01} = 0.4$ are the probabilities of "no audit" and "audit," respectively, in the next period. The second row has a similar interpretation.

To determine all conditional probabilities two periods hence, we simply square $\mathbf{P}$, as follows.

$$\mathbf{P}^{(2)} = \mathbf{P} \times \mathbf{P}$$

$$= \begin{bmatrix} 0.6 & 0.4 \\ 0.5 & 0.5 \end{bmatrix} \times \begin{bmatrix} 0.6 & 0.4 \\ 0.5 & 0.5 \end{bmatrix}$$

$$= \begin{bmatrix} 0.56 & 0.44 \\ 0.55 & 0.45 \end{bmatrix}$$

The resultant matrix indicates, for example, that the probability of no audit 2 years from now given that there was no audit in the current year is $p_{00}^{(2)} = 0.56$, and that the probability of an audit 2 years from now given that there was an audit in the current year is $p_{11}^{(2)} = 0.45$. How would you interpret $p_{01}^{(2)}$ and $p_{10}^{(2)}$?

Three years hence, the conditional probabilities are given by

$$\mathbf{P}^{(3)} = \mathbf{P}^{(2)} \times \mathbf{P}$$

$$= \begin{bmatrix} 0.56 & 0.44 \\ 0.55 & 0.45 \end{bmatrix} \times \begin{bmatrix} 0.6 & 0.4 \\ 0.5 & 0.5 \end{bmatrix}$$

$$= \begin{bmatrix} 0.556 & 0.444 \\ 0.555 & 0.454 \end{bmatrix}$$

Thus, $p_{01}^{(3)} = 0.444$ is the probability that there will be an audit in 3 years given that there was no audit in the current year. Similar calculations can be made for future years. In general, the *n*-step transition probability $p_{ij}^{(n)}$ for *n* periods from now is found by calculating $\mathbf{P}^{(n)}$.

Unconditional versus Conditional Probabilities

Just as in the case of the one-step transition probability p_{ij} between states i and j, the *n*-step transition probability $p_{ij}^{(n)}$ is a conditional value. The probability of being in state j after the Markov chain goes through *n* steps is statistically dependent on the initial state i. If the unconditional or absolute probability of being in state j after *n* steps, which we denote by $q_j(n)$, is desired, then we must perform the following calculation

$$\mathbf{q}(n) = \mathbf{q}(0)\mathbf{P}^{(n)} \text{ or } \mathbf{q}(n) = \mathbf{q}(n-1)\mathbf{P} \tag{3}$$

where $\mathbf{q}(n) = (q_0(n), q_1(n), \ldots, q_{m-1}(n))$ is the row vector of unconditional probabilities for all *m* states after *n* transitions, and $\mathbf{q}(0)$ is the row vector of *initial* unconditional probabilities. The components of $\mathbf{q}(n)$ are sometimes referred to as *transient* probabilities. The equations used in this section are described in Section 13.4.

In the brand switching example, the column totals in Table 12.4 represent market shares for the current week (week 27). Alternatively, we can say that the absolute probability is 125/500 (or 0.25) that a consumer selected at random from among the 500 in the sample has purchased brand 1 in the current week. Similarly, unconditional probabilities that brands 2 and 3 were purchased in the current week are 0.46 and 0.29, respectively. These values represent the initial unconditional probabilities, so

$$\mathbf{q}(0) = (0.25, 0.46, 0.29)$$

To predict market shares for, say, week 29 (that is, 2 weeks into the future), we simply apply Equation (3) with $n = 2$, $\mathbf{q}(0)$ as in the preceding equation, and $\mathbf{P}$ as given in Table 12.5.

$$\mathbf{q}^{(2)} = (0.25, 0.46, 0.29) \begin{bmatrix} 0.90 & 0.07 & 0.03 \\ 0.02 & 0.82 & 0.16 \\ 0.20 & 0.12 & 0.68 \end{bmatrix}^2$$

$$= (0.25, 0.46, 0.29) \begin{bmatrix} 0.8174 & 0.1240 & 0.0586 \\ 0.0664 & 0.6930 & 0.2406 \\ 0.3184 & 0.1940 & 0.4876 \end{bmatrix}$$

$$= (0.327, 0.406, 0.267)$$

Thus, the expected market shares 2 weeks hence are 32.7% for brand 1, 40.6% for brand 2, and 26.7% for brand 3.

With the help of Equation (3), we can determine the unconditional probability of being in any one of the m states after a fixed number of steps. The next question we wish to address concerns the long-run or limiting behavior of the system. We would like to know whether or not the system reaches a "steady state," and if so, how to compute the unconditional probability of being in any state $j \in S$ as $n \to \infty$.

12.3   STEADY-STATE SOLUTIONS

A natural question to ask is what happens to the transition probabilities $p_{ij}^{(n)}$ as n gets large? In fact, for many cases of interest, they approach a steady-state value. To illustrate the nature of the steady-state (time-independent) probabilities, consider the calculations in Table 12.8 for the IRS auditing problem. Let us first look at $p_{ij}^{(n)}$ as n increases. For example, the series given by p_{00} ($p_{00}^{(1)} = 0.6$, $p_{00}^{(2)} = 0.56$, $p_{00}^{(3)} = 0.556$, and so forth) indicates that these transition probabilities change by smaller and smaller increments at each successive step. This implies that $p_{ij}^{(n)}$ is asymptotically approaching a steady-state value as n gets large.

Now focus on the rows of $\mathbf{P}^{(n)}$ for n increasing. As we can see, they become identical. In particular, the rows of $\mathbf{P}^{(5)}$ are the same up to four decimal places. This illustrates the curious fact that the probability of being in any future state becomes independent of the initial state as time progresses. Moreover, from our previous observation, this probability is converging to its steady-state value, call it π_j, either from above (if $p_{ij} > \pi_j$) or from below (if $p_{ij} < \pi_j$). Figure 12.3 depicts the two cases. Here the curves are monotonic, but, in general, the transition probabilities may oscillate around their steady-state values.

For this example, the probability of finding the Markov chain in state j after a "large" number of transitions tends toward the value given by π_j. Because this tendency manifests itself regardless of the initial state or initial probability distribution, it follows that π_j is the *unconditional steady-state* probability for state j.

Table 12.8   The n-step Transition Matrix for IRS Example

Time, n	Transition matrix, $\mathbf{P}^{(n)}$
1	$\begin{bmatrix} 0.6 & 0.4 \\ 0.5 & 0.5 \end{bmatrix}$
2	$\begin{bmatrix} 0.56 & 0.44 \\ 0.55 & 0.45 \end{bmatrix}$
3	$\begin{bmatrix} 0.556 & 0.444 \\ 0.555 & 0.445 \end{bmatrix}$
4	$\begin{bmatrix} 0.5556 & 0.4444 \\ 0.5555 & 0.4445 \end{bmatrix}$
5	$\begin{bmatrix} 0.55556 & 0.44444 \\ 0.55555 & 0.44445 \end{bmatrix}$

Although the IRS and brand switching examples exhibit convergence, this is not always the case. For the finite-state case, steady-state probabilities will exist if every state can be reached from every other state in a finite number of transitions and if the process is aperiodic. When these conditions hold, the steady-state probabilities are independent of the initial state and are not very difficult to compute. Sufficient conditions for the existence of steady-state probabilities are discussed more extensively in Section 13.2.

The steady-state probabilities are the limits of the n-step probabilities.

$$\pi_j = \lim_{n \to \infty} \Pr\{X_n = j \mid X_0 = i\} = \lim_{n \to \infty} p_{ij}^{(n)} \text{ for all } i \text{ and } j$$

We use the m-dimensional row vector $\pi = (\pi_0, \pi_2, \dots, \pi_{m-1})$ to denote the steady-state probabilities. When the steady-state probabilities exist, they may be computed by solving the following system of linear equations. (These equations come from Property 2 in Section 13.5.)

$$\pi = \pi \mathbf{P} \qquad (4)$$

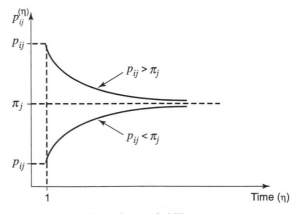

Figure 12.3   Asymptotic behavior of transient probability

$$\sum_{j=0}^{m-1} \pi_j = 1 \tag{5}$$

To see how this is done, let us return to the brand switching example. Since the brands are identified by the numbers 1, 2, and 3, we define the steady-state probabilities by the vector, $\pi = (\pi_1, \pi_2, \pi_3)$. From Equation (4), we get

$$(\pi_1, \pi_2, \pi_3) = (\pi_1, \pi_2, \pi_3) \begin{bmatrix} 0.90 & 0.07 & 0.03 \\ 0.02 & 0.82 & 0.16 \\ 0.20 & 0.12 & 0.68 \end{bmatrix}$$

Performing the matrix multiplication, we obtain three equations in three unknowns.

$$\pi_1 = 0.90\pi_1 + 0.02\pi_2 + 0.20\pi_3$$
$$\pi_2 = 0.07\pi_1 + 0.82\pi_2 + 0.12\pi_3$$
$$\pi_3 = 0.03\pi_1 + 0.16\pi_2 + 0.68\pi_3$$

Furthermore, from Equation (5), the following must hold.

$$\pi_1 + \pi_2 + \pi_3 = 1$$

This gives us a total of four equations in three unknowns. Since $\pi = (0, 0, 0)$ represents a trivial solution to the first three equations that is invalidated by the fourth, it follows that there is one redundant equation among the first three. Arbitrarily discarding the third equation and solving the remaining three simultaneously yields (to three significant digits) $\pi_1 = 0.474$, $\pi_2 = 0.321$, and $\pi_3 = 0.205$.

These results indicate that, over time, the market share of brand 1 will increase from its present value of 0.25 to its long-run stable value of 0.474. The market shares of brands 2 and 3 will erode, respectively, from 0.46 to 0.321 and from 0.29 to 0.205. Therefore, if present conditions continue, we can expect market share gains for brand 1 at the expense of losses for the other two brands. Note, however, that under steady-state conditions customers will continue to switch according to the stationary transition matrix. It is the absolute probabilities (market shares) that will change over time and finally stabilize.

Two additional points need to be made at this time. First, steady-state predictions are never achieved in actuality because of a combination of (1) errors in estimating **P**, (2) changes in **P** over time, and (3) changes in the nature of dependence relationships among the states. As in many areas where uncertainty is present, however, the use of steady-state values is an important diagnostic tool for the decision maker. Second, not all transition matrices lend themselves to the analysis of steady-state properties as presented here. In fact, steady-state probabilities may not exist for certain types of Markov chains. We explain this in more detail in Section 13.5.

As a second illustration of the calculations, consider the stockbroker example. The steady-state probabilities are the solutions to the following linear equations.

$$
\begin{array}{rcrcrcl}
 & & 0.75\,\pi_b & + & 0.75\,\pi_c & = & \pi_a \\
0.5\,\pi_a & & & + & 0.25\,\pi_c & = & \pi_b \\
0.5\,\pi_a & + & 0.25\,\pi_b & & & = & \pi_c \\
\pi_a & + & \pi_b & + & \pi_c & = & 1
\end{array}
$$

Once again, noting that the first three constraints are linearly dependent, one can be dropped. Solving the remaining three equations yields

$$\pi_a = 0.428, \quad \pi_b = 0.286, \quad \pi_c = 0.286$$

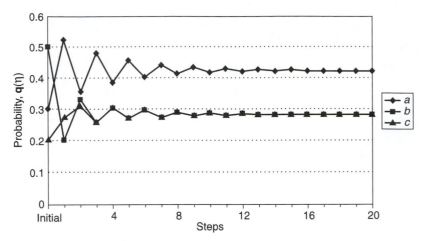

Figure 12.4 Transient probabilities as time progresses

Figure 12.4 plots the transient probabilities $\mathbf{q}(n)$ when the process starts with $\mathbf{q}(0) = (0.3, 0.5, 0.2)$. In this case, each component oscillates about its steady-state value.

12.4 ECONOMIC ANALYSIS

For modeling purposes, we consider two kinds of economic effects: (1) those incurred when the system is in a specified state and (2) those incurred when the system makes a transition from one state to another. Depending on the application, it may be appropriate to include one or both of these measures in the model. In this section, we use the general results from Section 13.6.

Revenue (or Cost) as a Function of State

To illustrate the revenue case, consider the stockbroker example. We repeat previously computed results for reference.

$$
\mathbf{P} = \begin{array}{c} \\ \\ \\ \end{array}\begin{array}{ccc} a & b & c \end{array} \\
\mathbf{P} = \begin{bmatrix} 0 & 0.50 & 0.50 \\ 0.75 & 0 & 0.25 \\ 0.75 & 0.25 & 0 \end{bmatrix}, \quad \mathbf{P}^{(2)} = \begin{bmatrix} 0.75 & 0.125 & 0.125 \\ 0.1875 & 0.4375 & 0.375 \\ 0.1875 & 0.375 & 0.4375 \end{bmatrix}
$$

$$
\boldsymbol{\pi} = (0.428, \ 0.286, \ 0.286)
$$

Suppose that every week spent pursuing potential clients on list a yields net revenues of $1700, every week spent on list b yields $1450, and every week spent on list c yields $1600. We group this information into a state revenue vector.

$$
\mathbf{C} = (c_a, c_b, c_c)^{\mathrm{T}} = (1700, 1450, 1600)^{\mathrm{T}}
$$

In different contexts, we may use $\mathbf{C}$ to denote the cost or some other economic measure associated with a state, but here we use it to describe revenue.

One might then ask, what would the stockbroker's expected revenue be during the nth week if he started with list a in the first week? The answer to this question comes from the definition of expected value. In the following discussion, we identify $f(X_n)$ as the net revenue if the system is in state X_n. Since X_n is a random variable, $f(X_n)$ is also a random variable. Assuming that the initial state is specified by the known probability vector $\mathbf{q}(0)$, the expected return in week n is

$$E[f(X_n)] = c_a q_a(n) + c_b q_b(n) + c_c q_c(n)$$

In more general terms, this is

$$E[f(X_n) \mid \mathbf{q}(0)] = \mathbf{q}(n)\mathbf{C} \tag{6}$$

Replacing $\mathbf{q}(n)$ by its equivalent from Equation (3), we have

$$E[f(X_n) \mid \mathbf{q}(0)] = \mathbf{q}(0)\mathbf{P}^{(n)}\mathbf{C}$$

For our example, state a is the initial state, so the expected return during week n is

$$E[f(X_n)] = p_{aa}^{(n)} \times 1700 + p_{ab}^{(n)} \times 1450 + p_{ac}^{(n)} \times 1600$$

The expected return during, say, the second week, given that the stockbroker starts with list a, is obtained by substituting the probability values from the first row of $\mathbf{P}^{(2)}$.

$$E[f(X_2) \mid X_0 = a] = 0.75 \times 1700 + 0.125 \times 1450 + 0.125 \times 1600$$
$$= 1656.25$$

When steady-state probabilities exist, we can use them to compute the expected revenue per week (see Property 6 in Section 13.6).

$$\lim_{n\to\infty} \frac{1}{n} \sum_{s=0}^{n-1} f(X_s) = \sum_{i\in S} \pi_i C_i = \pi\mathbf{C} \tag{7}$$

The steady-state return per week for the stockbroker is

$$\pi\mathbf{C} = (0.428, 0.286, 0.286)(1700, 1450, 1600)^T = 1599.9.$$

Revenue (or Cost) as a Function of Transition

As an extension of this example, suppose that the stockbroker incurs a setup cost each time he switches from one list to another. These costs are arrayed in the matrix $\mathbf{C}^R = (c_{ij}^R)$, where i and j range from 0 to $m - 1$. To include both state and transition effects in the analysis, we redefine the state revenue vector as $\mathbf{C}^S = (c_0^S, c_1^S, \ldots, c_{m-1}^S)^T$. For our example, the two matrices appear as follows.

$$\mathbf{C}^S = \begin{bmatrix} 1700 \\ 1450 \\ 1600 \end{bmatrix}, \quad \mathbf{C}^R = \begin{bmatrix} 0 & -200 & -300 \\ -250 & 0 & -300 \\ -200 & -250 & 0 \end{bmatrix}$$

The $\mathbf{C}^S$ vector shows the revenues for each state, and the $\mathbf{C}^R$ matrix shows the costs of transition as negative revenues. Since a transition occurs during each step of the Markov chain, the two matrices can be combined to obtain a net revenue vector $\mathbf{C} = (c_0, \ldots, c_{m-1})^T$. Given that the system is in state i, the one-step probability of next being in state j is given by the ith row of matrix $\mathbf{P}$. The expected cost of being in state i is then

$$c_i = c_i^S + \sum_{j=0}^{m-1} c_{ij}^R p_{ij} \tag{8}$$

We call $\mathbf{C}$ the expected state revenue vector. The summation in Equation (8) represents the scalar product of the ith row of $\mathbf{C}^R$ and the ith row of $\mathbf{P}$. Because $\mathbf{C}$ can always be computed from its two constituent matrices and $\mathbf{P}$, it is only necessary to refer to $\mathbf{C}$ in the analysis.

Using Equation (8), the new expected state revenue vector $C = (c_a, c_b, c_c)^T$ is

$c_a = \$1700 - \$0 \times 0 - \$200 \times 0.5 - \$300 \times 0.5 = \$1450$
$c_b = \$1450 - \$250 \times 0.75 - \$0 \times 0 - \$300 \times 0.25 = \$1187.5$
$c_c = \$1600 - \$200 \times 0.75 - \$250 \times 0.25 - \$0 \times 0 = \$1387.5$

Using the new value of C, the expected return during the second week given that the stockbroker starts with list a is

$$E[f(X_2) \mid X_0 = a] = 0.75 \times 1450 + 0.125 \times 1187.5 + 0.125 \times 1387.5$$
$$= 1409.4$$

Considering both revenue and setup costs, the long-run average revenue per week is

$$\pi C = (0.428, 0.286, 0.286)(1450, 1187.5, 1387.5)^T = 1357$$

Discounted Cash Flow

In many situations it is necessary for the decision maker to take into account the time value of money. Letting r be the interest rate per period, the discount factor $\alpha = 1/(1 + r)$ gives the present value of one dollar received one period from now.

To illustrate, consider a country road that degrades over time. The road can be classified into any of four states depending on the quality of the ride experienced by users: new, degraded, poor, and very poor. The quality of the road is measured at the beginning of each year. The road surface degrades over time because of weather and traffic, factors that cannot be known with certainty; however, a rule of the county government is that any road that has a very poor quality measure at the beginning of one year must be returned to the new state by the beginning of the next year. Based on models of road surface degradation, we obtain the following transition matrix for the four states. We number the states 0 for new, 1 for degraded, 2 for poor, and 3 for very poor. Note that the road quality at best decreases from year to year, but when the road reaches state 3 (very poor), it is returned to state 0 (new) at the beginning of the next year.

$$P = \begin{bmatrix} 0.3 & 0.4 & 0.2 & 0.1 \\ 0 & 0.2 & 0.4 & 0.4 \\ 0 & 0 & 0.2 & 0.8 \\ 1 & 0 & 0 & 0 \end{bmatrix}$$

For our example, we identify two cost components. The first is the state cost C^S, which represents the out-of-pocket expenses incurred annually by users owing to degraded road quality. The second is the transition cost C^R, which has a single nonzero element, the cost of repairing the road when it goes from state 3 to state 0. The two matrices are shown here together with the combined matrix C obtained from Equation (8). The interest rate for the county is 10%.

$$C^S = \begin{bmatrix} 0 \\ 200 \\ 600 \\ 2000 \end{bmatrix}, \quad C^R = \begin{bmatrix} 0 & 0 & 0 & 0 \\ 0 & 0 & 0 & 0 \\ 0 & 0 & 0 & 0 \\ 10{,}000 & 0 & 0 & 0 \end{bmatrix}, \quad C = \begin{bmatrix} 0 \\ 200 \\ 600 \\ 12{,}000 \end{bmatrix}$$

To determine the total expected discounted cost (revenue) for a Markov chain, we make use of the following equation stated in Property 7 in Section 13.6. The total expected discounted return is given by

$$E\left[\sum_{n=0}^{\infty}\alpha^n\left(f(X_n)\mid X_0=i\right)\right]=([\mathbf{I}-\alpha\mathbf{P}]^{-1}\mathbf{C})_i \tag{9}$$

where the subscript on the right-hand side refers to the ith component of the vector obtained after performing the matrix operations. In applying Equation (9), it will be assumed that all cash flows take place at the beginning of the period. If the cash flows were at the end, it would be necessary to multiply Equation (9) by α.

For an interest rate of 10%, we compute $\alpha = 1/(1 + 0.1) \approx 0.909$ and

$$[\mathbf{I}-\alpha\mathbf{P}]^{-1}\mathbf{C} = \left(\begin{bmatrix} 1 & 0 & 0 & 0 \\ 0 & 1 & 0 & 0 \\ 0 & 0 & 1 & 0 \\ 0 & 0 & 0 & 1 \end{bmatrix} - 0.909 \begin{bmatrix} 0.3 & 0.4 & 0.2 & 0.1 \\ 0 & 0.2 & 0.4 & 0.4 \\ 0 & 0 & 0.2 & 0.8 \\ 1 & 0 & 0 & 0 \end{bmatrix}\right)^{-1} \begin{bmatrix} 0 \\ 200 \\ 600 \\ 12{,}000 \end{bmatrix}$$

$$= \begin{bmatrix} 0.727 & -0.364 & -0.182 & -0.091 \\ 0 & 0.818 & -0.364 & -0.364 \\ 0 & 0 & 0.818 & -0.727 \\ -0.909 & 0 & 0 & 1 \end{bmatrix}^{-1} \begin{bmatrix} 0 \\ 200 \\ 600 \\ 12{,}000 \end{bmatrix} = \begin{bmatrix} 31{,}948 \\ 35{,}027 \\ 37{,}127 \\ 41{,}044 \end{bmatrix}$$

Thus, the present value of all future costs is $31,948 when the initial condition of the road under consideration is "new" and $41,044 when its surface is "very poor" and ready to be replaced.

12.5 APPLICATIONS

Light Bulb Replacement

A bowling center has just purchased a new outdoor sign containing 1000 light bulbs. Although business might increase as a result of better visibility in the neighborhood, the manager is worried about maintenance and cost. She wonders how many of the bulbs on average will have to be replaced each month and how much the budget for the repairs should be. The bowling center is owned by an entertainment company with an OR group, so she decides to give it a try.

You work for the OR group, and your boss asks you to provide technical help to the bowling center manager. This problem is similar to replacement-type problems you have analyzed in the past in which the time of failure is uncertain and there are a variety of maintenance options. You call the manager and learn that there is quite a bit of data concerning bulb failure. She provides you with maintenance reports for the old sign from which you construct an estimate of the probability distribution of the time to failure. Since failed bulbs are replaced monthly, you use monthly time intervals for the distribution. The results are summarized in Table 12.9. Columns correspond to the age of a bulb at the beginning of a month, so when $t = 0$, the bulb has just been placed in service and has a 0.5 probability of failing in the first month. Similarly, when $t = 3$, the bulb is 3 months old and has a 0.1 probability of failing in the fourth month.

The failure probability and cumulative probability rows are derived directly from the failure data. The last row of the table is the probability of failure in month t, given that the bulb had not previously failed. The formula for this probability is

$$f(0 \mid n \ge 0) = p(0), \quad p(t \mid n \ge t) = p(t)/(1 - F(t - 1)) \text{ for } t > 1$$

The data suggest that the bulbs have an infant mortality problem. There is an even chance that a bulb will out in the first month. The likelihood of a failure decreases in the

Table 12.9   Probability Data for Light Bulb Example

Age in months, t	0	1	2	3	4
Failure probability, $p(t)$	0.5	0.1	0.1	0.1	0.2
Cumulative probability, $F(t)$	0.5	0.6	0.7	0.8	1.0
Conditional probability of failure, $f(t \mid n \geq t)$	0.5	0.2	0.25	0.333	1.0

second, third, and fourth months. Nevertheless, a bulb is sure to fail before it is 5 months old.

To analyze this problem, you assume that age is the only variable affecting the likelihood of bulb failure. You define the state of the process to be the age of the bulb, which can take a value of 0, 1, 2, 3, or 4 months. Accordingly, you let $S = \{0, 1, 2, 3, 4\}$. Now consider a particular location holding a bulb of age t. At the beginning of the next month (after maintenance), the location will hold a bulb of age $t + 1$ if the bulb has not failed, or a bulb of age 0 if the bulb failed and was replaced with a new one. This information is described by the transition probabilities for the location. The probability of a transition from state t to state 0 is p_{t0}, and the probability of a transition from state t to state $t + 1$ is $p_{t, t+1}$. For this case

$$p_{t0} = f(t \mid n \geq t), \ p_{t, t+1} = 1 - f(t \mid n \geq t)$$

Because a bulb can only be replaced or increase in age by 1 month, all other transition probabilities from state t are zero. The following transition matrix $\mathbf{P}$ gives the transition probabilities p_{ij} for all states i and j. The corresponding state-transition network is shown in Figure 12.5.

$$\mathbf{P} = \begin{bmatrix} 0.5 & 0.5 & 0 & 0 & 0 \\ 0.2 & 0 & 0.8 & 0 & 0 \\ 0.25 & 0 & 0 & 0.75 & 0 \\ 0.333 & 0 & 0 & 0 & 0.667 \\ 1 & 0 & 0 & 0 & 0 \end{bmatrix} \begin{matrix} \text{New} \\ \text{1 mo.} \\ \text{2 mo.} \\ \text{3 mo.} \\ \text{4 mo.} \end{matrix}$$

New 1 mo. 2 mo. 3 mo. 4 mo.

This transition matrix together with the state definition comprise the Markov chain model. The example illustrates several characteristics that, as we have seen, must be true for all transition matrices. The elements of the matrix are probabilities, so they must all fall between 0 to 1. Because the columns of the matrix represent all possible states of the system, row

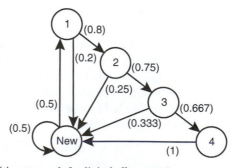

Figure 12.5   State-transition network for light bulb example

probabilities must sum to 1. Implicit in the construction of the matrix is the requirement that a transition probability depends only on the current state and the next state and not on the path used to reach the current state. Once again, this is called the Markovian property.

Economic Issues

There is a cost for performing the maintenance task of replacing light bulbs that depends on the number of bulbs that fail during the month. How can we represent the costs or benefits of this situation?

You learn that it costs $0.10 to inspect a bulb and $2 to replace it if it has failed. You incorporate these data into the model by specifying two economic components: the state cost vector $\mathbf{C}^S$, which specifies the cost per period the system is resident in a particular state, and the transition cost matrix $\mathbf{C}^R$, which specifies the cost for every transition. For our example, each component of the state cost matrix is $0.10, because every bulb is inspected each month regardless of its age. In the transition cost matrix you specify a cost of $2 every time the system moves into state 0. This is the event that requires a bulb replacement. The following matrices show the data for the example. State 0 represents a new bulb, state 1 a bulb 1 month old, and so on. The $\mathbf{C}$ matrix combines the other two matrices using $\mathbf{P}$ and Equation (8).

$$\mathbf{C}^S = \begin{bmatrix} 0.1 \\ 0.1 \\ 0.1 \\ 0.1 \\ 0.1 \end{bmatrix}, \quad \mathbf{C}^R = \begin{bmatrix} 2 & 0 & 0 & 0 & 0 \\ 2 & 0 & 0 & 0 & 0 \\ 2 & 0 & 0 & 0 & 0 \\ 2 & 0 & 0 & 0 & 0 \\ 2 & 0 & 0 & 0 & 0 \end{bmatrix}, \quad \mathbf{C} = \begin{bmatrix} 1.1 \\ 0.5 \\ 0.6 \\ 0.767 \\ 2.1 \end{bmatrix}$$

Transient Probabilities

When the sign is new, all bulbs are at age 0. The manager asks for an estimate of how many bulbs will be replaced during each of the first 12 months of operation.

You must advise the manager that since failures are a statistical phenomenon, the best you can provide is a probability distribution for each month that a bulb is in a given state. The first step is to derive this transient probability distribution for the age of the bulb in a single location and then use the results to answer the manager's question.

Because a bulb can be any of five ages at the beginning of a month, we use the state probability vector $\mathbf{q}(n) = (q_0(n), q_1(n), q_2(n), q_3(n), q_4(n))$ to describe the probability distribution for the age of a bulb at time n. Each component of this vector is a function of time. The initial probability vector is

$$\mathbf{q}(0) = (1, 0, 0, 0, 0)$$

indicating that each bulb starts out as new.

After 1 month, a bulb is replaced with probability 0.5, and will be 1 month old with probability 0.5, so the probabilities after 1 month are

$$\mathbf{q}(1) = (0.5, 0.5, 0, 0, 0)$$

The second month is more complicated, making simple reasoning insufficient. We take advantage of the theory of Markov chains and use Equation (3) for the computations to derive further results. In particular, the probability vector for month n can be obtained by multiplying the probability vector for the previous month, $n - 1$, by the transition matrix—i.e., $\mathbf{q}(n) = \mathbf{q}(n - 1)\mathbf{P}$.

Using Equation (3), we obtain the probability distributions shown in Table 12.10 through the first 12 months. The data for month 0 indicate that the location starts with a new bulb.

The rows labeled 0 through 11 represent the situation at the beginning of the first 12 months. Row 12 defines the situation at the beginning of year 2. At that time, for example, there is an almost 42% chance that a bulb in a given location is new. Other probabilities for the life distribution can similarly be read from the last row of the table.

The last column of the table is the expected cost associated with a single bulb during month n. These values were determined using the expected state cost vector $\mathbf{C}$ and Equation (6). The sum of rows 1 through 12 gives the replacement cost for the year: $10.91. This does not include the initial cost of the bulb. Assuming that the failure and repair of each of the bulbs is an independent event, the expected cost of maintaining the sign for the first year is then $10.91 \times 1000 = \$10,910$.

But how many bulbs are replaced in the first 12 months? The expected number in month n is $1000 \times q_0(n)$. Over the year, the expected number of replacements is the sum of the column labeled "New" multiplied by 1000. For the given data, each location receives, on average, 5.94 bulbs a year, or 5940 bulbs altogether. These numbers do not include the new bulbs installed at time 0. The actual number of bulbs replaced is a random variable, and the numbers computed here are expected values.

Steady-State Probabilities

The manager notices that although the numbers vary from month to month, they appear to be converging to some asymptotic values as time progresses. The manager wants an estimate of the distribution over the long run.

For discrete-time Markov chains, as long as each state can be reached from every other state, the transient probabilities will approach equilibrium. What is meant here is not that the status of a bulb becomes stable—it is dynamic by definition—but that after enough time has elapsed the probabilities will not change with time. Figure 12.6 illustrates this type of limiting behavior for the first 20 months of the process. After month 9, the oscillations begin to level off, with only minute differences being observed from month to month. Using Equa-

Table 12.10   Transient State Probabilities and Expected Cost

Month, n	New	1 month	2 months	3 months	4 months	Expected cost
0	1	0	0	0	0	
1	0.5	0.5	0	0	0	0.8
2	0.35	0.25	0.4	0	0	0.75
3	0.325	0.175	0.2	0.3	0	0.795
4	0.3475	0.1625	0.14	0.15	0.2	1.0825
5	0.4913	0.1738	0.13	0.105	0.1	0.9958
6	0.4479	0.2456	0.139	0.0975	0.07	0.9206
7	0.4103	0.2239	0.1965	0.1043	0.065	0.8976
8	0.3988	0.2052	0.1792	0.1474	0.0695	0.9077
9	0.4039	0.1994	0.1641	0.1344	0.0983	0.9518
10	0.4259	0.2019	0.1595	0.1231	0.0896	0.9476
11	0.4238	0.2129	0.1615	0.1196	0.0821	0.9336
12	0.4168	0.2119	0.1704	0.1212	0.0798	0.9271

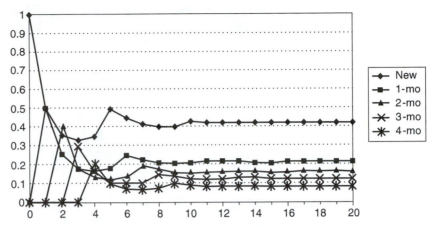

Figure 12.6   Transient probabilities for the light bulb example

tions (4) and (5), we compute the limiting values of $\mathbf{q}(n)$ to get the steady-state probability vector $\boldsymbol{\pi}$. For our example, we have

	New	1 mo.	2 mo.	3 mo.	4 mo.
$\boldsymbol{\pi} = ($	0.4167	0.2083	0.1667	0.125	0.0833)

with expected cost per period = $0.9333 per bulb—that is, the average cost of maintaining the sign is about $0.93 per bulb per month.

As we look further and further into the future, the steady-state probability vector $\boldsymbol{\pi}$ provides the probability distribution for the life of the bulb at a single location. When bulb failures are independent, the components of $\boldsymbol{\pi}$ give the proportion of the bulbs that have the same lives.

Evaluation of Alternatives

The fact that about 42% of the bulbs burn out in the first month means that the sign will not be very attractive or readable. The manager feels that this is a barely tolerable situation and is considering two other options. Each is discussed below.

Burn-in the bulbs: If you pay $2.5 for each bulb, the manufacturer will burn them for one month at the plant. The more expensive bulb will have one less month of life, but that first precarious month is eliminated. You wish to determine if the additional cost is worth it.

The absence of new bulbs implies that the state "New" should be deleted from the model, so the state space is reduced by one. All failure transitions now go into state 1 rather than state 0 since all replacement bulbs are 1 month old. If a bulb is in state 1, for example, there is a 0.2 probability that it will fail in the first month. In this case, the process returns to state 1; otherwise, it moves to state 2 with a 0.8 probability. The corresponding transition matrix, derived from the original transition matrix, is as follows.

$$
\begin{array}{cccc}
\text{1 mo.} & \text{2 mo.} & \text{3 mo.} & \text{4 mo.}
\end{array}
$$

$$
\mathbf{P} = \begin{bmatrix}
0.2 & 0.8 & 0 & 0 \\
0.25 & 0 & 0.75 & 0 \\
0.3333 & 0 & 0 & 0.6667 \\
1 & 0 & 0 & 0
\end{bmatrix}
\begin{array}{l}
\text{1 mo.} \\
\text{2 mo.} \\
\text{3 mo.} \\
\text{4 mo.}
\end{array}
$$

The cost of a transition into state 1 is now $2.50; all other transition costs remain at zero. The expected cost per bulb is now $0.993 per month. The updated cost matrix, along with the steady-state probability vector, is as follows.

$$\mathbf{C}^R = \begin{bmatrix} 2.5 & 0 & 0 & 0 \\ 2.5 & 0 & 0 & 0 \\ 2.5 & 0 & 0 & 0 \\ 2.5 & 0 & 0 & 0 \end{bmatrix} \begin{matrix} 1 \text{ mo.} \\ 2 \text{ mo.} \\ 3 \text{ mo.} \\ 4 \text{ mo.} \end{matrix}$$

$$\pi = (0.3571, 0.2857, 0.2143, 0.1429)$$

This option reduces the number of failed bulbs each month but increases the cost of maintenance by almost $0.06 per bulb per month, or $720 per year for the sign. It is up to the decision maker whether the additional expense is compensated by the benefit of having a better functioning sign.

Buy super bulbs: A new scientific development promises a bulb with a fixed life. Each one of these bulbs lasts exactly five months and costs $5. Now the bowling center can have a perfect sign.

The following transition matrix indicates that the process returns to the "New" state only at the end of the fourth month. Otherwise, bulbs invariably age 1 month. The transition cost matrix $\mathbf{C}^R$ also changes in this situation to reflect the $5 charge in the event of a purchase of a bulb. The state cost matrix $\mathbf{C}^S$ no longer needs to be included, because no inspection is required during the intermediate lives of the bulbs.

$$\begin{matrix} & \text{New} & 1 \text{ mo.} & 2 \text{ mo.} & 3 \text{ mo.} & 4 \text{ mo.} \end{matrix}$$

$$\mathbf{P} = \begin{bmatrix} 0 & 1 & 0 & 0 & 0 \\ 0 & 0 & 1 & 0 & 0 \\ 0 & 0 & 0 & 1 & 0 \\ 0 & 0 & 0 & 0 & 1 \\ 1 & 0 & 0 & 0 & 0 \end{bmatrix} \begin{matrix} \text{New} \\ 1 \text{ mo.} \\ 2 \text{ mo.} \\ 3 \text{ mo.} \\ 4 \text{ mo.} \end{matrix}$$

Plots of the transient probabilities for this case show that they are all cyclical. Figure 12.7 gives the plot for $q_0(n)$, the first component of $\mathbf{q}(n)$. Because the system is periodic, it returns to each state every five periods. Solving for the steady-state probability vector π confirms that there is an equal probability of being in any of the five states.

	New	1 mo.	2 mo.	3 mo.	4 mo.
$\pi = ($	0.2	0.2	0.2	0.2	0.2)

Thus, the system is in each state exactly 20% of the time. For periodic systems, π does not represent the limiting value of the state probability vector $\mathbf{q}(n)$—that is, $\mathbf{q}(n)$ does not converge to π or to any other vector as $n \rightarrow \infty$. Another way of saying this is that the long-run proportion of time the system is in state j does not equal the limiting probability of being in state j because the latter probability does not exist. More specifically,

$$\pi_j \neq \lim_{n \to \infty} \Pr(X_n = j \mid X_0 = i)$$

because the term on the right does not converge. For periodic systems, however, we have

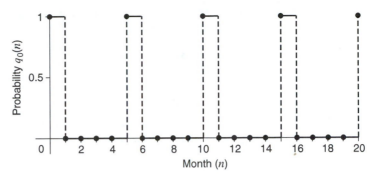

Figure 12.7   Transient probability $q_0(n)$ for periodic process

$$\pi_j = \lim_{n \to \infty} \frac{1}{n} \sum_{s=0}^{n-1} I(X_s = j) \tag{10}$$

where $I(X_s = j) = 1$ if the process visits state j at step s, 0 otherwise. Equation (10) indicates the proportion of time that the system is in state j over the long run. Since there is no need to inspect perfect bulbs, the expected cost per bulb is $1.00 per month.

The Game of Craps

As described in Section 11.3, in craps[2] the player rolls a pair of dice and sums the numbers showing. A total of 7 or 11 on the first roll wins for the player, whereas a total of 2, 3, or 12 loses. Any other number is called the point. The player then rolls the dice again. If she rolls the point number, she wins. If she throws a 7, she loses. Any other number requires another roll. The process continues until either a 7 or the point is thrown. The game is pure chance, devoid of strategy, so it should be straightforward to calculate the probability of winning and losing on each turn. Given that the payoff is one for one, the closer the probability of winning is to 0.5, the "fairer" the game is said to be.

This situation is a good example of a Markov chain for which the period is not measured in units of time but in throws of the dice. The discrete probability distribution for all possible outcomes associated with two dice is presented in Table 12.11. For example, the probability of throwing a 7 is approximately 0.167 (or 1 out of 6) whereas the probability of throwing an 11 is 0.056 (or 1 out of 18).

The state-transition network for this problem is shown in Figure 12.8. An activity is a roll of the dice culminating in an event represented by arcs in the diagram. The gambler begins at the node labeled "Start." The first roll leads to "Win," "Lose," or one of the states representing a point (P4, P5, P6, P8, P9, P10). The process stops when the system enters either state Win or state Lose. If the game were to begin again after reaching one of these terminal states, transitions back to the Start node would be included.

---

[2] For each person at the craps table, many different bets can be made, but our analysis will cover only the player throwing the dice. (For a complete treatment, the interested reader is referred to Patterson and Jaye, 1982.)

Table 12.11 Probabilities for the Throw of Two Dice

Sum	2	3	4	5	6	7	8	9	10	11	12
Probability	0.028	0.056	0.083	0.111	0.139	0.167	0.139	0.111	0.083	0.056	0.028

Using simple probability analysis concerning the throw of a pair of dice, we can determine the probabilities of the events shown in Figure 12.8. These are the transition probabilities p_{ij} from state i to state j, and are constant for each throw. For example, there are six ways of obtaining 7 and two ways of obtaining 11 out of 36 total possibilities (six for each die). Therefore, the probability of winning on the first roll is 8/36 = 2/9 $\cong$ 0.222.

To compute the components p_{ij} of the **P** matrix, we use the symbols S, W, L, 4, 5, 6, 8, 9, 10 to refer to the nine states. Let r_k be the probability of rolling the number k, as indicated in Table 12.11 for $k = 2, \ldots, 12$. On the first roll the player can win by throwing a 7 or an 11 ($p_{SW} = r_7 + r_{11}$), lose by throwing a 2, 3, or 12 ($p_{SL} = r_2 + r_3 + r_{12}$), or establish a point by throwing any other number, say 4, with $p_{S4} = r_4$. In subsequent rolls, she wins by throwing her point ($p_{4W} = r_4$), loses by throwing a 7 ($p_{4L} = r_7$), or rolls again with probability $1 - r_4 - r_7$.

The collection of the transition probabilities forms the transition matrix **P**, which, along with the state definitions, completely describes the Markov chain.

$$
\mathbf{P} = \begin{array}{c} \\ \text{Start} \\ \text{Win} \\ \text{Lose} \\ \text{P4} \\ \text{P5} \\ \text{P6} \\ \text{P8} \\ \text{P9} \\ \text{P10} \end{array}
\begin{array}{ccccccccc}
\text{Start} & \text{Win} & \text{Lose} & \text{P4} & \text{P5} & \text{P6} & \text{P8} & \text{P9} & \text{P10} \\
0 & 0.222 & 0.111 & 0.083 & 0.111 & 0.139 & 0.139 & 0.111 & 0.083 \\
0 & 1 & 0 & 0 & 0 & 0 & 0 & 0 & 0 \\
0 & 0 & 1 & 0 & 0 & 0 & 0 & 0 & 0 \\
0 & 0.083 & 0.167 & 0.75 & 0 & 0 & 0 & 0 & 0 \\
0 & 0.111 & 0.167 & 0 & 0.722 & 0 & 0 & 0 & 0 \\
0 & 0.139 & 0.167 & 0 & 0 & 0.694 & 0 & 0 & 0 \\
0 & 0.139 & 0.167 & 0 & 0 & 0 & 0.694 & 0 & 0 \\
0 & 0.111 & 0.167 & 0 & 0 & 0 & 0 & 0.722 & 0 \\
0 & 0.083 & 0.167 & 0 & 0 & 0 & 0 & 0 & 0.75
\end{array}
$$

"Win" and "Lose" are *absorbing* states signaling that the game is over. Once either state is entered, no further transitions are possible. This is apparent from **P**, which shows a probability of 1 on the main diagonal for these two states. One begins the game in "Start." The states P4, P5, P6, P8, P9, P10 correspond to the particular point the shooter is seeking.

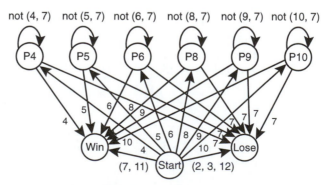

Figure 12.8 States and events for craps

Computing the transient probabilities for the first five rolls, there is a 0.333 chance that the game will end in a single roll (Win or Lose), as shown in Table 12.12. After five rolls, there is a 0.822 probability that the game is over. Note that the player has a slightly better chance of winning (0.422 versus 0.400) at this stage, but in the long run the advantage is reversed.

Of most concern to the casino are the long-run chances of a gambler winning. This information is provided by the absorbing state probabilities, which are computed from the formulas given in Section 13.8. Table 12.13 gives the probabilities of finishing in the two absorbing states as a function of the initial state. The most important line for the gambler as well as the casino is the row labeled "Start," which specifies the odds of winning and losing. In the long run, the gambler has a slightly less than even chance of winning. The other lines show the absorbing state probabilities if a point is thrown on the first roll.

Markov chain analysis can also be used to compute first passage probabilities. Here one wishes to determine the probability distribution associated with the number of steps required to make the first passage from some initial state s_I to some final state s_F. Once this distribution is known, it is possible to calculate the expected number of steps to go from s_I to s_F, as well as answer other common statistical questions. The details are provided in Section 13.7. For now, we simply enumerate two first passage distributions for 10 rolls of the dice, one from the Start state to the Win state, and one from the Start state to the Lose state. In either case, the first passage ends the game. The results are given in Table 12.14.

Omar's Barber Shop

You sit down to get a haircut and remark to Omar, "That was a long wait. If you weren't the best barber in town, I would have gone elsewhere."

Omar replies, "Well, you should come some other day than Saturday. Everybody seems to wait until the last minute and then expects to be taken right away. About an hour ago

Table 12.12 Transient Probabilities for Game of Craps

Roll number	Start	Win	Lose	P4	P5	P6	P8	P9	P10
0	1	0	0	0	0	0	0	0	0
1	0	0.222	0.111	0.083	0.111	0.139	0.139	0.111	0.083
2	0	0.299	0.222	0.063	0.08	0.096	0.096	0.080	0.063
3	0	0.354	0.302	0.047	0.058	0.067	0.067	0.058	0.047
4	0	0.394	0.359	0.035	0.042	0.047	0.047	0.042	0.035
5	0	0.422	0.400	0.026	0.030	0.032	0.032	0.030	0.026

Table 12.13 Absorbing State Probabilities

Initial state	Win	Lose
Start	0.493	0.507
P4	0.333	0.667
P5	0.400	0.600
P6	0.455	0.545
P8	0.455	0.545
P9	0.400	0.600
P10	0.333	0.667

Table 12.14   First Passage Probabilities

Roll	Start–Win	Start–Lose	Sum	Cumulative
1	0.222	0.111	0.333	0.333
2	0.077	0.111	0.188	0.522
3	0.055	0.080	0.135	0.656
4	0.039	0.057	0.097	0.753
5	0.028	0.041	0.069	0.822
6	0.020	0.030	0.050	0.872
7	0.014	0.021	0.036	0.908
8	0.010	0.015	0.026	0.933
9	0.007	0.011	0.018	0.952
10	0.005	0.008	0.013	0.965

there was no business at all, but I'll bet the shop will be full at closing time. Then I'll spend an hour just finishing up."

You are an OR student and need a term project for your stochastic processes class, so you offer Omar some free consulting. The next Saturday you watch the arrivals from across the street and notice a good deal of variability. By the nature of the business, you expect that customers arrive independently. This seems to be the case; sometimes there are long intervals between arrivals, and at other times several customers arrive very close together. You decide to model the number of arrivals during an interval with a Poisson distribution. Let $N(t)$ be a random variable that counts the number of arrivals through time t. The formula for the Poisson distribution is

$$\Pr\{N(t) = k\} = \frac{(\lambda t)^k e^{-\lambda t}}{k!} \quad \text{for } k = 0, 1, \ldots$$

where $\lambda > 0$ is a parameter (average number of customers per period). Theory tells us that $E[N(t)] = \lambda t$ and $Var[N(t)] = \lambda t$. A sample indicates that the average number of arrivals is four per hour. Before we can determine λ, however, we must specify the time interval or period to be used in the analysis. Omar has been cutting hair for 30 years and has the job down to a science. Every cut takes very close to 15 minutes, so this might be a good time interval to use.

Of course, customers do not like to wait. Omar has four chairs to accommodate the queue, but you observe that arrivals will enter the shop only when there is at least one empty chair. When all the chairs are occupied, customers *balk*. Omar suspects that they go to the new barber down the street.

To develop a Markov chain model, you divide the day into 15-minute intervals, implying that the Poisson parameter $\lambda = 1$ customer/period—i.e., $\lambda = (4 \text{ customers/hr})/(4 \text{ periods/hr})$. Omar opens at 9 A.M. and works until 5 P.M. for a total of 32 periods. A natural but partial definition of the states is the number in the shop, 0 through 5. Omar is idle when in state 0 and busy when in states 1 through 5.

To determine the transition matrix, you must rather carefully specify when the system is to be observed and the events that might occur during the 15-minute interval. There are two possibilities: (1) customers may enter and (2) customers may have their haircuts completed and leave. We define the states as the number of customers in the shop at the *beginning* of the interval, and specify that the number of persons entering is limited by the number of empty chairs at the beginning of the interval. With no customers in the shop, five customers can enter; with one in the shop, four can enter; and so on for each state. The number of customers that leave during the interval also depends on the number of customers in the shop at the beginning of the interval. With none in the shop, none will leave. With one or more in the shop, exactly one will leave. By defining both entries and departures in terms of the number of customers in the shop, we satisfy the Markovian property while ensuring that the two events represent independent random variables.

The transition matrix is best constructed one row at a time by considering each state in turn. For state 0 there are no customers in the shop at the beginning of the interval. Because customers arrive at random according to a Poisson distribution with $\lambda = 1$ (the expected number of arrivals in a 15-minute interval is 1 customer), the corresponding probabilities for $t = 1$ and $k = 0, \ldots, 6$ are as indicated in Table 12.15.

The number of arrivals is not the same as the number of entries, however. When more than five persons arrive during an interval, only five will enter while the others go elsewhere. Thus, the first row $\mathbf{p}_0 = (p_{00}, p_{01}, \ldots, p_{05})$ to three-decimal-place accuracy is

$$\mathbf{p}_0 = (0.368, 0.368, 0.184, 0.061, 0.015, 0.004)$$

The last component, $p_{05} = 0.004$, is the accumulation of the Poisson distribution for five or more arrivals (or $1 - \Sigma_{k=0}^{4} p_{0k}$). This and every other row of the transition matrix must sum to 1.

When one customer is in the shop at the beginning of the interval, at most four more can enter. Exactly one haircut will be completed and one customer will leave. This leads to the second row of the transition matrix $\mathbf{p}_1 = (p_{10}, p_{11}, \ldots, p_{15})$, or

$$\mathbf{p}_1 = (0.368, 0.368, 0.184, 0.061, 0.019, 0)$$

The value $p_{14} = 0.019$ is the probability that four or more persons will arrive. The value p_{15} is 0 because one customer is being served during the entire 15-minute interval, thus limiting the number that can enter to at most four. These results are a direct consequence of the assumption that both arrival and service processes are based on the number of customers in the shop at the beginning of the interval. The remainder of the transition matrix is computed using similar arguments.

$$
\mathbf{P} = \begin{array}{c} \\ 0 \\ 1 \\ 2 \\ 3 \\ 4 \\ 5 \end{array}
\begin{array}{c} \begin{array}{cccccc} 0 & 1 & 2 & 3 & 4 & 5 \end{array} \\
\left[\begin{array}{cccccc}
0.368 & 0.368 & 0.184 & 0.061 & 0.015 & 0.004 \\
0.368 & 0.368 & 0.184 & 0.061 & 0.019 & 0 \\
0 & 0.368 & 0.368 & 0.184 & 0.080 & 0 \\
0 & 0 & 0.368 & 0.368 & 0.264 & 0 \\
0 & 0 & 0 & 0.368 & 0.632 & 0 \\
0 & 0 & 0 & 0 & 1 & 0
\end{array} \right]
\end{array}
\qquad
\mathbf{C}^S = \left[\begin{array}{c} 0 \\ 10 \\ 10 \\ 10 \\ 10 \\ 10 \end{array} \right]
$$

The matrix $\mathbf{C}^S$ specifies revenue of \$10 for all states except the first. As long as someone is in the shop, Omar takes in \$10 every 15 minutes.

This completes the development of the Markov chain model, so you can now analyze Omar's operations over the 8-hour day (32 15-minute periods). You begin with a transient

Table 12.15   Poisson Probabilities for $\lambda = 1$ and $t = 1$

Arrivals, k	0	1	2	3	4	5	6
$\Pr\{N(1) = k\}$	0.3679	0.3679	0.1839	0.0613	0.0153	0.0031	0.0005

analysis. Assuming that customers will queue up at the beginning of the day, you start counting 15 minutes (1 period) before the shop opens. The probability distribution of the number of customers in the shop at the beginning of the first period is the Poisson distribution with mean equal to 1 (adjusted for balking). Transient probabilities are given in Table 12.16. The last column indicates the expected revenue per period.

After an initial transient period, the state probabilities begin to converge. The long-run values are

$$
\begin{array}{cccccc}
0 & 1 & 2 & 3 & 4 & 5
\end{array}
$$
$$\pi = (0.1153,\ 0.1982,\ 0.2252,\ 0.2302,\ 0.2307,\ 0.0004\ )$$

The expected revenue in the first 15 minutes is $6.32, whereas the expected revenue in the last 15 minutes is $8.85. Summing the last column of the table indicates that Omar can expect to make $274.61 in the course of a day. On the other hand, the steady-state expected revenue is $8.85 per period. Multiplying this value by 32 yields the steady-state revenue per day of $283.09, not far from the revenue predicted by the transient analysis. You conclude that a steady-state analysis will be sufficient for comparing alternative proposals for improvement.

The results show that although there are enough customers to keep Omar busy as the day progresses, he actually is idle about 12% of the time. Using the steady-state probabilities, you compute the expected number of customers in the shop at closing time to be

$$\sum_{k=0}^{5} k\pi_k = 2.26$$

This implies that Omar should expect to spend $2.26 \times 15 = 34$ minutes at the end of the day providing haircuts to the remaining customers. Of course, the number of customers is a random variable whose distribution is approximately the steady-state probability vector.

Table 12.16 Transient Probabilities for Omar's Barber Shop[a]

| Period, n | State, k | | | | | | Expected revenue |
	0	1	2	3	4	5	
0	1	0	0	0	0	0	
1	0.3679	0.3679	0.1839	0.0613	0.0153	0.0037	$6.32
2	0.2707	0.3384	0.2256	0.1071	0.0569	0.0013	$7.29
3	0.2240	0.3070	0.2344	0.1392	0.0943	0.0010	$7.76
4	0.1954	0.2816	0.2351	0.1616	0.1255	0.0008	$8.05
5	0.1755	0.2620	0.2337	0.1781	0.1501	0.0007	$8.25
6	0.1609	0.2469	0.2319	0.1905	0.1691	0.0006	$8.39
7	0.1500	0.2354	0.2304	0.2000	0.1836	0.0006	$8.50
.	.	.	.	.	.	.	.
.	.	.	.	.	.	.	.
20	0.1164	0.1993	0.2254	0.2292	0.2292	0.0004	$8.84
.	.	.	.	.	.	.	.
.	.	.	.	.	.	.	.
32	0.1154	0.1982	0.2252	0.2301	0.2306	0.0004	$8.85

[a]Some rows might not sum to 1 because of rounding.

EXERCISES

1. Louise Ciccone, a dealer in luxury cars, faces the following weekly demand distribution.

Demand	0	1	2
Probability	0.1	0.5	0.4

   She adopts the policy of placing an order for three cars whenever the inventory level drops to two or fewer cars at the end of a week. Assume that the order is placed just after taking inventory. If a customer arrives and there is no car available, the sale is lost. Show the transition matrix for the Markov chain that describes the inventory level at the end of each week under the following conditions.

   (a) The order takes 1 week to arrive.

   (b) The order takes 2 weeks to arrive. (*Hint*: To determine the state space, consider the possible number of cars on hand and cars on order just after taking inventory.)

   Also, compute the steady-state probabilities and the expected number of lost sales per week for each case.

2. Do both parts of Exercise 1 under the condition that the demand is assumed to follow a Poisson distribution with a mean of 1.5 cars per week.

3. A military maintenance depot overhauls tanks. There is room for three tanks in the facility and one tank in an overflow area outside. At most four tanks can be at the depot at one time. Every morning a tank arrives for an overhaul. If the depot is full, however, it is turned away, so no arrivals occur under these circumstances. When the depot is full, the entire overhaul schedule is delayed 1 day. On any given day, the following probabilities govern the completion of overhauls.

Number of tanks completed	0	1	2	3
Probability	0.2	0.4	0.3	0.1

   These values are independent of the number of tanks in the depot, but obviously no more tanks than are waiting at the start of the day can be completed.

   (a) Develop a Markov chain model for this situation. Begin by defining the state to be the number of tanks in the depot at the start of each day (after the scheduled arrival). Draw the network diagram and write the state-transition matrix.

   (b) Do the same when the state is defined as the number of tanks in the depot at the end of each day.

4. For the situation described in Exercise 3, use the Stochastic Analysis Excel add-in that accompanies this book to answer the following questions.

   (a) Assume that on Monday morning (after the arrival of a tank) there are three tanks in the maintenance facility but the overflow area is empty. List the state probabilities for the next 5 days.

   (b) Compute the steady-state probabilities. Use these probabilities to estimate the number of days in the year (assume 200 working days) on which a delay in the schedule is necessary.

   (c) From the steady-state probabilities, estimate the utilization of the depot in terms of percentage of its capacity that is used. Assume that the maintenance capacity is three tanks.

   Exercises 5 to 10 refer to the brand switching example in Section 12.1.

5. With respect to the data in Table 12.5:

   (a) Interpret the meanings of $p_{11} = 0.90$, $p_{22} = 0.82$, and $p_{33} = 0.68$.

   (b) To which brand is brand 2 most likely to lose customers?

   (c) Which brand has the most "loyal" customers?

   (d) Which brand has the least "loyal" customers?

6. Add a fourth state to the state space that represents "no purchase of brands 1, 2, or 3 in week n." Construct a new transition matrix given the following contingency table.

Brand (*j*)	1	2	3	4	Total
(*i*)					
1	90	5	3	2	100
2	5	200	40	5	250
3	25	15	80	10	130
4	5	10	5	0	20
Total	125	230	128	17	500

Is this table more comprehensive than Table 12.4? Interpret the meaning of p_{44}. Identify two explanations for the occurrence of $X_n = 4$. (*Hint:* When $X_n = 4$, does this necessarily mean that the consumer did not purchase this product in week n? Explain.)

7. Consider the three-brand market presented in the brand-switching example. Define a state space that includes the possibilities of zero, one, or two purchases of the product in a given week.

8. Calculate by hand the expected market shares three weeks into the future using both formulas given in Equation (3). Assume $\mathbf{q}(0) = (0.25, 0.46, 0.29)$.

9. (*Marketing Strategies*) To illustrate how price adjustments can affect consumer switching behavior in a Markov chain model, suppose that the following relationship is empirically valid.

$$\hat{p}_{ij} = p_{ij} + \alpha(s_{it} - \overline{s}_{it})$$

where
  p_{ij} = current (stationary) transition probability
  $\hat{p}_{ij}$ = new (stationary) unadjusted transition probability
  s_{in} = selling price of brand i in week n
  $\overline{s}_{in}$ = mean selling price of all brands other than brand i in week n
  α = parameter ($\alpha = 0.01$ when $i \neq j$ and $\alpha = -0.05$ when $i = j$)

Thus, when $i \neq j$, for example, a price for brand i in week n that is above the market average of all other prices (that is, $s_{in} > \overline{s}_{in}$) will have the effect of increasing the attraction power of other brands (that is, $\hat{p}_{ij}$ will be greater than p_{ij}). Given expected prices in week n, it follows that the preceding equation can be used to construct a new transition matrix. Note, however, that row sums will not be unity, hence the $\hat{p}_{ij}$ values for each row i must be adjusted to yield a sum of unity. For example, if unadjusted $\hat{p}_{11} = 0.9$, $\hat{p}_{12} = 0.5$, and $\hat{p}_{13} = 0.6$, the adjusted values for the first row would be determined as follows (noting that unadjusted $\hat{p}_{11} + \hat{p}_{12} + \hat{p}_{13} = 2$): $\hat{p}_{11} = 0.9/2 = 0.45$, $\hat{p}_{12} = 0.5/2 = 0.25$, and $\hat{p}_{13} = 0.6/2 = 0.3$. Also, if unadjusted $\hat{p}_{ij} < 0$, it should be set equal to zero before normalizing.

(a) Reason out changes in p_{ij} for the following situations: $s_{in} > \overline{s}_{in}$ and $i = j$; $s_{in} < \overline{s}_{in}$ and $i \neq j$; $s_{in} < \overline{s}_{in}$ and $i = j$; $s_{in} = \overline{s}_{in}$. Does the logic of this model reflect the phenomenon of "price snobbery"?

(b) Construct the new transition matrix (the equivalent of Table 12.5) given that expected market prices in week 28 are $1, $3, and $2 for brands 1, 2, and 3, respectively. Comment on the changes.

10. (*Homogeneity Assumption*) Table 12.4 and the resulting matrix of transition probabilities represent the aggregate switching behavior in a sample of the defined population. They do not represent individual consumer behavior in the sense of predicting what a particular person will do. The purchasing pattern for a particular consumer represents a stochastic process that could be modeled by a Markov chain. To gain a better understanding, construct $\mathbf{P}$ for the following 9-week purchasing pattern of a specific consumer: {2, 2, 2, 1, 1, 3, 1, 1, 1}. For practical reasons, firms are interested in aggregate rather than individual behavior. The act of aggregating consumers, however, implicitly assumes that they are *homogeneous* with respect to their transition matrices. Put another way, the construction of empirical probabilities as point estimates for true probabilities requires that the experiments (that is,

consumers in the sample) be performed under "identical" conditions (that is, homogeneous consumers). Can you think of any sampling procedures that would better guarantee homogeneity?

11. Suppose that the state-transition matrix for the IRS example discussed in Section 12.1 was based on a sample of homogeneous taxpayers. Calculate the proportion of taxpayers that will be audited (a) next year, (b) 2 years from now, and (c) 4 years from now, given that 10% were audited in the current year.

12. Consider a process in which a single worker must perform five stages of a manufacturing process, as indicated in the figure. For purposes of analysis, divide time into 1-hour segments. When the worker is idle, raw material enters the system during the next hour with probability p_A. At each stage i, the probability of completing the stage and moving on to the next stage during the current hour is $p_{C(i)}$. The goal is to find the efficiency of the worker and the throughput of the system. Each part should be done independently of the others.

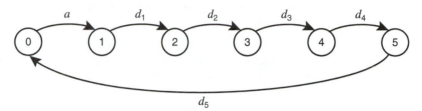

(a) Write the transition matrix for this situation. Use six states.

(b) Change the scenario so that the time between arrivals is exactly 2 hours. Counting begins when the worker becomes idle. Completion probabilities are still based on a 1-hour period.

(c) Change the scenario so that there is a space for one waiting job. Now define p_{A1} as the probability that an arrival occurs while the worker is busy. An arrival will wait if the space is empty; otherwise, it will not enter the system.

13. Allen Konigsberg has four daughters who are of marriageable age. He estimates that each daughter, operating independently, can find a spouse during a year with probability p. Because of budget considerations Mr. Konigsberg will allow only one daughter to be married in any given year. Unfortunately, if not allowed to marry, the daughter will lose her beau and will be forced to look again the following year. Find the probability distribution for the number of unmarried daughters as a function of time. Let the time interval be 1 year. Use $p = 0.6$. For each case, define a Markov chain model and find the transient probabilities for 10 years. If the system has a steady state, find the corresponding probabilities. Assume that the probabilities do not change over time.

(a) Solve the situation as given.

(b) How will the results change if Mr. Konigsberg drops the restriction of one marriage per year?

(c) How will the results change if we add the possibility of divorce? A married daughter divorces during a year with probability p_D. For computational purposes, assume that $p_D = 0.1$.

(d) Use the conditions in parts (b) and (c) and assume that each marriage costs $1000 (there is no cost for a divorce). How much should Mr. Konigsberg budget per year for marriages? Also assume that a daughter can, at most, either marry or divorce one time in any given year.

14. The figure shows a two-stage production flow line. At each stage, an operation is performed on the part being processed. Parts are introduced into the system at stage 1. Processing is serial, and each stage can hold only one part at a time.

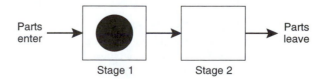

For purposes of analysis, discretize time into 1-minute intervals. At the beginning of an interval, stage 1 is empty, working, or blocked whereas stage 2 is either working, or empty. The system operates under the following rules.

- If stage 1 is empty at the beginning of an interval, a new part is introduced into stage 1 with probability 0.9. Work begins on the part; however, it cannot be completed during the minute it is introduced. If stage 1 is working or blocked, no part enters.
- If stage 2 is working at the beginning of an interval, the part will be completed and will leave the system with probability 0.8. Alternatively, it will remain in stage 2 with probability 0.2.
- If stage 1 is working at the beginning of an interval, the part will be completed with probability 0.6. A completed part will move to stage 2 if stage 2 is empty. Otherwise, it is blocked and remains in stage 1 until stage 2 becomes empty.

Develop a Markov chain model that describes this situation. Draw the corresponding network and find the steady-state probabilities for the system. Estimate the idle time for each stage and the throughput in parts per minute.

15. In a simplified Monopoly game there are four places on the board numbered 0 through 3. The matrix shows the transition probabilities between each of them on any one play of the game. Let the places be states.

$$
\mathbf{P} = \begin{array}{c} \\ 0 \\ 1 \\ 2 \\ 3 \end{array}
\begin{array}{c} \begin{array}{cccc} 0 & 1 & 2 & 3 \end{array} \\
\left[\begin{array}{cccc}
0 & 1 & 0 & 0 \\
0 & 0 & 1 & 0 \\
1/3 & 1/3 & 0 & 1/3 \\
1 & 0 & 0 & 0
\end{array} \right]
\end{array}
$$

(a) Find the steady-state probabilities for the corresponding Markov chain.

(b) If you receive $3 every time you reach state 2 and pay $1 when you reach any of the other states, what is your expected profit per play (in steady state).

(c) If you start in state 0, what is your long-run expected return? How would this result change if there were a discount rate of 1% per play due to fatigue?

(d) A modification of the preceding transition matrix is

$$
\mathbf{P} = \begin{array}{c} \\ 0 \\ 1 \\ 2 \\ 3 \end{array}
\begin{array}{c} \begin{array}{cccc} 0 & 1 & 2 & 3 \end{array} \\
\left[\begin{array}{cccc}
0 & 1 & 0 & 0 \\
0 & 0 & 1 & 0 \\
1/3 & 1/3 & 0 & 1/3 \\
0 & 0 & 0 & 1
\end{array} \right]
\end{array}
$$

If the system starts in state 0, what can you say about the transient probabilities after a large number of iterations?

16. The word processing center at Papers-R-Us.Com has three printers. The probability that a printer will fail during a given week is 0.1. Assume that failures are independent events. When one or more printers are in the shop at the beginning of the week, exactly one is repaired during the week. In each case, construct the transition matrix for this situation when the time interval is 1 week and the states describe the number of printers in the repair shop at the beginning of the week. The parts are not cumulative.

(a) Solve the problem as given.

(b) Change the situation so that each printer in the shop at the beginning of the week will be repaired during the week with a probability of 0.5. Assume that the repair operation for each printer is independent.

(c) Now assume that only one printer can be repaired at a time and it takes exactly 2 weeks for the repair. Also, a failed printer must be in the shop at the beginning of the week in order for the repair work to begin. Printers still fail according to the distribution given.

17. Set up and solve the following variations on Omar's Barber Shop problem presented in Section 12.5. Each part should be done independently of the others.

    (a) You note that sometimes Omar's haircutting time is not really deterministic as originally assumed. Actually, there is a probability d that he completes no haircuts in a 15-minute period, a probability d that he completes two haircuts, and a probability $1 - 2d$ that he completes one. Although the mean value of this distribution is 1, there is noticeable variability. How much does variability affect Omar's steady-state income? Plot a curve of d versus expected daily revenue.

    (b) Omar is thinking about removing one of the waiting chairs and putting in another barber chair. His plan is to hire a retired barber who he will pay $5 per cut, which he believes will also take 15 minutes. As storeowner, Omar will receive the remaining $5. Whenever there is a single customer in the shop, Omar will get the business. For this scenario, customers enter whenever a barber chair is empty, but the arrival rate will be reduced by 1/3 for every waiting chair that is full. How will this affect Omar's net revenue? How much will the retired barber earn per day on average? What proportion of the time will Omar be idle? What proportion of the time will the retired barber be idle? What proportion of the potential arriving customers will be served?

18. Enrollment at Big State's College of Engineering historically has ranged from 4500 to 6000 students. The lower bound represents a reliable pool of students within the community who have a hardcore interest in engineering. The upper bound represents a limit on capacity, which is enforced through admission controls. To gain a better understanding of the numbers, the dean's office has commissioned a study to assess the variability in fall enrollment each year. An analysis of past data indicates that the fluctuations are primarily a function of the economic climate that can be described by two states: *good* and *bad*.

    In an effort to develop a Markov chain model of the underlying process, assume that the enrollment in any 1 year will increase by 500 students, stay the same, or decrease by 500 students. As such, divide the range 4500 to 6000 into intervals of 500 students each. In a year in which the economic climate is *good*, enrollment will go up with probability 0.5, stay the same with probability 0.2, or go down by 500 with probability 0.3. In a year with a *bad* economic climate, enrollment will go up with probability 0.2, stay the same with probability 0.4, or go down with probability 0.4. These probabilities are based on the economic condition at the beginning of the year. The only exception occurs when the enrollment reaches 6000 students. At that point, the control measures will kick in to reduce enrollment to 5500 students at the beginning of the next year. When the enrollment is 4500 at the beginning of one year, it may not decrease further. In this case, the probability of decrease is combined with the probability that the enrollment doesn't change.

    The Bureau of Business Research has determined that if the climate is good at the beginning of a year, there is a 0.7 chance that it will be good at the beginning of the next year. If the climate is bad at the beginning of a year, there is a 0.6 chance that it will be bad again at the beginning of the next year.

    Begin by defining the states. Then draw the network diagram and construct the state-transition matrix for this problem. Find the steady-state probability distribution for the number of students in the College of Engineering.

19. Archy Leach, a recent graduate of Madame Bouffant's Beauty College, has opened his first hair salon. The facility has one stylist's chair and a waiting sofa that seats two. In any given minute of the day, only one of three things may happen.

    • A customer may arrive at the salon.
    • A customer may leave the salon.
    • There are no arrivals or departures.

The probability that a customer arrives in a particular minute is 0.01. If she arrives and finds the sofa full, she will not stay. Clearly, a customer may depart from the salon only if there is a customer in it. In this case, the probability that a customer departs in a particular minute is 0.02. Assume that during a 1-minute interval it is impossible for two or more events to occur.

(a) Construct the transition matrix that describes a 1-minute interval. The states are the number of people in the salon (not counting Archy).

(b) Compute steady-state probabilities and use them to answer the following questions.

(c) If Archy earns $5 each time a served customer leaves the salon, what are his expected daily earnings? Assume an 8-hour day.

(d) What percent of the time would you expect the salon to be full?

(e) What percent of the time would you expect the salon to have no customers?

20. A diamond merchant has determined that the following probability distribution describes the number of a particular type of customer who visits his store each day.

Number of customers	0	1	2
Probability	0.6	0.3	0.1

Assume that each of these customers will buy a Peruvian diamond if one is available. Otherwise, the sale is lost. Each morning the merchant looks at his inventory. If there are two or fewer Peruvian diamonds (and none on order), he orders three more from his supplier. They arrive 2 days later (on the second morning after the order).

(a) Model the inventory level as a Markov chain. Define the states for the system and write the corresponding transition matrix.

(b) If the store has three Peruvian diamonds on Monday morning, what is the probability distribution for the number on hand 5 days later? What is the probability that some Peruvian diamonds are on order?

(c) What is the steady-state probability that the store will have no Peruvian diamonds in the morning?

Use steady-state probabilities for parts (d) through (g).

(d) If the inventory holding cost is $10 per day per diamond, what is the expected annual inventory cost? Assume 260 workdays per year.

(e) What is the expected number of orders that must be placed each year? If the cost of placing an order is $50, what is the expected annual ordering cost?

(f) What is the expected annual cost of lost sales if the per unit cost is $200?

(g) Based on the results of parts (d), (e), and (f), experiment with different ordering policies with the aim of finding a policy that minimizes the sum of the inventory holding cost plus the ordering cost plus the lost sales cost.

21. The office manager in the computer repair example in Section 12.1 has the option of buying an additional machine to be kept as a spare. It will not be used in normal operations but will be placed into service when one of the originals fails. The effect will be to reduce the cost of sending out work to the typing service. Costs are: computer repair, $80; typing service with one failed computer, $75/day; typing service with two failed computers, $125/day.

The spare machine will fail only when in use and has the same failure rate as the original machines. The cost of ownership of the spare is $10 per day. Should the manager buy the spare?

22. Find the expected lifetime income of the stockbroker discussed in Section 12.2 as a function of the list he starts with. Assume a discount rate of 15% per year. What observations can you make about the results?

23. A manufacturing facility uses two identical machines that are attended by a single operator. Each machine requires the operator's attention at random points in time. The probability that a machine requires service in a period of 5 minutes is $p = 0.4$. The operator is able to service a machine in 5 minutes. Let us approximate this situation by assuming that when service is required, it is always initiated at the beginning of a 5-minute period. Construct a state-transition network for this problem and find the transition matrix associated with it. Define all terms. Answer the following numerical questions.

   (a) If both machines are operating properly at 8 A.M., find the state probabilities after 5 minutes, 10 minutes, and 40 minutes.

   (b) Find the steady-state probabilities for this process. What is the long-run probability that the operator is idle for a 5-minute period? For what fraction of all 5-minute periods is the operator busy? What is the long-run average number of machines that require service within a 5-minute period?

   (c) Assume that the opportunity cost in terms of lost production, if a machine is down or being serviced for a 5-minute period, is $5. What is the long-run average opportunity cost for an 8-hour shift (96 5-minute periods)?

24. Referring to Exercise 12 in Chapter 11, answer the following questions.

   (a) What fraction of the population is infected at any time if no drugs are available?

   (b) The cost of the drug program per week for each individual in the population is $5. If the cost of lost work as a result of the disease is assessed at $50 per week per person, will the program pay for itself?

25. A clothing wholesaler has 700 accounts that are past due. These accounts are classified as 0–30 days, 30–60 days, and 60–90 days overdue. Currently, there are 400 accounts in the 0–30-day category, 200 accounts in the 30–60-day category, and 100 accounts in the 60–90-day category. After 90 days, accounts are written off as being uncollectable. From past experience, the accounts receivable manager knows that the payment process behaves as a Markov chain. His estimate of the 1-month transition matrix is given in the table. Find the (approximate) number of accounts that will have to be written off as bad debts.

One-Month Transition Matrix

	0–30 days	30–60 days	60–90 days	Paid	Uncollectable
0–30 days overdue	0	0.7	0	0.3	0
30–60 days overdue	0	0	0.8	0.2	0
60–90 days overdue	0	0	0	0.5	0.5
Paid	0	0	0	1	0
Uncollectable	0	0	0	0	1

26. Heart patients at a local hospital can be found in one of two places: the coronary care unit or a regular room. Historical data indicate that of all heart patients, 83 percent leave the hospital alive whereas the other 17 percent die at some point during their treatment.

   (a) If we assume that the number of heart patients remains constant and that the 1-day transition probabilities are as shown in the table, what are the steady-state probabilities for an individual patient?

One-Day Transition Probabilities—Heart Patients

	CCU	Hospital rehabilitation	Discharged or deceased
Coronary care unit (CCU)	0.700	0.200	0.100
Hospital rehabilitation	0.050	0.800	0.150
Discharged or deceased	0.015	0.005	0.980

(b) If all the new heart patients went to a competing hospital and there was a 25 percent chance for a patient to leave the competing hospital each day, how would you change the 1-day transition matrix?

(c) Referring to part (b), what would the steady-state probabilities be?

27. (*Photocopier Maintenance*) Every office worker knows that photocopiers continue to suffer from a certain amount of downtime despite the great strides that have been made in technology. Assume that downtime can be modeled as a Markov chain with a transition period of 1 hour. For a particular office, the one-step transition matrix is as shown.

$$P = \begin{array}{c} \\ \text{Up} \\ \text{Down} \end{array} \begin{array}{c} \text{Up} \quad \text{Down} \\ \begin{bmatrix} 0.90 & 0.10 \\ 0.60 & 0.40 \end{bmatrix} \end{array}$$

(a) If the photocopier is currently up (running), what is the probability that it will be up after 3 hours of operation?

(b) What are the steady-state probabilities for the copier being up or down?

(c) The office manager is contemplating the replacement of the current copier with a newer, more reliable model whose one-step transition matrix is as follows.

$$P = \begin{array}{c} \\ \text{Up} \\ \text{Down} \end{array} \begin{array}{c} \text{Up} \quad \text{Down} \\ \begin{bmatrix} 0.95 & 0.05 \\ 0.85 & 0.15 \end{bmatrix} \end{array}$$

If the cost of system downtime is estimated at $600 per hour, what is the monthly break-even cost for the new machine? Assume that it will be in use 720 hours per month.

28. A delicate precision instrument has a component that is subject to random failure. In fact, if the instrument is operating properly at a given moment in time, then with probability 0.1 it will fail within the next 10-minute period. If the component fails, it can be replaced by a new one—an operation that also takes 10 minutes. The present supplier of replacement components does not guarantee that all replacement components are in proper working condition. The present quality standards are such that about 1% of the components supplied are defective. However, this can be discovered only after the defective component has been installed. If the component is defective, the instrument has to go through a new replacement operation. Assume that when a failure occurs, it always occurs at the end of a 10-minute period.

(a) Find the transition matrix associated with this process.

(b) Given that it was working properly initially, what are the probabilities that the instrument is not in proper working condition after 30 minutes and after 60 minutes?

(c) Find the steady-state probabilities. For what fraction of time is the instrument being repaired?

(d) Assume that each replacement component has a cost of $0.30, and that the opportunity cost in terms of losing profit during the time the instrument is not working is $10.80 per hour. What is the average cost per 10-minute period?

29. Past records indicate that the survival function for light bulbs in a traffic signal has the following pattern.

Age of bulbs in months, n	0	1	2	3	4	5
Number surviving to age n	1000	950	874	769	615	0

(a) If each light bulb is replaced after failure, find the transition matrix associated with this process. Assume that a replacement during the month is equivalent to a replacement at the end of the month.

(b) Determine the steady-state probabilities. What is the long-run average length of time that a bulb is used prior to being replaced? If an intersection has 40 bulbs, how many bulbs fail on the average per month? If an individual replacement has a cost of $2, what is the long-run average cost per month?

30. (*Group Replacement*) Consider the survival function in Exercise 29.

   (a) Assume now that all bulbs—regardless of age—are replaced every 3 months and that individual bulbs are replaced at failure, including those that fail during the month prior to a scheduled group replacement. On the basis of the transition matrix for the failure pattern of an individual bulb, determine the number of bulbs that are replaced at failure between group replacements. Note that failures in the month just prior to a group replacement are replaced also at failure. If group replacements cost $0.20 per bulb plus a $5.00 fixed cost per intersection, find the average cost per month for this policy.

   (b) Find the least-cost group replacement policy. Is its average monthly cost lower than that of a policy of individual replacements at failure only?

BIBLIOGRAPHY   Cochran, J.K., A. Murugan, and V. Krishnamurthy, "Generic Markov Models for Availability Estimation and Failure Characterization in Petroleum Refineries," *Computers & Operations Research*, Vol. 28, No. 1, pp. 1–12, 2000.

Feldman, R.M. and C. Valdez-Flores, *Applied Probability & Stochastic Processes*, PWS, Boston, 1996.

Karlin, S. and H.M. Taylor, *A First Course in Stochastic Processes*, Second Edition, Academic Press, New York, 1975.

Patterson, J.L. and W. Jaye, *Casino Gambling: Winning Techniques for Craps, Roulette, Baccarat, and Blackjack*, Perigee Books, New York, 1982.

Puterman, M.L., *Markov Decision Processes*, Wiley, New York, 1991.

Taylor, H.M. and S. Karlin, *An Introduction to Stochastic Modeling*, Revised Edition, Academic Press, New York, 1994.

Chapter 13

Mathematics of Discrete-Time Markov Chains

This chapter provides a description and justification of important concepts associated with discrete-time Markov chains (DTMCs), as well as the mathematical basis for computing their statistical properties. The developments are also relevant for continuous-time Markov chains (CTMCs), the subject of Chapters 14 and 15, because every CTMC contains an embedded DTMC. By implication, many statistical properties of CTMCs can be found using the same analytic procedures described here.

13.1 STATE-TRANSITION PROBABILITIES

A discrete-time Markov chain is a special case of a stochastic process. Rather than changing state at the moment of occurrence of some single event, the event for a Markov chain is the completion of a period. The transition to a new state is governed entirely by transition probabilities. When these probabilities depend only on the current state, the stochastic process has the Markovian property. In this case, we denote the transition probability of going from i to j by p_{ij}.

In what follows, X_n and $\{X_n\}$ are used to represent a random variable and the associated stochastic process, respectively. The subscript n is the time or step index and takes the values in the set $N = \{0, 1, 2, \ldots, \}$, which can be measured in any appropriate unit such as minutes or hours. The general time interval is called the period. To obtain a Markov chain, X_n must be restricted to discrete values. In addition, the transition probabilities p_{ij} must be time invariant and satisfy the Markovian property for all states i and j. For a stochastic process $\{X_n\}$ with discrete state space defined by the index set $S = \{0, 1, \ldots, m - 1\}$, the Markovian property can be expressed as

$$\Pr\{X_{n+1} = j \mid X_0 = k_0, X_1 = k_1, \ldots, X_n = i\} = \Pr\{X_{n+1} = j \mid X_n = i\} \tag{1}$$

for $n = 0, 1, \ldots,$ and all possible states $i, j, k_0, k_1, \ldots, k_{n-1}$ in S. Time invariance is equivalent to stationarity and can be stated symbolically as

$$p_{ij} = \Pr\{X_1 = j \mid X_0 = i\} = \Pr\{X_{n+1} = j \mid X_n = i\} \text{ for all } n = 1, 2, \ldots \tag{2}$$

To interpret Equation (1), think of time n as the present. The left-hand side is the probability of going to state j next, given the history of all past states. The right-hand side is the probability of going to state j next, given only the present state. The equality of the two terms implies that the past history of states provides no additional information to help predict the future if the current state is known. Equation (2) indicates that the one-step transition probabilities do not change as time progresses. They are the same in January as in August.

As we indicated in Chapter 12, the $m \times m$ transition matrix is denoted by $\mathbf{P} = (p_{ij})$, where the rows and columns of $\mathbf{P}$ are labeled 0 through $m - 1$ and $0 \le p_{ij} \le 1$, $i, j = 0, 1, \ldots, m-1$.

$$\mathbf{P} = \begin{bmatrix} p_{00} & p_{01} & p_{02} & \cdots & p_{0,m-1} \\ p_{10} & p_{11} & p_{12} & \cdots & p_{1,m-1} \\ \vdots & \vdots & \vdots & \cdots & \vdots \\ p_{m-1,0} & p_{m-1,1} & p_{m-1,2} & \cdots & p_{m-1,m-1} \end{bmatrix}$$

Because the system must always be in one of the m states, every row of the transition matrix must sum to 1.

$$\sum_{j=0}^{m-1} p_{ij} = 1, \quad i = 0, 1, \ldots, m-1 \tag{3}$$

When modeling with the DTMC, we first identify the states. To compute the transition probabilities for row i, representing state $\mathbf{s}_i$, we list the events that can occur while the system is in the state. Say we discover that event x occurs with probability $p(x)$ and that it leads to state $\mathbf{s}_j$ with the probability $p(i, j \mid x)$. If the event is the only one that leads from $\mathbf{s}_i$ to $\mathbf{s}_j$, then the transition probability is

$$p_{ij} = p(x)\, p(i, j \mid x)$$

It is often true for a DTMC that several events lead from $\mathbf{s}_i$ to $\mathbf{s}_j$. To compute p_{ij}, we must sum over the collection of events that result in the same transition. Recall that $Y(\mathbf{s}_i)$ is the set of events that can possibly occur in state $\mathbf{s}_i$. The transition probability is then

$$p_{ij} = \Sigma_{k \in Y(\mathbf{s}_i)} p(i, j \mid x_k) p(x_k)$$

where $p(x_k)$ is the unconditional probability that event x_k will occur, and $p(i, j \mid x_k)$ is the probability of the transition from $\mathbf{s}_i$ to $\mathbf{s}_j$ given event k. We saw an illustration of this in the game of craps example in Chapter 12. From the Start state several different throws of the dice yield the Win state. For this well-defined game, the conditional probabilities, $p(i, j \mid x_k)$, all equal 1, so the transition probability from Start to Win is the sum of the event probabilities.

The states and transition probabilities are graphically illustrated by a state-transition network, which is shown for the general case in Figure 13.1. In many applications, the corresponding network will not be totally connected, so it may not be possible to go from some state i to every other state j.

Several examples from Chapter 12 will be used in this and subsequent sections to illustrate basic concepts. To begin, consider the computer repair example described in Section 12.1 with states $S = \{0, 1, 2\}$ and the following transition matrix.

$$\mathbf{P} = \begin{array}{c} \\ \\ \end{array} \begin{matrix} 0 & \quad 1 & \quad 2 \end{matrix} \\ \begin{bmatrix} 0.6 & 0.3 & 0.1 \\ 0.8 & 0.2 & 0 \\ 1 & 0 & 0 \end{bmatrix} \begin{matrix} 0 \\ 1 \\ 2 \end{matrix}$$

To construct the network, recall that it is necessary to introduce a node for each state and an arc for each transition that has a nonzero probability. The state-transition network corresponding to $\mathbf{P}$ is repeated in Figure 13.2.

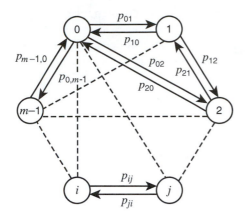

Figure 13.1   General state-transition network

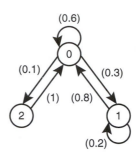

Figure 13.2   State-transition network for computer repair example

13.2   CLASSIFICATION OF STATES

In this section, we identify several classes to which a state may belong. Each class has special properties that affect the analysis, especially with regard to the long-run behavior of the stochastic process under investigation. Classification is done with reference to the state-transition network. For the most part, the discussion applies to all stochastic processes rather than just DTMCs.

Definitions

One state j is said to be *accessible* from another state i if it is possible to pass from state i to state j. This requires that a directed path exist in the state-transition network between node i and node j. The network for a birth–death process shown in Figure 13.3 will be used for illustrative purposes. We discuss birth–death models in Section 14.6.

It is possible to pass from some state i to another state j if one can trace a directed path following the arcs of the network from i to j. In Figure 13.3 we see that there is a path from state 0 to state 4, so state 4 is accessible from state 0. In fact, a path is present between each pair of states, so every state is accessible from every other state.

Alternatively, if state j is inaccessible from state i, there is no path from i to j. This case can be illustrated using the pure birth process shown in Figure 13.4. Clearly, state j is accessible from state i if $i < j$ but it is inaccessible from state i if $i > j$.

Two states *communicate* if both are accessible from each other. All states communicate in Figure 13.3, whereas no states communicate in Figure 13.4. A system is said to be *irreducible* if all states communicate, so Figure 13.3 describes an irreducible stochastic process.

Suppose a system is in state i at time n. If the system leaves state i but is sure to return to it some time in the future, the state is said to be *recurrent*. If a state is not recurrent, it

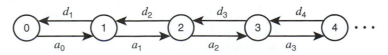

Figure 13.3   Birth–death process

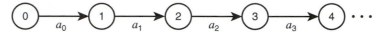

Figure 13.4 Pure birth process

is *transient*. The states in Figure 13.4 are transient. For the infinite-state case in Figure 13.3, the states may or may not be recurrent depending on the parameters of the system. Tests to determine if states are recurrent involve *probabilities of first passage* and are described later in the chapter.

The next concept we introduce is that of periodicity, but it is only appropriate for a stochastic process that changes state at fixed time intervals, such as the DTMC. A state is *periodic* if it can return to itself only after a fixed number of transitions greater than 1. The number of transitions is the period of the state. The states of the DTMC depicted in Figure 13.5a fall into this category and have period 3. Starting from any state, the system will return to that state after three transitions. The states in Figure 13.5b are also periodic with a period of 3. Starting in state 1, for example, the system can return to that state only after a multiple of three iterations. It will not necessarily return to state 1 every three iterations, but it will certainly not return except in multiples of three. A state that is not periodic is *aperiodic*. Continuous-time systems fall into this category even if the states repeat in a sequence. The randomness of the transition times makes the process aperiodic.

An *absorbing* state is one that locks in the system once it enters. Figure 13.6 shows a system with two absorbing states, 0 and 4. An event arc enters these states but none leaves. This system might represent the wealth of a gambler who begins with $2 and makes a series of wagers for $1 each. Let a_i be the event of winning in state i and let d_i be the event of losing in state i. The gambler quits when his wealth reaches $4 or when he has no money to wager. An absorbing state is said to be recurrent.

A *class* of states is defined as a set of states that communicate with each other. States in a class are either all recurrent or all transient and may be either all periodic or aperiodic.

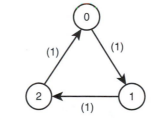

a. Each state is visited every three iterations

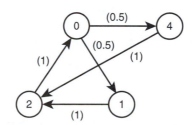

b. Each state is visited in multiples of three iterations

Figure 13.5 Periodic states

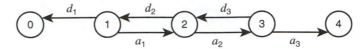

Figure 13.6   System with two absorbing states

The states in a transient class communicate only with each other, so no arcs enter a node in a transient class from a node not in the class. Arcs may leave, however, passing from a node in the class to one outside. In Figure 13.6, three classes exist. The first class consists of state 0 and is recurrent. The second class consists of state 4 and is similarly recurrent. The third class comprises the remaining nodes {1, 2, 3} and is transient. Figure 13.7 illustrates a system with two classes. The class consisting of nodes {0, 1, 2} is transient, because once the system has passed from 0 to 3, it will never return to the first class. The class consisting of nodes {3, 4, 5, 6} is recurrent, because once the system enters the class it never leaves. Both classes are periodic.

If the number of states is finite, the system is called finite; otherwise it is called infinite. In a finite system, there must be some class of states that is recurrent. If all the states communicate in a finite system, the system is recurrent and irreducible. For instance, the queuing system with a finite input source shown in Figure 13.8 has all recurrent states.

Illustration of Concepts

The diagrams in this discussion represent state-transition networks of stochastic processes. The accompanying matrices contain an X in entry (i, j) if a one-step transition is possible from state i to state j, and contain a 0 in entry (i, j) if the transition is not permitted. The problem is to classify each of the states, as well as the entire system. This requires only the connectivity properties of the networks, so no events or probabilities are shown.

Example 1

In the network depicted in Figure 13.9, we can see that a path can be traced from each state to every other state. Thus, every pair of states communicates, forming a single recurrent

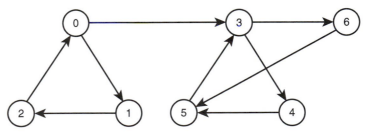

Figure 13.7   System with two classes of states

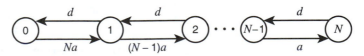

Figure 13.8   Queuing system with a finite input source

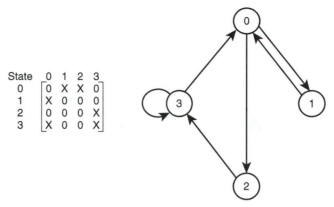

State	0	1	2	3
0	0	X	X	0
1	X	0	0	0
2	0	0	0	X
3	X	0	0	X

Figure 13.9   An aperiodic, irreducible Markov chain

class; however, the states are not periodic. Thus, the stochastic process is aperiodic and irreducible.

Example 2

Looking at the network in Figure 13.10, we see that states 0 and 1 communicate. No arcs leave this set, so states 0 and 1 comprise a recurrent class. State 2 communicates only with itself. Also, because no arcs leave state 2, it is absorbing and must be a recurrent class by itself. The opposite is true for state 3, which has one leaving arc but no entering arcs, so it must be a transient class. State 4 also has no entering arcs, making it inaccessible. Because an arc leaves state 4, it must comprise a unitary transient class. In summary, four classes of states have been defined.

Example 3

The graph in Figure 13.11 indicates that every state communicates with every other state, so it must represent an irreducible stochastic process. More careful review will show that

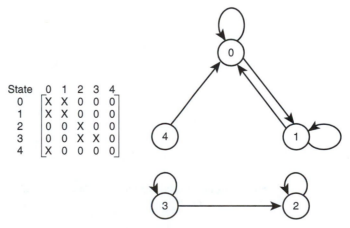

State	0	1	2	3	4
0	X	X	0	0	0
1	X	X	0	0	0
2	0	0	X	0	0
3	0	0	X	X	0
4	X	0	0	0	0

Figure 13.10   Markov chain with multiple classes

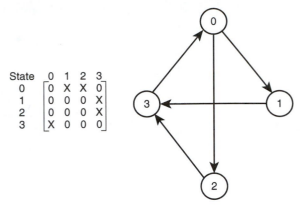

$$
\begin{array}{c c}
\text{State} & \begin{array}{c c c c}
0 & 1 & 2 & 3
\end{array} \\
\begin{array}{c}
0 \\ 1 \\ 2 \\ 3
\end{array} &
\left[
\begin{array}{c c c c}
0 & X & X & 0 \\
0 & 0 & 0 & X \\
0 & 0 & 0 & X \\
X & 0 & 0 & 0
\end{array}
\right]
\end{array}
$$

Figure 13.11   An irreducible, periodic Markov chain

every path from a given state back to itself must pass through a multiple of three arcs. For instance, starting at state 0, there are two possible paths that return to state 0: (0, 1, 3, 0) and (0, 2, 3, 0). Assuming the network represents a Markov chain with fixed length intervals, state 0 is periodic with period 3. Starting from state 1, paths returning to the state are: (1, 3, 0, 1), (1, 3, 0, 2, 3, 0, 1), and others that repeat the cycle (0, 2, 3, 0) more than once. Although the system may not return to state 1 every three steps, all paths are multiples of three arcs. State 1, therefore, is periodic with a period of 3. States 2 and 3 are also periodic with a period of 3. Consequently, the Markov chain is irreducible and periodic.

Example 4

In the network in Figure 13.12, states 0 and 3 communicate, but since arcs leave the set, the two states form a transient class. State 1 communicates only with itself, so it is absorbing and forms a recurrent class. Similarly, state 2 communicates only with itself, so it is a recurrent class and an absorbing state. Thus, the stochastic process has three classes of states, one transient and two recurrent.

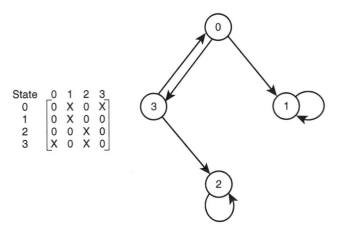

$$
\begin{array}{c c}
\text{State} & \begin{array}{c c c c}
0 & 1 & 2 & 3
\end{array} \\
\begin{array}{c}
0 \\ 1 \\ 2 \\ 3
\end{array} &
\left[
\begin{array}{c c c c}
0 & X & 0 & X \\
0 & X & 0 & 0 \\
0 & 0 & X & 0 \\
X & 0 & X & 0
\end{array}
\right]
\end{array}
$$

Figure 13.12   Markov chain with transient and recurrent states

13.3 MULTISTEP TRANSITION MATRIX

The problem we now consider concerns the computation of probability information after a specified number of periods or steps. The matrix **P** gives the transition probability from each state to every other state in a single period. The n-step transition probability $p_{ij}^{(n)}$ is the probability that, given that the system is in state i at some time τ, the system will be in state j at time $\tau + n$. The n-step transition matrix $\mathbf{P}^{(n)}$ gives the n-step transition probability from each state to every other state. If not otherwise stated, it is tacitly assumed that $\tau = 0$.

Property 1: Let $\{X_n : n = 0, 1, \ldots\}$ be a Markov chain with state space S and state-transition matrix **P**. Then, for i and $j \in S$, and $n = 1, 2, \ldots,$

$$\Pr\{X_n = j \mid X_0 = i\} = p_{ij}^{(n)}$$

where the right-hand side represents the ijth element of the matrix $\mathbf{P}^{(n)}$.

To compute the probability $p_{ij}^{(n)}$, we use the Chapman–Kolmogorov (C–K) equations.

$$p_{ij}^{(n)} = \sum_{r=0}^{\infty} p_{ir}^{(s)} p_{rj}^{(n-s)} \text{ for all } s = 1, \ldots, n-1 \text{ and for all } i, j$$

where, of course, $p_{ij}^{(1)} = p_{ij}$. These equations follow directly from the Markovian property and are most easily understood by noting that $p_{ir}^{(s)} p_{rj}^{(n-s)}$ represents the probability that, starting in state i, the process goes to state j in n transitions through a path that takes it into state r after s transitions. Hence, summing over all intermediate states r yields the probability that the process will be in state j after n transitions.

Identifying the transition probabilities as elements of the corresponding transition matrices, the C–K equations can be written in matrix notation.

$$\mathbf{P}^{(n)} = \mathbf{P}^{(s)}\mathbf{P}^{(n-s)} = \mathbf{P}^{(n-s)}\mathbf{P}^{(s)} \text{ for all } s = 1, \ldots, n-1 \text{ and for all } i, j$$

The order of multiplication is arbitrary. The choice of s is also arbitrary. The matrix $\mathbf{P}^{(n)}$ is found by multiplying together any two multistep transition matrices whose step indices sum to n.

Setting $s = 1$, we obtain a sequential expression from which the n-step transition matrix may be computed.

$$\mathbf{P}^{(n)} = \mathbf{P}\mathbf{P}^{(n-1)} \tag{4}$$

This is illustrated with a Markov chain that has two states labeled 0 and 1 and the transition matrix

$$\mathbf{P} = \begin{bmatrix} 0.2 & 0.8 \\ 0.7 & 0.3 \end{bmatrix}$$

To find $\mathbf{P}^{(2)}$, we use Equation (4) with $n = 2$.

$$\mathbf{P}^{(2)} = \begin{bmatrix} 0.2 & 0.8 \\ 0.7 & 0.3 \end{bmatrix}\begin{bmatrix} 0.2 & 0.8 \\ 0.7 & 0.3 \end{bmatrix} = \begin{bmatrix} 0.6 & 0.4 \\ 0.35 & 0.65 \end{bmatrix}$$

To find $\mathbf{P}^{(4)}$, compute $\mathbf{P}^{(4)} = \mathbf{P}\mathbf{P}\mathbf{P}\mathbf{P} = \mathbf{P}^{(2)}\mathbf{P}^{(2)}$.

$$\mathbf{P}^{(4)} = \begin{bmatrix} 0.6 & 0.4 \\ 0.35 & 0.65 \end{bmatrix}\begin{bmatrix} 0.6 & 0.4 \\ 0.35 & 0.65 \end{bmatrix} = \begin{bmatrix} 0.5 & 0.5 \\ 0.4375 & 0.5625 \end{bmatrix}$$

With regard to $p_{ij}^{(4)}$, the probability that the system starts in state i and arrives in state j after four transitions, we have

$$\Pr\{X_4 = 0 \mid X_0 = 0\} = 0.5, \quad \Pr\{X_4 = 1 \mid X_0 = 0\} = 0.5$$
$$\Pr\{X_4 = 0 \mid X_0 = 1\} = 0.4375, \quad \Pr\{X_4 = 1 \mid X_0 = 1\} = 0.5625$$

13.4   TRANSIENT PROBABILITY VECTOR

The transient probability vector $\mathbf{q}(n)$ gives the probability that the system is in each state at time n. For state space $S = \{0, 1, \ldots, m - 1\}$, we have

$$\mathbf{q}(n) = (q_0(n), q_1(n), \ldots, q_{m-1}(n))$$

The value of this vector depends on the initial state of the system given by $\mathbf{q}(0)$. If the system resides in some state i at time 0, one constructs the initial state probability vector $\mathbf{q}(0)$ by placing a 1 in position i and zeros in the other $m - 1$ positions. Then the probability vector for some later time n is computed by sequentially evaluating $\mathbf{q}(1), \mathbf{q}(2), \ldots, \mathbf{q}(n)$ using the following equation.

$$\mathbf{q}(n) = \mathbf{q}(n - 1)\mathbf{P} \tag{5}$$

Algebraically, the probability of being in state j at time n is

$$q_j(n) = \sum_{r=0}^{m-1} q_r(n-1)p_{rj}, \quad j = 0, 1, \ldots, m-1$$

The jth component of $\mathbf{q}(n)$ is the product of $\mathbf{q}(n - 1)$ and the jth column of $\mathbf{P}$. It is often easier to compute the vector $\mathbf{q}(n)$ rather than the entire matrix $\mathbf{P}^{(n)}$.

For the two-state example, given $\mathbf{q}(0) = (0, 1)$, we have

$$\mathbf{q}(1) = \mathbf{q}(0)\mathbf{P} = (0,1)\begin{bmatrix} 0.2 & 0.8 \\ 0.7 & 0.3 \end{bmatrix} = (0.7, 0.3)$$

$$\mathbf{q}(2) = \mathbf{q}(1)\mathbf{P} = (0.7, 0.3)\begin{bmatrix} 0.2 & 0.8 \\ 0.7 & 0.3 \end{bmatrix} = (0.35, 0.65)$$

$$\mathbf{q}(3) = \mathbf{q}(2)\mathbf{P} = (0.35, 0.65)\begin{bmatrix} 0.2 & 0.8 \\ 0.7 & 0.3 \end{bmatrix} = (0.525, 0.475)$$

$$\mathbf{q}(4) = \mathbf{q}(3)\mathbf{P} = (0.525, 0.475)\begin{bmatrix} 0.2 & 0.8 \\ 0.7 & 0.3 \end{bmatrix} = (0.4375, 0.5625)$$

The vector $\mathbf{q}(4)$ corresponds to the bottom row of $\mathbf{P}^{(4)}$ computed in Section 13.3, because the initial state of the system was state 1. If the system had started in state 0 [i.e., $\mathbf{q}(0) = (1, 0)$], we would have obtained the top row of $\mathbf{P}^{(4)}$. In general, when the initial state is i, $\mathbf{q}(0) = \mathbf{e}_i$, and the vector $\mathbf{q}(n)$ is equal to the ith row of $\mathbf{P}^{(n)}$. The components of $\mathbf{q}(n)$ must always sum to 1.

13.5   STEADY-STATE PROBABILITIES

In many cases, the n-step transient probabilities approach a limit as the number of periods increases. The limiting vector of state probabilities is called the steady-state vector and is denoted by $\boldsymbol{\pi}$.

In Section 13.2, much discussion was devoted to the classification of states and sets of states. Understanding those classifications is important, because the analysis of limiting behavior of a Markov chain is dependent on them. Sufficient conditions for convergence of the transient probabilities to steady-state probabilities are that the entire state space be finite, irreducible, and aperiodic. When these conditions are in effect, the steady-state probabilities are independent of the initial state and are not very difficult to compute, as we now formalize.

Equilibrium Conditions

Property 2: Let $\{X_n: n = 0, 1, \ldots \}$ be a Markov chain with finite state space S and state-transition matrix $\mathbf{P}$. Moreover, assume that S forms an irreducible, aperiodic set (recall that the former condition implies that all states are recurrent). We denote the steady-state probabilities by the vector $\boldsymbol{\pi}$, where

$$\pi_j = \lim_{n \to \infty} q_j(n), j = 0, 1, \ldots, m - 1 \text{ and any } \mathbf{q}(0)$$

or, equivalently,

$$\pi_j = \lim_{n \to \infty} p_{ij}^{(n)}, i, j = 0, 1, \ldots, m - 1$$

The solution to the following linear system gives the desired values.

$$\boldsymbol{\pi} = \boldsymbol{\pi} \mathbf{P} \tag{6}$$

$$\sum_{j \in S} \pi_j = 1 \tag{7}$$

We justify Equation (6) by taking the limits of both sides of Equation (5).

$$\lim_{n \to \infty} \mathbf{q}(n) = \lim_{n \to \infty} \mathbf{q}(n - 1)\mathbf{P}$$

When the limits exist, both $\lim_{n \to \infty} \mathbf{q}(n)$ and $\lim_{n \to \infty} \mathbf{q}(n - 1)$ equal $\boldsymbol{\pi}$, and we obtain Equation (6), which can be written as

$$\boldsymbol{\pi}(\mathbf{P} - \mathbf{I}) = \mathbf{0} \tag{8}$$

where $\mathbf{I}$ is the $m \times m$ identity matrix. Equation (7) must be true, since S includes all possible states. Writing Equation (8) explicitly, we have the following system of m equations in m unknowns.

$$
\begin{array}{cccccccc}
(p_{00} - 1)\pi_0 & + & p_{10}\pi_1 & + & p_{20}\pi_2 & \cdots & + & p_{m-1,0}\pi_{m-1} & = & 0 \\
p_{01}\pi_0 & + & (p_{11} - 1)\pi_1 & + & p_{21}\pi_2 & \cdots & + & p_{m-1,1}\pi_{m-1} & = & 0 \\
& & & & & & & \\
p_{0,m-1}\pi_0 & + & p_{1,m-1}\pi_1 & + & p_{2,m-1}\pi_2 & \cdots & + & (p_{m-1,m-1} - 1)\pi_{m-1} & = & 0
\end{array}
$$

Exactly one of these equations is redundant and must be replaced by Equation (7). When the conditions for the existence of the steady-state probabilities are satisfied, the resultant equations are independent and have a unique solution, $\boldsymbol{\pi}$. This vector will always be nonnegative.

To write a matrix expression for the steady-state probabilities, we form the matrix $\mathbf{A}_a$ by replacing the first column of $(\mathbf{P} - \mathbf{I})$ with all 1's, as follows.

$$\mathbf{A}_a = \begin{bmatrix} 1 & p_{01} & p_{02} & \cdots & p_{0,m-1} \\ 1 & p_{11}-1 & p_{12} & \cdots & p_{1,m-1} \\ 1 & p_{21} & p_{22}-1 & \cdots & p_{2,m-1} \\ \vdots & \vdots & \vdots & \cdots & \vdots \\ 1 & p_{m-1,1} & p_{m-1,2} & \cdots & p_{m-1,m-1}-1 \end{bmatrix}$$

Now the steady-state probabilities must satisfy

$$\boldsymbol{\pi}\mathbf{A}_a = \mathbf{e}_1 \tag{9}$$

where $\mathbf{e}_1 = (1, 0, \ldots, 0, 0)^{\mathrm{T}}$.

Solving Linear Equations to Find $\boldsymbol{\pi}$

For the computer repair example, we have

$$\mathbf{P} = \begin{bmatrix} 0.6 & 0.3 & 0.1 \\ 0.8 & 0.2 & 0 \\ 1 & 0 & 0 \end{bmatrix}, \mathbf{P}-\mathbf{I} = \begin{bmatrix} -0.4 & 0.3 & 0.1 \\ 0.8 & -0.8 & 0 \\ 1 & 0 & -1 \end{bmatrix}, \mathbf{A_a} = \begin{bmatrix} 1 & 0.3 & 0.1 \\ 1 & -0.8 & 0 \\ 1 & 0 & -1 \end{bmatrix}$$

which yields

$$\begin{aligned} \pi_0 + \quad \pi_1 + \quad \pi_2 &= 1 \\ 0.3\pi_0 - 0.8\pi_1 \qquad &= 0 \\ 0.1\pi_0 - \qquad\quad \pi_2 &= 0 \end{aligned}$$

Solving, we find $\boldsymbol{\pi} = (0.6780, 0.2542, 0.0678)$.

Flow Balance Approach

An alternative way of deriving the set of linear equations that does not depend on matrix manipulations and is often useful for hand computations is based on the principle of probability flow balance. Figure 13.13a depicts a general state-transition diagram for a three-state Markov chain. The transition diagram for the computer repair example is shown in Figure 13.13b.

The principle of probability flow balance means that when the system is in steady state, the flow of probability into a node representing a state must equal the flow of probability out. The flow from node i to node j is $p_{ij}\pi_i$. Referring to Figure 13.13b, we have

	Flow in	=	Flow out
State 0:	$0.8\pi_1 + \pi_2$	=	$0.4\pi_0$
State 1:	$0.3\pi_0$	=	$0.8\pi_1$
State 2:	$0.1\pi_0$	=	π_2

with normalization equation $\pi_0 + \pi_1 + \pi_2 = 1$.

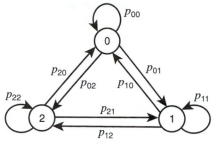

a. General flows

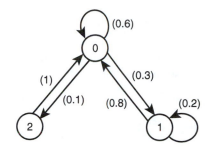

b. Computer repair example

Figure 13.13 State-transition diagram for three-state Markov chain

- From the equation for state 2: $\pi_2 = 0.1\pi_0$
- From the equation for state 1: $\pi_1 = 0.375\pi_0$
- From the normalizing equation: $\pi_0(1 + 0.375 + 0.1) = 1$
- Thus, $\pi_0 = 0.6780$, $\pi_1 = 0.2542$, $\pi_2 = 0.0678$.

The transition from a state to itself does not affect the balance equations, because this transition represents both flow in and flow out. The flow out of state i is simply π_i multiplied by the total of the transition probabilities out of the state less p_{ii}. Because the sum of the transition probabilities leaving a state must be 1, the coefficient on the right-hand side of each balance equation is $1 - p_{ii}$. In general, for the three-state chain, we have the following.

$$
\begin{array}{rcl}
\text{Flow in} & = & \text{Flow out} \\
\text{State 0:}\quad p_{10}\pi_1 + p_{20}\pi_2 & = & (1 - p_{00})\pi_0 \\
\text{State 1:}\quad p_{01}\pi_0 + p_{21}\pi_2 & = & (1 - p_{11})\pi_1 \\
\text{State 2:}\quad p_{02}\pi_0 + p_{12}\pi_1 & = & (1 - p_{22})\pi_2
\end{array}
$$

Since one of these equations is redundant, only two need to be constructed and then combined with the normalizing equation to obtain the desired linear system.

Multiple State Classes

Property 2 cannot be applied directly to a Markov chain whose state space is not irreducible—that is, has multiple classes of recurrent and/or transient state classes. Nevertheless, there is a way around this limitation, as Feldman and Valdez-Flores [1996] point out, when S is composed of recurrent and transient states. For such a case, all irreducible, recurrent sets are grouped together and the state-transition matrix **P** is rearranged so that the irreducible sets appear first and the transient states appear afterward. The new matrix will have the form

$$\mathbf{P} = \begin{bmatrix} \mathbf{P}_1 & & & & \\ & \mathbf{P}_2 & & & \\ & & \mathbf{P}_3 & & \\ & & & \ddots & \\ \mathbf{B}_1 & \mathbf{B}_2 & \mathbf{B}_3 & \cdots & \mathbf{B}_k \end{bmatrix}$$

where each submatrix $\mathbf{P}_l$ is a state-transition matrix for an irreducible subset of states indexed by C_l, $l = 1, \ldots, k - 1$, and the matrices $\mathbf{B}_1, \ldots, \mathbf{B}_k$ correspond to the one-step transitions for the transient states. It is now possible to apply Property 2 to each $\mathbf{P}_l$ with the following consequences.

- If state j is transient, $\lim_{n\to\infty} \Pr\{X_n = j \mid X_0 = i\} = 0$.
- If states i and j both belong to irreducible set C_l,

$$\lim_{n\to\infty} \Pr\{X_n = j \mid X_0 = i\} = \pi_j$$

where

$$\boldsymbol{\pi} = \boldsymbol{\pi}\mathbf{P}_l \text{ and } \sum_{j\in C_l}\pi_j = 1$$

- If state j is recurrent and X_0 is in the same irreducible set as j,

$$\lim_{n\to\infty} \frac{1}{n}\sum_{s=0}^{n-1}I(X_s = 1) = \pi_j \tag{10}$$

where $I(X_s = j) = 1$ if the process visited state j at step s, and 0 otherwise.

Equation (10) is quite general and should be intuitive. The notation $I(X_s = j)$ is used as a counter so that the summation on the left-hand side records the total number of times that the process is in state j. Thus, the equality indicates that the fraction of time spent in state j equals the steady-state probability of being in state j. As we have mentioned, this is the interpretation given to π_j for periodic systems. When

$$\lim_{n\to\infty} p_{ij}^{(n)} = \lim_{n\to\infty} \frac{1}{n}\sum_{s=0}^{n-1}I(X_s = 1) = \pi_j$$

the process is *ergodic*. All finite, irreducible, aperiodic Markov chains have this property.

13.6   ECONOMIC ANALYSIS

Markov chains are often used to analyze the net cost or revenue of an operation. The problem context will determine whether the analysis should be in terms of cost or revenue—but, of course, one can always be viewed as the negative of the other. For modeling purposes, we consider two kinds of economic effects: (1) those incurred when the system is in a specified state and (2) those incurred when the system makes a transition from one state to another.

Depending on the application, it may be appropriate to include one or both of these measures in the model.

To illustrate the use of the formulas presented in this section, we consider the stockbroker example described in Section 12.1. Recall that the broker can be in one of three states: $a, b,$ and c. The one- and two-step transition matrices and the steady-state solution are repeated here.

$$
\begin{array}{ccc} & a & b & c \end{array} \qquad\qquad \begin{array}{ccc} & a & b & c \end{array}
$$

$$
\mathbf{P} = \begin{bmatrix} 0 & 0.50 & 0.50 \\ 0.75 & 0 & 0.25 \\ 0.75 & 0.25 & 0 \end{bmatrix}, \quad \mathbf{P}^{(2)} = \begin{bmatrix} 0.75 & 0.125 & 0.125 \\ 0.1875 & 0.4375 & 0.375 \\ 0.1875 & 0.375 & 0.4375 \end{bmatrix}
$$

$$
\pi = (0.428, 0.286, 0.286)
$$

Expected Cost

In this development, the cost (revenue) of being in a particular state is represented by the m-dimensional column vector $\mathbf{C}^S = (c_0^S, c_1^S, \ldots, c_{m-1}^S)^T$, where c_i^S is the cost associated with state i. The cost of a transition is embodied in the $m \times m$ matrix $\mathbf{C}^R = (c_{ij}^R)$. Each component c_{ij}^R specifies the cost of going from state i to state j in a single step. We now define a third vector, $\mathbf{C} = (c_0, \ldots, c_{m-1})^T$, which is a combination of the two. Given that the system is in state i, the one-step probability of next being in state j is given by the ith row of the matrix $\mathbf{P}$. The expected cost of being in state i is then

$$
c_i = c_i^S + \sum_{j=0}^{m-1} c_{ij}^R p_{ij} \tag{11}
$$

We call $\mathbf{C}$ the expected state cost vector. The summation in Equation (11) represents the scalar product of the ith row of $\mathbf{C}^R$ and the ith row of $\mathbf{P}$. Because $\mathbf{C}$ can always be computed from its two constituent matrices and $\mathbf{P}$, it is only necessary to refer to $\mathbf{C}$ in the analysis.

For the stockbroker example, we have the following data.

$$
\mathbf{C}^S = \begin{bmatrix} -1700 \\ -1450 \\ -1600 \end{bmatrix}, \quad \mathbf{C}^R = \begin{bmatrix} 0 & 200 & 300 \\ 250 & 0 & 300 \\ 200 & 250 & 0 \end{bmatrix}, \quad \mathbf{C} = \begin{bmatrix} -1450 \\ -1187.5 \\ -1387.5 \end{bmatrix}
$$

The vector $\mathbf{C}^S$ represents the revenue earned by the broker while in each state. The transition cost matrix $\mathbf{C}^R$ shows the setup cost associated with switching from one list to another. The expected state cost $\mathbf{C}$ is computed using Equation (11). In contrast to the presentation in Chapter 12, we have expressed all matrices as costs rather than as revenues, so the negative entries in $\mathbf{C}^S$ and $\mathbf{C}$ represent net revenues.

To compute the expected cost during the nth step of the Markov chain, we make use of the result given in Property 3. Once again, X_n is the state of the system at step n and $f(X_n)$ is the cost. Both are random variables. The expected value operator is denoted by $E[\cdot]$.

Property 3: Let $\{X_n : n = 0, 1, \ldots \}$ be a Markov chain with finite state space S, state-transition matrix $\mathbf{P}$, and expected state cost vector $\mathbf{C}$. Assuming that the process starts in state i, the expected cost at the nth step is given by

$$
E[f(X_n) \mid X_0 = i] = \mathbf{e}_i \mathbf{P}^{(n)} \mathbf{C} \tag{12}
$$

where $\mathbf{e}_i$ is a new vector of 0's, except column i which is 1.

We illustrate the computations for $n = 2$. Using the previously defined matrices, we have

$$E[f(X_2) \mid X_0 = i] = \mathbf{e}_i \mathbf{P}^{(2)} \mathbf{C} = \mathbf{e}_i \begin{bmatrix} -1409 \\ -1312 \\ -1324 \end{bmatrix}$$

The vector $\mathbf{P}^{(2)}\mathbf{C}$ gives the expected cost for each starting state. Multiplying it by $\mathbf{e}_i$ selects the ith component. For example, for $i = 1$ for state a, we have

$$E[f(X_2) \mid X_0 = a] = \mathbf{e}_0 \mathbf{P}^{(2)} \mathbf{C} = -1409$$

Note that to determine the cumulative costs or returns through k transitions, it would be necessary to sum the results of Equation (12) for $n = 1, \ldots, k$.

Equation (12) assumes that the initial state is given. It is easily adjusted to accommodate a probability distribution, $\mathbf{q}(0)$, describing the state at $n = 0$. To compute the probability distribution of the state for step n, we can use Equation (5), which is now restated more formally.

Property 4: Let $\{X_n: n = 0, 1, \ldots\}$ be a Markov chain with finite state space S, state-transition matrix $\mathbf{P}$, and initial probability vector $\mathbf{q}(0)$ [i.e., $q_i(0) = \Pr\{X_0 = i\}$ for all $i \in S$]. Then

$$\Pr\{X_n = j \mid \mathbf{q}(0)\} = \sum_{i \in S} q_i(0) p_{ij}^{(n)}$$

or, in vector form,

$$\mathbf{q}(n) = \mathbf{q}(0)\mathbf{P}^{(n)}$$

which is equivalent to Equation (5).

The previous two properties can be combined into one statement.

Property 5: Let $\{X_n: n = 0, 1, \ldots\}$ be a Markov chain with finite state space S, state-transition matrix $\mathbf{P}$, initial probability vector $\mathbf{q}(0)$, and expected state cost vector $\mathbf{C}$. The expected economic return at the nth step is given by

$$E[f(X_n) \mid \mathbf{q}(0)] = \mathbf{q}(0)\mathbf{P}^{(n)}\mathbf{C} \tag{13}$$

For example, suppose that we do not know for sure which list the broker started with but know that there is a 30% chance that it was list a, a 50% chance that it was list b, and a 20% chance that it was list c. The expected cost in the second week is calculated to be

$$E[f(X_2) \mid \mathbf{q}(0)] = \mathbf{q}(0)\mathbf{P}^{(2)}\mathbf{C} = (0.30, 0.50, 0.20) \begin{bmatrix} -1409 \\ -1312 \\ -1324 \end{bmatrix} = -1344$$

When the goal is to determine long-run average cost per period, we have the following result.

Property 6: Let $\{X_n: n = 0, 1, \ldots\}$ be an irreducible Markov chain with finite state space S, state-state probabilities given by the vector $\boldsymbol{\pi}$, and expected state cost vector $\mathbf{C}$. Then the long-run average return per unit time is given by

$$\lim_{n\to\infty} \frac{1}{n} \sum_{s=0}^{n-1} f(X_s) = \sum_{i\in S} \pi_i c_i = \boldsymbol{\pi}\mathbf{C} \tag{14}$$

For the stockbroker, considering both revenue and setup costs,

$$\boldsymbol{\pi}\mathbf{C} = (0.428, 0.286, 0.286) \begin{bmatrix} -1450 \\ -1187.5 \\ -1387.5 \end{bmatrix} = -1357$$

Thus, his long-run average weekly profit is $1357.

Discounted Cash Flow

In many situations it is necessary for the decision maker to take into account the time value of money by computing the net present value of the cash flows generated by the Markov chain. Letting r be the interest rate per period, the discount factor $\alpha = 1/(1 + r)$ gives the present value of $1 received one period from now. For computational purposes, it will be assumed that all cash flows take place at the beginning of the period.

Suppose that a Markov chain generates the sequence of states $X_0, X_1, X_2, \ldots$ which continues indefinitely. The sequence of cash flows generated by the states is $f(X_0), f(X_1), f(X_2), \ldots$ The discounted cash flow is the net present value (NPV) which is computed in the following equation.

$$\text{NPV} = \sum_{n=0}^{\infty} \alpha^n f(X_n)$$

In this expression, α^n is the nth power of the discount factor. Because the individual cash flows are random variables, we would like to compute the expected value of the discounted cash flow when, say, the process starts in state i.

$$E[\text{NPV}] = E\left[\sum_{n=0}^{\infty} \alpha^n f(X_n) \mid X_0 = i \right]$$

The following result indicates how to perform these calculations.

Property 7: Let $\{X_n: n = 0, 1, \ldots \}$ be a Markov chain with finite state space S, state-transition matrix $\mathbf{P}$, expected state cost vector $\mathbf{C}$, and discount factor α, where $0 < \alpha < 1$. The total expected discounted cost is given by

$$E\left[\sum_{n=0}^{\infty} \alpha^n f(X_n) \mid X_0 = i \right] = ([\mathbf{I} - \alpha\mathbf{P}]^{-1}\mathbf{C})_i$$

where the subscript on the right-hand side refers to the ith component of the vector obtained after performing the matrix operations.

Note that the preceding formula is a function of the initial state i. If this value at $n = 0$ is not known but the unconditional probability distribution, $\mathbf{q}(0)$, is known, then the total expected discounted return is given by $\mathbf{q}(0)[\mathbf{I} - \alpha\mathbf{P}]^{-1}\mathbf{C}$, where $\mathbf{q}(0)$ is an m-dimensional row vector.

To illustrate the computations, let us consider the stockbroker with a rate of return for his time of 15% annually. This translates into a weekly interest rate of 0.29% with discount factor $\alpha = 1/(1 + 0.0029) = 0.99711$. For the given data, we have

$$[\mathbf{I} - \alpha\mathbf{P}]^{-1}\mathbf{C} = \left(\begin{bmatrix} 1 & 0 & 0 \\ 0 & 1 & 0 \\ 0 & 0 & 1 \end{bmatrix} - 0.99711 \begin{bmatrix} 0 & 0.50 & 0.50 \\ 0.75 & 0 & 0.25 \\ 0.75 & 0.25 & 0 \end{bmatrix} \right)^{-1} \begin{bmatrix} 1409 \\ 1311 \\ 1324 \end{bmatrix}$$

$$= \begin{bmatrix} 1 & -0.499 & -0.499 \\ -0.7478 & 1 & -0.249 \\ -0.7478 & -0.249 & 1 \end{bmatrix}^{-1} \begin{bmatrix} 1409 \\ 1311 \\ 1324 \end{bmatrix}$$

$$= \begin{bmatrix} 148.62 & 98.7 & 98.7 \\ 148.05 & 99.39 & 98.59 \\ 148.05 & 98.59 & 99.39 \end{bmatrix} \begin{bmatrix} 1409 \\ 1311 \\ 1324 \end{bmatrix} = \begin{bmatrix} 471{,}863 \\ 471{,}806 \\ 471{,}816 \end{bmatrix}$$

Thus, if the broker starts with list *a*, his long-run expected return is \$471,863. The other values are similar. The fact that the three final values are very close suggests that the initial state is not very important, at least in this case.

13.7   FIRST PASSAGE TIMES

In many applications, an important statistic is the number of steps (periods) required to reach some state *j* for the first time when the system begins in some arbitrary state *i*. This number is called the *first passage time* from state *i* to state *j*. Because the state transitions for a Markov chain are governed by probabilities, the first passage time is a discrete random variable taking on positive integer values. The smallest value is 1, because it will take at least one step to make the passage.

This section is concerned with computing the probability distribution of the first passage time along with its mean. In the developments, we make use of the following terms.

First passage time: The number of steps (time intervals) of the Markov chain required for the first passage from some state *i* to some other state *j*. This is a random variable denoted by T_j. Mathematically, we have

$$T_j = \min\{n \ge 1: X_n = j\}$$

where the minimum of the empty set is taken to be $+\infty$.

Probability distribution of the first passage time, $f_{ij}^{(n)}$: The probability that the first passage from *i* to *j* will take *n* steps, where *n* ranges from 1 to ∞. The probability that the process will reach every state *j* from state *i* is given by the equation

$$F_{ij} = \sum_{n=1}^{\infty} f_{ij}^{(n)} = \Pr\{T_j < \infty \mid X_0 = i\}$$

As we will see, the values of $f_{ij}^{(n)}$ are computed by a recursive procedure.

Expected first passage time, μ_{ij}: The expected value of the first passage time from state *i* to state *j* . This expectation is defined by

$$E[T_j \mid X_0 = i] \equiv \mu_{ij} = \begin{cases} \infty, & \text{if } \sum_{n=1}^{\infty} f_{ij}^{(n)} < 1 \\ \sum_{n=1}^{\infty} n f_{ij}^{(n)}, & \text{if } \sum_{n=1}^{\infty} f_{ij}^{(n)} = 1 \end{cases}$$

and can be computed by solving a set of linear equations.

Recurrent state: A state of a Markov chain that has the characteristic that if the system ever enters the state, it will eventually return to it. For a *positive* recurrent state i, we have

$$F_{ii} = \sum_{n=1}^{\infty} f_{ii}^{(n)} = 1 \text{ and } \mu_{ii} < \infty$$

and for a *null* recurrent state i we have

$$F_{ii} = \sum_{n=1}^{\infty} f_{ii}^{(n)} = 1 \text{ and } \mu_{ii} = \infty$$

Absorbing state: A state that has the characteristic that once the process enters the state it never leaves it. An absorbing state is also recurrent. If state i is an absorbing state, then $p_{ii} = 1$.

Transient state: A state that has the characteristic that once the process leaves the state, there is a finite probability that it will never return to it. For a transient state i, we have

$$F_{ii} = \sum_{n=1}^{\infty} f_{ii}^{(n)} < 1$$

Number of visits to a state: The total number of visits that the system makes to a fixed state j throughout the life of the Markov chain. This number is a random variable denoted by N_j and defined mathematically as

$$N_j = \sum_{n=0}^{\infty} I(X_n = j)$$

where the counter function $I(X_n = j) = 1$ if $X_t = j$, and 0 otherwise.

Expected number of visits: The expected number of visits to state j given that the initial state was i is denoted by R_{ij} and is defined by

$$R_{ij} = E[N_j \mid X_0 = i]$$

As might be expected, there is a close relationship between F_{ij} and R_{ij}, as we now state.

Property 8: Let F_{ij} and R_{ij} be as previously defined. Then

$$R_{ij} = \begin{cases} 1 / (1 - F_{jj}) & \text{for } i = j \\ F_{ij} / (1 - F_{jj}) & \text{for } i \neq j \end{cases}$$

where the convention 0/0 = 0 is used.

First Passage Time Distribution

Consider any two states i and j and assume that the system is in state i. The probabilities that the first passage time requires only one step is the probability of a one-step transition to state j, p_{ij}, so

$$f_{ij}^{(1)} = p_{ij}$$

The probability that the first passage requires exactly two steps is the probability of a two-step transition from state i to state j, $p_{ij}^{(2)}$, less the probability that the first passage time was one step *and* the system remained in state j.

$$f_{ij}^{(2)} = p_{ij}^{(2)} - f_{ij}^{(1)} p_{ij}$$

The remaining probabilities are derived in a similar manner, with the probability of an n-step transition reduced by the probability that the first passage from i to j occurred in fewer than n steps. Continuing for three-step, four-step, and finally the general n-step first passage probabilities, we have

$$f_{ij}^{(3)} = p_{ij}^{(3)} - f_{ij}^{(1)} p_{jj}^{(2)} - f_{ij}^{(2)} p_{jj}$$

$$f_{ij}^{(4)} = p_{ij}^{(4)} - f_{ij}^{(1)} p_{jj}^{(3)} - f_{ij}^{(2)} p_{jj}^{(2)} - f_{ij}^{(3)} p_{jj}$$

$$\cdot \qquad \cdot$$
$$\cdot \qquad \cdot$$
$$\cdot \qquad \cdot$$

$$f_{ij}^{(n)} = p_{jj}^{(n)} - f_{ij}^{(1)} p_{jj}^{(n-1)} - f_{ij}^{(2)} p_{jj}^{(n-2)} - \cdots - f_{ij}^{(n-1)} p_{jj}$$

To use these equations to evaluate the probabilities from 1 to n, all the multistep transition matrices from 1 to n must be available. The equations are evaluated recursively, first finding $f_{ij}^{(1)}$ then finding $f_{ij}^{(2)}$, and so on.

Alternatively, the first passage probabilities can be derived using a conditional first-step argument. The corresponding recursion is

$$f_{ij}^{(n)} = \begin{cases} \displaystyle\sum_{\substack{k=0 \\ k \neq j}}^{m-1} p_{ik}\, f_{kj}^{(n-1)}, & n > 1 \\[4mm] p_{ij}, & n = 1 \end{cases}$$

For $n > 1$, we can write this expression as

$$f_{ij}^{(n)} = \sum_{k=0}^{m-1} p_{ik} f_{kj}^{(n-1)} - p_{ij} f_{jj}^{(n-1)} = \sum_{\substack{k=0 \\ k \neq j}}^{m-1} p_{ik}\, f_{kj}^{(n-1)}$$

In matrix form with $\mathbf{F}^{(n)} = (f_{ij}^{(n)})$, we have $\mathbf{F}^{(1)} = \mathbf{P}$ and, for $n > 1$, $\mathbf{F}^{(n)} = \mathbf{PF}^{(n-1)} - \mathbf{PF}_d^{(n-1)}$, where $\mathbf{F}_d^{(n-1)}$ is the diagonal matrix formed by the diagonal elements of $\mathbf{F}^{(n-1)}$.

To illustrate the computations using the first set of recursive equations, consider again the computer repair example, where

$$\mathbf{P} = \begin{bmatrix} 0.6 & 0.3 & 0.1 \\ 0.8 & 0.2 & 0 \\ 1 & 0 & 0 \end{bmatrix}, \quad \mathbf{P}^{(2)} = \begin{bmatrix} 0.7 & 0.24 & 0.06 \\ 0.64 & 0.28 & 0.08 \\ 0.6 & 0.3 & 0.1 \end{bmatrix},$$

$$\mathbf{P}^{(3)} = \begin{bmatrix} 0.672 & 0.258 & 0.07 \\ 0.688 & 0.248 & 0.064 \\ 0.7 & 0.24 & 0.06 \end{bmatrix}, \text{ and } \mathbf{P}^{(4)} = \begin{bmatrix} 0.6796 & 0.2532 & 0.0672 \\ 0.6752 & 0.256 & 0.0688 \\ 0.672 & 0.258 & 0.07 \end{bmatrix}$$

We will find the first four components of the probability distribution of the first passage time from state 0 to state 0. The first passage time from a state to itself is called the *recurrence time*. From the preceding equations, we obtain

$$f_{00}^{(1)} = p_{00} = 0.6$$

$$f_{00}^{(2)} = p_{00}^{(2)} - f_{00}^{(1)} p_{00} = 0.7 - (0.6)(0.6) = 0.34$$

$$f_{00}^{(3)} = p_{00}^{(3)} - f_{00}^{(1)} p_{00}^{(2)} - f_{00}^{(2)} p_{00}$$
$$= 0.672 - (0.6)(0.7) - (0.34)(0.6) = 0.048$$

$$f_{00}^{(4)} = p_{00}^{(4)} - f_{00}^{(1)} p_{00}^{(3)} - f_{00}^{(2)} p_{00}^{(2)} - f_{00}^{(3)} p_{00}$$
$$= 0.6796 - (0.6)(0.672) - (0.34)(0.7) - (0.048)(0.6)$$
$$= 0.0096$$

If we call Ψ_0 the recurrence time for state 0, the probability distribution of Ψ_0 is

n	1	2	3	4	≥ 5
$\Pr\{\Psi_0 = n\}$	0.6	0.34	0.048	0.0096	0.0024

Expected First Passage Time

When all the states are recurrent, the following set of equations can be used to find the expected first passage times from each state i into some given state j.

$$\mu_{ij} = 1 + \sum_{r \neq j} p_{ir} \mu_{rj}, \quad i = 0, 1, \ldots, m - 1 \tag{15}$$

To justify Equation (15), we begin by noting that the expected first passage time is at least 1. This occurs when the system enters state j at the first step. In general, the system will be in state r after the first step with probability p_{ir}, and the expected *additional* time from r to j is μ_{rj}. The second term in Equation (15) sums over all possible values of the next state.

Given state j, Equation (15) corresponds to a set of m linear equations in m unknowns, μ_{ij}, $i = 0, 1, \ldots, m - 1$. The coefficients in these equations are the elements of the one-step transition matrix.

Continuing with the example, we illustrate the calculations by finding the expected value of the recurrence time for state $j = 0$. To do this, we must construct the set of equations for all first passages terminating at state 0.

$$\mu_{i0} = 1 + \sum_{r \neq 0} p_{ir} \mu_{r0} \text{ for } i = 0, 1, 2$$

$i = 0$: $\mu_{00} = 1 + p_{01}\mu_{10} + p_{02}\mu_{20}$ or $\mu_{00} = 1 + (0.3)\mu_{10} + (0.1)\mu_{20}$

$i = 1$: $\mu_{10} = 1 + p_{11}\mu_{10} + p_{12}\mu_{20}$ or $\mu_{10} = 1 + (0.2)\mu_{10} + (0.0)\mu_{20}$

$i = 2$: $\mu_{20} = 1 + p_{21}\mu_{10} + p_{22}\mu_{20}$ or $\mu_{20} = 1 + (0.0)\mu_{10} + (0.0)\mu_{20}$

The third equation indicates that $\mu_{20} = 1$ (i.e., one step). The second equation tells us that $\mu_{10} = 1/0.8 = 1.25$ (steps). Substituting these values into the first equation yields the final value $\mu_{00} = 1 + (0.3)(1.25) + (0.1)(1) = 1.475$.

Equation (15) also has a matrix description that may be preferable for computer implementation. Let $\mathbf{m}_j$ be the column vector of expected first passage times into state j, let $\mathbf{P}_j$ be the matrix $\mathbf{P}$ with column j replaced by zeros, and let $\mathbf{e}$ be an m-dimensional column vector of 1's. Then the equation defining the expected first passage times is

$$\mathbf{m}_j = \mathbf{e} + \mathbf{P}_j \mathbf{m}_j$$

Solving for $\mathbf{m}_j$ yields

$$\mathbf{m}_j = [\mathbf{I} - \mathbf{P}_j]^{-1}\mathbf{e}$$

For the example with $j = 0$, we have

$$\mathbf{P}_j = \begin{bmatrix} 0 & 0.3 & 0.1 \\ 0 & 0.2 & 0 \\ 0 & 0 & 0 \end{bmatrix}, \quad \mathbf{I} - \mathbf{P}_j = \begin{bmatrix} 1 & -0.3 & -0.1 \\ 0 & 0.8 & 0 \\ 0 & 0 & 1 \end{bmatrix},$$

$$[\mathbf{I} - \mathbf{P}_j]^{-1} = \begin{bmatrix} 1 & 0.375 & 0.1 \\ 0 & 1.25 & 0 \\ 0 & 0 & 1 \end{bmatrix} \text{ and } \mathbf{m}_j = [\mathbf{I} - \mathbf{P}_j]^{-1}\mathbf{e} = \begin{pmatrix} 1.475 \\ 1.25 \\ 1 \end{pmatrix}$$

Expected Recurrence Time

When $j = i$, the first passage time is simply the number of steps until the system returns to the initial state i. It is the recurrence time—the first passage time from a state back to itself. The expected recurrence time, μ_{ii}, can be computed using either Equation (12) or Equation (13) with $i = j$, but this is unnecessarily cumbersome. A more efficient method is based on the recognition that μ_{ii} is the inverse of the steady-state probability for state i—that is,

$$\mu_{ii} = 1/\pi_i, \ i = 0, 1, \ldots, m - 1$$

For our example, we have

$$\mu_{00} = 1/0.678 = 1.475, \ \mu_{11} = 1/0.2542 = 3.933, \ \mu_{22} = 1/0.0678 = 14.75$$

13.8   ABSORBING STATE PROBABILITIES

When a Markov chain has one or more absorbing states, the others must be transient. In many situations, we would like to know the probability that a particular absorbing state will ultimately be reached.

To analyze this problem, assume that the transient states are numbered 0 through k and that the absorbing states are numbered $k + 1$ through $m - 1$. The $\mathbf{P}$ matrix will be partitioned accordingly into four submatrices

$$\mathbf{P} = \begin{bmatrix} \mathbf{S} & \mathbf{T} \\ \mathbf{0} & \mathbf{I} \end{bmatrix}$$

where

$\mathbf{S}$ is the $(k + 1) \times (k + 1)$ matrix of transition probabilities between transient states,

$\mathbf{T}$ is the $(k + 1) \times (m - k - 1)$ matrix of transition probabilities from transient to absorbing states,

$\mathbf{0}$ is an $(m - k - 1) \times (k + 1)$ matrix of zeros (there can be no transitions from absorbing states to transient states), and

$\mathbf{I}$ is the order $m - k - 1$ identity matrix indicating that once the system enters an absorbing state it never leaves it.

As an example, consider the game of craps example discussed in Section 12.4 with $m = 9$. The seven states Start, P4, P5, P6, P8, P9, P10 are transient, and the two states Win and Lose are absorbing, so $k = 6$. Dividing the matrix **P** into its four components yields the following nontrivial submatrices.

	Win	Lose		Start	P4	P5	P6	P8	P9	P10	
	0.222	0.111		0	0.083	0.111	0.139	0.139	0.111	0.083	Start
	0.083	0.167		0	0.75	0	0	0	0	0	P4
	0.111	0.167		0	0	0.722	0	0	0	0	P5
$\mathbf{T} =$	0.139	0.167	$\mathbf{S} =$	0	0	0	0.694	0	0	0	P6
	0.139	0.167		0	0	0	0	0.694	0	0	P8
	0.111	0.167		0	0	0	0	0	0.722	0	P9
	0.083	0.167		0	0	0	0	0	0	0.75	P10

Our goal is to compute the probability, call it q_{ij}, that the system will pass to absorbing state j if it begins in transient state i. The matrix of these probabilities will be denoted by **Q** and has $k + 1$ rows and $m - k - 1$ columns, the same dimensions as **T**. Logically, a transition from i to j can result from two out of $k + 1$ mutually exclusive events. Either state j is reached directly with probability p_{ij}, or j is reached from some other transient state $r \neq j = 0, \ldots, k$. In the latter case, the joint probability of the independent events of passing to state r and then proceeding to state j is the product $p_{ir}q_{rj}$. Writing the equations for each absorbing state j for $j = k + 1, \ldots, m - 1$, we obtain

$$q_{ij} = p_{ij} + \sum_{r=0}^{k} p_{ir}q_{rj}, \quad i = 0, \ldots, k$$

For j fixed, we get $k + 1$ linear equations in $k + 1$ unknowns, q_{ij}, $i = 0, \ldots, k$. In matrix notation, this is equivalent to

$$\mathbf{Q} = \mathbf{T} + \mathbf{SQ}$$

or

$$\mathbf{Q} = [\mathbf{I} - \mathbf{S}]^{-1}\mathbf{T}$$

Using the Excel add-in on the accompanying CD, we find the absorbing probabilities for the game of craps as follows.

	Win	Lose	
	0.493	0.507	Start
	0.333	0.667	P10
	0.333	0.667	P4
$\mathbf{Q} =$	0.400	0.600	P5
	0.455	0.545	P6
	0.455	0.545	P8
	0.400	0.600	P9

The most interesting results are the win–lose probabilities for the transient state Start. As can be seen, the house has a 1.4% advantage at the outset [$100 \times (0.507 - 0.493)\%$]. This advantage increases for subsequent rolls, as indicated by the other absorbing probabilities. The corresponding values define the shooter's chances if she throws a point on her first turn.

EXERCISES

1. For each transition matrix, identify and classify the states of the corresponding stochastic process. Also, indicate whether or not the process is irreducible. The symbol "$\times$" in the matrices indicates that a transition is possible.

(a)

$$
\begin{array}{c}
0\ \ 1\ \ 2\ \ 3\ \ 4 \\
\begin{bmatrix}
0 & 0 & 0 & \times & \times \\
0 & \times & \times & 0 & 0 \\
0 & \times & \times & 0 & 0 \\
\times & 0 & 0 & 0 & 0 \\
\times & 0 & 0 & 0 & 0
\end{bmatrix}
\begin{array}{c} 0 \\ 1 \\ 2 \\ 3 \\ 4 \end{array}
\end{array}
$$

(b)

$$
\begin{array}{c}
0\ \ 1\ \ 2\ \ 3 \\
\begin{bmatrix}
\times & 0 & 0 & 0 \\
0 & \times & \times & 0 \\
0 & \times & \times & 0 \\
\times & 0 & 0 & \times
\end{bmatrix}
\begin{array}{c} 0 \\ 1 \\ 2 \\ 3 \end{array}
\end{array}
$$

(c)

$$
\begin{array}{c}
0\ \ 1\ \ 2\ \ 3 \\
\begin{bmatrix}
\times & \times & \times & 0 \\
0 & \times & 0 & \times \\
0 & 0 & \times & 0 \\
0 & \times & 0 & 0
\end{bmatrix}
\begin{array}{c} 0 \\ 1 \\ 2 \\ 3 \end{array}
\end{array}
$$

(d)

$$
\begin{array}{c}
0\ \ 1\ \ 2\ \ 3 \\
\begin{bmatrix}
0 & 0 & \times & \times \\
0 & 0 & \times & 0 \\
0 & \times & 0 & 0 \\
\times & \times & 0 & 0
\end{bmatrix}
\begin{array}{c} 0 \\ 1 \\ 2 \\ 3 \end{array}
\end{array}
$$

(e)

$$
\begin{array}{c}
0\ \ 1\ \ 2\ \ 3 \\
\begin{bmatrix}
0 & 0 & \times & \times \\
0 & \times & \times & 0 \\
\times & \times & 0 & 0 \\
\times & 0 & 0 & 0
\end{bmatrix}
\begin{array}{c} 0 \\ 1 \\ 2 \\ 3 \end{array}
\end{array}
$$

(f)

$$
\begin{array}{c}
0\ \ 1\ \ 2\ \ 3 \\
\begin{bmatrix}
0 & \times & 0 & 0 \\
0 & 0 & \times & 0 \\
0 & 0 & 0 & \times \\
\times & 0 & 0 & 0
\end{bmatrix}
\begin{array}{c} 0 \\ 1 \\ 2 \\ 3 \end{array}
\end{array}
$$

2. The following matrices give the state-transition probabilities for a Markov chain. Classify the states for each matrix. Also, indicate whether the Markov chain is irreducible.

(a)

$$
\begin{bmatrix}
0 & 0 & 0 & 0.5 & 0.5 & 0 \\
0 & 0 & 0 & 0 & 0 & 1 \\
0 & 0 & 0 & 0.1 & 0.9 & 0 \\
0.1 & 0.01 & 0.89 & 0 & 0 & 0 \\
0.8 & 0 & 0.2 & 0 & 0 & 0 \\
0 & 1 & 0 & 0 & 0 & 0
\end{bmatrix}
$$

(b)

$$
\begin{bmatrix}
0 & 0 & 0 & 0.5 & 0.5 & 0 \\
0 & 0 & 0 & 0 & 0 & 1 \\
0 & 0 & 0 & 0.1 & 0.9 & 0 \\
0.1 & 0.01 & 0.89 & 0 & 0 & 0 \\
0.8 & 0 & 0.2 & 0 & 0 & 0 \\
0.1 & 0.9 & 0 & 0 & 0 & 0
\end{bmatrix}
$$

(c)

$$
\begin{bmatrix}
0 & 1 & 0 & 0 \\
0 & 0 & 1 & 0 \\
0.3 & 0 & 0 & 0.7 \\
0 & 1 & 0 & 0
\end{bmatrix}
$$

(d)

$$
\begin{bmatrix}
0.2 & 0.8 & 0 & 0 \\
0 & 0.9 & 0.1 & 0 \\
0 & 0 & 1 & 0 \\
0 & 0 & 0.1 & 0.9
\end{bmatrix}
$$

(e)

$$
\begin{bmatrix}
0.2 & 0.8 & 0 & 0 \\
0 & 0.9 & 0.1 & 0 \\
0.3 & 0 & 0.2 & 0.5 \\
0 & 0 & 0.1 & 0.9
\end{bmatrix}
$$

(f)

$$
\begin{bmatrix}
0.2 & 0 & 0.8 & 0 \\
0 & 0.9 & 0.1 & 0 \\
0.3 & 0 & 0.7 & 0 \\
0 & 0 & 0.1 & 0.9
\end{bmatrix}
$$

3. The following problems are based on a single employee performing a series of tasks on a job. At any time, she may be either idle or working. If she is idle at the beginning of an hour, she receives a job

with probability p_A during the hour. If she is working at the beginning of the hour, she completes the job with probability p_C. Jobs do not arrive while she is working.

Using a DTMC model, we would like to estimate the efficiency of the employee measured by the proportion of time she is busy and the throughput rate of the system in terms of jobs completed per hour. For each case, define the states, construct the transition matrix, and identify the states as recurrent, transient, or absorbing. Time is divided into 1-hour segments. Solve each problem independently of the others.

(a) Model the situation as given. Use two states that describe the condition of the employee: idle and working.

(b) Change the conditions such that the probability of a job being completed depends on how many hours the job has been in progress. After the employee has worked on the job for 1 hour, the probability that it will be completed is p_{C1}, and the probability that another hour will be required is $(1 - p_{C1})$. In the second and subsequent hours, the probability that it will be completed is p_{C2}, and the probability that it will take another hour is $(1 - p_{C2})$. Define three states: idle, working during the first hour, and working during the second (and subsequent) hour(s). Note that only three states are needed if the probability of working the second and subsequent hours remains the same.

(c) Change the situation such that the time between arrivals is exactly 2 hours. Counting the time begins when the employee becomes idle. Completion probabilities are still based on a 1-hour period. Use three states: idle/just started waiting for an arrival, idle/waiting 1 hour for an arrival, and working.

(d) Now assume that there is space for one waiting job and define p_{A1} as the probability that an arrival will occur while the employee is busy. An arriving job will wait if the space is empty; otherwise, it will balk. Use three states: idle, working with no jobs waiting, and working with one job waiting.

4. Consider a DTMC with the states numbered 0, 1, and 2. The state-transition matrix is

$$\mathbf{P} = \begin{bmatrix} 0 & 0.3 & 0.7 \\ 1 & 0 & 0 \\ 1 & 0 & 0 \end{bmatrix}$$

(a) Given that the system begins in state 0, compute the state probabilities for four periods.

(b) Compute the n-step transition matrix for $n = 2$ through $n = 4$.

(c) Find the steady-state probabilities.

(d) Comment on the meaning of the steady-state probabilities in this case.

5. Find the steady-state probabilities for the DTMC associated with the following state-transition matrix. Do the calculations by hand.

$$\mathbf{P} = \begin{bmatrix} 1 & 0 & 0 \\ 0.1 & 0.7 & 0.2 \\ 0 & 0 & 1 \end{bmatrix}$$

Since this example contains absorbing states, comment on the meaning of the steady-state probabilities. Compute the probabilities of the absorbing states given that the system starts in state 1 (number the states 0, 1, and 2).

6. On a trip to Las Vegas, big-time gambler Barbara Kitchell is down to her last $20. Determined to continuing gambling, she finds a game that allows a bet of $10 on each play and for which the probability of winning is p_W. If she wins, she adds $10 to her cash supply; if she loses, her supply drops by $10. Because she has only a short time left, she adopts a strategy by which she will quit when her cash supply reaches $40 or when all her money is gone.

Develop a Markov chain model for each of the following situations. Assume that the process is observed after each play and that $p_w = 0.4$. Find the transient probabilities for 10 plays as well as the steady-state and absorbing state probabilities when appropriate.

(a) For the given situation, let the states be the cash supply: $0, 10, 20, 30, and 40. In addition, find the first passage probabilities from the initial state to the state $0, and also to the state $40.

(b) Change the model such that whenever Ms. Kitchell reaches $0, she borrows another $20 from her distraught daughter.

(c) Change the model as in part (b) but with the provision that the most Ms. Kitchell can borrow is $40.

7. For the computer repair example in Section 12.1, compute $f_{00}^{(5)}$ and $f_{00}^{(6)}$.

8. For the computer repair example in Section 12.1, find the first four values of the first passage probabilities as well as the expected first passage times for the following.

(a) Transition from state 0 to state 1
(b) Transition from state 1 to state 0
(c) Transition from state 1 to state 1

9. For a Markov chain with states numbered 0, 1, and 2, use the following transition matrix to answer the questions. In each case, write out all equations.

$$\mathbf{P} = \begin{bmatrix} 0 & 0.3 & 0.7 \\ 0.9 & 0.1 & 0 \\ 0.2 & 0 & 0.8 \end{bmatrix}$$

(a) Find the transient state probability vector for the first five steps, assuming that the system starts in state 2.

(b) Given that the system is in state 2, find the probabilities that the first passage to state 0 will require one step, two steps, and three steps.

(c) Find the expected first passage time from state 2 to state 0.

(d) Find the steady-state probability vector.

(e) Let the expected state cost vector $\mathbf{C} = (121, 47, 98)^T$ and the discount rate $r = 5\%$ per period. Find the total expected discounted cost when the process starts in each of the three states. What is the long-run average discounted cost when the initial probability distribution over the three states is $\mathbf{q}(0) = (0.2, 0.5, 0.3)$?

10. The transition matrix for a Markov chain with states numbered 0 through 3 is given. The time increment in the model is 1 week. In answering each part of this exercise, write out the relevant equations.

$$\mathbf{P} = \begin{bmatrix} 0.2 & 0.8 & 0 & 0 \\ 0 & 0.9 & 0.1 & 0 \\ 0.3 & 0 & 0.2 & 0.5 \\ 0 & 0 & 0.1 & 0.9 \end{bmatrix}$$

(a) Find the expected first passage time from state 3 to state 0.
(b) Find the expected first passage time from state 0 to state 3.
(c) Find the steady-state probabilities.
(d) Find the expected recurrence times for all states.
(e) For an expected state cost vector $\mathbf{C} = (1250, 1400, 900)^T$, find the long-run average cost of the process per week. For a nominal (annual) interest rate of 20%, find the total expected discounted cost of the process when it starts in each of the four states.

11. A submarine, operating clandestinely in the Atlantic Ocean, is resupplied at yearly intervals. At that time, failed parts are replaced. One critical part fails at random according to a Poisson distribution with a mean time between failures of 1/2 year. As a hedge against failure of this part, the ship carries two spares on board (three in all). Two replacements are delivered at the time of resupply. If the delivery results in more than three working parts, the excess is returned. Develop a Markov chain model for this situation. Define the states as the number of parts operating at the end of the year just before the delivery. Construct the transition matrix and analyze the system. Compute all relevant statistics, such as the probability that all parts will fail, various steady-state probabilities, and first passage times.

12. Referrring to Exercise 13 in Chapter 11, answer the following questions.

    (a) What is the probability that Ms. Hodler will sell the stock within 4 days? What is the expected net gain or loss per share when she sells within 4 days if the selling conditions are met or if she sells at the current price at the end of the fourth day?

    (b) What is the mean time before Ms. Hodler sells the stock?

13. The market for a product is shared by four brands. The table gives the present share of the market distribution and the percentage of people who switch from each brand to the other brands for consecutive purchases.

		To brand				
		1	2	3	4	Market share
	1	60	8	20	12	40%
From brand	2	15	40	25	20	20%
	3	25	16	50	9	30%
	4	28	12	20	40	10%

    (a) If one purchase is made every 2 months on average, predict the distribution of the share of the market after 6 months.

    (b) What is the long-run average share of the market for each brand if current purchase patterns are not altered?

    (d) What is the expected first passage time from brand 3 to brand 1?

    (c) What are the expected recurrence times for each brand?

14. Let $\{X\}$ be a DTMC with state space $\{a, b, c, d\}$ and transition probabilities given by

$$P = \begin{bmatrix} 0.1 & 0.3 & 0.6 & 0.0 \\ 0.0 & 0.2 & 0.5 & 0.3 \\ 0.5 & 0.0 & 0.0 & 0.5 \\ 0.0 & 1.0 & 0.0 & 0.0 \end{bmatrix}$$

Each time the chain is in state a, a profit of $20 is made; each visit to state b yields a $5 profit; each visit to state c yields $15; and each visit to state d costs $10. Letting $f(X_n)$ be a random variable associated with the economic return of the stochastic process at time n, find the following.

    (a) $E[f(X_5) \mid X_3 = c, X_4 = d]$
    (b) $E[f(X_1) \mid X_0 = b]$
    (c) $E[f(X_1)^2 \mid X_0 = b]$
    (d) $Var[f(X_1) \mid X_0 = b]$
    (e) $Var[f(X_1)]$ given that $\Pr\{X_0 = a\} = 0.2$, $\Pr\{X_0 = b\} = 0.4$, $\Pr\{X_0 = c\} = 0.3$, and $\Pr\{X_0 = d\} = 0.1$
    (f) The long-run average return per period

15. Consider the following transition matrix, which represents a DTMC with state space $\{a, b, c, d, e\}$.

$$\mathbf{P} = \begin{bmatrix} 0.3 & 0.0 & 0.0 & 0.7 & 0.0 \\ 0.0 & 1.0 & 0.0 & 0.0 & 0.0 \\ 1.0 & 0.0 & 0.0 & 0.0 & 0.0 \\ 0.0 & 0.0 & 0.5 & 0.5 & 0.0 \\ 0.0 & 0.2 & 0.4 & 0.0 & 0.4 \end{bmatrix}$$

(a) Draw the state diagram.

(b) List the transient states.

(c) List the irreducible set(s).

(d) Let F_{ij} denote the first passage probabilities of reaching (or returning to) state j given that $X_0 = i$. Calculate the $\mathbf{F}$ matrix.

(e) Let R_{ij} denote the expected number of visits to state j given that $X_0 = i$. Calculate the $\mathbf{R}$ matrix.

(f) Calculate the $\lim_{n \to \infty} \mathbf{P}^{(n)}$ matrix.

16. Let $\{X\}$ be a DTMC with state space $\{a, b, c, d, e, f\}$ and transition probabilities given by

$$\mathbf{P} = \begin{bmatrix} 0.3 & 0.5 & 0.0 & 0.0 & 0.0 & 0.2 \\ 0.0 & 0.5 & 0.0 & 0.5 & 0.0 & 0.0 \\ 0.0 & 0.0 & 1.0 & 0.0 & 0.0 & 0.0 \\ 0.0 & 0.3 & 0.0 & 0.0 & 0.0 & 0.7 \\ 0.1 & 0.0 & 0.1 & 0.0 & 0.8 & 0.0 \\ 0.0 & 1.0 & 0.0 & 0.0 & 0.0 & 0.0 \end{bmatrix}$$

(a) Draw the state diagram.

(b) List the recurrent states.

(c) List the irreducible set(s).

(d) List the transient states.

(e) Calculate the $\mathbf{F}$ matrix.

(f) Calculate the $\mathbf{R}$ matrix

(g) Calculate the $\lim_{n \to \infty} \mathbf{P}^{(n)}$ matrix.

BIBLIOGRAPHY

Broadie, M. and D. Joneja, "An Application of Markov Chain Analysis to the Game of Squash," *Decision Sciences*, Vol. 24, No. 5, pp. 1023–1035, 1993.

Feldman, R.M. and C. Valdez-Flores, *Applied Probability & Stochastic Processes*, PWS, Boston, 1996.

Kao, E., *An Introduction to Stochastic Processes*, Duxbury, Belmont, CA, 1997.

Ross, S.M., *Introduction to Probability Models*, Seventh Edition, Academic Press, San Diego, 2000.

Stewart, W.J. (editor), *Numerical Solutions of Markov Chains*, Dekker, New York, 1991.

Wolff, R.W., *Stochastic Modeling and the Theory of Queues*, Prentice-Hall, Englewood Cliffs, NJ, 1989.

Chapter **14**

Continuous-Time Markov Chains

A natural extension of a discrete-time Markov chain (DTMC) occurs when time is treated as a continuous parameter. In this chapter, we consider continuous-time, discrete-state stochastic processes but limit our attention to the case in which the Markovian property still holds—that is, the future realization of a system depends only on the current state and the random events that proceed from it. It happens that this property is satisfied in a continuous-time stochastic process only if all activity durations are exponentially distributed. Although this may sound somewhat restrictive, many practical situations can be modeled as continuous-time Markov chains (CTMCs), as they are called, and many powerful analytical results can be obtained. A primary example is an *M/M/s* queuing system, in which customer arrivals and service times follow an exponential distribution. Because it is possible to compute the steady-state probabilities for such systems, it is also possible to compute many performance-related statistics such as the average wait and the average number of customers in the queue. In addition, many critical design and operational questions can be answered with little computational effort.

We begin this chapter with some basic definitions and several examples that highlight various aspects of the model. Design and cost issues are discussed later to show how decision making can be introduced into the analysis. Chapter 15 provides a detailed look at several of the formulas used in the computations. Given the key role that Markov processes play in queuing systems, we devote Chapters 16 and 17 solely to that subject.

14.1 MARKOVIAN PROPERTY

Suppose now that we have a continuous-time stochastic process $\{Y_t : t \geq 0\}$ taking on values in the set of nonnegative integers. The definition of a CTMC is similar to that of a DTMC except that "one-step transition matrix" has no meaning in continuous time and so the Markovian property must hold for all future times instead of just for one step.

The Stochastic Process

Definition 1: The process $Y = \{Y_t : t \geq 0\}$ with state space S is a CTMC if the following condition holds for all $j \in S$, and $t, s \geq 0$

$$\Pr\{Y_{t+s} = j \mid Y_u, 0 \leq u \leq s\} = \Pr\{Y_{t+s} = j \mid Y_s\}$$

In addition, the chain is said to have *stationary* transitions if

$$\Pr\{Y_{t+s} = j \mid Y_s = i\} = \Pr\{Y_t = j \mid Y_0 = i\}$$

The first equation says that the conditional distribution of the future Y_{t+s} given the present Y_s and the past Y_u, $0 \le u \le s$, depends only on the present and is independent of the past. The second equation says that $\Pr\{Y_{t+s} = j \mid Y_s = i\}$ is independent of s. This is the time-homogeneous condition indicating that the probability law governing the process does not change during the life of the process. All Markov chains considered in this chapter are assumed to have stationary transition probabilities.

Note that when a CTMC is generalized by allowing the state to be continuous, we have what is referred to as a Markov process. Brownian motion is an example of a stochastic process in which both time and state are modeled as continuous parameters, and the Markovian property holds. Occasionally, we will refer to a CTMC as a Markov process.

Suppose that a CTMC enters state i at, say, time 0 and does not leave during the next 15 minutes—i.e., a transition does not occur. What is the probability that a transition will not occur in the next 5 minutes? To answer this question, we refer to the Markovian property, which tells us that the probability that the process will remain in state i during the interval [15, 20] is just the unconditional probability that it will stay in state i for at least 5 minutes. If we let T_i denote the amount of time that the process stays in state i before making a transition into a different state, then

$$\Pr\{T_i > 20 \mid T_i > 15\} = \Pr\{T_i > 5\}$$

or, in general,

$$\Pr\{T_i > s + t \mid T_i > s\} = \Pr\{T_i > t\}$$

for all $s, t \ge 0$. Hence, the random variable T_i is memoryless and thus is exponentially distributed. In fact, this analysis gives us another way of defining a CMTC—namely, a stochastic process that, each time it enters state i, has the following properties.

1. The amount of time it spends in state i before making a transition to a different state is exponentially distributed with a mean of, say, $1/\lambda_i$.

2. When the process leaves state i, it next enters state j with a probability of, say, p_{ij}, where p_{ij} must satisfy

$$p_{ii} = 0, \text{ for all } i \in S$$
$$\sum_{j \in S} p_{ij} = 1, \text{ for all } i \in S$$

The implication of the memoryless property is that it is a waste of effort to keep historical records on a process if it is Markovian.

Stockbroker Example

Consider a variation of the stockbroker situation originally presented in Section 12.2 in the form of a DTMC. The broker solicits business from client lists prepared by his firm's marketing department. Assume now that the amount of consecutive time he spends working from each of the three lists—a, b, and c—is a random variable with an exponential distribution. For a, the mean time is 2 weeks, for b it is 1 week, and for c it is $1\frac{1}{2}$ weeks. When the stockbroker finishes with list a, he flips a coin to determine which list to work from next; when he finishes with either b or c, he flips two coins. In this situation, and if he gets at least one head he returns to list a (75% chance), and if he gets two tails he picks up the other list (25% chance). Let Y_t be a random variable denoting the list from which the broker is working at time t. The process $Y = \{Y_t : t \ge 0\}$ is a CTMC because of the lack of memory inherent in the exponential distribution.

This example demonstrates a major characteristic of any CTMC: the process remains in each state for an exponentially distributed length of time and then, when it jumps, it jumps as though it were DTMC. This is illustrated in Figure 14.1. The time between jumps, $T_{n+1} - T_n$, is called the *sojourn time*. The Markovian property implies that the only information that can be used for predicting the sojourn time is the current state. Mathematically, this can be stated as

$$\Pr\{T_{n+1} - T_n \le t \mid X_0, \ldots, X_n, T_0, \ldots, T_n\} = \Pr\{T_{n+1} - T_n \le t \mid X_n\}$$

If we focus our attention on the process $X_1, X_2, \ldots$ for a moment, we will discover that we have what is called an *embedded DTMC*.

Definition 2: Let $Y = \{Y_t : t \ge 0\}$ be a CTMC with state space S and with jump times denoted by $T_1, T_2, \ldots$. Also let the embedded process at the jump times be $X_1, X_2, \ldots$ Then there is a collection of scalars λ_i for all $i \in S$, called the *mean sojourn rates*, and a state-transition matrix $\mathbf{P}$ called the *embedded discrete-time Markov matrix*, that satisfy:

$$\Pr\{T_{n+1} - T_n \le t \mid X_n = i\} = 1 - e^{-\lambda_i t}$$

and

$$\Pr\{X_{n+1} = j \mid X_n = i\} = p_{ij}$$

where each λ_i is positive and the diagonal elements of $\mathbf{P}$ are zero.

For the stockbroker, the embedded DTMC that results from Definition 2 is

$$\mathbf{P} = \begin{array}{c} a \\ b \\ c \end{array} \begin{bmatrix} 0 & 1/2 & 1/2 \\ 3/4 & 0 & 1/4 \\ 3/4 & 1/4 & 0 \end{bmatrix}$$

and the mean sojourn rates are $\lambda_a = 1/2$, $\lambda_b = 1$, and $\lambda_c = 2/3$. In general, the value of λ_i is the reciprocal of the mean time spent in state i before a transition to another state occurs.

Definition 3: Let $Y = \{Y_t : t \ge 0\}$ be a CTMC with state space S such that $0 < \lambda_i < \infty$ for all $i \in S$. A state is called *recurrent* or *transient* according to whether it is recurrent or transient in the embedded DTMC. A set of states is irreducible for the process if it is irreducible for the embedded chain.

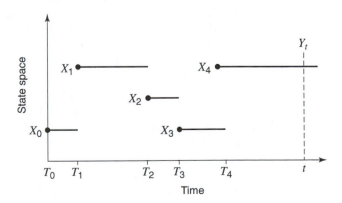

Figure 14.1   Hypothetical realization of a CTMC

Note the stipulation that $0 < \lambda_i < \infty$ means that the time spent in any state i is finite and nonzero. This eliminates any pathological cases. Moreover, we will see that when the states of a CTMC are positive recurrent, it will eventually reach a steady state.

As described in Section 14.2, a CTMC has an alternative representation in terms of a rate matrix that we denote by $\mathbf{R}$. This matrix has the same dimensions as $\mathbf{P}$, but its components are transition rates r_{ij}, where a transition rate is related to the transition probabilities and sojourn rates according to the equation

$$r_{ij} = \lambda_i p_{ij}$$

Using this expression, we find the rate matrix for the stockbroker example.

$$\mathbf{R} = \begin{bmatrix} 0 & 1/4 & 1/4 \\ 3/4 & 0 & 1/4 \\ 1/2 & 1/6 & 0 \end{bmatrix}$$

Either representation is valid for a CTMC. In many situations, it is more natural to use the rate matrix. We provide a variety of examples in subsequent sections.

14.2 MODEL COMPONENTS

For Markov processes, all the information concerning the model is entirely determined by the current state. It is said to have no memory because the future realization depends only on the current state and in no way on the past. The advantage that results when a process can be classified as Markovian is that its cost and performance can be easily calculated.

ATM Example

For a further illustration of the components of a Markov model, we return to the example introduced in Chapter 11 concerning a single automated teller machine (ATM) located in the foyer of a bank. Only one person can use the machine at a time, so a queue forms when two or more customers are present. Following the rule of first-come-first-served, a person in the queue will eventually receive service and leave. The number in the system is the sum of the number in service plus the number in the queue. The foyer is limited in size and can hold only five people. Since the weather is generally bad in the town where the bank is located, arriving customers balk when the foyer is full.

Statistics on the usage of this ATM indicate that the average time between arrivals is 30 seconds (or 0.5 minute) whereas the time for service averages 24 seconds (or 0.4 minute). Although the ATM has sufficient capacity to meet all demands, queues are frequently observed and customers are occasionally lost.

We want to perform an analysis to determine on average the number of people in the system, the waiting time for customers, the efficiency of the ATM, and the number of customers not served because there is no room in the foyer. These statistics can be used to provide guidance to managers in answering design questions such as whether or not another ATM should be installed and whether or not the size of the foyer should be expanded.

In Chapter 11, we described the process associated with this situation with the help of the state-transition network, which is repeated in Figure 14.2. The state is the number of customers in the foyer. A change in state occurs when there is an arrival (a) or a departure (d). When the foyer opens in the morning, the system is empty—in state 0. As customers arrive, the state increases. Because there is only one machine, customers must wait in a

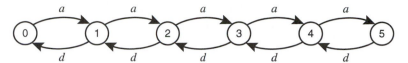

Figure 14.2   State-transition network for the ATM example

queue when the state is greater than 1. The state index increases and decreases as customers arrive and depart. When the foyer is full, the state reaches 5, and we assume that no other arrivals will occur until there is a departure.

The dynamic features of the model are provided by its *activities*. In this example, we identify service and arrival activities. The service activity begins when a customer first appears in front of the ATM and ends when all the requested banking operations have been completed. The culmination of the service activity is the event *d* in Figure 14.2. The duration of the activity that leads to this event is the time for service.

The arrival activity is the process that generates customers for the ATM. The arrival activity begins immediately after an arrival occurs and ends with the event of the next arrival. The event is an arrival (*a* in Figure 14.2), and the duration of this activity is the time between arrivals.

The Exponential Distribution

Suppose that arrivals at the ATM are governed by an exponential distribution with a mean of 0.5 minute. When an arrival occurs, the expected time to the next arrival is 0.5 minute. Suppose 1 minute goes by and no customer arrives. We might be tempted to say that the arrival is late and that one should occur at any time. For the exponential distribution, this would be wrong. In fact, the expected time to the next arrival is still 0.5 minute because of the memoryless property of the distribution. No matter how long the wait without an arrival, the remaining time still has an exponential distribution with the same expected value as at the start.

The exponential distribution is a continuous probability distribution with a single positive parameter λ called the rate. It models the random variable that is the time between events t, where the rate of occurrence of the events is λ. The density function and cumulative distribution for the general exponential distribution are

$$f(t) = \lambda e^{-\lambda t} \text{ for } t \geq 0$$
$$F(t) = 1 - e^{-\lambda t} \text{ for } t \geq 0$$

with parameters $Mean = 1/\lambda$ and $Var = (1/\lambda)^2$.

Figure 14.3 plots $f(t)$ for $\lambda = 2$. The function begins at λ and decreases asymptotically toward 0. Because the random variable is time, $f(t) = 0$ for $t < 0$. The duration of an activity represented by this distribution will most often be quite short, but occasionally it will assume large values. The probability that the activity will take 1 minute or less, for example, is roughly 86%. The standard deviation of the exponential distribution is equal to its mean. This is relatively large for most physical activities, so one might consider the exponential distribution to be somewhat extreme in terms of variability.

When the duration of the time between events is exponentially distributed, the number of occurrences of the event in a given time interval has a Poisson distribution. For the ATM, the Poisson parameter $\lambda = 2$ (customers per minute), so the expected number of arrivals in an interval t, expressed in minutes, is $2t$. The distribution of the number of arrivals is

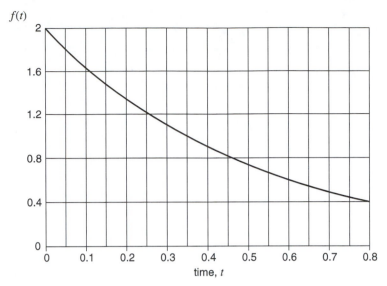

Figure 14.3   Exponential distribution with mean 0.5 ($\lambda = 2$)

$$\Pr\{k \text{ arrivals in time } t\} = \frac{(2t)^k e^{-2t}}{k!} \text{ for } k = 0, 1, \ldots$$

An activity whose duration has an exponential distribution is called a *Poisson process*.

When all activity durations of a stochastic process have exponential distributions, all activity completion times depend only on the current state. They do not depend on how that state was reached or how long the system has resided there. Thus, a stochastic process in which all activities are exponentially distributed is a Markov process. The exponential distribution is the *only* continuous distribution that has the memoryless property.

Validity of Assumptions

With the assumption of exponentially distributed activity times, we gain mathematical tractability, whereas without it, simulation is usually required for any analysis. As always, we must ask if the assumption is appropriate for the problem at hand.

When the arrival of customers at a system is considered, the exponential assumption for interarrival times can be defended on practical and empirical grounds. If the arrivals are independent events, use of the exponential distribution can be theoretically justified as well. The main requirement is that an arrival must not be affected in any way by the presence of any other customer. A classical example is the arrival of calls at a telephone switchboard. Practically speaking, the arrivals cannot influence each other. A second example derives from reliability theory and concerns the failure of complex devices consisting of a large number of components, say n, operating in series. Assume that in any small time period Δ (e.g., 1 second) each component has a very small probability p of failing. Furthermore, assume that components fail independently of each other. In this case, the number of failures in any period Δ is a binomial random variable with parameters n and p. If n is large and p is small as expected, the binomial distribution can be approximated by a Poisson distribution. In particular, the "law of small numbers" says that if we let $n \to \infty$, $p \to 0$, and $np \to \lambda$, then, in fact, the binomial random variable converges to a Poisson random variable. Now, if we replace a component as soon as it fails, the foregoing implies that the number of device failures in a small time period Δ is approximately Poisson distributed (because of the serial assumption, each time a component fails, so does the device). Coupling this with the independence assumption, we can see that the failure process is approximately Poisson distributed, so the times between failures are approximately exponential.

Justifying an exponential distribution of service times is often more difficult. If all customers receive approximately the same service, this assumption is unlikely to hold. The time it takes to get a haircut probably does not have an exponential distribution. The time required to check out at a grocery store may be exponentially distributed if the number of items being purchased is approximately exponentially distributed (making allowances for the discrete nature of this random variable). Note that the memoryless property need not hold for the individual customer. Clearly, a grocery store shopper with many items can expect a long service time. The exponential distribution is a property of the service times for the population rather than for the individual.

Rate Matrix

For queuing problems such as the ATM example, we typically use λ as the arrival rate and $1/\lambda$ as the mean time between arrivals. For the service activity, we use μ as the service rate and $1/\mu$ as the mean time for the service activity. For the ATM example, $\lambda = 2$/min and $\mu = 2.5$/min.

For the CTMC, activities are well represented by their rates of occurrence, so rather than using a state-transition network such as the one in Figure 14.2, we use a *rate network,* as shown in Figure 14.4. The rate network is easily constructed from the state-transition network by replacing the activity designation by the rate for the activity.

A computationally more convenient alternative to the rate network is the *rate matrix* **R**, whose element r_{ij} is the transition rate from state i to state j. The general rate matrix shown below has maximum state index $m - 1$. For some models, the number of states may be infinite, so **R** is infinite; however, we will mostly be concerned with finite cases.

General rate matrix	Rate matrix for the ATM example

$$\mathbf{R} = \begin{bmatrix} 0 & r_{01} & r_{02} & \cdots & r_{0,m-1} \\ r_{10} & 0 & r_{12} & \cdots & r_{1,m-1} \\ \vdots & \vdots & \vdots & & \vdots \\ r_{m-1,0} & r_{m-1,1} & r_{m-1,2} & \cdots & 0 \end{bmatrix} \qquad \mathbf{R} = \begin{bmatrix} 0 & \lambda & 0 & 0 & 0 & 0 \\ \mu & 0 & \lambda & 0 & 0 & 0 \\ 0 & \mu & 0 & \lambda & 0 & 0 \\ 0 & 0 & \mu & 0 & \lambda & 0 \\ 0 & 0 & 0 & \mu & 0 & \lambda \\ 0 & 0 & 0 & 0 & \mu & 0 \end{bmatrix}$$

The rate matrix for the ATM example allows only arrivals for state 0 and so has one nonzero entry, $r_{01} = \lambda$. State 5 allows only service completions and similarly has one nonzero entry, $r_{54} = \mu$. Every other state allows both events.

Note that for some problems there may be events that cause the system to remain in its current state. In such cases, we would have a positive transition rate r_{ii}.

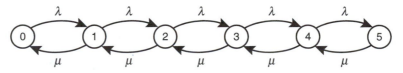

Figure 14.4   Rate network for the ATM example

14.3   TRANSIENT SOLUTIONS

Given that both interarrival and service times for the ATM are exponentially distributed with rates of 2 and 2.5 customers per minute, respectively, we can now begin to analyze the system using computational techniques available for CTMCs. Assume that the bank foyer opens at 8 A.M. Our first goal is to find the probability distribution of the number of customers in the system as a function of the time since opening. This is the transient solution.

The system begins in some initial state at time $t = 0$, the initial point of observation. In determining the probability of the system being in any particular state as a function of time, we note that soon after opening, the state of the system is influenced considerably by initial conditions—the number of customers queued up for service. During this so-called transient period, state probabilities may fluctuate widely.

As time passes, the probability distribution becomes less dependent on the initial conditions, so state probabilities approach constant values. When this happens, the system is said to be in steady state or equilibrium. There is no exact time at which the system leaves the transient period and enters equilibrium. In theory, if it reaches steady state at all, it does so only in the limit as time goes to infinity.

Closed-form representations that give the transient probabilities as functions of time are available for only very simple systems. In most cases, numerical approaches are necessary to approximate them. Here we use a numerical approach based on a DTMC approximation. Note that the DTMC used in this section is not the embedded DTMC mentioned in Section 14.1. To begin the analysis, let the unconditional probability that the system is in state i at time t be $q_i(t)$. The corresponding state vector is

$$\mathbf{q}(t) = (q_0(t), q_1(t), q_2(t), \ldots, q_{m-1}(t))$$

Since the system must be in one of the m mutually exclusive states, the components of this vector must sum to 1.

$$\sum_{i=0}^{m-1} q_i(t) = 1$$

Now, for some small interval of time Δ, let $n = t/\Delta$ be the number of steps or increments required to represent t. The transient solution of the process can be approximated at time $t = n\Delta$ with a DTMC by solving the following equation, the equivalent of Equation (3) in Chapter 12,

$$\mathbf{q}(n\Delta) = \mathbf{q}(0)\mathbf{P}^{(n)} \text{ or } \mathbf{q}(n\Delta + \Delta) = \mathbf{q}(n\Delta)\mathbf{P} \tag{1}$$

where $\mathbf{P}$ is a state-transition matrix determined from the rate matrix $\mathbf{R}$. For a general rate matrix, if we define α_i as the sum of all transition rates out of state i—that is,

$$\alpha_i = \sum_{j=0}^{m-1} r_{ij}$$

then

$$\mathbf{P} = \begin{bmatrix} 1 - \Delta\alpha_0 & \Delta r_{01} & \Delta r_{02} & \cdots & \Delta r_{0,m-1} \\ \Delta r_{10} & 1 - \Delta\alpha_1 & \Delta r_{12} & \cdots & \Delta r_{1,m-1} \\ \vdots & \vdots & \vdots & & \vdots \\ \Delta r_{m-1,0} & \Delta r_{m-1,1} & \Delta r_{m-1,2} & \cdots & 1 - \Delta\alpha_{m-1} \end{bmatrix}$$

For the ATM example, the Markov chain transition matrix is

$$\mathbf{P} = \begin{bmatrix} 1-\Delta\lambda & \Delta\lambda & 0 & 0 & 0 & 0 \\ \Delta\mu & 1-\Delta(\lambda+\mu) & \Delta\lambda & 0 & 0 & 0 \\ 0 & \Delta\mu & 1-\Delta(\lambda+\mu) & \Delta\lambda & 0 & 0 \\ 0 & 0 & \Delta\mu & 1-\Delta(\lambda+\mu) & \Delta\lambda & 0 \\ 0 & 0 & 0 & \Delta\mu & 1-\Delta(\lambda+\mu) & \Delta\lambda \\ 0 & 0 & 0 & 0 & \Delta\mu & 1-\Delta\mu \end{bmatrix}$$

$$= \begin{bmatrix} 1-2\Delta & 2\Delta & 0 & 0 & 0 & 0 \\ 2.5\Delta & 1-4.5\Delta & 2\Delta & 0 & 0 & 0 \\ 0 & 2.5\Delta & 1-4.5\Delta & 2\Delta & 0 & 0 \\ 0 & 0 & 2.5\Delta & 1-4.5\Delta & 2\Delta & 0 \\ 0 & 0 & 0 & 2.5\Delta & 1-4.5\Delta & 2\Delta \\ 0 & 0 & 0 & 0 & 2.5\Delta & 1-2.5\Delta \end{bmatrix}$$

Using an increment of $\Delta = 0.05$, if we wish to approximate the transient probabilities at $t = 1$ minute, then $n = 1/0.05 = 20$ steps. Assuming that the system starts empty at time 0, Figure 14.5 plots the transient response for $\Delta = 0.05$. The initial probability vector is

$$\mathbf{q}(0) = (1, 0, 0, 0, 0, 0)$$

and the probabilities for each step are computed using Equation (1). After 20 steps or 1 minute, the state probability is

$$\mathbf{q}(20\Delta) = \mathbf{q}(1) = (0.433, 0.291, 0.162, 0.075, 0.029, 0.011)$$

The accuracy of the approximation depends on Δ. For $\Delta = 0.025$ minute and $t = 1$ minute, $n = 40$ steps. After 40 computations using Equation (1) and the new Δ, we find

$$\mathbf{q}(40\Delta) = \mathbf{q}(1) = (0.435, 0.291, 0.160, 0.073, 0.029, 0.011)$$

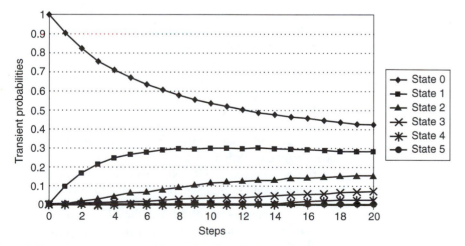

Figure 14.5   Transient solution of the ATM example for $\Delta = 0.05$

which is almost identical to the values obtained for $\Delta = 0.05$. Thus, doubling the computational effort produced little change in the results. Generally speaking, for any class of problems the analyst must determine when the point of diminishing returns has been reached. Increased accuracy (smaller Δ) can always be achieved at the expense of greater computational effort (larger n).

14.4   STEADY-STATE SOLUTIONS

The steady-state probability vector is the limit of $\mathbf{q}(t)$ as $t \to \infty$. One way to obtain an estimate of these values is to hold Δ fixed at, say, 0.05, allow n to increase to a very large number, and compute the limiting transient probability. This is unnecessary, however, because a simple set of equations is available for computing the steady-state probabilities directly. Their derivation will be presented in Section 15.2.

As in the discussion of DTMCs, let π_i^P be the steady-state probability for state i with corresponding vector $\boldsymbol{\pi}^P = (\pi_0^P, \pi_1^P, \ldots, \pi_{m-1}^P)$. The superscript "P" indicates that we are referring to a Markov process or a CTMC rather than a DTMC. When a steady state exists, we have

$$\pi_i^P = \lim_{t \to \infty} q_i(t), \quad i = 0, 1, \ldots, m-1$$

Of course, the term "steady state" does not mean that the process has reached a point where no more variability is observed. It simply means that as time increases our estimates of the state probabilities become uncoupled from the initial conditions, and so these estimates become constant.

In Section 15.2 we will show that the steady-state probabilities can be computed by solving a set of linear equations in m unknowns. Our point here is to illustrate modeling and the kinds of information that can be obtained from computational analysis, so we will wait until Chapter 15 to describe the mathematical details. The solution of the ATM example is

$$\boldsymbol{\pi}^P = (\ \pi_0^P, \quad \pi_1^P, \quad \pi_2^P, \quad \pi_3^P, \quad \pi_4^P, \quad \pi_5^P\ )$$
$$= (0.271, 0.217, 0.173, 0.139, 0.111, 0.089)$$

As we can see, these probabilities are not very close to the values obtained for $t = 1$ minute with the transient analysis. Therefore, we might ask how many steps are required (i.e., how big n should be) before the transient probabilities become good estimates of the steady-state values. To provide some insight, we set Δ equal to 0.025 minute and continued the computations for several hundred steps. The results are presented in Table 14.1. Although the transient probabilities do indeed approach the steady-state values, it appears that n must be quite large for a good approximation.

The steady-state probability vector provides management with critical information concerning the operations of a system. This information can be used in various ways to evaluate performance, quality of service, and design options. For the ATM example, it is straightforward to compute the following measures.

- The proportion of the time the ATM is idle: $\pi_0^P = 0.271$.
- The efficiency of the ATM or the proportion of the time the machine is busy: $1 - \pi_0^P = 0.729$.
- The proportion of customers obtaining immediate service: $\pi_0^P = 0.271$.

Table 14.1   Transient Computations for the ATM Example with $\Delta = 0.025$

Steps, n	Time (min)	q_0	q_1	q_2	q_3	q_4	q_5
0	0	1	0	0	0	0	0
40	1	0.435	0.291	0.160	0.073	0.029	0.011
80	2	0.348	0.258	0.175	0.110	0.066	0.042
120	3	0.311	0.239	0.175	0.124	0.087	0.063
160	4	0.292	0.228	0.175	0.131	0.098	0.075
200	5	0.282	0.223	0.174	0.135	0.104	0.082
240	6	0.277	0.220	0.174	0.137	0.107	0.085
.	.	.	.	.	.	.	.
.	.	.	.	.	.	.	.
.	.	.	.	.	.	.	.
Steady state	∞	0.271	0.217	0.173	0.139	0.111	0.089

- The proportion of customers who arrive and find the system full (the fraction not served): $\pi_5^P = 0.089$.

- The proportion of customers who wait: $1 - \pi_0^P - \pi_s^P = 0.640$

- The expected (average) number in the system:

$$\sum_{i=0}^{5} i\pi_i^P = 1.868 \text{ customers}$$

- The expected (average) number waiting in the queue:

$$\sum_{i=1}^{5} (i-1)\pi_i^P = 1.139 \text{ customers}$$

- The throughput rate of the system (or the average number of customers passing through the system):

Throughput rate $= \lambda(1 - \pi_5^P) = 1.822$ customers per minute

- The balking rate of the system (or the average number of customers lost to the system):

Balking rate $= \lambda\pi_5^P = 0.178$ customers per minute

- The average time in the system (for queuing systems, this is given by Little's law; see Section 16.1):

$$\text{Average time in system} = \frac{\text{average number in system}}{\text{throughput rate}}$$

$$= 1.868/1.822 = 1.025 \text{ min}$$

14.5 DESIGN ALTERNATIVES FOR THE ATM SYSTEM

On receiving these results, the bank manager notes that although the ATM is busy only 73% of the time and the space in the foyer is less than 40% utilized ($1.868/5 \times 100 = 37.36\%$), almost 9% of the customers are lost and those that are served must wait about 37 seconds on average (mean time in system less mean service time = $1.025 - 0.4 = 0.625$ min). Although these results may seem contradictory, such performance is common for systems with variability. To determine what it would take to improve service, the manager wishes to consider several alternative designs.

One of the benefits of good quantitative models is their ability to evaluate many candidate solutions rapidly and inexpensively. This can be quite valuable even when the results may only approximate actual system behavior. Because the modeling process requires a certain amount of abstraction, this is about as much as we can hope for.

Adding ATM Machines

For the ATM system, simple changes in the rate matrix will produce corresponding changes in performance that can easily be recalculated using available analysis techniques. One design option is to add more machines but leave the size of the foyer as is.

Suppose for simplicity that additional ATMs do not reduce the space for customers. The rate network for a system with three ATMs is shown in Figure 14.6. The service rate becomes 2μ with two customers in the system and 3μ with three customers.

The steady-state results for 1, 2, and 3 machines are shown in Table 14.2. With the addition of a second ATM (alternative ATM2), the proportion of customers lost and the average time in the queue both drop by almost an order of magnitude. Adding a third machine (alternative ATM3) nearly eliminates lost customers and waiting time.

Expanding the Size of the Foyer

Rather than adding more machines, the manager wishes to consider expanding the size of the foyer. Table 14.3 shows the results for two new alternatives, ATM4 and ATM5, which provide space for eight and 12 customers, respectively. As expected, the added space reduces the number of arrivals that balk, but this means that a higher proportion of customers must wait in line. More space increases the average number in the system as well as the time in the system, because more customers are actually being served—that is, fewer customers are lost.

Selecting the best alternative requires a tradeoff analysis between (1) machine acquisition and operating costs and (2) the costs associated with waiting and insufficient capacity. It is usually possible to improve system performance by adding more machines. These costs are easy to measure; the difficulty is obtaining reasonable estimates of the costs associated with lost or disgruntled customers.

ATM Teamed with a Human Teller

The bank is concerned that it is losing the personal touch by using ATMs. As an alternative to adding another machine or expanding the foyer, the manager is considering opening a window with a human teller to help relieve the congestion. The window will not reduce the space for customers nor will it increase the demand for service; however, the teller is

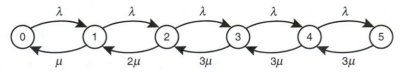

Figure 14.6 Rate network for three ATMs

Table 14.2   Alternative Designs with More ATMs

System measure	Alternative		
	ATM1	ATM2	ATM3
Number of machines	1	2	3
Capacity of foyer	5	5	5
Average number in system	1.8683	0.9187	0.8134
Average time in system	1.0252	0.4635	0.4078
Average number in queue	1.1394	0.1258	0.0156
Average time in queue	0.6252	0.0635	0.0078
Throughput rate	1.8224	1.9823	1.9946
Efficiency (utilization)	0.7289	0.3965	0.2659
Proportion who must wait	0.6401	0.2152	0.0484
Proportion of customers lost	0.0888	0.0088	0.0027

not as fast as the ATM for many transactions. When dealing with the teller, the average customer will require 1 minute for service. Recall that an ATM handles 2.5 customers per minute with an average service time of 24 seconds, and that customers arrive at an average rate of 2 per minute. Even though the ATM is faster, the manager feels that the human teller would be the first choice of customers even if both were available. For purposes of analysis, we again assume that the arrival and service processes are Poisson processes.

This is an example of a two-server queuing system. To develop a model, we need three state variables—two for the activities of the servers and one for the number of customers in the queue. We assign the index 1 to the human teller and 2 to the machine. The state vector is $\mathbf{s} = (s_1, s_2, s_3)$, where

$$s_i = \begin{cases} 0 & \text{if server } i \text{ is idle} \\ 1 & \text{if server } i \text{ is busy} \end{cases} \quad \text{for } i = 1, 2$$

s_3 = number in queue

Table 14.3   Alternative Designs with More Space

System measure	Alternative		
	ATM1	ATM4	ATM5
Number of machines	1	1	1
Capacity of foyer	5	8	12
Average number in system	1.8683	2.6048	3.2437
Average time in system	1.0252	1.3549	1.6458
Average number in queue	1.1394	1.8358	2.4554
Average time in queue	0.6252	0.9549	1.2458
Throughput rate	1.8224	1.9225	1.9709
Efficiency (utilization)	0.7289	0.7690	0.7884
Proportion who must wait	0.6401	0.7302	0.7738
Proportion of customers lost	0.0888	0.0388	0.0145

We define event a as an arrival and events d_1 and d_2 as service completions for the teller and the ATM, respectively. The state-transition network for the situation is depicted in Figure 14.7. The numbers in parentheses adjacent to the nodes represent particular states $\mathbf{s}_i$ of the system. For example, $\mathbf{s} = (110)$ positioned below node 3 indicates that both the ATM and the teller are busy and that no customers are waiting.

Note that we have included the event of an arrival while in state 6 as an arc that begins and ends at state 6. This represents the event of an arrival that balks because of the finite space available. For a CTMC, the steady-state probabilities are not affected by whether or not this arc is included. If there is a cost associated with the balk, however, the arc must be included to represent this cost. The economic evaluation of the process is considered in Section 15.3.

The rate diagram in Figure 14.8 is obtained by substituting the rates of occurrence for the event designations in Figure 14.7. We use λ for the arrival event, μ_1 for a service completion by the teller, and μ_2 for a service completion by the ATM. When two events lead to the same transition, their rates are summed. This is appropriate only when both activities are independent and exponentially distributed, as is the case here. Referring to Figure 14.7, the rate of occurrence of the event d_1 or d_2 is $\mu_1 + \mu_2$.

The accompanying rate matrix for $\lambda = 2$, $\mu_1 = 2.5$, and $\mu_2 = 1$ is as follows. Consider, for example, the transition from (111) to (110). The matrix element $r_{43} = \mu_1 + \mu_2 = 2.5 + 1 = 3.5$.

$$
\mathbf{R} = \begin{array}{c}
\\
(000) \\
(010) \\
(100) \\
(110) \\
(111) \\
(112) \\
(113)
\end{array}
\begin{array}{c}
\begin{array}{ccccccc}
(000) & (010) & (100) & (110) & (111) & (112) & (113)
\end{array} \\
\left[\begin{array}{ccccccc}
0 & 0 & 2 & 0 & 0 & 0 & 0 \\
2.5 & 0 & 0 & 2 & 0 & 0 & 0 \\
1 & 0 & 0 & 2 & 0 & 0 & 0 \\
0 & 1 & 2.5 & 0 & 2 & 0 & 0 \\
0 & 0 & 0 & 3.5 & 0 & 2 & 0 \\
0 & 0 & 0 & 0 & 3.5 & 0 & 2 \\
0 & 0 & 0 & 0 & 0 & 3.5 & 2
\end{array}\right]
\end{array}
$$

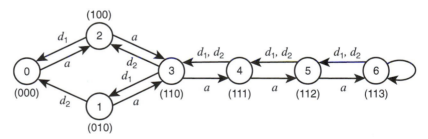

Figure 14.7   State-transition network for ATM with human teller

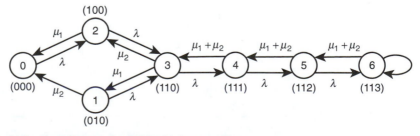

Figure 14.8   Rate diagram for ATM with human teller

Table 14.4   Comparison of ATM Alone and with Human Teller

System measure	Alternative		
	ATM1	ATM and human teller	ATM2
Average number in system	1.8683	1.580	0.9187
Average time in system	1.0252	0.8213	0.4635
Average number in queue	1.1394	0.3659	0.1258
Average time in queue	0.6252	0.1902	0.0635
Throughput rate	1.8224	1.9234	1.9823
ATM efficiency (utilization)	0.7289	0.4731	0.3965
Teller efficiency (utilization)	—	0.7408	—
Proportion who must wait	0.6401	0.3892	0.2152
Proportion of customers lost	0.0888	0.0383	0.0088

The steady-state solution for the preceding data is as follows.

$$\mathbf{s} = \quad (000) \quad (010) \quad (100) \quad (110) \quad (111) \quad (112) \quad (113)$$
$$\pi^{\mathrm{P}} = (0.214, \ 0.046, \ 0.313, \ 0.205, \ 0.117, \ 0.067, \ 0.038)$$

These values indicate, among other things, that there is now a 3.8% probability (π_6^{P}) that an arriving customer will find the system full and will be lost. This compares with 0.88% for the alternative with two ATMs in Table 14.2. Other statistics are presented in Table 14.4, which compares the ATM alone with the ATM and a human teller. System performance naturally improves when the teller window is added but is inferior to the performance of the ATM2 option (see Table 14.2; statistics repeated in Table 14.4) by a factor of more than 2 in all measures. Whether the human teller option is viable depends on the associated costs and perceived benefits.

14.6   BIRTH AND DEATH PROCESSES

A wide range of situations, such as warehousing and cross-docking operations, population infestations, and the servicing of customers on a hotline, can be described in terms of fairly simple stochastic processes. Moreover, it is usually straightforward to compute the steady-state behavior of the accompanying models. Our ability to obtain closed-form solutions is a direct consequence of the assumption that all activity durations are exponentially distributed. If this were not the case, more computationally intensive methods such as simulation would be needed because the Markov model would no longer be appropriate. In this section, we concentrate on an important special case characterized as birth–death processes.

Pure Birth Process

Hurricanes and tropical storms in the Atlantic Ocean are most prevalent during an 8-week period beginning in late August. It is customary to assign a name to each storm, starting with the letter A, such as Amy, and continuing on through the alphabet. For each year, 21 names have been specified with five letters not used. If it is assumed that weather forecasters know the probability distribution of the time between hurricanes and that these times are independent, we might ask what the probability is that we will run out of hurricanes before we run out of names—that is, the probability that the 21 names will be sufficient for a given season.

The number of hurricanes that have appeared up to some point in time can be viewed as a realization of a pure birth process. For this general class, populations only grow with time. The state of the system is the number in the population, and the only event is an arrival *a*. Figure 14.9 depicts the state-transition network for the class. The hurricane count starts in state 0, and the 21 names will be sufficient if the number of hurricanes in the system does not exceed 21 during the 8-week season.

When we assume an exponential distribution for the time between occurrences, this situation is described by a Markov process. The following rate matrix has infinite dimensions because the pure birth process has no finite upper bound. For the general case, the arrival rates have subscripts to indicate that their values may depend on the state.

$$\mathbf{R} = \begin{bmatrix} 0 & \lambda_0 & 0 & 0 & \cdots \\ 0 & 0 & \lambda_1 & 0 & \cdots \\ 0 & 0 & 0 & \lambda_2 & \cdots \\ 0 & 0 & 0 & 0 & \cdots \\ \vdots & \vdots & \vdots & \vdots & \ddots \end{bmatrix} \qquad \begin{array}{l} \text{Rate matrix for} \\ \text{pure birth process} \end{array}$$

Pure birth processes have no steady state, because the number in the system is increasing with time—i.e., the embedded Markov chain is not positive recurrent. However, when it is assumed that the mean sojourn rate is independent of the state and equal to λ, the transient probabilities are governed by a Poisson distribution whose parameter is proportional to time.

$$p_0(t) = e^{-\lambda t},\ p_1(t) = \lambda t e^{-\lambda t},\ \ldots,\ p_k(t) = \frac{(\lambda t)^k e^{-\lambda t}}{k!},\ \ldots$$

This is one of the few cases for which the transient probabilities are analytic functions. Note that we are using $p_k(t)$ instead of $q_k(t)$ in this section to conform to the notation in the queuing literature.

From the problem statement, we know that hurricanes arrive at an average rate of two per week, so $\lambda = 2$. Letting $N(t)$ be the random variable that counts the number of arrivals in t weeks, we wish to find the probability that $N(8)$ is less than or equal to 21; in other words, the probability that the 21 available names are sufficient. This can be written as

$$\Pr\{N(t) \le n\} = \sum_{k=0}^{n} \frac{(\lambda t)^k e^{-\lambda t}}{k!}$$

or

$$\Pr\{N(8) \le 21\} = e^{-16} \sum_{k=0}^{21} \frac{(16)^k}{k!} = 0.911$$

The computations were performed with the Random Variables Excel add-in on the accompanying CD. The results indicate that it is unlikely that we will run out of names.

Pure Death Process

A student is taking a self-paced class and must complete 10 study units in fixed sequence. After talking with other students who have previously taken the class, he estimates that the

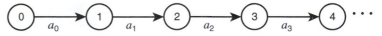

Figure 14.9   Pure birth process

average time to complete one unit is 1 week. This raises the question as to whether he will complete all units during the 14-week semester. To provide some guidance, he wishes to compute the probability of that occurrence, as well as the distribution of the number of unfinished units if he does not complete the class.

If we define the initial population as the 10 study units and observe the number of units remaining as the random variable of interest, we have described a pure death process. Such a process has a population that only declines with time. The only event, d, is a departure. Figure 14.10 depicts the state-transition network. For our example, the student starts in state 10 and will finish when and if he reaches state 0 during the 14-week period. The state space for the problem is $S = \{0, 1, \ldots, 10\}$.

The rate matrix for a pure death process has a single nonzero entry for row k, column $k - 1$ ($k = 1, \ldots, 10$), indicating the average number of transitions per unit time—a week for this problem. State 0 is an absorbing state because once the system has entered state 0, it can never leave it. For computational purposes, we assign a rate of 1 to r_{00}. This is the common way of dealing with absorbing states.

$$P = \begin{bmatrix} 1 & 0 & 0 & \cdots & 0 & 0 \\ \mu_1 & 0 & 0 & \cdots & 0 & 0 \\ 0 & \mu_2 & 0 & \cdots & 0 & 0 \\ 0 & 0 & \mu_3 & \cdots & 0 & 0 \\ \vdots & \vdots & \vdots & & \vdots & \vdots \\ 0 & 0 & 0 & \cdots & \mu_{10} & 0 \end{bmatrix} \quad \begin{array}{l} \text{Rate matrix for} \\ \text{pure death process} \end{array}$$

In the long run, the system will eventually reach state 0, so the steady-state probability vector is $\boldsymbol{\pi}^P = (1, 0, \ldots, 0)$—i.e., the probability of being in state 0 is 1. All other states are transient. Assume that the states have the same departure rate,

$$r_{k,k-1} = \mu_k = \mu, \quad k = 1, \ldots, 10$$

Then the transient probabilities are again based on the Poisson distribution. In particular, we have

$$p_{10-k}(t) = \frac{(\mu t)^k e^{-\mu t}}{k!}, \quad 0 \le k < 10$$

and

$$p_0(t) = 1 - \sum_{k=1}^{10} p_k(t)$$

Using a rate of one unit per week for μ, we compute the probability of completing k units in $t = 14$ weeks with the preceding distribution function. The results shown in Table 14.5 indicate that the chance of completing the course in 14 weeks is almost 90%. The remaining probabilities are less then 5%, shrinking to near zero for $k > 3$.

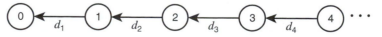

Figure 14.10   Pure death process

Table 14.5   Probability of Not Completing a Specific Number of Units

Incomplete units, k	0	1	2	3	4	5	6	≥ 7
Probability, $p_k(14)$	0.891	0.047	0.03	0.017	0.009	0.004	0.001	0.0005

We now add the condition that a unit is not complete until the student passes a competency exam. If he fails the exam, he must wait a week before being allowed to retake it. The student estimates that he has an 80% chance of passing any given exam and wishes to determine his overall probability of success.

This situation is easily handled by adjusting the departure rate by the probability that the student will pass an exam; in this case, $\mu_{new} = 0.8\mu_{old}$, where μ_{old} = one per week. The relevant Poisson distribution has the mean value of $\mu_{new} \times t = 0.8 \times 14 = 11.2$ weeks. The new results are shown in Table 14.6. The chance of completing all the units in 14 weeks is now 68%, a considerable reduction.

General Birth–Death Processes

A small rural hospital has a need for a staff of four full-time nurses. Unfortunately there is a staff turnover problem. Based on statistics gathered over the last several years, a nurse will remain on the staff for an average of 100 weeks before resigning. Because of turnovers, the number of full-time nurses ranges between 0 and 6. To fill vacancies, the hospital has a recruiting program. Under normal circumstances, the time between new hires averages 40 weeks. When the staff drops to 0 or 1, special incentives reduce the time between hires to an average of 20 weeks. There are no hires when the staff is at 6. A full-time nurse is paid $800 per week. Whenever the number of full-time nurses falls below 4, a temporary service fills the need up to 4 with a cost of $2000 per week per nurse. The hospital is interested in the long-run probability distribution of full-time nurses and the cost of staffing.

Combining the pure birth and pure death processes to allow both arrivals and departures, give a birth–death process as depicted in the state-transition network in Figure 14.11. For this example, an arrival (or birth) is a new hire and a departure (or death) is a resignation. The state measures the number of full-time nurses on the staff and the state space is finite, $S = \{0, 1, \ldots, 6\}$.

In order to obtain numerical results for this system, we assume that resignations and new hires can be modeled as Poisson processes. The general rate matrix for the problem is given below. Essentially, it is a composite of the rate matrices for the pure birth process and the pure death process.

$$\mathbf{R} = \begin{bmatrix} 0 & \lambda_0 & 0 & 0 & 0 & 0 & 0 \\ \mu_1 & 0 & \lambda_1 & 0 & 0 & 0 & 0 \\ 0 & \mu_2 & 0 & \lambda_2 & 0 & 0 & 0 \\ 0 & 0 & \mu_3 & 0 & \lambda_3 & 0 & 0 \\ 0 & 0 & 0 & \mu_4 & 0 & \lambda_4 & 0 \\ 0 & 0 & 0 & 0 & \mu_5 & 0 & \lambda_5 \\ 0 & 0 & 0 & 0 & 0 & \mu_6 & 0 \end{bmatrix} \quad \text{Rate matrix for a birth–death process}$$

Table 14.6   Updated Probability of Not Completing a Specific Number of Units

Incomplete units, k	0	1	2	3	4	5	6	≥ 7
Probability, $\pi_k(14)$	0.681	0.104	0.084	0.06	0.037	0.02	0.009	0.0042

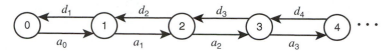

Figure 14.11   Birth–death process

A simple expression for the transient solution is not available but the steady-state probabilities can be found when the system is finite, as it is here. The departure rate is proportional to the inverse of the mean time to resignation and the number of full-time nurses currently on the staff, k, unless k is 0.

$$\mu_0 = 0, \quad \mu_k = k/100 = 0.01k \text{ for } 1 \le k \le 6$$

The arrival rate is the inverse of the time between hires and depends on the state as below.

$$\lambda_0 = 0.05, \quad \lambda_1 = 0.05, \quad \lambda_k = 0.025 \text{ for } 2 \le k \le 5, \quad \lambda_6 = 0$$

Substituting these values into the rate matrix, we compute the steady-state probabilities π^P with the Markov Process Excel add-in. The results are presented in Table 14.7 along with the input data. The table also includes a row that gives the cost per week as a function of the number of full-time nurses. Using these values, the expected cost per week is $4950. If there were no resignations, the cost of maintaining a staff of 4 full-time nurses would be $3200. The $1750 weekly difference between the expected cost and the minimum staffing cost is the cost due to variability. We have not included other relevant costs such as training costs. Note that the analysis evaluates the current situation, but does not indicate a solution to the problem. Other alternatives must be evaluated on a case-by-case basis.

Queuing Systems

A queuing system is an important special case of the birth–death process that has many practical applications. The basic queuing process is illustrated in Figure 14.12. Customers arrive from some input source. If the service mechanism is busy, they must wait in a line until a server is available. Some time is required for service, after which the customer departs. The details of the process depend on various assumptions and parameters adopted for the components of the system. For example, the input source may be finite, the customers might arrive in batches, several lines may form in front of the service mechanism, multiple servers may work together or separately, and the order in which waiting customers are serviced may depend on their priority ratings.

Table 14.7   Input and Output Data for Nurse Example

System measure	State, k						
	0	1	2	3	4	5	6
λ_k (per week)	0.05	0.05	0.025	0.025	0.025	0.025	0
μ_k (per week)	0	0.01	0.02	0.03	0.04	0.05	0.06
Cost/week	8000	6800	5600	4400	3200	4000	4800
π_k^P	0.025	0.125	0.312	0.260	0.163	0.081	0.034

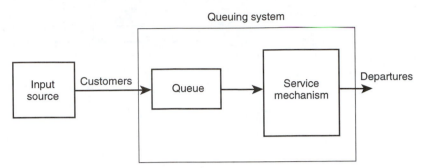

Figure 14.12 Schematic of a queuing system

The input source is also known as the calling population. It is the collection of customers that may have a need to use the system and is described by its size and interarrival time characteristics. The size is measured by the number of potential customers and may be finite or infinite. The interarrival time is usually a random variable with a specified probability distribution that informs the time between successive arrivals.

The queue is the place where customers reside while waiting for service. In the simplest case, all customers wait in a single line. The queue discipline defines the rules by which customers are selected for service. A popular discipline is first-in-first-out (FIFO), but last-in-first-out (LIFO), priority ordering, and random ordering are other possibilities. FIFO is also referred to as first-come-first-served (FCFS).

Service is provided by one or more servers (or channels) operating in parallel. For now, we will assume that they are identical. The characteristics of this component of the system are the number of servers and the service time distribution. The latter is a probability distribution for the time required to perform the service operation.

The following five-field notation is commonly used to identify most of the important characteristics of the queuing system.

Arrival distribution/Service distribution/Number of servers/

Maximum number in the system/Number in the calling population

The first and second fields specify the probability distributions of the durations of the arrival and service activities, respectively. The following symbols are used to identify prominent distributions.

M = exponential distribution

D = constant time

E_k = Erlang distribution with parameter k

G = general or arbitrary distribution

The third field indicates the number of servers, s. The fourth field, when present, indicates the maximum number of customers, K, allowed in the system. When there is no practical limit on this number, it is omitted. The fifth field specifies the size of the calling population, designated by the letter N. Similarly, when the potential number of customers is unlimited, this term is omitted. Of course, the fourth field cannot be omitted if the fifth is included.

To illustrate, a queuing system with two servers, an exponential distribution for the time between arrivals, a constant time for service, and unlimited numbers in the system and calling population would be described as an $M/D/2$ system. If the number in the system were limited to five, this would be an *M/D/2/5* system. If, in addition, we stated that the population had 20 customers, this would be an *M/D/2/5/20* system.

Infinite Queue

A software company operates a customer support hotline 24 hours a day. The company has just introduced a beta version of a product update, and their telephone lines are being flooded with calls as a result of unanticipated problems with installation. To ensure that all customers receive help, the company has installed a large number of holding lines. Calls are always answered, but customers may have to wait to talk to a representative. To make the situation more pleasurable, the company provides recorded country-western music to entertain waiting customers. Nevertheless, the quality assurance group has received numerous complaints and wants some guidance on how many service representatives to hire. Calls arrive every 10 seconds on average and are completed at the approximate rate of one every 3 minutes. For modeling purposes, it is assumed that both the arrival and service processes are Poisson.

When there is no upper limit on the number of customers that may reside in the queue, we say that the queue is infinite (we are referring to the potential size rather than the number of customers in line). Assuming that interarrival time and service time distributions are not exponential or Erlang, and that the calling population is very large, we have a *G/G/s* queuing system. It is common to designate this system as *GI/G/s*, where the *I* indicates that the times for the activities are independent random variables. We will not always include the *I* in our notation, but we do assume independence throughout the discussion.

In this model, customers arrive as a result of activity *a* (they have a problem with the software). We assume that there are *s* servers available to provide support in the form of an activity that we denote by *d*. If a customer arrives and finds all the servers busy, he or she joins a queue. The customers in the queue are served in the order in which they arrive (FCFS). The state-transition network for the queuing system is shown in Figure 14.13. Events labeled *kd* imply that there are *k* service activities proceeding simultaneously. Thus, we have *k* type *d* events on the calendar. We use *n*, the number of customers in the system, to define the state. When $n \leq s$, all the customers in the system are receiving service. When $n > s$, all the servers are busy and the number of service activities is at the maximum value *s*.

To provide guidance on hiring, it would be necessary to know something about the rate at which customers call, the time it takes to provide service, the cost of an additional employee, and the cost of waiting. The latter is the most difficult to estimate because it has several components, a subset of which, such as loss of good will, are impossible to measure. Therefore, rather than trying to assign dollar values to waiting time, most companies in this situation set a service standard that says something like "the probability that a customer will have to wait more than 5 minutes on average should be less than 0.01." A sufficient number of support personnel are then hired to realize this goal.

When exponential probability distributions are assumed for both interarrival times and service times, this system can be modeled as an *M/M/s* queue. The rate matrix for the infinite state birth–death process has nonzero transition rates.

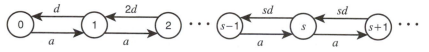

Figure 14.13   Queue with infinite capacity

$$r_{k,k+1} = \lambda \quad \text{for } k \geq 0$$

$$r_{k,k-1} = \begin{cases} k\mu & \text{for } 1 \leq k \leq s \\ s\mu & \text{for } k \geq s \end{cases}$$

With this in mind, we first address the problem of choosing the number of servers s. The absence of a limit on the number in the system means that the rate matrix has infinite dimensions. A property of an infinite queuing system is that for a steady state to exist, the total service capacity must exceed the arrival rate. This condition can be expressed mathematically as

$$\rho \equiv \frac{\lambda}{s\mu} < 1$$

where ρ is known as the *traffic intensity*. For the given data, $\lambda = 6$ calls per minute and $\mu = 1/3$ completions per minute, so

$$s > \frac{\lambda}{\mu} = \frac{6}{1/3} = 18$$

In other words, a minimum of 19 representatives are needed to ensure that the queue does not grow indefinitely. If 18 or fewer are provided, the number of customers on hold will get larger and larger with a corresponding increase in wait time. This assumes that neither the arrival rate nor the service rate changes over the course of the day. If the system is shut down at, say, midnight, then all customers on hold would be disconnected and would have to call back the next day.

The statistics for several design alternatives are given in Table 14.8. These results were obtained with the Queuing add-in rather than the Markov Process add-in because the latter cannot handle an infinite number of states. As expected, the proportion of time all representatives are busy (efficiency) decreases as their number increases; however, the quality of service, as measured by the proportion of customers who are placed on hold, also increases. With 25 representatives, only about 8% of the customers will wait and fewer than 1% will wait more than 1 minute. If management's objective is to keep operating costs as low as possible while ensuring system stability, the best choice is to hire 19 representatives, the fewest number that can handle the load. The average waiting time for this alternative is $W - \mu = 5.2493 - 3 \cong 2.25$ min. More than half of the callers will wait more than 1 minute.

Table 14.8   Results for the Hotline Example with Infinite Queue

Performance measure	Alternative		
	M/M/19	*M/M/22*	*M/M/25*
Average number in system, L	31.496	19.251	18.214
Average time in system, W	5.249	3.209	3.036
Average number in service, L_s	18	18	18
Efficiency, L_s/s	0.947	0.818	0.720
Proportion of customers that wait	0.750	0.278	0.083
Proportion that wait more than 1 minute	0.537	0.073	0.008

Finite Queue

After a detailed study, the quality assurance group decides that the large number of incoming lines is a mixed blessing. Since most customers are paying a long distance telephone charge for this "free" advice, they would rather receive a busy signal than be placed on hold. Moreover, it is not clear that everyone enjoys country-western music. As such, the company decides to reduce drastically the number of incoming lines but is unsure of exactly how many to provide. In the new environment, there will still be more lines than representatives, but when all holding lines are full, customers will receive a busy signal. The expectation is that they will call back later when the system is not so busy, or perhaps they will be motivated to solve whatever problem they are having themselves. Before making a decision on how many lines to maintain, the group wants a better understanding of the costs of the various alternatives.

When the number of customers in the system (in service plus waiting) is limited to some upper bound $K < \infty$, we have a situation characterized by a finite queue. The case under discussion is referred to as a $GI/G/s/K$ system. Recall that the fourth term indicates the maximum number permitted in the system.

This model is the same as the infinite queue case except arrivals cannot occur in state K. The state-transition network has $K + 1$ states with the maximum state index K. One can identify two possibilities for arrivals while in state K. The first possibility is that an arrival can occur but is turned away because the system is full. This situation is illustrated in Figure 14.14a. In general, a customer is said to *balk* when he or she does not enter the system. A full parking lot is a typical example where balking occurs. Cars trying to enter are turned away and must go elsewhere for service. In Figure 14.14a we see that the arrival event causes a transition from state K back to state K. Note that a customer may balk even in the $GI/G/s$ case. It often happens that arrivals become discouraged when the queue is long and decide not to wait.

A second possibility is that the arrival process is terminated when the system enters state K. Figure 14.14b shows the state-transition diagram where no arrival is on the calendar in state K. This might be the case when an inventory control manager has a standing order for some scarce part which is delivered one at a time. The order is canceled, however, whenever the inventory level reaches the maximum value K. Although the two cases are very similar and have the same steady-state solution for the $M/M/s/K$ system, they may have different economic consequences.

For the customer support example, we have a system equivalent to the one in Figure 14.14a, with s equal to the number of representatives and K equal to the number of incoming lines. The queue is provided by the holding lines, with the maximum number in the queue being equal to $K - s$.

a. Customer arrives but then leaves   b. No more arrivals after K

Figure 14.14   Possible treatment of arrival events in a finite queue

For a *GI/G/s/s* system, the number of servers equals the maximum number of customers in the system. In the case of the example, $K = s$, so there are no holding lines. All customers who do not receive a busy signal enter the system and receive immediate service.

If we continue to assume that arrivals and service completions are Poisson processes, we have an *M/M/s/K* queuing system. This means that the rates are now given by the equations

$$r_{k,k+1} = \lambda \quad \text{for } 0 \le k \le K - 1$$

$$r_{k,k-1} = \begin{cases} k\mu & \text{for } 1 \le k \le s \\ s\mu & \text{for } s < k \le K \end{cases}$$

where s is the number of representatives and K is the total number of lines. The number of holding lines is $K - s$. Thus, a maximum of $K - s$ customers will be waiting for service at any given time.

Table 14.9 highlights several performance measures for an *M/M/19/K* system with three different values of K. Limiting the number of lines has the desired effect of reducing the waiting time but at the cost of lost customers. When $K = 19$, there are no holding lines so there is no waiting, but more than 13% of the customers are lost. Increasing the holding lines to six ($K = 25$) decreases the percentage of lost customers to about 6%. Many other combinations of "number of representatives" and "number of holding lines" are possible and should be part of the full analysis. Before an informed decision can be made, however, it would be necessary to assess all costs, including the cost of lost customers and the cost of waiting. Unfortunately, both of these measures are extremely difficult to quantify.

Finite Input Source

Consider a taxi company with a fleet of 6 cabs and a repair shop to handle breakdowns. We now define the system as the repair shop and the state as the number of cabs in the shop. A server for the shop is a repair bay with a mechanic. Each server costs a fixed amount per day to operate. The calling population is the collection of taxis not in the shop. We assume that the taxis are identical with respect to breakdown rate and repair requirements, and that the servers are also identical with respect to skills and efficiency. The revenue for the company is a function of the number of cabs on the street in working order. Given this information, management wishes to determine the optimal number of repair bays to operate—i.e., they wish to maximize the difference between expected revenue and the expected cost of operating the bays.

Figure 14.15 shows the general case for a finite calling population N. For this example, $N = 6$, implying that the maximum number of cabs in the system is also 6. We associate the arrival process with an individual cab, so when the system is empty ($n = 0$), N different

Table 14.9 Results for the Hotline Example with Finite Queue

Performance measure	Alternative		
	M/M/19/19	*M/M/19/22*	*M/M/19/25*
Average number in system, L	15.548	17.002	18.297
Average time in system, W	3.000	3.096	3.239
Average number in service, L_s	15.542	16.473	16.948
Efficiency, L_s/s	0.818	0.867	0.892
Proportion of customers that wait, π_K^P	0	0.283	0.427
Proportion of customers not served, $\sum_{k=s}^{K-1} \pi_K^P$	0.136	0.085	0.059
Proportion that wait more than 1 minute	0	0.006	0.056

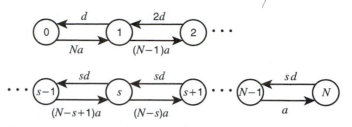

Figure 14.15 Network diagram for finite calling population

arrival events can occur. This is indicated by Na on the arc leaving node 0. For the general state n, $N - n$ units remain in the calling population, so $N - n$ arrival events are on the calendar. When the system is full, the calling population is empty, precluding any arrivals.

A departure from the system is triggered by a service completion. When the number of taxis in the system, n, is smaller than or equal to s (for this example we are to choose s), n service events are on the calendar. For n greater than s, all servers are busy and only s service events are on the calendar. Recall that a system with a finite input source and general service and arrival processes is denoted by $GI/G/s/K/N$. For this example we examine the system for $s = 1$ and 2.

Assuming that exponential distributions govern the time between failures of the taxis and the time to repair, our case can be modeled as an $M/M/s/N/N$ queuing system. Here, N is the number in the calling population (6 for the example) and provides a limit on the maximum number in the system. We identify the arrival process with an individual cab and specify λ as the failure rate for a cab.

$$r_{k,k+1} = (N - k)\lambda \quad \text{for } 0 \le k \le N$$

$$r_{k,k-1} = \begin{cases} k\mu & \text{for } 1 \le k \le s \\ s\mu & \text{for } s < k \le N \end{cases}$$

To compute numerical results we assume that the failure rate of each taxi is 1/3 per month and that the service rate of a bay is 4 per month. We analyze the system with one and two bays. The steady-state results obtained with the Queuing add-in are given in Table 14.10.

Table 14.10 Results for the Taxi Example with Finite Calling Population

	Alternative	
Performance measure	$M/M/1/6/6$	$M/M/2/6/6$
Average number in system, L	0.673	0.474
Average time in system, W	0.379	0.257
Average number in service, L_s	0.444	0.460
Efficiency, L_s/s	0.444	0.232
π_0^P	0.556	0.616
π_1^P	0.278	0.308
π_2^P	0.116	0.064
π_3^P	0.039	0.011
π_4^P	0.010	0.001
π_5^P	0.002	0
π_6^P	0	0
Expected revenue, $\bar{R}$	$5326	$5525

Assuming each working taxi yields a revenue of $1000 per day, Table 14.10 shows the expected revenues of the two plans. We see that the overall improvement in performance with two bays comes at a price. The extra bay is justified only if it costs less than $199 per day to operate.

14.7 PROBABILISTIC TRANSITIONS

Our discussion so far has been limited to the case in which the transition from one state to the next is predetermined given a current state and an event. For example, a service completion in an *M/M/s* queuing system reduces the number in the system by one with certainty. This situation can be generalized by defining multiple successor states along with the probability of going from the current state **s** to some successor state **s′**. We now show how to incorporate these transition probabilities into the model when the Markovian property holds. The discussion is limited to the case in which the number of successor states is countable and finite.

Transition Probabilities

Given that the system is in state **s** and event x takes place, we have previously indicated that the next state in the process is determined by the transition function

$$\mathbf{s}' = T(\mathbf{s}, x)$$

When the transition is not certain, it will now be supposed that the probability that the process will move to state **s′** is governed by the transition probability

$$p(\mathbf{s}, \mathbf{s}' \mid x)$$

This term is read as the probability of a transition from state **s** to state **s′** given that event x has occurred. Because of the nature of probabilities and assuming that events are mutually exclusive, it must be true that

$$0 \le p(\mathbf{s}, \mathbf{s}' \mid x) \le 1 \text{ and } \sum_{\mathbf{s}' \in S} p(\mathbf{s}, \mathbf{s}' \mid x) = 1$$

Figure 14.16 depicts a modified version of the state-transition network that incorporates these new data for two specific states $\mathbf{s}_i$ and $\mathbf{s}_j$. As we can see, an additional parameter is included on the arc corresponding to the transition probability distribution.

For a CTMC analysis, we must compute the rate of transition from $\mathbf{s}_i$ and $\mathbf{s}_j$. If event x occurs with rate λ, the rate assigned to the transition would be

$$r_{ij} = \lambda p(i, j \mid x)$$

Examples

Single Channel Queue with Two Kinds of Service

Consider a single bank teller providing service to arriving customers. All customers receive the normal banking service described by the activity d. Some proportion p, however, will

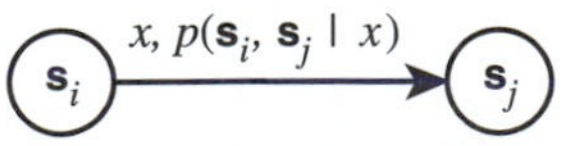

Figure 14.16   An event with corresponding transition probabilities

ask to buy traveler's checks after completing d. The activity of providing traveler's checks is denoted by c. Thus, the teller is either idle (i), providing service (d), or arranging for traveler's checks (c). Given data on arrival rates, service rates, and customer transition probabilities, management wants to estimate the proportion of the time the teller is idle as well as compile statistics on the waiting time of customers.

A model that can be used to address these concerns requires the following two-dimensional state vector $\mathbf{s} = (s_1, s_2)$ such that

$$s_1 = \text{number in system}$$
$$s_2 = \text{status of teller, where } s_2 \in \{i, d, c\}$$

The easiest way to represent this process is with the state-transition network diagram in Figure 14.17. If the teller is idle ($\mathbf{s} = (0, i)$), the only possible transition is the arrival of a customer. If the system is in state $\mathbf{s} = (j, d)$, $j = 1, 2, \ldots$, which means that the teller is providing regular service, the next state will be either $\mathbf{s}' = (j - 1, d)$, indicating that the customer departs (this occurs with probability $1 - p$), or $\mathbf{s}' = (j, c)$, indicating that the customer wants traveler's checks (this occurs with probability p). Thus, at the completion of activity d, there is either a transition with probability p to the state associated with the processing of traveler's checks or a transition with probability $(1 - p)$ to the state corresponding to one less customer in the system.

To illustrate the construction of the rate matrix, assume that the rates associated with arrivals (a), regular service (d), and traveler's checks are, respectively, λ, μ_1, and μ_2. Assume a limit of two customers at the teller. The rate matrix is then

$$\mathbf{R} = \begin{bmatrix} 0 & \lambda & 0 & 0 & 0 \\ (1-p)\mu_1 & 0 & p\mu_1 & \lambda & 0 \\ \mu_2 & 0 & 0 & 0 & \lambda \\ 0 & (1-p)\mu_1 & 0 & \lambda & p\mu_1 \\ 0 & 0 & \mu_2 & 0 & \lambda \end{bmatrix} \begin{matrix} (0,i) \\ (1,d) \\ (1,c) \\ (2,d) \\ (2,c) \end{matrix}$$

The state corresponding to each row is shown on the right of the matrix. The columns have the same designation as the rows. Note that the service rate μ_1 is multiplied by either p or $1 - p$ in the matrix. We place λ on the diagonals of rows 4 and 5 to indicate that an arrival does not result in a change of state.

A similar construction is appropriate for the remainder of the examples in this section, so it is sufficient to show the state-transition networks.

Service with Rework

Consider a machining operation in which there is a 0.4 probability that on completion a processed part will not be within tolerance. If the part is unacceptable, the operation is repeated

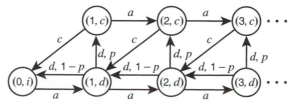

Figure 14.17   Queuing system with two types of service activities

immediately (this is called *rework*). Let us assume that the second try is always successful and that arrivals can occur only when the machine is idle. A simple extension would be to allow arriving parts to form a queue, and perhaps to have out-of-tolerance parts join the queue rather than be reworked immediately.

The basic situation can be represented with one state variable that assumes one of three values. When in state 0, the system is said to be empty and the machine is idle. Arrivals occur only in this state. When in state 1, the machine is working on the part for the first time. Rework, if necessary, occurs in state 2. The state-transition network for the process is shown in Figure 14.18. The event of a service completion (d_1) in state 1 gives rise to a transition to state 2 with a 0.4 probability and a transition to state 0 with a 0.6 probability.

A Multistage Process with a Single Worker and Rework

As an extension of the preceding example, assume that a part must undergo five successive operations performed by a single worker. The process begins with the arrival of a part at the first workstation and continues on through the remaining workstations in fixed order. When each specific operation has been completed, the part is inspected. If deemed to be within tolerance, it passes to the next operation; otherwise, it must be reworked by the worker. The probability of passing inspection on any try after completing any operation is p. Rework then occurs with probability $1 - p$. When the rework activity has been completed, the part returns to the *same* operation for processing again. When the part passes the final inspection, the worker becomes idle and waits for another arrival. Arrivals can occur only when the system is empty.

A two-dimensional state vector can be used to describe this situation. The first component s_1 identifies the workstation at which the part is being processed, and the second component s_2 indicates whether the worker is idle (i), processing the part (d), or performing rework (r). The state-transition network is shown in Figure 14.19. The vector $\mathbf{s} = (s_1,$

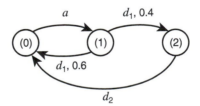

Figure 14.18 Rework diagram for single operation

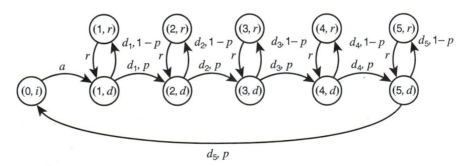

Figure 14.19 Multistage process with rework

s_2) within each node identifies the operation that is in progress and the status of the worker. The transition probabilities account for the uncertain results of the inspection.

EXERCISES

Use the Stochastic Analysis Excel add-in provided with this book to obtain numerical answers for the following exercises when required.

1. Consider Omar's barbershop described in Section 12.5, with one chair for cutting hair and three for waiting (four chairs in all). Assume that the mean time between customer arrivals is 15 minutes and the mean time for a haircut is 12 minutes. Also assume that both times are exponentially distributed and that customers who arrive and find the shop full, balk. If each customer pays $10 for a haircut, use the steady-state probabilities to compute Omar's revenue in an 8-hour day.

    Now make the following changes and compute Omar's average earnings. The changes are not cumulative.

    (a) Add a fourth waiting chair.

    (b) Change one of the waiting chairs to a barber's chair and hire another barber. Omar earns $5 for each haircut the second barber does. When both barbers are idle, an arrival goes to Omar.

    (c) When two or more of the waiting chairs are full, Omar works faster and reduces his average cut time to 10 minutes.

    (d) Change the arrival process. When all the waiting chairs are empty, all the arrivals enter. When one waiting chair is occupied, 1/3 of the arrivals balk. When two of the waiting chairs are occupied, 2/3 of the arrivals balk. When all three chairs are occupied, all the arrivals balk.

2. A remote Air Force base in Alaska has a surveillance system consisting of three radar units that are designed to operate continuously. The system can provide proper coverage as long as two of the three radar units are in working order. The time between failures is an exponentially distributed random variable with a mean of 1000 hours. Whenever a radar unit fails, a technician tries to fix it. The repair time is also exponentially distributed, but with a mean of 500 hours. When either two or three radar units are down, extra help is provided, reducing the repair time to an average of 200 hours. However, only one radar unit can be repaired at a time. Model this system as a CTMC and compute the steady-state probabilities.

    What is the probability that the system is operating with either two or three radar units? What is the probability that all radar units are down? What is the proportion of the time the repair shop is idle? What is the expected time a radar unit will spend in the shop (either waiting or in repair)? Show the rate network with your solution.

3. A gas station has three self-service pumps. The time required for a customer to pump her gas has an exponential distribution with a mean of 3 minutes. In addition to the space at the pumps, there is room for two more cars to wait. Cars arrive at random according to a Poisson process at an average rate of 60 per hour when there is room at the pumps. When all pumps are in use, the arrival rate drops to 40 per hour. When one customer is waiting, the arrival rate drops to 20 per hour. When both waiting spaces are full, no arrivals occur.

    (a) Compute the steady-state probabilities, the average number of customers either waiting or in service, and proportion of customers who are lost.

    (b) Redo the calculations for the case in which another pump is added. The number of waiting spaces remains at two.

4. A service system is composed of three sequential stations. Customers arrive according to a Poisson process at an average rate of four per hour. The service time at each station is exponentially distributed with a mean of 10 minutes. There is room in the system for three customers, one at each station. No waiting is permitted—i.e., customers cannot queue between stations or before the first station. A customer who completes service at one station and finds the next one busy must wait until it is available

before moving on. A customer arriving to find the first station busy is turned away. Compute the steady-state probabilities and the throughput rate for the system.

5. The figure shows the rate network for a CTMC. On the basis of this figure, determine the steady-state probabilities.

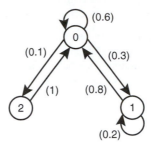

6. The figure shows the state-transition network for a Markov system with two absorbing states. The rates for the activities are as follows.

Activity	d_1	d_2	d_3	a_1	a_2	a_3
Rate	0.1	0.5	1	0.75	0.5	0.2

Construct the rate matrix and the embedded Markov chain matrix. Find the absorbing state probabilities when the system starts in state 2.

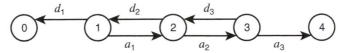

7. The figure shows the rate network for a CTMC. On the basis of this figure, determine the steady-state probabilities.

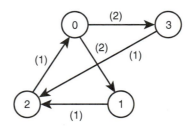

8. The figure shows the rate network for a CTMC. On the basis of this figure, determine the steady-state probabilities.

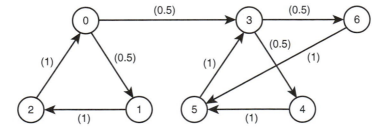

9. A military air base has three planes, each with its own repair crew. The planes fail independently according to an exponential distribution with a mean time to failure of 100 hours. The time to repair each plane is 30 hours on average and is also exponentially distributed.

   (a) Find the steady-state probabilities for the number of available planes. What is the average number of planes in working condition at any point in time?

   (b) Change the situation so that all three repair crews work together on one plane at a time. The mean repair time is reduced to 10 hours, but planes must now queue up for service. How does this change affect the average number of planes available? Is it better to have one crew dedicated to each plane or to combine the crews?

10. A truck running on four tires carries two spares in reserve. The tires in use are identical and each may fail while the truck is operating. The mean time to failure of a tire is 1 week (5 working days), and the failure time has an exponential distribution. When a tire fails, it is replaced with a spare if one is available. The truck cannot operate when it has less than four good tires.

    Several alternatives for repairing tires are given below. Consider each one separately and determine the proportion of time the truck is in working condition. Assume that failures are independent of each other. Repair times are exponentially distributed.

    (a) Whenever a tire fails, it is sent to the shop immediately. The average time to perform the repair and return the tire to the truck is 2 days.

    (b) When two tires are no longer in working condition, failed tires are sent to the shop. Tires are repaired and returned in sets of two. This takes 3 days on average.

    (c) When three tires have failed, the truck is sent to the shop and all tires are replaced. The repair time in this case takes an average of 4 days. Is there any benefit to replacing tires that have not failed when the failure time has an exponential distribution?

    (d) No repair is possible. The truck continues to operate until there are fewer than four good tires. What is the probability distribution for the time the truck can be used?

11. An assembly line consists of three stations in series. Each station is in one of three states: idle with no product, idle holding a product, and working on a product. There are sufficient subassemblies available so that whenever station 1 has no product it is immediately provided with a subassembly to work on. When station 3 completes a product, it leaves the system. Stations 1 and 2 cannot pass a unit to the following station unless that station is idle and empty. This represents a case of blocking. Assume the processing time of each station has an exponential distribution with a mean of 1 minute. Construct the rate matrix for this problem and find the steady-state solution. What is the average production rate of the assembly line?

12. A self-paced course consists of six units, numbered 1 through 6, that must be taken in sequence. The time it takes to study a unit can be modeled as an exponential random variable with a mean of 1 week. After completing a unit, the student takes a test and passes with probability p. Passing the test allows the student to go on to the next unit. The student must repeat the unit if he fails the test. When he has passed all six units, the course is complete. There is no limit to the number of times the student may take a test.

    Draw the rate network as a function of p. Write an expression for the cumulative distribution function associated with the time to finish the course. If $p = 0.5$, what is the probability that the student will complete the course in a 12-week semester?

13. Three placement machines work in parallel populating printed circuit boards. Raw boards are always available, so the machines can run continuously. In the absence of any misalignments or other faults, the production rate for a machine averages λ boards per minute. Each machine operates independently of the others so the production rate of the system is 3λ. The time it takes to populate a board can be modeled as an exponential random variable.

After all the components have been placed (event a), the board is inspected. If no faults are detected, it leaves the system and a new board is transferred to the machine. The probability of a fault is p. The following rules govern the operation of the system.

- When a fault is detected, the board is sent to a repair station.
- Only one technician is available, so boards needing repair form a queue.
- The time for repair is exponentially distributed with a mean of $1/\mu$.
- A placement machine cannot begin work on a new board until the board being repaired is returned for a touch-up operation. Consequently, it remains idle during board repair.
- Assume that all transfer and touch-up operations take negligible time.

(a) Draw the rate network for the problem.

(b) Construct the rate matrix for $\lambda = 1/\text{min}$, $\mu = 2/\text{min}$ and $p = 0.2$.

(c) Find the steady-state probabilities. What is the production rate of this system?

14. Consider the various production situations described in Section 11.4. For each case, let the arrival and service processes be Poisson. When queue lengths are limited, assume that arrivals balk when the system is full. For each of the following situations, compute the proportion of time the system will reside in each state and the production rate for each system.

(a) Consider the single server system depicted in Figure 11.9. The mean time between arrivals is 2 minutes. The mean time for service is 10 minutes.

(b) Consider the process represented by Figure 11.10. The mean time for arrivals is 2 minutes. The mean times for the five operations are shown in the table.

Activity	d_1	d_2	d_3	d_4	d_5
Mean time	1	2	3	4	5

(c) Consider the multistage process illustrated in Figure 11.11, but limit the number of customers in the system to five and the number of operations to three. Let the mean time between arrivals be 10 minutes, and let the three operations have the mean times given in the following table.

Activity	d_1	d_2	d_3
Mean time	2	3	4

(d) Consider the process shown in Figure 11.12, but limit the number of customers in the system to four. The mean times for the activities are given in the table below.

Activity	a	d_1	d_2
Mean time	3	4	8

(e) Consider the process depicted in Figure 11.14 with two classes of customers. Limit the number of customers in each queue to three (as shown in the figure). The mean times for the activities are given in the following table.

Activity	a_1	a_2	d
Mean time	10	10	8

15. Consider the various production situations described in Section 14.7. For each case, let the arrival and service processes be Poisson. When queue lengths are limited, assume that arrivals balk when the system is full. For each of the following situations, compute the proportion of time the system will reside in each state and the production rate for each system.

(a) Consider the queuing system depicted in Figure 14.17. Let $p = 0.1$ and limit the queue length to 5. The mean times for the activities are given in the following table.

Activity	a	d	c	
Mean time	6		4	10

(b) Consider the rework example illustrated in Figure 14.18. The mean times for the activities are given in the following table. Let the rework probability be 0.1.

Activity	a	s_1	s_2	
Mean time	2		4	10

(c) Consider the multistage rework example shown in Figure 14.19. Let the rework probability be 0.1, and let the activities have the mean times given in the following table.

Activity	a	d_1	d_2	d_3	d_4	d_5	r
Mean time	2	1	2	3	4	5	0.5

BIBLIOGRAPHY

Buzacott, J.A. and J.G. Shanthikumar, *Stochastic Models of Manufacturing Systems*, Prentice-Hall, Englewood Cliffs, NJ, 1993.

Feldman, R.M. and C. Valdez-Flores, *Applied Probability & Stochastic Processes*, PWS, Boston, 1996.

Gross, D. and C.M. Harris, *Fundamentals of Queueing Theory*, Third Edition, Wiley, New York, 1998.

Kao, E.P.C., *Introduction to Stochastic Processes*, Duxbury, Belmont, CA, 1997.

Lawler, G.F., *Introduction to Stochastic Processes*, Chapman & Hall, New York, 1995.

Mehdi, J., *Stochastic Models in Queueing Theory*, Academic Press, Boston, 1991.

Puterman, M.L., *Markov Decision Processes,* Wiley, New York, 1991.

Resnick, S.I., *Adventures in Stochastic Processes*, Birkhäuser, Boston, 1992.

Ross, S.M., *Introduction to Probability Models*, Seventh Edition, Academic Press, San Diego, 2000.

Chapter **15**

Mathematics of Continuous-Time Markov Chains

In this chapter, we derive several formulas used in Chapter 14 to analyze continuous-time Markov chains (CTMCs). We also discuss alternative or more streamlined approaches to the analysis that involve simple matrix algebra. In fact, the basic idea of a CTMC is quite simple, as you may recall. If we are in state i, our sojourn in that state will follow an exponential distribution with parameter λ_i. At the end of each sojourn, the process makes a transition to another state. The probability that the transition will be made to state j, p_{ij}, is the ratio of the transition rate from i to j, r_{ij}, to the total rate out of state i, λ_i.

In modeling a problem as a CTMC, we start with a definition of the state space and then construct the rate matrix $\mathbf{R}$, which contains the transition rates as described in Section 15.1. From the rate matrix, we can construct the generator matrix, as shown in Section 15.2. The time-dependent probabilities that the process is in a particular state at time t are characterized by the Kolmogorov equations, also presented in Section 15.2. These equations are used to derive the limiting probabilities $\boldsymbol{\pi}^{\mathrm{P}}$. A second way to derive $\boldsymbol{\pi}^{\mathrm{P}}$ using flow balance ideas is also discussed. In Section 15.3, we extend the economic methodology originally derived for discrete-time Markov chains (DTMCs). The focus is on expected value analysis and discounted cash flows. We next discuss first passage times in Section 15.4 and conclude with an in-depth look at birth-death processes in Section 15.5. The results provide the foundation needed for analyzing the steady-state properties of queuing systems, the most popular application of CTMCs.

15.1 EMBEDDED DISCRETE-TIME MARKOV CHAIN

As described in Section 14.1, every CTMC can be viewed as a combination of an embedded Markov chain and a state-dependent Poisson process. The first thing we show is that the transition matrix of the embedded DTMC is readily computed from the rate matrix of the CTMC, and can be analyzed with the same techniques discussed in Chapters 12 and 13. There are many other uses of the embedded DTMC that parallel those described previously. For example, when a CTMC includes absorbing states, the steady-state probabilities do not exist; however, one can compute absorbing probabilities using the same techniques developed for DTMCs.

ATM Teamed with Human Teller

To illustrate the primary components of a CTMC, we use the ATM plus human teller example presented in Section 14.5. Recall that a bank is considering providing service in its foyer with a combination of an ATM and a human teller. On average, two customers per minute arrive at the facility. The capacity of the foyer is five customers, either waiting or in service. When dealing with the teller, customers will require an average of 1 minute of service. Alternatively, the average service time at the ATM is 24 seconds. A customer entering the foyer will choose the first available server, whether human or ATM. If both are idle, the customer will choose the human teller. For purposes of analysis, we again assume that the arrival and service processes are Poisson processes.

The model of this system has three state variables—two for the activities of the servers and one for the number of customers in the queue. We assign the index 1 to the human teller and 2 to the ATM machine. The state vector is $\mathbf{s} = (s_1, s_2, s_3)$, where

$$s_i = \begin{cases} 0 & \text{if server i is idle} \\ 1 & \text{if server i is busy} \end{cases} \quad \text{for } i = 1, 2$$

$s_3 = $ number in queue

The rate diagram for this example is repeated in Figure 15.1. We use λ to represent the arrival rate (2/min), μ_1 is the service rate of the human teller (1/min), and μ_2 is the service rate of the ATM (2.5/min).

Markov Chain Representation

To obtain a realization of the process over some period of time, we simulated the operations of the corresponding *M/M/2/5* queuing system for 20 steps. The results are shown in Table 15.1, where each step represents either an arrival or a departure. The state numbers are identified with the network in Figure 15.1. At time zero, the system is in state 0. As time passes, it moves from state to state as customers arrive and are served. Each transition to a new state is a step of the process. For each step n, the system resides in a state X_n for an interval of time $T_n - T_{n-1}$ (recall that this value is known as the *sojourn time* or *residence time*). Each row in the table gives the step number, the time at which the transition occurs, the state index, and the interval between two successive steps.

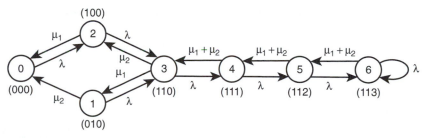

Figure 15.1   Rate diagram for ATM with human teller

Table 15.1 Simulation of ATM with Human Teller

Step, n	Time, T_n	State, X_n	Interval, $T_n - T_{n-1}$
0	0	0	
1	0.61	2	0.61
2	0.64	3	0.03
3	0.87	2	0.23
4	0.95	3	0.08
5	0.99	2	0.04
6	1.03	3	0.04
7	1.16	2	0.13
8	1.19	3	0.03
9	1.25	4	0.05
10	1.28	3	0.04
11	1.51	4	0.23
12	1.54	5	0.03
13	1.60	4	0.06
14	3.21	3	1.62
15	3.29	2	0.08
16	3.59	3	0.30
17	3.65	1	0.05
18	3.91	0	0.26
19	4.64	2	0.73
20	5.47	3	0.83

For any CTMC, its embedded DTMC determines the transitions between states and the associated Poisson process determines the sojourn time at each step. The DTMC is described by a state-transition matrix **P** and the Poisson process by a rate vector λ that specifies the transition rate at each step.

The Embedded DTMC from the Rate Matrix

Let the state space $S = \{0, 1, \ldots, m-1\}$ and consider state $i \in S$ of the CTMC. The rate of transitions out of state is λ_i, where

$$\lambda_i = \sum_{j=0}^{m-1} r_{ij}$$

and r_{ij} is the average rate at which the process goes from state i to state j. The embedded Markov chain is defined by the transition matrix **P**, whose elements p_{ij} represent the state-transition probabilities. In terms of the system parameters,

$$\Pr\{\text{transition to } j \mid \text{system is in state } i\} = p_{ij} = \frac{r_{ij}}{\lambda_i} \tag{1}$$

The general transition matrix is then

$$\mathbf{P} = \begin{bmatrix} r_{00}/\lambda_1 & r_{01}/\lambda_0 & r_{02}/\lambda_0 & \cdots & r_{0,m-1}/\lambda_0 \\ r_{10}/\lambda_1 & r_{11}/\lambda_1 & r_{12}/\lambda_1 & \cdots & r_{1,m-1}/\lambda_1 \\ \vdots & \vdots & \vdots & & \vdots \\ r_{m-1,0}/\lambda_{m-11} & r_{m-1,1}/\lambda_{m-1} & r_{m-1,2}/\lambda_{m-1} & \cdots & r_{m-1,m-1}/\lambda_{m-1} \end{bmatrix}$$

For the ATM plus human teller example, we have

	(000)	(010)	(100)	(110)	(111)	(112)	(113)	i	λ_i
(000)	0	0	2	0	0	0	0	0	2
(010)	2.5	0	0	2	0	0	0	1	4.5
(100)	1	0	0	2	0	0	0	2	3
R = (110)	0	1	2.5	0	2	0	0	3	5.5
(111)	0	0	0	3.5	0	2	0	4	5.5
(112)	0	0	0	0	3.5	0	2	5	5.5
(113)	0	0	0	0	0	3.5	2	6	5.5

and

	(000)	(010)	(100)	(110)	(111)	(112)	(113)
(000)	0	0	1	0	0	0	0
(010)	0.556	0	0	0.444	0	0	0
(100)	0.333	0	0	0.667	0	0	0
P = (110)	0	0.182	0.455	0	0.364	0	0
(111)	0	0	0	0.636	0	0.364	0
(112)	0	0	0	0	0.636	0	0.364
(113)	0	0	0	0	0	0.636	0.364

The embedded DTMC does not include the time information for the problem; it only describes the dynamics of the transitions between states. The rate vector λ, shown with the preceding matrix for **R**, gives the rate of transition out of the states $i \in S$, whereas $1/\lambda_i$ indicates the mean sojourn time in state i.

In the construction of the **P** and **R** matrices, we allow nonzero entries on the main diagonal, although in many situations these values (p_{ii} and r_{ii}) are irrelevant to the analysis. Consequently, they can be set equal to 0 to simplify the calculations. In some situations, however, there are events that result in the system remaining in a particular state, as the preceding example demonstrates. When the foyer is full, the arrival of a customer does not affect the number in the system because the customer balks. This may have important economic implications that would be lost if the rates on the main diagonal were all set equal to 0. For this reason, we allow nonzero entries on the main diagonal but note when they can be neglected.

Rate Matrix from the Embedded DTMC

In some cases, it may be natural to express a model in terms of a DTMC together with the mean sojourn times. Consider again the stockbroker example in Section 14.1. This example has three states—a, b, and c—representing different lists of customers, with sojourn times of 2 weeks, 1 week, and 1.5 weeks, respectively. When the current list is exhausted, a new list is determined by flipping two coins, resulting in the transition matrix

$$\mathbf{P} = \begin{array}{c} a \\ b \\ c \end{array} \begin{bmatrix} 0 & 1/2 & 1/2 \\ 3/4 & 0 & 1/4 \\ 3/4 & 1/4 & 0 \end{bmatrix}$$

The sojourn rates are the inverses of the mean sojourn times and are

$$\lambda_a = 1/2, \ \lambda_b = 1 \text{ and } \lambda_c = 2/3$$

We compute the components of the rate matrix by solving Equation (1) for r_{ij}.

$$\text{Rate}\{\text{transition to } j \mid \text{system is in state } i\} = r_{ij} = \lambda_i p_{ij}$$

For the stockbroker, we find

$$\mathbf{R} = \begin{bmatrix} 0 & 1/4 & 1/4 \\ 3/4 & 0 & 1/4 \\ 1/2 & 1/6 & 0 \end{bmatrix}$$

The rate matrix allows a different interpretation of the problem. The nonzero entries represent activities whose mean durations are the inverses of the rates. For our example, while the stockbroker is working on the current list, say list c, the office staff is preparing the other two lists a and b. The mean time to prepare list a is 2 weeks, and the mean time to prepare list b is 6 weeks. These two activities are governed by exponential distributions. As soon as one of the new lists is finished, it is given to the broker. Thus, rather than view the situation as selecting lists by flipping coins as originally described, we can view the situation in terms of selecting lists on a first-come-first-served basis. Both viewpoints have the same solution.

15.2   STEADY-STATE PROBABILITIES

We now address the problem of finding the steady-state probabilities for both the CTMC and the embedded DTMC. The CTMC probabilities, $\boldsymbol{\pi}^P$, are found by solving a set of m linear equations derived from the rate matrix. The embedded DTMC probabilities, $\boldsymbol{\pi}^D$, are proportional to the CTMC probabilities.

The Generator Matrix

The information contained in the rate matrix can be used to construct the *generator matrix* $\mathbf{G}$, to facilitate the computations of steady-state probabilities. For purposes of this section, we assume that the rate matrix has all zeros along the main diagonal.

Definition 1: Let $Y = \{Y_t : t \geq 0\}$ be a CTMC with the rate matrix $\mathbf{R} = (r_{ij})$ for the state space S. The generator matrix $\mathbf{G} = (g_{ij})$ for the CTMC is given by

$$g_{ij} = \begin{cases} -\lambda_i & \text{for } i = j \\ r_{ij} & \text{for } i \neq j \end{cases} \tag{2}$$

where

$$\lambda_i = \Sigma_{j \in S} \, r_{ij}$$

Equivalently, the generator matrix can be obtained directly from the embedded DTMC matrix $\mathbf{P} = (p_{ij})$ and the mean sojourn rate λ_i for each state i in the state space S. The generator matrix $\mathbf{G} = (g_{ij})$ is

$$g_{ij} = \begin{cases} -\lambda_i & \text{for } i = j \\ \lambda_i p_{ij} & \text{for } i \neq j \end{cases}$$

From Definition 1, we see that **G** has two properties: (1) the elements of each row sum to zero, and (2) the off-diagonal elements are nonnegative. This follows from the fact that transition rates are nonnegative and the diagonal element is the negative of the sum of the other elements in the row.

To illustrate the construction of **G** from the embedded DTMC, we use the stockbroker example with $\lambda = (1/2, 1, 2/3)$ and transition matrix **P**.

$$\text{From } \mathbf{P} = \begin{bmatrix} 0 & 1/2 & 1/2 \\ 3/4 & 0 & 1/4 \\ 3/4 & 1/4 & 0 \end{bmatrix}, \text{ we get } \mathbf{G} = \begin{bmatrix} -1/2 & 1/4 & 1/4 \\ 3/4 & -1 & 1/4 \\ 1/2 & 1/6 & -2/3 \end{bmatrix}$$

If we start from the rate matrix, the same results are obtained.

The primary use of the generator matrix is in computing the steady-state probabilities $\pi^P = (\pi_0^P, \pi_1^P, \ldots, \pi_{m-1}^P)$, of the CTMC. To derive the necessary equations, let

$$q_{ij}(t) = \Pr\{Y_{t+s} = j \mid Y_s = i\}$$

be the probability that a process currently in state i will be in state j after t units of time. The quantities $q_{ij}(t)$ are called the *transition probabilities* of a CTMC. We now make use of Kolmogorov's forward equations, which follow from the Chapman–Kolmogorov equations given in Chapter 13 (see Ross [2000] for details):

$$\frac{dq_{ij}(t)}{dt} \equiv q_{ij}'(t) = \sum_{k \neq j} r_{kj} q_{ik}(t) - \lambda_j q_{ij}(t) \tag{3}$$

where $r_{kj} = \lambda_k p_{kj}$ is the instantaneous transition rate and $\lambda_k = \Sigma_j r_{kj}$ is the mean sojourn rate. Equation (3) says that the rate of change of the transition probabilities at time t equals the probability of going from i to an intermediate state k times the rate at which the process goes from k to j summed over all k, minus the probability of going from i to j times the rate at which the process leaves j. If we let $t \to \infty$, the transition probabilities become independent of the initial state, so

$$\pi_j^P = \lim_{t \to \infty} \Pr\{Y_t = j \mid Y_0 = i\} = \lim_{t \to \infty} q_{ij}(t) \tag{4}$$

and we have

$$\lim_{t \to \infty} q_{ij}'(t) = \lim_{t \to \infty} \left[\sum_{k \neq j} r_{kj} q_{ik}(t) - \lambda_j q_{ij}(t) \right]$$

$$= \sum_{k \neq j} r_{kj} \pi_k^P - \lambda_j \pi_j^P$$

$$= \sum_{k \neq j} \lambda_k p_{kj} \pi_k^P - \lambda_j \pi_j^P$$

where it was implicitly assumed that the limit and summation in the first equation can be interchanged. This is permissible for most applications, including all birth and death processes and all finite state models. Now, because $q_{ij}(t)$ is a probability, it is bounded between 0 and 1, so it follows that if a steady-state solution exists, $q_{ij}'(t)$ converges as t approaches ∞. In the limit, then, $q_{ij}(t)$ does not change with time, so its derivative must go to zero, yielding

$$0 = \sum_{k \neq j} \lambda_k p_{kj} \pi_k^P - \lambda_j \pi_j^P \quad \text{for all states } j \tag{5}$$

In matrix form, Equation (5) is equivalent to

$$\pi^{\mathrm{P}}\mathbf{G} = \mathbf{0} \tag{6}$$

For state space $S = \{0, 1, \ldots, m-1\}$, Equation (6) yields m linear equations. If we also consider the fact that the components of π^{P} must sum to 1

$$\sum_{j=0}^{m-1} \pi_j^{\mathrm{P}} = 1 \tag{7}$$

we have $m + 1$ equations in m unknowns; however, one of the equations in Equation (6) is redundant. Arbitrarily replacing the first one with Equation (7) leads to a linearly independent system with an equal number of equations and unknowns. To present this result in compact form, we define the *augmented* generator matrix $\mathbf{G}_a$ as

$$\mathbf{G}_a = \begin{bmatrix} 1 & r_{01} & r_{02} & \cdots & r_{0,m-1} \\ 1 & -\lambda_1 & r_{12} & \cdots & r_{1,m-1} \\ \vdots & \vdots & \vdots & & \vdots \\ 1 & r_{m-1,1} & r_{m-1,2} & \cdots & -\lambda_{m-1} \end{bmatrix}$$

and the first unit vector as $\mathbf{e}_1 = (1, 0, \ldots, 0, 0)^{\mathrm{T}}$. Thus, Equations (6) and (7) are equivalent to

$$\pi^{\mathrm{P}}\mathbf{G}_a = \mathbf{e}_1 \tag{8}$$

Solving this set of equations yields the steady-state probabilities when they exist. In formal terms, we have the following.

Property 1: Let $Y = \{Y_t : t \geq 0\}$ be a CTMC with an irreducible, recurrent state space S and augmented generator matrix $\mathbf{G}_a$. If π^{P} is the vector of steady-state probabilities, then π^{P} is a solution to Equation (8).

For the stockbroker example, the augmented generator matrix is

$$\mathbf{G}_a = \begin{bmatrix} 1 & 1/4 & 1/4 \\ 1 & -1 & 1/4 \\ 1 & 1/6 & -2/3 \end{bmatrix}$$

so Equation (8) becomes

$$\pi_a^{\mathrm{P}} + \pi_b^{\mathrm{P}} + \pi_c^{\mathrm{P}} = 1$$

$$\tfrac{1}{4}\pi_a^{\mathrm{P}} - \pi_b^{\mathrm{P}} + \tfrac{1}{6}\pi_c^{\mathrm{P}} = 0$$

$$\tfrac{1}{4}\pi_a^{\mathrm{P}} + \tfrac{1}{4}\pi_b^{\mathrm{P}} - \tfrac{2}{3}\pi_c^{\mathrm{P}} = 0$$

with solution $\pi^{\mathrm{P}} = (6/11, 2/11, 3/11)$.

In the preceding developments, we have assumed that the steady-state probabilities π^{P} exist. Sufficient conditions for this to be true are as follows.

1. All states of the Markov chain communicate—i.e., the chain is irreducible: Starting in state i, there is a positive probability of eventually visiting state j, for all i, j.

2. The Markov chain is positive recurrent (see Section 13.7)—i.e., starting in any state, the mean time to return to that state is finite.

If these conditions hold, π_j^P can be viewed as being the long-run proportion of the time that the process is in state j.

Flow Balance

A second way to determine the set of equations defining the steady-state probabilities $\boldsymbol{\pi}^P$ is through the principle of *probability flow balance*. Since the system must be either in some state j or not, in any interval $(0, t)$ the number of transitions into state j must be within 1 of the number of transitions out of state j. Hence, in the long run, the rate at which transitions into j occur must equal the rate at which transitions out of j occur.

When the process is in state j, it leaves at rate λ_j, and since π_j^P is the proportion of time it is in state j, it follows that

$$\lambda_j \pi_j^P = \text{rate at which the process leaves state } j$$

Similarly, when the process is in state k, it enters j at a rate r_{kj}. Given that π_k^P is the proportion of time in state k, we see that the rate at which transitions from k to j occur is just $r_{kj}\pi_k^P$, so

$$\sum_{k \neq j} r_{kj}\pi_k^P = \text{rate at which the process enters state } j$$

Equation (5) implies that the rates in and out of each state must be equal, or balanced, for the steady-state probabilities.

Probability flow in = probability flow out

$$\sum_{k \neq j} r_{kj}\pi_k^P = \lambda_j\pi_j^P \text{ for } j \in S$$

To construct the flow balance equations, it is convenient to use the rate diagram. For the stockbroker example with state space $S = \{a, b, c\}$, the corresponding diagram is shown in Figure 15.2, where the numbers along the arcs represent the rates r_{ij}.

The equations defining steady state can be written directly from the diagram.

Probability flow in = probability flow out

State a $\frac{3}{4}\pi_b^P + \frac{1}{2}\pi_c^P = \left(\frac{1}{4} + \frac{1}{4}\right)\pi_a^P$

State b $\frac{1}{4}\pi_a^P + \frac{1}{6}\pi_c^P = \left(\frac{3}{4} + \frac{1}{4}\right)\pi_b^P$

State c $\frac{1}{4}\pi_a^P + \frac{1}{4}\pi_b^P = \left(\frac{1}{2} + \frac{1}{6}\right)\pi_c^P$

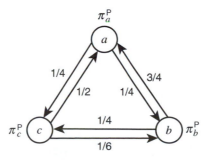

Figure 15.2   Rate diagram for stockbroker example

Moving the terms on the right-hand sides of these equations to the left and summing yields $0 = 0$, confirming that they are linearly dependent. As before, we replace one of these equations with Equation (7) and solve to get $\pi^P = (6/11, 2/11, 3/11)$. This, of course, is the same solution obtained directly with the augmented generator matrix and Equation (8).

Steady State for the Embedded DTMC

Because every CTMC has an associated DTMC, we can compute the steady-state probabilities for the DTMC using Equations (7) and (8) in Chapter 13. We now repeat these equations using π^D to represent the steady-state probabilities for the embedded DTMC.

$$\pi^D(\mathbf{P} - \mathbf{I}) = \mathbf{0} \tag{9}$$

and

$$\sum_{j \in S} \pi_j^D = 1 \tag{10}$$

Dropping one equation from (9) and solving the remaining system, we obtain the steady-state probabilities for the DTMC.

$$\pi^D = (0.428, 0.286, 0.286)$$

One might ask what this solution has to do with the previously computed steady-state vector computed for the CTMC.

$$\pi^P = (6/11, 2/11, 3/11) = (0.545, 0.182, 0.273)$$

The components of π^P describe the proportion of time the system spends in each state, whereas the components of π^D describe the proportion of steps the system spends in each state. To show the relationship between the two vectors mathematically, let $Y = \{Y_t : t \geq 0\}$ be a CTMC with an irreducible, recurrent state space, and let $X = \{X_n : n = 0, 1, \ldots\}$ be an embedded DTMC with steady-state probabilities π^D

$$\pi_j^D = \lim_{t \to \infty} \Pr\{X_n = j \mid X_0 = i\}$$

For the CTMC,

$$\pi_j^P = \lim_{t \to \infty} \Pr\{Y_t = j \mid Y_0 = i\}$$

In both cases, the limiting values are independent of the initial state, because the state space is assumed to be irreducible and recurrent. Now, given one of the two vectors π^D and π^P and the mean transition rates λ, the other vector can be computed as follows.

$$\pi_j^P = \frac{\pi_j^D / \lambda_j}{\sum_{k=0}^{m-1} \pi_k^D / \lambda_k}$$

or

$$\pi_j^D = \frac{\pi_j^P \lambda_j}{\sum_{k=0}^{m-1} \pi_k^P \lambda_k} \quad \text{for } j = 0, 1, \ldots, m-1 \tag{11}$$

Because the components of π^D and π^P are proportional, one can always find one when given the other.

In the stockbroker example, we conclude from π^P that the stockbroker spends roughly 55% of his time soliciting clients from list a, 18% of his time soliciting from list b, and 27% of his time with soliciting from list c. If we look at the process in terms of the number of times the broker was working from a particular list, we conclude from π^D that 43% of the lists were of type a, 29% were of type b, and 29% were of type c. The proportions given by π^D are different from those given by π^P, because in the first case we are concerned only with the transition from one list to another, whereas in the second case we are interested in the time a list is held.

When $p_{ii} \neq 0$

A common assumption in the modeling of systems as CTMCs is that a transition must take the system from its current state i to some other state j. By implication, then, $p_{ii} = 0$ for all $i \in S$, and hence the diagonal elements of both the transition matrix $\mathbf{P}$ and the rate matrix $\mathbf{R}$ are zero. One might ask what happens when this is not the case. To illustrate, we change the transition matrix $\mathbf{P}$ and state transition rates λ in the stockbroker example as follows.

$$\mathbf{P} = \begin{bmatrix} 0 & 1/2 & 1/2 \\ 3/4 & 0 & 1/4 \\ 9/16 & 3/16 & 1/4 \end{bmatrix} \quad \lambda = \begin{bmatrix} 1/2 & 1 & 8/9 \end{bmatrix}$$

When the broker has list c during some time interval, there is a 1/4 chance that he will be given list c at the next update. The other probabilities in row 3 have been proportionally adjusted so that the elements of the row still sum to 1. From λ and $\mathbf{P}$, we compute the associated rate matrix.

$$\mathbf{R} = \begin{bmatrix} 0 & 1/4 & 1/4 \\ 3/4 & 0 & 1/4 \\ 1/2 & 1/6 & 2/9 \end{bmatrix}$$

For the case of nonzero rates on the diagonal, we slightly modify the formula given by Equation (2) for determining the generator matrix. The new formula is

$$g_{ij} = \begin{cases} -\lambda_i + r_{ii} & \text{for } i = j \\ r_{ij} & \text{for } i \neq j \end{cases} \tag{12}$$

where

$$\lambda_i = \Sigma_{j \in S} r_{ij}$$

Using Equation (12), we compute

$$\mathbf{G} = \begin{bmatrix} -1/2 & 1/4 & 1/4 \\ 3/4 & -1 & 1/4 \\ 1/2 & 1/6 & -2/3 \end{bmatrix}$$

The generator matrix has not changed with the nonzero entry on the diagonal, so the steady-state probabilities for the CTMC do not change.

$$\pi^P = (0.545, 0.182, 0.273)$$

Using Equation (11) to compute the corresponding steady-state probabilities of the DTMC, we find, however, that π^D is now

$$\pi^D = (0.391, 0.261, 0.348)$$

The reason for the change in π^D is that we have modified the conditions of the problem and allowed list c to follow itself. The proportion of time the broker holds list c has not changed, but the proportion of lists that are of type c has changed because we view consecutive lists of the same type as different. This result is important only when there is an economic effect associated with changing from one list to another. Related issues are discussed in the next section.

15.3 ECONOMIC ANALYSIS

In systems design, a procedure is needed for evaluating the costs of the different alternatives. As in the case of DTMCs, we identify two economic factors. The first is the state cost vector $\mathbf{C}^S$, whose components c_i^S describe the *rate* of cost accumulation while in state i. If this is the only economic consideration, then the long-run cost per unit time is simply the scalar product of the steady-state probabilities and $\mathbf{C}^S$.

Property 2: Let $Y = \{Y_t : t \geq 0\}$ be a CTMC with an irreducible, recurrent state space S and a cost rate vector $\mathbf{C}^S$ (i.e., c_i^S is the rate at which cost is accumulated whenever the process is in state i). Also, let π^P denote the steady-state probabilities as defined by Equation (4). Then the long-run cost per unit time is given by

$$\bar{c}^S = \lim_{t \to \infty} \frac{1}{t} E\left[\int_0^t c_{Y_u}^S \, du \right] = \sum_{i \in S} c_i^S \pi_i^P \tag{13}$$

independent of the initial state.

In Equation (13), $E[\cdot]$ is the expected value operator and $c_{Y_u}^S$ is the rate at which cost is accumulated when the CTMC is in state Y_u. As an example, consider again the bank foyer with an ATM and human teller operating in parallel. One possible assignment of costs when the system capacity is 5 is

$$\mathbf{C}^S = (0, 1, 1, 2, 3, 4, 5)$$

where the unit of measurement is dollars per minute. The components of $\mathbf{C}^S$ are intended to reflect the cost of customers using the system and are simply $\$1 \times$ (number of customers in the system). For state $(0, 0, 0)$, the system is empty, so no cost is being incurred. For states $(0, 1, 0)$ and $(1, 0, 0)$, there is one person in the system, so the rate is $\$1$/min—i.e., costs are accumulating for that one person at a rate of $\$1$/min. For state $(1, 1, 0)$, there are two persons in the system, so costs are accumulating at a rate of $\$2$/min.

The state cost could be a more complicated function of the number of customers in the system. Perhaps customer discomfort increases as the foyer becomes more crowded. Then the state cost would be a convex nonlinear function of the number in the system.

The second matrix used in the economic analysis is $\mathbf{C}^R$, whose components c_{ij}^R denote the cost of a transition from state i to state j. For the ATM example, let us assume that each time a customer is served, the bank profits by $\$2$. This is reflected in the following matrix.

$$\mathbf{C}^R = \begin{array}{c} \begin{array}{ccccccc} (000) & (010) & (100) & (110) & (111) & (112) & (113) \end{array} \\ \begin{bmatrix} 0 & 0 & 0 & 0 & 0 & 0 & 0 \\ -2 & 0 & 0 & 0 & 0 & 0 & 0 \\ -2 & 0 & 0 & 0 & 0 & 0 & 0 \\ 0 & -2 & -2 & 0 & 0 & 0 & 0 \\ 0 & 0 & 0 & -2 & 0 & 0 & 0 \\ 0 & 0 & 0 & 0 & -2 & 0 & 0 \\ 0 & 0 & 0 & 0 & 0 & -2 & 0 \end{bmatrix} \end{array} \begin{array}{cc} \text{State} & \text{Index} \\ (000) & 0 \\ (010) & 1 \\ (100) & 2 \\ (110) & 3 \\ (111) & 4 \\ (112) & 5 \\ (113) & 6 \end{array}$$

Since each departure represents revenue and the problem is stated in terms of cost, we insert "–2" whenever a transition results in a customer leaving the system. For consistency, the units are –\$2 per customer.

We are now in a position to compute the expected cost of transitions out of each state. Based on the transition probabilities of the embedded Markov chain, the expected cost of a transition out of state i is

$$\sum_{j=0}^{m-1} c_{ij}^R p_{ij}, \quad i = 0, 1, \ldots, m-1$$

The expected transition time out of state i is $1/\lambda_i$, so we convert this cost into a rate by dividing it by the expected transition time.

$$c_i^R = \sum_{j=0}^{m-1} \frac{c_{ij}^R p_{ij}}{1/\lambda_i} = \sum_{j=0}^{m-1} c_{ij}^R p_{ij} \lambda_i = \sum_{j=0}^{m-1} c_{ij}^R r_{ij}$$

This expression gives the cost rate c_i^R for a transition out of state i. Adding this to the state cost c_i^S, we find the combined cost rate for state i to be

$$c_i = c_i^S + c_i^R = c_i^S + \sum_{j=0}^{m-1} c_{ij}^R r_{ij} = c_i^S + \sum_{j=0}^{m-1} c_{ij}^R p_{ij} \lambda_i \tag{14}$$

With the steady-state probability distribution for the states $\boldsymbol{\pi}^P$, we can compute the expected cost rate for the process as follows.

$$\bar{c} = \bar{c}^S + \bar{c}^R \equiv \sum_{i=0}^{m-1} c_i^S \pi_i^P + \sum_{i=0}^{m-1} c_i^R \pi_i^P = \sum_{i=0}^{m-1} c_i \pi_i^P \tag{15}$$

To illustrate, we calculate c_1 and c_2 for the ATM example.

$$c_1 = c_1^S + c_1^R = c_1^S + c_{1,0}^R r_{1,0} = 1 + (-2)2.5 = -4$$

$$c_2 = c_2^S + c_2^R = c_2^S + c_{2,0}^R r_{2,0} = 1 + (-2)1 = -1$$

These values indicate profits for states $(0, 1, 0)$ and $(1, 0, 0)$ that correspond to $i = 1$ and $i = 2$, respectively. Similarly, we derive the entire cost rate vector $\mathbf{c}_1$ as follows.

$$\mathbf{c} = (0, -4, -1, -5, -4, -3, -2)$$

Using this vector and the steady-state probabilities computed in Chapter 14 ($\pi^P = (0.214,$ $0.046, 0.313, 0.205, 0.117, 0.067, 0.038)$), we have

$$\bar{c} = \sum_{i=0}^{6} c_i \pi_i^P = -2.267$$

Thus, the system is profitable, because negative costs represent a positive return. At steady state, the profit is \$2.267 per minute of operation.

The following statement summarizes these results.

Property 3: Let $Y = \{Y_t : t \geq 0\}$ be a CTMC with an irreducible, recurrent state space S, a cost rate vector $\mathbf{C}^S$, and a transition cost matrix $\mathbf{C}^R$ (i.e., c_{ij}^R is the cost incurred whenever the process jumps from state i to state j). Also, let π^P denote the steady-state probabilities. Then the long-run cost per unit time is given by

$$\bar{c} = \sum_{i=0}^{m-1} \left[c_i^S + \sum_{j=0}^{m-1} c_{ij}^R r_{ij} \right] \pi_i^P = \sum_{i=0}^{m-1} \left[c_i^S + \sum_{j=0}^{m-1} c_{ij}^R p_{ij} \lambda_i \right] \pi_i^P$$

Note that we provide two forms for the result, one involving the transition rates and one involving the transition probabilities of the embedded Markov chain. Either may be used.

We should also point out that this development does not assume the rates on the diagonal, r_{ii}, are all zero. For our example, we have $r_{66} = 2$, the rate of arrivals while the ATM foyer is full. This plays no part in the aforementioned profit because the cost $c_{66}^R = 0$. If we assign a cost of 10 for each customer lost because of the foyer being full, the value of c_6 becomes

$$c_6 = 5 + (3.5)(-2) + (2)(10) = 18$$

From Equation (15), the steady-state cost, including the cost, of lost customers, is $\bar{c} = -1.50$. The charge for lost customers reduces the profit to \$1.50 per minute.

Stockbroker Example (Continued)

It was stated in Section 12.1 that the stockbroker's revenue (expressed as a cost) per week was

$$\mathbf{C}^S = (-1700, -1450, -1600)$$

and his transition cost matrix and state-transition matrix, respectively, were

$$\mathbf{C}^R = \begin{bmatrix} 0 & 200 & 300 \\ 250 & 0 & 300 \\ 200 & 250 & 0 \end{bmatrix} \text{ and } \mathbf{P} = \begin{bmatrix} 0 & 0.5 & 0.5 \\ 0.75 & 0 & 0.25 \\ 0.75 & 0.25 & 0 \end{bmatrix}$$

for state space $S = \{a, b, c\}$. Substituting these values into Equation (14) with $\pi^P = (6/11,$ $2/11, 3/11)$ and $\lambda = (1/2, 1, 2/3)$ yields

$$\begin{aligned} \bar{c} &= [-1700 + 200(0.5)(0.5) + 300(0.5)(0.5)]\pi_1 \\ &\quad + [-1450 + 250(0.75)(1) + 300(0.25)(1)]\pi_2 \\ &\quad + [-1600 + 200(0.75)(2/3) + 250(0.25)(2/3)]\pi_3 \\ &= -1575\pi_1 - 1187.5\pi_2 - 1458.33\pi_3 = -1472.73 \end{aligned}$$

Thus, his expected revenue per week is \$1472.73.

The cost model can be used to compute a variety of characteristics by suitably adjusting the components of $\mathbf{C}^S$ and $\mathbf{C}^R$. For the ATM example, by setting $\mathbf{C}^R$ equal to zero and using the current value of $\mathbf{C}^S$, Equation (14) gives the expected number in the system. Alternatively, setting $\mathbf{C}^S$ equal to zero and replacing each –2 with 1 in $\mathbf{C}^R$ yields the throughput rate of the system in terms of customers per minute.

Discounted Cash Flow

In many applications, it is important to consider the time value of money. In the discrete case, we specify a rate of return or interest rate per period, r, with corresponding discount factor α, where $\alpha = 1/(1 + r)$. This means that the present value of one dollar obtained one period from the present is equal to α. For a continuous-time problem, we let β denote the *discount rate*, which in this case would be the same as the rate of return except that we assume that compounding occurs continuously. Accordingly, the present value of one dollar obtained one period from the present equals $e^{-\beta}$. Unfortunately, it is difficult to include the jump time costs in this type of analysis, but it is easy to account for the costs incurred whenever the process is in state i. The following result gives the formula used to compute the discounted expected cost (revenue) for a continuous-time process. The formula makes use of the generator matrix $\mathbf{G}$ given in Definition 1.

Property 4: Let Y = $\{Y_t : t \geq 0\}$ be a CTMC with generator matrix $\mathbf{G}$, cost rate vector $\mathbf{C}^S$, and discount rate β. Then, the present value of the total expected discounted cost (over an infinite planning horizon) is given by

$$E\left[\int_0^\infty e^{-\beta u}\mathbf{C}^S_{Y_u}\,du \mid Y_0 = 1\right] = \left([\beta\mathbf{I} - \mathbf{G}]^{-1}\mathbf{C}^S\right)_i \tag{16}$$

where the subscript i on the right-hand side of the equation refers to the ith component of the vector that results from the matrix operations.

To illustrate these computations, consider an entrepreneur who wants to invest in the housing market in a suburban community. Historical trends indicate that the market is cyclical with respect to profits and can be classified into four states: excellent, good, moderate, and poor. Based on long-term statistics, the expected times in the four categories are shown in Table 15.2. Also shown are the state that results when a transition occurs and the annual profit rate. The entrepreneur wants to know if her investment will have an expected return of at least 10%. The answer depends in part on the state of the market when the initial investment is made.

This example is a little odd in that an excellent market always follows a poor one, but these numbers serve to emphasize the variability of the expected discounted return with the initial state. Assuming that the actual transitions occur according to an exponential distribution, the rate matrix describing this situation is as follows. The transition rates are the inverses of the expected times.

Table 15.2   Data for the Housing Example

State	0	1	2	3
Category	Excellent	Good	Moderate	Poor
Expected time (years)	0.5	1	2	2
Next state	Good	Moderate	Poor	Excellent
Profit rate (per year)	1000	500	0	–500

$$\mathbf{R} = \begin{bmatrix} 0 & 2 & 0 & 0 \\ 0 & 0 & 1 & 0 \\ 0 & 0 & 0 & 0.5 \\ 0.5 & 0 & 0 & 0 \end{bmatrix}$$

Using these data, the problem is to find the expected discounted present value of the cash flow for an investment in the housing market based on the current state of the business cycle. We associate the state vector $\mathbf{C}^S$ with the annual profit rates of the categories. The generator matrix is constructed as described in Section 15.2.

$$\mathbf{C}^S = \begin{bmatrix} 1000 \\ 500 \\ 0 \\ -500 \end{bmatrix}, \quad \mathbf{G} = \begin{bmatrix} -2 & 2 & 0 & 0 \\ 0 & -1 & 1 & 0 \\ 0 & 0 & -0.5 & 0.5 \\ 0.5 & 0 & 0 & -0.5 \end{bmatrix}$$

To determine the total expected discounted cost (revenue) for the CTMC, we make use of Equation (16) with a discount rate of $\beta = 0.1$.

$$[\beta \mathbf{I} - \mathbf{G}]^{-1}\mathbf{C}^S = \left(0.1 \begin{bmatrix} 1 & 0 & 0 & 0 \\ 0 & 1 & 0 & 0 \\ 0 & 0 & 1 & 0 \\ 0 & 0 & 0 & 1 \end{bmatrix} - \begin{bmatrix} -2 & 2 & 0 & 0 \\ 0 & -1 & 1 & 1 \\ 0 & 0 & -0.5 & 0.5 \\ 0.5 & 0 & 0 & 0 \end{bmatrix} \right)^{-1} \begin{bmatrix} 1000 \\ 500 \\ 0 \\ -500 \end{bmatrix}$$

$$= \begin{bmatrix} 2.1 & -2 & 0 & 0 \\ 0 & 1.1 & -1 & 0 \\ 0 & 0 & 0.6 & -0.5 \\ -0.5 & 0 & 0 & 0.6 \end{bmatrix}^{-1} \begin{bmatrix} 1000 \\ 500 \\ 0 \\ -500 \end{bmatrix} = \begin{bmatrix} 772 \\ 311 \\ -158 \\ -189 \end{bmatrix}$$

Thus, if the initial market condition is either excellent or good, the net present value of the cash flow is positive, indicating that the rate of return will be greater than 10%. On the other hand, if the market is in the moderate or poor state, the rate of return will be less than 10%,

It is interesting to compute the steady-state annual profit for this example. Using Equation (8), we first compute the steady-state probabilities.

$$\pi^P = (0.0909, 0.1818, 0.3636, 0.3636)$$

Combining these values with the expected profits listed in Table 15.2 yields an expected annual return of

$$\pi^P \mathbf{C}^S = 0$$

Thus, in the long run, this investment yields no profit, but it still may be a good opportunity if the entrepreneur chooses the right moment to enter the market.

15.4 FIRST PASSAGE TIMES

For a DTMC, the first passage time from one state i to another state j is a discrete random variable measured in the number of steps. In contrast, the first passage time from i to j for a CTMC is a continuous random variable. We will not attempt to find an explicit representation of the first passage time distributions, but we will provide a procedure for determining the expected values of these random variables.

Expected First Passage Time

We use the ATM and human teller system described in Section 15.1 for an example. The accompanying transition matrix of the embedded DTMC is used in the calculations.

The expected first passage times from each state i into some given state j has two parts, as shown in the following equation

$$\mu_{ij} = \frac{1}{\lambda_i} + \sum_{r \neq j} p_{ir}\mu_{rj}, \quad i = 0, 1, \ldots, m-1, j = 0, 1, \ldots, m-1 \tag{17}$$

The first term is the expected time the system resides in state i. No matter what the realization of the next state, the system remains in state i for a time whose expected value is $1/\lambda_i$. The terms in the summation involve the probabilities from the embedded DTMC transition matrix. The system will be in state r after the first step with probability p_{ir}, and the expected *additional* time from r to j is μ_{rj}. The summation in Equation (17) is over all possible values of the next state, excluding state j. The time for passage directly into state j is accounted for by the first term of Equation (17).

For any given final state j, Equation (17) is used to construct a set of m linear equations in m unknowns, μ_{ij}, $i = 0, 1, \ldots, m - 1$. The coefficients in these equations are the elements of the one-step transition matrix of the embedded DTMC. The constant terms in the equation are the values of $1/\lambda_i$.

We illustrate the calculations by finding the expected value of the first passage times into state $j = 0$. This would be interesting for the ATM example because state 0 corresponds to no customers in the system. A reasonable question is: What is the expected time between successive observations of an empty foyer? Although we desire only the value of μ_{00}, we write and solve the equations for the expected time from every state to state 0. The ATM example has seven states, so we must construct seven equations in seven unknowns.

$$\mu_{i0} = \frac{1}{\lambda_i} + \sum_{r \neq j} p_{ir}\mu_{r0}, \quad i = 0, 1, \ldots, 6$$

We include in the following equations only those terms that have nonzero transition probabilities.

$$i = 0: \quad \mu_{00} = \frac{1}{\lambda_0} + p_{02}\mu_{20} = 0.5 + \mu_{20}$$

$$i = 1: \quad \mu_{10} = \frac{1}{\lambda_1} + p_{13}\mu_{30} = \frac{1}{4.5} + 0.444\mu_{30}$$

$$i = 2: \quad \mu_{20} = \frac{1}{\lambda_2} + p_{23}\mu_{30} = \frac{1}{3} + 0.667\mu_{30}$$

$$\vdots \qquad \vdots \qquad\qquad\qquad\qquad \vdots$$

$$i = 6: \quad \mu_{60} = \frac{1}{\lambda_6} + p_{65}\mu_{50} + p_{66}\mu_{60} = \frac{1}{5.5} + 0.636\mu_{50} + 0.364\mu_{60}$$

Equation (17) also has a matrix description, which may be preferable for computer implementation. Let $\mathbf{m}_j$ be the column vector of expected first passage times into state j, and let $\mathbf{P}_j$ be the matrix $\mathbf{P}$ with column j replaced by zeros, and

$$\tau = \left(\frac{1}{\lambda_0}, \frac{1}{\lambda_1}, \ldots, \frac{1}{\lambda_{m-1}} \right)^{\mathrm{T}}$$

where $\boldsymbol{\tau}$ is an m-dimensional column vector giving the expected time for the system to leave each state. The vector equation defining the expected first passage times is

$$\mathbf{m}_j = \boldsymbol{\tau} + \mathbf{P}_j \mathbf{m}_j \tag{18}$$

Solving Equation (18) for $\mathbf{m}_j$ gives

$$\mathbf{m}_j = [\mathbf{I} - \mathbf{P}_j]^{-1} \boldsymbol{\tau} \tag{19}$$

For our example with $j = 0$, we have

$$\mathbf{I} - \mathbf{P}_0 = \begin{bmatrix} 1 & 0 & -1 & 0 & 0 & 0 & 0 \\ 0 & 1 & 0 & -0.444 & 0 & 0 & 0 \\ 0 & 0 & 1 & -0.667 & 0 & 0 & 0 \\ 0 & -0.812 & -0.455 & 1 & -0.364 & 0 & 0 \\ 0 & 0 & 0 & -0.636 & 1 & -0.364 & 0 \\ 0 & 0 & 0 & 0 & -0.636 & 1 & -0.364 \\ 0 & 0 & 0 & 0 & 0 & -0.636 & 0.636 \end{bmatrix}$$

$$\boldsymbol{\tau} = (0.5,\ 0.222,\ 0.333,\ 0.182,\ 0.182,\ 0.182,\ 0.182)^{\mathrm{T}}$$

Finding the inverse of $\mathbf{I} - \mathbf{P}_0$ and substituting it into Equation (19), we obtain

$$\mathbf{m}_0 = (2.34,\ 1.23,\ 1.84,\ 2.26,\ 2.80,\ 3.25,\ 3.54)^{\mathrm{T}}$$

Thus, the expected first passage time from state 0 back to state 0 is 2.34 minutes, whereas the expected first passage time from the full state ($i = 6$) to the empty state ($j = 0$) is 3.54 minutes.

Expected Number of Steps to First Passage

Section 13.7 develops the formulas for the first passage time for the DTMC. The number of steps to first passage into state j is determined by the solution of

$$\mathbf{m}_j^{\mathrm{s}} = [\mathbf{I} - \mathbf{P}_j]^{-1} \mathbf{e} \tag{20}$$

We have added the superscript "s" to the vector $\mathbf{m}_j$ to indicate that the measure is steps rather than clock time. Equation (20) can be evaluated for a CTMC using the embedded DTMC. For our example, we obtain

$$\mathbf{m}_j^{\mathrm{s}} = (9.18,\ 5.79,\ 8.18,\ 10.78,\ 13.76,\ 16.23,\ 17.80)^{\mathrm{T}}$$

From the solution, we note that 9.18 changes of state occur on the average when passing from state 0 back to state 0. This kind of information may be useful, for example, if we want to know how much activity transpires between idle periods.

15.5 BIRTH-DEATH PROCESSES

An important class of CTMC is the birth-death process in which a population is observed over time. Since we most often deal with a population consisting of customers in a queuing system, we equate a birth with an arrival and a death with a service completion or departure. The state of the system at any time is the number of customers in the population, so $S = \{0, 1, \ldots\}$. This number, denoted by $N(t)$, is a random variable when the arrivals and/or departures occur according to some random mechanism. In addition, if we assume that the stochastic process $\{N(t): t \geq 0\}$ is Markovian, all activity durations must be exponentially distributed. We will use the results obtained in this section to derive steady-state probabilities for the queuing models in Chapter 16.

An unbounded birth-death process is a model characterized by a population that takes on integer values 0, 1, 2, ... When the system is in some state k at a given time, it will move to state $k + 1$ if an arrival occurs or to state $k - 1$ if a departure occurs. These are the only possible transitions from all but the first state, as illustrated in Figure 15.3. At state 0, the only possible transition is to state 1.

When the system is in state k, the time to the next arrival is governed by an exponential distribution with rate parameter λ_k. Similarly, the time to the next departure is governed by an exponential distribution with rate parameter μ_k. Although we can formulate a set of differential equations that describe the transient solution, it is not possible—at least for the general case—to solve them in closed form. Therefore, we will leave this issue to more advanced texts.

Steady-State Solutions

The steady-state probabilities can be computed using the principle of flow balance. The procedure for determining the set of equations whose solution yields π^P is to set the probability flow into a state equal to the probability flow out of that state. Referring to Figure 15.3 as a guide, these equations are as follows.

$$\text{probability flow in} = \text{probability flow out}$$

State 0:
$$\mu_1 \pi_1^P = \lambda_0 \pi_0^P$$

State 1:
$$\lambda_0 \pi_0^P + \mu_2 \pi_2^P = (\mu_1 + \lambda_1)\pi_1^P$$

State 2:
$$\lambda_1 \pi_1^P + \mu_3 \pi_3^P = (\mu_2 + \lambda_2)\pi_2^P$$

$$\vdots \qquad\qquad \vdots$$

State k:
$$\lambda_{k-1} \pi_{k-1}^P + \mu_{k+1} \pi_{k+1}^P = (\mu_k + \lambda_k)\pi_k^P, \quad k \geq 1$$

By adding to each equation the one preceding it, we get

$$\mu_1 \pi_1^P = \lambda_0 \pi_0^P$$

$$\mu_2 \pi_2^P = \lambda_1 \pi_1^P$$

$$\mu_3 \pi_3^P = \lambda_2 \pi_2^P$$

$$\vdots$$

$$\mu_{k+1} \pi_{k+1}^P = \lambda_k \pi_k^P, \quad k \geq 0$$

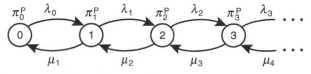

Figure 15.3   Rate network for the birth–death process

Solving in terms of π_0^P yields

$$\pi_1^P = \left(\frac{\lambda_0}{\mu_1}\right)\pi_0^P$$

$$\pi_2^P = \left(\frac{\lambda_1}{\mu_2}\right)\pi_1^P = \left(\frac{\lambda_1\lambda_0}{\mu_2\mu_1}\right)\pi_0^P$$

$$\pi_3^P = \left(\frac{\lambda_2}{\mu_3}\right)\pi_2^P = \left(\frac{\lambda_2\lambda_1\lambda_0}{\mu_3\mu_2\mu_1}\right)\pi_0^P$$

$$\vdots$$

$$\pi_k^P = \left(\frac{\lambda_{k-1}}{\mu_k}\right)\pi_{k-1}^P = \left(\frac{\lambda_{k-1}\lambda_{k-2}\ \cdots\ \lambda_1\lambda_0}{\mu_k\mu_{k-1}\ \cdots\ \mu_2\mu_1}\right)\pi_0^P$$

To simplify, we write these equations as

$$\pi_k^P = C_k\pi_0^P \tag{21}$$

where $C_k = \lambda_0\lambda_1\lambda_2\ \cdots\ \lambda_{k-1}/\mu_1\mu_2\mu_3\ \cdots\ \mu_k$ is called the *state factor*. The numerator of this expression is the product of all arrival rates for states less that k, whereas the denominator is the product of all departure rates for states less than or equal to k.

The value of π_0^P is determined by noting that the sum of the state probabilities must be 1.

$$\pi_0^P + \pi_1^P + \pi_2^P + \cdots = 1$$

so

$$\pi_0^P\left(1 + C_1 + C_2 + \cdots\right) = 1$$

and

$$\pi_0^P = \frac{1}{\left(1 + \Sigma_{k=1}^{\infty}C_k\right)} \tag{22}$$

All other probabilities are computed from Equation (21). It can be shown that a necessary and sufficient condition for the steady-state probabilities to exist for a birth–death process is that

$$\sum_{k=1}^{\infty}C_k < \infty \tag{23}$$

For the multiserver queuing system *M/M/s*, for example, Condition (23) reduces to

$$\sum_{k=s+1}^{\infty}\frac{\lambda^k}{(s\mu)^k} < \infty$$

which is equivalent to $\lambda/s\mu < 1$.

Relation among Time in System, Number in System, and Arrival Rate

With the system in statistical equilibrium (having reached steady state), we can identify three important measures that are useful for assessing performance and making design decisions: (1) the expected number in the system, (2) the expected arrival rate, and (3) the expected residence time or time in the system. Given the state of the system, the number in the system and arrival rate are known, so their expected values are computed directly from the steady-state probabilities.

$$\text{Expected number in the system: } L = \sum_{k=1}^{\infty} k\pi_k^{\text{P}} = \left(\sum_{k=1}^{\infty} kC_k \right)\pi_0^{\text{P}}$$

$$\text{Expected arrival rate: } \lambda_a = \sum_{k=0}^{\infty} \lambda_k \pi_k^{\text{P}} = \left(\lambda_0 + \sum_{k=1}^{\infty} \lambda_k C_k \right)\pi_0^{\text{P}}$$

The time interval between the arrival and departure of a customer is the residence time. A very useful result known as Little's law provides the expected value of the residence time distribution W.

$$\text{Average residence time: } W = \frac{L}{\lambda_a}$$

This formula underpins much of queuing theory and, although certainly valid for birth-death processes, has much wider application than the Markov models considered in this chapter.

Expected Transition Time and Variance

Another interesting measure for a birth-death process with birth rates $\{\lambda_n\}$ and death rates $\{\mu_n\}$ is the expected time it takes for the system to go from some state i to some other state j. Let T_i denote the time, starting from state i, required for the process to enter state $i + 1$, $i \geq 0$. We will compute $E[T_i]$ recursively starting with $i = 0$. Now, since T_0 is exponential with rate λ_0, we have

$$E[T_0] = 1/\lambda_0$$

For $i > 0$, it is necessary to consider whether the first transition takes the process into state $i - 1$ or state $i + 1$. Let

$$I_i = \begin{cases} 1, & \text{if the first transition is from } i \text{ to } i+1 \\ 0, & \text{if the first transition is from } i \text{ to } i-1 \end{cases}$$

and note that

$$E\big[T_i \mid I_i = 1\big] = \frac{1}{\lambda_i + \mu_i}$$

$$E\big[T_i \mid I_i = 0\big] = \frac{1}{\lambda_i + \mu_i} + E\big[T_{i-1}\big] + E\big[T_i\big]$$

This follows because, independent of whether the first transition is the result of a birth or a death, the time until it occurs is exponential, with rate $\lambda_i + \mu_i$. If the first transition is a birth, the population size is $i + 1$, so no additional time is needed; if it is a death, the population size becomes $i - 1$ and the additional time needed to reach $i + 1$ is equal to the time it takes to return to state i (this time has a mean of $E[T_{i-1}]$) plus the additional time it then takes to reach $i + 1$ (this time has a mean of $E[T_i]$). Thus, given that the probability of the first transition being a birth is $\lambda_i/(\lambda_i + \mu_i)$, a little arithmetic yields

$$E\big[T_i\big] = \frac{1}{\lambda_i + \mu_i} + \frac{\mu_i}{\lambda_i + \mu_i}\Big(E\big[T_{i-1}\big] + E\big[T_i\big]\Big)$$

or, equivalently,

$$E[T_i] = \frac{1}{\lambda_i} + \frac{\mu_i}{\lambda_i} E[T_{i-1}], \quad i \geq 1 \tag{24}$$

where it is assumed that $\mu_0 = 0$. Starting with $E[T_0] = 1/\lambda_0$, this yields an efficient method of successively computing $E[T_1]$, $E[T_2]$, and so on.

Suppose that we want to determine the expected time for the system to go from state i to state j where $i < j$. This can be found by using the foregoing recursive procedure and by noting that the desired quantity is equal to $E[T_i] + E[T_{i+1}] + \cdots + E[T_{j-1}]$.

Example

For the birth–death process having parameters $\lambda_i = \lambda$ and $\mu_i = \mu$ for all i,

$$E[T_i] = \frac{1}{\lambda} + \frac{\mu}{\lambda} E[T_{i-1}] = \frac{1}{\lambda}\left(1 + \mu E[T_{i-1}]\right)$$

Starting with $E[T_0] = 1/\lambda$, we see that

$$E[T_1] = \frac{1}{\lambda}\left(1 + \frac{\mu}{\lambda}\right)$$

$$E[T_2] = \frac{1}{\lambda}\left(1 + \frac{\mu}{\lambda} + \left(\frac{\mu}{\lambda}\right)^2\right)$$

and, in general,

$$E[T_i] = \frac{1}{\lambda}\left(1 + \frac{\mu}{\lambda} + \left(\frac{\mu}{\lambda}\right)^2 + \cdots + \left(\frac{\mu}{\lambda}\right)^i\right)$$

$$= \frac{1 - (\mu/\lambda)^{i+1}}{\lambda - \mu}, \quad i \geq 0$$

The expected time to reach state j starting in state i, $i < j$, is

$$E[\text{time to go from } i \text{ to } j] = \sum_{k=i}^{j-1} E[T_k]$$

$$= \frac{j-i}{\lambda - \mu} - \frac{(\mu/\lambda)^{i+1}(1 - (\mu/\lambda)^{j-i})}{(\lambda - \mu)(1 - \mu/\lambda)}$$

In the preceding developments, we assumed that $\lambda \neq \mu$. If $\lambda = \mu$, then

$$E[T_i] = \frac{i+1}{\lambda}$$

and

$$E[\text{time to go from } i \text{ to } j] = \frac{j(j+1) - i(i+1)}{2\lambda}$$

It is also possible to compute the variance of the time to go from state 0 to state $i + 1$ using the concept of conditional variance. We will forego the derivation and simply state the result. The interested reader is referred to Ross [2000]. For a general birth-death process, we have

$$Var[T_i] = \frac{1}{\lambda_i(\lambda_i + \mu_i)} + \frac{\mu_i}{\lambda_i} Var[T_{i-1}] + \frac{\mu_i}{\lambda_i + \mu_i}\left(E[T_{i-1}] + E[T_i]\right)^2$$

Starting with $Var[T_0] = 1/(\lambda_0)^2$ and using Equation (24) to obtain the expectations, we can recursively compute $Var[T_i]$. Moreover, if we want the variance of the time to reach state j starting from state i, $i < j$, this value can be expressed as the time to go from i to $i + 1$ plus the additional time to go from $i + 1$ to $i + 2$, and so on. Because the Markovian property implies that these successive random variables are independent, it follows that

$$Var[\text{time to go from } i \text{ to } j] = \sum_{k=i}^{j-1} Var[T_k]$$

EXERCISES

1. Fill in the missing entries in the following generator matrix for a CTMC. What are the mean sojourn rates and the state-transition matrix for the embedded Markov chain?

$$G = \begin{bmatrix} -8 & 3 & ? & 1 \\ 4 & ? & 0 & 5 \\ 7 & ? & -10 & 1 \\ 4 & 5 & 1 & ? \end{bmatrix}$$

2. Elaine Johnson runs a hot dog stand at the corner of Broadway and 42nd Street in New York City. Depending on the local economy, her business is in one of four states: high income, medium income, low income, deficit. Over the years she has observed that her business moves among these states according to a CTMC $\{Y_t\}$ with state space $S = \{h, m, l, d\}$ and generator matrix

$$G = \begin{matrix} h \\ m \\ l \\ d \end{matrix} \begin{bmatrix} -0.03 & 0.02 & 0.01 & 0.0 \\ 0.0 & -0.04 & 0.03 & 0.1 \\ 0.0 & 0.0 & -0.05 & 0.05 \\ 0.15 & 0.0 & 0.0 & -0.15 \end{bmatrix}$$

The hot dog stand produces profit at rates of $900, $550, $300, and –$700 per week when the business is in states h, m, l, or d, respectively. Elaine would like to reduce the time spent in the fourth state and has determined that doubling the cost (i.e., from $700 to $1400) of advertising in the subway while in state d will cut in half the mean time spent in state d. Is the additional expense worthwhile?

(a) Use the long-run average profit for the criterion.

(b) Use a total discounted profit criterion, assuming an annual discount rate β of 20%.

(c) Model the problem as a DTMC with 1-week increments. Use a total discounted profit criterion assuming that the company expects a 20% annual rate of return.

3. A small convenience store off a major highway has room for only five customers inside. Cars arrive according to an exponential distribution, with a mean interarrival time of 10 cars per hour. The number of people in each car is a random variable N, with distribution $Pr\{N = 1\} = 0.1$, $Pr\{N = 2\} = 0.7$, and $Pr\{N = 3\} = 0.2$. Arrivals enter the store and stay an exponential length of time when possible. The mean length of time in the store is 10 minutes, and each person acts independently of the others, leaving individually and waiting in their cars for the others in their parties. If a car arrives and the store is too full for everyone in the car to enter, it will leave with all its passengers. Model the store as a CTMC $\{Y_t\}$, where Y_t denotes the number of individuals in the store at time t. Write the generator matrix (Resnick [1992]).

4. A piece of machinery consists of two major components. The time to failure for component A is exponentially distributed with a mean of 100 hours. The time to failure for component B is also exponentially distributed and has a mean of 200 hours. When one component fails, the machine is turned off and maintenance is performed. The time required to fix either component is exponentially distributed with means of 5 hours for A and 4 hours for B. Let Y be a CTMC with state space $S = \{0, 1, 2\}$, where state 0 denotes that the machine is working, state 1 denotes that component A has failed, and state 2 denotes that component B has failed.

(a) Give the generator matrix for Y.

(b) What is the long-run probability that the machine is working?

(c) An outside contractor does the repair work on the components and charges $100 per hour plus travel expenses, which amount to an additional $500 for each visit. The company has determined that they can hire and train their own repairperson. This option will cost $40 per hour regardless of whether the machine is running or down. Ignoring the initial training cost and the possibility that the employee who is hired to perform the repairs can do other things while the machine is running, is it economically worthwhile to pursue this option?

5. Let Y be a CTMC with state space $S = \{0, 1, 2, 3\}$ and rate matrix

$$\mathbf{R} = \begin{bmatrix} 0 & 4 & 0 & 1 \\ 6 & 0 & 4 & 0 \\ 0 & 0 & 0 & 1 \\ 2 & 0 & 0 & 0 \end{bmatrix}$$

Costs of $250, $450, $800, and $1500 per unit time are incurred while the process is in states 0, 1, 2, and 3, respectively. Furthermore, each time a transition from state 3 to state 0 occurs, an additional cost of $2000 is incurred. (No other transitions result in additional costs.) For a maintenance fee of $600 per unit time, all state costs can be cut in half and the "transition" cost can be eliminated. Based on long-run averages, is the maintenance cost worthwhile?

6. An electronic relay in a communications device works as follows. Electric impulses arrive at the relay according to a Poisson process with a mean interarrival rate of 90 impulses per minute. An impulse is stored until a third one arrives, at which time the relay "fires" and enters a recovery phase. If an impulse arrives during the recovery phase, it is ignored. The length of time that the relay resides in the recovery phase follows an exponential random variable with a mean of 1 second. After the recovery phase is over, the cycle is repeated—that is, the third arriving impulse will instantaneously fire the relay (Resnick [1992]).

(a) Write the generator matrix for a CTMC model of the dynamics of this relay.

(b) What is the long-run probability that the relay is in the recovery phase?

(c) How many times would you expect the relay to fire each minute?

7. Using the data in Table 15.1 as a guide, simulate the CTMC defined by the following rate matrix with state space $S = \{0, 1, 2\}$. Begin by drawing the rate diagram.

$$\mathbf{R} = \begin{bmatrix} 0 & 0.3 & 0.7 \\ 0.1 & 0 & 0.1 \\ 0.3 & 0.2 & 0 \end{bmatrix}$$

From the simulation, estimate the probability that the process will be in state 1 at time $t = 10$ given that the process starts in state 0.

8. For Exercise 7, write a computer program to determine the long-run probability of the process being in each of the three states. In developing your program, remember that the long-run probability for a particular state j can be estimated by the fraction of time that the process spends in that state. The estimate can be improved slightly if the initial conditions are ignored. In other words,

$$\pi_j^P \cong \left(\int_{t_0}^t I(Y_u = j)du\right)\bigg/(t - t_0)$$

where t_0 is small and t is large. Also, the function $I(Y_u = j)$ counts the number of times the process visits state j. Part of the problem is to determine quantitative values for "small" and "large."

9. Suppose that a one-celled organism can be in either of two states, A or B. An organism in state A will change to state B at an exponential rate of α; an organism in state B will divide into two new organisms of type A at the exponential rate of β. Define an appropriate CTMC model for a population of such organisms and determine the corresponding parameter values. Also, draw the rate diagram.

10. Consider two machines that are maintained by a single repair crew. Statistics indicate that the time between failures for machine i is exponentially distributed, with an average rate of λ_i, $i = 1, 2$. The repair times for both machines are exponentially distributed, with rate μ. Can this situation be modeled as a birth–death process? If so, what are the model parameters? If not, how can it analyzed?

11. There are N persons in a population, some of whom have a certain infection that spreads as follows. Contacts between two members of the population occur in accordance with a Poisson process with rate λ. When a contact occurs, it is equally likely to involve any of the $\binom{N}{2}$ pairs of persons in the population. If a contact involves an infected and a noninfected person, then with probability p the latter gets the infection. No cure exists, so once infected person remains so forever.

    Let Y_t denote the number of infected members of the population at time t.

    (a) Draw the rate diagram for this situation. Is $\{Y_t : t \geq 0\}$ a CTMC? Explain.

    (b) Starting with a single infected individual, what is the expected time until all members are infected?

12. Consider a birth–death process with birth rates $\lambda_i = (i + 1)\lambda$ and death rates $\mu_i = i\mu$ for all $i \geq 0$.

    (a) Determine the expected time for the process to go from state 0 to state 3.

    (b) Determine the expected time for the process to go from state 2 to state 5.

    (c) Discuss how to find the variances in parts (a) and (b).

13. The Club A-Go-Go has two sound boards used to control the music it plays nightly. Because only one board is needed to run the equipment, the other serves as a spare. The times between failures are exponentially distributed, with a mean of $1/\lambda$. When a board fails, it is replaced by the spare, if available, and is sent immediately to a nearby repair shop. A single technician runs the repair shop and takes an exponential amount of time, with a mean of $1/\mu$, to effect repairs. A newly failed board enters service if the technician is free; otherwise, it waits until the technician finishes with the other board. When finished, the technician begins working on the second board. Starting with both boards operational, find the following.

    (a) The expected value of the time until both boards are in the repair facility.

    (b) The variance of the time until both boards are in the repair facility.

    (c) The proportion of time, over the long run, that a working board is available.

14. Suppose in Exercise 13 that when both sound boards are down a second technician is called in to work on the one that has failed most recently. Assume that all repair times remain exponential, with rate μ. Now find the proportion of time that at least one board is operational and compare your answer with that obtained in Exercise 13.

BIBLIOGRAPHY Heyman, D.P. and M.J. Sobel, *Stochastic Models*, Vol. 2, *Handbook in Operations Research*, North-Holland, Amsterdam, 1990.

Resnick, S.I., *Adventures in Stochastic Processes*, Birkhäuser, Boston, 1992.

Ross, S.M., *Introduction to Probability Models*, Seventh Edition, Academic Press, San Diego, 2000.

Tijms, H.C., *Stochastic Models: An Algorithmic Approach*, Wiley, New York, 1994.

Chapter 16

Queuing Models

A situation that is familiar to everyone is waiting in a line. A typical example might be the line of customers that forms in front of the service windows at a post office. The number of customers in the line grows and shrinks with time, and, as anyone who has had the experience knows, the wait can be highly unpredictable. Because the number of customers in line is a random variable that changes with time, the system of customers and servers fits the definition of a stochastic process.

Other familiar situations that involve lines or queues include ticket booths at theaters, conference registration desks, red lights at traffic signals, buffer storage on assembly lines, e-mail on servers, and taxis outside airports. Basically, a queue results whenever existing demand temporarily exceeds the capacity of the service facility—i.e., whenever an arriving customer cannot receive immediate attention because all servers are busy. This situation is almost always guaranteed to occur at some time in any system that has probabilistic arrival and service patterns. Tradeoffs between the costs of increasing service capacity and the costs incurred by customers having to wait prevent an easy solution to this design problem. If the cost of expanding a service facility were no object, then, theoretically, enough servers could be provided to handle all arriving customers without delay. In reality, however, a reduction in service capacity results in a concurrent increase in the cost associated with waiting. For example, consider the costs of operating a fire station. The amortized cost of a truck along with a driver and crew is about $400,000 per year, and must be compared with the hard-to-measure cost of burning property having to wait for the arrival of the fire-fighting equipment. The basic objective in most queuing problems is to achieve a balance between these costs.

The study of waiting lines was pioneered by the Danish mathematician A. K. Erlang, who published *The Theory of Probabilities and Telephone Conversations* in 1908. In later work he observed that a telephone system could often be characterized by either (1) Poisson inputs, exponential holding (service) times, and multiple channels (servers) or (2) Poisson inputs, constant holding times, and a single channel. Much of the theory he developed was soon put into practice by the engineers at Bell Laboratories. As with many OR models, progress was initially rapid but began to slow when computational barriers appeared. A resurgence of interest in queuing systems in the last 10 years has seen many of those barriers fall, and we are now at the point where the gap between theory and practice is a lot less daunting.

In this chapter, we are primarily concerned with a single queuing station having one or more servers operating in steady state. A variety of analytical results are presented for such systems, mostly under the assumption that the arrival and service processes are Markovian. Our chief objective is to highlight the formulas that can be used to compute statistical estimates for such measures as the average number in the queue, the average waiting time for a customer, and the probability that the service mechanism is busy. Near the end of this chapter we consider models that are not Markovian.

For the single queuing station, there are several different models that represent different operating conditions. Our presentation of material is geared toward an understanding of the use and limitations of each model. One of the key insights that can be gained from the examples presented here is that queuing systems may not be very efficient in terms of resource utilization. Queues form and customers wait even though servers may be idle much of the time. The fault is not in the model or the underlying assumptions; it is a direct consequence of the variability of the arrival and service processes. If this variability could be eliminated, systems could be designed economically so that there would be little or no waiting and hence no need for queuing theory.

In many applications, such as discrete parts manufacturing and emergency room operations, the system under study may contain several queues, and customers or products flow from one queue to another in a random fashion. In recent years, most of the theoretical work done on waiting lines has involved "networks" of queues and decision making under uncertainty. These topics are taken up in Chapter 17.

16.1   SYSTEM CHARACTERISTICS

Figure 16.1 shows the basic components of a queuing system. Potential and actual customers are represented by the small circles and may be persons, parts, machines, or almost anything else. The servers are represented by the numbered rectangles and may be any resource, such as a person, a machine, or a repair shop, that performs a function. Customers arrive at the system from some input source and enter service immediately if any one of the servers is idle. If all the servers are busy, the customer waits in a queue until one is available. After some finite amount of time, the customer departs. The details of the process depend on the various assumptions and parameter values adopted for the components of the system.

The input source, also known as the *calling population*, is the collection of potential customers that might have need for the services offered by the system. It is characterized by its size, N, which is often assumed to be infinite for modeling purposes, and by the probability distribution governing the interarrival times. The queue is the number of customers waiting for service, and may be concentrated at a fixed location such as a bank foyer or

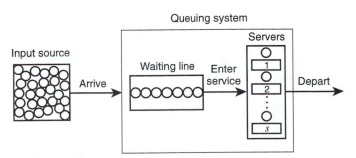

Figure 16.1   Schematic of a general queuing system

may be distributed in time and space such as airplanes approaching a runway. The queue discipline defines the rules by which customers are selected for service. A common discipline is first-come-first-served (FCFS), otherwise known as first-in-first-out (FIFO), but other possibilities are last-in-first-out (LIFO), priority schemes, and random selection. Most of the results presented in this chapter, with the exception of the formulas associated with the waiting time distribution, are valid regardless of the discipline. When preemption is allowed or priority rules are used, additional formulas are needed to compute waiting time probabilities. Unless otherwise stated, then, an FCFS discipline will be assumed.

The service mechanism is the process by which customers are served. The assumption in most of this chapter is that service is provided by one or more identical servers (channels) operating in parallel. For systems that involve a network of queues, however, various configurations will be considered. The characteristics of service are the number of channels, s, and the service time probability distribution. The queuing system is the combination of the waiting line and the service channels. The parameters used to define a system and the performance measures used to analyze it are summarized at the end of this section.

States, State Probabilities, and Expected Values

The first measure of interest is the expected number of customers in the system, denoted by L. To get a better understanding of how the actual number in the system, $N(t)$, varies with time, we focus on birth-death processes as described in previous chapters. The state-transition network for a single-server system with infinite calling population is repeated in Figure 16.2. The states, indicated by the nodes, show the number of customers in the system. The arcs represent events: an arrival, denoted by a, causes the number in the system to increase by 1 whereas a departure, denoted by d, causes the number in the system to decrease by 1. Since both the interarrival and service times are random variables, $N(t)$ is also a random variable. For stable systems, there exist steady-state probabilities for the number of customers in the system, which is independent of time. We call the steady-state probability of n customers π_n, where n is a nonnegative integer.

For a queuing system with s channels, the states identify important characteristics with respect to the number of customers in the queue and the number in service. These characteristics, which are illustrated in Figure 16.3, are as follows.

- When $n = 0$, the system is empty, all servers are idle, and there is no queue. An arriving customer immediately enters service.

- When $1 \leq n \leq s - 1$, all customers are being served, there is no queue, and the number of busy servers is n. An arriving customer immediately enters service.

- When $n = s$, all customers are being served, there is no queue, and all servers are busy. An arriving customer enters the queue.

- When $n \geq s + 1$, a queue with $n - s$ customers is present and all servers are busy. An arriving customer enters the queue.

If an equilibrium exists, the steady-state probabilities have two meanings:

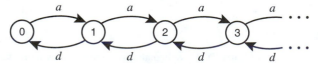

Figure 16.2   Birth–death process

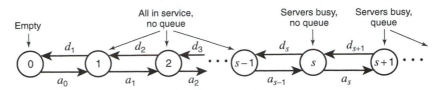

Figure 16.3 Queuing system with s servers

π_n = probability of finding the system in state n at some randomly selected time,

or

π_n = proportion of time the system is in state n

The expected value formulas given below are a direct application of the definition of expected value when the system is in steady state. Note that by "efficiency" we mean the proportion of the time the servers are busy. This is sometimes referred to as the *system utilization*.

Definition of Steady-State Terms

State of the system: Total number of customers in the system, including the number in the queue and in service

State space: $S = \{0, 1, 2, \dots\}$

Expected number in service: $L_s = \sum_{n=0}^{s-1} n\pi_n + s\left[1 - \sum_{n=0}^{s-1} \pi_n\right]$

Expected number in the queue: $L_q = \sum_{n=s+1}^{\infty} (n-s)\pi_n$

Expected number in the system: $L = L_s + L_q = \sum_{n=0}^{\infty} n\pi_n$

Efficiency of the system: $E = L_s/s$

Computing System Characteristics from State Probabilities

To illustrate how steady-state measures are computed, suppose that a queuing system with three servers is observed for a long period of time and data are collected on the proportion of time the system is in each of the states. We assume an infinite calling population so that the probability of an arrival in any interval of time is independent of the number already in the system. Capacity is limited, so whenever there is an arrival when six customers ($K = 6$) are present, the arriving customer balks and goes elsewhere for service. Estimates of the steady-state probabilities are given in Table 16.1.

Using these estimates, we can now answer several interesting questions about the system.

Table 16.1 Steady-State Probabilities for System with $K = 6$

State, n	0	1	2	3	4	5	6
Probability, π_n	0.068	0.170	0.212	0.177	0.147	0.123	0.103

1. What is the probability that all servers are idle?

$$\text{Pr\{all servers idle\}} = \pi_0 = 0.068$$

2. What is the probability that a customer will not have to wait?

$$\text{Pr\{no wait\}} = \pi_0 + \pi_1 + \pi_2 = 0.450$$

3. What is the probability that a customer will have to wait in the queue?

$$\text{Pr\{wait\}} = 1 - \text{Pr\{no wait\}} - \text{Pr\{full\}} = 1(\pi_0 + \pi_1 + \pi_2 + \pi_6) = \; = 0.447$$

4. What is the probability that an arriving customer will be lost?

$$\text{Pr\{lost customer\}} = \pi_6 = 0.103$$

5. What is the expected number of customers in the queue?

$$L_q = 1\pi_4 + 2\pi_5 + 3\pi_6 = 0.702$$

6. What is the expected number of customers in service?

$$L_s = \pi_1 + 2\pi_2 + 3(1 - \pi_0 - \pi_1 - \pi_2) = 2.244$$

7. What is the expected number of customers in the system?

$$L = L_q + L_s = 2.946$$

8. What is the efficiency (utilization) of the servers?

$$E = L_s/s = 2.244/3 \text{ or } 74.8\%$$

Relationship between Time and Number in System

There is a strong relationship between the expected number in the system and the expected time required to pass through the system. The basic result, derived by John Little, is that in expected value terms, the number of items residing in a queuing system is proportional to the ratio of the rate of flow through the system to the residence time. The general expression is known as Little's law and is typically stated in the following form.

$$\text{Expected number in system} = \text{flow rate} \times \text{residence time} \qquad (1)$$

This formula is extremely robust, describing any situation in which items enter a facility, remain for some time, and then leave. There are no assumptions about arrival or service time distributions, the size of the calling population, or limits on the queue. In most applications, the flow rate is equivalent to the average or effective arrival rate, which we denote by $\bar{\lambda}$. If any two terms in Equation (1) are known, the third can be calculated. In addition, the relationship holds when "queue" or "service" is substituted for "system," assuming that the other terms are appropriately defined.

To apply Equation (1) to queuing systems, we reintroduce the notation for arrival and service rates and the expected times in service, in the queue, and in the system. Of course, since we are dealing with a stochastic process, the time a customer spends in the system is a continuous random variable with some probability distribution. Whether or not the exact form of this distribution is known, Little's law can be used to compute its expected value. As indicated in the accompanying sidebar, Little's law relates expected number to expected time by means of the arrival and service rates. It is individually applicable to the service mechanism, the queue, and the system.

The expected time a customer spends actually receiving service, W_s, is simply the expected value of the service time distribution $1/\mu$. The expected number in service is then $\bar{\lambda}/\mu$, independent of other parameters of the queuing system. When the effective arrival rate

$\bar{\lambda}$ equals the mean arrival rate λ, as is the case with an uncapacitated system, we use the simpler notation λ instead of $\bar{\lambda}$.

Definition of Time Parameters

Mean arrival rate: λ

Mean time between arrivals: $1/\lambda$

Effective arrival rate: $\bar{\lambda}$

Mean service rate: μ

Mean time for service: $1/\mu$

Expected time in service: $W_s = 1/\mu$

Expected time in queue: W_q

Expected time in system: $W = W_s + W_q$

Little's Law for Queuing Systems

$L_s = \bar{\lambda} W_s = \bar{\lambda}/\mu$

$L_q = \bar{\lambda} W_q$ or $W_q = L_q/\bar{\lambda}$

$L = \bar{\lambda} W$ or $W = L/\bar{\lambda} = W_s + W_q$

Example 1

Earlier in this section, we computed the following measures: $L_s = 2.244$, $L_q = 0.702$, and $L = 2.946$. On further analysis of the data, we learned that, on average, five customers per hour try the use the system ($\lambda = 5$), but some are turned away because of limited capacity. With this additional information, we now wish to estimate the mean time a customer spends in the system.

Applying Little's law, we must first compute the actual rate of flow through the system to compute the expected times. The preceding calculations showed that $\pi_6 = 0.103$, so 10.3% of the customers find the system full, so the actual throughput rate $\bar{\lambda} = \lambda \times$ (Probability system not full) $= 5(1 - 0.103) = 4.485$ per hour. This rate yields the following results.

$$\bar{\lambda} = 5(1 - \pi_6) = 4.485/\text{hour} \qquad W_q = L_q/\bar{\lambda} = 0.157 \text{ hour}$$
$$W_s = L_s/\bar{\lambda} = 0.5 \text{ hour} \qquad W = L/\bar{\lambda} = 0.657 \text{ hour}$$

In the more general case in which the arrival rate is a function of the number in the system, n, we have

$$\bar{\lambda} = \sum_{n=0}^{\infty} \lambda_n \pi_n$$

For this example, $\lambda_n = 5$, $n = 1, \ldots, 5$, and 0 otherwise, which similarly yields $\bar{\lambda} = 4.485$. When $\lambda_n = \lambda$ is constant for all n, the preceding expression reduces to $\bar{\lambda} = \lambda$, and Little's law can be stated as $L = \lambda W$. In such cases, we will always use λ instead of $\bar{\lambda}$ in the formulas.

Example 2

A post office has three windows staffed by clerks. Customers arrive every 30 seconds, on average, and require approximately 1.25 minutes of service. They remain in the system for an average time of 3 minutes. What is the expected number of customers in the system, and what proportion of time are the clerks idle?

Analysis

$s = 3$ (given)	$L = \lambda W = 6$ customers
$\lambda = 2$ customers/min (given)	$W_q = W - W_s = 1.75$ min
$W = 3$ min (given)	$L_q = \lambda W_q = 3.5$ customers
$W_s = 1.25$ min (given)	$L_s = \lambda W_s = 2.5$ customers
$\mu = 1/W_s = 0.8$ customers/min	$E = L_s/s = 2.5/3 \rightarrow 83.3\%$

Using the preceding formulas with $\overline{\lambda} = \lambda$ and the given information, we can easily compute a variety of system performance measures. In general, the unit for the arrival rate is customers per unit of time. The word "customers" is used in a generic sense and refers to the items in the system. The specific unit of time (seconds, minutes, hours, or days) is not critical, but the same units must be used for all time and rate measures associated with the problem. Minutes are used here.

Input and output data obtained from the analysis are shown above. The results, once again, illustrate the difficulties facing designers of queuing systems. The clerks are idle almost 17% of the time whereas the queue length averages 3.5 customers. These apparently contradictory results are attributable to the stochastic nature of the process.

Notation Describing the System

Characteristics of the System

State Total number of customers in the system—i.e., the number in the queue plus the number in service

K The maximum number in the system. When the maximum number in the system is finite and this number has been reached, an arriving customer "balks" and does not enter.

π_n Probability of n customers in the system at steady state

L Expected number of customers in the system at steady state

W Expected time customer spends in the system at steady state

Characteristics of the Arrival Process

N Size of the input source

λ_n Arrival rate or the expected number of arrivals per unit time when n customers are in the system

λ Arrival rate when the state of the system does not affect the rate of customer arrivals. The expected time between arrivals is $1/\lambda$

$\overline{\lambda}$ Average or effective arrival rate when the state of the system affects the rate of arrival

Characteristics of the Queue

Maximum number in the queue The maximum number in the system less the number of servers; $K - s$

L_q Expected number of customers in the queue at steady state

W_q Expected customer waiting time in the queue at steady state

Queue discipline The rule by which customers are chosen from the queue to receive service. Possible rules are first-come-first-served, last-come-first-served, or selection by some priority rule. Unless otherwise stated, it is assumed that customers form a single queue even if there are multiple service channels.

Characteristics of the Service Process

s Number of service channels. All are assumed to be identical

L_s Expected number in service at steady state

μ_n Mean service rate for the system when *n* customers are present

μ Mean service rate for a single busy server when the number in the system does not affect the service rate. The expected time to complete a single service activity is $1/\mu$.

ρ Traffic intensity. This is the ratio between the rate at which arrivals attempt to use the system and the maximum service rate of the system.

E Efficiency or utilization. This is the ratio between the average number of customers in service and the number of servers.

Specification of System Characteristics

The five-field notation that follows is commonly used to identify some of the important characteristics of a queuing system.

Arrival distribution / Service distribution / Number of servers / Maximum number in the system / Number in the calling population

The following symbols are used to identify interarrival and service time distributions.

M = exponential distribution

D = constant time distribution

E_k = Erlang distribution with parameter *k*

GI = general independent distribution

G = general or arbitrary distribution

16.2 MARKOV QUEUING SYSTEMS

For various queuing models, when the interarrival and interservice times are assumed to be exponentially distributed, closed-form expressions are available for the steady-state probabilities $\pi = (\pi_0, \pi_1, \ldots, \pi_n, \ldots)$. These probabilities permit the derivation of formulas for the expected number of customers in the queue, the expected number in service, and the expected number in the system, as well as the expected time customers spend waiting. Using the standard five-field notation, the systems considered in this section all have the designation

$$M/M/s/K/N$$

where M indicates that the arrival and service processes are Poisson processes, s is the number of servers, K is the maximum number of customers allowed in the system, and N is the size of the calling population. When K and N are infinite, only the first three fields are used. When K is finite but N is infinite, the first four fields are used. The case in which K is infinite and N is finite is logically void.

Some authors include a sixth field to identify the queue discipline. Because we are considering FCFS only, this field is omitted. Nevertheless, there is nothing in the derivation of the steady-state probabilities for any of the Markov queuing systems that depends on queue discipline. Indeed, it can be shown that as long as the selection for service is not a function of a customer's service time, the $\{\pi_n\}$ are independent of queue discipline. The proof of Little's law also remains unchanged, and since the average number in the system is unaltered, so is the average waiting time. In other words, the steady-state probabilities and measures of effectiveness hold for arbitrary queue disciplines provided that (1) all arrivals stay in the queue until served (no reneging), (2) the mean service time is the same for all customers, (3) the server completes service before it starts on the next customer (no preemption), and (4) a service channel always admits a waiting customer immediately on completion of another.

Single Channel Queue, *M/M*/1

The simplest queuing model has a single server operating at an average rate of μ with customers arriving at an average rate of λ. The diagram in Figure 16.4 shows the queue as open ended—i.e., there is no upper limit on the number of customers that are permitted to wait. This contrasts with models presented later, in which the number in the system is restricted.

An important measure for queuing systems is the traffic intensity ρ, where for a single channel model

$$\rho = \lambda/\mu$$

If the system is to provide service to all customers and undergoes some variability in either the arrival or service process, ρ must be strictly less than 1. This condition is necessary for the existence of a steady-state solution and the validity of the expressions in Figure 16.4. These simple closed form expressions can be derived from basic probability concepts. We show how this is done for the *M/M/s* model in the subsection Multichannel Queue, starting with the derivation of $\{\pi_n\}$. In addition to the measures already discussed, we include P_B, the probability that an arriving customer will find all the servers busy or, in other words, the probability that a customer must wait in the queue.

It is interesting to observe that all measures in Figure 16.4 except those involving time depend only on the traffic intensity ρ—the ratio of the arrival rate to the service rate. Measures that refer to time are all proportional to the expected service time $1/\mu$.

Although it is useful to know the expected amount of time a customer spends in the queue, it is also useful to be able to compute waiting time probabilities. Of course, when the system is empty, an arriving customer will receive immediate service and will not have to wait. When the system is not empty and n customers are present, an arrival must wait for n service completions before entering service. Each service activity has an exponential probability distribution with a rate of μ. From probability theory, we know that the sum of n exponentially distributed random variables is a random variable having an Erlang distribution with parameter n. Using the cumulative Erlang distribution, one can derive an expression describing the probability that the wait in the queue will exceed a specified value $t \geq 0$. Let T_q denote the random variable "time spent waiting in the queue" and let $W_q(t)$ denote its cumulative probability distribution. The appropriate formulas are as follows.

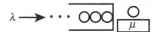

Measure	Formula
ρ	λ/μ
π_0	$1-\rho$
$\pi_n,\ n \geq 1$	$\rho^n \pi_0$
P_B	ρ
L_q	$\dfrac{\rho^2}{1-\rho}$
L_s	ρ
L	$\dfrac{\rho}{1-\rho}$
W_q	$\dfrac{\rho}{\mu(1-\rho)}$
W_s	$1/\mu$
W	$\dfrac{1}{\mu(1-\rho)}$
E	ρ

Figure 16.4 $M/M/1$ queuing system

$$W_q(t) = \begin{cases} \Pr\{T_q = 0\} = \pi_0 = 1-\rho & \text{for } t = 0 \\ \Pr\{T_q \leq t\} = 1 - \rho e^{-\mu(1-\rho)t} & \text{for } t \geq 0 \end{cases}$$

The probability that an arriving customer will not have to wait, $\Pr\{T_q = 0\}$, is the probability that the customer will find the system empty. The formula for $\Pr\{T_q \leq t\}$ allows the computation of probabilities for times greater than zero. Its complement, $\Pr\{T_q > t\} \equiv 1 - \Pr\{T_q \leq t\}$, is often more useful, however, so the formulas for this expression, rather than the expression for $W_q(t)$, will be given in the remainder of this chapter.

A final point to make about $W_q(t)$ is that although its derivative for $t > 0$ does not integrate to 1 over the range $(0, \infty)$ and thus is not a true density, the addition of the point $t = 0$ with its probability $1 - \rho$ yields a valid composite density function, $W_q(t)$.

Workstation Congestion

Different parts that are forged by a metal processing company must all pass through a single machine. On average, it takes 30 seconds for the machine to perform the required operations, although there is much variability owing to the diverse nature of the parts. The time between arrivals at the workstation is about 40 seconds. There seems to be plenty of time for the machine to meet the demand, but frequently a queue can be observed that stretches beyond the space allowed for work in process. In fact, there is actually room for only three parts to wait in the immediate area of the machine. Additional arrivals back up on the material handling system.

Management feels that the congestion at this workstation is affecting the efficiency of the entire facility. Consequently, you have been asked to analyze the situation and propose a solution. As a goal, management has stated that the queue must remain in the available space next to the machine 90% of the time and that no more than 10% of the parts should have to wait more than 1 minute.

For purposes of analysis, let us assume that the arrivals occur independently and that the time between arrivals has an exponential distribution. We also assume that the service time is exponentially distributed. This is reasonable because of the high variability associated with the processing requirements of the different parts.

The next issue that needs to be addressed is the permissible length of the queue. Although it is limited in practice to the storage space for the material handling system, there has never been a situation in which this capacity has been reached, so we will assume that the queue can be of infinite length. Collectively, then, these assumptions lead to a single channel queuing model. The steady-state performance measures for this $M/M/1$ system are as follows.

System: $M/M/1$

$\lambda = 1/40 = 0.025/\text{sec} = 1.5/\text{min}$

$\mu = 1/30 = 0.333/\text{sec} = 2/\text{min}$

$\rho = \lambda/\mu = 0.75$

$\boldsymbol{\pi} = (\pi_0, \pi_1, \pi_2, \pi_3, \pi_4, \ldots)$

$\quad = (0.25, 0.188, 0.141, 0.105, 0.079, \ldots)$

Proportion of time space exceeded:

$\quad \Pr\{n > 4\} = 1 - \Sigma_{i=0}^{4} \pi_i = 0.237$

$L = \rho/(1 - \rho) = 3.0$

$W = L/\lambda = 2 \text{ min}$

$L_q = L - \rho = 2.25$

$W_q = W - 1/\mu = 1.5 \text{ min}$

Proportion waiting too long:

$\quad \Pr\{T_q > 1\} = \rho e^{-\mu(1-\rho)} = 0.455$

These measures indicate that the present configuration does not satisfy management's goals. The line will exceed the available space almost one-quarter of the time ($\Pr\{n > 4\} = 0.237$), and almost half of the items must wait more than 1 minute ($\Pr\{T_q > 1\} = 0.455$). One way to improve this situation is to increase the number of machines, as discussed in the next two subsections.

Multichannel Queue, $M/M/s$

For this system, we again assume an unlimited queue and an infinite calling population. With service rate μ per channel, the maximum service rate is $s\mu$ and the traffic intensity is

$$\rho = \frac{\lambda}{s\mu}$$

For a stable system, we require that $\rho < 1$.

When $s > 1$, the formulas for the $M/M/s$ model are a bit more complicated than those in Figure 16.4, but still can be derived in a straightforward manner given λ, μ, and s, as we now show. In fact, all the formulas except those related to time depend only on the parameters ρ and s. We begin with the derivation of $\{\pi_n\}$.

Derivation of Steady-State Probabilities

Since the queuing model is a special case of the birth–death model with $\lambda_n = \lambda$ and $\mu = \mu_n$ for all n, we can find the steady-state probabilities using the state factor method described in Section 15.5. The state factors for the $M/M/s$ model are

$$C_n = \frac{\lambda^n}{n! \mu^n} = \frac{(s\rho)^n}{n!} \quad \text{for } 0 \leq n < s$$

$$C_n = \frac{\lambda^n}{s! s^{n-s} \mu^n} = \frac{s^s \rho^n}{s!} \quad \text{for } n \geq s$$

For zero customers in the system, π_0 is the inverse of the sum of the state factors.

$$\pi_0 = \left[\sum_{n=0}^{\infty} C_n\right]^{-1} = \left[\sum_{n=0}^{s-1} \frac{(s\rho)^n}{n!} + \frac{s^s}{s!}\sum_{n=s}^{\infty}\rho^n\right]^{-1}$$

$$= \left[\sum_{n=0}^{s-1} \frac{(s\rho)^n}{n!} + \frac{(s\rho)^s}{s!(1-\rho)}\right]^{-1}$$

The infinite geometric series on the right in the first equation has been replaced by its closed-form equivalent.[1] The resultant expression given in the second equation has s terms in the denominator for states 0 through $s-1$ and one additional term representing the remaining states. As ρ approaches 1, the second term becomes very large, indicating that the probability that the system is empty goes to zero as the arrival rate approaches the maximum service rate. The expression is meaningless for $\rho \geq 1$, because the system has no steady-state solution in this range. Finally, when $s = 1$, $\pi_0 = 1 - \rho$, as expected.

The remaining steady-state probabilities can be computed as a function of π_0 using the principle of flow balance as applied to birth-death processes (see Section 15.5).

$$\pi_n = \frac{(s\rho)^n \pi_0}{n!}, \quad 1 \leq n \leq s$$

$$\pi_n = \frac{s^s \rho^n \pi_0}{s!}, \quad n \geq s$$

Note that both expressions are identical when $n = s$.

Probability That All Servers Are Busy

An arriving customer must wait in the queue when all the servers are busy. This situation occurs when the system is in state s or greater, so the corresponding probability is computed as follows.

$$P_B = \sum_{n=s}^{\infty}\pi_n = \frac{\pi_0 s^s}{s!}\sum_{n=s}^{\infty}\rho^n = \frac{\pi_0(s\rho)^s}{s!(1-\rho)} = \frac{\pi_s}{1-\rho}$$

Once again, this result was obtained by replacing the infinite series by its equivalent after making the substitution $n = k + s$.

Average Number and Time in the System and Queue

First we derive the expression for the average number of customers in the queue at steady state. Using the general formula for expected value given in Section 16.1, we have

$$L_q = \sum_{n=s}^{\infty}(n-s)\pi_n = \pi_s\sum_{n=s}^{\infty}(n-s)\rho^{n-s} = \pi_s\sum_{k=0}^{\infty}k\rho^k$$

The second summation is obtained by noting that $\pi_n = \pi_s\rho^{n-s}$, whereas the third summation follows from the substitution $n = k + s$. Replacing the infinite series with its closed-form equivalent,[2] yields

---

[1]The infinite geometric series $\sum_{k=0}^{\infty}\rho^k = \frac{1}{(1-\rho)^2}$ for $\rho < 1$. To obtain this form where the index starts at 0, we make the substitution $n = k + s$.

[2] The infinite series $\sum_{k=0}^{\infty} k\rho^k = \frac{\rho}{(1-\rho)^2}$ for $\rho < 1$.

$$L_q = \frac{s^s \rho^{s+1} \pi_0}{s!(1-\rho)^2}$$

The expected time in the queue can now be found using Little's law.

$$W_q = L_q / \lambda$$

The average time in service is, by definition,

$$W_s = 1/\mu$$

and the average number of customers in service is

$$L_s = \lambda W_s = s\rho = \lambda/\mu$$

Interestingly, all of these measures are independent of the number of channels.

To find the average number of customers in the system and the expected amount of time that each customer is in the system, we simply sum the corresponding queue and service values already computed.

$$L = L_q + L_s, \ W = W_q + W_s$$

The efficiency of the service process is the proportion of the time the servers are busy

$$E = L_s / s = \rho$$

which is nothing more than the traffic intensity of the system.

Probabilities for Time in the System and Time in the Queue

Let T_q be a random variable denoting the time a customer spends in the queue, and let T_{sys} be the time the customer spends in the system. We use $T_{q:n}$ and $T_{sys:n}$ as the corresponding times conditioned by the fact that n customers are already in the system when the customer arrives. $T_{sys:n}$ differs from $T_{q:n}$ by the service time t_s, which is also a random variable. If the mean service time is $1/\mu$, we have

$$T_{sys:n} = T_{q:n} + t_s \ \text{ and } \ E[T_{sys:n}] = E[T_{q:n}] + 1/\mu$$

Suppose a customer enters the system and finds n customers already in service or in the queue. When n is less than s, service begins immediately, so

$$T_{q:n} = 0, \ n < s$$

With s or more customers present, the arriving customer must wait in the queue. Service begins after $n - s + 1$ service completions. Therefore, the time in the queue is

$$T_{q:n} = \sum_{k=1}^{n-s+1} \tau_k, \ \ n \geq s$$

where τ_k is a random variable corresponding to the time of the kth service completion. The mean of τ_k is $1/s\mu$, since the rate of service is $s\mu$ when all the channels are busy. The simplest result that can be obtained from this situation is that the expected times that the customer must be in the queue and the system, respectively, are

$$E[T_{q:n}] = 0 \ \text{ and } \ E[T_{sys:n}] = 1/\mu, \ \text{if } n < s$$
$$E[T_{q:n}] = (n - s + 1)(1/s\mu) \ \text{ and } \ E[T_{sys:n}] = E[T_{q:n}] + 1/\mu, \ \text{if } n \geq s$$

More can be said about the distribution of the time in the queue by noting once again that the sum of n independently distributed exponential random variables has an Erlang distribution with parameters n and $s\mu$. Using this fact, it can be shown that for the $M/M/s$ model the time in the queue (not dependent on the number in the system when entering) has an exponential distribution with mean $1/s\mu(1 - \rho)t$. In particular,

$$\Pr\{T_q = 0\} = \sum_{n=0}^{s-1} \pi_n = 1 - \frac{(s\rho)^s \pi_0}{s!(1-\rho)}$$

and

$$\Pr\{T_q > t\} = \frac{(s\rho)^s \pi_0}{s!(1-\rho)} e^{-s\mu(1-\rho)t}, \quad t > 0$$

Thus, when there are s servers, there will be no wait if an arriving customer finds fewer than s customers in the system. When the number in the system is s or more, the customer must wait for $n - s + 1$ service completions.

Using probability theory, it is also possible to derive an expression for the probability that a customer will spend more than t units of time in the system. Without going through the details, this result is

$$\Pr\{T_{sys} > t\} = e^{-\mu t}\left[1 + \frac{(s\rho)^s \pi_0}{s!(1-\rho)}\left(\frac{1 - e^{-\mu t(s-1-s\rho)}}{s-1-s\rho}\right)\right], \quad t \geq 0$$

Figure 16.5 presents the formulas for the $M/M/s$ model. To reduce the number of entries, the expression for L, which is the sum of L_s and L_q, is omitted, as are the time measures since they are readily available from Little's law.

$$W_q = L_q/\lambda, \quad W_s = 1/\mu, \quad W = L/\lambda$$

Selecting the Number of Servers for the Congested Workstation

In an attempt to meet the goals specified in the preceding example, a proposal is made to add a second machine at the workstation. The new machine will operate at the same rate as the first, and waiting parts will continue to form a single line with no upper limit. To analyze this new situation, we use the formulas for the multichannel queue with $s = 2$.

System: *M/M/2*

$\rho = \lambda/s\mu = E = 0.375$

$\pi = (\pi_0, \pi_1, \pi_2, \pi_3, \pi_4, \pi_5, \ldots)$
   $= (0.455, 0.341, 0.128, 0.048, 0.018, 0.007, \ldots)$

$\Pr\{n > 4\} = 0.011$

$L_q = (\pi_0(\lambda/\mu)^2\rho) / (2!(1 - \rho)^2) = 0.1227$

$W_q = L_q/\lambda = 0.0818$ min

$L = L_q + \lambda/\mu = 0.8727$

$W = L/\lambda = 0.5818$ min

$\Pr\{T_q = 0\} = \pi_0 + \pi_1 = 0.796$

$\Pr\{T_q > 1\} = (1 - \Pr\{T_q = 0\})e^{-2\mu(1-\rho t)} = 0.0167$

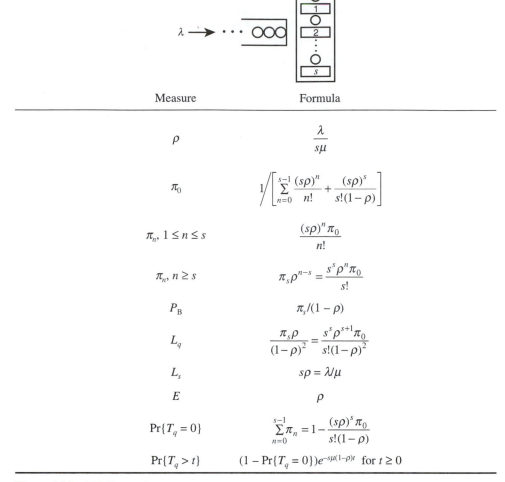

Measure	Formula
ρ	$\dfrac{\lambda}{s\mu}$
π_0	$1\left/\left[\displaystyle\sum_{n=0}^{s-1}\dfrac{(s\rho)^n}{n!}+\dfrac{(s\rho)^s}{s!(1-\rho)}\right]\right.$
$\pi_n,\ 1\le n\le s$	$\dfrac{(s\rho)^n\pi_0}{n!}$
$\pi_n,\ n\ge s$	$\pi_s\rho^{n-s}=\dfrac{s^s\rho^n\pi_0}{s!}$
P_B	$\pi_s/(1-\rho)$
L_q	$\dfrac{\pi_s\rho}{(1-\rho)^2}=\dfrac{s^s\rho^{s+1}\pi_0}{s!(1-\rho)^2}$
L_s	$s\rho=\lambda/\mu$
E	ρ
$\Pr\{T_q=0\}$	$\displaystyle\sum_{n=0}^{s-1}\pi_n=1-\dfrac{(s\rho)^s\pi_0}{s!(1-\rho)}$
$\Pr\{T_q>t\}$	$(1-\Pr\{T_q=0\})e^{-s\mu(1-\rho)t}$ for $t\ge 0$

Figure 16.5 *M/M/s* queuing system

This solution meets management's goals, with the waiting parts exceeding the available space by only about 1% ($\Pr\{n>4\}=0.011$) of the time and with fewer than 2% of the items waiting more than 1 minute ($\Pr\{T_q>1\}=0.0167$). We are assuming here that the added machine used one of the spaces for work in process. Efficiency, however, falls from 75% to 37.5%. The variability of the system coupled with management's requirements force the machines to be idle 62.5% of the time. This value is simply $1-\rho$ but can be calculated directly from the steady-state probabilities—i.e., $\Pr\{$a specified server is idle$\}=1-\rho=\pi_0+0.5\pi_1=0.455+0.5(0.341)\cong 0.625$.

Telephone Answering System

A city-owned utility company wants to determine a staffing plan for its customer representatives. During a specific time of the day, calls arrive at an average rate of 10 per minute, and it takes an average of 1 minute to respond to each inquiry. You have been asked to recommend the number of operators that would provide a "satisfactory" level of service to the calling population. Assume that both arrival and service processes are Poisson processes.

With an arrival rate λ of 10 per minute and a service time $1/\mu$ of 1 minute, the traffic intensity $\rho=\lambda/s\mu=10/s$, so at least 11 operators are required to obtain stability. Ten, for example, would yield a value of $\rho = 1$, which is too high. Table 16.2 enumerates several measures related to the performance of the system associated with 11, 12, and 13 operators. Clearly, the number s has a great effect on the queue characteristics and the level of customer service. Management must resolve the tradeoff between the cost of additional operators and the costs incurred by customers having to wait.

Finite System with a Single Channel, *M/M/1/K*

As a further generalization of the Markovian queuing process, we now consider the case of a single-server system that can handle only a finite number of customers K at any point in time. With this qualification, the maximum number of customers in the queue is limited to $K - 1$. When all available spaces are occupied, the system is said to be full and arriving customers are turned away—i.e., they balk. To complete the model, we assume that service times have an exponential distribution with service rate μ and that customers arrive according to a Poisson process with rate λ. This leads to an *M/M/1/K* system, as sketched in Figure 16.6.

For models with limited capacity, stability is not an issue. In particular, the traffic intensity ρ need not be less than 1 since the queue cannot grow indefinitely. It is useful, however, to define several new measures, the first being P_F, the probability that the system is full. This is the probability that an arriving customer will find the system full, so it is the probability of a balk.

For a finite system, some flow is lost, so the actual flow rate entering the system $\bar{\lambda}$ is less than the flow rate λ arriving at the system. Recalling that Little's law uses the flow rate passing through the system, we have

$$W_q = L_q/\bar{\lambda}, \; W = L/\bar{\lambda}, \; W_s = L_s/\bar{\lambda} = 1/\mu$$

These results are for a single channel *M/M/1/K* system. The remaining parameters and measures for this system are shown in Figure 16.6 along with the corresponding values for an *M/M/1* system. Once observed, the differences should be easy to explain. For the allowable states, the state probabilities for the finite system are greater than those for the infinite system. Also, the expected numbers of customers in the queue, in service, and in the system are all lower for the finite system. Both effects are attributable to balking.

Overburdened Server

An electronics technician working for a firm that designs printed circuit boards is constantly behind schedule. Jobs submitted by the engineering department pile up at his bench and never seem to get finished on time. As a consequence, the technician often works weekends to catch up.

A quick review of the work records suggests the reason for the problem. Jobs arrive at an average rate of 1.1 per day, whereas it takes an average of 1 day to complete a job.

Table 16.2 Comparison of Multiserver Systems

Measure	*M/M/*11	*M/M/*12	*M/M/*13
L_q	6.821	2.247	0.951
W_q	0.682	0.225	0.095
E	0.909	0.833	0.769
$\Pr\{T_q = 0\}$	0.318	0.551	0.715
$\Pr\{T_q > 1\}$	0.251	0.061	0.014

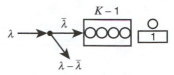

Measure	$M/M/1/K$	$M/M/1$
ρ	λ/μ	λ/μ
π_0	$\dfrac{1-\rho}{1-\rho^{K+1}}$	$1-\rho$
$\pi_n,\ 1 \le n \le K$	$\pi_0 \rho^n$	$\pi_0 \rho^n$
P_{F}	$\pi_K = \pi_0 \rho^K$	0
P_{B}	$\dfrac{\rho(1-\rho^K)}{1-\rho^{K+1}}$	ρ
L_q	$\dfrac{\rho^2[(K-1)\rho^K - K\rho^{K-1} + 1]}{(1-\rho)(1-\rho^{K+1})}$	$\dfrac{\rho^2}{1-\rho}$
L_s	$\rho(1 - P_{\mathrm{F}})$	ρ
L	$\dfrac{\rho}{1-\rho} - \dfrac{(K+1)\rho^{K+1}}{1-\rho^{K+1}}$	$\dfrac{\rho}{1-\rho}$
$\bar{\lambda}$	$\lambda(1 - P_{\mathrm{F}})$	λ
E	$\rho(1 - P_{\mathrm{F}})$	ρ

Figure 16.6 $M/M/1/K$ queuing system for $\rho \ne 1$

Because the arrival rate is greater than the processing rate, the situation is unstable, and without the weekend to reduce the load, the queue would eventually grow without bound. To unburden the technician, management proposes to install a fixed number of bins to hold waiting jobs. When the bins are full, the policy will be to send the overflow to an outside contractor. Two measures will be used to evaluate this approach: (1) the average time in the queue for a job worked on by the technician and (2) the proportion of jobs sent to the contractor. It is desirable to keep both of these measures as small as possible.

The suggested policy leads to a finite queuing system. Assuming that interarrival and service times are exponentially distributed, we can use the formulas for the $M/M/1/K$ system with $\rho = 1.1$. Table 16.3 shows the values of P_{F}, the probability of a balk, and of W_q, the expected queue time as a function of K, the number of bins provided.

As we can see, there is a clear tradeoff between the two measures. More bins mean that a larger portion of the work will be done by the technician, but they also mean a longer wait. To determine the optimal number of bins, we must identify the costs associated with the competing alternatives, including overtime for the technician and the use of outside services as well as the cost resulting from job delays. Thus, more data and analyses are needed to arrive at a solution.

Finite System with Multiple Channels, $M/M/s/K$

The general formulas for a finite queuing system with s identical service channels each operating at a rate μ are given in Figure 16.7. Comparing them with the formulas given in Figure 16.5 for the case where no limit is placed on the queue size, we see the same effects

Table 16.3 Parametric Results for *M/M/1/K* Example

K	1	2	3	4	5	6	7	8	9	10
W_q (days)	0.00	0.52	1.06	1.62	2.19	2.78	3.38	4.00	4.63	5.27
P_F	0.52	0.37	0.29	0.24	0.21	0.19	0.17	0.16	0.15	0.14

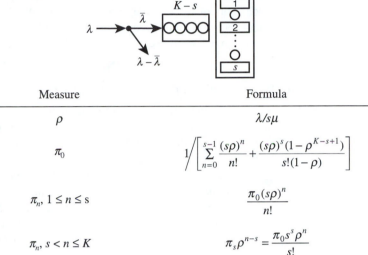

Measure	Formula
ρ	$\lambda/s\mu$
π_0	$1 \left/ \left[\displaystyle\sum_{n=0}^{s-1} \frac{(s\rho)^n}{n!} + \frac{(s\rho)^s(1-\rho^{K-s+1})}{s!(1-\rho)} \right] \right.$
$\pi_n, 1 \leq n \leq s$	$\dfrac{\pi_0(s\rho)^n}{n!}$
$\pi_n, s < n \leq K$	$\pi_s\rho^{n-s} = \dfrac{\pi_0 s^s \rho^n}{s!}$
P_F	$\pi_K = \pi_s\rho^{K-s} = \dfrac{\pi_0 s^s \rho^K}{s!}$
P_B	$\dfrac{\pi_s(1-\rho^{K-s+1})}{1-\rho}$
L_q	$\dfrac{\pi_s\rho}{(1-\rho)^2}\left[1 - \rho^{K-s} - (K-s)(1-\rho)\rho^{K-s}\right]$
L_s	$s\rho(1-P_F) = \bar{\lambda}/\mu$
$\bar{\lambda}$	$\lambda(1-P_F)$
E	$\rho(1-P_F)$
$q_n, 0 \leq n \leq K-1$	$\pi_n/(1-P_F)$
$\Pr\{T_q = 0\}$	$\displaystyle\sum_{n=0}^{s-1} q_n$
$\Pr\{T_q > t\}$ for $t \geq 0$	$\displaystyle\sum_{n=s}^{K-1} q_n\left[\sum_{i=0}^{n-s}\frac{(s\mu t)^i e^{-s\mu t}}{i!}\right]$

Figure 16.7 *M/M/s/K* queuing system when $\rho \neq 1$

we observed for the single channel, finite queue case—i.e., balking reduces the average number of customers in the system. Also, given that the state probabilities must sum to 1, the values of π_n for the finite system are greater than the corresponding values for the infinite system when $n \leq K$.

Near the bottom of the tabular portion of Figure 16.7, we provide an expression for computing the waiting time probability of a customer in the queue, $\Pr\{T_q > t\}$. This expression makes use of q_n, the probability that a customer entering the system will find n customers present, rather than the state probability π_n. When an arriving customer finds K customers in the system, he or she balks and does not enter. To account for this situation, it is necessary to divide π_n by the probability that an arriving customer will enter the system, $1 - \pi_K$. Thus, $q_n = \pi_n/(1 - P_B)$, $0 \le n \le K - 1$.

Note that the formulas in Figure 16.7 are specialized to the case when the arrival rate does not equal the maximum service rate, $\rho \ne 1$. Such a system would be exactly in balance, but the formulas for π_0 would be undefined in this case. The formulas for the case in which $\rho = 1$ are presented in Figure 16.8 in a subsequent subsection.

Machine Processing with Limited Space for Work in Process

Parts arrive at a machine station at an average rate of 1.5 per minute. The mean time for service is 30 seconds. When the machine is busy, parts queue up until there are three waiting in line. At that point, balking occurs and any new arrivals are sent for alternative processing. We wish to analyze this situation under the criteria that no more than 5% of arriving parts receive alternative processing and no more than 10% of the parts that are serviced directly spend more than 1 minute in the queue.

Assuming that both the arrival and service processes are Poisson processes, we have an $M/M/s/K$ model, where $K = 4$ is the maximum number of parts in the system (including service). Numerical results for the single machine case are as follows.

System: $M/M/1/4$

$\lambda = 1.5$/min

$\mu = 2$/min

$\rho = \lambda/\mu = 0.75$

Balking probability: $P_F = 0.104$

$\bar{\lambda} = \lambda(1 - P_F) = 1.344$/min

$L = 1.444$

$W = 1.074$ min

$E = 67.22\%$

$\Pr\{T_q > 1\} = 0.225$

We observe that more than 10% of the arriving parts balk (P_F in Table 16.4). Because of balking, the flow rate through the system is 1.344/min, which is less than the arrival rate of 1.5/min.

To find the probability that the waiting time is greater than 1 minute, we must compute π_n and then q_n for $n = 0, \ldots, 4$. We also need the cumulative value of the Poisson distribution for $n > s$ with parameter $s\mu t$ for $n = 1, \ldots, 4$. The results are shown in Table 16.4 for the case in which $s = 1$, $K = 4$, $t = 1$, and $s\mu t = 2$. Using these results, we have

$$\Pr\{T_q > 1\} = \sum_{n=s}^{K-1} q_n \left[\sum_{i=0}^{n-s} \frac{(s\mu t)^i e^{-s\mu t}}{i!} \right] = 0.225$$

so there is a 22.5% probability that a part will have to wait 1 minute or more.

Table 16.4 Results for $M/M/1/4$ System

State index, n	0	1	2	3	4
π_n	0.328	0.246	0.184	0.138	0.104
q_n	0.366	0.274	0.206	0.154	0
$\sum_{i=0}^{n-s} \dfrac{(s\mu t)^i e^{-s\mu t}}{i!}$	0	0.135	0.406	0.677	0.857

System: $M/M/2/5$

$\pi_K = 0.0068$

$\bar{\lambda} = \lambda(1 - \pi_K) = 1.49$

$L = 0.85$

$W = 0.57$ min

$E = 37.2\%$

$\Pr\{T_q > 1\} = 0.011$

We now propose to add another machine with the same production rate as the first. Let us assume that the new machine does not reduce the space for waiting parts, so there is still room for three parts. To analyze the new situation, we use the formulas for an $M/M/2/5$ system. The accompanying solution meets our original goals: the percentage of balking parts is less than 1% and the probability of a wait time greater than 1 minute is about 1%.

Finite System with Traffic Intensity Equal to 1, $M/M/s/K$ and $\rho = 1$

In Figures 16.6 and 16.7, we see that several of the expressions for a finite queuing system contain $1 - \rho$ in the denominator and so cannot be applied when the traffic intensity $\rho = 1$. This case is important because it represents the situation in which the capacity of a system is just equal to the demand for service. New expressions must be derived. The full set of formulas for the $M/M/s/K$ model is given in Table 16.5 for general s and the special case where $s = 1$. As we can see, the state probabilities now have the interesting characteristic that they are all equal when there are s or more in the system—i.e., $\pi_n = \pi_s$, $s \leq n \leq K$. Thus, with one server and $\rho = 1$, all state probabilities are equal to $1/(K + 1)$.

To get a better understanding of the behavior of finite queuing systems that operate at or near the threshold value $\rho = 1$, consider a company whose policy requires that service capacity never exceed demand. For the case in which $K = 5$ and $\mu = 5$, the results are given in Figure 16.8. As demand grows (λ increases), the efficiencies of the servers decline and the probability of balking customers (π_5) increases. Obviously, the policy is not a good one when there is variability.

System with No Queue Allowed

When the number of servers is equal to the maximum number of customers in the system, we have the case in which no queue is permitted to form. The designation is $M/M/s/s$ for Markovian arrival and service processes. An example might be a telephone switchboard with no holding lines, or a parking lot with no space on the street to wait if the lot is full.

Measure	$M/M/s/K$	$M/M/1/K$
ρ	$\lambda/s\mu$	λ/μ
π_0	$1\Bigg/\left[\displaystyle\sum_{n=0}^{s-1}\frac{(s\rho)^n}{n!}+\frac{(s\rho)^s(K-s+1)}{s!}\right]$	$\dfrac{1}{K+1}$
$\pi_n,\ 1\le n\le s$	$\dfrac{\pi_0 s^n}{n!}$	$\dfrac{1}{K+1}$
$\pi_n,\ s<n\le K$	π_s	$\dfrac{1}{K+1}$
P_F	π_s	$\dfrac{1}{K+1}$
P_B	$(K-s+1)\pi_s$	$\dfrac{K}{K+1}$
L_q	$\pi_s(K-s)(K-s+1)/2$	$\dfrac{K(K-1)}{2(K+1)}$
L_s	$s(1-\pi_s)$	$\dfrac{K}{K+1}$
$\bar{\lambda}$	$\lambda(1-\pi_s)$	$\dfrac{\lambda K}{K+1}$
E	$1-\pi_s$	$\dfrac{K}{K+1}$
$q_n,\ 0\le n\le K-1$	$\dfrac{\pi_n}{1-\pi_s}$	$\dfrac{1}{K}$
$\Pr\{T_q=0\}$	$\displaystyle\sum_{n=0}^{s-1}q_n$	$\dfrac{1}{K}$
$\Pr\{T_q>t\}$ for $t\ge 0$	$\displaystyle\sum_{n=s}^{K-1}q_n\left[\sum_{i=0}^{n-s}\frac{(s\mu t)^i e^{-s\mu t}}{i!}\right]$	$\dfrac{1}{K}\displaystyle\sum_{n=1}^{K-1}\left[\sum_{i=0}^{n-1}\frac{(\mu t)^i e^{-\mu t}}{i!}\right]$

Figure 16.8 $M/M/s/K$ queuing system with $\rho=1$

This situation is illustrated in the top portion of Figure 16.9. Customers arrive at a rate λ and either enter service immediately at a rate $\bar{\lambda}$ or balk at a rate $\lambda-\bar{\lambda}$. The steady-state probabilities shown in the accompanying table hold for any value of ρ and, somewhat surprisingly, for any service distribution with mean $1/\mu$. The probability of exactly s customers in the system, π_s, is computed by what is called Erlang's loss formula. This value gives the proportion of customers that find the system full and hence balk; thus, $P_F=\pi_s$. Finally, the waiting time probability $\Pr\{T_q\ge t\}=0$ for all t since queues are not allowed. As a consequence, there is no longer a need to compute q_n.

Interestingly, the formulas in the table that accompanies Figure 16.9 are valid for both $M/M/s/s$ and $M/G/s/s$ systems. When $s\to\infty$, we have the ample service model $M/G/\infty$.

Table 16.5 Results for Different $M/M/s/K$ System with $\rho = 1$

Measure	$M/M/1/5$	$M/M/2/5$	$M/M/3/5$	$M/M/4/5$	$M/M/5/5$
λ	5	10	15	20	25
μ	5	5	5	5	5
$\bar{\lambda}$	4.167	8.182	11.932	15.259	17.878
L_q	1.667	1.097	0.614	0.237	0
W_q	0.4	0.133	0.051	0.016	0
L	2.5	2.727	3.0	3.289	3.576
W	0.6	0.333	0.251	0.216	0.2
E	0.833	0.818	0.795	0.763	0.715
P_B	0.833	0.727	0.614	0.474	0.285
π_0	0.167	0.091	0.045	0.022	0.011
π_1	0.167	0.182	0.136	0.089	0.055
π_2	0.167	0.182	0.205	0.178	0.137
π_3	0.167	0.182	0.205	0.237	0.228
π_4	0.167	0.182	0.205	0.237	0.285
π_5	0.167	0.182	0.205	0.237	0.285

Telephone Answering System with No Queue Allowed

We return to the utility company switchboard example considered earlier in this chapter. During the busy time of the day, customers call at an average rate of 10 per minute and require 1 minute on average to speak to a company representative. No holding lines are provided. A customer who calls when all lines are full receives a busy signal. We assume

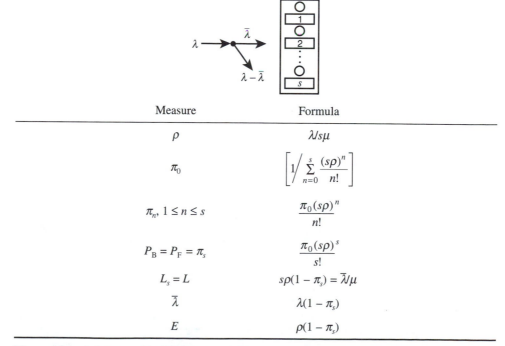

Measure	Formula
ρ	$\lambda/s\mu$
π_0	$\left[1 \middle/ \sum_{n=0}^{s} \dfrac{(s\rho)^n}{n!} \right]$
$\pi_n, \ 1 \le n \le s$	$\dfrac{\pi_0(s\rho)^n}{n!}$
$P_B = P_F = \pi_s$	$\dfrac{\pi_0(s\rho)^s}{s!}$
$L_s = L$	$s\rho(1 - \pi_s) = \bar{\lambda}/\mu$
$\bar{\lambda}$	$\lambda(1 - \pi_s)$
E	$\rho(1 - \pi_s)$

Figure 16.9 $M/M/s/s$ or $M/G/s/s$ queuing system

that he or she does not call back that day. Again, the question is how many lines should be provided.

The analyses for 10, 11, and 12 lines are summarized in Table 16.6. As more lines are added, the throughput rate $\bar{\lambda}$ goes up, and so the number of frustrated customers goes down; however, efficiency decreases with more capacity. Recall that the throughput rate is the equilibrium rate at which customers enter and leave the system. When there is no balking, $\bar{\lambda} = \lambda$. Because the average service time is 1, the average number in the system is the same as the throughput rate.

The bottom row of Table 16.6 gives the probability that a caller will receive a busy signal. Once again, a tradeoff exists between the level of service and the cost of providing additional lines—either lines for holding or lines to be answered directly by a company representative.

Finite Input Source System, *M/M/s/K/N*

When the size of the calling population N is small enough so that the number of customers in the system affects the arrival rate, we use the finite input source model shown in the top portion of Figure 16.10. Thus, when the number of customers in the queuing system is n ($n = 0, 1, \ldots, K$), there are only $N - n$ potential customers remaining in the input source. To analyze this situation for s servers, we assume that the service times are identical exponential random variables with mean $1/\mu$. When a member of the population is not already in the queuing system, the time until the member will enter is governed by an exponential distribution with mean $1/\lambda$. Thus, the arrival rate depends on the system state. When the system is empty, the arrival rate is $N\lambda$, and when one member of the population is in the system, the arrival rate is $(N-1)\lambda$. The arrival rate continues to decrease with the number in the system until, when all members of the population are being served or are waiting for service, the arrival rate is 0. We assue arrivals balk when $n = K$ and $K < N$.

Using the birth–death model discussed in Section 15.5, we can determine the arrival and service rates as functions of the state n.

$$\lambda_n = \begin{cases} (N-n)\lambda, & 0 \leq n < K \\ 0, & n \geq K \end{cases}$$

$$\mu_n = \begin{cases} n\mu, & 0 \leq n < s \\ s\mu, & n \geq s \end{cases}$$

These values lead to the system measures shown in Figure 16.10. The probability that an arrival finds n customers in the system, q_n, is equal to

$$q_n = \frac{(N-n)\pi_n}{N-L} \quad \text{for } n = 0, \ldots, K-1$$

Table 16.6   Parametric Results for $s = K$

System measure	*M/M/10/10*	*M/M/11/11*	*M/M/12/12*
Throughput rate, $\bar{\lambda}$	7.854	8.368	8.803
Efficiency, $E = L_s/s$	0.785	0.761	0.734
Average number in service, L_s	7.854	8.338	8.803
Average time in service, W_s	1	1	1
Balking probability, $P_F = \pi_s$	0.2146	0.1632	0.1197

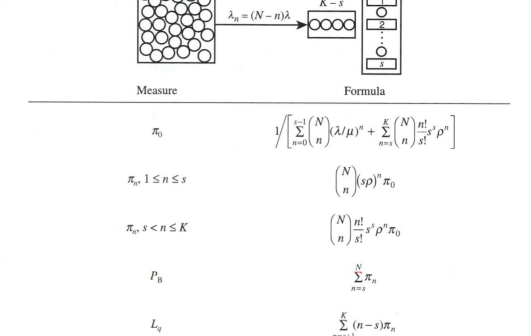

Measure	Formula
π_0	$1 \Big/ \left[\displaystyle\sum_{n=0}^{s-1} \binom{N}{n}(\lambda/\mu)^n + \sum_{n=s}^{K} \binom{N}{n}\frac{n!}{s!}s^s\rho^n \right]$
$\pi_n,\ 1 \le n \le s$	$\binom{N}{n}(s\rho)^n \pi_0$
$\pi_n,\ s < n \le K$	$\binom{N}{n}\frac{n!}{s!}s^s\rho^n\pi_0$
P_B	$\displaystyle\sum_{n=s}^{N}\pi_n$
L_q	$\displaystyle\sum_{n=s+1}^{K}(n-s)\pi_n$
L_s	$L_s = \displaystyle\sum_{n=1}^{s-1} n\pi_n + s\Big(1-\sum_{n=0}^{s-1}\pi_n\Big)$
$\overline{\lambda}$	$\lambda(N-L)$
E	L_s/s

Figure 16.10 *M/M/s/K/N* queuing system (finite input source)

where the denominator of this expression was selected to ensure that $\sum_{n=0}^{K-1} q_n = 1$. The formula for computing $\Pr\{T_q > t\}$, the waiting time probability in the queue, is given by

$$\Pr\{T_q > t\} = \sum_{n=s}^{K-1} q_n \sum_{i=0}^{n-s} \frac{(s\mu t)^i e^{-s\mu t}}{i!}$$

To illustrate the use of the model, we consider a repair facility for a work center with N machines. The repair facility has s identical servers. The time for service is exponentially distributed, with mean $1/\mu$. The inputs to the system are provided by failures among the operating machines, each of which alternates between being in and being out of the repair system. When a machine leaves the system, the time until its next failure is assumed to have an exponential distribution, with rate λ. In other words, $1/\lambda$ is the mean time to failure.

This situation fits the *M/M/s/K/N* queuing model. The last two components of the notation indicate that the number in the system is limited to K and that there are N customers in the calling population. Arrivals that occur when the system is full (in state K) balk and return to the calling population. When K equals N, as is the case here, the results in Figure 16.10 hold even if the arrival process is not exponential.

Note that it is unrealistic to suppose that $K < N$ for this example. When a machine is failed and cannot enter the repair facility because of a limited queue size, it remains failed and does not return to the calling population. To make the model appropriate, one must assume that the failed machine that balks is somehow instantly repaired outside of the repair facility. Since this is not realistic for many examples, one most often has $K = N$ for finite population models.

Interestingly, the model can be generalized to include the use of spares with a minimal amount of effort. Assume that there are N operating machines with Y spares on hand. Each operating machine fails at a rate λ, but spare machines in excess of N do not fail. When an operating machine fails it is sent to the repair facility, and if a spare is available it replaces the failed machine. A machine that has been repaired returns to operation if fewer than N are currently operating, or otherwise becomes a spare.

For this model, the arrival rates in the birth-death model are slightly different and are given by

$$\lambda_n = \begin{cases} N\lambda & 0 \leq n < Y \\ (N - n + Y)\lambda, & Y \leq n < Y + N \\ 0, & n = Y + N \end{cases}$$

With s servers, we have

$$\mu_n = \begin{cases} n\mu, & 0 \leq n < s \\ s\mu, & n \geq s \end{cases}$$

This situation can be solved as a general Markov process, but formulas similar to those in Figure 16.10 are available for computing the steady-state measures. If Y is very large, we essentially have an infinite calling population with arrival rate $N\lambda$, which reduces to an $M/M/s/\infty$ system with λ replaced by $N\lambda$. For details, see Gross and Harris [1998].

Repair Shop

Consider a repair shop that handles all the repair work for the lathes of a tool and die maker. There are five lathes, each of which fails once every 50 hours on average according to an exponential distribution. The time required to repair a machine also has an exponential distribution, with an average time of 20 hours. All the lathes are operating when they are not in the repair shop.

An analysis of the system for one and two repair technicians is presented in Table 16.7. For a single technician, we see that more than half of the five machines are in the repair shop at any time ($L = 2.67$) and that there is about an 8.6% chance (π_5) that no machines are operating at all. This solution may not be acceptable for the company because so many machines are in the repair shop.

The system with two technicians reduces the average number of lathes in the shop to 1.66, leaving 3.33 in operation. All five machines are in the repair shop only a little more than 1% of the time. This improvement is at the cost of an additional technician. On the average, the technicians are idle about one-third of the time ($1 - E = 1 - 0.6678 = 0.3322$).

Table 16.7   Results for Repair Shop with Finite Input Source

Measure	M/M/1/5/5	M/M/2/5/5
λ	0.02/hour	0.02/hour
μ	0.05/hour	0.05/hour
$(\pi_0, \pi_1, \pi_2,$	(0.0697, 0.1395, 0.2231,	(0.1661, 0.3322, 0.2657,
$\pi_3, \pi_4, \pi_5)$	0.2678, 0.2142, 0.0857)	0.1594, 0.0638, 0.0128)
$\bar{\lambda}$	0.0465/hour	0.0668/hour
L_q	1.74	0.32
W_q	37.50 hours	4.871 hours
L_s	0.93	1.34
L	2.67	1.66
W	57.5 hours	24.87 hours
E	93%	66.78%

16.3   NON-MARKOV SYSTEMS

Analytical formulas are available for a few situations that do not require the Markov assumption for either the input or service process. These formulas are useful because, in practical instances, interarrival and service time distributions may be not be reasonably approximated by exponential distributions. The formulas also illustrate the effects that different probability distributions have on the system measures.

M/G/1 System

One very useful result available for single-server systems is called the Pollaczek-Khintchine (P-K) formula. For the results to be valid, service times may follow a general distribution with mean m_s and standard deviation σ_s, but interarrival times are still required to be exponentially distributed. In terms of the service rate, $m_s = 1/\mu$. The top portion of Figure 16.11 illustrates the components of the system. To ensure stability in steady state, the traffic intensity ρ must be less than 1.

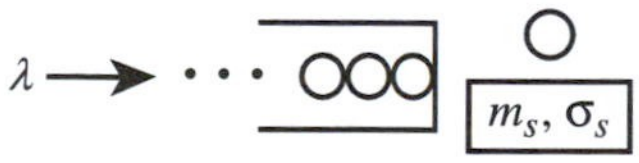

$$\lambda \rightarrow \cdots \; \fbox{OOO} \; \underset{m_s, \sigma_s}{\fbox{$\bigcirc$}}$$

$$m_s = 1/\mu, \; \sigma_s \text{ given}$$

$$\rho = \lambda/\mu = \lambda m_s$$

$$\pi_0 = 1 - \rho$$

$$L_q = \frac{\lambda^2(\sigma_s)^2 + \rho^2}{2(1-\rho)}$$

$$W_q = \left[\frac{1 + (\sigma_s/m_s)^2}{2}\right]\left[\frac{m_s\rho}{(1-\rho)}\right]$$

$$L_s = \lambda/\mu, \; L = L_q + L_s, \; W = L/\lambda$$

Figure 16.11 $M/G/1$ system ($\rho < 1$)

The P-K formula is associated with L_q and is shown in Figure 16.11. Once L_q has been determined, Little's law can be used to find W_q, although in Figure 16.11 it is written as a function of the coefficient of variation, σ_s / μ_s. The other measures are computed in the usual way.

The formula for L_q illustrates the effect of variability of service times on queue lengths. For the exponential distribution, for example, $\sigma_s = m_s$, whereas for constant service time (same for each customer), $\sigma_s = 0$. The expected numbers of customers in the queue for the two cases are as follows.

$$\text{Exponential service time distribution: } L_q = \frac{2\rho^2}{2(1-\rho)}$$

$$\text{Constant service distribution: } L_q = \frac{\rho^2}{2(1-\rho)}$$

Thus, the average queue length doubles when the standard deviation of the service time increases from 0 to m_s.

Value of Reduction in Service Process Variability

Stocking personnel in a warehouse get calls for orders at an average rate of 8.5 per hour. The mean time to fill an order is 6 minutes or 0.1 hour. For modeling purposes, assume that both times are exponentially distributed. This leads to an $M/M/1$ system with an average time in the queue, W_q, of 34 min or 0.567 hour. An opportunity arises to reduce the variability of the process of filling orders by updating the inventory control software. The warehouse manager wonders if the updating would be worth the cost.

Part of the answer depends on how the benefits of reducing variability would be used and how the overall performance of the system would change. In this case, the manager wants to increase the throughput rate of the system without reducing the average waiting time in the queue, W_q. In Table 16.8, we compare the current system with two $M/G/1$ systems whose variabilities, as measured by σ_s, are reduced significantly. For each case, the average time to fill an order is held fixed at $m_s = 0.1$ hour. Using the equation for W_q in Figure 16.11 with $W_q = 0.567$, we solve for ρ and then determine the corresponding arrival rate λ, the throughput of the system. It is clear from the results that throughput increases when variability is reduced. Notice also that the queue length L_q increases as a result of the greater throughput rate, but, by design, the wait is the same in all cases.

Table 16.8   Comparison of Systems with Different Service Variances

Measure	Case 1	Case 2	Case 3	
Model	$M/M/1$	$M/G/1$	$M/G/1$	
W_q	0.567	0.567	0.567	} Fixed values
m_s	0.1	0.1	0.1	
σ_s	0.1	0.05	0	} Parameter
ρ	0.85	0.9007	0.9189	
λ	8.500	9.007	9.189	} Computed values
L_q	4.817	5.104	5.207	

Non-Poisson System Approximations, *GI/G/*1

We now consider a single-server system characterized by general input and service processes. That is, neither the interarrival time distribution nor the service time distribution need be exponential. It will be assumed, however, that the means and variances of both these distributions are known, and that the random variables associated with arrivals are independent and identically distributed (i.i.d.). Such a system is denoted by *GI/G/*1, where *GI* stands for "general independent" and *G* stands for "general" although the service time random variables are also assumed to be i.i.d. The components of the system are shown in the top portion of Figure 16.12.

As we have stated, the coefficient of variation of a probability distribution is the ratio of its standard deviation to its mean—i.e., $c = \sigma/m$. The value of c is small when the standard deviation is small relative to the mean. In the extreme when $\sigma = 0$, there is no variability, so $c = 0$. This would be the case for a deterministic process. For the exponential distribution, the mean and standard deviation are equal, so the coefficient of variation is 1.

Letting the subscripts a and s refer to the arrival and service processes, respectively, an approximation for the mean time in the queue for a general single-server system is

$$W_q(GI/G/1) \cong \left[\frac{c_a^2 + c_s^2}{2}\right] W_q(M/M/1) \tag{2}$$

In this equation, c_a and c_s are coefficients of variation, as defined in Figure 16.12, and $W_q(M/M/1)$ denotes the mean waiting time for an *M/M/*1 system. Explicit formulas for approximations of W_q and L_q are also given in Figure 16.12. Their derivation is credited to Ward Whitt, whose 1983 paper provides a comprehensive survey of the subject.

Equation (2) implies that W_q for the general system is approximately equal to W_q for the Poisson system multiplied by a factor that depends on the coefficients of variation of the arrival and service processes squared. For Poisson arrivals and service, this factor equals 1, as expected. The implication of the formula is that the time in the queue is an increasing function of the variability of both the arrival and service processes. For a Poisson arrival process, $c_a = 1$, the approximation provides the same result as the P-K formula.

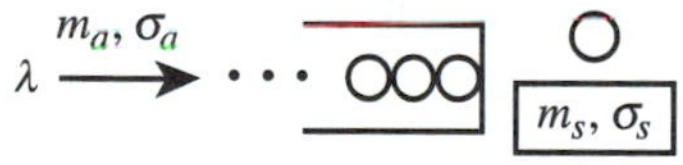

$$\lambda \xrightarrow{m_a, \sigma_a} \cdots \boxed{OOO} \; \underset{\boxed{m_s, \sigma_s}}{O}$$

$$m_s = 1/\mu, \; \sigma_s \text{ given}$$
$$m_a = 1/\lambda, \; \sigma_a \text{ given}$$
$$\rho = \lambda/\mu$$

$$c_a = \frac{\sigma_a}{m_a}, \quad c_s = \frac{\sigma_s}{m_s}$$

$$\pi_0 = 1 - \rho$$

$$L_q \cong \left[\frac{c_a^2 + c_s^2}{2}\right]\left[\frac{\rho^2}{(1-\rho)}\right]$$

$$W_q \cong \left[\frac{c_a^2 + c_s^2}{2}\right]\left[\frac{m_s\rho}{(1-\rho)}\right]$$

$$L_s = m_s$$
$$L = L_q + L_s, \; W = L/\lambda$$

Figure 16.12   *GI/G/*1 system $(\rho < 1)$

Value of Reduction in Arrival Process Variability

The inventory manager is impressed with the results of the reduction in the variability of the service process, so he wants to consider the effects of reducing the variability of the arrival process. He believes that it is possible to schedule orders so that the arrival process has a coefficient of variation less than 1.

Using the approximation equation for W_q in Figure 16.12, with $W_q = 0.567$ and $m_s = 0.1$, we solve for ρ for several values of c_a and c_s and then determine the corresponding arrival rate λ (and hence m_a) as a function of these parameters. The variance associated with arrivals, σ_a, follows from the definition of c_a.

Table 16.9 shows the effect of reducing the values of c_a and c_s. Case 1 refers to the baseline *M/M/*1 system whereas cases 2 and 3 are for *GI/G/*1 models. As the coefficients of variation of both service times and interarrival times are reduced, the throughput λ approaches its theoretical maximum of 10. Recall that the arrival rate cannot exceed the service rate if the system is to remain stable. At the point where the arrival rate is exactly equal to the service rate, the Poisson formulas fail, but with zero variability the manager should be able to schedule arrivals so there is no wait in the queue.

Multichannel Approximation, *GI/G/s*

The equivalent of Equation (2) for a multichannel system with s servers is

$$W_q(GI \,/\, G \,/\, s) \cong \left[\frac{c_a^2 + c_s^2}{2}\right] W_q(M \,/\, M \,/\, s)$$

Once again, no assumptions are made about the input and service processes other than the fact that their means and variances are known and the corresponding random variables are i.i.d. The full set of approximations for a *GI/G/s* system is given in Figure 16.13. Their derivation is based on the formulas associated with a multichannel Poisson system. Note, however, that the computed value of π_0 does not approximate the probability that the system is empty, but is simply one of the components needed for the calculation of W_q.

An additional approximation has proven quite useful when analyzing a series of queuing stations that might arise in a manufacturing system or transportation network. In par-

Table 16.9 Comparison of Systems with Different Arrival Variances

Measure	Case			
	1	2	3	
Model	*M/M/*1	*GI/G/*1	*GI/G/*1	
W_q	0.567	0.567	0.567	Fixed values
m_s	0.1	0.1	0.1	
c_a	1	0.5	0.1	
c_s	1	0.5	0.1	Parameters
σ_s	0.1	0.05	0.01	
ρ	0.85	0.9577	0.9982	
m_a	0.118	0.104	0.100	
σ_a	0.118	0.052	0.010	Computed values
λ	8.500	9.577	9.982	
L_q	4.817	5.427	5.662	

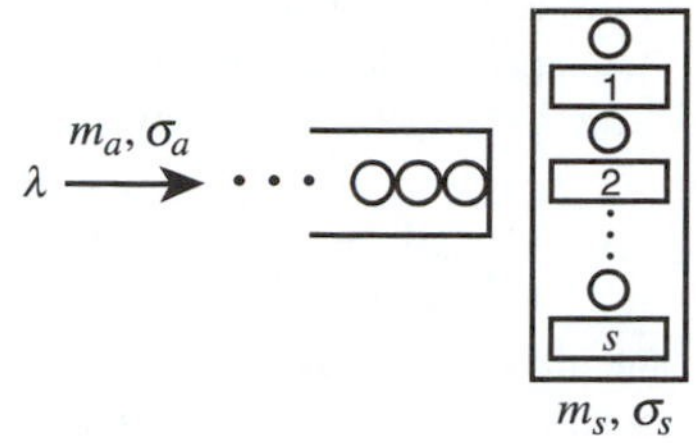

$$m_s = 1/\mu, \ \sigma_s \text{ given}$$

$$m_a = 1/\lambda, \ \sigma_a \text{ given}$$

$$\rho = \lambda/s\mu$$

$$c_a = \frac{\sigma_a}{m_a}, \quad c_s = \frac{\sigma_s}{m_s}$$

$$\pi_0 = 1 \Big/ \left[\sum_{n=0}^{s-1} \frac{(s\rho)^n}{n!} + \frac{(s\rho)^s}{s!(1-\rho)} \right]$$

$$W_q \cong \left[\frac{c_a^2 + c_s^2}{2} \right] \frac{\pi_0 m_s s^s \rho^s}{s!(1-\rho)^2}$$

$$L_q = \lambda W_q = \left[\frac{c_a^2 + c_s^2}{2} \right] \frac{\pi_0 s^s \rho^{s+1}}{s!(1-\rho)^2}$$

$$L_s = \lambda m_s$$

$$E = \rho$$

Figure 16.13 *GI/G/s* queuing system

ticular, it is possible to get a reliable estimate of the coefficient of variation of the output of a queuing station as a function of its input and service processes. Letting $c_d = \sigma_d/m_d$ be the coefficient of variation of the departure process, we have the following approximation.

$$c_d^2 \cong \rho^2 c_s^2 + (1-\rho^2) c_a^2$$

For a stable system, customers leave the system at the same rate at which they enter, so $m_d = m_a$ and the expression for c_d^2 can be used to determine σ_d. (Note that, in general, m_d is not the same as m_s.)

The output coefficient of variation, c_d, depends on c_a and c_s as well as on the value of ρ. When ρ is small, the system is likely to be idle much of the time, so the value of c_d is primarily a function of the arrival process. When ρ is close to 1, the system is busy and the value of c_d depends primarily on the service process. This is intuitively satisfying. For intermediate values of ρ, the results are surely approximate. The formulas in Figure 16.13, however, extend considerably the range of systems for which analytical solutions can be found. We will illustrate the use of the non-Poisson approximations when we discuss queuing networks in the next chapter.

EXERCISES

1. For each situation, identify the type of queuing system described, the arrival rate, and the service rate. Use the relationships given in Sections 16.1 to 16.3 to compute the system performance measures.

   (a) The time between arrivals of patients at a medical clinic has a normal distribution with a mean of 12 minutes and a standard deviation of 5 minutes. Three doctors treat the patients. The treatment time is exponentially distributed with a mean of 30 minutes. The secretary who has been collecting the relevant statistics has found that the average time a patient must wait before being seen by a doctor is 20 minutes.

(b) Cars arrive at a stop sign on a two-lane road at an average rate of 220 per hour. The arrival follows a Poisson process (i.e., the time between arrivals has an exponential distribution). The time it takes a car to clear the stop sign once it reaches the front of one of the lanes also has an exponential distribution with a mean of 10 seconds. The average number of cars at the stop sign (either waiting in the queue or preparing to cross) is 8.4.

(c) The situation is the same as in part (b) except that whenever there are six cars at the corner, any additional arriving traffic takes a convenient side street and bypasses the stop sign.

(d) An overseas airline counter has five open positions. The average time required for check-in of an international passenger is 8 minutes and is normally distributed, with a standard deviation of 2 minutes. The time between arrivals has an exponential distribution with a mean rate of 0.6 passenger per minute. The average number of passengers either being served or waiting in line is 13.

(e) Four robots have been installed in a work cell to manufacture parts. Each robot takes exactly 4 minutes to make one part, whose average demand is 50 per hour. The time between orders for a part has an exponential distribution. The average number of parts in the queue is 4.1.

2. Three drive-in windows have been installed at a branch office of a bank. Besides the spaces adjacent to the windows, there is room for three additional cars to wait. Arriving cars balk when there is no waiting space available and do not join the queue. Work sampling has established the following percentages.

Number of busy windows	0	1	2	3
Percent of the time	10	15	20	55
Number of waiting cars	0	1	2	3
Percent of the time	60	20	15	5

What type of queuing model would be appropriate for this system, and what performance measures can you compute? Show your computations.

3. A finite queue has three servers and two additional spaces for waiting customers. There is an infinite input source, and the arrival rate is 50/hour. When arrivals find the queue full, they balk. The following steady-state probabilities have been determined for the number of customers in the system.

$$\pi_0 = 0.05, \ \pi_1 = 0.12, \ \pi_2 = 0.23, \ \pi_3 = 0.25, \ \pi_4 = 0.20, \ \pi_5 = 0.15$$

Find each of the following values.

(a) Expected number of customers in the queue
(b) Expected time in the queue
(c) Expected number of customers in the system
(d) Expected time in the system
(e) Expected number of customers in service
(f) Expected time in service
(g) Proportion of customers who balk
(h) Percent utilization of the servers
(i) Probability that there will be fewer than two in the system
(j) Probability that there will be four or more in the system
(k) Probability that an arriving customer will obtain immediate service (rather than having to wait in a queue)

4. Let π_0, π_1, π_2, . . . stand for the steady-state probabilities associated with a queuing system. In general, π_n is the probability that the system is in state n. In each case, answer the question in terms of these probabilities and any other symbols needed. Also indicate what kind of queuing model fits the situation described with respect to the five-field notation. All terms introduced in your answer should be explained—that is, you should explain how the calculations would be performed if more data were available.

   (a) The system has an infinite queue with three servers. What is the probability that an arriving customer will spend no time in the queue?

   (b) The system has a finite queue with two servers and three spaces in the queue. Customers finding the queue full will balk. The arrival rate is 20 per hour. What is the average time spent in the queue?

   (c) The queuing system represents a repair shop for a taxicab company. The shop services only the five cabs owned by the company, and there is a single mechanic who does the work. The state of the system represents the number of cabs in the shop. What is the average number of cabs not in the repair shop?

5. A multichannel queuing system has four servers. The input source is infinite and the average arrival rate is three customers per hour. The mean time required to complete a service operation is 1 hour. Both arrival and service processes are Poisson processes. There is no limit on the number in the queue.

   (a) What is the probability that more than one customer will arrive in a 15-minute period?

   (b) What is the probability that the next service operation will require less than 1 hour?

   (c) What is the probability that each of the next two service operations will require more than 30 minutes?

   (d) Say you are a customer in this system. You have waited for 1 hour in the queue and you have been in service for 25 minutes. How long should you expect to remain in the system before your service is completed?

   (e) What is the expected number of arrivals in an 8-hour period?

   (f) An individual customer arrives at the system and finds one other person in the queue. If the queue discipline is first-come-first-served, how long should the customer expect to spend in the system?

   (g) Assuming that all servers remain busy, what is the probability that exactly four customers will be served in the next hour?

   (h) Assuming that all servers remain busy, what is the expected number of customers served in a 4-hour period?

6. For the system described in Exercise 5, find the following values.

   (a) Expected number of customers in the queue

   (b) Expected time in the queue

   (c) Expected number of customers in the system

   (d) Expected time in the system

   (e) Expected number of customers in service

   (f) Expected time in service

   (g) Proportion of customers who balk

   (h) Percent utilization of the servers

   (i) Probability that there will be fewer than two in the system

   (j) Probability that there will be four or more in the system

   (k) Probability that an arriving customer will obtain immediate service (rather than having to wait in a queue)

7. Customers arrive at random at a self-service gasoline station at an average rate of 12 per hour. The station has a single pump. The time spent by a motorist at the pump has an exponential distribution with a mean of 3 minutes. In parts (a) to (e), assume that there is no limit on the number of cars that can wait.

   (a) What proportion of the time is the pump busy?

   (b) How many spaces for waiting cars should be made available on the station property to ensure that there is sufficient room to wait 85% of the time?

   (c) What is the average time required for a customer to get gas?

   (d) If the arrival rate changes to 50 per hour, what is the minimum number of pumps required to handle the load? Answer parts (a) to (c) for this number of pumps.

   (e) Repeat part (d) for one more than the minimum number of pumps.

   (f) Analyze this system if the arrival rate is 45 per hour and four pumps are used. Assume that only four cars can wait in the queue. How many customers will be lost because of this limit?

   (g) Analyze the system described in part (f) given that the service time is normally distributed. This will make it more uniform. Specifically, assume that the average service time is still 3 minutes but that the standard deviation is 1 minute.

8. Carver Interior Designs plans to advertise a new product and to take orders by phone. Currently, Carver, Inc. has two phone lines. Using data collected during a dry run of the system, it is determined that the average time to take a customer's order is 1.5 minutes. If the advertising campaign has the desired effect, orders will arrive at a rate of 2 per minute on average. If a customer receives a busy signal, the order is assumed to be lost. For each case, assume that the arrival and service processes are Poisson and compute the average time in the system, the percent utilization of the phones, and the proportion of customers lost.

   (a) Leave the phone system as it is.

   (b) Install an additional phone with an operator.

   (c) Have three phones with operators and one holding line.

   (d) Have three phones but add a call holding facility that can handle any number of waiting customers.

   (e) Have four phones but add a call holding facility that can handle any number of waiting customers.

9. A microcomputer laboratory in a mechanical engineering department has six Super-X computers for use by students. During the busy time of the day, students arrive at the lab according to a Poisson process at an average rate of 12 per hour. Unfortunately, many times a student finds all the computers busy. In such circumstances, she will not wait but will rush off to one of the other computer labs on campus. The average time a student spends on a computer is 1/2 hour, but the time is actually exponentially distributed.

   (a) Sketch the rate network for this situation.

   (b) What is the average number of students waiting?

   (c) What proportion of the arriving students will be served by this computer lab?

   (d) What is the utilization of the computers in the lab—i.e., what proportion of the time are the computers in use?

10. A family with four children has a single bathroom. Each morning during a 1-hour period before school, there is a congestion problem related to its use. The parents minimize the problem by remaining in bed. While lying in bed, the mother, who is studying queuing theory, makes the following observations. The requirement to use the bathroom seems to be a random process. The time between the moment when one child leaves the bathroom and the moment when the same child returns averages 15 minutes, but actually the time is a random variable with an exponential distribution. All children are assumed to have the same bathroom needs and hence the same arrival rate. The time spent in the bathroom in any one instance is also exponentially distributed, with a mean of 5 minutes.

A child arriving at the bathroom and finding it occupied will wait in a queue, all the while taunting the person inside in a loud voice and thus disturbing the parents.

(a) Although this is clearly a transient situation, determine the steady-state probabilities associated with the number in the system.

(b) When there is no queue, the house is relatively calm. Using the results from part (a), how many minutes in the 1-hour period should the parents expect to have peace and quiet?

11. Faculty members at Big State University must submit annual reports to their department chairpersons showing their accomplishments during the previous year. The reports are usually handwritten and must be retyped by a secretary in the format required by the dean. During the late spring, the reports arrive at the chairpersons' offices at an average rate of one per day; however, the actual arrival process is Poisson. The time required to retype the information is about 1.5 days but in fact is highly variable owing to the professors' bad writing and different levels of accomplishment. For purposes of analysis, assume that this time is exponentially distributed. In order to finish the reports promptly, two secretaries are assigned full time for this job.

The associate chairperson of the industrial engineering department is proposing a different procedure that will require all professors to enter their data on printed forms. The secretary will then enter the data into a personal computer. It is estimated that this procedure will not change the arrival process, but the time required to enter the data into the computer and print out the report will now be normally distributed, with a mean of 2/3 day and a standard deviation of 1/3 day. Under this proposal, only one secretary will be required.

Compare the present method with the proposed method in terms of the utilization of the secretaries and the mean interval between the time when the professor submits the report and the time when the secretary finishes it.

12. A surveillance system requires three working radar units to scan the horizon. When three or more units are up, the system is said to be in the "operational" mode. If only two are working, it is in the "reduced effectiveness" mode. When fewer than two are working, the system is in the "failed" mode. In order to maintain effectiveness, five radar units have been installed, two of which are redundant if all are in good repair.

Each unit fails randomly and at the same average rate—once every 500 hours. When a failure occurs, the unit is sent to a repair shop where one unit is worked on at a time. If more than one unit has failed, the others must wait in a queue. The repair operation takes an average of 40 hours, but the actual repair time has an exponential distribution. Compute the proportion of time that the system will be in each of the three modes.

13. A Broadway theater box office receives telephone calls at the average rate of one call every 10 seconds. Each call takes an average of 3 minutes to complete. If an operator is free, the call is taken immediately; otherwise, it is placed on hold if a holding line is available. Both the arrival and service processes are Poisson.

The theater wishes to design the system so that the average time a customer is on hold is no more than 1 minute and that no more than 1% of the calls are lost. Determine the minimum numbers of operators and holding lines that are required to satisfy this service standard.

14. The schematic below represents an inspection area for finished products that arrive at the average rate of 12 per hour according to a Poisson process. The mean inspection time is 12 minutes, and the variance is 360 minutes squared.

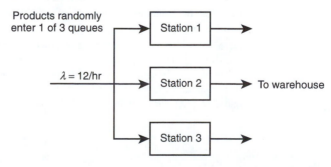

(a) Determine the expected waiting time for a product, the expected length of each queue, and the expected number of products in the *overall* system.

(b) Same as part (a) except for a variance of 36 minutes squared.

(c) Same as part (a) for an *M/D/*1 model with a 12-minute inspection time. Does the variance effect appear significant?

15. A power plant operating 24 hours per day has four identical turbine-generators, each capable of producing 3 megawatts of power. The demand at any time is 6 megawatts, so that when all four turbines are in working condition, one is kept on "warm-standby," one on "cold-standby," and the other two in full operation. If one turbine is down, then two are operating and one is on warm-standby. If two are down, both working turbines are operating. If only one turbine is working, then the company must purchase 3 megawatts of power from another source. If all turbines are down, the company must purchase 6 megawatts.

    If a turbine is in the operating mode, its mean time to failure is 3 weeks. If a turbine is in warm-standby, its mean time to failure is 9 weeks. The company has two technicians that can repair failed turbines, but it only takes one technician to affect a repair. On average, a repair can be done in half a week. If a turbine is in cold-standby, it cannot fail. (We assume that all switchovers from warm-standby to working or from cold-standby to warm-standby are instantaneous.) The company has two technicians who can repair failed turbines, but it takes only one technician to effect a repair. Assuming that all times are exponentially distributed, determine the expected megawatt hours that must be purchased each year. (*Hint:* use the birth-death equations to determine steady-state probabilities.)

16. Martha Stark, M.D. runs a family medical practice in a suburb of Boston. Although she schedules patients every half-hour, she has noticed that many come a bit late and a handful arrive much earlier than necessary. In the last few months, she has also received several subtle complaints about the waiting time.

    Having an undergraduate degree in statistics, Dr. Stark decided to collect data on arrival and service times in an effort to characterize the corresponding distributions. Using a Chi-square goodness-of-fit test, she was able to confirm that the interarrival times were exponentially distributed with a mean of 30 minutes. Unfortunately, no closed-form distribution provided a good fit for the time she spends with patients diagnosing and treating their illnesses. The only thing that she can say for sure is that the sample mean is 25 minutes and the sample standard deviation is 10 minutes.

    (a) Assuming steady-state conditions, determine the amount of time a patient can expect to wait before seeing Dr. Stark and the average number of patients in the office at any given time.

    (b) Develop a simulation model that can be used to verify your results in part (a). Run the model to determine the probability distribution for the number of patients in the office. If the waiting room has two chairs, what is the probability that an arriving patient will have to stand?

17. Simulate an *M/M/*1/4 queuing system with a mean arrival rate of eight customers per hour and a mean service rate of six customers per hour, either by hand or (better), by using a spreadsheet program such as Excel. Run the simulation for the equivalent of about 10 hours. Start with the system empty and organize the results in a table. A separate random number must be used to generate each event.

    (a) How many customers are turned away in the first 2 hours?

    (b) Discard the output from the first hour. Now, using the output from the remaining 9 hours, calculate the following steady-state measures; L, L_q, W, W_q, π_0, and π_4.

    (c) Calculate the same measures listed in part (b) using the formulas in Figure 16.7. Compare the two sets of results.

       Note that one way to generate a realization of an exponential random variable is to use the inverse transformation technique. As described in Section 18.3, let λ be the parameter of the exponential random variable X with cumulative distribution function $F(X) = 1 - e^{-\lambda X}$. Also, let R be a uniform random variable defined on the interval [0, 1]. To obtain the required expression, solve the equation $F(X) = R$ for X in terms of R. This leads to

$$X = \frac{-1}{\lambda} \ln(1 - R)$$

Now, using a uniform random number generator, find a value R_i and plug it into the preceding expression to obtain the corresponding realization X_i for $i = 1, 2, \ldots$ Alternatively, a value of R_i can be obtained from a published table of uniform random numbers. For this application, it will be necessary to map each random number into $[0, 1]$ if it is generated over some other range.

18. Simulate a *GI/G/1/4* system for 10 hours. Let the interarrival times and service times have continuous uniform distributions between 0 and 15 minutes and between 0 and 20 minutes, respectively, and assume that the system starts empty.

    (a) How many customers are turned away in the first 2 hours?

    (b) Discard the output from the first hour. Now, using the output from the remaining 9 hours, calculate the following steady-state measures; L, L_q, W, W_q, π_0, and π_4.

    (c) Compare the results in part (b) with those obtained in part (b) of Exercise 17.

BIBLIOGRAPHY

Feldman, R.M. and C. Valdez-Flores, *Applied Probability & Stochastic Processes*, PWS, Boston, 1996.

Gross, D. and C.M. Harris, *Fundamentals of Queueing Theory*, Third Edition, Wiley, New York, 1998.

Jang, J., J. Suh, and C.R. Liu, "A New Procedure to Estimate Waiting Time in *GI/G/2* System by Server Observation," *Computers & Operations Research*, Vol. 28, No. 6, pp. 597–611, 2001.

Kao, E.P.C., *Introduction to Stochastic Processes*, Duxbury, Belmont, CA, 1997.

Kleinrock, L., *Queueing Systems*, Vol. I, Wiley, New York, 1975.

Larson, R.C., "The Queue Inference Engine: Deducing Queue Statistics from Transactional Data," *Management Science*, Vol. 36, No. 5, pp. 586–601, 1990.

Little, J.D.C., "A Proof for the Formula $L = \lambda W$," *Operations Research*, Vol. 9, pp. 383–387, 1961.

Mehdi, J., *Stochastic Models in Queueing Theory*, Academic Press, Boston, 1991.

Ross, S.M., *Introduction to Probability Models*, Fifth Edition, Academic Press, San Diego, 1993.

Wolff, R.W., *Stochastic Models and the Theory of Queues*, Prentice-Hall, Engelwood Cliffs, NJ, 1989.

Chapter 17

Queuing Networks and Decision Models

In many applications, rather than passing through a single queue, an arrival will pass through a series of queues arranged in a network structure. The flow may be serial, as in an assembly line, or it may be much more general, with feedback loops allowing for reentry as units flow from one station to the next. The first part of this chapter is devoted to investigation of network configurations in which each node is modeled as a queue. One example is the registration procedure at a college, where students visit several stations as required for approval of their current programs. Each station involves one or more servers and a queue of waiting students. Another example is a job shop in which machines are grouped by function. A collection of machines of the same type can be modeled as a queuing station. A job entering the shop is processed at one or more stations before it leaves. A job may visit a station more than once for processing or rework.

In the second part of this chapter we will consider various decision problems that arise in single queuing systems as well as networks of queues. In some cases the parameters of a queuing model may be variable and we may want to investigate alternative solutions using cost or some other measure of performance before fixing the design. The models presented in this chapter and in Chapters 11–16 allow the analyst to evaluate different solutions but not necessarily to determine optimal parameter values. These models are descriptive rather than prescriptive in nature and so reveal the implications of a decision rather than providing the optimal decision. As a consequence, some trial and error is usually required to arrive at an optimal design.

17.1 JACKSON NETWORKS

The analysis of networks of queues can be quite difficult unless they have some special structure. In this section we investigate such a case, in which all external arrivals are Poisson processes and each queue within the network has unlimited capacity and exponential service times. This type of system is called a *Jackson network*, and some of its steady-state behavior can be described using the same formulas developed for single station Markovian queues. As we have seen, these formulas are very easy to apply and yield numerical results that are critical to the design of a system.

When all the stations in a network can be modeled as *M/M/s* queues, we have the very important result that each station can be analyzed independently using the formulas given in Section 16.2. This was proved by J. R. Jackson [1957], thus giving rise to the name. In such networks, a station can receive input from external sources, from other stations in the network, or even from its own output. The last case is called *rework, feedback,* or *reentrant flow,* depending on the application.

Definition 1: A network of queues is called a Jackson network if the following conditions are satisfied.

1. All outside arrivals at each queuing station in the network must follow a Poisson process.

2. All service times must be exponentially distributed.

3. All queues must have unlimited capacity.

4. When a job leaves one station, the probability that it will go to another station is independent of its past history and of the location of any other job.

The first three conditions parallel the standard assumptions associated with an *M/M/s* queue. The fourth condition is not quite as straightforward, but nevertheless can be viewed as a Markovian assumption. Its effect is that the movement of units through the network can be described as a Markov chain. Note that the fourth condition rules out real-time decision making. For example, if a job in a machine shop can be processed at one of three stations after its first operation, it would not necessarily be sent to the station with the shortest queue. Rather, the station assignment would be determined by random selection based on some probability distribution.

To illustrate some of the more general ideas, consider a system with v parallel processing stations that feed a final packaging station j, as shown in Figure 17.1. For the moment, we assume that each of the v stations receives input from an external source only and that a fraction of its output goes to station j while the remainder leaves the system—that is, there is no flow from station i to station k ($k = 1, \ldots, v$ and $k \neq j$). Station j receives inputs from outside the network at a rate of γ_j, as well as from the v processing stations with probability ϕ_{ij}, $i = 1, \ldots, v$. Now, if the flow rate through station i is λ_i, then the flow rate into station j is the sum of all these sources.

$$\lambda_j = \gamma_j + \sum_{i=1}^{v} \phi_{ij} \lambda_i \tag{1}$$

We assume that all the external arrival processes and all the service processes are Poisson. The critical characteristic of a Jackson network, as relevant for this example, is that the station can be analyzed as if the interarrival times for station j had an exponential distribution with parameter given by Equation (1). Note that the interarrival times at station j do not necessarily have an exponential distribution, but the station can be analyzed as if they did.

The implication of this assertion is that the steady-state probability distribution for the number at each station follows the formulas for the *M/M/s* model. Coupled with Little's law, this allows us to analyze the two-stage queuing network in Figure 17.1 and, by extension, any open queuing network (customers are allowed to enter and leave the system) of arbitrary topology. It should be stressed that such an analysis is appropriate only for queues with unlimited capacity, Poisson input and service processes, and independent transfer (switching) probabilities ϕ_{ij}. Under these conditions, the steady-state results are valid.

Because the *M/M/s* formulas can be used to compute the station statistics, the only difficult part of the analysis is determining the input rate to each queuing station. As in Figure 17.1, let λ_i denote the total input to station i, and observe that for a system in steady state, the input must equal the output and thus the total output from station i is λ_i. Next observe that the input to any station i must equal the input from outside the system plus

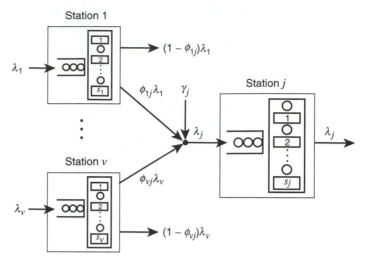

Figure 17.1   Two-stage queuing network

any output from the other stations routed to i. Thus, for an arbitrary network with m stations, we have the general relationship

$$\lambda_i = \gamma_i + \sum_{k=1}^{m} \phi_{ki}\lambda_k, \quad i = 1, \ldots, m \tag{2}$$

where γ_i is the rate of arrival at station i from outside the network and ϕ_{ki} is the probability that output from station k will be routed to station i. The system of m linear equations given by Equation (2) can be solved to determine the net input rate λ_i for each station. This result is stated formally as follows.

Property 1: Let $\boldsymbol{\Phi}$ be the $m \times m$ probability matrix that describes the routing of units within a Jackson network, and let γ_i denote the mean arrival rate of units going directly to station i from outside the system. Then,

$$\boldsymbol{\lambda} = \boldsymbol{\gamma}(\mathbf{I} - \boldsymbol{\Phi})^{-1}$$

where $\boldsymbol{\gamma} = (\gamma_1, \ldots, \gamma_m)$ and the components of the vector $\boldsymbol{\lambda}$ denote the arrival rates into the various stations—that is, λ_i is the net rate into station i.

Note that, unlike the state-transition matrix used for Markov chains, the rows of the $\boldsymbol{\Phi}$ matrix need not sum to 1; that is, $\Sigma_i\,\phi_{ki} \leq 1$. This follows because it is not necessary to account for the flow that leaves the system when we compute $\boldsymbol{\lambda}$.

After the net rate into each node is known, the network can be decomposed and each node treated as if it were an independent queuing system with Poisson input. This is in spite of the fact that if there is any feedback within the network, not only will the input streams to the nodes be dependent on each other, but the input processes will not be Poisson. These complications, however, do not affect the analysis described in the remainder of this section, because the stationary distributions at each node are the same as they would be if each were an independent $M/M/s$ queue.

Property 2: Consider a Jackson network comprising m nodes. Let N_i denote a random variable indicating the number of jobs at node i (the number in the queue plus the number in service). Then,

$$\Pr\{N_1 = n_1, \ldots, N_m = n_m\} = \Pr\{N_1 = n_1\} \times \cdots \times \Pr\{N_m = n_m\}$$

and the probabilities $\Pr\{N_i = n_i\}$ for $n_i = 0, 1, \ldots$ can be calculated using the equations in Section 16.2 for π_i, $i = 0, 1, \ldots$ associated with the *M/M/s* queue.

17.2   NETWORK EXAMPLES

As we have described, a Jackson network consists of a collection of connected *M/M/s* queues with known parameters. Flow enters the network at one or more of the stations, leaves at others, and may pass from one station to another in accordance with known probabilities. Whenever there are two or more flow paths leaving a station, the path taken by an entity is randomly determined from these distributions. Several examples will now be given to illustrate how the computations are performed and what kinds of results can be expected.

Computer Center

All batch jobs submitted to a computer center must first pass through an input processor before moving on to the central processor station where the bulk of the work is performed. Because of errors, only 80% of the jobs go through the central processor; the remaining 20% are rejected. Of the jobs that pass through the central processor successfully, 40% are routed to a printer station where a hard copy is produced.

Jobs arrive randomly at the computer center at an average rate of 10 per minute. To handle the load, each station may have several processors operating in parallel. The times for the three steps have exponential distributions with means as follows: 10 seconds for an input processor, 5 seconds for a central processor, and 70 seconds for a printer. When all the processors at a station are in use, an arriving job must wait in a queue. All queues are assumed to have unlimited capacity. Our goal is to find the minimum number of processors of each type and compute the average time required for a job to pass through the system.

This is a queuing network with three stations in series comprising, respectively, input processors, central processors, and printers. We first compute the rates into each processor. Writing Equation (2) for $m = 3$, we get the following linear system.

$$\lambda_1 = \gamma_1 + \phi_{11}\lambda_1 + \phi_{21}\lambda_2 + \phi_{31}\lambda_3$$
$$\lambda_2 = \gamma_2 + \phi_{12}\lambda_1 + \phi_{22}\lambda_2 + \phi_{32}\lambda_3$$
$$\lambda_3 = \gamma_3 + \phi_{13}\lambda_1 + \phi_{23}\lambda_2 + \phi_{33}\lambda_3$$

The only nonzero parameters are $\gamma_1 = 10$ per minute, $\phi_{12} = 0.8$, and $\phi_{23} = 0.4$, so it is an easy matter to compute the arrival rates λ_i. Their values are given in Table 17.1.

The next step is to find the minimum number of channels at each station i that will ensure that $\rho_i < 1$, the requirement for stability. Recalling that $\rho = \lambda/s\mu$, we set s_i equal to the smallest integer greater than λ_i/μ_i. The results are given in Table 17.1.

With the processors modeled as independent *M/M/s* systems, each is analyzed using the formulas in Section 16.2 to determine steady-state characteristics. Table 17.2 presents the important system measures at each station and in total. Since a completed job must pass through all three stations, the expected time that a job will spend waiting in queues is 4.169 minutes and the average time it will spend in processing is 1.384 minutes. Summing these two values, we get the total time in the system—5.553 minutes.

From Table 17.2, we can see that the printer station is the bottleneck, because it has the longest queues and waiting times. Improving this component of the system by obtaining

Table 17.1   Data for the Computer Center

System measure	Input processor	Central processor	Printer
External arrival rate, γ_i	10/min	0	0
Total arrival rate, λ_i	10/min	8/min	3.2/min
Service rate, μ_i	6/min	20/min	0.857/min
Minimum channels, s_i	2	1	4
Traffic intensity, ρ_i	0.833	0.400	0.933

Table 17.2   Results for the Various Stations in the Computer Center

System measure	Input processor	Central processor	Printer station	Total
Model	$M/M/2$	$M/M/1$	$M/M/4$	
L_q	3.788	0.267	12.023	16.077
W_q	0.379	0.033	3.757	4.169
L_s	1.667	0.400	3.734	5.801
W_s	0.167	0.050	1.167	1.384

another printer (increasing s_3) or by increasing the printing rate (increasing μ_3) will reduce the queuing time and the total system time.

Job Shop

An electronics company assembles three classes of products in a job shop environment that is represented by the directed network shown in Figure 17.2. The rectangles identify workstations that comprise a specific machine type. The other data in the rectangles indicate the number of parallel machines (s) at the station and the average service rate (μ) of each. The order rate for each class of products along with the routing information is given in Table 17.3. For example, products in class 1 must visit machines A, B, D, and F in that order. It will be assumed that orders for the products arrive according to Poisson processes characterized by the given rates, and that processing times at the machines are exponentially distributed.

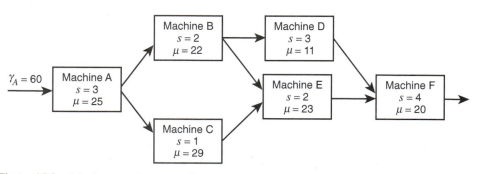

Figure 17.2   Job shop queuing network

Table 17.3   Data for the Job Shop Example

Product class	Order rate	Route
1	30/month	A B D F
2	10/month	A B E F
3	20/month	A C E F

Table 17.4 gives the workstation parameters, the arrival rates at each station derived from Equation (2), and related performance measures obtained from the $M/M/s$ formulas. When evaluating the results, it is most revealing to look at the L_q row. It is apparent that the stations with the largest average queues are B and D. This is also indicated by the high traffic intensities ρ for these stations. Although both stations have the same value of ρ, they are not of equal importance; station B has a larger queue than station D. If the products are equally profitable, then station B is the most critical in the network. Regarding the average number of products in service, we note that $L_s = \lambda/\mu$ and so is unaffected by s.

For problems of this type, the number of units in the system constitutes the work-in-process (WIP) inventory. It is the goal of most manufacturers to reduce WIP, since it represents a significant capital investment. It is also interesting to look at the average time through the system (the manufacturing lead time) for each product. This information is given in Table 17.5. The times were determined by summing the times that each product spends at the various stations. The WIP was determined by Little's law, multiplying the lead time by the order rate.

Note that in Table 17.4 stations B and D have the highest utilization (0.909) and the greatest time in queue. As shown in Table 17.5, because product 1 passes through both station B and station D, it has a much higher lead time than the other two products. It also has a higher order rate. Because the WIP is determined by multiplying the order rate (flow rate) by the lead time (residence time), its value for product 1 is much higher than for the other two products.

Printed Circuit Board Assembly

Facilities for populating printed circuit boards are generally configured as flow lines. A simplified version of a printed circuit board facility is shown in Figure 17.3. The first station

Table 17.4   Results for the Job Shop Example

Measure	Machine					
	A	B	C	D	E	F
Model	$M/M/3$	$M/M/2$	$M/M/1$	$M/M/3$	$M/M/2$	$M/M/4$
γ	60	0	0	0	0	0
μ	25	22	29	11	23	20
s	3	2	1	3	2	4
λ	60	40	20	30	30	60
ρ	0.800	0.909	0.690	0.909	0.652	0.750
L	4.989	10.476	2.222	11.059	2.270	4.528
W	0.083	0.262	0.111	0.369	0.076	0.075
L_q	2.589	8.658	1.533	8.332	0.965	1.528
W_q	0.043	0.216	0.077	0.278	0.032	0.025

Table 17.5   System Performance Measures for the Job Shop Example

Product	Order rate (per month)	Route	Lead time (months)	Queue time (months)	WIP (units)
1	30	A B D F	0.789	0.563	23.67
2	10	A B E F	0.496	0.317	4.96
3	20	A C E F	0.345	0.177	6.91

is where raw boards are inspected and marked and component kits are prepared. Let us assume that the orders arrive at the rate of 60 boards per hour and that, on average, 80 boards can be processed per hour. A certain fraction, 5%, must be recycled at this stage because of defects, whereas 75% move on to the surface mount technology (SMT) center and the remaining 20% go to the through-hole center. Station 2 consists of three SMT robotic placement machines that can each process an average of 30 boards per hour. It receives input from both the preparation area and outside customers. Station 3 has a single through-hole (T-H) machine that operates at an average rate of 20 boards per hour.

After the small components have been placed on the boards, the boards all go to station 4, which is manually staffed by eight operators who perform a preliminary inspection and attach large components such as heat sinks and batteries. Each operator can process 15 boards per hour on average. Next, at station 5, the boards flow through a wave solder machine that has a capacity of 100 boards per hour. There are two solder machines operated in parallel. Boards are also cleaned and touched up at this stage. At station 6, testing on one of five parallel machines is conducted to look for opens and shorts, and to determine if the boards function correctly. The testing rate is 30 per hour. On average, 6% of the boards must return to station 5 for additional soldering and 12% to station 4 for rework. The remaining 82% pass all tests and are shipped. Table 17.6 summarizes the operational data. The switching probability matrix is as follows.

$$\Phi = \begin{bmatrix} 0.05 & 0.75 & 0.2 & 0 & 0 & 0 \\ 0 & 0 & 0 & 1 & 0 & 0 \\ 0 & 0 & 0 & 1 & 0 & 0 \\ 0 & 0 & 0 & 0 & 1 & 0 \\ 0 & 0 & 0 & 0 & 0 & 1 \\ 0 & 0 & 0 & 0.12 & 0.06 & 0 \end{bmatrix} \begin{matrix} 1 \\ 2 \\ 3 \\ 4 \\ 5 \\ 6 \end{matrix}$$

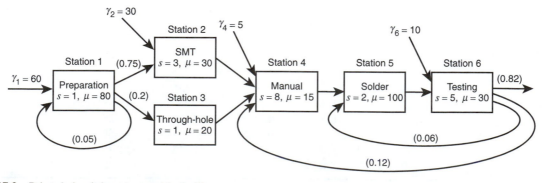

Figure 17.3   Printed circuit board assembly facility

Table 17.6 Operational Data for Printed Circuit Board Assembly

Measure	Station 1 (preparation)	Station 2 (SMT)	Station 3 (T-H)	Station 4 (manual)	Station 5 (solder)	Station 6 (testing)
γ	60	30	0	5	0	10
μ	80	30	20	15	100	30
s	1	3	1	8	2	5
Model	$M/M/1$	$M/M/3$	$M/M/1$	$M/M/8$	$M/M/2$	$M/M/5$

For purposes of analysis, it will be assumed that all processes are Markovian. This allows us to compute the various system performance measures as listed in Table 17.7. The first step in the computations is to solve Equation (2) to get the values of λ_i, $i = 1, \ldots, 6$.

$$\lambda_1 = 60 + 0.05\lambda_1$$
$$\lambda_2 = 30 + 0.75\lambda_1$$
$$\lambda_3 = 0 + 0.2\lambda_1$$
$$\lambda_4 = 5 + \lambda_2 + \lambda_3 + 0.12\lambda_6$$
$$\lambda_5 = 0 + \lambda_4 + 0.06\lambda_6$$
$$\lambda_6 = 10 + \lambda_5$$

Once these arrival rates are known, we can use the formulas for the $M/M/s$ model to find the expected number of boards at each station, their expected waiting times, and the station efficiencies. The results are presented in Table 17.7 and indicate that the system is very well balanced. Station 4 is the most efficient (as indicated by the values of ρ) and also the one with the largest average queue. It is not unusual for these two measures to be highly correlated. The queue time at station 4 is shorter than that at station 3 because the flow rate at station 4 is much greater ($W_q = L_q/\lambda$). The long system time at station 4, W, is attributable primarily to the relatively large service time at that station. The value of W is unaffected by the number of parallel servers.

The results for the full system are summarized in Table 17.8. The expected level of WIP is 38.58 boards, and the cycle time (W) is about 1/2 hour. These values reflect a simplified situation with no machine failures or jams. Also omitted from the description of the process are several intermediate steps and the burn-in process in which boards are heat treated to trigger temperature-sensitive failures.

Although we report the mean residence times across all stations in the network, these values (W, W_q, and W_s) are not useful measures and may be misconstrued as the expected flow times. To compute the desired expected values, we can use a system-wide version of

Table 17.7 Performance Results for Printed Circuit Board Assembly

Measure	Station 1 (preparation)	Station 2 (SMT)	Station 3 (T-H)	Station 4 (manual)	Station 5 (solder)	Station 6 (testing)
λ	63.16	77.38	12.63	110.37	118.05	128.05
ρ	0.79	0.86	0.63	0.92	0.59	0.85
L	3.75	7.15	1.71	16.02	1.81	8.13
W	0.059	0.092	0.136	0.145	0.015	0.063
L_q	2.96	4.57	1.08	8.66	0.63	3.86
W_q	0.047	0.059	0.086	0.078	0.005	0.030

Table 17.8   System Results for Printed Circuit Board Example

Performance measure	Value
Mean number in system, L	38.57
Mean time in system, W	0.510
Mean number in system queues, L_q	21.76
Mean time in system queues, W_q	0.305
Mean number in system service, L_s	16.81
Mean time in system service, W_s	0.206

Little's law. Letting W' be the mean flow time through the network, L the mean number in the system, and γ the rate of arrival at the system, we have $W' = L/\gamma$, where $\gamma = \Sigma_i \, \gamma_i$. Similar results apply to W_q' and W_s', the mean total queue time and service time, respectively, of an arbitrary customer who passes through the network. For this example, $\gamma = 105$, which yields $W' = 0.367$, $W_q' = 0.207$, and $W_s' = 0.160$. An alternative approach to computing these measures is described in Section 17.3.

17.3   EXPECTED FLOW TIME THROUGH THE NETWORK

A major reason for developing a network model for a multistation queuing system is to estimate the length of time needed for a job to travel through the network. The basic mathematical tool for analyzing this flow time involves the sum of random variables.

Theory

Suppose that $X_1, X_2, \ldots, X_N$ are identical and independently distributed random variables with mean and variance given by μ and σ^2, respectively (here we are using μ for the mean of the distribution, not the service rate). Furthermore, let N be a random variable having a nonnegative integer value independent of the X random variables. Now define the random sum $S = X_1 + \cdots + X_N$, where if $N = 0$, then $S = 0$. In other words, S is the sum of N random variables, where N is also a random variable. This construction is required to account for reentrant flow in the network. Given these conditions, it is possible to show that

$$E[S] = \mu E[N]$$
$$Var[S] = \sigma^2 E[N] + \mu^2 Var[N]$$

The two important components needed for determining the flow time through a network are (1) the time spent at each node and (2) the number of times each node is visited. We denote the mean time a job spends at node i per visit by W_i, where the value of W_i (mean station time in a queuing system) can be determined with the formulas in Section 16.2 for the $M/M/s$ model. The mean number of visits to each node can be calculated by Property 8 in Chapter 13 after the initial probability vector has been obtained. This vector can be interpreted as the probability mass function for the initial node of an arriving job.

Property 3: Consider a Jackson network containing m nodes with T_{net} being a random variable denoting the time that a job spends in the network. Let Φ be the switching probability matrix and let γ be the vector giving the external arrival rates. Define the potential matrix Ω as

$$\mathbf{\Omega} = (\mathbf{I} - \mathbf{\Phi})^{-1}$$

and the initial m-dimensional probability vector $\mathbf{v}$ with components

$$v_i = \frac{\gamma_i}{\sum_{j=1}^{m} \gamma_j}, \; i = 1, \ldots, m$$

Then the mean flow time through the network is given by

$$E[T_{\text{net}}] = \sum_{i=1}^{m} W_i \mathbf{v} \mathbf{\Omega} \mathbf{e}_i \tag{3}$$

where W_i is the mean waiting time at node i and $\mathbf{e}_i$ is the ith unit vector from the $m \times m$ identity matrix.

It is also possible to give the variance of the flow time through a network, but the calculations are more involved. The main reason for the additional complication is that the number of visits to one node is not independent of the number of visits to another node. In particular, it is possible to derive the covariance relationship

$$\text{cov}[N_j, N_k] = \begin{cases} \mathbf{v}\mathbf{\Omega}\mathbf{e}_j \left(2\omega_{jj} - \mathbf{v}\mathbf{\Omega}\mathbf{e}_j - 1 \right) & \text{for } j = k \\ \mathbf{v}\mathbf{\Omega}\mathbf{e}_j \omega_{jk} + \mathbf{v}\mathbf{\Omega}\mathbf{e}_k \omega_{kj} - \mathbf{v}\mathbf{\Omega}\mathbf{e}_j \mathbf{v}\mathbf{\Omega}\mathbf{e}_k & \text{for } j \neq k \end{cases}$$

where N_j is a random variable denoting the number of visits to node j, and ω_{kj} is the kjth component of $\mathbf{\Omega}$ This relationship among the various nodes allows us to compute the variance of T_{net}.

Property 4: Consider a queuing network as defined in Property 3. The variance of the flow time through the network is given by

$$Var[T_{\text{net}}] = \sum_{i=1}^{m} Var[T_i] \mathbf{v}\mathbf{\Omega}\mathbf{e}_i + \sum_{j=1}^{m} \sum_{k=1}^{m} W_j W_k \, \text{cov}[N_j, N_k]$$

where $Var[T_i]$ is the variance of the waiting time at node i.

Flow Time Example

To illustrate the computation of the expected flow time, consider a three-station Jackson network with external arrivals at station 1 only. Let $\gamma_1 = 8$ per hour and let $s_1 = 1$, $s_2 = 2$, and $s_3 = 1$, with corresponding service rates $\mu_1 = 10$ per hour, $\mu_2 = 4$ per hour, and $\mu_3 = 12$ per hour. Also, the switching probability matrix is

$$\mathbf{\Phi} = \begin{bmatrix} 0 & 0.75 & 0.25 \\ 0.10 & 0 & 0.90 \\ 0.05 & 0 & 0 \end{bmatrix}$$

indicating that there are reentrant flows in an otherwise serial configuration.

The first step is to compute the value of λ_i for each station i. Using Equation (2), we find that $\lambda = (9.104, 6.828, 8.421)$. Next we compute W_1, W_2, and W_3. Using the $M/M/1$ formula for W with $\lambda_1 = 9.104$ per hour and $\mu_1 = 10$ per hour, we obtain $W_1 = 1.12$ hours. Using the $M/M/2$ formula for W with $\lambda_2 = 6.828$ per hour and $\mu_2 = 4$ per hour yields $W_2 = 0.92$ hour. Finally, returning to the $M/M/1$ formula with $\lambda_3 = 8.421$ per hour and $\mu_3 = 12$

per hour, we get $W_3 = 0.28$ hour. Now we must find the number of visits to each node; specifically, we determine the matrix

$$\Omega = \begin{bmatrix} 1.00 & -0.75 & -0.25 \\ -0.10 & 1.0 & -0.90 \\ -0.05 & 0 & 1.0 \end{bmatrix}^{-1} = \begin{bmatrix} 1.38 & 0.853 & 1.053 \\ 0.165 & 1.124 & 1.053 \\ 0.057 & 0.043 & 1.053 \end{bmatrix}$$

and apply Equation (3). For this example, the initial probability vector is $\mathbf{v} = (1, 0, 0)$, indicating that all jobs enter the network through node 1; therefore,

$$E[T_{net}] = 1.12 \times 1.138 + 0.92 \times 0.853 + 0.28 \times 1.053 \cong 2.35 \text{ hours}$$

Alternatively, we can use the system-wide version of Little's law to find W, the equivalent of $E[T_{net}]$. For our example, $L = L_1 + L_2 + L_3 = 10.196 + 6.282 + 2.357 = 18.835$ and $\gamma = 8$. Therefore, $W = L/\gamma = 18.835/8 \cong 2.35$, which is the same result and evidently much easier to compute.

17.4   NON-POISSON NETWORKS

In this section we consider networks for which the Jackson assumptions do not hold. We use the equations presented in Section 16.3 for non-Markov systems to analyze individual stations. Because the formulas require the values of the standard deviations for interarrival and service times, additional approximations are necessary to estimate these values. Although the results of this section are clearly not exact, they are important because so many systems do not satisfy the Markovian assumption.

Approximation for Serial Systems

Figure 17.4 shows a three-station assembly line in which arriving units pass through each station in sequence. Associated with station i is a mean processing time m_{si} and standard deviation σ_{si} for $i = 1, 2, 3$. Arrival and service data are given in Table 17.9. Each station is assumed to be a $GI/G/1$ queue. The means and standard deviations are measured in minutes. We would like to estimate the expected number of units in the queues and the expected waiting times.

In this example, the mean time between arrivals is $m_a = 5$ minutes, so $\lambda = 0.2$ per minute. The mean time between departures at station 1, and equivalently the mean time between arrivals at station 2, is the same as m_a, because all units are processed and there are no losses. Similarly, the departures from stations 2 and 3 all have the same mean, m_a. The standard deviation of the time between departures at the three stations, σ_{d1}, σ_{d2}, and σ_{d3}, will differ, however, because of the joint effects of arrival and service variabilities on departure variability. The approximate relation is

$$c_d^2 = \rho^2 c_s^2 + (1 - \rho^2) c_a^2, \text{ where } c_d = \sigma_d / m_a$$

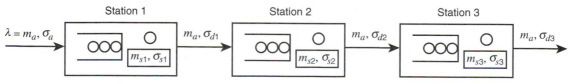

Figure 17.4   A series of non-Poisson stations

Table 17.9   Data for Non-Poisson Network

External arrivals	Station 1	Service process	Station 1	Station 2	Station 3
m_a	5	m_s	4	4	4
σ_a	2	σ_s	2	2	2
c_a	0.4	c_s	0.5	0.5	0.5

The departure coefficient of variation c_d^2 is the same as the arrival coefficient of variation of the next stage.

For the network in Figure 17.4, the queues can be analyzed sequentially starting with station 1 using the formula

$$W_q(GI / G / 1) = \left[\frac{c_a^2 + c_s^2}{2} \right] W_q(M / M / 1)$$

The results in Table 17.10 indicate that the average total time a unit spends in the three queues is 10.93 minutes, which is much shorter than the time in a system with Poisson arrival and service processes. This is not unexpected, because an exponential distribution (used for Poisson systems) has a coefficient of variation of 1, which is quite high. Most other distributions (with reasonable parameters) have coefficients of variation less than 1, so they have less variability. The foregoing approximation expresses the mean time in the queue as the product of the mean time in queue for a Poisson process times a factor related to the coefficients of variation of the arrival and service processes. The smaller the coefficients of variation, the smaller the factor. If the arrival and service processes both have $c = 1/2$ (half of exponential), the mean time in queue is reduced by a factor of 4 in comparison with the exponential. Thus, one would expect the results for a non-Poisson system (with less variability) to be smaller than those for a Poisson system. In the extreme, when the arrival and service processes are constant ($c = 0$), there is no queue.

To obtain the average total time spent at each station, we use the fact that $W = W_q + m_s$ and then Little's law to get L and L_q. Because all stations have the same mean time between arrivals, they all have the same arrival rate $1/m_a$ or λ, which for this case is 0.2. As previously mentioned, for three stations in series and no losses, the arrival rate at each station must be the same in equilibrium. For station 1, then, $W = 3.28 + 4 = 7.28$ minutes, $L = \lambda W = (0.2)(7.28) = 1.456$, and $L_q = \lambda W_q = 0.656$.

To analyze this system, we assumed that the queues between stations were unlimited in size. This is not usually the case for assembly lines, where the amount of buffer space is finite. The analysis of such systems is much more complicated, and the equations used above are not valid.

Table 17.10   Results for Non-Markov Assembly Line

Measure	Station 1	Station 2	Station 3	Total
c_a^2	0.16	0.218	0.238	—
c_s^2	0.25	0.25	0.25	—
ρ	0.8	0.8	0.8	—
$W_q(M/M/1)$	16	16	16	48
$W_q(GI/G/1)$	3.28	3.74	3.90	10.92

Approximations for Fixed Routings

Jackson networks can be used to approximate the behavior of some queuing networks even when the four conditions of Definition 1 do not all hold. A very common scenario is that instead of each job being routed through the network according to a Markov chain, there are classes of jobs for which each job within a class has a fixed route. For example, consider an assembly line with two workstations. Further suppose that jobs arrive at a rate of 10 per hour, where arriving jobs begin processing at station 1, proceed to station 2, then return to station 1 for a final processing step. This flow is illustrated in Figure 17.5a. Such a route is deterministic and so does not satisfy the Markov assumption (condition 4) of Definition 1.

Nevertheless, a Jackson network can be used to approximate a network with deterministic routing by fixing the switching probabilities so that the mean flow rates in the Jackson network are identical to the flow rates in the network with deterministic routes. In the assembly line example, we could assume that all jobs start at station 1 and, when finished, leave the system with a 50% probability or proceed to station 2 with a 50% probability. The switching probability matrix governing the flow of jobs within the two-station network is

$$\mathbf{P} = \begin{bmatrix} 0 & 0.5 \\ 1 & 0 \end{bmatrix}$$

The schematic in Figure 17.5b represents the stochastic interpretation. Notice that in both networks an input rate of $\gamma = 10$ per hour results in net flows of 20 per hour into station 1 and 10 per hour into station 2. The difficulty with the Jackson network approximation is that the variances of the two networks will not be the same.

A Jackson network model will always produce a high variability in flow time. This means that when flow times are deterministic, as in the preceding example, the approximation approach will not be appropriate. However, consider a network system with several distinct classes. Even if each class has deterministic routes, the actual flow times through the network would be random because of the variability in processing. Therefore, when a system contains several classes, it may be reasonable to model the aggregate as a Jackson network.

Feldman and Valdez-Flores [1996] examine this case for fixed routings. The only condition from Definition 1 that they enforce ahead of time is the unlimited capacity assumption at each node. The assumption of Poisson input is used as an approximation, although the arrival rates of the individual classes need not be Poisson rates. Other authors have shown that when many independent arrival processes are merged, the resultant arrival process tends to be a Poisson process even if the individual processes are not. The two main issues that must be addressed before an analysis can be conducted are how to obtain each node's arrival

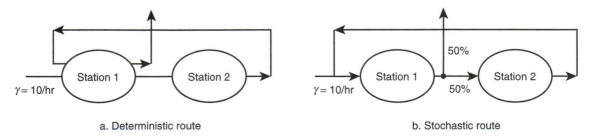

a. Deterministic route   b. Stochastic route

Figure 17.5   Comparison of networks with deterministic and stochastic routing

rate and how to obtain the mean service time. The interested reader is referred to more advanced texts for details.

17.5   OPTIMAL DESIGN OF QUEUES

The techniques and formulas developed in the preceding sections allow us to analyze the behavior of many types of stochastic systems. Given a model structure and values for the various parameters, one can compute measures of effectiveness such as expected number and time in the system. This is in contrast with the material presented in the early chapters of the book, which, given the model structure and parameter values, allows us to determine optimal solutions. The procedures presented in the later chapters of the book have thus far been descriptive—i.e., they have provided normative information on how the system will behave under certain assumptions. This is in contrast with prescriptive information, which tells us how to achieve optimal results. In this section, we introduce some typical decision problems involving queuing systems and we provide analytic guidance. Powerful techniques for optimizing general stochastic models, however, are unavailable for most situations. Enumerating a set of candidates and then evaluating their performance is the primary means of finding good solutions.

Design of a Single Facility

Suppose customers arrive at random at an $M/M/s$ facility at an average rate of 110 per hour. Each channel can serve roughly 30 customers per hour. The waiting cost depends linearly on the number of customers in the system and is equal to $20 per hour per customer. The cost of providing a service channel is $50 per hour. What is the optimal number of channels?

The optimal design must balance the cost of providing the service channels against the waiting costs that the system imposes on its customers. To analyze the situation, let $E[SC]$ be the expected service cost and let $E[WC]$ be the expected waiting cost. The expected total cost is

$$E[TC] = E[SC] + E[WC]$$

Both components on the right-hand side of this equation are measured in cost per unit time and must be expressed in the same time units.

In general terms, the waiting cost depends on how many customers are in the system. If we denote by $g(n)$ the cost per unit time with n customers in the system, then the expected waiting cost is computed from the steady-state probabilities.

$$E[WC] = \sum_{n=1}^{\infty} g(n)\pi_n$$

For the case in which $g(n)$ is a linear function of the number of customers in the system, the unit cost assigned to an individual, call it C_w (in dollars per unit time), is constant. The expected waiting cost is then proportional to the average number in the system.

$$E[WC] = \sum_{n=1}^{\infty} nC_w\pi_n = C_w L$$

When the cost C_s of each server is also a constant, the service cost is proportional to s and the total cost is

$$E[TC] = sC_s + C_w L$$

For this expression to be dimensionally consistent, C_s must be specified in dollars per unit time. For our example, we have a facility cost of \$50 per hour and a linear waiting cost of \$20 per hour, so the total cost as a function of the number of channels is

$$E[TC] = 50s + 20L$$

The decision variable is the integer s. The first term is linear in s, but because L is a complicated function of s, as described in Section 16.2, the second term is highly nonlinear.

As the number of channels in a queuing system increases, the average number and time in the system both decrease. There is a tradeoff between the cost of waiting and the cost of service. For an $M/M/s$ system in equilibrium, the minimum value of s is equal to the smallest integer greater than λ/μ. The optimal number of channels is found by starting from this minimum and evaluating $E[TC]$ for increasing values of s. When the total cost stops decreasing and begins to increase, the optimal solution has been found. This follows, because L is a convex (decreasing) function of s.

For the preceding data, $\lambda = 110$ and $\mu = 30$, so the minimum number of channels is $\lfloor 110/30 \rfloor + 1 = 4$. Using the steady-state formulas for the $M/M/s$ model, we compute the values of L for several values of s. The resultant costs are shown in Table 17.11. The optimal number of channels $s^* = 5$.

Waiting Cost Only for Those in the Queue

It is often the case that a customer does not mind the time spent in service, but only the time spent waiting for service. Consider the preceding example with the qualification that the only cost incurred by a customer is \$20 per hour for the time waiting in the queue—that is, there is no cost for the time spent in service.

To analyze this case, we use the same total cost function as before but separate the number in the queue L_q and the number in service L_s. Let C_w and C_w' be the hourly cost in the queue and service, respectively.

$$E[TC] = sC_s + C_w L_q + C_w' L_s$$

The number in service is λ/μ, so

$$E[TC] = sC_s + C_w L_q + C_w'(\lambda/\mu)$$

The last term on the right does not depend on s, so the solution that minimizes the expected total cost is the same whether or not the times in service and queue have the same cost. The optimal number of channels s^* remains 5.

Division of the Load between Two Service Facilities

Assume that two separate service facilities are now available to handle the demand described in the example. Dividing the arrivals equally, each facility has an arrival rate of 55 customers per hour. Waiting and service costs are as previously stated. The goal is to find the optimal

Table 17.11   Parametric Results for a Single Facility

s	L	$E[SC]$	$E[WC]$	$E[TC]$	
4	12.71	200	254	454	
5	4.86	250	97	347	← Minimum
6	4.00	300	80	380	

number of channels for each facility and compare the resulting system cost with that of the single facility with five channels.

For $\lambda/\mu = 55/30$, the minimum number of channels for each facility is 2. The optimal number is three, as shown in Table 17.12. Each station costs $198 per hour, so the cost of the two stations is $396 per hour, compared with a cost of $347 per hour for a single station handling the entire load.

When facility and waiting costs are linear, it is always best to provide a single facility with multiple channels rather than create a separate facility and waiting line for each server. It is for this reason that many real-world queuing systems with multiple servers have only one queue. This is evident in post offices, banks, and amusement parks, where ropes or metal bars organize waiting customers into long snake-like lines. Individual lines behind each server are used only in situations in which it is impractical to channel arrivals into a single queue, such as at supermarket check-out lines and highway toll booths.

In a later example, we will investigate a situation in which the locations of the facilities are also decision variables. Here the time spent in travel is relevant, so the optimal solution is not necessarily a single facility.

Variable Service Rate

Consider again the preceding situation but assume that the service rate is continuously adjustable. Increasing the rate, however, imposes additional costs. In particular, assume that the cost of a service channel is a linearly increasing function of μ of the form

$$f(\mu) = c_f + c_v\mu$$

where c_f is the fixed cost of establishing a channel and c_v is the unit cost associated with the service rate. Our goal is to find the optimal design by specifying μ and s.

Under these conditions, the cost model can be written as

$$E[TC] = sf(\mu) + E[WC]$$

which is a nonlinear function of μ and s, even when s is fixed. The corresponding optimization problem is an integer, nonlinear program in nonnegative variables.

When the alternatives for μ are continuous and $f(\mu)$ is either a linear or a concave function of μ, it is always optimal to use a single channel rather than multiple channels for the queuing system. Where it is applicable, this result is extremely powerful, because the optimal solution can be found without searching over alternative values of s.

For our example, the cost of providing service is linear in μ, so it is optimal to use a single channel. For an $M/M/1$ model, the average number in the system is

$$L = \frac{\rho}{1-\rho} = \frac{\lambda}{\mu-\lambda}$$

Table 17.12   Parametric Results for One Facility with Half the Load

s	L	$E[SC]$	$E[WC]$	$E[TC]$
2	11.48	100	230	330
3	2.41	150	48	198
4	1.95	200	39	239

resulting in a total cost function of

$$E[\text{TC}] = c_f + c_v\mu + \frac{C_w\lambda}{(\mu - \lambda)}$$

Taking the derivative of this function with respect to μ, setting it equal to zero, and solving for μ yields the optimal value

$$\mu^* = \lambda + \sqrt{\frac{C_w\lambda}{c_v}}$$

For $c_f = 20$, $c_v = 1$, and $C_w = 20$, we get an optimal service rate of

$$\mu^* = 156.9/\text{hr}$$

using the preceding formula. To determine the expected cost, we must first find the expected number of customers in an $M/M/1$ system with $\lambda = 110$ and $\mu = 156.9$. The analysis indicates that $L = 2.345$ customers, so the expected total cost of the system is $E[\text{TC}] = (20 + 156.9) + 20(2.35) = \$224$. This compares favorably with a cost of \$347 for a facility with five servers, each with a service rate of 30 per hour. Thus, for the given data, it is better to provide one, very fast server rather than a facility with five relatively slow servers.

Other Decision Problems for a Single Facility

When the scope of the analysis is expanded to include choices for the maximum number of customers allowed in the system and the size of the calling population, a variety of new decision problems can be posed, even for a single-server system. For finite queues, balking is a concern. If a cost function for balking can be specified on the basis of, say, lost sales or loss of good will, it is possible to extend the analysis to include the cost of providing service, the cost of waiting, and the cost of balking. For problems with finite populations, one might be interested in the selection of the optimal population size.

In general, the analytic approach is to construct an objective function that embodies the cost components of all variable factors. As in the deterministic case, an optimal solution is one that minimizes this function. The variables may be continuous, as was the service rate μ in the preceding example, or discrete, as in the case of the number of servers s. Other possible discrete variables include the maximum number of customers in the system K and the size of the population N. Since we have closed-form expressions for the expected number in the system, the expected time in the system, and related measures of interest, functions can be devised that explicitly describe the effects of the decision variables on cost.

When the discrete features of the system are fixed, nonlinear programming can often be used to select optimal values of continuous parameters. This often allows solution by partial enumeration. Of course, the validity of the results depends on the convexity characteristics of the model.

When the continuous features of the system are fixed, discrete enumeration of alternatives may be necessary. Although this is not a powerful optimization method, the number of alternatives is often small, and efficient evaluation techniques may provide solutions without excessive amounts of computation. When it makes more sense to address the continuous and discrete features of a decision problem simultaneously, a combination of nonlinear programming and discrete search may be necessary. This is the case, for example, when complete enumeration is impractical.

17.6   SYSTEMS WITH TRAVEL TIMES

As indicated in the preceding section, it is rarely beneficial to divide the resources of a single service facility into separate systems. When customer travel time is important, however, it may be cost effective to establish several facilities at appropriate distances from each other. In such cases, the locations of facilities, as well as the number of facilities, are decision variables. Typical examples include fire stations in a city, restrooms in an office complex, and tool cribs in a manufacturing plant.

Travel Cost

To develop a modeling framework, we consider a two-dimensional plane in which movement is limited to directions parallel to the orthogonal axes x and y. If a route from one point to another follows a path that is always parallel to the axes, the travel is said to be rectilinear. This assumption is appropriate for manufacturing plants and office buildings in which aisles and/or hallways form rectangular arrays, and for cities with streets arranged in grid patterns. The rectilinear distance between two points (x_1, y_1) and (x_2, y_2) is

$$D = |x_1 - x_2| + |y_1 - y_2|$$

Besides being justified in many practical situations, this metric simplifies the calculations of some solution methods.

   In the following discussion, we will define the coordinate system so that the service facility is at the origin, as in Figure 17.6. The distance D from point (x, y) to the facility is then $D = |x| + |y|$.

   We also define v as the speed of travel for a customer going to and from a service facility. Consequently, the time T to travel the distance D is D/v. For the rectangle in Figure 17.6, the point $(-a, -b)$ identifies the lower left corner of the service region and the point (c, d) identifies the upper right corner (a, b, c, and d are nonnegative). Assuming that the customers are spread uniformly throughout the rectangle, it can be shown that the expected rectilinear travel time for a customer traveling to and from the facility (round trip) is

$$E[T] = \frac{1}{v}\left[\frac{a^2 + c^2}{a + c} + \frac{b^2 + d^2}{b + d}\right] \qquad (4)$$

This value is the travel time averaged over all customers in the region.

   Now let C_t be the cost per unit time for customers traveling to and from the service facility. The cost per trip is then

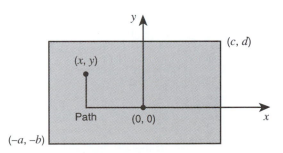

Figure 17.6   Rectangular region with service facility at the origin

$$C_t E[T]$$

and the cost per unit time (or the cost rate) for travel is

$$E[\text{TV}] = \lambda C_t E[T]$$

where λ is the rate of arrival at the facility.

System Design

We will use the results of this section to solve problems regarding the dispersion of service facilities in a geographical region. For the simplest case, let us assume that there are n facilities each having the same number of channels, serving homogeneous regions, and operating from the same relative locations within their respective regions. The parameters of interest are C_f, the cost of a facility; C_s, the cost of a server; and λ_T, the total arrival rate. The total expected cost is

$$E[\text{TC}] = n\Big(C_f + sC_s + E[\text{WC}] + \lambda C_t E[T]\Big)$$

where $\lambda = \lambda_T/n$.

When it is important to differentiate among the facilities, the notation in the preceding expression should be expanded to include subscripts for the individual parameters and variables. These terms must then be summed over all facilities to get the total expected cost. The relevant formula is

$$E[\text{TC}] = \sum_{i=1}^{n}\Big(C_{fi} + s_i C_{si} + E[\text{WC}_i] + \lambda_i C_{ti} E[\text{T}_i]\Big)$$

In this case, the total arrival rate is $\lambda_T = \lambda_1 + \lambda_2 + \cdots + \lambda_n$.

Locating Service Facilities within a Plant

The floor plan of a manufacturing plant is shown in Figure 17.7, where all dimensions are in feet. To provide food service for its employees, management is planning to install one or two snack bars (stations) in the plant. Two plans are under consideration:

1. Install stations at locations A and B.
2. Install a single station at C.

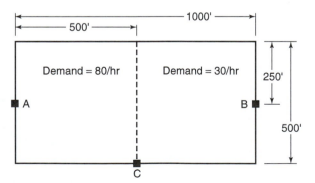

Figure 17.7   Plant with nonuniform demand

A statistical analysis has shown that the demand for the food service can be modeled as a Poisson process, but in the left half of the plant the arrival rate is expected to be 80 customers per hour whereas in the right half it is expected to be 30 per hour. The demand is uniformly distributed within each sector. The design calls for each facility to have one or more channels that will each serve customers at an average rate of 40 per hour. Previous experience indicates that the service time should be exponentially distributed.

The cost associated with employees who are waiting, being served, or traveling to or from the facility is $20/hour. They travel at a rate of 10,000 feet per hour. All travel follows a rectilinear path, and a two-way trip is required for each visit.

The cost of establishing a station at any location is $C_f = \$20,000$ per year independent of the number of channels, with an added cost of $C_s = \$15,000$ per year per channel. If plan 1 is chosen, the demand in the left side of the plant will be handled at station A and the demand in the right side will be handled by station B. If plan 2 is chosen, all the demand(s) will be handled at station C.

We wish to determine the better of the two plans. To do this, it is necessary to consider the number of channels at each station as well as the location(s) of the station(s). For the analysis we assume that there are 2000 work hours per year.

Recognizing that each station can be modeled as an independent *M/M/s* queue, we can analyze each station separately and then add the resultant costs. To begin, consider plan 1 with stations at A and B. Evaluating the left half of the region first, we define a coordinate system with station A located at the origin, as shown in Figure 17.8.

For purposes of estimating travel time, we have the coordinates

$$a = 0, b = 250, c = 500, d = 250$$

These values allow us to compute the expected two-way travel time using Equation (4).

$$E[T] = 0.075 \text{ hr}$$

Because most of the data are given in hours, we will express the fixed and variable facility costs in per-hour equivalents for a year.

$$C_f = 20,000/2000 = \$10/\text{hr}$$
$$C_s = 15,000/2000 = \$7.50/\text{hr}$$

Continuing, the service cost is a linear function of the number of channels

$$E[SC] = 10 + 7.5s \text{ (\$/hr)}$$

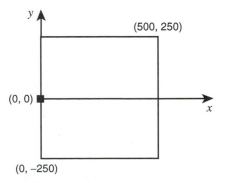

Figure 17.8 Coordinate system for station A

and the waiting cost varies directly with the number of employees in the system

$$E[WC] = 20L \text{ (\$/hr)}$$

The final component is the travel cost

$$E[TV] = \lambda C_t E[T] = (80)(20)(0.0750) = \$120/hr$$

Summing these costs, the expected total hourly cost for station A is

$$E[TC_A] = 10 + 7.5s + 20L + 120$$

which reduces to a function of s and L. However, L is really a function of s, so to determine the optimal number of channels we can enumerate over s. For $\lambda = 80$ and $\mu = 40$, the number of channels must be greater than 2 to ensure stability, so the enumeration begins with $s = 3$. The results are presented in Table 17.13. The optimal number of channels is 4, and the cost of the corresponding system at A is \$203.

Station B is similar to station A with respect to travel time, but it has an arrival rate of 30 per hour. The total cost function for station B is

$$E[TC_B] = 10 + 7.5s + 20L + 45$$

Enumerating over s for $\lambda = 30$ and $\mu = 40$ yields the results shown in Table 17.14. The optimal number of channels is 2, and the corresponding cost is \$87. Combining this solution with the solution for station A gives us a total cost of \$290 per hour for plan 1.

We now consider plan 2, a single station at location C. The first thing to note is that the rate of arrival at station C is equal to the sum of the rates for the two sides of the plant, so $\lambda = 110$ per hour. Travel costs must be calculated separately for each side. Relative to the right side with the location of C at $(0, 0)$, we have $a = 0$, $b = 0$, $c = 500$, and $d = 500$. Using Equation (4) yields an average two-way travel distance of 1000 feet. The two-way travel time is 0.1 hour. Because of symmetry, the same time is appropriate for the left side of the plant. The expected travel cost is the sum of the two components, so

$$E[TV] = (80)(20)(0.1) + (30)(20)(0.1) = \$220/hr$$

Therefore, the expected total cost for station C is

$$E[TC_C] = 10 + 7.5s + 20L + 220$$

Table 17.13 Parametric Results for Station A

s	L	$E[SC]$	$E[WC]$	$E[TV]$	$E[TC]$
3	2.888	32.5	57.76	120	210
4	2.174	40.0	43.48	120	203
5	2.040	47.5	40.80	120	208

Table 17.14 Parametric Results for Station B

s	L	$E[SC]$	$E[WC]$	$E[TV]$	$E[TC]$
1	3.000	17.5	60.0	45	123
2	0.873	25.0	17.45	45	87
3	0.765	32.5	15.29	45	93

Table 17.15   Parametric Results for Station C

s	L	$E[SC]$	$E[WC]$	$E[TV]$	$E[TC]$
3	12.06	32.5	240.1	220	494
4	3.65	40.0	73.02	220	333
5	2.97	47.5	59.37	220	327
6	2.81	55.0	56.19	220	331

For $\lambda_T = 110$ and $\mu = 40$, s must be an integer greater than $110/40 = 2.75$, so the enumeration starts at $s = 3$. The computational results are shown in Table 17.15; the optimal number of channels is 5, and the corresponding cost is \$327 per hour. Because this value is greater than the cost of a system with stations at A and B, we conclude that plan 1 is the better alternative.

EXERCISES

1. An assembly line has five stations through which every unit being produced must pass in the same fixed order. The arrival of units at the line is random, following a Poisson process. The mean time between arrivals is 20 minutes. The stations operate independently, and queues are allowed to form in front of each station. All stations have the same mean processing time of 15 minutes. For analysis purposes, we assume that arrivals follow a Poisson process for all stations. [This assumption is correct for part (a) below, but not for part (b).] Determine the total average number of units in the five queues for each case.

   (a) The processing time at each station has an exponential distribution.
   (b) The processing time at each station is a constant, exactly 15 minute.

2. An alternative to the assembly line in Exercise 1 is a single machine that performs all five operations. The machine can work on only one unit at a time, so all five operations must be completed before processing can begin on the next unit. Determine the average queue length, the average number of units in the system, the average waiting time in the queue, and the average waiting time in the system for each case.

   (a) Assume that the time required to perform each operation has an exponential distribution, with a mean time of 3 minutes.
   (b) Assume that the time required to perform each operation has a constant value of 3 minutes.

3. For the computer center example in Section 17.2, determine the minimum number of channels required at each station when the arrival rate is increased to 15 per minutes.

   (a) Display the results for this situation in a table similar to Table 17.1.
   (b) For which station would the addition of another channel most reduce the average time spent on a job passing through all three stations.

4. For the job shop example in Section 17.2, show the effect of adding an additional channel for station B. Rework Tables 17.4 and 17.5 for the revised system.

5. For the job shop example in Section 17.2, change the order rate of product 2 to 20 per month. If necessary, add additional channels to the stations to allow a steady-state solution. Compute the steady-state performance measures and display them in tables similar to Tables 17.4 and 17.5. Identify the bottleneck station.

6. For the food service location example in Section 17.6, find the cost of a single facility placed at the center of the manufacturing plant.

7. Two operations are required to repair a roof: the old roof must be removed, and then the new one must be installed. The time between the beginning of the removal of the old roof and the completion of the installation of the new roof is important because the interior of the house may be damaged if it rains during this period.

   Customers call at random to a roofing contractor at an average rate of 12 per month (4 weeks). It takes on average 0.5 week to remove a roof and 1 week to install a new one. Both times are exponentially distributed. The contractor has two kinds of crews, a removal crew and an installation crew. When an order arrives, the customer must wait until a removal crew is available. After the roof has been removed, the customer enters a queue to wait for an installation crew to become available.

   (a) Model this situation as a queuing network and draw the corresponding diagram. Show the relevant arrival and service rates for each station. Compute the minimum number of crews required for the two operations.

   (b) Compute the average danger period using the minimum number of crews. The danger period is the time between the moment when the removal crew begins to remove the old roof and the moment when the installation crew completes the installation of the new roof.

8. Two alternative designs are being considered for the service facility of a finite queuing system. Alternative A has a single server and alternative B has two servers. The cost of the service facility has two components: a fixed cost of $50,000 per year, which is independent of the number of servers, and a variable cost of $30,000 per server per year, which is proportional to the number of servers. Simulation analysis has determined the following steady-state probability distributions for the number of customers in the system.

	Design A						Design B				
n	0	1	2	3	4	n	0	1	2	3	4
π_n	0.12	0.25	0.38	0.15	0.10	π_n	0.38	0.30	0.21	0.06	0.05

   In addition, the simulation has determined that 10% of the customers balk with design A whereas 5% balk with design B. The arrival rate (including balks) is 1500 customers per year.

   Management estimates that a customer who balks costs the company $100 in lost profits. The cost of customers waiting in the system is $20 per hour (per customer). Determine which design alternative minimizes the total expected cost. Assume that there are 2000 work hours in a year.

9. A pizza company is initiating a delivery service in a growing suburb of a city. For simplicity, assume that the suburb has a square shape, as indicated in the figure (all distances are in miles). Two alternatives are being considered: either a single shop is to be located at site A, or four shops are to be located at sites B, C, D, and E. In the latter case, the suburb will be divided into quadrants, with each customer being served by the closest shop.

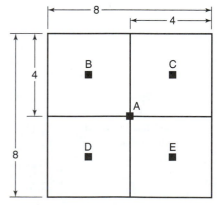

Layout for pizza delivery problem

Calls for pizza arrive at random at an average rate of 20 per hour and are distributed uniformly throughout the entire suburb. When a call is received, the pizza is prepared and then delivered to the customer by a driver. The preparation time is constant and is independent of the location of the order. Since it will not affect the location decision, we consider only the delivery process, which is assumed to be a Poisson process.

The delivery time is the round-trip time from the shop to the customer and back to the shop. Assume that all travel is rectilinear with respect to the orientation of the city boundaries, and that the average speed of travel is 20 miles per hour.

(a) Compute the average travel time from the given geometry for both alternatives. The delivery process is a queuing system, because orders must wait for the availability of a driver.

(b) There is a fixed cost of $15,000 per year for establishing a shop. The cost of hiring a driver and providing a delivery vehicle is also $15,000 per year. Determine which of the two alternatives is better. The solution must specify the location(s) of the shop(s) and the number of drivers to hire for each location. To account for service quality, assign a cost of $20 per hour for the time required for a customer to receive his or her pizza. Because this cost varies linearly with time, the fixed preparation time can be neglected in the total cost computation. Assume that there are 2000 work hours in a year.

10. A blood bank has two kinds of customers, those that donate their blood and those that sell it. Donors arrive at the blood bank at an average rate of 6 per hour whereas sellers arrive at an average rate of 10 per hour. Both arrival processes are random and independent. There are two receptionists for the two types of customers owing to the need to fill out different forms. The "donation" receptionist can handle an average of 10 customers per hour, and the "selling" receptionist can handle an average of 15 customers per hour. The service times for both receptionists are exponentially distributed.

After filling out the forms, both types of customers go to the blood-donation room and lie on tables where the required amount of blood is drawn. Because of the different reactions experienced, the time a person spends on the table is exponentially distributed, with a mean of 10 minutes. A sufficient number of tables are available to handle the total load of the two types of customers without the queue becoming infinitely large.

(a) Draw a diagram that represents this situation, showing all relevant queuing system parameters. What is the minimum number of tables required?

(b) What is the average time required for a person to finish the blood-donation process? Counting begins after processing by the receptionist and includes the time waiting for a table and the time for giving blood.

(c) Both types of customers share a single waiting room when they are not being processed. What is the average number of persons using the waiting room?

(d) What is the probability that the waiting room is empty?

11. (*Typing Pool Model*) Suppose that documents to be typed at an office arrive at the rate of 100 per day and secretaries can type at the rate of 50 documents per secretary per day. Assume that both processes are Poisson. Each secretary earns $80 per day, and the cost of holding documents in the system is estimated at $25 per document per day.

(a) Let C_s be the cost per secretary per day, C_w the cost of waiting per day, s the number of secretaries, L the expected number of documents in the system, and z the expected total cost. Write the expected total cost equation for z using this notation. Find the optimal number of secretaries s^* that minimizes the expected total cost per day if each secretary works with his or her own queue. Make use of the fact that

$$z(s^*) \leq z(s^* - 1) \text{ and } z(s^*) \leq z(s^* + 1)$$

(b) Suppose that secretaries are to be combined into a typing pool with one queue. Determine the optimal number of secretaries and the associated cost by enumeration—that is, by evaluating the total cost equation for $s = 1, 2, 3, \ldots$.

(c) Let $L(s^*)$ be the expected number of documents in the system in steady state with s^* secretaries, and let $z(s^*)$ be the expected minimum total cost. Show that

$$L(s^*) - L(s^* + 1) \leq C_s / C_w \leq L(s^* - 1) - L(s^*)$$

(d) Is it better to pool the secretaries? Do you have any reservations about the validity of the analysis?

12. Given the schematic in Exercise 14 in Chapter 16, determine the optimal number of inspection stations if each station costs $10 per hour to operate and the cost of holding each product in the system is $2 per hour.

13. (*Pollution Control Model*) A manufacturing company continuously operates five independent waste treatment facilities for cleaning effluents that are discharged into a river. The operation of each facility, however, is not completely reliable as a result of equipment failures. The time between breakdowns in each facility is exponentially distributed, with a mean of 14 days. On average, it takes a repair crew one day to get a facility back up and running. Repair times are similarly exponential, and each crew costs the company $1000 per day. The EPA does not allow the company to discharge untreated waste, so whenever a facility is "down" the plant operates at less than full capacity. This downtime is costed at $50,000 per day per facility. Determine the optimal number of repair crews to hire by enumerating total cost per day for 1, 2, 3, 4, and 5 crews. Is the equation given in Exercise 11c satisfied?

14. (*Airline Reservation System*) Bonzi airlines has four telephone lines at their reservation desk. During the late shift, incoming phone calls arrive every 6 minutes on average according to a Poisson process. The time required to service a caller is exponentially distributed, with a mean of 5 minutes. If a call comes in and all operators are busy, the caller is placed on hold, provided that a line is available. Otherwise, the caller receives a busy signal. Callers encountering a busy signal are assumed to place reservations elsewhere. It is further assumed that calls on hold never renege.

(a) Determine the minimum number of operators (s) needed to ensure that the probability of rejection is less than 10 percent.

(b) Assuming that one operator is to be used, determine the minimum number of phone lines (K) needed to ensure a probability of rejection below 10 percent.

(c) Discuss the tradeoffs between the two types of decisions indicated in parts (a) and (b). Specifically, determine the total cost per hour for each of these cases if $20 per hour is the cost of an operator, $2 per hour is the cost of an additional phone line, $25 is the cost of a rejected call, and $10 per hour per call in the system is the cost of waiting.

(d) Describe how the optimal joint decision for s and K could be determined.

15. Consider the following tandem queuing system for a product that requires two steps to assemble.

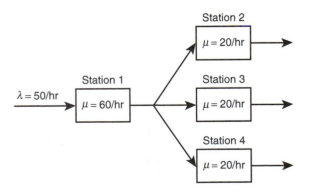

Given that the assumptions for the standard $M/M/s$ model are satisfied, determine the mean length of each queue and the expected time a product spends on the assembly line. Assume that stations 2, 3, and 4 form an $M/M/3$ system.

16. (*State Unemployment Compensation*) A state agency that handles compensation for those who are unemployed is considering two options for processing applications. Option 1: Four clerks process applications in parallel from a single queue. Each clerk fills out the required form in the presence of the applicant based on information that is verbally related to the clerk. Processing time is exponentially distributed with a mean of 45 minutes. Option 2: Each applicant first fills out the form without the help of the clerk. The time required to accomplish this is exponentially distributed, with a mean of 65 minutes. When the applicant finishes the form, he or she joins a single queue to await a review by one of the four clerks. The time required to review a form is exponentially distributed with a mean of 5 minutes.

Given that the arrival of applicants is a Poisson process with a mean rate of 4.8 per hour, compare the two options with respect to the expected number of applicants in the system and the expected time in the system.

17. The simplest type of queuing network is an *M/M/*1 system with Bernoulli feedback. Answer the following questions for the system shown in the diagram.

(a) What is the probability that the system is empty?

(b) What is the expected number of customers in the system?

(c) What is the probability that a customer will pass through the server exactly twice?

(d) What is the expected number of times that a customer will pass through the server?

(e) What is the expected amount of time that a customer will spend in the system?

18. Consider the queuing network shown in the diagram. The first node is an *M/M/*1 system and the second is an *M/M/*2 system. Answer the following questions.

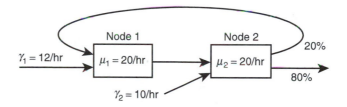

(a) What is the probability that node 1 is empty?

(b) What is the probability that the entire network is empty?

(c) What is the probability that there is exactly one job in the queue at node 1?

(d) What is the probability that there is exactly one job within the network?

(e) What is the expected number of jobs within the network?

(f) What is the expected flow time through the system for an arbitrarily selected job? (Answer this question in two different ways—first using Property 3, then using Little's law.)

(g) What is the expected flow time through the system for a job that starts at node 1? Can Little's law be used for this determination?

(h) A cost of $100 per hour is associated with each job within the system. A proposal has been made that a third node be added after the second. The third node would be an *M/M/s* system and would eliminate the existing feedback. Each server added to the third node would operate at a rate of 20 per hour and would cost $200 per hour. Would the third node be worthwhile? If so, how many servers should it contain?

19. Patients arrive at a hospital emergency room according to a Poisson process at a mean rate of 24 per hour. Twenty percent of them first go to the receptionist to fill out paperwork, 60% are seen by a doctor immediately, and the rest go directly to surgery. After finishing with the receptionist, 75% are sent to a doctor, 20% to surgery, and 5% to another facility within the hospital for treatment. After seeing a doctor, 90% of the patients leave the hospital and the rest go into surgery for minor operations (patients suffering major injuries are sent to the operating room in the hospital). When the surgery is finished, the patient leaves the hospital.

The one receptionist takes an exponential amount of time to gather the necessary information, averaging five minutes per patient. The noncritical patients are seen by one of three doctors; time of service is exponentially distributed, with a mean of 6 minutes. Two surgeons are available to perform minor operations. Each surgeon can treat five patients per hour on average, but the actual time is exponentially distributed. Assuming that the underlying system can be modeled as a Jackson network, respond to the following.

(a) Draw a block diagram to represent the emergency room network.

(b) What is the corresponding switching probability matrix?

(c) For a patient who sees a doctor, what is the average amount of time he or she spends in the queue?

(d) What is the average number of patients in the emergency room?

(e) Compute the mean flow time through the network—that is, compute the average time that a patient spends inside the emergency room.

(f) Compute the variance of the flow time through the network.

(g) Surgeons often have to perform other duties outside the emergency room during their shifts. The chief administrator wants to know, on the average, the amount of time per hour that no more than one surgeon will be absent from the emergency room.

(h) Answer parts (a) through (d) again with one change in the patient flow dynamics. Assume that after spending time in surgery, 15% of the patients must see a doctor whereas the remaining 85% leave without further consultations. (As such, a small fraction of patients who first see a doctor are sent into surgery and then return to see a doctor.)

20. After graduating from Big State University with a bachelor's degree in operations research, Norma Jeane Mortenson was hired as a manager of a local Barnes & Noble book store. With her training in queuing theory, she was eager to apply the techniques she had learned in the classroom. She was told that one of the most frequent customer complaints was that it took too long to go through the checkout line. To get a better understanding of the waiting time, she asked one of her employees to record customer arrival times at the cashier between 8:00 A.M. and noon. The following data were recorded: 8:05, 8:07, 8:17, 8:18, 8:19, 8:25, 8:27, 8:32, 8:35, 8:40, 8:45, 8:47, 8:48, 8:48, 9:00, 9:02, 9:14, 9:15, 9:17, 9:23, 9:27, 9:29, 9:35, 9:37, 9:45, 9:55, 10:01, 10:12, 10:15, 10:30, 10:32, 10:39, 10:47, 10:50, 11:05, 11:07, 11:25, 11:27, 11:31, 11:33, 11:43, 11:49, 12:05.

Another employee was asked to measure how long it took the cashier to check out a customer. On average, the service time was found to be 3.5 minutes with a corresponding standard deviation of 5.0 minutes. The store had two cash registers but currently employed only one cashier.

(a) Using the approximation formulas in Section 17.3, estimate the expected length of time that a customer has to wait before receiving service and the expected number of customers in front of the cashier.

(b) Suppose that Norma Jeane's knowledge of queuing theory was marginal and that an $M/M/1$ model was incorrectly used under the assumption that the arrival process was Poisson with a mean estimated from the data and that the service times were exponentially distributed with a mean of 3.5 minutes. Determine the approximate difference between the estimates obtained in part (a) and the estimates that would be obtained using the (incorrect) Markovian assumptions.

(c) Now there are two alternatives available to Norma Jeane to reduce the waiting time at the cashier. The first is to buy a bar code reader, which would reduce the standard deviation of the service time to 1.7 minutes but would cost $50 per month for a service contract. The second is to hire another cashier to operate the second cash register. The new cashier would work at the same rate as the original one and would cost $500 per month. Assuming that the cost of waiting for a customer is $0.25 per minute on average, Norma Jeane wants to know which is the better alternative. Assume that the congestion problem exists for only 4 hours per day, 5 days per week.

BIBLIOGRAPHY

Bard, J.F., K. Srinivasan, and D. Tirupati, "An Optimization Approach to Capital Expansion in Semiconductor Manufacturing," *International Journal of Production Research*, Vol. 37, No. 15, pp. 3359–3382, 1999.

Feldman, R.M. and C. Valdez-Flores, *Applied Probability & Stochastic Processes*, PWS, Boston, 1996.

Gross, D. and C.M. Harris, *Fundamentals of Queueing Theory*, Third Edition, Wiley, New York, 1998.

Jackson, J.R. "Networks of Waiting Lines," *Operations Research*, Vol. 5, pp. 518–521, 1957.

Kao, E.P.C., *Introduction to Stochastic Processes*, Duxbury, Belmont, CA, 1997.

Mehdi, J., *Stochastic Models in Queueing Theory*, Academic Press, Boston, 1991.

Ross, S.M., *Introduction to Probability Models*, Fifth Edition, Academic Press, San Diego, 1993.

Ross, S.M., *Stochastic Processes*, Second Edition, Wiley, New York, 1996.

Whitt, W., "The Queuing Network Analyzer," *Bell System Technical Journal*, Vol. 62, No. 9, pp. 2779–2815, 1983.

Whitt, W., "Open and Closed Models for Networks of Queues," *AT&T Bell Laboratories Technical Journal*, Vol. 63, No. 9, pp. 1911–1979, 1984.

Chapter 18

Simulation

Most of us have played video games that simulate both real and imaginary adventures we can only dream about experiencing. Examples include driving a racecar in the Indianapolis 500, searching for precious stones in the Andes, and fending off attacks from invading aliens. More common are movies, TV programs, and theatrical productions that provide the simulated drama and sensationalism that are normally absent in our everyday lives. At a practical level, airplane pilots train using equipment that simulates flight, the military fights mock battles that simulate real combat, and politicians preparing for debates face staff members who simulate their opponents. In each case, an environment is created that is as close to reality as possible.

Thus, we can think of simulation as an alternative representation of reality—a certain type of model. For our purposes, we will be using a computer rather than specialized equipment, actors, and props for the simulation. In some cases it will be possible to develop a graphical representation of the system under study and observe its simulated behavior. This will require the use of a special-purpose language that allows the user to manipulate symbols and objects to construct the system model. A common example might involve the simulation of an assembly line in which electronic components, chassis, and wire harnesses are joined together at various workstations. Additional system components might include a warehouse for inventory, movement of raw materials, and subassemblies on the shop floor, testing, rework, packing, and shipping. Many simulation languages are available from commercial vendors. In this chapter, we do not concentrate on languages per se but on the methodology underlying their use. Our primary goal is to convey an understanding of their power, appropriateness, and limitations, as well as the effort involved in their development, testing, and implementation.

Given that simulation is based on a mathematical model, further distinctions can be cited, such as deterministic versus stochastic and static versus dynamic. Deterministic simulations involve variables and parameters that are fixed and known with certainty, whereas stochastic simulations assign probability distributions to some or all variables and parameters. In practice, both types of simulations are important, although many analysts are inclined to narrow their definition to those experiments based only on stochastic processes.

For the most part, this chapter will highlight *discrete-event dynamic simulation* (DEDS). Of interest are events that occur at discrete points in time, such as the arrival of a customer, the completion of a transaction, the breakdown of a machine, or the shipment of a product. The systems to be analyzed will consist of discrete entities that are processed over time—that is, individual items that flow through the system asynchronously. Games of chance and board games can also be characterized in these terms, although in a game the length of the activity or time between events is not particularly relevant to the outcome. We can also speak of a *static* simulation in which there is no notion of a system evolving over time. One example might be the kinds of experiments performed by statisticians to derive the null distributions of complicated test statistics. To compute the statistic of interest, it might be necessary to generate realizations of many random variables and then manipulate the results in some complicated way at each iteration. In this situation, there is no system clock running in the background as there would be in a dynamic simulation. All this is in contrast to continuous-type systems in which differential or difference equations are used to describe behavior. Examples include electronic circuits, waste water treatment, chemical processing, and the movement of a robotic arm. In each of these cases, the simulation is performed in synchronous time steps with the state of the system changing continuously.

Many authors treat discrete-event simulation, whether static or dynamic, as a subset of *Monte Carlo methods* (MCMs). In broad terms, MCMs fit into the branch of applied mathematics concerned with experiments involving random numbers—i.e., random samples from the uniform distribution on the unit interval [0, 1]. Examples include sampling from known distributions, coin flips, card games, and the like. In each of these cases, MCMs can be used to estimate the probability distributions of all possible outcomes and all conditional strategies.

In Section 18.1, we describe where simulation fits in the catalog of quantitative techniques and provide some motivation for its use. We then introduce the components of the discrete-event model. This is followed by a presentation of the inverse transformation method, which can be used to generate samples for any known probability distribution. Next we describe how a model is run and what data need to be recorded. The computational procedures arc illustrated with queuing and inventory examples. We also provide a blueprint for conducting an analysis along with methods for analyzing output data.

18.1   NATURE AND PURPOSE OF SIMULATION

Of the analytic methods that comprise operations research, simulation stands in sharp contrast to the mathematical programming algorithms and stochastic models presented earlier. With simulation, the analyst creates a model of a system that describes some process involving individual entities such as persons, products, or messages. The components of the model try to reproduce, with varying degrees of accuracy, the actual operations of the real components of the process. Most likely the system will have time-varying inputs and time-varying outputs that are affected by random events. The components of the simulation are interconnected and can often be viewed as a network with complex input–output relationships. Moreover, the flow of entities through the system is controlled by logic rules that derive from the operating rules and policies associated with the process being modeled.

Because the model takes the form of a computer program, which operates as a facsimile of its real-world counterpart, it is much less restricted than analytical models such as those encountered in Chapters 14 through 17. Within the limitations of the input and output interfaces, a skilled programmer can duplicate, with a high level of accuracy, most systems that can be observed and rationalized. Because of this capacity for detail, simulation has become a very popular method of analysis. Particularly appealing is its ability to model random variables with arbitrary probability distributions and systems that comprise a variety of interacting random processes. Modern simulation languages are very powerful tools, allowing even a beginner to create representations of complex systems.

Several factors that motivate the use of simulation are as follows.

1. Simulation may be the only alternative that can provide solutions to the problem under study. For example, it is not possible to obtain transient (time-dependent) solutions for complex queuing models in closed form or by solving a set of equations, but they are readily obtained using simulation methods.

2. Models to be simulated can represent real-world processes more realistically because fewer restrictive assumptions are required. Examples include the use of nondeterministic lead times in an inventory model, non-Poisson arrivals or service times in a queuing process, and nondeterministic parameters in a multiperiod production scheduling and inventory control problem. Each of these situations results in analytic models that are intractable.

3. Changes in configuration or structure can be easily implemented to answer "What happens if . . . ?" questions. For example, various decision rules can be tested for altering the number of servers in a network of queues.

4. In most cases, simulation is less costly than actual experimentation; in other cases, it may be the only reasonable initial approach, such as when a system does not yet exist but its theoretical relationships are well known. For example, solar energy thermal collection systems for homes have been tested by simulation prior to manufacture to help solve site-specific problems or to explore new design issues.

5. Simulation can be used for pedagogical (teaching) purposes either to illustrate a model (as in Section 18.9, where we simulate an $M/M/1$ queue) or to enhance comprehension of a process, such as revenue management policies used by airlines to price tickets over time.

6. For many dynamic processes, simulation provides the only means of direct and detailed observation within specified time limits. The simulation approach also allows time compression, whereby a simulation accomplishes in minutes what might require years of actual experimentation.

With these advantages and others, one might ask "Why not approach all modeling through simulation?" First, simulation is time consuming and costly compared with many analytic approaches. For example, a simulation to estimate optimal reorder levels and quantities for an inventory problem requires an extensive search for optimal values of the controllable variables, whereas an analytic solution would not. This issue is explored in Sections 18.6 and 18.9. Second, certain issues associated with design, validation, and estimation are complex at best and unresolvable at worst. Because it attempts to reproduce significant amounts of detail, a simulation model may require a large programming effort, its accuracy may be difficult to verify, and the computational burden it imposes may be extensive compared with other approaches. Like queuing analysis, simulation is a tool that requires the enumeration of alternatives to determine an optimal design. Unlike queuing analysis, simulation does not yield expected values. Rather, making a simulation run is like observing a system in the real world and recording the relevant statistics. Since statistics are themselves random variables, the interpretation of the results must be done carefully with procedures based on the appropriate theory. This requirement is often neglected in practice, giving rise to the possibility that the results will be either misinterpreted or misused.

18.2   MODEL COMPONENTS

To explain how simulation works, it is first necessary to introduce some new terminology and to redefine several terms used in previous chapters. The basic concept is that of a *system,* which can be thought of as a collection of entities, rules, and objects that interact with each other to achieve some stated goal. From a modeling point of view, it is necessary to draw a boundary around the system under investigation and treat everything outside the boundary either as fixed in time or as exogenous inputs. Where the boundary is drawn is often arbitrary but depends to some extent on the scope of the analysis and the availability of data. In the system, an object of interest is called an *entity* and has properties called *attributes*. For example, in a mail processing and distribution center the pieces of mail would be the entities and would be characterized by source, class, size, quality of address, current level of processing, destination, and so on. The system boundary might be the building itself or perhaps that portion of the facility used to handle letters and flats if these pieces are the focus of the study. Parcels, bundles, and retail operations might be outside the system boundary.

When developing a discrete event dynamic simulation model, we do not necessarily use equations to represent the system. More often, we identify the objects that perform the operations, the rules and logic that govern transactions, the possible paths that the entities can take, and the circumstances under which the attributes of an entity will change. Much of this can be represented in a flow diagram similar to the ones used in the two preceding chapters on queuing. Thinking in these terms, we say that an entity enters the system, undergoes various transactions, resides in queues, is routed to different locations, and then leaves the system. As the entity experiences an event such as the completion of a service operation, a new activity is initiated. The simulation jumps through time, simulating one activity and its terminal event after another, collecting statistics as the clock moves forward. When the run is complete, the path that each entity followed is known. This allows us to compute many descriptive statistics such as average queue lengths and waiting times, probability distributions for the numbers of entities in the various queues, and component efficiencies. Through proper interpretation, much can be learned about how the system performed and how it is likely to perform under similar conditions.

When an analysis of a complex system is being conducted, the basic building blocks of a simulation model can be pieced together in novel and intricate ways to represent almost any dynamic situation. These building blocks are as follows.

1. The *state of the system:* the collection of variables necessary to describe the status of the system at any given time. Examples include the number of patients being treated in an emergency room and the number of pieces of mail waiting to be sequenced by a bar code sorter.

2. *List of possible states* of the system that are feasible. In previous chapters, we called this the state space.

3. *List of possible events* that would change the state of the system (e.g., the diagnosis of a patient's illness or the landing of an aircraft).

4. A *simulation clock* that records the passage of (simulated) time in the software that implements the model.

5. A method for *randomly generating events* from the list of possible events, and an accounting scheme for keeping track of when they occur in simulated time.

6. A set of rules or formulas that govern the *state transactions* occasioned by the various type of events.

Although building block 5 emphasizes the need to deal with the randomness in a system, a simulation model, as we have noted, can be either stochastic or deterministic. Our interest centers mostly on stochastic models in which randomness plays a major role. When all activities and events are deterministic, it is still possible and useful to use simulation to study the behavior of a system. In these cases, the model can be viewed simply as an accounting scheme for keeping track of events and gathering statistics.

Process

Whether it is in manufacturing, transportation, patient care, or dealing with a bureaucracy, it is natural to characterize the dynamics of the system in terms of entities (e.g., persons or items) entering, passing through, and leaving various processes. Conceptually, a process is a series of steps at which one or more operations are performed on the entity. When all the steps have been completed, the person or item leaves the process, hopefully better off for the experience. Examples are patients passing through a medical facility, raw material passing through a manufacturing plant, and messages passing through an information system. Of course, not all situations involve a series of discrete activities on discrete entities; however, simulation has been particularly successful in modeling these kinds of systems. One of the simplest processes is the queuing system illustrated in Figure 18.1, which might represent the check-in counter at an airport. Passengers or customers arrive randomly, with the time between arrivals governed by some probability distribution. The time for service is also a random variable governed by a different probability distribution. When a passenger arrives and finds all the clerks busy, she joins a queue. Typically, passengers in the queue are served in the order in which they arrive unless there is a disruption in the flight schedule. We simplify the following discussion by assuming that there is only one server.

For simulation purposes, a customer in our example receives the general designation of *entity*. An entity is the item passing through the system being simulated. In different situations, entities might be customers, messages, products, or documents, among others.

Time, Events, and State

When the movement of entities through a dynamic process is being simulated, *time* is a critical variable. The simulation starts at time zero, runs for some specified period of time during which entities move through the process, and then terminates. Of primary interest are the times at which *events* occur. In the airline check-in example, there are two kinds of events: arrivals and service completions (we give the latter the shorter designation *service*). We move through time by jumping from one event to the next. For this reason, the procedure is called a *next-event simulation*.

Alternatively, *fixed time increment* simulations proceed by equal increments of the clock. At the end of each interval of time, the system is scanned in order to record the occurrence

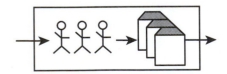

Figure 18.1   Simple queuing system

or nonoccurrence of events. The Dynamo simulation language, developed for modeling continuous-time systems that can be represented by difference equations, uses fixed time increments. Many interesting systems, including several that have arisen in production and inventory control, fit this category.

The response of the system to an event depends on the system *state*. The state indicates information about the system. For the check-in example, it is sufficient to describe the state using three variables: time, the number of entities in the queue, and the number of entities in service.

Logic

The logic of the system specifies the changes in state caused by the occurrence of an event. A single-server queuing system operates according to two rules.

- When the *arrival* event occurs and the entity finds the server idle, the entity directly enters service, and the number in service increases by 1. Otherwise, the entity enters the queue, and the number in the queue increases by 1.

- When the *service* event occurs and the queue is empty, the server becomes idle and the number in service decreases by 1. Otherwise, the first entity in the queue enters service activity, and the number in the queue is reduced by 1.

Events also play the role of triggering future events as specified by the logic of the simulation. Again, we find a rule for each of the two events. The rules may depend on the current state of the system.

- When the *arrival* event occurs, the next arrival event may be scheduled. The time of the next arrival is the current time plus the time between arrivals (t_a). The value of t_a is a simulated random observation from a specified probability distribution for interarrival times. If the server is idle when the event occurs, a service event must also be scheduled. This will be illustrated presently.

- When the *service* event occurs and an entity enters the service activity from the queue, the next service event may be scheduled. The time of this event is the current time plus the time for service (t_s). The value of t_s is a simulated random observation from a specified probability distribution for service times.

New events triggered in this fashion are put on a *calendar*. There are two characteristics of a calendar event—its type and its time. In this simple example, the type is either *arrival* or *service*. For more complex systems there would be many event types, but the logic described above would carry over in an analogous manner. A final point about generating events is that the interarrival and service times can all be generated at the beginning of the simulation run. In other words, we do not have to wait for an arrival before generating the next interarrival time. From a computational point of view, there is considerable efficiency in generating them all at once.

18.3 INVERSE TRANSFORMATION METHOD FOR GENERATING RANDOM VARIATES

The inverse transformation method (ITM) is a popular technique that can be used to generate random variates (realizations of random variables) from many common probability distributions such as the exponential distribution, the Weibull distribution, the uniform distribution, and all empirical continuous distributions. Moreover, it embodies the underlying principle for sampling from a wide variety of discrete distributions. This method will be explained in detail, first for the exponential and Poisson distributions and then for an empirical distribution. In the presentation, a specific value of a random variable (r.v.) will be represented by a lowercase letter (say, x), and the random variable itself will be denoted by an uppercase letter (say, X).

Exponential Distribution

From elementary knowledge of probability theory, recall that the probability density function (PDF) for an exponential random variable is given by

$$f(x) = \begin{cases} \lambda e^{-\lambda x}, & x \geq 0 \\ 0, & x < 0 \end{cases}$$

and that the cumulative distribution function (CDF) is given by

$$F(x) = \Pr\{X \leq x\} = \int_{-\infty}^{x} f(t)dt = \begin{cases} 1 - e^{-\lambda x}, & x \geq 0 \\ 0, & x < 0 \end{cases}$$

The parameter λ is interpreted as the mean number of occurrences per unit time, and the expected value of the random variable X is $E[X] = 1/\lambda$. The goal is to develop a procedure for generating realizations of $X_1, X_2, X_3, \ldots$ that come from an exponential distribution.

The ITM can be used for this purpose whenever the CDF, $F(x)$, has a form such that its inverse F^{-1} can be computed explicitly by analytic means. The idea is to solve

$$x = F^{-1}(u) \tag{1}$$

for different realizations of the uniform random variable U. We now give a step-by-step procedure for doing this.

Step 1: Identify the CDF of the desired random variable, X. For the exponential distribution, $F(x) = 1 - e^{-\lambda x}$, $x \geq 0$.

Step 2: Set $F(x) = u$ on the range over which X is defined. Here, U is a uniform random variable defined on the interval from 0 to 1, written $U(0, 1)$. Approaches to generating values of U will be discussed presently. For the exponential distribution, we have $1 - e^{-\lambda x} = u$ over the range $x \geq 0$.

Step 3: Solve the equation $F(x) = u$ for x in terms of u. For the exponential distribution, the solution proceeds as follows.

$$1 - e^{-\lambda x} = u$$
$$-\lambda x = \ln(1 - u)$$
$$x = \frac{-1}{\lambda} \ln(1 - u) \tag{2}$$

Equation (2) is called the *random variate generator* for the exponential distribution. In general, Equation (2) is written as $x = F^{-1}(u)$. The final step involves generation of the required realizations.

Step 4: Generate (as needed) uniform random variates $u_1, u_2, u_3, \ldots$ over [0, 1] and compute the desired random variates by solving

$$x_i = F^{-1}(u_i), \quad i = 1, 2, \ldots$$

For the exponential case, we have

$$x_i = \frac{-1}{\lambda}\ln(1 - u_i), \quad i = 1, 2, \ldots \tag{3}$$

Figure 18.2 gives a graphical interpretation of the ITM. The CDF shown is $F(x) = 1 - e^{-x}$, an exponential distribution with rate $\lambda = 1$. To generate a realization of X, first a random number u is generated between 0 and 1, then a horizontal line is drawn from u to the graph of the CDF, and finally, a vertical line is dropped to the horizontal axis to obtain x, the desired result. In Figure 18.2, $u = 0.74$ and, using Equation (3), we get $x = 1.34$.

In practice, Equation (2) is usually simplified by replacing $1 - u$ with u, since both are uniformly distributed on $[0, 1]$. The new equation is

$$x = \frac{-1}{\lambda}\ln(u)$$

The ITM gets its name from the fact that a transformation is made from the u scale to the x scale. Note that this procedure is the inverse of the usual procedure. Typically, we map a specific x into a specific u through $F(x)$—that is, $F(x) = u$.

Poisson Distribution

If $N(t)$ is a Poisson process with parameter λ, it can be proven, as stated in previous chapters, that the random variable representing the time between any two occurrences of $N(t)$ is exponential with the same parameter λ. Now, if k Poisson arrivals occur over some arbitrary time t, then the sum of the times between the k arrivals *must* be less than or equal to t. If x_i represents the time between arrivals $i - 1$ and i, then the following relationship must hold.

$$\sum_{i=1}^{k} x_i \leq t < \sum_{i=1}^{k+1} x_i$$

Thus, Poisson random variates with parameter λ can be generated by summing exponential random variates (x_i) until the expression above is satisfied. The simulated Poisson random variate is the value of k. Note that x_i is found from Equation (3).

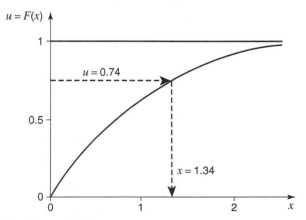

Figure 18.2 Graphical view of ITM

Normal Distribution

Many methods have been developed for generating normally distributed random variables, but the ITM is not one of them because the inverse CDF cannot be computed analytically. The standard normal CDF with $\mu = 0$ and $\sigma^2 = 1$ is given by

$$F(x) = \int_{-\infty}^{x} \frac{1}{\sqrt{2\pi}} e^{-t^2/2} dt, \quad -\infty < x < \infty$$

To use the ITM, it is necessary to be able to solve (in closed form) $F(x) = u$ for x in terms of u, which is not possible for the normal distribution. As an alternative, we present a technique based on the *central limit theorem* (CLT). The CLT asserts that the sum of n independent and identically distributed random variables $X_1, X_2, \ldots, X_n$, each with mean μ and finite variance σ^2, is approximately normally distributed with mean $n\mu$ and variance $n\sigma^2$. Applying this to n uniform random variables, $U_i, i = 1, \ldots, n$, on the interval [0, 1] with mean 0.5 and variance 1/12, it follows that

$$Z = \frac{\sum_{i=1}^{n} U_i - 0.5n}{(n/12)^{1/2}}$$

is approximately normally distributed with mean 0 and variance 1, written $N(0, 1)$. The approximation becomes better the larger the value of n, although it is generally thought that $n = 12$ is sufficiently large for an adequate result. Moreover, the use of $n = 12$ is the most efficient for computational purposes because the square root and a division operation are avoided, as can be seen by substituting $n = 12$ into the preceding equation to get

$$Z = \sum_{i=1}^{12} U_i - 6$$

for an approximate normal random variable with mean 0 and variance 1. If it is desirable to generate a normal random variable Y with mean μ_Y and variance σ_Y^2, we would first generate Z and then use the transformation

$$Y = \mu_Y + \sigma_Y Z$$

To see how to use this technique, assume that service times at a supermarket checkout counter are normally distributed with mean $\mu = 7.3$ min and variance $\sigma^2 = 11.7$ min^2. To generate a typical service time, we first need to obtain 12 uniformly distributed random numbers between 0 and 1. Say,

0.1489,	0.1758,	0.2774,	0.6033,	0.9813,	0.1052
0.7484,	0.1816,	0.1699,	0.7350,	0.6430,	0.8803

Now, we use the equation for Y to obtain the sample

$$y = 7.3 + \sqrt{11.7}\left(\sum_{i=1}^{12} u_i - 6\right) = 6.10 \text{ min}$$

where u_i is the ith sample from $U(0, 1)$.

Many authors recommend this technique for generating approximate normal random variates, but an exact approach is always preferable. For more detail, see any of the simulation texts referenced at the end of this chapter.

Empirical and Discrete Distribution

To illustrate the use of the ITM for discrete distributions, consider the random variable X, "number of emergency room arrivals per hour," and the data in Table 18.1. We use the notation, once again, where a specific value of a random variable is represented by a lowercase letter and the random variable itself is denoted by an uppercase letter.

According to the relative frequency definition of probability, the probability of zero arrivals, or $\Pr\{X = 0\}$, is 0.10, $\Pr\{X = 1\} = 0.28$, ..., and $\Pr\{X = 6\} = 0.01$. Now, if U is a second random variable that is uniformly distributed over the range [0, 1], then, by definition, $\Pr\{0.00 \leq U \leq 0.10\} = 0.10$, $\Pr\{0.10 < U \leq 0.38\} = 0.28$, ..., and $\Pr\{0.990 < U \leq 1.00\} = 0.01$. Since the probabilities for the given range of U are, respectively, identical to the probabilities for the given values of X, it follows that occurrences of U can be used to simulate or "artificially reconstruct" occurrences of X as we did in the continuous case. An examination of Table 18.1 reveals that random variates of U within the specified intervals are equivalent to corresponding random variates of X. In other words, U can be used for artificial generation of X precisely because the probabilities associated with the specified ranges of U are exactly equivalent to the probabilities associated with the corresponding values of X.

As we can see in Table 18.1, the probability mass function (pmf), $f(x)$, is simply the relative frequency of each occurrence, and the CDF, $F(x)$, is equivalent to specified values of U. Once again, we use Equation (3) to determine a specific realization x given a sample u from the uniform distribution—that is, $x = F^{-1}(u)$. Table 18.2 illustrates a "hand" simulation for six 1-hour periods based on the data in Table 18.1. The simulated number of arrivals for each period first requires the generation of a random variate (u) from the distribution of U. The first realization in Table 18.2, $u_1 = 0.246$, corresponds to one simulated arrival ($x = 1$) between 10 A.M. and 11 A.M., since 0.246 falls between 0.10 and 0.38 on the scale given in Table 18.1. Continuing down the second column, we find that $u_2 = 0.514$ falls in the interval [0.38, 0.67], which corresponds to $x = 2$ (two arrivals between 10 A.M. and noon). To check your understanding, you should confirm the remaining entries in Table 18.2.

The transformation of u to x for our present example is shown graphically in Figure 18.3. In general, the simulation of a *discrete* random variable can be represented by the graph of its CDF. Conceptually, the simulation of a continuous random variable is treated in the same manner.

Table 18.1   Distribution of Emergency Room Arrivals

Number of arrivals per hour, x	Frequency of occurrence	pmf, $f(x)$	CDF, $F(x)$	Interval along 0-1 scale
0	10	0.10	0.10	0.00 – 0.10
1	28	0.28	0.38	0.10 – 0.38
2	29	0.29	0.67	0.38 – 0.67
3	16	0.16	0.83	0.67 – 0.83
4	10	0.10	0.93	0.83 – 0.93
5	6	0.06	0.99	0.93 – 0.99
6	1	0.01	1.00	0.99 – 1.00
Total	100	1.00		

Table 18.2   Sample Simulation of Emergency Room Arrivals

Time period (24-hour clock)	Uniform random variate on [0, 1], u	Number of arrivals during specified period, x
10 – 11	0.246	1
11 – 12	0.514	2
12 – 13	0.898	4
13 – 14	0.030	0
14 – 15	0.152	1
15 – 16	0.573	2

Actual simulations of *empirical* probability distributions can be implemented in the manner of Tables 18.1 and 18.2. This is referred to as the *tabular method*. In general, let $x - 1$ and x be two consecutive values of a discrete random variable X, and let $F(x - 1) = \Pr\{X \leq x - 1\}$. Given a uniform random decimal fraction $u \in U(0, 1)$, the associated random variate is found as that value of x such that $F(x - 1) \leq u < F(x)$. Thus, with each value of x we associate a range of uniform random decimal fractions of size $\Pr\{X = x\}$. If $F(x)$ is specified up to k significant digits after the decimal point, we use uniform random decimal fractions with up to k significant digits. The range of u for each x is $[F(x - 1), F(x) - 10^{-k}]$.

For theoretical distributions, such as the Poisson and exponential, random variates of X are generated through the use of algorithms or formulas for $F^{-1}(u)$, as described previously. Nevertheless, the tabular method can always be used whether or not algebraic expressions are available. For a continuous random variable defined over a finite range, we would select a grid of, say, $N + 1$ points covering its range: $x_0, x_1, \ldots, x_n, \ldots, x_N$. For each grid point x_n, we find $F(x_n)$. The corresponding interval of uniform random decimal fractions associated with the interval $[x_{n-1} < x \leq x_n]$ is $[F(x - 1), F(x) - 10^{-k}]$ for random numbers

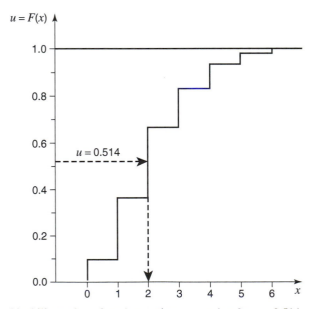

Figure 18.3   Graphical illustration of random variate generation for $u = 0.514$

with k-digit accuracy. To determine x for a given value of u, we would then use linear interpolation between x_{n-1} and x_n.

$$x = x_{n-1} + \left(x_n - x_{n-1}\right)\frac{u - F(x_{n-1}) - 10^{-k}}{F(x_n) - F(x_{n-1})}$$

When N is large, implying that the grid size is small, it is probably sufficient to approximate x as the midpoint of each interval.

For continuous random variables defined over an infinite range, it is permissible to truncate the tails at a point where the probability of the random variable falling outside the finite range becomes smaller than 10^{-k}. This approximation is similarly valid for discrete random variables.

Generation of 0-1 Uniformly Distributed Random Variates

All Monte Carlo methods, such as discrete-event simulation, require the generation of a sequence of random numbers (variates of U) uniformly distributed over the interval [0, 1]. The process that generates u must satisfy the following conditions: (1) the parent population must be *uniformly distributed*, (2) the sampled observations (variates) from that population must be *independent* (no correlation), (3) the process should be fast (or cheap), and (4) the process should have a long period if it repeats itself.

The most common methods for generating u include (1) use of special electronic devices, (2) physical phenomena (for example, fluctuations in some crystalline structure), and (3) the application of a numerical process. Of the three, numerical processes based on so-called *congruential methods* most nearly satisfy the four criteria outlined above. Suffice it to say that all computer manufacturers and software vendors selling analytic products make available computational procedures (usually in the form of subroutines or functions) for generating random numbers. Therefore, a simulation model programmed for a computer need only incorporate the appropriate subroutine for generating u.

Random numbers for U based on numerical methods are called *pseudorandom numbers*, as opposed to *true* random numbers, which are generated by some physical process. Believe it or not, pseudorandom numbers are preferable because the exact same sequence of random numbers is capable of being repeated at will by the analyst. This has control value when the analyst wishes to alter a policy or decision variable in answering "What if . . .?" questions. Without the capability of repeating the same sequence of random variates, experimental results would differ because of both conscious changes by the analyst and random changes owing to a different random number sequence. The latter effect would represent a confounding factor that would have to be isolated by replications of the experiment. Pseudorandom numbers, therefore, uniquely isolate the effects of changes in controllable conditions. Furthermore, purely random effects can be estimated easily by rerunning experiments with a new *random number seed*—i.e., an arbitrary number provided by the analyst that initializes the random number subroutine.

18.4   SIMULATING COMPLEX RANDOM VARIABLES

For relatively simple random events such as the arrival of patients at an emergency room or the repair of a malfunctioning piece of equipment, it is often possible to fit either an empirical or a known distribution to the underlying process. For more complex systems, events are composites of multiple random phenomena, so more effort is required to derive the corresponding distributions. In particular, it is necessary to simulate the constituent random variables that define the process and observe their effects with the help of statistical methods. This approach is widely used in operations research, as illustrated throughout this section.

Production Process Example

Consider an assembly operation fed by three independent workstations. Each station produces one of three types of parts, which are then assembled to yield a finished product. The stations have been balanced in terms of work load so that the average production rate of each is 10 units per day. The actual output, however, is a discrete random variable uniformly distributed over the interval [8, 12]. Work in process cannot be kept from one day to the next, so the production of the system is equal to the output of the least productive station. Let X_1, X_2, and X_3 be random variables corresponding to the production rates of the three stations. The daily production of the system is a random variable Y defined as

$$Y = \min\{X_1, X_2, X_3\}$$

Our interest is in learning about the distribution of Y.

Students with advanced training in mathematical statistics may be able to derive the cumulative distribution function of Y given this information using probability theory. We will take the conceptually easier approach of simulation. In particular, Table 18.3 presents 10 days of randomly generated production data obtained with the equation

$$x_i = \text{ROUND}(7.5 + 5u), \quad i = 1, 2, 3$$

where u is a 0-1 uniformly distributed random number. The ROUND function rounds the summation to the nearest integer thus, providing a uniformly distributed integer-valued number from 8 to 12, as desired. With these observations, we compute the simulated production of the system as

$$y = \min\{x_1, x_2, x_3\}$$

The statistical mean and standard deviation of the system production y are computed as follows for $n = 10$.

The *mean* or *average* of the data is: $\bar{y} = \sum_{i=1}^{n} y_i/n = 8.7$

The *sample variance* of the data is: $s^2 = \sum_{i=1}^{n}(y_i - \bar{y})^2/(n-1) = 0.4556$

The *standard deviation* of the data is: $s = \sqrt{s^2} = 0.6749$

Table 18.3   Simulated Values for Workstation Output

r.v.					Day					
	1	2	3	4	5	6	7	8	9	10
x_1	11	10	8	11	9	12	8	11	12	9
x_2	12	10	10	12	8	9	8	9	10	9
x_3	9	8	8	9	9	12	12	9	10	10
y	9	8	8	9	8	9	8	9	10	9

We conclude that the average daily production of the system is well below the average of the individual stations. The cause of this reduction in capacity is the variability in station output.

Because the statistics depend on the specific simulated observations, it is advisable to replicate the experiment several times to learn more about the variability of the simulated results. Table 18.4 gives the values of $\bar{y}$ for six replications. Although the results vary, they stay within a narrow range. Our conclusion about the reduction in the system capacity is certainly justified.

The grand average and the standard deviation of the six observations in Table 18.4 are 8.833 and 0.186, respectively. Since each of the replication averages are determined by the sum of 10 numbers, we can infer from the central limit theorem that the average observations are approximately normally distributed. This allows us to perform several additional statistical tests on the results, such as determining confidence limits.

Project Planning

A particular project consists of three tasks, A, B, and C. Tasks A and B can be done in parallel, whereas task C can begin only after tasks A and B have been completed. The times required for the tasks are T_A, T_B, and T_C, respectively, and are random variables; T_A has an exponential distribution with a mean of 10 hours, T_B has a uniform distribution that ranges from 6 to 14 hours, and T_C has a normal distribution with a mean of 10 hours and a standard deviation of 3 hours. The time required to complete the project, Y, is similarly a random variable because it depends on T_A, T_B, and T_C in the following manner.

$$Y = \max\{T_A, T_B\} + T_C$$

Our goal is to complete the project in 20 hours. What is the probability that we will be successful?

Solving this problem analytically would be difficult, because the completion time is not a simple function of the three random variables. Using simulation, however, we can get very good results. Table 18.5 gives the outputs for 10 realizations of the project. The rows labeled u_A, u_B, and u_C provide the random numbers used to simulate the corresponding task time random variables, and the rows labeled t_A, t_B, and t_C show the associated observations obtained using the generating procedures discussed in Section 18.3. For each of the 10 observations, one value of y was computed by substituting the random variates t_A, t_B, and t_C into the preceding equation.

It is clear from the bottom row in Table 18.5 that there is quite a bit of variability in the estimates of the time required to complete the project, y. Given that there are three values below 20, an initial estimate of the probability of completion in less than 20 hours would be 0.3. However, a much larger sample is needed for an accurate determination.

Table 18.4   Expect Workstation Output for Several Experiments

Replication	1	2	3	4	5	6
$\bar{y}$	8.7	8.6	9.1	9	8.8	8.8

Table 18.5   Simulated Results for Time Required to Complete Project

	Observation									
r.v.	1	2	3	4	5	6	7	8	9	10
u_A	0.663	0.979	0.256	0.312	0.141	0.092	0.311	0.251	0.139	0.983
t_A	10.89	38.41	2.954	3.745	1.52	0.968	3.722	2.89	1.497	40.76
u_B	0.953	0.33	0.139	0.326	0.512	0.379	0.964	0.335	0.578	0.432
t_B	13.63	8.638	7.111	8.609	10.1	9.034	13.71	8.681	10.62	9.459
u_C	0.731	0.615	0.557	0.18	0.729	0.305	0.869	0.861	0.554	0.488
t_C	11.85	10.88	10.43	7.253	11.83	8.466	13.36	13.25	10.41	9.912
y	25.48	49.28	17.54	15.86	21.93	17.5	27.07	21.94	21.03	50.67

18.5   DISCRETE-EVENT DYNAMIC SIMULATION

Perhaps the easiest way to understand how DEDS works is to simulate the simplest of systems, a single channel queue. Performing the computations involves following the logic of the process as customers arrive, receive service, and then depart. As simulated time progresses, we will gather statistics on the state of the system. For a single channel queue, the state has two components: the number of customers in service and the number of customers in the queue.

Simulating Activity Times

We illustrate the single channel queue with Poisson arrival and service processes. The simulation process is the same for any distribution, but the Poisson assumption allows us to compare the simulated result with the actual expected values computed using exponential queuing models.

An initial state must be specified before the simulation can begin. For this example, we assume that at time zero one customer is in service and one is in the queue. In general, there are three ways to specify the length of a run. The first involves a limit on the amount of simulated time that is allowed to pass, the second involves a limit on the number of entities that are to be processed, and the third involves a limit on CPU time.

Using the second approach, we now simulate the arrival of 10 additional customers, giving a total of 12 in all to be processed. Let the mean time between arrivals be 5 minutes and let the mean time for service be 4 minutes. Both random variables are assumed to be exponentially distributed. To simulate 12 customers, we need to generate 10 interarrival times and 12 service times from the respective distributions. The first step is to generate 22 random numbers in the interval [0, 1] and use Equation (3) to compute the corresponding realizations. The results are presented in Table 18.6, where $r_{a(i)}$ and $r_{s(i)}$ denote the uniform random variates, and $t_{a(i)}$ and $t_{s(i)}$ denote the corresponding randomly generated interarrival and service times, for customer i.

Except for the first two customers (entities) identified in the table, each has been assigned two random numbers to simulate their interarrival and service times. The first two customers are assumed to arrive at time zero, so they each need only one random number. Since the mean time between arrivals is the reciprocal of the arrival rate, we have $1/\lambda = m_a = 5$ minutes in this case. The interarrival time for the third customer, for example, is

$$t_{a(3)} = -m_a \ln (1 - r_{a(3)}) = -5 \ln(1 - 0.932) = -5 \ln(0.068) = 13.44$$

Table 18.6 Simulated Interarrival and Service Times

Customer number, i	Random number, $r_{a(i)}$	Interarrival time, $t_{a(i)}$	Random number, $r_{s(i)}$	Service time, $t_{s(i)}$
1	—	—	0.475	2.58
2	—	—	0.457	2.44
3	0.932	13.44	0.851	7.62
4	0.708	6.15	0.022	0.09
5	0.517	3.64	0.906	9.45
6	0.357	2.21	0.158	0.69
7	0.033	0.17	0.966	13.58
8	0.633	5.01	0.518	2.92
9	0.024	0.12	0.187	0.83
10	0.300	1.78	0.395	2.01
11	0.513	3.60	0.828	7.04
12	0.776	7.47	0.097	0.41
Total		43.59		49.66

This is the simulated time interval between the arrival of the second customer (time 0) and the arrival of the third customer. The service time for the third customer is simulated in a similar fashion using the mean service time, $m_s = 4$, as follows.

$$t_{s(3)} = -m_s \ln(1 - r_{s(3)}) = -4 \ln(1 - 0.851) = -4 \ln(0.149) = 7.62$$

The remaining data in Table 18.6 were generated with the random numbers provided. Times are rounded to two significant digits. We will use these data in our step-by-step illustration of the simulation process.

Next-Event Simulation for the *M*/*M*/1 Queue

In this example, the calendar will consist of at most two events, the next arrival and the next service completion, along with the time at which each occurs. We write the calendar as a list using the letters A and S to denote an arrival and a service, respectively. By arranging the calendar in order of decreasing time, the last event on the list will be the next event. The initial calendar is then

$$\{(A, 13.44), (S, 2.58)\}$$

and the next event is a service at time 2.58. When the service event occurs, the entity in the queue begins service, so its service event must be added to the calendar. As we will discuss shortly, it will occur 2.44 minutes after the first service event. The logic of the process being modeled as well as the realizations of interarrival and service times given in Table 18.6 are used to update the calendar at each iteration of the simulation.

Table 18.7 provides a common format for describing the progress of the computations. Each row corresponds to the state of the system at the time of the referenced event. Starting with event 1 (initial conditions), the information in each row can be used to generate the next event (row). Regarding the table headings, the first column contains a sequence number used for accounting purposes. The second and third columns give the event type and the time of its occurrence. For example, there are two service completions, $S1$ and $S2$, before the third arrival $A3$. This follows from the interarrival and service times given in

Table 18.6. Together with the time, the fourth and fifth columns describe the state of the system. The last column shows the current calendar sorted so that the next event is the last in the list.

The first row of Table 18.7 reflects the initial conditions, whereas the remaining rows correspond to the simulated output. Given entries for the first k rows, the next event is the last element on the current calendar. This event now defines row $k + 1$, and the time entry in column 3 is the associated event time. The numbers in service and in the queue are adjusted according to the logic of the simulation, and a new event is added to the calendar, if appropriate, as specified by the logic. For example, the next event for row 1 is $(S1, 2.58)$ and is given in row 2. Since this event is a service and an entity is in the queue, this entity enters the service activity and its completion time is added to the calendar. The completion time is the current simulated clock time plus the simulated service time from Table 18.6.

$$\text{Time for } S2 = 2.58 + 2.44 = 5.02$$

The number in the queue is reduced by 1.

When an arrival event occurs, as in row 4, the next arrival is added to the calendar. The time of this event is the current time plus the simulated interarrival time from Table 18.6.

$$\text{Time for } A4 = 13.44 + 6.15 = 19.59$$

The results of the simulation for all 12 customers are shown in Table 18.7. Note that the simulation moves in discrete time steps from one event to the next in an asynchronous

Table 18.7 Steps in Next-Event Simulation for $M/M/1$ Queue

Event number	Event	Time	Number in service	Number in queue	Calendar
1	A1, A2	0	1	1	$\{(A3, 13.44), (S1, 2.58)\}$
2	S1	2.58	1	0	$\{(A3, 13.44), (S2, 5.02)\}$
3	S2	5.02	0	0	$\{(A3, 13.44)\}$
4	A3	13.44	1	0	$\{(S3, 21.06), (A4, 19.59)\}$
5	A4	19.59	1	1	$\{(A5, 23.23), (S3, 21.06)\}$
6	S3	21.06	1	0	$\{(A5, 23.23), (S4, 21.15)\}$
7	S4	21.15	0	0	$\{(A5, 23.23)\}$
8	A5	23.23	1	0	$\{(S5, 32.68), (A6, 25.44)\}$
9	A6	25.44	1	1	$\{(S5, 32.68), (A7, 25.61)\}$
10	A7	25.61	1	2	$\{(S5, 32.68), (A8, 30.62)\}$
11	A8	30.62	1	3	$\{(S5, 32.68), (A9, 30.74)\}$
12	A9	30.74	1	4	$\{(S5, 32.68), (A10, 32.52)\}$
13	A10	32.52	1	5	$\{(A11, 36.12), (S5, 32.68)\}$
14	S5	32.68	1	4	$\{(A11, 36.12), (S6, 33.37)\}$
15	S6	33.37	1	3	$\{(S7, 46.95), (A11, 36.12)\}$
16	A11	36.12	1	4	$\{(S7, 46.95), (A12, 43.59)\}$
17	A12	43.59	1	5	$\{(S7, 46.95)\}$
18	S7	46.95	1	4	$\{(S8, 49.87)\}$
19	S8	49.87	1	3	$\{(S9, 50.07)\}$
20	S9	50.07	1	2	$\{(S10, 52.71)\}$
21	S10	52.71	1	1	$\{(S11, 59.75)\}$
22	S11	59.75	1	0	$\{(S12, 60.16)\}$
23	S12	60.16	0	0	

manner. Figure 18.4 provides a graphical view of the state variables associated with the simulation run. Starting with two customers at time 0, the number in the system (number in service plus number in the queue) first drops to zero and then rises to a maximum of six. The number drops back to zero after the last arrival, as the remaining customers are serviced.

The results of the simulation run depend on the particular set of random numbers used in generating the interarrival and service times in Table 18.6. A different set of realizations will yield different results. A simulation run is like an observation of the actual system in that each run provides a different result. Of course, simulating 12 customers is wholly inadequate if the goal is to determine steady-state behavior. Hundreds or perhaps thousands of customers must be simulated before any meaningful conclusions can be drawn, even for this simple system.

Statistics from the Simulation

The purpose of a simulation is to provide insight into how the real system behaves. With regard to queuing systems, the primary measures of interest are the average number of entities in the queue, the average number of entities in service, the average time an entity spends in the queue, the average time an entity spends in service, and the proportion of time that the server is idle. Each of these performance measures can be estimated from the simulation output.

Recall that the average value of any function of time $f(t)$ over the interval $[T_S, T_F]$ is the area under the graph of the function over this interval divided by its length—i.e.,

$$\bar{f} = \frac{1}{(T_S - T_F)} \int_{T_S}^{T_F} f(t)\,dt$$

We use this expression with the information in Figure 18.4 and Table 18.7 to compute the average numbers of customers in the queue and in service. For this table we have chosen $T_S = 0$ and $T_F = T$, where $T = 60.16$ is the time of the last event. In practice, one would usually choose $T_S > 0$ and $T_F < T$ in order to minimize the effects of the initial conditions and the last few events of the simulation. The interval $[0, T_S]$ is called the *warm-up period*, whereas

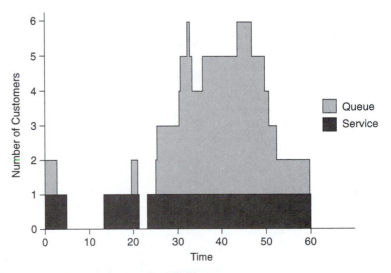

Figure 18.4   Output of simulation run for *M/M/*1 queue

the interval $[T_F, T]$ is called the *cool-down period*. The interval $[T_S, T_F]$ is called the *data collection period*.

The approach is to compute the areas under the respective curves between each two consecutive event and then sum the results that occur during the data collection period. Table 18.8 provides a convenient format for arraying the necessary data.

Let t_k be the time of event k, t_{k+1} the time of the next event, q_k the number in the queue after event k, and s_k the number in service after event k. Then, during the time interval $[t_k, t_{k+1}]$, the contributions to the service and queue areas are

$$\text{Service area } k = (t_{k+1} - t_k)s_k$$
$$\text{Queue area } k = (t_{k+1} - t_k)q_k$$

These quantities are presented in the last two columns of Table 18.8. To compute the total area under the service curve and queue curve, we sum the respective columns. Here, we compute the results from 0 to T, the time when the last customer leaves the system. For this realization, $T = 60.16$. The relevant areas are presented at the bottom of Table 18.8.

Once these areas have been computed, the performance measures can be estimated. The following notation will be used to describe the computations. To retain generality, the results will be given for a system with s servers, realizing that $s = 1$ for the example. We

Table 18.8   Computation of Areas under Output Curves

Event number, k	Time	Number in service	Number in queue	Service area k	Queue area k
1	0	1	1	2.58	2.58
2	2.58	1	0	2.44	0.00
3	5.02	0	0	0.00	0.00
4	13.44	1	0	6.15	0.00
5	19.59	1	1	1.47	1.47
6	21.06	1	0	0.09	0.00
7	21.15	0	0	0.00	0.00
8	23.23	1	0	2.21	0.00
9	25.44	1	1	0.17	0.17
10	25.61	1	2	5.01	10.02
11	30.62	1	3	0.12	0.36
12	30.74	1	4	1.79	7.12
13	32.52	1	5	0.15	0.80
14	32.68	1	4	0.69	2.76
15	33.37	1	3	2.76	8.25
16	36.12	1	4	7.47	29.88
17	43.59	1	5	3.35	16.80
18	46.95	1	4	2.92	11.68
19	49.87	1	3	0.13	2.49
20	50.07	1	2	2.64	4.02
21	52.71	1	1	7.04	7.04
22	59.75	1	0	0.41	0.00
23	60.16	0	0		
			Summation from 0 to T:	49.59	105.44

also include the effects of balking, although no balking will occur in the current example because the queue size is unlimited.

Input parameters:

s = number of servers

K = maximum number in the system

T_S = end of warm-up period

T_F = beginning of cool-down period

n_0 = initial number in the system

Statistics from the simulation:

T = time when the last customer leaves the system

n_a = number of arrivals during the data collection period (When $T_S = 0$, we include the initial number n_0.)

n_b = number of balks during the data collection period

A_s = area under the curve that would result by plotting the number in service versus time during the data collection period

A_q = area under the curve that would result by plotting the number in the queue versus time during the data collection period

The performance measures are calculated from the total areas under the curves associated with the following formulas. We are using the steady-state notation from queuing theory to identify the measures, but we note that the measures are statistical estimates of the steady-state values. For this example, we are using the total simulation time as the data collection period, so $T_S = 0$ and $T_F = T$.

Average number in service: $L_s = A_s/(T_F - T_S) = 49.59/60.16 = 0.82$

Average number in the queue: $L_q = A_q/(T_F - T_S) = 105/60.16 = 1.75$

Average number in the system: $L = L_s + L_q = 0.82 + 1.75 = 2.57$

Efficiency (proportion of time servers are busy): $\rho = L_s/s = 0.82$

Arrival and throughput rates are computed from the number of arrivals during the data collection period. The throughput rate is the rate of the arrivals that do not balk. For this example, there are no balks because the queue is unlimited. Also, when $T_S > 0$, there is some approximation involved because there may be departures from the system during the data collection period that are not counted. Similarly, when $T_F < T$, some arrivals may be counted that do not depart during the data collection period. These effects are minimized when the data collection period is long.

Arrival rate: $\lambda = n_a/(T_F - T_S) = 0.20$

Throughput rate: $\bar{\lambda} = (n_a - n_b)/(T_F - T_S) = 0.20$

We compute time estimates using Little's law.

Average time in service: $W_s = L_s/\bar{\lambda} = 4.13$

Average time in queue: $W_q = L_q/\bar{\lambda} = 8.79$

Average time in the system: $W = W_s + W_q = 12.92$

These statistics can be used to evaluate the behavior of the system being modeled. Remember that the simulation will yield different results when different sets of random numbers are used to determine the interarrival and service times. Issues related to the experimental design and output analysis will be taken up in later sections.

An important question in general is the accuracy of the results. The answer depends on several factors, including the accuracy of the assumptions regarding the process distributions, the lengths of the warm-up, cool-down, and data collection periods, the fidelity with which the logic used to control the flow of entities matches actual operating procedures, and the demarcation of the system boundary. For our example, we can evaluate the solution quality with absolute precision, because exact formulas are available for the $M/M/1$ queue.

Table 18.9 presents a comparison of the theoretical and simulated results. All times are in minutes. What stands out is that when only 12 entities are considered, the estimated values are not very accurate. The last column of the table gives the simulated output for 1000 entities obtained with the Simulate Queue option of the Excel Queuing Add-in accompanying the book. These results are much better but still haven't converged to the theoretical values. In fact, the important lesson that should be drawn from these comparisons is that thousands of iterations are required before useful results can be obtained for even the simplest of systems. For more complex systems, such as those that arise in manufacturing and transportation, hundreds of thousands of iterations are required. This greatly increases the computation effort involved in the simulation.

One set of steady-state values that we have not yet discussed is the probability distribution for the number of entities in the system. For our example, it is a simple matter of looking at Table 18.7 or 18.8 and identifying the time intervals in which there are 0, 1, 2, ... entities in the system, summing their lengths, and then dividing the result by $T = 60.16$. To estimate π_0, for example, we observe that there are two intervals in which there are zero customers present: $[t_3, t_4]$ and $[t_7, t_8]$. Thus, our estimate is $\hat{\pi}_0 = [(13.44 - 5.02) + (23.23 - 21.15)] / 60.16 = 0.17$. This compares well with the theoretical value of $\pi_0 = 0.2$. For π_1, there are four intervals in which there is one customer present: $[t_2, t_3]$, $[t_4, t_5]$, $[t_6, t_7]$, and $[t_8, t_9]$, so $\hat{\pi}_1 = [(5.02 - 2.58) + (19.59 - 13.44) + (21.15 - 21.06) + (25.44 - 23.23)] / 60.16 = 0.18$. The theoretical value, however, is $\pi_1 = 0.126$, so once again the results indicate a need for many more iterations or a much larger value of T.

The final point about running a simulation concerns the warm-up period. In general, when the goal is to evaluate steady-state behavior, it is good practice to discard the first 5 to 10 percent of the output before computing any statistics. This policy is designed to remove any bias that might be introduced by the initial conditions. For simple queuing models, this

Table 18.9 Comparison of Results for $M/M/1$ Queue

Measure	Theoretical value	Simulated value, 12 entities	Simulated value, 1000 entities
Average number in system, L	4	2.59	3.84
Average number in queue, L_q	3.2	1.75	3.03
Average number in service, L_s	0.8	0.83	0.81
Average time in system, W	20	12.93	19.00
Average time in queue, W_q	16	8.79	15.01
Average time in service, W_s	4	4.14	3.99
Efficiency, ρ	0.8	0.84	0.81

may not be an important issue but for a manufacturing system with significant amounts of work in process, the initial conditions could have a measurable effect on the results.

$M/M/s$ Queue

The logic for the single channel queue readily extends to the multichannel case. The simulation becomes a bit more complicated, however, because we must now accommodate an expanded number of options. We include in our logic the possibility that balking might occur in a finite queuing system, although the example for this section is infinite. The finite queue case is illustrated in the next subsection. The logic for the simulation is independent of the interarrival and service distributions, but again we assume exponential distributions so that we can compare the simulation results with the actual expected values.

- If an arrival occurs when all servers are busy but the queue is not full (for a finite queue), the number in the queue increases by 1.
- If an arrival occurs when the queue is full (for a finite queue), the arrival balks and the numbers in service and in the queue do not change.
- If an arrival occurs when a server is idle, the customer goes directly to a service channel; the number in service increases by 1, and the number in the queue remains at zero.
- If a service occurs when the queue is empty, the number in service decreases by 1, and the number in the queue remains at zero.
- If a service occurs when the queue is not empty, the number in the queue decreases by 1, and the number in service remains at s.

For two channels, we simply modify the calendar in Table 18.7 to account for the second server. The simulation is shown in Table 18.10. The next event for a given row is the event taken from the Arrival calendar, the Server 1 calendar, or the Server 2 calendar with the earliest time. The event and the logical rules listed above determine the change in state.

The performance measures are calculated in the same way as they were for the single channel queue by first finding the total areas under the curves during the data collection period. For this case, we use $T_S = 0$ and $T_F = T = 45.48$, the time the last customer leaves the system.

Average number in service: $L_s = A_s/(T_F - T_S) = 1.09$

Average number in the queue: $L_q = A_q/(T_F - T_S) = 0.30$

Average number in the system: $L = L_s + L_q = 1.39$

Efficiency (proportion of time servers are busy): $\rho = L_s/s = 0.55$

Arrival rate: $\lambda = n_a/(T_F - T_S) = 0.26$

Throughput rate: $\bar{\lambda} = (n_a - n_b)/(T_F - T_S) = 0.26$

We compute the time estimates using Little's law.

Average time in service: $W_s = L_s/\bar{\lambda} = 4.14$

Average time in queue: $W_q = L_q/\bar{\lambda} = 1.14$

Average time in the system: $W = W_s + W_q = 5.28$

A comparison of these statistics with those obtained for the $M/M/1$ queue shows that the addition of the second channel has reduced the average number (L_q) and time (W_q) in

Table 18.10   Steps in Next-Event Simulation for *M/M/2* Queue

Event number	Event	Time	Number in service	Number in queue	Arrival calendar	Server 1 calendar	Server 2 calendar
1	A1, A2	0	2	0	(A3, 13.44)	(S1, 2.58)	(S2, 2.44)
2	S2	2.44	1	0	(A3, 13.44)	(S1, 2.58)	
3	S1	2.58	0	0	(A3, 13.44)		
4	A3	13.44	1	0	(A4, 19.59)	(S3, 21.06)	
5	A4	19.59	2	0	(A5, 23.23)	(S3, 21.06)	(S4, 19.63)
6	S4	19.68	1	0	(A5, 23.23)	(S3, 21.06)	
7	S3	21.06	0	0	(A5, 23.23)		
8	A5	23.23	1	0	(A6, 25.44)	(S5, 32.68)	
9	A6	25.44	2	0	(A7, 25.61)	(S5, 32.68)	(S6, 26.13)
10	A7	25.61	2	1	(A8, 30.62)	(S5, 32.68)	(S6, 26.13)
11	S6	26.13	2	0	(A8, 30.62)	(S5, 32.68)	(S7, 39.71)
12	A8	30.62	2	1	(A9, 30.74)}	(S5, 32.68)	(S7, 39.71)
13	A9	30.74	2	2	(A10, 32.53)	(S5, 32.68)	(S7, 39.71)
14	A10	32.52	2	3	(A11, 36.13)	(S5, 32.68)	(S7, 39.71)
15	S5	32.68	2	2	(A11, 36.13)	(S8, 35.60)	(S7, 39.71)
16	S8	35.60	2	1	(A11, 36.13)	(S9, 36.43)	(S7, 39.71)
17	A11	36.12	2	2	(A12, 43.60)	(S9, 36.43)	(S7, 39.71)
18	S9	36.43	2	1	(A12, 43.60)	(S10, 38.44)	(S7, 39.71)
19	S10	38.44	2	0	(A12, 43.60)	(S11, 45.48)	(S7, 39.71)
20	S7	39.71	1	0	(A12, 43.60)	(S11, 45.48)	
21	A12	43.59	2	0		(S11, 45.48)	(S12, 44.00)
22	S12	44.00	1	0		(S11, 45.48)	
23	S11	45.48	0	0			

the queue. Of course, since there are two servers doing the same work as one in the previous case, overall efficiency is reduced.

A Finite Queue System with Balking

Now we consider a slightly more complex situation—a multichannel queue with *s* servers and a maximum of *K* customers allowed in the system. If the queue is full when a customer arrives, the customer balks. In other words, if an arrival occurs when the queue is full (the number in the queue is $K - s$), the entity balks and the numbers in the queue and in service do not change.

To illustrate this situation, we simulate a single channel queue, except now we assume that a maximum of two customers are allowed in the system. Using the same arrival and service data as for the preceding situation, the output for an *M/M/1/2* system is shown in Table 18.11. Arrivals 7, 8, 9, and 10 balk when they find the system full. The service times for these customers are not used, because they never enter the system. A valuable statistic for a simulation of a finite queue is the proportion of customers who balk. For this case, the proportion is $P_F = p_2 = 4/12 = 0.333$. This compares with the theoretical value of $P_F = 0.262$, computed with the Markov queuing formulas from Section 16.2. A simulation run with 1000 arrivals had a balking rate of 0.248.

Table 18.11   Steps in Simulation of *M/M/1/2* Queue with Balking

Event number	Event	Time	Number in service	Number in queue	Calendar
1	A1, A2	0	1	1	{(A, 13.44), (S, 2.58)}
2	S1	2.58	1	0	{(A, 13.44), (S, 5.02)}
3	S2	5.02	0	0	{(A, 13.44)}
4	A3	13.44	1	0	{(S, 21.06), (A, 19.59)}
5	A4	19.59	1	1	{(A, 23.23), (S, 21.06)}
6	S3	21.06	1	0	{(A, 23.23), (S, 21.15)}
7	S4	21.15	0	0	{(A, 23.23)}
8	A5	23.23	1	0	{(S, 32.68), (A, 25.44)}
9	A6	25.44	1	1	{(S, 32.68), (A, 25.61)}
10	A7-balk	25.61	1	1	{(S, 32.68), (A, 30.62)}
11	A8-balk	30.62	1	1	{(S, 32.68), (A, 30.74)}
12	A9-balk	30.74	1	1	{(S, 32.68), (A, 32.53)}
13	A10-balk	32.53	1	1	{(A, 36.13), (S, 32.68)}
14	S5	32.68	1	0	{(A, 36.13), (S, 33.37)}
15	S6	33.37	0	0	{(A, 36.12)}
16	A11	36.12	1	0	{(A, 43.59), (S, 43.16)}
17	S11	43.16	0	0	{(A, 43.59)}
18	A12	43.59	1	0	{(S, 44.00)}
19	S12	44.00	0	0	
20	End	50	0	0	

Statistical Variation and Simulation

The results of a simulation depend on the random numbers used to generate realizations from the input probability distributions. The random number generator requires a seed that determines the sequence obtained. Different seeds should be used for different replications of the model.

As mentioned, all random number generators are really pseudorandom number generators because the same seed will produce the same stream of numbers regardless of how many runs are made. This feature is important when one is trying to verify that the calculations are being performed correctly or to isolate the effects of small changes in parameter values and model structure.

For a given system, different results will be obtained depending on the initial state of the system, the random number seed, the simulation time, and the values of T_S and T_F that define the data collection period. For the best estimation of steady-state measures, the initial values for the system should be set to minimize the time required to reach steady state. Unfortunately, it is not possible to determine these values a priori, but sample runs will provide some guidance. A long simulation time with sufficient warm-up and cool-down periods will reduce to some extent the effects of the initial and final conditions; however, several independent runs with moderate simulation times are probably better than one very long run. Data from several independent runs provide an estimate of the statistical variability of the performance measures and allow the use of standard statistical methods.

For proper interpretation, it is necessary to treat the output of a simulation run as a single observation from a population of possible results. Standard statistical techniques, such as confidence limits and tests of hypotheses, should be used to analyze multiple runs of a given model.

18.6 SIMULATION OF AN INVENTORY SYSTEM

Now that we know the basics of setting up, running, and collecting output statistics for a queuing system, we investigate a slightly different situation. The subject of the analysis is a reorder point inventory system with probabilistic demands and lead times. Although the model is conceptually simple, analytic solutions are restrictive and not generally available. The following conditions inform a need to simulate.

1. Stochastic demands and lead times characterized by specific probability distributions that may be theoretical or empirical
2. Nonstationarities such as changing demand patterns or trends over time
3. Nonlinearities or discontinuities in the cost functions that negate analytic solutions
4. Complex system interactions and the presence, for example, of multiproducts, multiechelons, resource constraints, and queuing considerations

Reorder Point Inventory Model

Our goal is to estimate the reorder point (R) and reorder quantity (Q) that minimize the sum of average holding, ordering, and shortage costs per unit time. Tables 18.12 and 18.13 present the sample data and the empirically constructed probability distributions for demand per day (D) and days of lead time (T_L), respectively. For example, the second column in Table 18.12 breaks down 361 days of observations by frequency of occurrence. The probability mass function (pmf) is shown in the third column, and the cumulative distribution function (CDF) is shown in the fourth column. As before, the assignment of random number ranges are based on the CDFs. It is assumed that the given probability distributions typify the random behaviors of D and T_L and are stationary.

Figure 18.5 depicts the logic of the inventory system. At the outset of a study, it is always a good idea to draw a flow diagram that captures the logic, parameters, and major components of the system. As the operating rules and component interactions become better understood, more detail can be added. For the inventory model, the following rules govern its

Table 18.12   Frequency Distribution and Assignment of Random Numbers to Daily Demand

Daily demand, d	Demand frequency	pmf, $f(d)$	CDF, $F(d)$	Range of 0-1 uniform random number, u_1
6	11	0.03	0.03	0.00 – 0.03
7	18	0.05	0.08	0.03 – 0.08
8	22	0.06	0.14	0.08 – 0.14
9	29	0.08	0.22	0.14 – 0.22
10	36	0.10	0.32	0.22 – 0.32
11	43	0.12	0.44	0.32 – 0.44
12	50	0.14	0.58	0.44 – 0.58
13	58	0.16	0.74	0.58 – 0.74
14	43	0.12	0.86	0.74 – 0.86
15	29	0.08	0.94	0.86 – 0.94
16	22	0.06	1.00	0.94 – 1.00
Total	361	1.00		

Table 18.13   Frequency Distribution and Assignment of Random Numbers to Lead Time

Lead time, t_L	Lead time frequency	pmf, $f(t_L)$	CDF, $F(t_L)$	Range of 0–1 uniform random number, u_2
1	12	0.1	0.1	0.00 – 0.10
2	30	0.25	0.35	0.10 – 0.35
3	54	0.45	0.80	0.35 – 0.80
4	18	0.15	0.95	0.80 – 0.95
5	6	0.05	1.00	0.95 – 1.00
Total	120	1.00		

behavior. The discussion refers to Table 18.14, which presents the results of 20 days of simulation. For this table we are using a reorder point of $R = 35$, the approximate demand for the expected lead time, and an order quantity of $Q = 70$, the approximate optimal lot size for a deterministic inventory system with the parameters used for the illustration.

1. The holding cost H_t for a given day t is based on the ending inventory I_t for that day as long as $I_t > 0$—that is, $H_t = c_h \times I_t$, where c_h is the unit holding cost per day. If $I_t \leq 0$, then $H_t = 0$. Depending on the application, H_t could also be based on the beginning inventory or average inventory for the day.

2. The shortage cost S_t for a given day t is a function of the *lost sales* for that day— that is, $S_t = c_s \times |I_t|$ when $I_t < 0$, where c_s is the cost of a lost sale. If the ending inventory for a given day is negative ($I_t < 0$), then the beginning inventory for the next day is set at zero. (See day 11 in Table 18.14.) Thus, *back-ordering* is not allowed. Other models could base this cost on the number of outstanding back-orders per day, in which case c_s would be the unit shortage cost per day. Still other models could use a mixture of lost sales and back-orders, if appropriate.

3. If the ending inventory for a given day t is less than the reorder point (that is, if $I_t < R$), then an order for Q units is placed and an ordering cost c_o is incurred. At this point, a random number for the lead time variate is used to determine when the order is to be received. (See days 2, 8, and 15 in Table 18.14.) For example, the order placed on day 2 can be used on day 4, since the lead time is 2 days. Note that only one order can be outstanding at any one time for this example. Quantity discounts can be incorporated easily by making c_o a function of Q.

4. Beginning inventory is augmented by Q on the day an order is received. (See days 4, 12, and 18 in Table 18.14.)

The results presented in Table 18.14 are for initial inventory $I_0 = 50$ units; cost parameters $c_h = \$0.10$/day/unit, $c_o = \$20$/order, and $c_s = \$5$/unit; and policy variables $R = 35$ units and $Q = 70$ units. The table was created in Microsoft Excel using the Simulation add-in. The u_1 and u_2 values were randomly generated from a uniform distribution. To understand how the results were obtained, the reader should carefully follow the logic in Figure 18.5 and work out the first few rows in Table 18.14.

A measure of merit for the inventory system is the expected cost of operation per day. This is estimated by averaging the sum of the holding, ordering, and shortage costs over the simulation run. When we take n as the number of simulated days, the average daily cost is

$$\overline{C} = \left(\sum_{t=1}^{n} H_t + O_t + S_t \right) \Big/ n$$

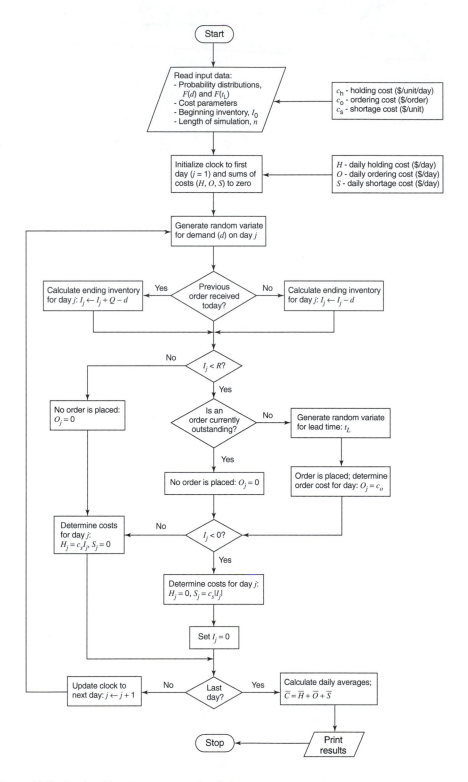

Figure 18.5   Logic of inventory system simulation

Table 18.14   Results of 20-Day Simulation of Inventory System

Day	Units of initial inventory	Units received, Q	u for demand, u_1	Units of demand, d	Final inventory position	u for lead time, u_2	Days of lead time, t_L	Holding cost in $, H$	Order cost in $, O$	Shortage cost in $, S$	Total cost in $, C$
1	50	0	0.084	8	42	0.322	—	4.2	0	0	4.2
2	42	0	0.768	14	28	0.855	2	2.8	20	0	22.8
3	28	0	0.045	7	21	0.488	—	2.1	0	0	2.1
4	21	70	0.237	10	81	0.108	—	8.1	0	0	8.1
5	81	0	0.786	14	67	0.786	—	6.7	0	0	6.7
6	67	0	0.511	12	55	0.229	—	5.5	0	0	5.5
7	55	0	0.262	10	45	0.894	—	4.5	0	0	4.5
8	45	0	0.997	16	29	0.652	4	2.9	20	0	22.9
9	29	0	0.199	9	20	0.939	—	2	0	0	2
10	20	0	0.800	14	6	0.521	—	0.6	0	0	0.6
11	6	0	0.071	7	−1	0.072	—	0	0	5	5
12	0	70	0.139	8	62	0.190	—	6.2	0	0	6.2
13	62	0	0.371	11	51	0.763	—	5.1	0	0	5.1
14	51	0	0.374	11	40	0.354	—	4	0	0	4
15	40	0	0.253	10	30	0.098	3	3	20	0	23
16	30	0	0.090	8	22	0.434	—	2.2	0	0	2.2
17	22	0	0.357	11	11	0.384	—	1.1	0	0	1.1
18	11	70	0.464	12	69	0.602	—	6.9	0	0	6.9
19	69	0	0.422	11	58	0.877	—	5.8	0	0	5.8
20	58	0	0.335	11	47	0.351	—	4.7	0	0	4.7
Total	787	—	—	214	783	—	—	78.3	60	5	143.4
Average	39.35	—	—	10.7	39.15	—	—	3.92	3	0.25	7.17

For the example, we obtain a mean cost per day of $7.17. In an actual simulation, a greater number of days (sample size, n) would be needed to estimate more accurately the mean total cost per day. Also, it should be realized that the beginning inventory for day 1 will affect the cost calculations; however, this effect diminishes as n becomes larger. We discuss the tradeoff between sampling error and cost of implementation (sample size) in Section 18.8.

Finding Optimal Values of R and Q

One of the goals of the analysis is to determine the "optimal" combination of R and Q. The mean total cost ($\overline{C}$) is a discrete, nonlinear, nonanalytic function of the decision variables R and Q, so it is impossible to use traditional optimization methods. Rather, we will evaluate $\overline{C}$ over a grid of possible values. Such an experimental design is called a grid search.

To be effective, the user must specify appropriate ranges for R and Q—in particular, ranges that are likely to contain the unknown optimal R/Q combination. For simple inventory models such as ours, good point estimates of the optimal values are provided by the basic EOQ formula for Q

$$Q^* = \left[\frac{2c_o E(D)}{c_h}\right]^{1/2} \tag{4}$$

and by the expected demand between placement and receipt of an order for R

$$R = E(D) \times E(T_L) \tag{5}$$

where $E(D)$ is mean demand per unit time and $E(T_L)$ is mean time between placement and receipt of an order. Note that the value provided by Equation (4) is likely to be a bit low because shortage costs are ignored. The estimate given by Equation (5) should make sense in that the computation provides the expected demand between the placement and receipt of an order; hence, on the average, inventory will be zero when an order arrives.

Using the standard formulas for computing averages in conjunction with the data in Tables 18.12 and 18.13, we get $E(D) = 11.65$ units per day and $E(T_L) = 2.8$ days. Equations (4) and (5) yield $Q^* = 68$ units and $R = 32$ units, respectively. Therefore, ranges of 25–60 for R and 40–120 for Q should be reasonable first approximations within which we might expect to find the optimal R and Q values. We choose a grid size of 5 for R and 10 for Q, resulting in $8 \times 9 = 72$ instances to be evaluated.

The mean total cost values for these 72 runs are shown in Table 18.15 for $n = 1000$. Based on the output, we see that $R = 40$ units and $Q = 70$ units give the best results. This policy leads to a cost of $8.37 per day on average. At this point, we can be reasonably certain that the "global" minimum rather than some "local" minimum of $\bar{C}$ has been bracketed, because 8.37 is well embedded in the table. That is, if you conceptualize the cost table as a topographical map with "ridges" outlining a "valley," you can see that 8.37 is on the valley floor with high ridges all around the perimeter of the table. Note that all these runs were made with a fixed random number seed for the random numbers used to generate the demand quantities and the lead times, so every run has the same demand sequence. Since the random numbers are associated with days, the lead time values depend on the particular days when orders are placed. These days depend on the specific values of R and Q being simulated.

Now that we have located the approximate floor of the valley, we would like to "zoom in" to get a better fix on optimal R and Q. This is done by narrowing the ranges for R and Q and reducing the grid size to 1 for both variables. The results, presented in Table 18.16, show the cells that surround the apparent optimal solution at $R = 41$ and $Q = 67$. The results indicate that the valley floor is fairly flat but there is a good deal of up-and-down variation near the optimal solution. This is a result of the discrete nature of the problem and the specific random numbers used for the simulation.

Table 18.15   Mean Total Cost $\bar{C}$ for First Experiment

		Quantity ordered (Q)[a]								
		40	50	60	70	80	90	100	110	120
	25	14.31	11.94	10.56	10.32	10.38	9.87	10.28	9.85	9.95
	30	11.64	10.20	9.50	9.29	9.21	8.62	9.35	9.26	9.27
	35	10.46	9.47	8.84	8.77	8.43	8.70	8.86	8.90	9.20
Reorder	40	10.33	9.07	8.62	**8.37**	8.48	8.52	8.73	8.98	9.08
point (R)	45	10.32	9.12	8.57	8.59	8.51	8.83	9.06	9.32	9.58
	50	9.94	9.15	8.84	9.01	9.06	9.17	9.39	9.70	10.06
	55	10.12	9.52	9.35	9.23	9.33	9.54	9.65	10.06	10.42
	60	10.35	10.03	9.76	9.74	9.85	9.95	10.09	10.44	10.81

[a] The lowest average total cost is $8.37 when $R = 40$ and $Q = 70$.

Table 18.16   Mean Total Cost $\bar{C}$ for Second Experiment

		Quantity ordered (Q)[a]							
		64	65	66	67	68	69	70	71
	35	8.93	8.72	8.47	8.73	8.36	8.75	8.77	8.98
	36	8.38	8.78	8.28	8.52	8.22	8.89	8.61	8.99
	37	8.52	8.85	8.40	8.54	8.23	8.60	8.59	8.87
	38	8.58	8.98	8.52	8.39	8.32	8.53	8.78	8.96
Reorder	39	8.66	9.04	8.56	8.28	8.39	8.58	8.52	8.81
point (R)	40	8.71	8.54	8.68	8.34	8.43	8.44	8.37	8.23
	41	8.48	8.37	8.73	**8.17**	8.53	8.43	8.43	8.51
	42	8.54	8.51	8.70	8.20	8.35	8.40	8.42	8.51
	43	8.50	8.48	8.80	8.33	8.47	8.47	8.51	8.56
	44	8.60	8.37	8.75	8.41	8.62	8.57	8.50	8.56

[a]The lowest average total cost is $8.17 when $R = 41$ and $Q = 67$.

We must stress that the optimal values are for specific random number seeds and length of run. Changing the run length, demand seed, or lead time seed will result in different optimal values. One would expect that a longer run would return better results than a shorter one because the simulation would have more opportunity to represent rare events.

18.7   STEPS IN A SIMULATION STUDY

So far we have taken an ad hoc approach to performing a simulation study. We now wish to provide a template that parallels the steps described in Chapter 1 for any operations research analysis. However, because simulation often involves more experimentation than other methods, it is worth defining those steps in greater detail. Figure 18.6 depicts the major phases that a model builder typically goes through in a simulation study.

Problem Formulation

Every study should begin with a statement of the problem that is agreed to by all stakeholders involved. It is difficult to arrive at the right answer if you are working on the wrong problem. Therefore, the first step is to formulate the problem properly. Often, the manager has only a vague idea of what the problem is. It is the job of the analyst to translate this vague idea into an explicit, written statement that captures the goals of the study. It should be realized, however, that this statement is likely to change over time. As the work progresses, the analyst becomes more knowledgeable about the system being simulated and about the goals of the organization. It may be necessary to modify the model accordingly. Usually, the statement of goals takes the form of questions to be answered, hypotheses to be tested, and effects to be estimated. In this phase, it is also necessary to identify the criteria to be used to evaluate these questions.

Model Building

The formulation or construction of a model is probably as much an art as it is a science. The art of modeling is enhanced by an ability to abstract the essential features of a problem, to select and modify basic assumptions that characterize the system, and then to enrich and elab-

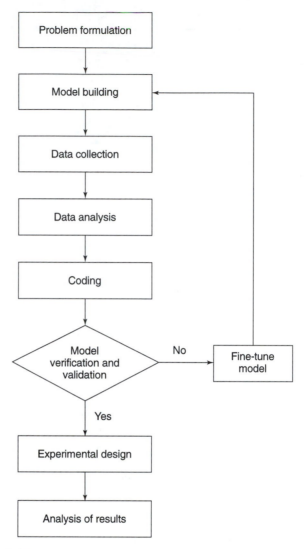

Figure 18.6   Simulation steps

orate the model until a useful approximation results. Thus, it is best to start with a simple formulation and build toward complexity but without ever going beyond what is required to accomplish the study's goals. In fact, the analyst must strike a balance between model realism and the cost of developing the model. If a crucial variable or functional relationship is omitted, the model will not accurately predict the behavior of the real system. If the model contains excessive detail, it may be too expensive to collect the required data, or to code and run it.

Data Collection

The next step, and possibly the most time-consuming step, is the job of collecting data. There is a constant interplay between the construction of the model and the collection of the needed input data. Quantitative data are necessary for several reasons. First, data are required to describe the system being simulated. If you do not understand the real system thoroughly, it is not very likely that you will simulate it properly. Second, data must be gathered as the foundation for generating the various stochastic variables in the system. For example, in the study of a bank, if the goal is to learn about the length of the waiting lines as the number of tellers changes, the types of data needed will be the distributions of interarrival times (at different times of the day), the service time distributions for the tellers, and historical distributions of the lengths of the waiting lines under varying conditions. The last item would be used to validate the model. Thus, real data concerning arrivals and service times must be gathered and analyzed to determine the proper probability distributions and their parameters. Also, data are necessary to test and validate the model. To use a model to make decisions, the decision maker must be confident that real-world phenomena have been adequately and accurately represented. Often, the best way to accomplish this validation is to compare simulated output with historical data.

Data Analysis

Once the data have been collected, they must be organized and then analyzed. Often it is necessary to discard outliers and other questionable elements. The data are generally used to estimate parameter values and, for random phenomena, to construct cumulative distribution functions. These distributions may be treated as empirical, or tests may be done to see if they fit any known forms. Although it is possible to use subjective probability distributions, this is somewhat atypical; probability distributions, based on empirical data usually yield more reliable results and are thus preferable.

As noted in Section 18.3, two basic tasks must be performed to generate samples of random variables. First, the raw data of a stochastic phenomenon must be analyzed to determine how that random variable is distributed. Then a function must be derived or the inverse transformation method applied to generate realizations of the variable using 0-1 uniformly distributed random numbers. The steps for doing this are as follows.

1. Group the data into intervals based on frequency of occurrence.
2. Depict this frequency distribution graphically as a histogram.
3. Hypothesize a probability distribution from the shape of the histogram.
4. Estimate the probability distribution parameters using sample statistics.
5. Test the hypothesis using one of several statistical tests such as the *chi-square* or *Kolmogorov–Smirnov* goodness-of-fit test (see Appendices A.2 and A.3 for this chapter on the CD accompanying the book).
6. If the hypothesis is rejected, distribution parameters can be perturbed or changed slightly, and a new hypothesis tested.
7. If no known probability distribution can be found to fit the data, use the empirical frequency distribution constructed in Step (1).

To illustrate this procedure, consider the following 35 service times (in minutes) at a fast food drive-up window collected during one day: 3.4, 6.2, 0.5, 3.9, 9.6, 7.9, 6.9, 4.2, 2.7, 5.5, 4.8, 1.8, 9.0, 3.4, 3.9, 6.5, 0.5, 5.8, 3.6, 8.9, 2.4, 0.4, 4.7, 0.8, 3.4, 5.4, 4.2, 5.5, 7.9, 0.6, 9.5, 0.0, 9.5, 5.1, 6.7. The frequency distribution for these raw data is reflected in Table 18.17 and depicted in Figure 18.7.

From the shape of the histogram in Figure 18.7, let us hypothesize that the random variable (service time for fast food order) is uniformly distributed between 0 and 10. Therefore,

H_0: Sample is drawn from a uniformly distributed population with parameters $a = 0$ and $b = 10$.

Table 18.17   Frequency Distribution for Drive-up Window Service Times

Class	Frequency of observation
0 – 2	7
2 – 4	8
4 – 6	9
6 – 8	6
8 – 10	5

H_A: Population is not uniformly distributed with parameters $a = 0$ and $b = 10$.

To test the null hypothesis, two approaches are available—either a standard chi-square test or the generally more powerful Kolmogorov–Smirnov test. The latter entails the following steps.

1. Formulate the null hypothesis and the alternative hypothesis.

2. Define a set of class intervals that covers the range over which the observed data fall. Calculate the relative frequency for each class by dividing the number of observations in the class by the sample size n.

3. Compute the cumulative probability distribution of the sample data by successively adding the relative frequencies of all classes. [This is also called the *observed cumulative distribution*, or $S_n(x)$, where x is the cumulative sum.]

4. Establish the theoretical probability for each class by taking the definite integral of the hypothesized probability density function over the class interval or by using the appropriate table or formula when available.

5. Establish the cumulative probability distribution of the theoretical distribution that has been hypothesized by successively adding the theoretical probabilities of all classes. [This is also called the *theoretical cumulative distribution*, or $F(x)$.]

6. Compute the absolute difference between the observed cumulative distribution and the theoretical cumulative distribution for each class, $|S_n(x) - F(x)|$. This operation gives you the absolute difference.

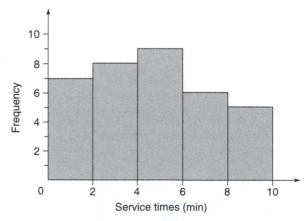

Figure 18.7   Histogram for drive-up window service times

7. Find the maximum absolute difference over all classes. Compare this value with the critical value $d_\alpha(n)$ found in a standard Kolmogorov–Smirnov table for a specified level of significance α (see Appendix A.3 on the CD). If the critical value exceeds the maximum absolute difference, then the null hypothesis cannot be rejected.

To see how this procedure can be applied, let n be the sample size, let α be the level of significance of the test, and let $d_\alpha(n)$ be the critical Kolmogorov–Smirnov value. Using the drive-up window data, we get the results shown in Table 18.18 for Steps (1) to (6). For $n = 35$ and a level of significance of $\alpha = 0.05$, we have $d_{0.05}(35) = 0.23$, implying that the null hypothesis cannot be rejected at Step (7).

Coding

Virtually all operations research studies require the use of a computer, but for simulation there is no alternative. The vast amounts of data that need to be stored and the heavy computational burden require computer implementation. From several points of view, the selection of a language becomes a matter of some consequence. Fortunately, there are many choices, with more than 100 vendors providing commercial products and about half as many researchers and academicians making available their experimental codes for nominal fees. The Society of Computer Simulation publishes an annual *Directory of Simulation Software* that lists almost all vendors and their products. Many operations research and industrial engineering trade journals also publish annual surveys of simulation software (e.g., see Elliot [2000] and Swain [2001]).

Despite this proliferation, a more basic language may still be preferable. At the outset of a study, the analyst must decide whether to program the model in a general-purpose language such as C++, Fortran, or Visual Basic, in a general-purpose simulation language such as Arena, SLAM, or Witness, or in a special-purpose simulation language intended specifically for, say, manufacturing facilities or telecommunication networks. General-purpose languages require a much longer development time, but usually run much faster than simulation languages because they have less overhead. On the other hand, the use of a simulation language greatly speeds up the coding and verification steps by offering much greater efficiency. Many routine functions such as the simulation clock, the collection of statistics, the next-event logic, and report generation are built into these languages. In particular, the programmer has far less detail to be concerned with, so the coding effort is reduced. An additional benefit is that the probability of creating a valid program is increased. Moreover, most full-scale simulation languages now have friendly graphical user interfaces that relieve

Table 18.18 Kolmogorov–Smirnov Test Applied to Drive-up Window Data

Class	Theoretical probability	Relative frequency	Theoretical CDF, $F(x)$	Observed CDF, $S_n(x)$	Absolute difference, $\lvert S_n(x) - F(x) \rvert$	
$0 - 2$	0.2	0.200	0.2	0.200	0.0	
$2 - 4$	0.2	0.229	0.4	0.429	0.029	
$4 - 6$	0.2	0.257	0.6	0.686	0.086	← Largest
$6 - 8$	0.2	0.171	0.8	0.857	0.057	value
$8 - 10$	0.2	0.143	1.0	1.000	0.0	
Total	1.0	1.000				

the analyst of the task of learning a new programming language. Almost all the development and testing can be done on the screen by simply manipulating objects.

Model Verification and Validation

Once the model is up and running, it is necessary to determine whether or not the logic has been implemented correctly and whether or not the calculations are being performed as intended. This step is referred to as *verification*. With complex models, extensive debugging is the rule. It is possible that a simulation model is valid as designed but invalid as implemented. Standard testing techniques such as manual calculations and program traces should be employed to ensure concurrence between design and output. It is necessary to verify the absence of programming errors before conducting any analysis. When it has been determined that the input parameters and the logic of the model are correctly represented in the code, verification has been completed.

The next step is to determine whether or not the model is an accurate representation of the real system. The program can be perfect, but the implemented model may still be invalid. *Validation* is usually achieved through model calibration, an iterative process of comparing model output with actual system behavior and using the discrepancies between the two, and the insights gained at this stage, to narrow the differences. Although absolute accuracy is probably unattainable, this process is repeated until the analyst and the decision maker achieve sufficient confidence in the output. The steps involved in validation include the following.

Variable generation test: The data analysis process involves the application of nonparametric goodness-of-fit tests to hypotheses concerning the distributions of the various stochastic variables. These same tests should be applied to the outputs from the various process generators to ensure that the real-world variables and simulated variables are distributed in the same manner. For example, if the interarrival time in a real queuing system is normally distributed with a mean of 10 minutes and a standard deviation of 3 minutes, then the random variable of interarrival times being generated in the simulation program should also be normally distributed (or very nearly normally distributed) with $\mu = 10$ and $\sigma = 3$.

Subjective validation: The design, as well as the output, of the simulation model should be reviewed by the people who are most familiar with the real system. To minimize bias, it is best that the subjective validation be done by persons not directly involved in the simulation study.

Historical validation: If the goal is to simulate an existing system, it is often possible to simulate the system as it is presently configured and then compare historical data with simulation output. For example, if the real system is a hospital emergency room, vital statistics such as the average wait time of a patient and the average time a patient spends with the medical staff should be compared with the distributions of various output variables. The absence of significant differences between simulated results and historical results may tend to validate the model, but it does not guarantee that the program will accurately predict the behavior of the real system under different conditions.

Confidence in the validity of a simulation model is crucial to its use. Management will not adopt any tool whose output does not conform to known behavior under normal operating conditions. For this reason, the analyst should make every effort to substantiate the validity of the model.

Experimental Design

Once a simulation model has been implemented and validated, it can be used for its original purpose—experimentation. Simulation is a means of providing information necessary for decision making when a real-world system cannot be sufficiently manipulated. A simulation model synthetically gathers the information necessary to describe the system. A critical aim is to do this at the lowest possible cost. Because real-world experiments are more costly than simulation experiments, the analyst can explore a greater number of alternatives when using a simulation model. For example, if the system under study is a technical support hotline and the decision variables are the number and type of hardware specialists, the number and type of software specialists, as well as the referral policies, simulation provides a ready means of evaluating many combinations of the decision variables in order to derive a cost-effective staffing plan. If, however, experiments were made on the real system, far fewer alternatives could be evaluated.

Defining the alternative scenarios that are to be run is the joint responsibility of the analyst and the relevant decision makers who are relying on the simulation results. Typically, these scenarios allow management to evaluate a variety of system configurations and operating policies. For example, in the design of an automated mail processing and distribution center, the Postal Service and model builders would specify a host of equipment configurations, material handling systems, operationing policies, and labor rules to be tested in conjunction with a variety of service standards and mail arrival profiles at different points in time.

Once the scenarios have been identified, the analyst must deal with questions concerning the length of the simulation run, the initialization period, the sample size, and how to design an optimal system. Many of the answers can be found in the traditional literature concerning experimental design and in most simulation texts. The question of deriving optimal settings was partially addressed in Section 18.6. In Section 18.9, we will touch on the other issues.

Analysis of Results

If the simulation model is valid and the experiments have been designed properly, analysis of simulation output may proceed in a fairly straightforward manner. It is the function of the analyst to interpret simulation results and make the appropriate inferences necessary for rational decision making. Often, certain statistical techniques, such as analysis of variance, can be helpful in evaluating the results.

18.8   IMPLEMENTATION ISSUES

When simulating dynamic phenomena, we are usually interested in their long-run or average behavior after a steady state has been reached. For Markov queuing processes (Chapter 16), we saw that after a sufficiently large number of transitions the system settled into a near steady state, where it became independent of the initial conditions. The same ideas can be applied in stochastic simulation, provided the structure of the system is such that it approaches a steady state. For the $M/M/1$ queue examined in Section 18.5, for example, steady state implies that if we were to observe the number of customers in the queue repeatedly at random points in time, after a large number of samples we would expect to see 3.2 customers on average. In other words, as the length of the simulation run increases, the effects of the initial conditions under which the simulation was started are washed out.

The speed at which this happens may also be affected by the choice of the initial conditions. Therefore, care should be taken in choosing initial values for the states. Starting in an empty state—e.g., with no initial workload and empty queues—may be a bad choice except for dynamic phenomena where the empty state is a natural occurrence at the beginning of every period, such as waiting line situations that go through a daily cycle. For the $M/M/1$ queue, starting out with four customers in the system would be the best choice since this corresponds to the average number in the system in steady state.

As previously mentioned, a convenient way to remove any bias in the results that may arise from the initial conditions is to exclude the initial portion of each simulation run from the analysis and begin accumulating operating statistics only after this initial period. Alternatively, the ending conditions of each run can be used as the initial conditions of the next run.

Many simulation projects involve comparisons of several different modes of operation. For example, a company might want to compare its present inventory policy with a proposed policy and determine the differences in average annual costs and other performance measures. In such instances, it is advantageous to use the same sequence of random events—e.g., the same sequence of daily demands and production lead times. Although corresponding runs are no longer independent, this approach reduces the variability of the differences observed and is a common practice in statistical inference. For consistency, all runs should also be started with the same initial conditions.

Sample Size and Simulated Probability Distributions

Monte Carlo methods seek to reproduce underlying probability distributions artificially by the techniques described earlier. A critical design question concerns the sample size (n) that should be selected in order to ensure a given degree of accuracy between simulated and theoretical observations. Prior to the experiment, this question is usually answered by determining the sample size (number of simulated observations) that will provide sample statistics for estimating population parameters within specified maximum errors and confidence levels according to the central limit theorem.

Table 18.19 illustrates calculations for means of the demand and lead time distributions used in the inventory simulation in Section 18.6 as well as several related error calculations. Note that n increases exponentially as the error decreases, which clearly illustrates the cost incurred (in increased computer time) for greater accuracy in estimating population parameters with sample statistics.

The values $E(D) = 11.65$ units and $E(T_L) = 2.8$ days were calculated from the empirical distributions in Tables 18.12 and 18.13, respectively, using the definition of the mean. Similarly, $\sigma_D = 2.594$ units and $\sigma_{T_L} = 0.980$ day were calculated using the definition of the standard deviation with the same distributions. The sample size n was calculated using Equation (A.8) (see Appendix A.1 on the CD) for the stated maximum error ε and desired level of confidence α.

For the inventory simulation, the number of random demands generated in a run of the simulation is much greater than the number of lead times. In a run of 1000 days, 1000 random demands are generated, whereas the number of lead time variates generated depends on the reorder quantity. For a reorder quantity of 70, an order is placed on average every 6.00 days, so in 1000 days we would expect only 166 orders. For a given simulation run, the demand process is more accurately represented than the lead time process. We should also note that the formulas in Appendix A.1 do not really hold for the daily outputs of the

Table 18.19   Error as a Function of Sample Size for Inventory Simulation

Measure	Demand, D	Lead time, T_L
Estimated mean	11.65	2.8
Estimated standard deviation	2.594	0.980
Sample size (n) for 1% error	1904	4704
Sample size (n) for 5% error	76	188
Sample size (n) for 10% error	19	47

inventory simulation because the observations are not independent. For example, the inventory level on day t is highly correlated with the inventory level on day $t + 1$, and the formulas in Appendix A.1 assume independence.

Variance Reduction Techniques

The outcome of a simulation run can be thought of as representing one observation of a random variable. If we wish to get sufficiently reliable estimates (in terms of small standard errors), we need either one very long run with respect to simulated time or a large number of smaller runs that we then average. There are several techniques for improving the efficiency of parameter estimates obtained from simulation known as *variance reduction techniques*. Although only a cursory discussion of this topic is suitable for this text, we will attempt to provide some insight into one approach referred to as the *antithetic variate method*.

Suppose that the simulation involves generation of daily demands from a given probability distribution. Rather than generating a total of n independent demands, we generate two sequences of $n/2$ demands that exhibit a high negative correlation. This can easily be achieved by using u_i, the actual random decimal fractions generated for the first sequence ($i = 1, \ldots, n/2$), and ($1 - u_i$), also random decimal fractions, for the second sequence. Whenever u_i produces a demand value above the median, ($1 - u_i$) produces a value below the median, and vice versa. As a consequence, the average demand over both sequences tends to be closer to the expected value of the demand than that of a single sequence of n independently generated demands, thus reducing the variability of the simulation results.

Most variance reduction techniques were developed for simulations in the physical sciences. The complexity of the systems encountered in operations research problems renders their use much more difficult. The improvements gained often may not justify the additional modeling and programming costs incurred. Improvements in reliability of a similar magnitude may sometimes be achieved less expensively by increasing either the length of each simulation run or the number of runs. For more discussion on this topic, see Wilson and Pritsker [1984].

Production Runs

There is no universal theory that tells the analyst how long a simulation run should be or how long the warm-up period should be. With regard to the former, the longer the better, but practical considerations, such as time required per iteration, the experimental design, budget limitations, and the scope of the study, are often the determining factors. We will presently identify one statistical approach that has a theoretical justification. With regard to the length of the warm-up period, a general rule might be to discard the first 5 to 10% of the output before collecting statistics. Alternatively, a better approach would be to determine when a steady state is reached, but the effort involved in doing this could be prohibitive. A related issue concerns the number of runs—say, one long run versus several shorter runs. Although the amount of data collected in either case may be about the same, it might take longer and be more cumbersome to conduct several shorter runs.

The difficulty with one long, uninterrupted run, however, is that it provides no information about the variance of the processes being studied. Thus, it is not possible to compute confidence intervals on the performance measures, as discussed in the next subsection. In addition, a single run may highlight unusual process interactions or correlations that occur infrequently (and that just happened to occur on this run), leading the analyst to conclude incorrectly that these events and the accompanying behavior are normal.

In most cases, the random variables being modeled, whether they represent the number of patients in an emergency room or the number of items back-ordered in an inventory system, are highly correlated. This means that if we examine the state of the system at short intervals, we are not likely to see much change. One way to avoid this problem and still use one long run is to divide the output into fixed-size blocks. To reduce the effects of autocorrelation and to allow for a warm-up period, it would be necessary to discard a fraction of the initial output between every two blocks. Unfortunately, most commercial codes do not have provisions for collecting data in this manner.

Performance Estimation and Number of Replications

When simulation is used, care must be taken in estimating steady-state system performance. The analytic path can be strewn with pitfalls for the statistically unwary. To illustrate, consider the results in Table 18.20 for estimating the criterion "mean total cost per day" for the inventory model in Section 18.6. Five sets of simulation runs (*replications*) were made for each level of n by simply changing the random number seed and restarting the simulation under otherwise identical conditions ($R = 41$, $Q = 67$, $c_s = 5$, $c_h = 0.1$, and $c_o = 20$). The seven samples in each column of the table use the same random number seed. The average and standard deviation of the mean daily costs are computed for each sample size and shown in the last two columns of the table. For example, five separate runs for 1000 simulated days have a grand mean, $\bar{\bar{C}}$, of 8.433, and the unbiased estimate of the standard error of $\bar{\bar{C}}$, $\hat{\sigma}_{\bar{C}}$, is 0.170. Note that $\hat{\sigma}_{\bar{C}}$ is the square root of the value computed from Equation (A.5) in Appendix A.1.

The principle of *stochastic convergence* governs the accuracy of the statistical results and is well illustrated by the plot of the grand mean and the grand mean increased or decreased by one standard deviation in Figure 18.8. As the sample size increases, the grand mean ($\bar{\bar{C}}$) converges to the steady-state value, which appears to be in the neighborhood of $8.45 per day. Furthermore, as predicted by Equation (A.3) and verified by $\hat{\sigma}_{\bar{C}}$ in Table 18.20, variation about the expected value decreases as n increases.

Also note that the replication mean ($\bar{C}$) approaches the steady-state value for each replication number. Thus, we have the choice of running one long simulation or running several smaller ones and averaging their results. The former has the advantage of requiring only one warm-up period, whereas the latter has the advantage of producing independent observations of the sample means. Because the sample means are independent and approximately normally distributed, they can be used for creating confidence limits.

Table 18.20   Mean Total Cost per Day ($\bar{C}$) as a Function of Sample Size (n)

Sample size, n	Replication mean, $\bar{C}$					Grand average, $\bar{\bar{C}}$	Standard deviation, $\hat{\sigma}_{\bar{C}}$
	1	2	3	4	5		
25	7.668	7.264	7.468	7.592	9.760	7.950	1.023
100	9.815	7.111	8.102	8.510	8.925	8.493	1.000
500	8.869	8.090	8.765	8.317	8.766	8.561	0.339
1000	8.812	8.577	8.468	8.535	8.795	8.637	0.157
2000	8.733	8.749	8.479	8.698	8.836	8.699	0.133
5000	8.688	8.649	8.554	8.666	8.681	8.648	0.054
10,000	8.604	8.587	8.602	8.637	8.633	8.613	0.022

For steady-state estimates, we must be sure that transient effects do not confound the calculations. For instance, the size of the initial inventory affects the calculation of $\bar{C}$ in the inventory model, and the starting length of a queue affects the estimation of the steady-state waiting time in a queuing simulation. By definition, in the steady state, succeeding observations of the performance measure are time dependent; hence, the observations used to estimate the steady-state value should not exhibit correlation over time (termed *serial correlation*). One approach to this problem is to calculate a serial correlation coefficient for a block of the first k observations and test for significance. If significant, assume that the first k observations are part of the transient phase and repeat the procedure on the next k observations by continuing the simulation at the point where it previously ended. This procedure is continued until the first block of k observations is found for which the null hypothesis of no correlation is accepted. At this point, the k observations in that block can be used to estimate the steady-state criterion.

A more straightforward (but no less costly) approach is "reasonably" initializing values for the necessary parameters and variables and simulating until stochastic convergence is achieved. The usual difficulty with this approach is that one may not know what reasonable values are (theoretically, they should be near steady state). For the inventory problem, this approach works well because we can derive good estimates based on simple analytic formulas. For example, using Equations (4) and (5) we get $Q = 68$ and $R = 32$, suggesting that a beginning inventory between 40 and 90 might lead to quick convergence. In the examples in this section, we used an initial inventory of 50. In any case, a long simulation will wash out transient effects. A glance at Table 18.20 or Figure 18.8 will show that a run of 1000 simulated days provides a reasonable estimate of the steady-state average cost.

Because stochastic convergence may be slow, its realization is costly in terms of computer time. For this reason, much research has been centered on variance reduction techniques, as previously mentioned, which provide lower standard errors than the conventional approach for a given value of n. At present, however, there is no definitive answer to the problem of estimating steady-state criteria.

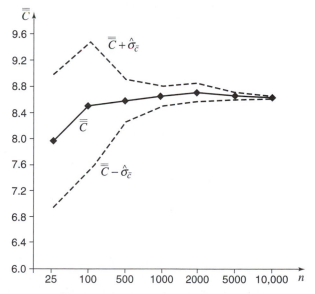

Figure 18.8   Illustration of stochastic convergence for inventory problem

Assuming that the mean steady-state value of the criterion has been estimated in a manner that is free of transient effects, it is possible to establish a *confidence interval* based on Equation (A.6). Let μ_C be the true mean of the average cost per day for the inventory simulation. Using the results from Table 18.20 for $n = 1000$, we get

$$\text{C.I.} = 8.433 \pm (1.96)(0.170)$$
$$= 8.433 \pm 0.333$$

at the 95% level of significance. In general, 95% of the confidence intervals that are constructed using this procedure will contain the true mean.

Validation Revisited

Simulation models typically comprise several separate components (or subroutines) representing the various subsystems of the process being simulated. Each of these components should be tested separately. However, once these components have been put together as one interacting package, the model as a whole must be tested. This validation consists of two steps.

First, it is necessary to determine whether the logical connections between the various parts are correct. This is best done by running the whole simulation model over several events suitably chosen to test the various paths through the system. The events chosen should include exceptional and extreme circumstances, and intermediate status reports should be printed out at various crucial points, such as all decision forks along each path. These status reports should be carefully checked against the results obtained by duplicating at least some portion of the simulation by hand or with an auxiliary code. Once the internal logic has been checked, these intermediate status report printouts can be eliminated from the program.

The second step is to test whether or not the model as a whole can properly reproduce the real-world process. This validation is considerably more difficult. If the simulation describes an existing system, it can presumably be tested against past data and its performance compared with the results actually experienced. Many simulation models, however, describe hypothetical or planned systems for which no past performance data are available. In such instances, the analyst has no alternative but to perform carefully the first step of the validation process and then make a value judgment about the reasonableness of the results obtained. In addition, the analyst should be constantly on the watch for possible anomalies or counterintuitive and unusual results, and attempt to find logical explanations for any discrepancies. This level of scrutiny should continue until a sufficient degree of confidence has been gained in the correctness of the model as a whole.

Documentation

In the course of any study, documenting the methodology and the program is critical. If the software is to be used by several different analysts, a road map is needed for specifying input, executing the code, compiling output, and interpreting the results. Most studies involve extensive analyses to determine relationships between decision variables and output measures, optimal parameter settings, and the sensitivity of system performance to certain design features. Without an understanding of how the code works or what is required to specify or alter input values, the analyst is mostly operating in the dark.

There is a myth that documentation and reporting are important only if the client requests them. Nothing could be farther from the truth. You never know who will question or challenge the model assumptions, logic, scope, or results down the road. Documenting what was done and agreed to (and by whom) is critical, because the analyst may not remember or may not be available for consultation. Moreover, the client may disagree with the analyst's recollection. Documentation and reporting are often needed at some future time when the client asks for additional explanation of the results, for more analysis, or for further development of the model. This can happen months or even years later. It can be difficult for even the same analyst to pick up a model a year later, and it will be much more difficult for a newcomer. With proper documentation and reporting however, the learning curve will not be very steep.

User Acceptance

The successful use of a simulation depends, to a large extent, on how well the steps outlined in Figure 18.6 have been carried out. It is also contingent on how thoroughly the analyst has involved the end-user during the developmental process. If the user has been an active participant at each step, and understands the nature of the model and its output, the likelihood of a full implementation is greatly enhanced. Conversely, if the model and its underlying assumptions have not been properly communicated and approved, implementation will probably suffer regardless of the model's validity and potential usefulness. If we were to single out the most crucial step in the entire process, it would be model verification and validation. An invalid model will be worthless at best, and at worst could be costly and dangerous.

Simulation as a Power Tool

We are used to thinking of simulation as a powerful tool that allows us both to quantify and to observe the behavior of new systems. Whether the system is a production line, a distribution center, or a communications network, simulation can be used to study and compare alternative designs or to troubleshoot existing systems. With simulation models, we are free to imagine how an existing system might perform if altered, or to imagine and explicitly visualize how a new system might behave even before the prototype is completed. The ability to construct and execute models and to generate statistics and animate behavior with ease are some of the main attractions of modern software (Swain [2001]).

Decision Making

Building a model is rarely an end in itself; instead, the goal of most analyses is to make a decision. To assist in analysis, simulation software is increasingly able to exchange information with other software tools in an integrated way, and hence is being positioned within a general decision-making framework. This integration supports both analysis and the inclusion of simulation within larger contexts. For instance, simulation may be integrated with presentation software to document and report on findings or to facilitate analyses with spreadsheet or statistical software. Likewise, simulation models may be driven by solutions obtained from scheduling or optimization software, and then run to provide insight into the proposed solution or to aid in the evaluation of alternative solutions. Integration speeds the process of design and evaluation, making simulation a valuable component within the process.

Many simulation products also include tools for both input and output data analysis, such as Expertfit, Bestfit, and Stat::fit for distribution fitting (see Law [1998] and Swain [2001]). Support for run control and statistical analysis of results is also becoming more common. In the former case, there are mechanisms for allowing both replication of simulation runs and more complicated experimental designs. In the latter case, there is support for the storage of production runs together with the ability to produce summary graphics and analysis.

Simulation is only one modeling methodology. Because it operates by the construction of an explicit trajectory or the realization of a process, each replication provides only a single sample of the possible behavior of a system. This is in contrast to many analytical models that provide exact results over all possible outcomes. This limitation is often far outweighed by the flexibility of simulation and the direct representation of entities within the model. In addition, because simulation directly represents a realization of the system behavior dynamically, it can be joined with animated outputs and used for validation, analysis, and training.

Visualization

The interplay between simulation and visualization is a powerful combination. Visualization adds impact to simulation output and often enhances the credibility of the model—and not just for a nontechnical audience. Almost all simulation software has some level of animation, from basic two-dimensional geometric figures to products that make you feel that you are there. Developers are already taking the next step, combining simulation with virtual reality to provide the equivalent of what one might experience in a pilot training simulator.

In the corporate world, simulation is becoming increasingly commonplace. For example, Intel uses simulation at all stages of product and process development throughout the entire life cycle of their production facilities. Simulation studies parallel developments in the physical process. For most manufacturers, as the size and complexity of their systems increase, simulation is no longer a luxury but a necessity for proper analysis to support good decisions. Recognizing this, the Department of Defense has embraced the methodology with even more enthusiasm. Their vision is to use simulation at all stages of the acquisition cycle—design (including the operational aspects), manufacturing, maintenance, training, and even the evaluation of tactical doctrines.

18.9   ASSESSMENT OF SIMULATION METHODOLOGY

Nowhere in applied operations research does the concept of a system come forth as naturally as in simulation. For the analyst using mathematical optimization techniques, the complexities of the real world leave little choice but to make abstractions, approximations, and simplifications in the models. These limitations need not curtail the use of simulation. The mathematical complexities of simulation seldom go beyond simple numeric computations or logical operations. Hence, much more detail and more interactions among the various parts of a system can be taken into account. As a result, simulation models may be fairly true representations of the real world. This is one of the great attractions of the simulation methodology over other quantitative techniques.

With this advantage, however, come two potentially severe handicaps. Simulation is not an a priori optimizing tool, but rather a tool of analysis often used for evaluating the performance of decisions derived by other means. Each simulation run simply traces through the effects of decision rules that are specified in full detail as part of the input. Attempts at optimization have to be made by a slow process of trial and error. All we can usually strive for is "good" decision rules rather than optimal ones, and even good rules may be costly in terms of computer time.

To be successful, simulation models have to incorporate large amounts of detail. As a consequence, the effort and cost that go into building such models is usually much greater than for comparable optimization models. Nevertheless, the uses of simulation are almost unlimited, as the following list suggests.

- Waiting lines: evaluation of alternative proposed facilities or of alternative modes of operating existing facilities.

- Job shop scheduling: evaluation of alternative dispatching rules and forecasting of workloads at each machine center. Such forecasts may be used to initiate corrective action to eliminate potential bottlenecks.

- Process plant operations: the entire operation of a process plant, such as a refinery, is simulated, unit by unit, to determine output composition as a function of input mix and instrument settings (deterministic).

- Company-wide planning models and budget simulations: balance sheets, income statements, and cash flows are projected over time using the conventional accounting structure (deterministic).

- Evaluation of PERT (program evaluation and review technique) networks used in project management to determine more representative completion time distributions.

- Operation of transportation systems: ocean or river shipping, dispatching of rail freight, and airline scheduling.

- Assessment of energy use and production systems on a national or international level.

- Operational gaming that allows human intervention: financial planning, evaluation of marketing strategies, airline ticket pricing, and simulated space flights.

With minor exceptions, such as operational gaming, simulation is best used only as a means of last resort, when all else fails. Generally speaking, mathematical tools are much more efficient for evaluating and optimizing a system's performance. Only if such tools cannot adequately reproduce the complexities of a real system should the operations researcher take refuge in simulation. The apparent simplicity of simulation is deceptive and may lead the unsuspecting analyst into a quagmire that may prove expensive as well as calamitous. Before a simulation project is initiated, the objectives of the study and the likely outcomes should be clearly spelled out and rough cost estimates obtained. Properly used, simulation can be an effective tool in the hands of an experienced operations researcher.

EXERCISES

When random numbers are called for in any of the exercises, use the three-digit numbers in the following table unless otherwise directed.

Random number table

005	192	941	720	596	206	233	711	909	994
624	684	748	187	069	399	527	291	388	965
599	081	086	757	758	345	845	664	775	645
914	326	213	101	662	263	784	995	347	006
344	986	776	312	512	435	160	014	255	349
799	812	863	992	680	589	143	858	958	662

1. Simulate 12 service times from a normally distributed population with a mean of 4 minutes and a standard deviation of 1 minute. For each service time computation, use four random numbers in conjunction with the formula given in Section 18.3. Use the first 48 random numbers above, starting with the first row. If the simulated time is less than zero, replace it with zero.

   (a) Compute the mean and standard deviation of the sample of simulated service times. How do they compare with the population mean and standard deviation ($\mu = 4$ and $\sigma = 1$).

   (b) Use the Kolmogorov–Smirnov test with $\alpha = 0.05$ to determine if the numbers generated are from a normal population.

2. Using the same random numbers as in Exercise 1, simulate 12 observations from an Erlang distribution with $k = 4$ and $\lambda = 0.25$. Recall that an Erlang random variable X has the following probability density function

$$f(x) = \begin{cases} k\lambda(k\lambda x)^{k-1} e^{-k\lambda x} / (k-1)! & \text{for } x \geq 0 \\ 0 & \text{otherwise} \end{cases}$$

where k is a positive integer and λ is a positive number. In particular, X is the sum of k (independent) exponential random variables, each with rate parameter $k\lambda$. This means that $E[X] = 1/\lambda$ and $Var[X] = 1/k\lambda^2$. The usefulness of the Erlang distribution stems from its relationship with the exponential distribution. In modeling process times, the exponential distribution is often inappropriate because the standard deviation is as large as the mean. Engineers usually try to design systems whose process time standard deviations are significantly smaller than their means. Notice that for the Erlang distribution, the standard deviation decreases as the square root of the parameter k increases, so that process times with a small standard deviation can often be approximated by an Erlang random variable.

3. There are three horses in a race. Historical records show that the mean and standard deviation for the time it takes each horse to complete a race are as shown in the table. What proportion of the time will each horse win? Assume that each completion time random variable is normally distributed. Simulate 10 races.

Horse	Mean time (min)	Standard deviation (min)
1	10	1
2	9.5	0.5
3	10.5	3

4. Use the second column of the random number table preceding Exercise 1 to generate six observations for each of the following situations.

   (a) A Bernoulli distribution with $p = 0.4$. Use the range $(0 \leq u \leq 0.6)$ for 0 and $(0.6 < u \leq 1)$ for 1
   (b) A binomial distribution with $n = 5$ and $p = 0.4$
   (c) A geometric distribution with $p = 0.4$
   (d) A Poisson distribution with $\lambda = 2$
   (e) A Weibull distribution with location parameter $v = 0$, scale parameter $\alpha = 1$, and shape parameter $\beta = 2$

5. Use the columns of the random number table to generate 10 observations from each of the following. Use one column of 6 numbers for each observation.

   (a) A standard normal distribution
   (b) A normal distribution with $\mu = 100$ and $\sigma = 20$
   (c) A lognormal distribution whose underlying distribution is standard normal

6. Use the columns in the random number table to generate 10 observations from the following. Use one column of 6 numbers for each observation.

   (a) A binomial distribution with $n = 6$ and $p = 0.4$ by simulating Bernoulli trials with $p = 0.4$
   (b) An Erlang distribution with $k = 6$ and $\lambda = 2$

7. (*Triangular distribution*) Consider a random variable X with probability density function (pdf)

$$f(x) = \begin{cases} x, & 0 \leq x \leq 1 \\ 2 - x, & 1 < x \leq 2 \\ 0, & \text{otherwise} \end{cases}$$

This is known as a triangular distribution with endpoints [0, 2] and mode at 1.

(a) Derive the cumulative distribution function $F(x)$ for X.

(b) Determine analytically the mean and variance of X.

(c) Develop a step-by-step scheme that can be used to generate observations of X.

(d) Use the first row of the random number table to generate 10 observations.

8. Develop a step-by-step generation scheme for the triangular distribution with pdf

$$f(x) = \begin{cases} \frac{1}{2}(x-2), & 2 \leq x \leq 3 \\ \frac{1}{2}(2-\frac{x}{3}), & 3 < x \leq 6 \\ 0, & \text{otherwise} \end{cases}$$

Use the bottom row of the random number table to generate 10 observations of X. Compute the sample mean and compare it with the true mean of the distribution.

9. Develop a random variate generator for random variable X with the following pdf.

$$f(x) = \begin{cases} e^{2x}, & -\infty < x \leq 0 \\ e^{-2x}, & 0 < x < \infty \end{cases}$$

Use the second row of the random number table to generate 10 observations of X.

10. When evaluating investments with costs and benefits realized at different points in time, it is standard practice to compute the net present worth (NPW) of the cash flows. Consider an investment of amount P with annual returns R_k for $k = 1, \ldots, n$, where n is the life of the investment. The NPW is defined as

$$\text{NPW} = -P + \sum_{k=1}^{n} \frac{R_k}{(1+i)^k}$$

where i is the investor's minimum acceptable rate of return (MARR). If NPW ≥ 0, the investment provides a return at least as great as the MARR. If NPW < 0, the investment fails to provide the MARR. Your financial advisor tells you that a particular investment of $1000 will provide a return of $500 per year for 3 years. Your MARR is 20%, or 0.2.

(a) Is this investment acceptable according to your NPW criterion?

(b) Your advisor adds the information that the life of the investment is uncertain and is, in fact, uniformly distributed with $p_n = 0.2$ over the range $n = 1, \ldots, 5$. Use the following random numbers to simulate 10 replications of the life of the investment and compute the NPW in each case.

0.5758	0.4998	0.3290	0.3854	0.6784	0.8579
0.5095	0.6824	0.3762	0.1351	0.2555	0.9773

Based on your simulation, what is the probability that the investment will yield the MARR?

(c) Your advisor now tells you that the annual revenue is uncertain. Although it will be the same each year, the revenue could be negative, and is normally distributed with a mean of $500 and a standard deviation of $200. Use the following random numbers to simulate 10 observations of the annual revenue.

0.5153	0.3297	0.6807	0.0935	0.9872	0.6339
0.0858	0.3229	0.5285	0.4451	0.3177	0.1562

Combine these results with the results in part (b) to determine the probability that the investment will yield the MARR.

11. A company has 18, 000 employees at the beginning of the year, each of whom contributes $125 per month toward the cost of medical insurance. Historical data suggest that the cost of insurance claims is about $250 per month per employee. Because the company pays the excess above $125, management wants to estimate the cost to the company of providing coverage. The number of employees is expected to increase at the rate of 2% a month, and the average claim per employee is expected to increase by 1% a month. The spreadsheet that follows computes the annual cost, with these assumptions, to be a little over $37 million.

Modify this analysis so that the growth rates each month are random variables. Assume that the growth rate for the number of employees is uniformly distributed between 1 and 3%, and that the growth rate for claims is uniformly distributed between 0.5 and 1.5%. Both growth rates are independent random variables for each month of the year. Perform a simulation analysis to estimate the mean and standard deviation of the total contribution by the company. Simulate 1 year of operation 10 times and average the results.

Initial conditions

No. of employees = 18,000
Average claim = $250
Contribution per employee = $125

Rate of change

Increase per month = 2%
Increase per month = 1%

Medical Insurance Costs for Deterministic Data

Month	Number of employees	Employee contribution	Average claim per employee	Total claims	Company contribution
Initial	18,000	$2,250,000	$250	$4,500,000	$2,250,000
1	18,360	$2,295,000	$253	$4,635,900	$2,340,900
2	18,727	$2,340,900	$255	$4,775,904	$2,435,004
3	19,102	$2,387,718	$258	$4,920,136	$2,532,418
4	19,484	$2,435,472	$260	$5,068,725	$2,633,252
5	19,873	$2,484,182	$263	$5,221,800	$2,737,618
6	20,271	$2,533,865	$265	$5,379,498	$2,845,633
7	20,676	$2,584,543	$268	$5,541,959	$2,957,417
8	21,090	$2,636,234	$271	$5,709,326	$3,073,093
9	21,512	$2,688,958	$273	$5,881,748	$3,192,790
10	21,942	$2,742,737	$276	$6,059,377	$3,316,639
11	22,381	$2,797,592	$279	$6,242,370	$3,444,778
12	22,828	$2,853,544	$282	$6,430,890	$3,577,346
Total		$33,030,745		$70,367,633	$37,336,888

12. Use the service times found in Exercise 1 to replace those in Table 18.6. With these numbers, simulate the single channel queue described in Section 18.5 where now the service times are, of course, normally distributed. Compute the various statistics for this modified problem. Is the system better or worse when the service times have less variability? (Recall that the standard deviation of the exponential distribution is equal to its mean.)

13. Use the arrival and service data in Table 18.6 and simulate an $M/M/2/3$ queue with balking. Construct a table similar to Table 18.11. Compute the evaluation statistics for the two channel system. Compare the results with the steady-state values computed from the appropriate formulas in Chapter 16.

14. From the information presented in Table 18.11 for the $M/M/1/2$ queue, compute the areas under the curves associated with the number in service and the number in the queue, as was done for the single channel case in Table 18.8. From these data, compute the evaluation statistics and observe the effect of limiting the queue size to 1. Compare the results with the steady-state values computed from the appropriate formulas in Chapter 16.

15. Consider a network consisting of two queues in series. The first is the single channel Markov system ($M/M/1$) described in Section 16.2 and the second is a single channel system with a mean service time of 3 minutes following an exponential distribution. The second system receives all its inputs from the first.

   (a) Simulate the operation of the second system using the information in Table 18.7 to determine the arrival schedule. Use the last two columns of the random number table preceding Exercise 1 to generate the service times. Compute the evaluation statistics for the network.

   (b) Assume that 20% of the departures from the first system leave the network rather than enter the second system. Also assume that, in addition to the remaining 80% who enter the second system, there is a second process generating arrivals for the second system whose interarrival times are exponentially distributed with a mean of 10 minutes. Starting with the first column in the random number table, generate six arrivals for the second system and then simulate the two queues in series. Compute the evaluation statistics for the network.

16. Simulate a two-channel queuing system given the data in the table. The system has a finite queue with a maximum number of two waiting spaces. Customers arriving when the queue is full will balk. Of course, the time for service given is irrelevant for a balking customer. The queue discipline is first-come-first-served, and the first customer arrives at time zero. Compute the evaluation statistics, such as L_S, L_q, and L, described in Section 18.5 from the simulation output.

Customer	Time between arrivals	Service time
1	–	12
2	5	8
3	2	13
4	7	7
5	1	8
6	8	10
7	2	11
8	5	10
9	3	5
10	7	9

17. Consider a finite input source system with five machines ($M/M/1/5/5$), as described in Section 16.2. The failure rate of each machine when in operation is 0.1 per hour. When a machine fails, it is sent to the repair shop—a single channel queue with a service rate of 0.3 per hour. The rate of arrival at the shop is $0.1n$ per hour, where n is the number of machines currently operating.

   (a) Simulate this system for 50 hours starting with one machine already in the repair shop.

   (b) Compute the equivalent of the performance measures given in Figure 16.9 from the output of the simulation and compare them with their theoretical steady-state counterparts.

18. Use the Queue Simulate option in the Queuing Excel add-in that comes with the text (or any other program) to simulate the $M/M/1$ queue discussed in Section 18.5. Let the initial number in the system be 3, and perform 10 replications. Each replication should process a total of 200 customers. Record the average number in the system, L, and the average time in the system, W, in each case. For these statistics, compute the mean and standard deviation over the sample of 10 observations. Assuming the statistics come from normal populations, use the t-distribution to find confidence intervals for the true steady-state values of L and W. Compare the values found with the analytic steady-state solution computed with the queuing formulas in Chapter 16.

19. Repeat Exercise 18 for 1000 customers.

20. This exercise refers to the inventory system simulated in Section 18.6.

    (a) Calculate the mean $E[D]$ and the $Var[D]$ of the demand using the data in Table 18.12. Then calculate the sample mean ($\bar{d}$) and an unbiased estimator of the variance ($\hat{\sigma}_D^2$) using the data in Table 18.14. Compare the results. Based on the t-distribution, is $E[D]$ within a 95% confidence interval constructed from $\bar{d}$ and $\hat{\sigma}_D^2$? Note that $\hat{\sigma}_D^2$ should be calculated from Equation (A.5) in Appendix A.1 on the CD.

    (b) Repeat part (a) for the lead time random variable T_L.

21. Confirm the calculations in Table 18.20 for (a) demand and (b) lead time. Note that ε in Equation (A.8) is absolute error, not percent error—that is, $\varepsilon = $ (percent error/100) $\times$ (assumed theoretical mean).

22. Confirm the results for $\bar{\bar{C}}$ and $\hat{\sigma}_{\bar{c}}$ in Table 18.20 for $n = 1000$.

23. Calculate the 95% confidence interval for the true mean total cost when $n = 2000, 5000$, and 10,000 from the data in Table 18.20. Does it make sense that this interval for $n = 10,000$ is narrower than the interval calculated for $n = 5000$? Explain.

24. Use Equation (A.7) in Appendix A.1 on the CD to compute the 95% confidence interval for the true mean, call it $\bar{C}$, of the total inventory cost given in Table 18.14. Interpret the meaning of this confidence interval and explain the circumstances under which it is valid?

25. Test the null hypothesis that a simulated criterion is consistent with the actual criterion given the following data.

	Time period									
	1	2	3	4	5	6	7	8	9	10
Actual	10	12	15	20	18	17	24	28	30	25
Simulated	8	14	12	17	20	19	23	30	32	27

Use a 0.05 level of significance and the following statistical tests.

    (a) Mann–Whitney test

    (b) Matched t-test

    (c) Analysis of variance test

Exercises 26 to 30 are design problems. Use a simulation program to analyze the situations and arrive at some conclusion. There are no "right" answers to these problems, because there are several measures of effectiveness in each case. Construct curves of the performance measures computed as functions of the values of the decision variables. Run the program at least 10 times for each parameter setting to determine the variability of the simulation results.

26. Jeremy Sitzer is planning to open a restaurant that will operate from 12 noon to 2 P.M. each day. During this time, parties are expected to arrive at random at an average rate of six per hour. Each party will use one table and will spend an average of 0.5 hour at the table, with a standard deviation of 10 minutes (assume a normal distribution for this time). Customers already seated will be allowed to remain after 2 P.M. How many tables should Mr. Sitzer plan for? If a waiter spends an average of 10 minutes with a party (assume this time has an exponential distribution), how many waiters are required?

27. A bank wants to determine how many drive-up windows to open and how much room to provide for waiting cars. If the number of cars exceeds the allotted space, they will be forced to move on. Arrivals follow a Poisson process with an average rate of one car per minute. The average service time is 2.5 minutes, with a standard deviation of 1 minute. This time is normally distributed. Specify one or more performance measures and use them to determine how many drive-up windows the bank should open.

28. A shop currently has a machine with a highly variable processing time that is exponentially distributed, with a mean of 20 minutes. Jobs arrive at the machine irregularly at an average rate of 2.5 per hour (the time between arrivals has an exponential distribution), and long queues have been observed. Management has been considering the installation of a second machine with the same characteristics to reduce the length of the queue. The chief industrial engineer has proposed that, rather than purchase a second machine, a robot be installed with much less variability. In fact, the robot will require exactly 20 minutes for each processing operation. Analyze the two alternatives.

29. Dr. Stark, who we met in Chapter 16, can partially control the rate of arrivals to her office by keeping an appointment book. Experience indicates that the average time required to treat a patient is 15 minutes and follows an exponential distribution. On the other hand, whatever the schedule, the time between arrivals is normally distributed, with a standard deviation of 5 minutes. Discuss how Dr. Stark should schedule arrivals given that she wishes to spend 4 hours each morning seeing patients. To ensure efficient use of her time, she always arranges for two patients to be waiting when the office opens, but will work overtime, if necessary, until everyone scheduled for an appointment has been seen.

30. A consortium of wealthy dot-com executives is planning to build a private airport outside of Austin, Texas to better accommodate high-tech business travelers. It is expected that there will be two busy periods in the day, 1 hour in the morning and 1 hour in the evening. During these times, the arrival rate is estimated to be 12 planes per hour. Historical data show that arrivals occur at random and require the use of a dedicated runway for about 10 minutes, with a standard deviation of 3 minutes. How many runways should be provided?

31. Consider the following container port operation. The port has two berths, each with one container crane. Each crane can work only one ship at a time. Ships arrive at the port and wait for a tug to pilot them to a berth if one is free. There is only one tug. Once a ship has docked, it will be unloaded by a crane. The containers are lifted off the ship and placed directly onto flat-deck rail wagons, of which there is an unlimited supply. Once all containers have been removed, the ship is loaded with new containers, which are also brought to the dock by rail. The number of containers to be unloaded and loaded varies from ship to ship. Loaded ships clear their berths with the help of the tug.

(a) Identify the components of this system in terms of entities, attributes of the entities, activities and their associated events, and the calendar.

(b) Draw a flow diagram embodying the logic of a simulation of the port.

32. (*Deterministic model*) A soft drink manufacturer forecasts the following pattern of demand (in thousands of gallons) over the coming 12-month period.

Month	1	2	3	4	5	6	7	8	9	10	11	12
Demand	2500	1800	2000	2800	3500	4800	5600	6000	4500	3200	1200	3600

The manufacturing facilities can be operated with one or two shifts, with or without overtime. The following table identifies the maximum output capacities and associated costs.

	One shift		Two shifts	
Production setup	No overtime	With overtime	No overtime	With overtime
Maximum output	2000	2500	4000	5000
Cost ($\times$ $1000)	120	165	200	280

Goods produced in each month can be used to satisfy the demand in that month or any subsequent month. Any goods carried forward to later months incur a storage cost of $100 per 1000 gallons per month. An increase in the number of shifts results in a "hiring and training" cost of $18,000, whereas a decrease from two shifts to one shift costs the firm $20,000 in severance pay. At the beginning of the planning horizon, there are no goods in stock, and in the preceding period the firm was operating with only one shift. Overtime cost is proportional to the amount of labor used.

(a) Simulate this operation, assuming that the plant always runs at the maximum regular capacity for each production setup. Two shifts are used for months 5 through 10 only, and overtime is used as necessary. What is the cost of this schedule?

(b) By trial and error, use simulation to determine what you consider to be the optimal production schedule and its cost. Regular production may be at less than full capacity but at full-capacity cost.

33. (*Deterministic model*) Consider the operation of a single-track railway line connecting stations A to B, B to C, C to D, and D to E. No more than one train can be on the track between adjacent stations. A new train may enter a track segment only when the preceding train has cleared it. Trains may cross only at stations C and D; station B does not have sufficient siding for two trains. Hence, no train may enter the track from C to B if there is already a train traveling from A to B. However, a second train may enter the track from A to B if there is already a train traveling from B to C. The same rules also hold in the opposite direction. The sidings at each station are not considered part of the track between adjacent stations. No more than three trains can be in each of stations C and D simultaneously. The travel and switching times in minutes for regular and express trains are as follows.

	Travel times								Switching times		
	A–B	B–C	C–D	D–E	E–D	D–C	C–B	B–A	B	C	D
Regular	20	12	30	16	24	28	14	18	10	30	20
Express	15	8	15	12	16	24	10	14	0	5	5

The timetable provides the following departure times. At A in the direction of E: regular at 8:00 A.M., express at 10:00 A.M., regular at 10:20 A.M. At E in the direction of A: express at 9:10 A.M., regular at 9:30 A.M., regular at 11:00 A.M. Simulate this operation using next-event incrementation for the 4-hour period from 8 A.M. to 12 noon. At 8 A.M., there is one regular train at station D ready to depart toward station A.

34. Using the random number table, generate 10 random variates for the following probability mass function.

Value, x	0	1	2	3	4	5	6	7	8
pmf, $p(x)$	0.156	0.234	0.208	0.161	0.095	0.064	0.036	0.028	0.018

35. A firm wishes to investigate the profitability of a new product. Consumer tests on samples of the product made at a small test plant have produced favorable results. From these tests and other market surveys, it is estimated that the potential sales should range from 40 to 100 units in about 9 out of 10 quarters within 3 years of the product's introduction.

One of the alternatives considered involves the construction of a plant with a capacity of 100 units per quarter at a cost of $400,000. Such a plant would be in operation within 1 year of the decision to go ahead with the new product. Since the product can be stored only for very short periods, production would follow actual sales very closely. Hence, any potential sales above 100 units per quarter would be lost. The fixed production cost per quarter is estimated at $30,000, whereas the variable production costs per unit are highly nonlinear, as follows.

$$\text{For } 0 \le q \le 50, \quad \$(100 - q) \text{ per unit}$$
$$50 < q \le 80, \quad \$50 \text{ per unit}$$
$$80 < q \le 100, \quad \$0.625q \text{ per unit}$$

Sales predictions are as follows.

Operating year	Sales range	Sales price/unit
1	20 to 60 units per quarter	$880
2	30 to 90 units per quarter	$890
3, 4, . . .	40 to 100 units per quarter	$900

All ranges are quoted with odds of 9 out of 10, and actual sales are assumed to be approximately normally distributed. (Note that sales cannot be negative.) Management would like to determine by simulation the distribution of the cumulative net cash flow over the first 5 years of plant operation (after plant construction). Round sales figures to the nearest unit.

(a) Using the random number table starting in row 1 (or a random number generator of your choosing), simulate the cash flow quarter by quarter over a 5-year period. What is the net cash flow?

(b) (*More computations*) Redo the simulation 10 more times, continuing with the random number table where you left off in part (a). Once you reach the end of the table, go to the upper right corner and read the numbers from right to left (alternatively, you can use a computer-based random number generator). Find the average net cash flow and the standard deviation of the net cash flow.

(c) (*Computer implementation*) Write a computer program that will perform this simulation. Make 100 runs, each over a 10-year period, discounting the quarterly cash flows at a rate of 3% per quarter. Construct a frequency histogram of the net discounted cash flows containing about 10 class intervals. (If you need a refresher on discounting, see the material in Sections 5.2 and 13.6). What conclusion do you reach about the profitability of the project, assuming that 10 years is the productive life of the plant?

36. Canal A is connected to the lower-level canal B by a lock. Boats enter the lock from canal A, the gates are closed, the water level is lowered to the level of canal B, the gates leading to canal B are opened, and the boats leave. At this point, any boats in canal B wanting to be raised to canal A enter the lock, the gates are closed, the water level is raised to the level of canal A, the gates are opened, and the boats leave. The lock has a capacity of four boats, and it takes 4 minutes to lower or raise the water level. Only one boat may enter the lock at a time, and a boat takes 1 minute to be moored. However, all boats leave the lock one after the other, taking a total of 2 minutes (regardless of the number of boats).

Boats arrive at the lock in canal A at an average rate of 12 per hour, and in canal B at an average rate of 15 per hour. Arrivals in both canals follow a Poisson distribution. The current mode of operation during the busy hours, for which the above mentioned arrival rates hold, is to fill or empty the lock whenever four boats have been moored inside the lock.

At 11 A.M., the system is in the following state. The lock gates are open to traffic from canal A. Three boats are already moored in the lock, and the next arrival from canal A is scheduled at 11:02. Five boats are currently waiting in canal B to go through the lock to canal A. The next arrival from canal B is scheduled at 11:04. Simulate this system until 11:30 A.M. Keep enough detail to determine the number of boats passing through the lock in each direction, the average time boats spend in the lock between arrival and departure in each direction, and the number of times the lock is raised and lowered. Use the following sequence of two-digit random numbers to generate the arrivals: 56 03 09 78 38 47 01 98 03 16 14 56 17 11 98 82 51 97 93 04.

37. A job shop has three work centers, X, Y, and Z. Each job has to go through some or all centers in a prescribed unique sequence. A work center can process only one job at a time, but the next job can enter immediately after the preceding job departs. The current state of the system is as follows.

Jobs currently in the system	1	2	3	4	5	6	7	8	9	10
Sequence of centers left to be entered	Z	Z	X	XY	XYZ	XZ	YXZ	Y	YZ	Y

Work center X is currently processing job 1; the scheduled release time from work center X is at simulated time 124 minutes. Work center Y is currently processing job 2, with a scheduled release time of 180 minutes. Jobs are processed by each work center on a first-come-first-served basis. The current job order at work center X is 1, 3, 4, 5, 6, and the current job order at work center Y is 2, 7, 8, 9, 10. Note that work center Z is currently idle. Processing times in minutes at each work center are normally distributed as follows.

	Work center		
	X	Y	Z
Mean (min)	120	200	180
Standard deviation (min)	40	40	60

New jobs enter the system at the beginning of each day, at which time they are added to the file of jobs waiting for processing at each center. The current simulated time is $t = 100$ minutes.

(a) Identify the components of this system in terms of entities, attributes of the entities, activities and their associated events, and the calendar.

(b) Draw a flow diagram similar to Figure 18.5 for the inventory system that depicts the logic of the work center operations.

(c) Simulate the system with the given data until simulated time $t = 480$ minutes.

38. Draw two flow diagrams representing the activities and logic for the problem in Exercise 31, one for the movement of ships and one for the unloading and loading operations. The latter should contain the details associated with the loading and unloading activities of the former.

39. Bonzi Delivery operates a 24-hour parcel pick-up service for air freight delivery to East Coast cities. Customers make pick-up requests by phone to a dispatcher, who fills in an order form and then assigns the request to the first available driver. After completing a pick-up, the driver fills in a pick-up report, which he or she files with the dispatcher, and then waits for a new assignment. A dispatcher will interrupt assigning a job to a driver to take a customer phone call, and will continue with the job assignment only after having prepared the pick-up order form. There are two dispatchers who alternate taking customer calls except when a call comes in while a dispatcher is still engaged with the preceding customer. The number of incoming phone lines is sufficiently large so that no customer ever gets a busy signal. All pick-up orders are available to both dispatchers for assignment to drivers. Past records for a particular office indicate the following information.

• The average rate of pick-up requests is 24 per hour and follows a Poisson process.

• The time required to receive a pick-up request and fill in the order form is normally distributed with a mean of 120 seconds and a standard deviation of 20 seconds.

• The time to make a job assignment is a constant 60 seconds.

• The time to make a pick-up and file the report is normally distributed with a mean of 1500 seconds and a standard deviation of 400 seconds.

• There are 16 drivers.

(a) Identify entity classes and activities, and list all entities jointly engaged in each activity. Identify the calendar to be used for a next-event simulation.

(b) Draw a flow diagram similar to Figure 18.5 in the text for the inventory system depicting the logic of the dispatching operations.

(c) Assume that at 11:00 A.M. the system is in the following state: (1) no pending requests; (2) dispatcher 1 is idle; (3) dispatcher 2 is processing a pick-up request with 80 sconds to go; (4) the next pick-up request will be received at 100 seconds from now; (5) drivers 1, 2, 3 and 4 are idle; (6) busy drivers and pick-up completion times (including filling a pick-up report) in 1370, 14–510, 15–870, 16–1410. Simulate this system for 1 hour. Round all times to the nearest 10 seconds. Collect statistics on the total idle time of dispatchers and drivers, and on the total time pick-up orders wait in the office until they are assigned to a driver. Generate all activity times at the start of each activity only.

BIBLIOGRAPHY

Banks, J. and R. Gibson, "Selecting Simulation Software," *IIE Solutions*, Vol. 29, No. 5, pp. 29–32, 1997.

Banks, J. and R. Gibson, "10 Rules for Determining when Simulation Is Not Appropriate," *IIE Solutions*, Vol. 29, No. 9, pp. 30–32, 1997.

Banks, J., J.S. Carson II, B.L. Nelson, and D.M. Nicol, *Discrete Event System Simulation*, Third Edition, Prentice-Hall, Upper Saddle Ridge, NJ, 2001.

Bard, J.F., "Benchmarking Simulation Software for Use in Modeling Postal Operations," *Computers & Industrial Engineering*, Vol. 32, No. 3, pp. 607–625, 1997.

Bard, J.F., A. deSilva, and A. Bergevin, "Evaluating Simulation Software for Postal Service Use: Technique Versus Perception," *IEEE Transactions on Engineering Management*, Vol. 44, No. 1, pp. 31–42, 1997.

Bowden, R., "The Spectrum of Simulation Software," *IIE Solutions,* Vol. 30, No. 5, pp. 44–46, 1998.

Elliot, M., "Buyer's Guide, Simulation," *IIE Solutions*, Vol. 32, No. 5, pp. 55–64, 2000.

Evans, J.R. and D.L. Olsen, *Introduction to Simulation and Risk Analysis*, Prentice-Hall, Upper Saddle Ridge, NJ, 1998.

Fishman, G.S., *Monte Carlo: Concepts, Algorithms, and Applications*, Springer-Verlag, New York, 1996.

Forrester, J.W., S*ystem Dynamics*, Wright-Allen Press, Cambridge, MA, 1972.

Fu, M.C. and K.J. Healy, "Techniques for Optimization via Simulation: An Experimental Study on an (s, S) Inventory System," *IIE Transactions on Operations Engineering*, Vol. 29, No. 3, pp. 191–199, 1997.

Goldsman, D. and J.R. Wilson (editors), *Special Issue of IIE Transactions on Operations Engineering Honoring Alan Pritsker*, Vol. 33, No. 3, 2001.

Grossman, T.A. Jr., "Teacher's Forum: Spreadsheet Modeling and Simulation Improves Understanding of Queues," *Interfaces*, Vol. 29, No. 3, pp. 88–103, 1999.

Harrell, C., B.K. Ghosh, and R. Bowden, *Simulation Using ProModel*, McGraw-Hill, New York, 2000.

Kelton, W.D, R.P. Sadowski, and D.A. Sadowski, *Simulation with Arena*, McGraw-Hill, New York, 1998.

Kleijnen, J.P.C., *Statistical Tools for Simulation Practitioners*, Dekker, New York, 1987.

Kleijnen, J.P.C., "Experimental Design for Sensitivity Analysis, Optimization and Validation of Simulation Models," in *Handbook of Simulation*, J. Bands (editor), Wiley, New York, pp. 582–625, 1998.

Law, A.M., *ExpertFit*, Averill M. Law & Associates, Tucson, AZ, 1998.

Law, A.M. and W.D. Kelton, *Simulation Modeling and Analysis*, Third Edition, McGraw-Hill, New York, 2000.

Lawrence, J.A. Jr. and B.A. Pasternack, *Applied Management Science: A Computer-Integrated Approach for Decision Making*, Wiley, New York, 1998.

Lilegdon, W.R., D.L. Martin, and A.A.B. Pritsker, "FACTORY AIM: A Manufacturing Simulation System," *Simulation*, Vol. 62, No. 2, pp. 367–372, 1994.

Montgomery, D.C., *Design and Analysis of Experiments*, Fourth Edition, Wiley, New York, 1997.

Pegden, C.D., R.E. Shannon, and R.P. Sadowski, *Introduction to Simulation Using SIMAN*, Second Edition, McGraw-Hill, New York, 1995.

Pritsker, A.A.B. and J.J. O'Reilly, *Simulation with Visual SLAM and AweSim*, Second Edition, Wiley, New York, 1999.

Rohrer, M. and J. Banks, "Required Skills of a Simulation Analyst," *IIE Solutions*, Vol. 30, No. 5, pp. 7–23, 1998.

Rubinstein, R.Y., B. Melamed, and A. Shapiro, *Modern Simulation and Modeling*, Wiley, New York, 1998.

Schmeser, B.W., "Simulation Experiments," in *Handbook In Operations Research and Management Science,* Vol. 2, *Stochastic Models*, D.P. Heyman and M.J. Sobel (editors), North- Holland, Amsterdam, pp. 295–330, 1990.

Swain, J.J., "Simulation Software Survey: Power Tools for Visualization and Decision Making," *OR/MS Today*, Vol. 28, No. 1, pp. 52–63, 2001.

Thesen, A. and L.E. Travis, *Simulation for Decision Making*, West, Minneapolis, MN, 1993.

Wert, S.D., J.F. Bard, A. deSilva, and T.A. Feo, "A Simulation Analysis of Advanced Concepts for Semi-Automated Mail Processing," *Journal of the Operational Research Society*, Vol. 42, No. 12, pp. 1071–1086, 1991.

Wilson, J.R. and A.A.B. Pritsker, "Experimental Evaluation of Variance Reduction Techniques for Queuing Simulation Using Generalized Concomitant Variables," *Management Science*, Vol. 30, No. 12, pp. 1459–1472, 1984.

Index

LICENSING AND WARRANTY AGREEMENT
For MPL

Notice to Users: Do not install or use the CD-ROM until you have read and agreed to this agreement. You will be bound by the terms of this agreement if you install or use the CD-ROM or otherwise signify acceptance of this agreement. If you do not agree to the terms contained in this agreement, do not install or use any portion of this CD-ROM.

License: The material in the CD-ROM, MPL for Windows 4.2 with CPLEX and CONOPT solvers (the "Software") is owned by Maximal Software, Inc. ("Maximal") with mailing address 2111 Wilson Blvd., Arlington, VA, USA 22201 and is copyrighted and protected by United States copyright laws and international treaty provisions. All rights are reserved to the respective copyright holders. Maximal hereby grants the user a nonexclusive license to use the Software as set forth below. No part of the Software may be reproduced, stored in a retrieval system, distributed (including but not limited to over the www/Internet), decompiled, reverse engineered, reconfigured, transmitted, or transcribed, in any form or by any means – electronic, mechanical, photocopying, recording, or otherwise – without the prior written permission of Maximal. The Software may not, under any circumstances, be reproduced and/or downloaded for sale. Please note since this Software is provided free of charge with the book, there are some restrictions on how the Software can be used. Users can install the Software on their own computer and use it for their own personal optimization modeling projects. Using the Software for any commercial purposes other than evaluation and/or distributed it on multiple computers or computer networks is not allowed.

Warranty: Maximal provides no warranties, expressed or implied, including the implied warranties of merchantability or fitness for a particular purpose, and shall not be liable for any damages, including direct, special, indirect, incidental, consequential, or otherwise.

For Technical Support:
info@maximalsoftware.com

6. <u>NO WARRANTY</u>. IMAGINE THAT! AND ITS LICENSORS DO NOT AND CANNOT WARRANT THE PERFORMANCE OF OR THE RESULTS THAT MAY BE OBTAINED BY USING THE SOFTWARE. ACCORDINGLY, THE SOFTWARE IS LICENSED "AS IS" AND IMAGINE THAT! AND LICENSORS HEREBY SPECIFICALLY DISCLAIMS ANY AND ALL EXPRESS AND IMPLIED WARRANTIES WITH RESPECT TO THE SOFTWARE, INCLUDING, WITHOUT LIMITATION, THE WARRANTY OF MERCHANTABILITY AND OF FITNESS FOR A PARTICULAR PURPOSE. Some jurisdictions do not allow the exclusion of implied warranties, so the above exclusion may not apply to you. Imagine That! warrants the CD on which the Software is recorded to be free of defective material and workmanship under normal use for 30 days after the date of delivery. During the 30 day period, you may return the CD to Imagine That! and it will be replaced without charge. Replacement of the CD is your SOLE AND EXCLUSIVE REMEDY in the event of a defect. The above warranty gives you specific legal rights and you may also have other rights that vary from state to state.

7. <u>LIMITATION OF LIABILITY.</u> UNDER NO CIRCUMSTANCES SHALL IMAGINE THAT! BE LIABLE TO YOU OR ANY OTHER PERSON FOR ANY SPECIAL, INCIDENTAL OR CONSEQUENTIAL DAMAGES, INCLUDING, WITHOUT LIMITATION, LOST PROFITS OR LOST DATA, LOSS OF OTHER PROGRAMS, OR OTHERWISE, AND WHETHER ARISING OUT OF BREACH OF WARRANTY, BREACH OF CONTRACT, TORT (INCLUDING NEGLIGENCE) OR OTHERWISE, EVEN IF ADVISED OF SUCH DAMAGE OR IF SUCH DAMAGE COULD HAVE BEEN REASONABLY FORESEEN, EXCEPT ONLY IN CASE OF PERSONAL INJURY WHERE APPLICABLE LAW REQUIRES SUCH LIABILITY. Some states do not allow the exclusion or limitation of incidental or consequential damages, so the above limitation or exclusion may not apply to you. In no case shall the liability of Imagine That! exceed the purchase price for the Software.

8. <u>TERM AND TERMINATION.</u> This Agreement shall continue until terminated. This license terminates automatically if you violate any terms of the Agreement. You may terminate this agreement at any time. Upon termination you must promptly delete all copies of the Software. Sections 2, 5, and 6 shall survive termination.

9. <u>CONTRACTING PARTIES.</u> If the Software is installed on computers owned or operated by a corporation or other legal entity, then this Agreement is formed by and between Imagine That! and such entity. The individual executing this Agreement represents and warrants to Imagine That! that they have the authority to bind such entity to the terms and conditions of this Agreement.

10. <u>U.S. GOVERNMENT RESTRICTED PROVISIONS.</u> If this software was acquired by or on behalf of a unit or agency of the United States Government, this provision applies. This software: (a) was developed at private expense, and no part of it was developed with government funds, (b) is a trade secret of Imagine That! for all purposes of the Freedom of Information Act, (c) is "commercial computer software" subject to limited utilization as provided in the contract between the vendor and the government entity, and (d) in all respects is proprietary data belonging solely to Imagine That! For the Department of Defense (DOD), this Software is sold only with "Restricted Rights" as defined in the DOD Supplement to the Federal Acquisition Regulations, 52.227-7013(c)(1)(ii) and use, duplication, or disclosure is subject to restrictions as set forth in subdivision (c)(1)(ii) of the Rights in Technical Data and Computer Software clause at 52.227-7013.

11. <u>GOVERNING LAW.</u> This agreement shall be governed by and construed and enforced in accordance with the laws of the State of California, excluding conflict of law rules. No action arising under this Agreement may be brought more than one year after the cause of action has accrued.

12. <u>GENERAL.</u> Imagine That! is a registered trademark and Extend, Industry Suite, Extend Suite, Extend+Manufacturing, Extend+BPR, and Extend+Industry are trademarks of Imagine That, Inc. SDI Industry is a trademark of Simulation Dynamics, Inc. Proof Animation is a trademark of Wolverine Software Corporation. Stat::Fit is a registered trademark of Geer Mountain Software.

To contact Imagine That! or if you have any questions about this agreement:
Imagine That, Inc
6830 Via Del Oro, Suite 230
San Jose, CA 95119 USA
Phone 408-365-0305   • Fax 408-629-1251   • Email extend@imaginethatinc.com.